The Victorian City

Images and Realities

The

Edited by

H. J. Dyos and Michael Wolff

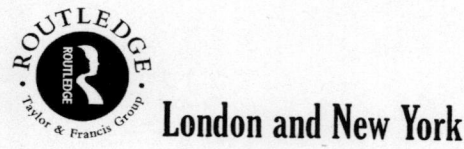

London and New York

Victorian City

Images and Realities

Volume 1

First published 1973
by Routledge Kegan and Paul Ltd
Reprinted in 1999, 2001 (twice) by Routledge
11 New Fetter Lane, London EC4P 4EE
Simultaneously published in the USA and Canada
by Routledge, 29 West 35th Street, New York, NY 10001
© 1973, 1999 H J Dyos and Michael Wolff
Routledge is an imprint of the Taylor & Francis Group
Printed and bound in Great Britain by
Antony Rowe Limited,Chippenham,Wiltshire

All rights reserved. No part of this book may be reprinted
or reproduced or utilised in any form or by any electronic,
mechanical, or other means, now known or hereafter
invented, including photocopying and recording, or in any
information storage or retrieval system, without
permission in writing from the publishers.

British Library Cataloguing in Publication Data

A catalogue record for this book is available from the
British Library

Library of Congress Cataloging in Publication Data

ISBN 0-415-19323-0

To H. L. BEALES in his eighty-fifth year

Pioneer and Exponent of the Victorian World

Contents

Volume One

Preface   xxv
Acknowledgments   xxxi

I Past and Present

1. The Urbanizing World   Eric E. Lampard   3
2. Voices from Within   Paul Thompson   59

II Numbers of People

3. The Human Aggregate   Asa Briggs   83
4. The Contagion of Numbers   J. A. Banks   105
5. Comers and Goers   Raphael Samuel   123
6. Pubs   Brian Harrison   161

Contents

7	The Literature of the Streets	Victor E. Neuburg	191
8	The Metropolis on Stage	Michael R. Booth	211

III Shapes on the Ground

9	The Camera's Eye	G. H. Martin and David Francis	227
10	The Face of the Industrial City	G. F. Chadwick	247
11	Reading the Illegible	Steven Marcus	257
12	The Power of the Railway	Jack Simmons	277
13	London, the Artifact	John Summerson	311
14	House upon House	Donald J. Olsen	333
15	Slums and Suburbs	H. J. Dyos and D. A. Reeder	359

IV A Change of Accent

16	Another Part of the Island	G. F. A. Best	389
17	Metropolitan Types	Lynn Lees	413

Volume Two

V Ideas in the Air

18	The Awful Sublimity of the Victorian City	Nicholas Taylor	431
19	Victorian Artists and the Urban Milieu	E. D. H. Johnson	449
20	The Frightened Poets	G. Robert Stange	475
21	From 'Know-not-Where' to 'Nowhere'	George Levine	495
22	The Novel between City and Country	U. C. Knoepflmacher	517
23	Dickens and London	Philip Collins	537
24	Pictures from the Magazines	Michael Wolff and Celina Fox	559

VI A Body of Troubles

25	Fact and Fiction in the East End	P. J. Keating	585
26	Unfit for Human Habitation	Anthony S. Wohl	603
27	Disease, Debility, and Death	George Rosen	625
28	Training Urban Man	Richard L. Schoenwald	669
29	Prostitution and Paterfamilias	Eric Trudgill	693
30	The Culture of Poverty	Gertrude Himmelfarb	707

VII A New Earth

31	Literary Voices of an Industrial Town	Martha Vicinus	739
32	Areas of Urban Politics	Derek Fraser	763
33	Orange and Green	Sybil E. Baker	789

Contents

34	Challenge to the Church	David E. H. Mole	815
35	Catholic Faith of the Irish Slums	Sheridan Gilley	837
36	Feelings and Festivals	John Kent	855
37	The Way Out	Stanley Pierson	873

VIII Epilogue

38	The Way We Live Now	H. J. Dyos and Michael Wolff	893

Notes on Contributors   909

Index   917

Illustrations

Volume One

Between pages 150 and 151

1 Sailors' Home in the East End
2 The Strangers' Home, Limehouse
3 The Bull's-Eye in Whitechapel
4 Gypsy encampment in Notting Dale
5 Gypsies on Epsom Downs
6 'Out of the parish'
7 'All the way from Manchester . . .'
8 Spring herrings at Yarmouth
9 Casual ward at Marylebone workhouse
10 Labour-yard at Bethnal Green
11 Frost fair in St James's Park
12 Thaw in Cheapside
13 London gypsies
14 Moved on by the police
15 On the tramp
16 Van-dwellers at the Agricultural Hall
17 Sandwich-board man
18 Italian street musicians

Illustrations

  19 Mobile circus
  20 Suffolk maltsters
  21 Navvies at the Crystal Palace
  22 Hop-pickers
  23 Whitby fishermen
  24 Dundee whaler

Between pages 182 and 183

  25 The King's Head, Southwark
  26 The White Horse, Fetter Lane
  27 The Bull and Mouth, St Martin's-le-Grand
  28 London street-corner on Sunday
  29 Ginshop interior
  30 The Old Oak, Hampstead
  31 Tom Spring's Parlour
  32 The Cyder Cellars, Covent Garden
  33 Cream ginshop
  34 'Father, don't go'
  35 The Rosemary Branch, Islington
  36 The Surrey Music-Hall
  37 Dinner at the London Tavern
  38 Discussion at the Belvedere
  39 Signing the pledge
  40 The Band of Hope

Between pages 198 and 199

  41 Catnach's shop
  42 A ballad-seller
  43 'Marriage of the Queen'
  44 'John Bull & his Party'
  45 'Mister Billy Roupell'
  46 'Strike of the London Cabmen'
  47 'Shocking Murder'

Between pages 214 and 215

  48 *Cross Roads of Life!*
  49 *After Dark*
  50 *The Great City*
  51 *Lost in London*
  52 The Surrey Theatre
  53–4 The Adelphi Theatre: Old and New
  55 The London Pavilion
  56 The Alhambra Theatre of Varieties

Illustrations

57 Theatre Royal, Haymarket
58 The Garrick Theatre, Whitechapel
59 The Gaiety Theatre
60 The Grecian Theatre
61 The Britannia Theatre
62 *The Streets of London*, Princess's Theatre
63 *The Great City*, Drury Lane Theatre
64 *Lost in London*, Adelphi Theatre
65 *The Long Strike*, Lyceum Theatre

Between pages 246 and 247

66 Temple Bar
67 St John's Gate, Clerkenwell
68 The Banqueting Hall, Whitehall
69 Butcher's Row, Aldgate
70 Butcher Row, Coventry
71 Hotel Cecil, forecourt
72 The Oxford Arms, Newgate
73 Whitehorse Close, Edinburgh
74 Sackville Street, Dublin
75 The Old Town, Edinburgh
76 The Victoria Embankment
77 Leyton, the first Town Hall and Free Library
78 The Sailors' Home, Great Yarmouth
79 Bristol General Hospital
80 Market place, Leicester
81 Corporation Street, Birmingham
82 Royal opening of the Crystal Palace
83 Wet-plate photographer
84 The Free Trade Hall, Manchester
85 The Chain Pier, Brighton
86 Street-seller of crockery-ware
87 London Stereoscopic Company
88 The *carte de visite*
89 Stereo card workshop
90 Stereoscope
91 *Slum life in our great cities*
92 Ragamuffins
93 Seven Dials
94 Liverpool fire engine
95 Street trader
96 Local patriarch
97 Round game

Illustrations

98 Kerbside game
99 Street vendor
100 Hot chestnuts
101 Second-hand shoes
102 Bootblack
103 Hokey-pokey ice-cream man
104 Ice cream, the native product
105 Hot food
106 Ragged boys
107 Street-cleaner
108 Bricklayers
109 London cart-horse
110 Collapse at work
111 Horse inspection
112 Household Cavalry
113 Laying road-blocks
114 Steamroller
115 The Rows, Chester
116 Borough High Street
117 Lime Street, Liverpool
118 Clapham Road on Derby Day
119 Whitehall
120 Kennington turnpike gate
121 Fleet Street
122 Holborn
123 The Trongate, Glasgow
124 Policeman on point-duty
125 Boarding a bus
126 Omnibus competition
127 Horse-tram
128 Bristol quays
129 West Hartlepool docks
130 Southampton pier
131 Emigrants
132 Horse-trough
133 Kingsbury & Neasden station
134 Street carrier
135 London Bridge
136 Sailing barges
137 Barges in the Pool
138 The frozen Thames
139 Market place, Aberdeen
140 Covent Garden Market

Illustrations

141 Flat-Iron Market, Salford
142 Butcher's shop
143 Pawnbroker's and off-licence
144 Back-street store
145 An ironmonger's shop
146 Butter and eggs
147 The counting house
148 Manchester telephone exchange
149 Liverpool Cotton Exchange
150 Kodak head office
151 Advertising float
152 Sandwich-board man
153 Rotten Row
154 Fun of the fair
155 Surrey Gardens
156 Pub entertainment
157 Cyclists
158 Race-course tipster
159 At the Derby
160 Greenwich pier
161 The *Skylark*
162 Hastings beach
163 Hastings pier
164 Scarborough
165 Blackpool Tower
166 Beach music
167 Opening the Manchester Ship Canal
168 Declaration of the poll
169 Cost of burial
170 Oakum-picking
171 In the park
172 Municipal gardening
173 Public benefaction
174 The Brown Library, Liverpool
175 Kingsway in the making
176 Sheffield Town Hall
177 Electrification of Manchester's trams

Between pages 256 and 257
(Nos 178–235 all relate to Manchester)

178 Victoria Bridge
179 Victoria Street Fish Market
180 Old Wellington Inn

Illustrations

181	The Vintners' Arms, Smithy Door
182	The Sun Inn, Long Millgate
183	Free Trade Hall
184	Newall's Buildings, Market Street
185	Watts's warehouse
186	King Street
187	Bank of England
188	Old Royal Exchange
189	Queen's Hotel
190	Albert Memorial
191	The Manchester & Salford Bank
192–3	Manchester and Salford, 1857 and 1890
194	Bridgewater Canal
195	River Irwell
196	London Road station
197	Oxford Road
198	Exchange Station
199–201	Market Street
202	The Town Hall
203	St Ann's Square
204	Smithy Door market
205	Smithy Door
206	The Shambles
207	The Royal Exchange
208	Victoria Buildings
209	Daniel Lee's warehouse
210	Piccadilly
211	Shudehill Poultry Market
212	Market Street
213	Cross Street
214–17	Rochdale Road
218	The Police Court
219	The Assize Courts
220	Belle Vue Gaol
221	Palace of Varieties
222	Reference Library
223	Folly Theatre
224	Jersey Dwellings, Ancoats
225	Strangeways Gaol
226	The Whit Walk
227	Botanical Gardens, Old Trafford
228	Crematorium
229–30	The south-eastern suburbs, 1848 and 1895

Illustrations

231–2 The central area, 1848 and 1895
233–4 The environs of the Exchange, 1850 and 1890
235 The Manchester city region, *c.* 1900

Between pages 288 and 289

236–7 Pearson's proposed Central Railway Terminus
238–9 Farringdon station
240 Metropolitan Railway: Clerkenwell tunnel
241 Metropolitan Railway: St John's Wood Junction
242 Thames Embankment
243 'Ruskin's Nut'
244 Euston and Marylebone Roads
245 Greenwich Railway
246 St Pancras station under construction
247 Bridge and station at Cannon Street
248 The old Fleet Prison and Ludgate station
249 Bridge over Ludgate Hill
250 Bridge over Friargate, Derby
251 Bridge over Church Street, Hackney
252 Railway entrance into Bath
253 Stoke-on-Trent station
254 New Street station, Birmingham
255 Temple Meads joint station, Bristol
256 Royal Border Bridge, Berwick-on-Tweed
257 Railway viaduct, Leicester

Between pages 320 and 321

258 Threadneedle Street
259 National Discount Company, Cornhill
260 City Bank, Threadneedle Street
261 Nos 59–61 Mark Lane
262 Design for General Credit Company, Lothbury
263 Flats, Victoria Street
264 The London School Board Offices
265 Southwark Street
266 New Zealand Chambers, Leadenhall Street
267 Northumberland Avenue
268 Albert Buildings, Queen Victoria Street
269 Hotel Cecil
270 The Red House, Bayswater Road
271 Lowther Lodge, Kensington Gore
272 Design for Chelsea Public Library
273 British Museum of Natural History

Illustrations

274	Business premises, Duke Street	
275	No. 1 Old Bond Street	
276	Grosvenor Hotel	
277	Albert Hall Mansions	
278	The Palace Theatre	
279	Royal Albert Hall	
280	The Imperial Institute	

Between pages 352 and 353

281–7	Plans of the Chalcots Estate	
288–91	Elevations of buildings thereon	

Between pages 396 and 397

292	St Leonard's Hill, Edinburgh	
293	Amphion Place, Calton Road	
294	Lower Viewcraig Row	
295	Tweeddale Court	

Volume Two

Between pages 448 and 449

296	Newgate Gaol, exercise yard	
297	Leicester Gaol	
298	Newgate Gaol, entrance	
299	Woolwich Arsenal	
300	No. 10 Gipsy Hill	
301	Wright: 'Arkwright's Cotton Mill'	
302	Cox's Stack, Dundee	
303	Edinburgh Gas Works	
304	Marshall's Mills, Leeds	
305	Euston Arch	
306	The Archway, Highgate	
307	Hungerford Market	
308	Chester General Station	
309	Great Western Arcade, Birmingham	
310	Great Wheel, Earls Court Exhibition	
311	Leeds Corn Exchange	
312	Barton's Arcade, Manchester	
313	Metropolitan Tabernacle, Elephant & Castle	
314	Particular Baptist Chapel, Leicester	
315	St Matthew's, Brixton	

Illustrations

316 St Thomas's, Leeds
317 Milner Square, Islington
318 Grand Hotel, Scarborough
319 Nos 129–169 Victoria Street, Westminster
320 Lodging houses, Saltaire
321 No. 107 Tulse Hill, Brixton
322 Bristol General Hospital
323 London & Westminster Bank, Lothbury
324 Halifax Town Hall
325 London warehouses
326 Stanley Dock, Liverpool
327 Watts's warehouse, Manchester
328 Egyptian Halls, Glasgow
329 Portland Street warehouse, Manchester
330 Scott: 'Iron and Coal'
331 No. 104 Stokes Croft, Bristol
332 St Martin's Northern Schools, Westminster
333 St Pancras station
334 Christ Church, Brixton
335 Tram transformer station, Islington
336 Design for County Hall, London
337 Thiepval Memorial

Between pages 464 and 465

338 Cruikshank: 'London going out of Town . . .'
339 Seymour: 'Heaven and Earth'
340 Egley: 'Omnibus Life in London'
341 Wyld: 'View of Manchester'
342 Frith: 'The Railway Station'
343 Hicks: 'The General Post Office . . .'
344 Houghton: 'Holborn in 1861'
345 Madox Brown: 'Work'
346 Crowe: 'The Dinner Hour: Wigan'
347 Doré: 'Dudley Street, Seven Dials'
348 Gavarni: 'Le Gin'
349 Solomon: 'I am Starving'
350 Fildes: 'Applicants . . . to a Casual Ward'
351 Holl: 'Deserted—A Foundling'
352 Holman Hunt: 'London Bridge . . .'
353 Watts: 'Found Drowned'
354 Shotter Boys: 'St Dunstan's, Fleet Street'
355 Bourne: 'Early Stages of the Excavation towards Euston'
356 Turner: 'Newcastle on Tyne'

Illustrations

    357   Turner: 'Burning of the Houses of Parliament'
    358   Whistler: 'Nocturne in Blue and Silver . . .'
    359   Grimshaw: 'Hull'
    360   Greaves: 'Chelsea Regatta'
    361   Sickert: 'The Old Bedford'

Between pages 560 and 561

    362   Prospect of Manchester
    363   London in 1842
    364   Mosley Street, Manchester
    365   The Fleet Street sewer
    366   Billingsgate Fish Market
    367   'Down Whitechapel Way'
    368   Dram-drinker
    369   Lord Mayor's procession
    370   'Familiarity breeds contempt'
371–2   Social Science Congress, Sheffield
    373   Opening of the Thames Embankment
    374   The Great Social Evil
    375   Workman's Home, Leather-lane
    376   The genuine West-End article
    377   'The Nearest Way Home'
    378   'Culture for the Million'
    379   'Substance and Shadow'
    380   'The Real Street Obstructions'
    381   Demonstration in Hyde Park
    382   Night in the fog
    383   Street encounter

Between pages 572 and 573

    384   'The Homeless Poor'
    385   'Houseless and Hungry'
    386   'Distress in London'
    387   Supper to Homeless Boys
    388   'The Young Ravens'
    389   Lincoln-court tenement
    390   Rent Day
    391   House of Correction, Coldbath-Fields
    392   Men's Casual Ward, West London
    393   Attic occupied by family of ten
    394   Dwellings of the Poor, Bethnal-Green
    395   'The Real Sufferers by the Money Pressure'
396–400   Taking the census

Illustrations

401	'Poor Law divorce'
402	'Poor Law exercise'
403	'Poor Law imprisonment'
404	'Transportation of the casual poor'
405	'Condition of the poor'

Between pages 640 and 641

406	'Court for King Cholera'
407	'Father Thames Introducing his Offspring'
408	'The Great Lozenge-Maker'

Between pages 672 and 673

409	'Sanitary and Insanitary Matters'
410	'The Water that John Drinks'
411	Edwin Chadwick
412	'The Alderman and the Apothecary'
413	Intramural interment
414	Burial space
415	Sanitary condition of Bethnal Green
416	'Picturesque Sketches of London'
417	'London with a Clean Front On'
418	'Cholera Prevented!'
419	Plan of Sewerage
420	Subway for sewage, gas, and water supply
421–2	House drainage
423	Doulton's best valve closet
424	Main drainage works, Crossness
425	Holborn Viaduct subway
426	Blackfriars Bridge Approach

Between pages 800 and 801

427–30	Belfast riots, 1864
431–2	Mob violence, 1886
433	Shankhill Road Barracks
434	Orange Arch, Belfast

Maps

Volume One

I	Licensed premises, Bethnal Green, 1899	164
II	Licensed premises, Strand, 1899	165
III	Licensed premises, dockland, 1899	166
IV	Licensed premises, Bloomsbury, 1899	167
V	Drink map of Oxford, 1883	177
VI	London theaters and music-halls, 1875–1901	214
VII	London terminal stations	282
VIII	Sheffield railway connections	286
IX	Nottingham railways within a five-mile radius	290
X	Leicester railways within a five-mile radius	291
XI	Eton College Estate, Hampstead	336
XII–XIII	Population densities: London and Paris	418

Volume Two

XIV–XV	Townships and wards of Leeds	767
XVI–XVII	The development of Belfast	791
XVIII	Distribution of Roman Catholics in Belfast	794
XIX–XX	Religious denominations in Birmingham, 1815 and 1865	824–5

See also Illustrations 192–3, 229–35, 281–7, 415

Preface

The idea for this book was planted, though we did not realize it at the time, in a conversation we had in London in the summer of 1965. We were looking for a theme for a symposium which the editors of *Victorian Studies* planned to hold at Indiana University (the home of the journal since its inception in 1957), under the auspices of the University and the American Council of Learned Societies.

The occasion demanded a theme which was capable of being handled in different ways by scholars from as wide a range of disciplines as possible. The choice eventually and, as it now seems, naturally fell on the Victorian City. Urban history, of course, was already beginning to make its mark, and the growth of cities in the nineteenth century was understandably one of its chief preoccupations. The ways in which the circumstances of urban life had influenced, or were influenced by, the ideas, the values, and the creative expressions of men and women living through that experience were being studied perhaps less explicitly but with increasing curiosity. It struck us that it might be particularly interesting to hear what some of the academic disciplines engaged on this common ground had to say to each other. Here, it seemed, was the most promising topic we could find to provide a focus for the immense range and variety of Victorian experience. Not only so, however, for the global process of urbanization we could see going on around us first gathered momentum in Victorian Britain. To study the inhabitants of Victorian Britain as city-dwellers was therefore to come upon them in their most telling role, the prototype of modern urban man. We could not conceivably have found a theme to touch the sinew of our respective subjects more directly. This book is an outcome of the dialogue that ensued.

The thirty or so scholars who faced each other in Bloomington in March 1967 mined no golden interdisciplinary nuggets and probably no one expected that they would. Indeed, it had become clear before they dispersed that a great synthesis of disciplines was as undesirable as it was unattainable. Yet we could not shrug off the fact that any man's experience of life—or any community's either—transgresses all conventional academic boundaries: that the various professional approaches to the study of man in his time are all more or less artificial as compared with the totality of his experience. However difficult it may be, it is vital to the ultimate purpose of the social sciences or the humanities that people should be recognized as having *lived* in the round. What was undeniably valuable, therefore, in our discussions was the pooling of specialized knowledge, the demonstration of unfamiliar ways into familiar problems, the atmosphere of freedom from arbitrary academic constraints. These things helped to put old thoughts into new contexts, brought new thoughts out of old settings, generated unexpected insights. It is our hope that we have not lost the momentum of those exciting exchanges in the even more ambitious programme of papers we have gathered into this book. For the underlying purpose remains—not only to encircle our subject more completely than any one of us might do alone but also to blaze trails of new ideas for each other. As far as we know, the combined pursuit of such a theme as ours by a team of scholars so large and multifarious is something new in itself.

Of the papers included in this book, not one is an untouched relic of the Bloomington conference, though four—those by Banks, Best, Briggs, and Stange—are more or less developed versions of papers first presented there and subsequently printed in two special numbers of *Victorian Studies*; four more—those by Dyos and Reeder, Keating, Himmelfarb, and Marcus—contain some material that has already appeared elsewhere. The remaining twenty-nine papers have not been published in any form before and have been written specially for this book. Sixteen of the present contributors, in fact, took part in the Bloomington symposium though the majority of them did so as discussants or chairmen of the discussions. Every contributor to this book, without exception, was invited to write on a specific topic and to keep in view the scope of the other papers and of the book as a whole. This entailed a very substantial amount of redrafting as the book took shape. In these respects it represents in its finished form a co-operative effort of an unusual order and, in a very real sense, a continuing discussion of its theme.

Exacting though some of this has been, we cannot pretend to have beaten all the bounds of our subject, nor should we like to believe that we have said the last word on any of its aspects. The study of the past cannot divest itself of the present, and the questions forced from us now about the past are unlikely to serve for an indefinite future. Despite the size of this book, we think of what we have tried to do, not (as the Victorians themselves might have done) as an encyclopedia to be consulted for verified fact, but as a connected series of essays, some weightily empirical but others deliberately speculative, all of them brief incursions into territory which they could not hope to subdue at a swoop. Every chapter has had to be pruned, some of them severely, and many topics which we would have liked to include have had to be left out. A full bibliography of the subject as treated would require a volume to itself, and we hope that we may later be able to attend to that, but the notes to each chapter are generally extensive and are a guide to the larger literature.

It may not be amiss to add that the book has been designed to be read straight through. There is hardly any cross-reference between chapters and no editorial superscriptions to them. We hope that the logic of its own structure will make the book's connections clear. One small but important point of presentation will be noticed in the spelling and other literary conventions. We have deliberately retained both United Kingdom and North American usages in the belief, formed by our own experience, that to force either into the mould—or mold—of the other does damage to what is being said. This having been such a transatlantic experience in the writing and editing, it seems to us entirely apt that the reader also might be alerted to its extent.

We should explain here that the main title to this book is purely descriptive and not generic. We are concerned with the city in the Victorian period but are not concerned to isolate it as a type in a whole declension of urban forms and, more particularly, to differentiate it from what might be inferred from it, namely the Georgian city or the Edwardian city. One reason for this is that much of the physical fabric and many of the underlying attitudes governing the way in which urban life was shaped in the first half or more of the period were an unexpended legacy, not only from the most recent, but from the remoter past. Nor was the vast investment of capital and creative energies in the Victorian city suddenly liquidated when Victoria died. Indeed, there are some respects in which a great many cities of the present day, wherever they may be, can fairly be described as Victorian without decrying them—a point to which we shall return at the end of the book. It would, therefore, seem unduly pedantic to be too precise about terminal dates, and this we have tried to avoid. It would be equally fatuous for us to contend for the distinctiveness of the Victorian experience without extending the comparison in a number of other directions, though we are inclined to think that some of the most lasting impressions of the cities dealt with in this book will be formed when the next logical step is taken of making more intercultural comparisons. For the time being we have had to be content for the most part with inter-city comparisons within the United Kingdom.

It makes sense to us, too, not to define 'city' with episcopal nicety. The English legal authorities of the seventeenth century maintained that a city was not such unless it was an incorporated town or borough which was or had been the see of a bishop. Whenever a borough was elevated to an episcopal see before that time, it generally became known as a city, though there were boroughs, like Sherborne and Dorchester, which had once had bishops but never became cities; there was also at least one place, Ely, which could boast a bishop but no borough charter; no doubt there were others, like Leicester, where the terms 'city' and 'borough' were mixed up promiscuously in their charters. It was, and remains, confusing. Though lexicographers went on repeating each other over these things, the first edition of the *Encyclopaedia Britannica* declared soothingly in 1771 that 'city' was used in England as little more than a synonym for 'town', 'while at the same time there is a kind of traditional feeling of dignity connected with it.' That understanding seems to have remained undisturbed for over a hundred years. When any 'mere village'—as Manchester was to Defoe—grew sufficiently in economic strength it aspired, especially after 1835, to be chartered as a borough, as Manchester itself was in 1838. By comparison London, encumbered in this respect by an ancient corporation presiding over the affairs of a mere particle of London proper, though popularly known as a whole as the largest city in the world, could never usurp the style of the City of London, and it eventually became a county instead. In the same year in which this

Preface

came about, 1889, a rather smaller industrial town, Birmingham, though not an episcopal see either, acquired the style of 'city' by royal charter—the first to have done so in these circumstances. The concept of city we prefer to use in this book, therefore, is what might be called the common-sense one. We mean by it simply any large centre of population which was generally regarded as such at the time. We use it in preference to 'town' just because it is the larger places that occupy us. No one yet knows for certain what the necessary and sufficient distinguishing marks of an urban as distinct from an agrarian society are in the modern world. It is not even clear that these things are a matter of scale. However, we suspect that, to some extent at least, they are, and in choosing to concentrate on the largest places we hope to illuminate some of their inherent characteristics.

There is little we need say here about the basic theme of the book itself. In discussing just now what a city meant at law and to its citizens we were palpably enough dealing at once with certain images and realities of the city in fact. We were referring to actual places and verifiable occurrences as well as to attitudes towards them both. We could reasonably speak of a local image of these municipal realities, and if we wished to know better what forces were actually controlling them in a particular place we would pay as much attention, not merely to what was said to have happened, nor even to what people thought about what was happening, as to what they thought about themselves or others as it did so. We can—mercifully perhaps—seldom listen to such things so closely. Almost all the evidence that can be admitted is indirect and woefully incomplete and fragile. We cannot make any ropes from such sand. But the tissue we have is generally more serviceable than is commonly supposed. The vocabulary people use, the conventions they obey or disregard, the actions they perform, the things they find funny, the beliefs they find acceptable, the circumstances they tolerate, the fantasies they project—all these leave some kind of observable residue. There are, we think, more valid documents to be read than historians sometimes like to believe.

In trying to show how to read some of them in this book we should perhaps add one word more about its arrangement. Just as we have avoided any territorial or thematic division of our editorial labour, so we have deliberately avoided polarizing the images and realities to be treated. None of the contributors deals exclusively with one category or the other. We are not in fact making any arbitrary or sustained distinction between opposing forces, so to speak. We do not see in one camp the historians of fact all brimming with realities and in another the historians of values all burnishing their images. The realities cannot sometimes be communicated without the appropriate images; they cannot sometimes be perceived because of them. Both the realities and the images are of many different orders and it becomes a matter of discovering just how much each is revealed by the other, and which is really which. That is what this book is all about.

We must here record our most heartfelt thanks to every contributor to this book. We thank them particularly for accepting our editorial criticisms, suggestions, and incessant demands, and we can but hope that in reading the book as a whole they will think their efforts worthwhile and ours justified. We are also grateful in due measure to all those friends and colleagues whose names appear among the more formal acknowledgments set out below. We thank them for their invaluable encouragement and for the opportunities, advice, and practical aid they gave us, sometimes perhaps without realizing it. Whatever weaknesses there may be in the finished work we accept unreservedly as our own; whatever strengths there are we are happy to share to the full with all who have been concerned.

Finally, we take pleasure in the opportunity this book affords of offering our salute to the man who, in his long years as Reader in Economic History at the London School of Economics, gave so freely and unselfishly of his unrivalled store of knowledge about the Victorians.

H. J. Dyos
Michael Wolff

Acknowledgments

To the Editor of *Victorian Studies* for permission to reprint, in amended or developed form, material first published in its pages and now forming part of chapters 3, 4, 15, 16, 20, and 30; to the Society of Authors, on behalf of the Bernard Shaw Estate, for the passage from *Widowers' Houses* in chapter 15; and to Penguin Books Ltd for the passage in chapter 23 from Robert Baldick's translation of J.-K. Huysmans' *A Rebours* (1884) under the title *Against Nature* (1959).

Her Majesty Queen Elizabeth II has graciously permitted us to reproduce Wyld's 'View of Manchester' (341). To each of the following for their courteous permission to reproduce paintings, drawings, or photographs in their possession, as indicated: Aberdeen University Library (56, 66, 70, 73, 74, 81, 116–18, 121–23, 128, 129, 139, 161–4, 176, 249, 255, 267, 309, 310, 313, 318); the Trustees of Sir Colin and Lady Anderson (344); the Editor, *Architectural Review* (304); Ashmolean Museum, Oxford (352); B. T. Batsford Ltd (277, 298); Mr M. Bernard (343); the Curators of the Bodleian Library (26, 40); the Trustees of the British Museum (35, 253, 338, 339, 356); Brown, Son & Ferguson Ltd (24); Country Life Ltd (337); Courtauld Institute of Art (351); Davis Galleries, New York City (349); Derby Museum & Art Gallery (250); the Editor, *Dundee Courier* (302); Edinburgh Central Public Library (293, 295); Sir Arthur Elton (355); the Provost and Fellows of Eton College (281–91); Faber & Faber Ltd (20); Ferrers Gallery, London (359); Mr David Francis (19, 23, 68, 74, 88, 90–2, 95–7, 99–101, 104–6, 115, 119, 126, 130, 134, 144, 151–3, 165, 166, 170, 172, 227, 305); Greater London Council Photograph Library (25, 67, 69, 72, 76, 135, 143, 155, 168, 300, 321, 334, 335); the Keeper of Prints and Pictures, Guildhall Library (13, 17, 18, 27, 31, 32, 37, 53, 55, 93, 112, 120, 159); Miss Caroline Herschel (40); Kodak Museum (84, 89, 102, 103, 107, 109, 113, 124, 125, 127, 132,

Acknowledgments

133, 136–8, 140, 142, 147, 150, 154, 157, 177); Liverpool Public Libraries (94, 98, 110, 114, 131, 145, 146, 149, 156, 158, 169, 171, 173, 174); the Trustees of the London Museum (36); Mr Howarth Loomes (77–80, 83, 85, 160, 185); Mr Adam H. Malcolm (294); Manchester City Art Galleries (345, 346, 360); Manchester Public Libraries (108, 111, 141, 148, 167, 177–84, 186–226, 228, 312, 327); Mansell Collection (21, 22); National Gallery of Canada, Ottawa (361); National Monuments Record (71, 256, 261, 266, 268–70, 275, 276, 278, 280, 299, 315, 333); National Trust (330); Mr James M. Oakes (301); the Director-General, Ordnance Survey (229–35); Philadelphia Museum of Art (357); Royal Commission on Ancient Monuments, Scotland (303, 328); Royal Holloway College, University of London (342, 350); Mr R. B. Sawrey-Cookson (30); Science Museum, London (245, 251); Scotsman Publications Ltd (292); Mr Leslie Shepard (43–7); Sir John Summerson (306, 316); Mr Timothy Summerson (259, 260, 279); Tate Gallery, London (340, 358); Ulster Museum (428, 429, 433, 434); Victoria & Albert Museum (48–51); the Trustees of the Watts Gallery, Compton (353); Mr Reece Winstone (331). Also, to Douglas Birch, Peter Boulton, William Burgan, G. Kitson Clark, John Clive, B. W. Coe, J. Mordaunt Crook, Ralph Davis, K. M. Dexter, Olive Dyos, Arthur Elton, Rupert Evans, Joan Farrell, Bill Forsyth, C. J. Francis, Colin Franklin, Kenneth Garfield, Terry Garfield, John Harrison, Ihab Hassan, Peter Hennock, S. W. F. Holloway, James Howgego, May Katzen, John Kellett, Paul Koda, Henry Kowalski, John Krause, Paul Laxton, Roy MacLeod, C. E. Makepeace, Standish Meacham, Martin Meisel, J. Hillis Miller, Mary Murray, Richard Noland, Robert O'Kell, David Orme, Ian Paterson, Morse Peckham, Thomas Pennant, Harold Perkin, Elaine Quaglini, John Henry Raleigh, Gareth Rees, Helene Roberts, Mark Roskill, Joseph Rykwert, Jerome Schneewind, Tania Senff, David Spring, Gareth Stedman Jones, Hazel Taylor, Iain Taylor, Edward Thompson, John Tobias, Alan Trachtenberg, Alexander Welsh, Tappan Wilder, Sara Wolff, Calvin Woodard, and Paul Zietlow.

I Past and Present

1 The Urbanizing World

Eric E. Lampard

Each day newspaper readers, radio listeners, and television viewers around the world are confronted with the torrent of events. If the headlines of 1973 were merely updating yesterday's unresolved problems, the public might well relax. Their parents also lived with these problems and the world is still there. Besides, the mass media provide many distractions, and the commercialization of anxiety is a notorious vice of our time. Only the increasing reference to population explosions, urban implosions, and an alleged crisis in human ecology, perhaps, need give rise to the uneasy feeling that, beneath the familiar epiphenomena of tumult and wars, the old frameworks of interpretation and understanding are no longer sufficient.

Certainly many parts of the Victorian framework are no longer there. Britain's queen is not an empress and some two dozen sovereignties have risen in place of her empire. The specter of communism that haunted the old Europe of Victoria's innumerable kin has, in the present century, conjured up a global counter-spook more potent and resourceful than the holy alliance against which Marx and Engels inveighed. The concert of great powers which, in Victoria's twilight years, still held a world in fee has given way before a nuclear deadlock of super-powers around which the proliferating nations of the earth must form as satellites or supplicants.

We have entered upon a new era of world history which, however obscure and shapeless it may be at present, will nevertheless come to be as conventional in the future as modern, medieval, and ancient history became to historians of Europe in Victorian times. That is the major premise of 'contemporary history,' which has been defined as beginning when the problems which are actual in the world today first took visible shape. The promise of contemporary history is to construct a framework within which the world can be interpreted and understood as a *unit*. One of the major tasks confronting contemporary historians is, therefore, 'to clarify the basic structural changes' which have been shaping the world scene and which, notwithstanding a certain overlap with the more recent past, will eventually give our era a definitive character and identity of its own.[1]

This essay is such an exploration in contemporary history. Its theme is the Victorian city in an urbanizing world. Its object is to review the urban climax of late-Victorian times from the global vantage point of the 1970s and delineate, however tentatively, those of its societal processes which have been pressing the world forward into the mold of the twentieth Christian century.

The First Urban Transformation

We look first at the urban transformation of the nineteenth century. During the first half of that century Victorian Britain became the world's first urbanized society. In 1801 one-fifth of the population lived in cities and towns with 10,000 or more inhabitants; one in every twelve persons was a Londoner. By the year of the Great Exhibition, the proportion of the greatly enlarged population resident in such places was 38 percent, and London's 2,362,000 inhabitants now represented a slightly smaller share of the national population. For the first time, the census in 1851 reported an aggregate 'urban' population which exceeded the 'rural' population in size, albeit by less than one percentage point. At the beginning of the century only the Dutch Netherlands had surpassed Britain in the proportion of her population resident in towns with 10,000 or more inhabitants, and the Dutch *gemeenten* of that size included a considerable segment of rural population. Elsewhere, with the exception of the nine provinces that would shortly unite as Belgium, fewer than one person in ten could be found residing in concentrations of even this modest size.[2]

By mid-century the Belgian level of concentration had risen to one person in every five, approximating the level of Great Britain in 1801. But in the Netherlands, meanwhile, the rural segment had grown faster than the town segment and, if anything, the Dutch population was slightly less 'urban' in 1849 than it had been at the end of the eighteenth century. Elsewhere, only France, Saxony, Prussia, and the United States could yet claim more than one person in ten residing in concentrations of 10,000 and over. Thus the early-Victorians who, following Robert Vaughan, had designated their era 'an age of great cities' had somewhat missed the point. Great cities there were, larger perhaps and more numerous than ever before, but in Great Britain this phenomenon was only an outward symptom of a more profound tendency: the urbanization of society. In the rapidity and extent of the urbanization of her growing population, Great Britain was without peer (see Table 1.1). If the distinctive achievement of Britain in the latter part of the eighteenth century was to inaugurate the Industrial Revolution, her no less remarkable feat in the first half of the nineteenth century was to accomplish the first urban transformation.[3]

Table 1.1 *Annual rates of urban concentration, cities 10,000 and over, selected countries, 1800–50 and 1850–90*

Country	1800–50	1850–90
England and Wales	0·36	0·56 (1)*
Scotland	0·30	0·44 (6)
Belgium	0·18	0·32 (8)
U.S.A.	0·16	0·39 (7)
Saxony	0·14	0·52 (2)
Canada	0·11c	0·22 (13)
Switzerland	0·11	0·24 (11)
Prussia	0·10	0·47 (4)
France	0·10	0·29 (9)
Hungary	0·07	0·21 (14)
Bavaria[a]	0·06	0·24 (12)
Ireland	0·05	0·20 (16)
Norway	0·05	0·20 (16)
Russian Empire (Europe)	0·04	0·14 (18)
Austria	0·03	0·21 (14)
Portugal	0·00	−0·01 (19)
Australasia	—	0·46 (5)
Argentina[b]	—	0·48 (3)
Chile	—	0·26 (10)

N.B.: Annual rate of concentration is percentage point shift in level of urbanization over intercensal years nearest terminal dates, $\delta(u/p)$.

* Figures in parentheses, rank order, 1850–90.

[a]Cities ≥ 20,000.   [b]Cities ≥ 2,000.   [c]1820–50.

For dates of census or estimates: A. F. Weber, *The Growth of Cities in the Nineteenth Century* (New York, 1899), Table cxii.

It is significant that, despite the antiquity of its Latin root, the word *urbanize* was scarcely used in English before the late nineteenth century. People commonly spoke of 'the growth of towns,' the life of townsmen was 'urban life,' but when the verb 'to urbanize' was employed it generally had the connotation 'to render urbane': courteous, refined in manner, elegant or suave. To urbanize was almost synonymous with civilize: to bring out of barbarism, to enlighten, and it is not surprising that as the word took on the more literal meaning, 'to render urban,' the lexicographers spoke simply of 'removing the rural character' of a place or population.

The reluctance of lexicographers to commit themselves to a more positive definition is understandable. No less an enthusiast for the factory system than William Cooke Taylor had suggested in 1840 that the northern industrial towns revealed 'a system of social life constructed on a wholly new principle, a principle yet vague and indefinite but developing itself by its own spontaneous force and daily producing effects which no human foresight had anticipated.' Two years later he conceded that a stranger passing through the towns 'cannot contemplate these crowded hives without feelings of anxiety and apprehension ... The population is hourly increasing in breadth and strength. It is an aggregate of masses.' In 1850 most of the manufacturing and mining aggregates were still small 'walking towns' made up of mills, chimneys, pitheads, and tips, narrow cobbled or dirt streets with grimy row (terraced) houses running out over hill and dale into the countryside. In the larger places the social landscape was already more varied, with wholesale and retail areas, warehouses, parks, church steeples, and public buildings, with cuttings and viaducts carrying canals or railways into the congested factory districts. The horse-drawn omnibus had appeared on improved roads, and, in order to escape the soot, odors, and noise, the proprietary classes were moving to the edge of town. For the masses the countryside lay further away.[4]

During the third quarter of the century, when the census classified more than half the population as urban, the loss of rural character could no longer even imply a passage to urbanity. Work, work places, and the working day were increasingly differentiated from the inherited routines of domestic life and livelihood. Provincial accents and manners still combined with local materials to give an outwardly variegated texture to a deepening uniformity of the industrial urban fabric. A growing segment of the town population was thought to be dangerously uncivilized and the prospect of 'rural depopulation' a social disaster. 'Civilized man,' de Tocqueville said, 'is turned back almost into a savage.' The social philosopher J. S. Mackenzie was uttering a commonplace by the 1890s in stigmatizing the growth of large cities as 'perhaps the greatest of all the problems of modern civilization.' Amidst a continuing celebration of material progress and the enlargement of individual opportunity and experience, some Victorian critics remembered the 'marks of weakness, marks of woe' which Blake had seen on the faces of eighteenth-century Londoners, or echoed Rilke's contemporary judgment on 'the cities' guilt.'[5]

There is no need to rehearse what Asa Briggs has termed 'the international debate' about cities. 'If it has to be a choice between Beverley and Jarrow,' declared J. B. Priestley in 1933, 'write me down a medievalist.' But neither pessimists nor optimists could win the debate, for the issue was never put to the question in either the nineteenth or the twentieth centuries. The population, meanwhile, was voting with its feet. It was rendering itself urban. By the year of Victoria's death, three-quarters of the population of Great Britain was classified by the census as urban. Whether the people were really better off in the grim industrial towns of the Midlands and the North, Priestley still could not decide in 1933 but 'they [had] all rushed into the towns and mills as soon as they could ... which suggests that the dear old quaint England they were escaping from could not have been very satisfying.' What was it about Merrie England, he queried, 'that kept the numbers down?'

The style of urban life in Britain was changing rapidly by 1933. Two years earlier the census of England and Wales had reported 79·9 percent of the population living in urban areas with 2,000 or more inhabitants. Of the twenty-five places with 20,000 or more residents which had increased their populations by more than 30 percent between 1921 and 1931, nineteen were situated in the Home Counties. With one exception, the others were city suburbs or seaside resorts. This fringe development had, in fact, become marked since the census of 1881, especially around London, and had only been temporarily arrested by the First World War. This was a Britain not of Garden Cities but of arterial roads and petrol pumps, of wireless sets and Woolworths, of suburban council estates and semi-detached bungalows, of new factories and cinemas that looked more like exhibition buildings. It was a style, Priestley thought, belonging more to the age itself than to 'this particular island ... America, I supposed, was its real birthplace.' In the course of his *English Journey*, Priestley had found three townscapes: the Old, the Nineteenth-Century, and the New, all coexisting, 'variously and most fascinatingly mingled in every part of the country.'[6] The urban transformation of Britain had gathered momentum in the latter part of the nineteenth century. By means of restrictive legislation, the reform of public administration, and prodigious feats of engineering, the late-Victorians had striven, however tardily and ineffectually, to make urban life livable. For better and for worse the economic system had done the rest. In the early 1930s Priestley had found his more numerous townspeople better off materially, even when living on the dole, but lacking a little of the spontaneity he could remember from his Bradford childhood at the turn of the century.

Only in the last quarter of the nineteenth century could people anywhere begin to get the urban transformation in perspective. Amidst the welter of blue books, white papers, census volumes, social surveys, tracts, and dissertations that poured from the printing-offices on one aspect or another of the Social Question, a new kind of 'scientific monograph' appeared, dealing with the subject of urban and rural populations and the relationships between them. The focus of the new studies was not so much the growth of great cities and their problems as population growth and redistribution and the related processes of concentration and agglomeration. By 1899 Adna F. Weber concluded that 'the concentration of population in cities' was now commonly recognized as 'the most remarkable social phenomenon of the present century.' For him also it was not of one particular country but of the age. Whereas in 1790 the United States, a virgin country with undeveloped resources, had little more than 3 percent of its 3·9 million inhabitants living in cities, Australia in 1891, still an undeveloped country, had more than 33 percent of its 3·8 million inhabitants doing so. 'But Australia is of the nineteenth century; and that is the vital fact which explains the difference in the distribution of population ... What is true of Australia,' Weber affirmed, 'is, in a greater or lesser degree, true of the other countries in the civilized world. The tendency towards concentration ... is all but universal in the Western world.'[7]

At no point in his monumental study of the growth of cities did Weber use the term 'urbanization.' He wrote instead of 'the tendency towards concentration,' a measure of urbanization in international and historical comparison which has not been improved upon to this day. He expressed the tendency simply as the share of a country's population resident in cities of a given size, usually 10,000 or more inhabitants. In some cases he was able to push his urban threshold down to 5,000 and even 2,000, but most demographers would nowadays agree that a minimum size of 20,000 for the 'agglomerated population' is the only one that can be used with any prudence in international studies. It is impossible, of course, to compare the variety of classifications of 'urban' population employed in national censuses, owing to the diversity of conditions of settlement and administration that prevail throughout the world at any time. By most national criteria, the size threshold of 20,000 is somewhat high and probably represents divergent combinations and degrees of urban characteristics in different countries.[8] Equally, not all population in places with less than 20,000 inhabitants can be sensibly classified as 'rural.' For many comparative purposes the size limit of 100,000 is quite satisfactory, but in most nineteenth-century situations, as well as in developing countries today, that threshold is unduly high.

If we adopt the measure of 20,000 and more for the agglomerated population *c*. 1890 and compare the level of concentration prevailing in the fifty-nine countries for which Weber furnishes data with that achieved in England and Wales in 1801—16·9 percent—then the universal tendency remarked by Weber at the end of the century does not appear to have gone very far—even in 'the Western world.' Hardly more than a dozen countries had yet surpassed the level attained by England and Wales at the beginning of the century. Of fifteen countries, outside England and Wales, with more than 16·9 percent of their population urbanized, all, with the exceptions of the Netherlands and Scotland, had achieved that level since midcentury. By 1891, Scotland with 42·4 and Australia (six colonies) with 40·9 percent of their populations urbanized were the only countries to have surpassed the level of England and Wales in 1851—35·0 percent. Of the sixteen countries which exceeded the 16·9 percent level *c*. 1890, ten were located in Europe, three in Latin America, two in Australasia, and one in North America. Only six of them could yet be characterized as 'industrializing societies' in which a rapidly expanding share of their labor forces had been committed to full-time employment in manufacturing, mining, and construction industries. Among the other ten countries, six had comparatively small populations resident within a confined national territory, while two had comparatively small populations in very extensive national territories. In several cases the aggregate level of concentration reflected the numerical predominance of one large 'primate' city rather than a pyramid of different-sized centers extending throughout the entire size-distribution. Thus, apart from Great Britain itself, what was true of Australia in 1891 was true of other 'civilized' countries, only to a much lesser degree.

At the close of the Victorian era it was still possible for Weber to carry out a revealing analysis of urban growth within the successive size-classes: 2,000–10,000; 10,000–20,000; 20,000–100,000; and 100,000 plus. Almost everywhere the largest cities were growing more rapidly than the smaller but, for all the talk about 'great cities', only Britain and the United States yet appeared to have been developing a class of cities over 500,000. There were about twenty such agglomerations in the early nineties—four in Great Britain; two in European Russia; five in the rest of

Europe (including Constantinople); three in the United States; two in East Asia, one in South Asia; and two in Latin America (see Table 1.2). Peking was almost certainly in this class, and, by 1900, Budapest, Warsaw, Madrid, Birmingham, Greater Brussels, Melbourne, Sydney, Madras, and possibly Shanghai were as well.[9]

Table 1.2 *World's largest agglomerations, 1890, 1920, 1960*
(000s)

	1890		1920		1960	
1	London	4,212	New York-N.J.	8,047	New York-N.J.	14,163
2	New York-N.J.	2,741	London	7,236	Tokyo-Yokohama	13,534
3	Paris	2,448	Paris	4,965	London	8,190
4	Berlin	1,579	Tokyo-Yokohama	4,168	Shanghai	7,500
5	Tokyo-Yokohama	1,390	Berlin	4,025	Paris	7,140
6	Vienna	1,342	Ruhrgebiet	3,730	Buenos Aires	6,775
7	Chicago	1,100	Chicago	3,315	Los Angeles	6,568
8	Philadelphia	1,047	Manchester	2,306	Moscow	6,150
9	St Petersburg	1,003	Philadelphia	2,302	Chicago	5,998
10	Constantinople	874	Buenos Aires	2,275	Calcutta	5,810
11	Moscow	822	Boston	2,210	Osaka	5,158
12	Bombay	822	Osaka	1,889	Ruhrgebiet	4,960
13	Rio de Janeiro	800	Vienna	1,845	Mexico City	4,825
14	Osaka	800	Calcutta	1,820	Rio de Janeiro	4,700
15	Calcutta	741	Shanghai	1,700	São Paulo	4,375
16	Hamburg-Altona	712	Birmingham	1,694	Bombay	4,040
17	Manchester	703	Glasgow	1,630	Philadelphia	3,655
18	Buenos Aires	677	Hamburg-Altona	1,545	Detroit	3,560
19	Glasgow	658	Leeds	1,445	Peking	3,500
20	Liverpool	518	Rio de Janeiro	1,325	Leningrad	3,400
21	Budapest	492	Bombay	1,275	Cairo	3,320
22	Melbourne	491	Pittsburgh	1,261	Berlin	3,274
23	Warsaw	485	Budapest	1,225	Djakarta	2,850
24	Birmingham	478	Liverpool	1,201	Tientsin	2,750
25	Madrid	470	Detroit	1,119	Boston	2,700
26	Brussels	465	Brussels	1,070	Hongkong	2,614
27	Naples	463	Istanbul	1,000	S. Francisco Bay	2,442
28	Madras	452	St Louis	978	Manchester	2,427
29	Boston	448	Moscow	950	Seoul	2,400
30	Baltimore	434	Cleveland	924	Birmingham	2,300

Source: Tokyo-Yokohama and Osaka populations based on populations resident 1890s reported in *Encyclopaedia Britannica*, 11th edn rather than on place of *legal* residence, *honseki*: see I. B. Taeuber, 'Urbanization and Population Change in the Development of Modern Japan,' *Economic Development and Cultural Change*, ix (1960), Part 1, 1–28. Other 1890 figures: A. F. Weber, *The Growth of Cities*, Table clxiii. 1920 and 1960: U.N. Dept of Economic and Social Affairs, 'Growth of the World's Urban and Rural Population, 1920–2000,' unpublished, corrected to 9 Dec. 1968.

These international comparisons suggest the singularity of the Victorian urban achievement. While Saxony, Prussia, Australasia (seven colonies), and perhaps Argentina and the United States, now approached Britain's rate of concentration around 1890, their degree of concentration still lagged. Apart from Australia, in so many respects a *rara avis*, no country yet had more than 30 percent of its population urbanized. Great Britain, with four or five of the world's largest conurbations, with more than half her population resident in cities of 20,000 or more, and with almost three-quarters of the population classified as urban, presented a situation at the end of Victoria's reign as 'unique' as that perceived by Cooke Taylor and others at its beginning. Britain had made its painful adjustment to the first urban transformation. This alteration profoundly affected the quantity and quality not only of town life and livelihood but also of rural life and livelihood. It is understandable that, amidst the pomp and circumstance of the Jubilees, the predominant mood was self-congratulatory and that, at the time of the Queen's death, the nation's genuine mourning was tinged with self-satisfaction.

The Demographic Transition in the Cities

The second peculiar achievement of Victorian Britain was to be the first to make what may be called the urban demographic transition from high to low birth- and death-rates at a comparatively late stage in the urban transformation we have been considering. This transition was completed between about 1890 and 1910. From this time the growth of the total population became largely dependent on the natural increase of the urbanized population itself.

During Victoria's reign the population of Great Britain had doubled. In 1901 the censuses reported almost 77 percent of thirty-six million inhabitants resident in urban areas. Notwithstanding the phenomenal growth of the towns, Britain's crude death-rate, which peaked around 24–25 per 1,000 in 1849, had fallen from 22–23 per 1,000 in the later sixties to about 17–18 per 1,000 at the close of the century. Unfortunately this change had not yet affected the very young. The numbers of deaths under one year per 1,000 live births was as high again in the late 1890s, 155–160 per 1,000, as it had been in the late sixties. Meanwhile the crude birth-rate, which had risen from about 30 per 1,000 in the late 1830s to 36 per 1,000 in the late sixties, had fallen steadily thereafter to less than 29 at the time of the Queen's death. In so far as these improvements had persisted over the two decades of highest incremental urbanization, 1870–90 (see Table 1.3), they reflected not only the advances in social medicine and vast expenditures on water systems and sanitation, but also very real gains in the material level of living brought by industrialization.[10] Average real wages in England and Wales, adjusted for unemployment, had fluctuated upwards over the second half of the century and by 1901 the index stood at more than eighty points above the level for 1850. By no means all of the increase can be attributed to price changes. Average retail prices had followed much the same pattern as crude

Table 1.3 *Population concentration and industrialization of labor force and product, England & Wales and the United States, by decades, 1800–1900*

Date	Urban shift	England & Wales				United States		
		Labor force shift		National income shift		Urban shift	Labor force shift		Commodity value-added shift
		N-A	MMC	N-A	MMC		N-A	MMC	N-A
				(percentage point shifts)					
	(1)	(2)	(3)	(4)	(5)	(6)	(7)	(8)	(9)
1800–10	1·2	2·9	0·5	−3·2	−2·6	0·9	−1·2	—	—
1810–20	1·6	4·6	8·2	9·6	11·1	0·0	5·0	—	—
1820–30	2·6	3·8	2·4	2·7	2·5	1·8	8·2	—	—
1830–40	2·5	2·4	−0·3	1·3	0·0	1·8	7·5	—	—
1840–50	3·3	0·5	2·4	1·8	−0·1	4·0	8·3	6·3	12·2
1850–60	2·8	3·0	0·7	2·5	2·2	3·6	1·9	−0·8	4·2
1860–70	2·1	3·6	−0·5	3·6	1·6	4·8	0·4	6·5	2·9
1870–80	5·9	2·5	0·4	3·8	−0·5	1·8	1·2	−0·8	3·6
1880–90	8·7	2·1	0·4	1·8	0·8	6·3	8·6	1·5	11·6
1890–1900	1·7	1·8	2·4	2·2	1·8	3·9	2·5	1·0	4·1

N.B.: Urban shift is intercensal δ(u/p) in percentage point terms. E & W concentrations ≥ 20,000; U.S. concentrations ≥ 8,000.
N-A labor force is non-agricultural; MMC is the manufacturing, mining, and construction segment of the N-A labor force. The same notation applies to national income in the case of England & Wales and to commodity value-added in the U.S.A. N-A labor force shift is intercensal δ(N-AL/LF) in percentage point terms and similarly for national income and commodity value-added.
Source: E. E. Lampard, 'Historical Contours of Contemporary Urban Society,' *Journal of Contemporary History*, iv (1969), Table II.

birth-rates, rising until about 1873 and then falling steadily to 1901 when the index stood fully ten points below the mid-century level. Over the entire length of Victoria's reign, gross national product had roughly quadrupled in size. Net national income *per capita*, adjusted for price changes, had more than doubled, rising from £18 in 1855 to £42 in 1901. These figures reveal little about changes in the distribution of income among different classes in the population. The numbers of 'the poor' may well have increased in this period, but that they ever formed a growing share of the labor force, except in periods of severe economic depression, seems unlikely.

What underlay these long-run achievements was, of course, the unfolding of the Industrial Revolution. While population multiplied in the countryside, it was in the manufacturing towns that industrial reorganization created the jobs. Industrialism soon put its stamp upon emerging urban society in every country and was itself stimulated and shaped by the related concentration of population. The coincidence of high rates of population growth with still higher rates of concentration identifies the demographic processes underlying the early industrial-urban transformation, while the later transition to low vital rates among the more highly-urbanized populations marks the completion of the transformation from the economic and demographic regimes of preindustrial society.

National populations may grow from an excess of births over deaths and/or from net migration. But population increases that are not accompanied by even greater *relative* increases in concentration cannot achieve higher levels of urbanization. Since there is no evidence before the late-nineteenth century that urban rates of natural increase were higher than rural rates, in Great Britain or elsewhere, the rapid and sustained rates of concentration experienced throughout much of that century could have resulted only from (1) a rural-to-urban shift brought about by net migration.[11] Unless concentration is arrested, the transformation enters another critical phase (2) when the magnitude of the migration comes to exceed that of the rural natural increase. At this juncture the *absolute* size of the rural population begins to fall and, barring immigration from abroad, the pool of potential migrants to the cities also declines. Thereafter the rate of concentration appears to increase and the rate of national population growth tends to slacken unless and until (3) the urban segment begins to reproduce itself at more than replacement levels. There is thus an interval, of longer or shorter duration in different countries, when the process of concentration portends 'race suicide.' Only when urban mortality falls, and more especially the mortality of infants, do the industrial cities cease to be 'devourers of population' and does their own excess of births assure the growth of total population.

The growth and redistribution of population between town and country will meanwhile have involved great changes in the relative numbers of children, adults, and old people in each place, as well as alterations in the demographic contributions of the different socio-economic classes. The lower, largely newer, strata in the urban population now provide the greater part of the increment, since the middle strata tend to have smaller families as their level of living improves. Thus, while the urban demographic transition allows the survival of 'the race,' it threatens a new regime in which the city proletariat outbreeds the higher and more economically productive strata. As the socio-economic and spatial structures of the urbanized population are further differentiated during (4) the metropolitan phase of the transformation, the fertility of the entire population subsides as people from the lower orders 'rise' to occupy either the new niches created by economic change or older ones vacated by the less fecund middling strata. This final turn to low fertility among *all* classes generally offsets the incremental effect of low mortality and removes the 'menace' to society from proletarian fecundity. The industrial-urban order thus achieves the low vital rates and structural stability that characterizes the high *per capita* income countries of the mid-twentieth century.

The Changing Balance of Migration and Natural Increase

Migration was the major component of the urban increase in Victorian Britain even after some of the towns had begun to show a slight excess of births over deaths. Assuming that all births and deaths occurring in the London area are attributable to London population, it would seem that the deficits of the eighteenth century were turned into a positive increase quite early in the nineteenth century. One set of figures based on pre-registration data suggests that the ratio of net migration to total London increase during the first decade was less than one quarter. Migration continued to grow in absolute terms and probably reached its maximum in 1841-51, when it formed 40·2 percent of the increment, but it fell off again during the sixties and, in 1881-91, finally turned negative with the 'export' of population to suburbs outside the registration district. Except for this suburban trend of the eighties, the other large towns in England and Wales followed London's demographic tendencies, although their ratio of net migration to total increase was higher. Between 1821 and 1871 that ratio was always above a third and it peaked during the forties at almost half.[12]

The diversity of experience among the larger towns is well illustrated, however, by the following examples which show the proportion that in-migration contributed to total increase per thousand in each instance over the decade 1881-91: Manchester 717, Belfast 654, Leeds 300, Birmingham 257, and Edinburgh 94. At this point in the century, too, some of the large towns were beginning to export population beyond their boundaries, while increasing in size. The following examples show the proportion which out-migration bore to total increase per thousand during the eighties: Sheffield 4, Dublin 120, Bristol 785, and London 1,289. Liverpool, with 2,481, actually experienced an absolute decline over the decade 1881-91, although, if the areas which were annexed before 1901 are included in the totals for these census years, the population of Liverpool had grown—by less than one percentage point.[13]

Thus, by the closing decades of the century, Victorian society was remarkable not only in the degree of concentration but also in the fecundity of its urban population. In no other European country for which data are available does the urban population come so near to reproducing itself. Of course, large cities were growing rapidly in most European countries at the time, but in few instances does the contribution of natural increase seem to have *outweighed* that of migration. Émile Levasseur cited a study reported in 1877 which showed that migration had furnished more than half the growth of twenty-three of Europe's thirty largest cities. Without this influx seven of the cities would have rapidly decreased in size, since even their natural increase was heavily dependent on the city-born children of migrant parents. Six of the twelve large cities of France were also found subject to an excess of deaths. In a survey of eighty-eight European cities, 200,000 and over, for the years 1880-1890, Richard Boeckh concluded that at least half of their population increase was attributable to net migration and the balance to a combination of natural increase and annexation. A. F. Weber summarized his evidence with the conclusion that migration had played 'the largest role in the growth of French and Italian cities, then in the German and Scandinavian, and finally in the English cities.' He might have added the Russian cities (St Petersburg, Odessa, and probably Moscow) to the French and Italian centers among the class of 'devourers of population' in Europe and, in the light of Irene Taeuber's researches, we may include the large cities of Japan before and after the Meiji restoration. The fact that Victoria's great capital had grown so conspicuously from natural increase was due, Weber thought, to 'London's precedence in the making of sanitary improvements.'[14]

There is no reason to suppose that the United States' experience differed very greatly from that of continental Europe. The phenomenal growth of such Middle and Far Western cities as Chicago, Minneapolis, Cleveland, and San Francisco, or Detroit, Los Angeles, and Seattle by the turn of the century, could only have stemmed from an influx of foreign as well as native-born migrants. Between 1865 and 1890 the population of Boston, for example, had risen by 133 percent, of which roughly half was attributable to net migration, about 20 percent to excess of births, and the residual 30 percent to political annexation. If Boston's natural increase had furnished 'a subordinate part' of its growth, as one observer reckoned, it is noteworthy that its crude birth-rate none the less was already higher than its death-rate. Taking the six-state New England region as a whole, in which the urban population ('towns' of 10,000 and over) almost equaled the population of the rural 'towns' (under 10,000) by 1890, the crude urban birth- and death-rates were 29·7 and 21·0 respectively, compared with 20·0 and 18·7 for the rural districts. The rate of urban increase was thus 8·7 per 1,000 and the rural rate only 1·3. As a consequence of both rural-to-urban and net interregional migration, this slight excess of births did not suffice to maintain the region's rural population; it had been in *absolute* decline since the decade of the Civil War. Hence the greater part of the region's urban growth was probably contributed by foreign-born migrants and their foreign- or native-born children.[15]

There is no evidence that the urban rate of natural increase nationwide equaled the rural rate; there are no grounds for believing that the urban population could yet reproduce its numbers, let alone grow, without sizeable in-flows of migrants. This was still true for Europe, although the natural increases of town and country in England and Wales were virtually equal by the late eighties. One estimate by Charles Booth gives the two segments natural increase-rates of 14·03 per 1,000 and 14·13 per 1,000 respectively. Only in Sweden, still a comparatively rural country, did the crude rates approach this degree of parity, 12·3 per 1,000 for rural population and 11·3 per 1,000 for urban over the decade 1881–90. Urban death-rates in Europe were already somewhat lower than in the United States but urban birth-rates in Europe were, in general, much lower than in the United States.[16]

The United States also differed from Europe in the relative contribution of international migration to its urban increment. No country, with the possible exception of Australia before 1890, had accepted such a large proportion of foreign-born into its population as the United States. By 1890, with the flood tide of immigration still more than a decade away, some 31·8 percent of the residents of the twenty-eight largest cities, 100,000 and over, were foreign-born. These cities contained 15·5 percent of all US population but only 12·4 percent of the native-born and 33·4 percent of the foreign-born. Canada and some of the Latin American countries also had higher proportions of foreigners living in their cities than any European country, but in both volume and proportion their levels of immigration fell far short of the United States. In Europe, Scotland and Saxony had the highest proportions of non-natives living in their cities, about 13 and 10 percent respectively, but their foreign shares include the natives of other parts of the British Isles or other German states whose presence, apart from sectarian differences, scarcely betokened cultural pluralism. Among the great European cities of 1890, Vienna, with 11 percent foreign-born, was the most cosmopolitan and almost three-quarters of her alien element were natives of the confederated state, Hungary.[17]

By the close of the Victorian era, therefore, internal migration was generally contributing less to the urban increment than it had during the third quarter of the century. Among larger, more developed European countries, only France and Italy, perhaps, lagged in this respect. In Great Britain, on the other hand, the towns came closest to reproducing themselves. In England and Wales the *absolute* decline in rural population, marked since 1871, was arrested by 1901 with the spread of population to suburbs which lay outside town boundaries. With rural population rising again, except for the war years 1914–18, and the population of cities of 20,000 and more continuing to grow, it is not surprising that, in the decades on either side of the First World War, the population of smaller towns and urban districts, 2,000–19,999, declined. In other words, population concentration was no longer a predominantly rural-to-urban movement but, over and beyond the suburban spread, a movement within the city-size distribution itself.[18]

If the growth of cities in the more urbanized countries was becoming less dependent on rural-to-urban migration by the 1890s, this does not mean that the volume of internal migration was subsiding. The most highly urbanized populations were generally still the ones with the most mobile populations. In 1890–1 England and Wales had less than 72 percent of total population residing in the county of birth (a smaller share if Scotland is included), Saxony, with only 69 percent, and Prussia, with 70 percent, dwelling in the *Ort/Kreis* of birth, were the most internally-mobile countries on the Continent and were, if anything, becoming more mobile. In ten other European countries, from 79 to 90 percent were enumerated in their native county or equivalent district and half to three-quarters of these people were actually living in their locality of birth. The United States, with 21·5 percent of its natives living outside their state of birth in 1890, probably contained the most internally mobile population in the world. Most internal movement seems to have been over relatively short distances. This was especially true for females. Only the very largest European cities exercised much international allure although, in general, the larger the city the more likely it was to attract migrants from remoter provinces. Younger males appear more frequently in the long-distance movement than females, and this difference may have reflected the influence of large industrial labor markets and, in some instances, the location of garrison towns.[19]

Sex and Age Structure of Urban Populations

By 1891 there were 1,064 women in the population of England and Wales for every 1,000 males. For London the figure was 1,116, for the urban sanitary districts 1,090, and for rural districts only 1,010. In Scotland the imbalance of females was slightly larger. In the German *Reichsgebiet* the *Frauenüberschuss* in 1890 was 1,040, but in small towns, 5,000–20,000, there were only 994 females; in cities 20,000–100,000, there were 1,004 females, and in the largest cities, over 100,000, the preponderance was 1,057. Evidence from a dozen other European countries broadly confirms the fact of the imbalance, the notable exceptions being Russia and Serbia, where the preponderance of men seems to have been greater in the cities than in the countryside. One explanation may be that many of the men in the larger Russian towns were there temporarily, returning later to their families in the country. In 1891 the colony of New South Wales had a female deficit of more than 8 percent in the rural areas, but in Sydney the sex ratio virtually balanced. Parts of the American West also resembled New South Wales, but in the United States as a whole a female deficit of 1·2 percent turned to a slight surplus in the urban population ($\geq 2,500$). Nevertheless, in the twenty-eight large cities of 1890 ($\geq 100,000$) there were 999 females for every 1,000 males. It is noteworthy that in countries which benefited so much from international migration, such as Australia and the United States, the female surplus did not occur nationwide.[20]

Registration data from a number of countries indicate that the surplus of women was largely the outcome of heavier mortality among infant males 'which within the first year usually effaces the superiority of male births.' Since infant mortality was usually higher in the cities, the excess of females had become characteristic of the urban population in many countries. The heavier mortality of men in working ages—owing to physical hazards of certain male occupations and, as many Victorians believed, to effects of 'vice and crime'—tended to confirm the male deficit in the towns.[21] The in-migration differential augmented an existing female surplus, while the greater out-migration of men from a number of large cities further distorted the sex ratio in that part of the city-size distribution.

At the close of the nineteenth century the age structure of most urban populations differed from that of most national populations. The latter resembled a pyramid; the more rapidly a population was growing from its own natural increase, the greater the proportion in the younger age groups, the broader the base of the pyramid, and the more gradual the slope of the structure to its apex formed by the oldest age groups. The age structure of the urban population departed from the national pyramid, however, largely as a consequence of the greater proportion of city folk in the productive adult years and the correspondingly smaller shares of dependent children and retired people.[22]

That migration contributed to this bulging of city age structures in the productive adult years is clear enough. However, most of the nineteenth-century evidence reveals little about age *at the time* of migration. Charles Booth reported that among

295 male and female migrants to London from English villages, 235 or 85 percent had been between the ages of fifteen and twenty-five; only 16 were under fifteen years and only 17 above the age of thirty. Over half the migrants resident in Frankfurt-am-Main in 1890 were of the age groups twenty to forty years, while in Berlin the number of migrants per thousand population in 1885 clustered in five-year groups between thirty and sixty years. Some 80 percent of the male labor force, aged thirty to sixty years, had been born *outside* the capital. Native Berliners constituted 42·3 percent of the population but the natives, of course, included the children of the newcomers. About half the entire adult population, twenty-five years and over, in this city of 1,315,287 had in 1885 lived there for less than fifteen years.[23]

As populations experienced later phases of the urban transformation, differences in the age composition and sex ratios of urban and national populations would be reduced. This would tend to erode the rural bulges in age groups under fifteen years, and thereby dry up the historic reservoir for migrants to the city. If the rate of natural increase also declined, the population would, barring foreign immigration, tend to 'age'; a smaller share would be younger persons and a growing proportion would be older people. Such changes could be expected to have important repercussions on both the production and consumption sides of the economy as the shares of households in different phases of the life cycle expanded or contracted.

Urban Fertility and the Socio-Economic Environment

The exceptional size and fecundity of the British urban population have already been noted in reference to the relative contribution of natural increase to city growth. A number of conditions incident to this first urban transformation—especially the growth of economic opportunities—should be noted. While the proportion of married adults of all ages and both sexes had risen in England and Wales since mid-century (less in Scotland, especially for adult males), the census of 1891 revealed a marked difference in the levels of nuptiality for urban and rural adults of each sex in the ages fifteen to forty years. In the highly urbanized county of Lancashire, for example, 469 males and 475 females in every thousand adults of each sex in these ages were married; in London the figures were 464 males and 456 females. The predominantly rural population of Herefordshire had 402 males and 422 females married and little Rutland only 388 males and 427 females.

Dr W. Ogle had already noted, however, that the marriage-rate among single men, twenty to forty-five years, was highest for the rural population of Bedfordshire where employment opportunities for women were exceptionally good owing to the localization of the lace and straw-plaiting industries in that county. His unromantic hypothesis, that men would propose more readily to girls who were themselves earning money, seemed to be confirmed by the proportion of girls, fifteen to twenty-five years, gainfully occupied in industry. This condition held generally for the larger towns in England and was especially notable where opportunities for employment were greatest, as in the textile factories of Lancashire.[24]

The effect of economic environment was also noteworthy on the high refined marriage-rate prevailing in the textile towns of Massachusetts. Nevertheless, the married state of the adult population in the twenty-eight largest cities of the United States in 1890 was, unlike that for England and Wales, well below the national average for both men and women. Contrary to public opinion, moreover, it was not the large foreign-born element which lowered their aggregate level of nuptiality; rather was it the native-born white population of native parentage. The Negro was also a depressing factor in the nuptiality of large cities (not for the U.S. population as a whole), but the black proportion of their population was only 4·3 percent nationwide and insignificant outside of the four large cities of the South. A marked tendency on the part of the native white population in large cities to postpone marriage, a function of their improved socio-economic status perhaps, reduced the nuptiality of the city populations below that of the rest of the country. In so far as the age and socio-economic selectivity of the suburban trend may already have carried a greater share of the married native adults out beyond the boundaries of the older and larger cities, it could have contributed to the same result.[25]

In England and Wales both the high marriage-rate and the level of nuptiality for all adult ages in the urban population exceeded those for the rural population. In the United States, where younger unmarried people were quitting the rural areas in large numbers, the same was only true for the marriage-rate. The refined marriage-rate in some European countries may also have been higher for the city populations, but this was not generally the case. Denmark provides a well-documented exception where the large city of Copenhagen encouraged marriage at an earlier age than was possible in the Danish countryside. But in Sweden, Austria, and such German states as provide data, marriage-rates and nuptiality were generally lower among the urban populations than among their rural counterparts, except in the heavy industrial centers of the Ruhr. In the higher age groups which included older migrants and natives, the difference in nuptiality was less marked.[26]

By the early 1890s, nevertheless, crude birth-rates were generally higher among urban populations than they were among rural. With the exceptions of Sweden and Prussia, J. E. Wappäus had found the same to be true for seven European countries around mid-century. To the extent that cities were the destinations of young adult migrants and had a *Frauenüberschuss* in the child-bearing age, the explanation of their fertility seemed obvious. In Sweden the higher rural fertility, as measured by crude rates, had disappeared during the decade 1861–70 but in Prussia it had persisted and by 1890–1 stood at 40 per 1,000 compared with 36·3 for the urban population. Refined birth-rates based on the same census failed to change the picture and showed, moreover, that female fertility in ages sixteen to fifty *decreased* with size of city, especially above the 100,000 level. Even the incidence of illegitimacy was slightly higher in the Prussian rural districts, except for the very largest cities. On the other hand, refined birth-rates for Saxony for 1879–83 indicated that in all but one of the five *Regierungsbezirke* legitimate births were higher in the urban districts than in rural districts, although this did not hold for the large cities of Dresden, Leipzig, and Chemnitz. Only in the Leipzig *Bezirk* did the rate of urban illegitimacy surpass the rural rate, but this did not hold for the great commercial city itself. The refined urban birth-rate in Denmark, outside of Copenhagen, was also higher than the rural rate, but illegitimacy in the capital was reported twice as frequently as in other urban and rural communities.

In regard to family size Wappäus had shown that, around mid-century, the numbers of children per marriage in six of seven European countries was greater in their rural districts than in the cities. By the 1890s, however, it was no longer true that the large city was the peculiar resort of relatively sterile families.[27] Marital fertility appeared to be falling in a number of countries which were more and less urbanized. In the United States, a country with falling fertility but with a comparatively high birth-rate, the number of persons per census family nationwide—4·93 in 1890—was slightly smaller than in the twenty-eight large cities—4·99. The average for the nation was raised considerably by the size of family units in the South which contained the overwhelming majority of large Negro families but only four of the large cities. Among the ten largest centers, ranging in size from New York to Cleveland, seven showed a greater number of persons per family than did the states in which they were situated; two had a smaller number; and one virtually the same number as its particular state. While the proportion of foreign-born residents may have been a variable affecting family size in some of the larger cities, data for the registration state of Massachusetts and its Boston metropolis showed that married women of native parentage in the cities also had more children than their counterparts in rural areas. Three of the four largest cities in Germany, on the other hand, Berlin, Leipzig, and Munich, had a refined birth-rate that was below the average for the Reich's twenty-six largest cities, whereas in Hamburg and the six smallest of this class the birth-rate was above the average.[28]

In a number of countries, therefore, there was no direct relation between size of agglomeration and fertility. The mere concentration of population did not account for either the tendency to marry or the fecundity of married life. Since cities of more or less equal size in the same country or province commonly exhibited wide variations in their birth-rates, A. F. Weber was right to conclude that the 'conditions affecting the fruitfulness of marriage are so numerous and complicated that statisticians and social philosophers are still in dispute.'[29]

Urban Mortality and the Physical Environment

The issue of mortality in cities posed no such complexities to understanding. At the end of the century it was almost universally true that rates of mortality were lowest in country districts and increased steadily with size and density of agglomerated population. Death expectancy was still highest in great cities. In 1893 the Registrar-

General, whose office coextended with Victoria's reign, apportioned the population of England and Wales into fifteen groups, increasing in density from 138 to 19,584 persons per square mile and in crude mortality from 14·75 to 30·70 per 1,000. He also grouped together certain 'selected healthy districts' with low mortality, 14–15 deaths per 1,000, embracing about one-sixth of the population, mostly rural, and contrasted their life expectancy at birth, 51·48 years, with that of the Manchester district, 27·78 years, and with that of the entire country, 43·66 years. Thus a Manchester-born man in Manchester might expect to live just over half as long as his country cousin.

In the United States the crude death-rate among the rural population of the census registration states in 1890 was 15·34—not much greater than that of the 'healthy districts' of England and Wales. The rate for their urban segments was 22·15 for all ages, but in the special 'metropolitan district' of New York it was 24·61. Almost without exception city mortality increased in severity for every period of life with the agglomerative size of a population and was at all ages heavier than rural mortality. The crude death-rate for the twenty-eight large cities of 1890 was 23·28 and, for similar cities of 100,000 and more residents in the registration states, 21·62 for all ages. Weber vividly summarized the cheapness of human life in cities at the time:[30]

> whereas the average person born in Massachusetts may expect to live 41·49 years, the average person born in Boston may expect to live only 34·89 years . . . while 426 out of 1000 men born in Prussia survive to the age of 50 years, only 318 native Berliners reach the same age . . . while the mean age at death is 42 years and 2 months in France, it is but 28 years and 19 days in Paris . . . while the average duration of life in the rural population of the Netherlands is 38·12 years, in the urban population it is only 30·31 years.

The decades on either side of 1890 mark a turn from the old mortality to the new. In 1891 the crude death-rate in Manchester was 26·0 per 1,000 but, over the years 1895–1904, the average rate was only 22·6. Among the large cities of England and Wales, Manchester was exceeded only by Liverpool, whose average was 23·2. London's average mortality was 18·2 per 1,000 and was surpassed by Newcastle, 20·9, Birmingham, 20·2, Sheffield, 19·6, Leeds, 18·7 (as well as Liverpool and Manchester). The average mortality in Bradford was 17·7, while Bristol, 16·9, and Leicester, 16·7, fell well below the rate for the capital. By the year of the Queen's death crude mortality in Glasgow had fallen to 21·0 per 1,000. Within little more than a quarter of a century after the Prince Consort's death, the life expectancy of the average Londoner had risen from twenty-five to thirty-seven years; over the middle years of the reign the crude death-rate for the capital had averaged 23·4 (1861–80), but by 1905 it had fallen to 15·6 when the rate for England and Wales stood at 15·3 per 1,000. In the latter year the crude rates for Bristol and Leicester were already below those of the lowest-density population groups reported by the Registrar-General as recently as 1893.[31]

In view of the death-rates obtaining in the industrial cities during the middle period of Victoria's reign the achievements of the closing years were remarkable. It is clear from Table 1.4 that by the early 1900s the mortality of England's great

Table 1.4 *Crude mortality in twelve selected cities, c. 1900*

	City	Average death-rate per 1,000, 1895–1904	Death-rate per 1,000 1905	Net change (permillage point)
1	Manchester	22·6	18·0	−4·6
2	Newcastle	20·9	16·8	−4·1
3	Birmingham	20·2	16·2	−4·0
4	Liverpool	23·2	19·6	−3·6
5	London	18·2	15·6	−2·6
6	Brussels	16·7	14·5	−2·2
7	New York	20·2	18·3	−1·9
8	Paris	19·2	17·4	−1·8
9	Vienna	20·0	19·0	−1·0
10	Berlin	17·8	17·2	−0·6
11	St Petersburg	25·9	25·3	−0·6
12	Rome	19·1	20·6	+1·5

Source: Data from 11th edn *Encyclopaedia Britannica*, XVI, p. 946.

manufacturing centers was on a par with that of the capitals of Europe and the two largest cities in the world. Nor was this improvement merely a consequence of the very high death-rates ruling in the older manufacturing towns as recently as 1890. The ancient woollen-city of Leicester, which in the nineteenth century had become a center for the boot and shoe industry and whose population had reached 212,000 by 1901, also achieved a net decline of 3·4 permillage points from its comparatively low average rate of 16·7 for the years 1895–1904. Newcastle's experience was particularly significant, since Tyneside had some of the most 'overcrowded' housing in Britain and, as late as 1933, struck Priestley as the grimmest part of England outside the dock districts of Liverpool.

The relationships between mortality and the socio-physical environment had been generally understood throughout Victoria's reign. Poor housing, impure water, lack of fresh air and sunlight, the prevalence of dirt and disease were the common lot of those living in the compact 'walking towns' of mid-century and especially of the poorer working classes. The *sine qua non* of the city, as in one way or another of every human habitation, was the water supply. Whether for sustenance, sanitation, fire-fighting, or industrial use, water was the original public utility and, historically, the first urban 'problem.' Londoners had been obliged to seek 'sweete waters abroad' in Hertfordshire even before the upsurge of the city's population in the early seventeenth century. By Victoria's time, when nine water companies competed to supply the capital, conditions had probably deteriorated and, regardless of whether it was obtained from fountains, dug wells, or the companies, 'a drop of London water' was not likely to be any purer than that depicted in the notorious *Punch* cartoon in 1850. Urban water systems were a natural breeding ground for typhoid and cholera and any real improvement had to await advances in hydraulic engineering and the development of bacteriological science, almost all of which belong to the last quarter of the nineteenth century.[32]

It was impossible for the nineteenth-century market-economy to house the growing, urbanizing, population in any but the most rudimentary way. Public and philanthropic efforts could do little more than advertise the 'problem.' In his *Housing Problem in England* Ernest R. Dewsnup cited twenty-eight major public Housing and Health Acts between 1851 and 1903. Their terms were largely permissive and depended for their implementation and enforcement mostly upon the initiatives and resources of local authorities. The authorities were generally more responsive to health regulation than to housing legislation, except when their influences were combined, as in the standards imposed upon builders for *new* construction. Although the Nuisances Removal Act of 1855, for example, had given statutory recognition to 'overcrowding' and had indicated its dire consequences to health, the abatement provisions were nugatory. An early report of the London County Council's Medical Officer of Health, nearly four decades later, showed that the death-rate increased in the same ratio as the proportion of the population residing more than two in a room in tenements with less than five rooms. Newsholme's *Vital Statistics* revealed that the death-rate in Glasgow was 27·7 for families living in one or two rooms, 19·4 for those in three and four rooms, and only 11·2 for those in five or more rooms. In one English town the death-rate was reported as 37·3 in districts containing 50 percent of 'back-to-back' houses compared with 26·1 in districts without them. Between 1885 and 1905 Manchester closed down nearly 10,000 'back-to-back' houses and allowed only about half to be reoccupied after thorough reconstruction.[33]

The English statutory definition of overcrowding, more than two persons per room, proves to be a far better measure of urban living conditions than the crude demographic criterion of density. In 1901 the census reported 8·2 percent of the population overcrowded, 8·9 percent in urban districts as against 5·8 for the rural. London with 16·0 percent overcrowding (Finsbury had 35·2 percent and a death-rate exceeding 19 per 1,000) was poorly housed by the national standard, but the situation ranged from Gateshead (35·5 percent) and Newcastle (30·5 percent) to places like Leicester (1·0 percent) and Bournemouth—'the Garden City of the South'— with only 0·6 percent overcrowded. Making no allowance for population changes, Huddersfield led the country over the decade 1891–1901 in reducing its overcrowding by seven percentage points and Nottingham appears to be unique among eighty-four towns of more than 50,000 population in showing no improvement.

North of the border the housing situation was far more grave. Compared with

8·2 percent overcrowding in England and Wales, the corresponding figure for Scotland was 50·6 percent. Dundee reported 63·0 percent of its residents in tenements of one or two rooms, Kilmarnock 62·2 percent, Glasgow 55·1, Edinburgh 41·3, and Aberdeen 39·3 percent. Even Ireland was better off in respect of overcrowding: 40·6 percent in Dublin (24·7 percent in one room), 31·7 percent in Limerick, but only 16·7 in Derry and 8·2 percent in Belfast. In England and Wales, at least, E. R. Dewsnup was convinced that 'gross overcrowding' probably affected no more than 2·8 percent of the population by 1901 and he implied that the 'housing shortage' was only gravely acute in a few highly publicized localities.[34]

Given the variety of local circumstances, the almost universal reduction in overcrowding during Victoria's last decade was extraordinary. One contemporary, Arthur Shadwell, insisted, however, that 'the most important' element in the recent improvement of English housing conditions was the electric tramcar, which from 1896 had made the outskirts of towns accessible to more people and presented private builders with more opportunities for profit. The declining birth-rate, he thought, had also reduced pressure on housing; many imprudent parents, such as those residing in the small antiquated miners' houses in Durham or the tenements of Finsbury, Shoreditch, and Bethnal Green, had been obliged to relocate their homes at great cost and inconvenience. Much overcrowding, he declared, was in any case 'voluntary' as evidenced by the many large families taking in lodgers, or the tendency of foreigners to 'herd together' in London, the ports, and a few of the manufacturing centers. Finally, and most gradual in its effect on housing, there was the improvement in public-health administration revealed, for example, in the accounts of building renovations given by the medical officers of London's metropolitan boroughs. In varying degrees, Shadwell concluded, these three tendencies indicated that 'the process of urbanization has been modified by one of suburbanization.'[35]

Infant Mortality: Key to the Urban Demographic Transition

However much the experts might still dispute the relative significance of housing and health regulations in reducing urban mortality, they were agreed that the stubborn predominance of city deaths over those of the countryside was a consequence of the heavier mortality of the city's children. The gradual approximation of urban and rural death-rates in England and Wales since the late 1860s was largely a result of the decline in deaths among people aged fifteen to forty-five and, to a lesser extent, five to fourteen. The widest divergence between the age-specific death-rates of town and country populations was still under five. As late as 1890 the towns had achieved a lower rate only for females between the ages of fifteen and thirty-five. Nationwide the infant mortality-rate (deaths under one year per 1,000 live births) was edging upwards again as more people concentrated in the towns.

Among nine functionally-grouped categories of socio-economic environment, only seaside resorts and pleasure places yet enjoyed age-specific death-rates approaching those of the rural districts. For females aged five to forty-five, their rates were already below the corresponding rural rates. The highest death-rates for children of both sexes under five years were reported from the Staffordshire potteries and the Lancashire manufacturing districts, while Manchester itself and the northern colliery districts were unique among the nine categories in having somewhat higher death-rates for children five to fifteen years than for those under five. Even so, mortality among the under-fives in Manchester was well above the national average for that age group during 1881–90 and far above the average for the Registrar-General's 'selected healthy districts.' Yet chances for survival beyond the age of fifteen in Manchester tended to approximate those for the corresponding age groups in the healthy districts, at least until the age of thirty-five.[36]

In his consideration of causes and remedies for infant mortality, the Registrar-General revealed in 1893 that the town rate exceeded the rural rate in the first week of life by 23 percent and that the differential increased to 97 percent by the fourth week, reaching its maximum of 273 percent in the sixth month; throughout the balance of the first year of life the two rates converged rapidly. By the late nineties death-rates in England and Wales were declining for both sexes at every age, except from sixty-five to seventy-four, and the parents of city-born infants who survived the first six months could reasonably expect all their children to live longer, fuller lives than if they had been born only five years earlier. The national infant mortality-rate reached its peak of 160–3 per 1,000 in 1898–9 and then declined rapidly. By 1905 it was down to 128; in 1910, when the population was more than 78 percent urbanized, the rate was 105. The average rate for 1899–1901 was 146, but for 1909–11 was 114—the greatest decline in any decade since registration began in 1837. After 1915 infant mortality never again exceeded 100 and after 1933 fell below 60 per 1,000. Thus, by 1910, the last great barrier to the cities of England and Wales reproducing themselves had been overcome and the urban demographic transition had been accomplished.[37]

Elsewhere in Europe infant death-rates mostly lagged far behind those of England and Wales. While the crude death-rate in London had fallen below 20 in the late 1880s and infant deaths had hovered around 164 per 1,000, Vienna's crude rate fell to 23 and the infant rate to 203. From 1881 to 1890 the infant death-rate in the sixteen largest cities of Prussia fell by more than twenty-five permillage points, standing at 241·7 in 1891 (250·5 in Berlin). Slums were less prevalent in Prussia than in England owing to the comparatively recent development of most German industrial cities; but during the 1890s living conditions in the working-class districts of German cities deteriorated with the influx of newcomers; there were no uniform standards and overcrowding tended to increase. In 1900 Königshütte, for example, one of the chief coal and iron centers of Silesia, had 70·4 percent of its rapidly growing population 'overcrowded' by the English standard of two or fewer rooms per dwelling unit; Breslau had 49·9 percent overcrowded; Chemnitz, with the highest infant mortality in Saxony, had 36·5 percent overcrowded. By the end of the century, however, crude death-rates in German cities generally were beginning to fall and children were among the principal beneficiaries. In Bavaria, for example, notably in the ancient city of Munich which had 28·7 percent of its population overcrowded in 1900, the infant death-rate in the towns had already fallen below the level prevailing in the countryside.[38] In the Netherlands an infant death-rate of 195 per 1,000 for the twelve largest cities in 1891–2 was already below the countrywide average of 203. In Switzerland, where infant mortality was almost fifty permillage points above the English rate in 1890–2, the rate for the fifteen largest cities, 157·5, had actually fallen below that for the thirty or so largest centers of England and Wales.[39]

There was less cause for congratulation in the United States. The crude death-rate in New York City, to be sure, had fallen from around 32·2 per 1,000 at mid-century to 21·5 in 1896 and to 16·8 in 1908. But the infant death-rate for the twenty-eight largest cities of 1890 was 236·8 per 1,000, exceeding that for the German *Grossstädte* outside Prussia, while the rate for the metropolitan district of New York, 264·4, surpassed even that for Berlin. The rate for the entire urban population of the registration states of 1890 was 243·3 compared with only 121·2 for their rural populations. The state of Massachusetts could take pride, perhaps, in the fact that by 1885 infant mortality in Boston and three densely populated suburbs was already down to 185·8 per 1,000.[40]

Making due allowance for shortcomings of American statistics outside of the census registration states, the situation is still hard to understand. Average population densities, outside lower Manhattan and a few blocks in other major centers, were much lower than in Continental cities, owing to the preference for home ownership and to the fact that, on the whole, boundaries were more flexible and not circumscribed by lines of ancient walls and fortifications. While the proportions of home ownership in English cities were generally lower than in the United States, the preference for individual houses was marked in both countries, but the average number of persons per census dwelling in the United States (7·64 in the twenty-eight largest cities of 1890) was considerably higher than in England and Wales. Yet public standards for sanitation and sewage disposal were much lower. The U.S. Commissioner of Labor in 1894, reported that in New York some 360,000 people lived in 'slum' conditions—162,000 in Chicago, 35,000 in Philadelphia, and 25,000 in Baltimore, where 530 families were living in single rooms—the highest proportion in the country. In New York 44·6 percent of families lived in two rooms—27·9 percent in Baltimore, 19·4 percent in Philadelphia, and 19·1 percent in the huge 'spread city' of Chicago, the older parts of which had been substantially rebuilt since the great fire of 1871. Although average earnings per family in all of these cities were high by European standards, the high proportions of families living in overcrowded housing showed, according to one observer, that the worst kind of Continental conditions 'reproduce themselves in American cities.'

If bad housing in the United States was nevertheless not so widespread as in Europe, conditions in lower Manhattan had been publicized into a 'world-wide scandal.' Density in New York City in 1890 at fifty-nine persons per acre was no greater than it was in London but whereas London's density was decreasing, that of

New York appeared to be increasing. The *Report* of the New York Tenement House Committee in 1894 revealed that density was increasing in ten of the city's twenty-four wards. Two of these, it is true, were residential suburbs north of 42nd Street, but three which lay on the lower East Side below 14th Street were the most densely populated in North America. A. F. Weber claimed that the 10th Ward, with a density of 523·6 persons per acre, was 'the densest district in the Western World, Josefstadt in Prague having 485·4, the *quartier* Bonne-Nouvelle in *arrondissement* Bourse in Paris 434·19, and Bethnal Green North in London 365·3.' By 1900 density in Manhattan was 149, on the lower East Side 382, and in the 10th Ward, 735 persons per acre. Between 1900 and 1905 density was generally increasing all over Manhattan and in 1905 there were reported to be twelve city blocks with from 1,000 to 1,400 persons to the acre. In 1901 London's most densely populated borough had 182 persons per acre, but its most densely settled district, much smaller in area than the lower East Side, was 396. Nevertheless, the crude death-rate in New York did not increase. The figure for the predominantly 'Russian Jewish' 10th Ward in 1894 was no more than 17·1 per 1,000, a rate which was bettered in only two wards of the city, one a business district and the other an up-town suburban ward. The mortality of children under five years in the 10th Ward was 58·3 per thousand compared with rates ranging up to 183 in some other wards, and 76·6 for the city as a whole.[41] The corresponding rate for London children in 1891 was 66·4 per thousand. Clearly the experience of the 10th Ward demonstrated that domestic and personal hygiene could do much to save the lives of children among a population that had many of the outward characteristics of a 'culture of poverty.' The Peabody and Octavia Hill ventures in London had likewise shown that in 'properly managed' buildings, selected poor at any rate could be trained to save themselves.

Child mortality in New York declined rapidly around the turn of the century even as housing conditions deteriorated in many working-class districts. Medical inspection of school children, isolation of contagious diseases, and more stringent inspection of milk, all helped. By 1906 the city had reduced its death-rate for all children under five years to 55 per 1,000. Infant mortality itself was more intractable. In Massachusetts the last third of the nineteenth century had witnessed death-rates for infants ranging from the 150s to the 220s. The rate for 1890–4 averaged 163·2 per 1,000 when more than 80 percent of its population was classified as urban. By 1910, nevertheless, the decline in infant deaths was unmistakable; the average for 1905–9 was 134·3, the lowest for any quinquennium in Massachusetts since the late 1850s. During the years 1915–19, which included the postwar influenza epidemic, the infant rate was averaging 100·2, only slightly above the level for England and Wales.[42]

The situation in the United States as a whole is less clear. The population of the country's death registration area, established in 1900, was heavily weighted by the inclusion of large urban populations from outside the regular registration states of the North East. Infant mortality in this population, which covered about 40 percent of the whole, remained somewhat above the rate for Massachusetts during the early 1900s. The impact of the secular fall in infant deaths was nevertheless strong enough to effect a marked improvement in crude mortality for most of the nation's larger cities. By 1905 the death-rate for New York (five boroughs) was already down to 18·3 per 1,000 and to 16·8 in 1908 when, by this measure, the city ranked ninth among the fifteen largest centers whose individual rates ranged from a high of 22·7 in the case of New Orleans and 19·3 in Washington, D.C., to lows of 14·1 in Chicago and 13·6 in Milwaukee. If it was true, as one English critic affirmed, that 'American cities have nothing to learn from other countries in regard to bad housing' and 'nothing to teach in the way of reform,' their crude mortality at least could compare with any but the most advanced industrial states such as Great Britain and Belgium and was already much better than in backward states such as Italy or European Russia.[43]

The mortality aspect of the demographic transition in the cities posed fewer difficulties of interpretation than the fertility side. The cumulative effects of improved diet, more sanitary dwellings, public health, and medical services were generally reflected in lower death-rates for all ages and classes, but only the most involved social and economic analysis could begin to 'explain' the peculiar turn in fertility. What emerged as a demonstrably negative correlation between family fertility and socio-economic status confirmed the observations of men such as A. F. Weber that 'the birth rate in large cities diminishes as one goes from the poor to the rich quarters, and that the age at marriage increases, and the size of the family diminishes as one passes from classes low in the social scale to the responsible mercantile and professional classes.'[44] Only with the steep decline in infant mortality after 1890, however, could the lower orders begin to develop a more prudential control over their own fertility. The demographic transition within the urbanized population was thus a behavioral correlate of the social and spatial structural-differentiation which marks the transformation of the Victorian industrial city into the twentieth-century metropolis.

Alternative Urban Futures: The Vision of 1900

The demographic transition produced a mood of *fin de siècle* among late nineteenth-century social thinkers no less pessimistic and confused than that which pervaded the literary world. Since the costs of raising children in the city tended to be higher than in the country (for both private households and public authorities), and inasmuch as city children were less of an economic asset than farm children (especially among the poorly-paid lower-middle classes), a growing number of urbanites was evidently substituting a higher level of consumption for the larger families of earlier generations. But, in so far as the absolute size as well as the fertility of the ageing rural population was also beginning to fall in more 'civilized' countries, while the lower urban strata tended to breed more prolifically than the middle strata, the decline in urban mortality could no longer be regarded as an unmixed blessing. However much pundits like Herbert Spencer, Francis Galton, Georg Hansen, or Alfred Marshall had differed on the principles and details of the change, they seemed to imply that life in the industrial cities was becoming at once essential and inimical to the further progress of civilization. The selective processes of the urban transformation had brought out the noblest and the worst in mankind. Such experts left little doubt that they identified themselves with the former tendency and the ignorant and imprudent masses with the latter.

At the very dawn of the industrial era Adam Smith had predicted some such outcome for the populations of manufacturing towns. The division of labor which was the source of enhanced productivity and material betterment in both town and country would reduce the ordinary tasks of manual workers 'to a very few simple operations.' Lacking any power of judgment concerning 'the ordinary duties of private life,' and incapable of judging 'the great and extensive interests of his country,' the town worker would eventually become 'as stupid and ignorant as it is possible for a human creature to become.' His industrial skill would be acquired at the cost of his 'intellectual, social, and martial virtues.' Smith insisted that 'in every improved and civilized society this is the state into which . . . the great body of the people must necessarily fall, unless government takes some pains to prevent it.' The Scot's judgment on the social character and outlook of the urban merchant or manufacturer was only less uncomplimentary than his view of the worker.[45]

More than a century after Smith, J. A. Hobson preserved many of the same doubts. Town life, as opposed to town work, was doubtless 'educative of certain intelligence and moral qualities.' The competitive stimulus of town life, Hobson thought, more than offset the deadening effect of town work, but the 'character of town education and intelligence' was none the less limited. Smartness, glibness, 'half-baked information' and prejudice among workers and businessmen alike had produced a shallow, self-seeking outlook which paid off on the job, perhaps, but remained essentially unscientific and anti-social. While he conceded Alfred Marshall's point that the progressive mechanization of work tended to substitute 'higher or more intellectual forms of skill' for the older manual skills, Hobson insisted that this description was 'not yet applicable to most factory trades.' More significant, there had been 'a progressive weakening of the bonds of moral cohesion between individuals' and he concluded that, the growth of production notwithstanding, 'the forces driving an increased proportion of our population into towns are bringing about a decadence of *morale*, which is the necessary counterpart of the deterioration of national physique.'[46]

The demographic transition in the cities had now added a note of urgency to the century-long debate concerning the effects of specialization of roles on citizens and their social environments. From across the Atlantic, W. Z. Ripley declared, in a somewhat overdrawn contrast between urban and rural populations, that marriage and family life, unless deferred, had become 'an expensive luxury' for the more enterprising people and 'every child . . . a handicap for further advancement.' Arthur T. Hadley, shortly to be elevated to the presidency of Yale, expressed the fear of many observers that the poorer classes, now relieved of both their hunger and their high mortality, would soon out-multiply the more ambitious and prudent members of society, the builders of industrial civilization. While still adhering to the tenets of Christian Darwinism, Hadley was no longer sure that those 'fittest for civilization' could necessarily survive: 'It is not that social ambition *in itself* constitutes a greater preventive check to population than the need of subsistence; but that the need of subsistence is felt by all men alike, emotional as well as intellectual, while social ambition stamps the man or race that possess it as having reached the level of intellectual morality.' The emotional masses, he opined, were incapable of 'ethical selection.'[47]

Against these confusions, other observers looked back to proposals such as Charles Kingsley's fusion of city and country modes of life or Andrew Jackson Downing's 'country cottage residences' as the best hope for the industrial city and its threatened civilization. For such people, the countryside had to be brought into the city, and a part of the city, at least, returned to the land. Charles H. Cooley, like Hadley a student of modern transportation, posed the late-Victorian dilemma most succinctly as 'a permanent conflict between the needs of industry and the needs of humanity. Industry says men must aggregate. Humanity says they must not, or if they must, let it be only during working hours and let the necessity not extend to their wives and children.' It was 'the office of the city railways,' Cooley declared, 'to reconcile these conflicting requirements.' Cooley's vision of the industrial city was of a new order of space and time: rapid transit and the residential suburb were conceived as the most practical and benign remedies for the physical and moral 'problems' afflicting the impacted populations of great cities.[48]

There was a certain myopic quality to the late-Victorian belief in the suburbs. Suburbs were as old as the cities themselves; the clustering of populations on the edges of towns was, if anything, an older form of city growth than their concentration at the core. The peripheral areas of early industrial towns had sometimes grown faster than their centers, even before the coming of steam railways in the 1830s. London, Manchester, Brussels, Philadelphia, Boston, Chicago, Berlin, and Sydney were essentially agglomerations of 'suburbs' and 'satellites,' some of which were later consolidated or annexed. The suburbs of one generation were often indistinguishable from the town of the next; sometimes yesterday's suburbs became today's slums. In *Sesame and Lilies* Ruskin had bitterly condemned the 'festering and wretched suburb.'

If the Victorians had studied the social morphology of city growth more closely, they might have been more guarded in their belief in the redemptive powers of suburban life. The fact that developments in technology and organization now permitted spatial differentiation on an unprecedented scale was surely as much cause for concern as for congratulation. While reducing the difference that had grown up between town and country environments, a much more intensive differentiation of land uses and social space would now be achieved over a greatly extended urbanized area. The city population might well become even more divided from itself.[49] The subjective gravity of 'the problem' was such that many otherwise informed observers grasped this particular panacea uncritically.

Here was the remedy for a deteriorating social environment that went beyond restrictive housing legislation or countrified neighborhoods. By increasing the speed of public transportation to ten miles per hour, the electrified street-car could more than double the journey-to-work distance attained by horse-drawn cars. Its promoters claimed that the electric tramway would be more flexible than cumbersome rackrailways or cable cars, and much cheaper to install than either tube or open-cut subway systems. New modes of electric power generation and transmission, moreover, could lead to a rapid dispersion of manufacturing jobs to the outskirts of towns, if not back to the countryside. Alfred Marshall had, in fact, been urging the decentralization of London's clothing industry and its poorly-housed working people on economic as well as civic grounds since the early 1880s. A. F. Weber agreed that, in so far as concentrated poverty and overcrowding had once been concomitants of agglomeration, the planned movement of people to new 'colonies' in the country, such as Marshall proposed, would certainly mitigate 'the evils of city life.'

Marshall had made many other suggestions for improvement which Weber could endorse: piped water, steam heat, compressed and 'ozonized' air in compact multi-tubular tunnels beneath city streets. Weber upheld Marshall's technical recommendations for reducing air pollution from household and industrial smoke and he applauded the economist's vision that 'every house might be in electric communication with the rest of town.' But the optimism of Weber, the most informed student of the city at the turn of the century, was grounded on something more substantial than a trendy confidence in quick technological solutions to intractable social and moral problems. He added Marshall's insights to his own hard evidence to show that, for all the real danger of the city, the general forces of economic and social progress were already in the ascendant.

Even before Weber, Marshall had maintained that advances in medical science, improvements in public health services, and the growth of material wealth 'all tend to lessen mortality and to increase health and strength and to lengthen life.' The very rapid growth of town populations, on the other hand, lowered vitality and tended to raise death-rates while 'the higher strain of the population' inclined 'to marry later and have fewer children than the lower' strain. The former impulses, unchecked by the latter, could, if overpopulation were avoided, bring man 'to a physical and mental excellence superior to any that the world had yet known; while if the latter act unchecked he would speedily degenerate.' Recent experience with the Victorian cities had convinced Marshall that the former set of impulses was 'slightly preponderating.' The population of England and Wales was still growing; 'those who are out of health in body and mind are certainly not an increasing part of the whole' while the rest were better fed and clothed and, with few exceptions, stronger than ever before.[50]

The qualified affirmation of the economist was carried over into the luminous vision and confident practice of the stenographer-turned-inventor, Ebenezer Howard. Published in 1898, Howard's seminal tract *Tomorrow: A Peaceful Path to Real Reform* had more citations to Marshall's writings than to any other single source. Howard's 'plan' for Garden Cities was an explicit critique of the sprawling peripheral development so familiar to late-Victorians and their descendants. For Howard the suburbs were a 'railway chaos' and should be replaced by towns linked with each other and with a central city by a 'railway system' including rapid public transit; the intervening spaces should be preserved for uses other than commerce or residence. Howard's 'town-country magnet' was, in Lewis Mumford's words, *'not a suburb but the antithesis of a suburb: not a more rural retreat, but a more integrated foundation for an effective urban life.'* He hoped that the Garden City experiment would not only relieve congestion and economize in the use of land but also serve as 'the stepping stone to a higher and better form of industrial life generally throughout the country.' Long before the private automobile, Howard foresaw the kind of 'alien' settlement and encompassing smog through which Priestley would pass on his English journey in 1933. The subsequent failure of the British town-planning movement to grasp more firmly Howard's design for 'social cities' is no reflexion on the practicality of his idea.[51]

No planner ever penetrated more deeply into the changing character of the late Victorian city than the indefatigable Scottish biologist and human ecologist, Patrick Geddes. Like Howard, Geddes recognized that the city of the future would differ as much from the mid-Victorian city of his childhood as that city, Perth, has done from the burghs and villages of preindustrial Scotland. If the existing city was to be reconciled with the needs of both industry and humanity, then its future development must be 'one capable of increasingly conscious evolution,' in course of which *quantitative* growth in wealth and population must be transformed into *qualitative* progress. Under a commission from Andrew Carnegie to his native Dunfermline in 1903, Geddes had opportunity to demonstrate his remarkable gift for mediating between a 'community's' past and future. Although Carnegie's initial interest had not gone far beyond introducing 'more sweetness and light into the lives of working people,' and despite the fact that Geddes' plan failed to induce the canny Dunfermline burghers to spend much more of Carnegie's money, *City Development. A Study of Parks, Gardens, and Culture Institutes*, published in 1904 at the author's expense, established Geddes as a master whose works and influence would be spread halfway round an urbanizing world.

Geddes predicted that 'the world is now rapidly entering upon a new era of civic development.' In the more urbanized countries 'the last generation has had to carry out great works of prime necessity, as of water supply, sanitation and the like; elementary education, too, has been begun; so that to some, even pioneers in their day, our city development may seem wellnigh complete.' In fact, city development had entered on an urgent new phase 'that of ensuring healthier conditions, of providing happier and nobler ones.' Coming generations would have to study the social history and needs of their particular communities and regions in order to relate their 'culture' to opportunities and limitations shown by a study of their physical environments. Meanwhile, technical education and technology, unless infused by an intimate knowledge of science and art and consciously directed to social and civic

The Victorian City: Past and Present

ends, would only tend to exacerbate the condition of man in cities. Thus, beginning with 'the fundamental problem of purifying our stream and cultivating our garden,' men would have to progress 'naturally and necessarily . . . towards the idea, first of bettered dwellings of the body, and then to that of higher palaces of the spirit.' Without a fusion of the 'material and intellectual, domestic and civic, scientific and artistic' into the 'fundamental basis of natural and industrial reality much of our present-day idealism but flutters in the void; while our would-be practical world . . . is continually sinking into material failure or stagnation, moral discouragement, or decay.' Though the would-be practical world would often decry Geddes' pronouncements as 'utopian' or 'anti-urban', he was consistently seeking to pose an alternative to the present, to awaken an urbanizing world to the necessity for *eu-technics* and to the promise of *Eu-topia*, the best of each place in its fitness and beauty.[52]

The Urbanizing World and the Social Question

The urbanizing world at the close of Victoria's reign was still a western world. High rates and levels of urban concentration had been achieved only in countries of predominantly European population and culture. By 1890 the very rapid concentration in Australia and the northeastern United States had already indicated that the transformation was no longer peculiar to Europe itself. The case of Australia was particularly significant since it had achieved both high levels of urbanization—with more than a third of the population living in four colonial capitals in 1891—and comparatively high levels of material living without benefit of industrialization on the pattern of either Britain or the United States.

By 1900 two-thirds of the world's rapidly growing urban population was located in Europe (excluding the Russian Empire), North America, and Australasia (see Table 1.5). The proportion in Asia continued to fall and the fraction in Africa barely held its own. In contrast to the first half of the nineteenth century when the fastest growing share of world urban population had been situated in Europe, it was the Americas, notably North America, which were registering the greatest share of the increment during the last quarter of the century. Over the first half of the twentieth century, the Americas' share of world urban population continued to grow steadily, but Europe's declined almost as rapidly as that of Asia in the latter part of the nineteenth century. Between 1900 and 1950 the absolute numbers of city dwellers in Asia grew by nearly 450 percent, and in the Soviet Union by 495 percent, compared with only 160 percent for the combined urban populations of Europe and both Americas.

Since 1950 many of these tendencies have persisted. As of 1970 the Soviet Union and Latin America appear to be undergoing the most rapid structural change from rural to urban, followed by Australasia, North America, and East Asia in that order (see Table 1.6). The decadal rate of urban concentration in South Asia and Africa does not appear to be accelerating, although it is somewhat above the rates experienced in those regions during the years between the two World Wars. Meanwhile, the rate of concentration in Europe is currently the slowest among the eight major world regions. By the mid-twentieth century the urbanizing world had long ceased to be an exclusively 'western' world.

Table 1.5 *Growth of world population and world urban population, 1800–1970*

	Population (millions)	Percent urban (≥ 20,000)	Percent urban (national definitions)	Share of world urban population (≥ 20,000) in Europe, N. America and Australasia*
1800	906	2·4	—	30·8
1850	1,171	4·3	—	48·3
1900	1,608	9·2	—	65·4
1920	1,860	14·3	19·3	61·6
1930	2,069	16·3	22·0	58·6
1940	2,295	18·9	24·8	51·1
1950	2,515	21·2	28·0	47·0*
1960	2,991	25·4	33·1	41·0
1970	3,584	28·1	37·1	36·3

*1900 and earlier, in cities ≥ 100,000; 1950, 43·5 percent in cities ≥ 100,000.
Source: 1800, 1850, 1900: K. Davis and H. H. Golden, cited by P. M. Hauser, ed., *Urbanization in Asia and the Far East* (UNESCO: Calcutta, 1957), pp. 55–60. Since 1920: U.N. Dept of Economic and Social Affairs, provisional data, 9 Dec. 1968.

Table 1.6 *Levels and rates of world urbanization, by regions, 1920–70*

Region	Percent of population agglomerated (≥ 20,000)						Rate of urban concentration (percentage point shift)	
	1920	1930	1940	1950	1960	1970	1920–50	1950–70
N. America	41	46	46	51	58	62	+10	+11
Australasia	37	38	41	46	53	58	+9	+12
Europe	35	37	40	41	44	45	+6	+4
U.S.S.R.	10	13	24	28	36	43	+18	+15
L. America	14	17	20	25	33	38	+11	+13
World	14	16	19	21	25	28	+7	+7
E. Asia	7	9	12	14	19	22	+7	+8
S. Asia	6	7	8	11	14	16	+5	+5
Africa	5	6	7	10	13	16	+5	+6

N.B.: Italicized figures indicate a rate of regional concentration, δ(u/p), above world rate in preceding decade. The last two columns present summary rates for three and two decades respectively.
Source: U.N. Dept of Economic and Social Affairs, provisional data, 9 Dec. 1968.

The world-wide process of urbanization since the 1920s is the outcome of much the same economic, technological, and demographic forces as were felt in parts of Europe and North America during the Victorian era. There, however, the metropolitan form of organization after 1890 had tended to mitigate some of the older central city problems, although it aggravated others by dispersing political and fiscal resources as well as population over a wider and more socially-fragmented urban area. By the 1920s the metropolitan ring populations of North America and some European countries were growing faster *nationwide* than the populations of their central cities and by 1970 the aggregate population residing in suburban and satellite rings of United States metropolitan areas actually outnumbered the aggregate population of the central cities themselves. Only in the interwar years did the pace and form of population concentration in a few of the less urbanized countries begin to assume a metropolitan pattern. The phase began early in Japan, and in certain Latin American countries such as Argentina, but the fuller impact of urbanism and metropolitanism has only been registered in most parts of the 'Third World' since the 1940s.[53]

For a period after the Second World War the average rate of population growth was rising in most parts of the world. During 1950–60, the less developed areas were growing at a faster annual rate, 1·9 percent, than the more advanced areas, 1·3 percent, with the single exception of Australasia, 2·3 percent, a major beneficiary of international migration. Latin America led the world in the 1950s with an annual rate of 2·8 percent, while its average rate of urban population growth (as distinct from its rate of urban concentration), 5·5 percent, far exceeded that of any other major continental region except Africa, 5·4 percent. United Nations' projections of population growth for the period 1960–80 likewise give Latin America the fastest rate, 2·9 percent per annum, for the immediate future. The projected rate of urban population growth for Latin America, 4·4 percent, however, will lag behind that for Africa, 4·6 percent, during the 1970s, but will far exceed that for any other region except South Asia, 4·1 percent.[54] Latin America is now therefore the most critical region for both population and urban growth.

Some of the recent effects of these changes are exhibited in Table 1.7 which also presents comparable data for high-income countries. Perhaps the most startling fact to emerge from it is that, by 1960, crude death-rates in India and many Latin American countries *already approximated those* of the more economically advanced countries. This accomplishment indicates the effectiveness of cheap health technologies, such as D.D.T., in the control of some forms of pandemic disease which had hitherto taken a heavy toll of life.[55] It also reveals, in light of high infant mortality-rates, the comparative youthfulness of the Indian and Latin American populations. The wide discrepancy between infant mortality-rates of high- and low-income countries, on the other hand, underlines the inadequacy of medical facilities, and relatively unhealthy conditions in the under-developed areas. Countries with comparatively high infant mortality-rates are, with the exception of Argentina, also countries with comparatively high birth-rates and, with the exceptions of Argentina and Chile, countries with comparatively low levels of urbanization. As of the early 1960s, Japan appears to be the only country outside of Europe, North America, and Australasia to have yet passed through anything resembling the urban demographic transition to low fertility and low mortality for infants as well as older age groups.

Table 1.7 *Level of urbanization, crude vital rates, infant mortality, numbers of persons per room, daily calorie intake per person, for selected countries, 1960–5*

Per capita output range ($)[a]	Country	Level of urbanization, percent ($\geq 20{,}000$)	Birth-rate	Death-rate	Infant mortality	Persons per room	Average daily calorie consumption
			Per 1,000				
Over 1,200	U.S.A.	59	18·5	9·5	23·4	0·7	3,140
	Canada	53	19·6	7·6	23·6	0·7	3,090
	Australia	66	19·6	8·8	18·5	0·7	3,160
	N. Zealand	59	22·5	8·9	17·7	0·7	3,410
	Sweden	40	15·8	10·0	13·3	0·8	2,950
601–1,200	U.K.	69	18·3	11·5	19·6	0·7	2,360
	German F.R.	52	17·9	11·2	23·8	0·9	2,900
	Belgium	52	16·4	12·1	24·1	0·6	3,150
	Netherlands	60	19·2	8·1	14·4	0·8	2,890
	France	48	17·7	11·1	22·0	1·0	3,050
301–600	Italy	47	18·9	9·5	35·6	1·1	2,810
	German D.R.	40	15·8	11·2	23·2	1·0	—
	U.S.S.R.	36	18·4	7·3	27·0	1·5	—
	Japan	46	18·6	7·1	18·5	1·2	2,320
	Argentina	55	22·5	8·2	60·7	1·4	3,040
	Chile	53	32·1	10·7	107·1	1·7	2,370
101–300	Mexico	35	44·2	9·5	60·7	2·9	2,660
	Peru	26	43–45	14–15	66·5	2·2	2,160
	Ecuador	25	46–50	15–18	93·0	2·5	1,970
Under 100	India	14	38·4	12·9	139·0	2·6	1,980
	Pakistan	10	43·4	15·4	145·6	3·1	2,260

[a]Conversion to U.S. dollars by means of current exchange rates: E. E. Hagen, 'Some Facts about Income Levels and Economic Growth,' *Review of Economics & Statistics*, xlii (Feb. 1960), 63.
Source: Statistical Office, United Nations, *Statistical Yearbook, Demographic Yearbook*, and *Population and Vital Statistics Report*.

It is clear from Table 1.7 that levels of *per capita* output and degree of urbanization among high-income countries do not in themselves explain differences in vital rates or quality of the socio-physical environment. It must be remembered that some of the more obvious differences in socio-economic and demographic characteristics among low-income countries may well reflect divergences in the relative size and condition of their respective urban and rural components. Data from Japan, for example, suggest that as recently as 1926–30, when the population was about 28 percent urbanized, mortality in cities was still very much higher than in the countryside. In the immediate postwar years this was narrowing rapidly and by 1950, when the country was nearly 40 percent urbanized, registered infant deaths in Tokyo, 42 per 1,000, and Osaka, 50 per 1,000, for example, were already below the national rate of 60 per 1,000. Some fairly reliable data from four countries in South Asia indicated that, as of 1950, the national rate of infant deaths had been declining since the 1930s but some less complete data for large cities, such as Calcutta and Bombay, showed that an increasing rate had prevailed up until the late 1940s and had only recently been reversed. In 1950, however, the rate of infant mortality among the national populations of Ceylon, 82 per 1,000, and India, 127 per 1,000, was still considerably below the rate for such large cities as Colombo, 112, and Bombay, 152. Even if the rate in the large cities was no longer rising, Table 1.7 shows that, for the Indian population as a whole, average infant mortality over the decade 1951–61 *was* higher than at the mid-century, with no more than a two percentage point shift in the interim level of urban concentration.[56]

A more reliable measure of effective fertility in a population is given by the child–woman (c–w) ratio, or the number of live births less childhood deaths under age five per 1,000 women of child-bearing ages. In most East and South Asian countries the c–w ratios around 1950 were still very high compared with the more industrial-urban countries of 'the West.' Nevertheless, the c–w ratios for large Asian cities were almost always lower than for their respective national populations. In Tokyo, 454 per 1,000, and Colombo, 467 per 1,000, at least, the ratio was no greater than in Canada, 498, Chile, 480, or Finland, 476, whose populations then showed the highest c–w ratios among the more developed non-Asian countries. Making due allowance for the quality of Asian data it is clear, nevertheless, that the registered rates of infant deaths in East and South Asian cities during the early 1950s were already well *below* the rates prevailing in the larger cities of Continental Europe and North America at the turn of the century. Thus, while infant mortality may possibly increase and c–w ratios decline somewhat, especially in South Asian countries, the further urbanization of their populations is not likely to have quite the same 'braking effect' on total population growth as it did in the more urbanized parts of Europe and North America in the late-Victorian era.

In regard to the urban demographic transition the Japanese model may be more relevant to Asia than was the earlier experience of Europe or the United States. The drastic steps taken by the Japanese after 1950 to curb their own population, however, occurred at a comparatively advanced stage in their process of industrialization. Table 1.7 shows that in the early 1960s Japan already compared quite favorably with the high-income countries by most measures other than *per capita* product. By 1970 *per capita* product was rising steadily and Japan appears to have accomplished its urban transformation—more than 51 percent of population resident in cities of 20,000 and over—very successfully.[57]

Meanwhile, fertility remains highest in Latin America. In tropical American countries crude birth-estimates ranged up to 44–50 per 1,000 in the early 1960s, comparable to the rates achieved in parts of North America early in the nineteenth century. The more reliable child–woman ratio allows us to make a more careful analysis of the urban-rural differentials in Latin America. It was highest in the Dominican Republic, 749 per 1,000 in 1950, and lowest in highly urbanized Argentina, 423 per 1,000 in 1947. All other countries, except Chile and Haiti, however, had c–w ratios above 500, but Chile was one of the most urbanized countries in the region and Haiti the very least; hence the relation between level of urbanization and the effective fertility of Latin populations does not seem to be a very systematic one. Nevertheless, a comparison of the c–w ratios for total and for non-single women between the principal city and national populations in six countries led U.N. observers to assert 'an inverse association of the level of fertility with the size of the locality.' In general, they concluded, 'the fertility of the urban population is uniformly below that of the total population.'[58]

Data for nine Latin American countries, based on national intercensal periods prior to mid-century, reveal that the greater part of incremental urban growth in Mexico (58 percent), Chile (53 percent), and Brazil (51 percent) already stemmed from urban natural increase. Venezuela (29 percent) and Colombia (32 percent) were the countries whose urban increments had benefited least from natural increase and most from the migration of native and foreign-born elements. The annual average percentage growth in major urban areas in the ensuing years was, not surprisingly, highest in Venezuela, 7·6 percent in 1950–5, and Colombia, 6·6 percent, 1951–5, compared with rates of less than 4·0 percent for Brazil, Mexico, and Chile. Buenos Aires appeared to be the only major urban area with a rate of population growth lower than its country's national rate. U.N. observers of Latin America in the late 1950s endorsed the view of their counterparts in East and South Asia at the time that 'while the volume of rural-urban migration is largely dependent on the economic situation, the demographic characteristics of the migrants are conditioned by long-established social patterns.'

The situation in many developing countries today is perhaps more complex than this conclusion suggests. Some low-income countries are said to be 'over-urbanized in relation to [their] degree of economic development.' The massive urban housing deficits reflected in the growth and size of Indian *bustees*, Brazilian *favelhas*, or Peruvian *barriadas* are cited as evidence of the recent concentration of rural poverty, the spread of underemployment, and of the general explosiveness of the resulting social situation. Such conditions are thought to preclude the achievement of 'political stability and economic growth' which is said to have characterized the development of most 'Western nations' in the past.[59]

Victims of modern warfare such as Korea or Vietnam, and countries such as Egypt, Iraq, Peru, or Colombia in the late 1950s did indeed appear to have reached quite high levels of urbanization and non-agricultural employment in relation to their national levels of *per capita* income. Middle East countries, with about 23 percent of population living in cities of 20,000 and over by 1960, had an average level of gross domestic product *per capita* of only $75 in 1958. Was India at the time, with but 14 percent of her population urbanized and a *per capita* G.D.P. only $5 below the level of the Middle East, achieving a more balanced development? Was the U.S.S.R. with 36 percent of its population urbanized, and a *per capita* G.D.P. above $700 more balanced than Japan with 46 percent of population urbanized and a G.D.P. of only $650 *per capita*? Was New Zealand, at the same level of urbanization as the United States in 1960, but with a *per capita* G.D.P. of only $1,454 compared with $2,324 for the U.S., overurbanized and underdeveloped?

Clearly, the formulation implies some criterion of *balanced* urban and economic growth. Data from 150 countries around mid-century showed a high correlation between the level of urbanization and the degree to which the labor force was employed outside of agriculture. There was, likewise, an inverse relation between dependence of the labor force on agriculture and the level of *per capita* product. At any time it is possible to derive a regression line that summarizes the strong statistical association between (a) the proportion of population found, say, in cities of 20,000 and over, and (b) the proportion of total labor force or product contributed by the non-agricultural sector. Such an operation would reveal some conspicuously deviant countries in which a higher than 'expected' portion of the population is found in cities on the basis of the country's industrial composition and, by the same measure, cases of 'underurbanization' as well. But it does not follow that the only 'justifiable' balance is one that approximates the figment of the regression line. One might well inquire whether in such a deviant case as the Middle East, for example, any substantial numbers in urban places could be justified in the context of extremely low rural *and* urban productivity? But we do not know whether conditions in the urban places are in fact worse than in the alternative rural places and, as was the case in early nineteenth-century Britain, 'those best in a position to judge are continuing their migration.' What we do know is that the situation in the low-income countries today is, to say the least, uncomfortable and often desperate. As Britton Harris has suggested with reference to Indian urbanization, 'if . . . the cities are overlarge in relation to opportunity, so too are the rural areas.' Overurbanization combined with 'overruralization' probably means no more and no less than overpopulation.[60]

Another feature of contemporary urbanization in low-income countries gives rise to a more reasonable concern. In some parts of the world, population concentration is increasing the predominance of one national center within the size-distribution of city populations or the national population as a whole. Buenos Aires, Santiago, and Caracas are prominent examples in Latin America, although the phenomenon is not peculiar to that part of the world either today or in the past. Metropolitan Sydney had developed its primacy in New South Wales by 1861, almost a decade before Buenos Aires had assumed a similar position in Argentina. By 1891 Glasgow contained nearly 20 percent of the population of Scotland and Glaswegians outnumbered the aggregate of its seven next largest cities. London had also been steadily enlarging its share of the population of England and Wales during Victoria's reign and by the 1890s contained just over 20 percent of the whole. Nevertheless, the aggregate for the next sixteen largest cities (of 1871) had grown more rapidly than London over every census decade. By 1901 some thirty cities in England and Wales outside of London contained more than 100,000 population and, in the aggregate, more than a quarter of the nation's population. On the Continent, however, Stockholm, Oslo, Paris, Brussels, Berlin (Prussia), Budapest (Hungary), and Athens (Greece) were at one time or another primate cities in the manner of Glasgow in Scotland in 1891.[61]

It appears from this that there is no close empirical connection between the fact of primacy and economic underdevelopment. There is evidence, nevertheless, that a progression from primacy to a log-normal type distribution of city sizes often does occur as a society develops toward greater complexity and as the patterning effect of any single socio-economic variable is obscured. The examples of London and New York in their respective city-size distributions appear to illustrate this progression at least since the early nineteenth century. But the several patterns of city-size distribution obtaining in the contemporary world do not appear closely related to either the level of economic development or the degree of urbanization of national populations.[62] Very large centers are often unable to provide either adequate public services or full-time employment for a sizeable portion of their residents; alternatively, they are thought to exercise a constraining or 'parasitical' influence over the development of their national or provincial populations. The long history of controversies concerning the optimum size of cities and the effects of 'overcentralization' suggests, however, that such problems, real or imagined, are not entirely novel.[63] Nor does the existence of the primacy phenomenon anywhere in the world necessarily constitute evidence of either 'internal' or 'external' colonialism.

In the contemporary world the term 'colonialism' is a politically-charged expression to denote, in Richard Morse's phrase, 'a relationship between two systems showing discontinuities of structure and inherent purpose.' But the dominance which large cities exert over smaller centers and intervening countryside is as much a cause of their preponderance as it is the consequence. They grow larger, if not always faster, than provincial and local centers because, historically, they have served to integrate the social systems of which they form a part. Their integral role with respect to transport and communications, information flows, credit resources, and so on induces the gravitation of many other functions which benefit from 'cost' or other friction-reducing access to the node. Inequalities and inequities inhere in such a structure, to be sure, but they arise as much from the reciprocal relationships involved as from any others. The subordination of provinces to a metropolis is not indisputable evidence of rival systems with contradictory purposes, though provincial feelings of being victimized can certainly subvert their essentially symbiotic relations with the metropolis.

Morse affirms that, in Latin America at least, the term 'colonialism' may be analytically useful when applied to certain 'inter-ethnic situations' and even to 'certain forms of international influence and manipulation,' but that 'applied to the workings of a total society the expression becomes tendentious, while applied to relations between "city" and "country" it merely obfuscates.' He concludes that in the larger realm of 'collective attitude and social action,' now even more than in the past perhaps, 'the city becomes a theater, not a player—a node of forces, not a quantum of energy.' The conception of cities as theaters of forces or, less elegantly, as foci of generalized nodality, is central to an understanding of recent alterations in the organization of human ecosystems and many related phenomena of social change.[64]

Cities have become the focus for the twin forces of specialization and differentiation which have contributed so greatly to the making of the contemporary world. During the Victorian era exceptional numbers of people were concentrated in impacted areas of different size, density, and function. Land-extensive activities, such as staple agriculture, were excluded from the vicinity of cities, whose internal spaces were meanwhile reorganized and reshaped into specialized districts of work and residence and whose populations were required to allocate particular time for labor and domesticity, switching from production to consumption roles at specific times. By the early twentieth century a metropolitan city-dweller in many parts of Europe and North America was already obliged to play increasingly differentiated roles at work, at home, on the lengthening journey between, and in his increasingly fragmented 'community.' During the span of an individual's career, the sequence and specificity of his roles were modulated, not by ageing alone, but by changes in socio-economic status as well. Over the generations the daily routine and unfolding life-histories of individuals departed quite radically from those of their ancestors, urban or rural. The growing specializations of cities and countrysides and the integral linking of small towns, cities, and metropolitan centers were also manifestations of the same specialization-differentiation-reintegration tendency which, as in the case of vertically-integrated factory systems and large business enterprises, yielded increasing returns to scale. Notwithstanding distinctive sets of institutional-legal constraints on individual production units and corporate bureaucracies there is, by analogy, a similarly centralized mode of organization and control over 'systems of cities'—metropolitan hierarchy—which inheres in technological progress.[65]

Perhaps the most striking behavioral manifestations of the industrial-urban transformation were: (1) the generally inverse relation between family fertility and socio-economic status, and (2) the age and status selectivity of population movements to different suburban amphitheaters. Thus the climax of the Victorian city inaugurated and institutionalized a new order of space and time that ultimately penetrated the most intimate realms of personal consciousness and private judgment.[66] By and large, a majority of urbanites adjusted to the demands and disciplines of the emergent order, reconciled to it because for them at least the benefits in real income and acquired status outweighed the heavy costs that mostly fell on other segments of society. In its most recent metropolitan phase, the process of transformation has involved—in aspiration, if not actuality—the *embourgeoisement* of virtually the entire population.

Until the 1940s the spread of urbanization to the world was a gradual and highly selective process. Temporal as well as cultural and contextual differences from late

Victorian Europe and North America almost certainly assured that the outcome would diverge from any nineteenth-century model of change and adjustment. Before the mid-twentieth century, the very phenomena of urbanization and economic growth had ceased, so to speak, to be historical happenings and had become the subjects of highly self-conscious sets of competing ideologies. During the inter-war years, however, no country outside of A. F. Weber's 'western world,' except Japan and the Soviet Union, appeared to be undergoing anything resembling a rapid industrial-urban transformation (see Table 1.8). But in the two decades after 1940,

Table 1.8 *Rates of urban concentration, leading countries, 1920–60*

(*percentage point shifts*)

1920–40				1940–60			
Japan	16	S. Africa	8	Venezuela	26	Finland	14
U.S.S.R.	14	Morocco	8	Uruguay	18	Brazil	13
Spain	10	Uruguay	8	Mexico	17	Peru	13
Denmark	9	Venezuela	8	Colombia	17	U.S.S.R.	12
Chile	9	France	8	Chile	16	Bulgaria	12
Korea	9	Greece	8	Korea	16	Egypt	12
New Zealand	8	Sweden	8	New Zealand	15	Canada	12
Iran	8	Finland	8	Argentina	14	Australia	12

Rate of urban concentration: $\delta(u/p)$, cities $\geq 20{,}000$.
Source: U.N. Dept of Economic and Social Affairs, provisional data, 9 Dec. 1968.

there were few countries which did not experience a marked surge in their rates of concentration. Only in Great Britain, the most urbanized country in the world, did the situation appear to have stabilized with about 70 percent of population concentrated in urban areas of 20,000 and over. The most awesome change occurred in Latin America, while in Korea and a few North African and Middle Eastern countries, levels of concentration rose by more than 10 percentage points. The absolute numbers of people moving to cities in India, Pakistan, or Indonesia could likewise no longer be ignored.[67] Even the United States, which had been conspicuously unsuccessful among developed nations in implementing any kind of 'urban' policy at home, became publicly concerned with the urban problems of low-income countries.

In view of population growth, the *per capita* domestic resources available in most low-income countries were judged to be inadequate if economic growth was to be achieved under conditions of 'political stability.' Resources that could be earned from the normal course of international trade, moreover, seemed uncertain and, depending on the terms of trade, comparatively expensive. Under Cold War conditions, moreover, neither the United States nor Soviet blocs could rely upon the United Nations to serve as the agency for devising and implementing assistance policy; they each preferred to make bilateral arrangements. Given the necessity for 'development,' as the problem is defined and measured by developers from high-income countries, the need for international assistance in the contemporary world is great, almost without regard to political-ideological strings attached by its donors.

The experience of Australia and New Zealand indicates, nevertheless, that on the product side at least, developing countries need not follow the form of Great Britain, Germany, the United States or Japan in their respective transformations. Certain tendencies in the structure of consumption in the economically-developed countries are more alike, however, although the proportion of actual goods and services comprising private and public expenditures differ greatly over time according to technologies and tastes. The proportions of disposable incomes spent on such basic consumption needs as food, clothing, furniture, or shelter do not necessarily increase over time.[68] Thus coefficients derived from cross-sectional structural associations between levels of urbanization, shares of manufacturing in total labor force or product, and additions to total real income and product in the advanced countries, even when they agree in broad direction, may differ greatly in *magnitudes* owing to the numerous variables involved: the size of populations, grades of raw materials, quality of the labor force relative to changing technologies, involvements in international trade, and, ultimately, to differences in social and cultural tradition.

All this implies a high degree of contingency in the social transformation required to accomplish economic growth. It also helps to account for the selective character of urbanization in different parts of the world. More important, it suggests that low-income countries are not doomed to recapitulate the 'modernizing' experiences of the more developed worlds, western or eastern. There is a margin for hope that, *if* certain acceptable minimum levels of material existence were reached, no social catastrophe need follow, nor even 'development.' Indeed, forced development, western or eastern style, might bring catastrophe. The outcome might depend very often on the extent to which the average number of persons 'per private household' can be lowered. High death-rates notwithstanding, the average size of 'private households' in countries with *per capita* G.D.P. under $200 in 1958, five members, was almost one-and-a-half times as large as in moderate and high-income countries with *per capita* G.D.P. of $575 and over. The proportion of total household population in the 'under $200' countries contributed by households of seven or more members was close to half at the time, compared with little more than 15 percent in the higher-income countries. One barely credible means of raising *per capita* incomes to some locally defined and tolerable minimum might be an effort to substitute such an acceptable level for the seventh, sixth, and fifth births per household by successive target dates before the *annus mirabilis vatum*, A.D. 2000. Only when one has contemplated the full complexity of factors, social and normative, affecting fertility in different cultures, perhaps, can one turn back to consider alternative, if scarcely less complex, strategies for stimulating economic growth![69]

The Human Eco-System in an Urbanizing World

While the reduction in fertility might lessen some of the more obvious pressures making for social catastrophe in newly-developing countries, a comparable achievement in many high-income countries might prove to be a greater boon to the entire planet. In a world of finite resources there are boundary conditions, albeit uncertain, which set limits to all human procreation. The *effective* population in the human ecosystem is not mere numbers, but rather the impact of numbers on physical and social environments. With less than 6 percent of world population in 1970, for example, the United States consumes 40 percent of the annual world production of raw materials. If, on a very rough calculation, one additional North American birth has an effect on natural resource consumption of approximately twenty-five births in India or Pakistan, the recent decline in U.S. fertility might be regarded as providing some ecological relief for the rest of mankind. During the 1950s the refined fertility of American women averaged 3·35 children but by 1970 the figure had dropped to 2·45; the U.S. population is currently growing by little more than one percent per annum (and in some high-income countries of Europe the rate is already less than one percent). A fertility-rate in the United States of 2·11 children would, barring immigration, eventually produce zero population growth. If the trend continues, the Director of the U.S. Bureau of the Census confidently predicts a near doubling of average family income to $15,000 (constant) by the year after 1984 when, he affirms, the nation will be transformed into 'a society of an affluent majority.' But the effect of near-zero population growth, without some radical change in consumption patterns, might actually involve even greater demands upon physical resources than are currently being made.[70]

Unfortunately the United States, like other industrial-urban countries, is also an effluent society. Even at the very low rate of population increase obtaining today, consumption of electric energy is growing at an annual rate of 8 percent and, according to René Dubos, the quantity of accumulated thermal pollutants is doubling each decade. Eventually the production of electric energy by any technological means, Dubos insists, will have to be halted because the heat pollution of the earth has no *technical* solution. At some point conversion to solar energy sources will become literally vital.[71]

There are two features of the ecological problem which have become critical. With increasing control over mortality, the industrial-urban environment has been unable to generate sufficiently powerful resistances to moderate the mounting abuse of the human ecosystem. Information has been signaled too slowly and resulting feedbacks have been too weak to bring about corrective action. As a consequence, we have no very dependable knowledge of the time remaining in which such actions can be instituted—whether, for example, the horizon is five, fifty, or five hundred years.

Climatic changes are already apparent. For example, between 1957 and 1963 the observed dust-content of the air over the United States, outside the cities, doubled. This had the effect of making the earth brighter and reducing the penetration of sunlight to its surface. From 1950 to the early 1960s a 5 to 10 percent rise in turbidity is estimated to have produced a fall of from 0·6 to 0·7 of a degree centigrade in average temperature—an amount equivalent to the warming achieved over the previous century. The effects of these climatic changes are felt differently, but in the more industrially urbanized areas certain tendencies are common. The built-up areas of cities form surfaces which are from 50 to 60 percent waterproof and as a consequence city surfaces become much drier and dustier than those of open country which has organic cover. Since industrial and household chimneys are also emitting vast quantities of solid pollutants every day, the combined effects of waterproof surfaces and air turbidity are unmistakable. By the mid-1960s there were from ten to fifteen times as many dust particles on average in the air over open country as in 1950, and 10,000 times as many particles over the more polluted parts of cities. Over the urbanized region that sprawls along the Atlantic seaboard from north of Boston to south of Washington, D.C., the dust load of the air increased in less than fifteen years by a factor of twenty. The possibility of adapting solar energy was thus to some extent impaired. There was said to be 30 percent less penetration of sunlight and 90 percent less ultra-violet light and a concurrent increase in the nuclei available for fog condensation (in addition to the rising level of chemical pollution reflected in measures of lead or mercury in human blood samples).[72]

The time horizon is part of the ecological dilemma. We do not know how long the industrial-urban populations have before their opportunities for exercising rational options will be foreclosed by, say, river and ocean pollution, air poisoning, or heat accumulation. We are no less ignorant of the time-telescoping effects, in newly urbanizing countries, of the introduction of new forms of transport, new sources and forms of heat energy, or of electronic modes of communication in advance of general literacy. Even if more residents of high-income countries become less mesmerized by G.N.P. as it grows grosser in its content and ecological aftermath, the population of developing countries will, short of some Gandhian conversion, still have to raise their own levels of output and income somewhat if they are to attain hypothetically acceptable standards in regard to nutrition, medical services, or housing. Moreover, in default of profound changes in the outlook and behavior of populations in high-income countries, it will not be easy to convince either governments or masses in the Third World that they must now forgo the greater part of the 'progress' which has belatedly been offered them in the form of economic growth. Yet the decisions made by people in low-income countries in regard to growth will also determine the range of options remaining open to their own children in the not too distant future. There is a real danger that, if they adopt the environmentally-destructive features of industrial-urban technology, they run the risk of accelerating that desert-breeding sequence of negative feedbacks on the ecosystem which has been rendering portions of the earth's surface uninhabitable since the second millennium B.C.

Underlying the social and ecological dilemmas of both high- and low-income countries is the contemporary formulation of the Victorian concept of progress. This idea took hold as a small island nation—with little more than one percent of world population—embarked upon an industrial transformation four or five decades in advance of any other country. Within little more than a century, the technological and organizational forms embodying the transformation had spread to include one-fifth of a greatly enlarged world population. Since the close of Victoria's reign advocates of progress have generally assumed that the economic and social benefits to be derived from this great alteration in occupational and residential patterns far outweighed any social costs entailed. Until recently the ecological costs were ignored altogether since most adherents of progress were unaware that human populations were part of a human ecosystem.

Thus the intellectual framework devised for the study of industrial-urban development tended to become extremely narrow. Certain concomitants of the transformation were adduced to explain and rationalize the material changes involved. Attention was focused on *measurable* inputs and outputs. Larger net differences in real values of inputs and outputs are conventionally reckoned a more efficient use of scarce resources; rising productivity in the output and utilization of resources is attributed to the more intangible, but no less benign, influences of technology and organization. To the extent that net product and income grow faster than population, almost without regard to their actual distribution, the average level of living is said to be raised, real incomes enlarged, and a country's welfare enhanced. By the same entelechy, sectional and class antagonisms are progressively resolved by urbanization, upward mobility, and increased consumption of goods and services. All but the technical aspects of social problems are thus solved by the process itself.

This intellectual framework is none the less a hastily improvised structure designed to meet the exigencies of contemporary history. It represents a rapid translation of the Victorian idea of economic progress into the mid-twentieth-century idea of economic growth. While the inherited liberal and socialist systems of thought still differ on the desirability and extent of conscious public intervention in the socio-economic transformation, they are alike convinced of the transformation's necessity and virtue. Despite a brief flurry of disquiet concerning the need 'to conserve' natural resources, and a throb of neo-Malthusian alarm about the technical elasticity of the food supply in a period of rising prices before the First World War, this faith in progressive economic solutions to what others had mistakenly conceived as political, social, or even moral questions has remained unabashed.

It is the contemporary idea of 'growth,' albeit managed growth, that inspires the 20 percent of world population in high-income countries to recommend a parallel and more expeditious transformation to the less affluent 80 percent of mankind. The identification of economic and social progress with rising G.N.P. *per capita*, and a corollary belief that problems of production and consumption are solvable for all but a minority of social misfits and intellectual intransigents, are common ground between liberal and socialist growth managers competing in today's urbanizing world. The faith that, on a firm foundation of economic growth, social and all but the most profound human problems are manageable is the meaning of the Victorian urban-industrial transformation projected into contemporary history.

From an ecological perspective, the growth framework is inadequate and potentially calamitous. Even from the economic standpoint the apparatus of theory and measurement must be judged incomplete in so far as it reduces the complexity of men's relationships with their social and physical environments to the formal simplicity of an ideal market-place. The utility functions employed by economists are by definition only associated with voluntary exchanges between two parties: buyers and sellers. Waste products and other side effects of normal transactions are not taken into account as goods and services. Yet 'bads' and 'disservices' arise in the course of such exchanges, regardless of whether either party wants them or not. Before the automobile, for example, gasoline or petrol was largely a 'waste' by-product of kerosene manufacture; it was customarily dumped into the rivers and streams adjacent to refineries. Thus neither the full product nor the full cost of business decisions is entered in a firm's ledgers or even in a country's income and product accounts. Such products and costs are absorbed by either the social or the physical environment.[73]

In respect of human environments, the rationality of the market has always been highly dubious. Economists have long recognized that population concentration gives rise to productivity increments which are independent of the operational decisions of firms which locate at such points. The economies of concentration and urbanization are essentially various types of scale economies. But just as part of the economies which are independent of the operations of firms nevertheless accrue to those firms, as well as to the larger economy, so do the *diseconomies* of concentration. Most of the diseconomies that arise, whether from concentration or from the operations of individual firms and industries, are by accounting-practice kept 'external' to those firms. Some of these costs fall on other firms, for example those connected with traffic congestion or water pollution, and are duly accounted as internal costs of production. Still others are borne by households in such forms as poisonous air, noxious drains, sooty window-sills, or insufferable noise. Householders must clean up such products at their own expense, adapt themselves to a degraded or unhealthy environment, which probably entails a capital loss, or complain to the local authority. If they can afford it, householders are free to remove to another site the costs of which, if not actually lower than at the present one, will at least be 'voluntarily' incurred. Thus optimizing behavior by one unit in the economy is almost always achieved by coerced sub-optimizing of other units and by uncompensated damage to the environment.

Society stands as a cushion between the economy and the ecosystem. The balance of externalized and unrequited costs redounds to society in the form of 'urban' or 'social' problems. Municipalities or other authorities eventually charge back a portion of these costs to households and enterprises in the form of tax bills; the residuum passes into the environment as an ecosystem cost. Ecological costs are virtually disregarded by all practical decision-making units except perhaps as they sometimes reappear in the rising prices of raw material inputs. Since the socialist countries have the same system of economic belief as liberal-capitalist countries and differ only in respect of 'ecclesiastical' control over assets, their awareness of the effects of externalities on the ecosystem is no more sensitive than that of their political antagonists. Indeed, in so far as state enterprises create external diseconomies and state authorities 'plan' to pass such wastes and side effects into the people's environment, their decisions might be judged, by outsiders at least, the more culpable and coercive.

Historically, political adjustments in the form of regulation and taxation have moderated some of the socially more dysfunctional effects of the urban-industrial transformation. More recently, governmental decisions affecting the social and physical environments have been subjected to a welfare-accounting technique which is thought to provide a more comprehensive 'cost-benefit' analysis than conventional accounting practice. The procedure involves itemizing all the foreseeable consequences of a proposal—such as building a motor highway or extending a rapid transit system—and estimating the full costs and benefits of alternative actions. Since costs and benefits are expressed in money values, it is comparatively easy to make the rational choice in favor of the particular proposal which maximizes benefit-values relative to costs. By such means, economic decisions made in the public sector are brought into conformity with the ultimate good of Gross National Product. In liberal societies, progress is further safeguarded by the requirement that no decision be taken in the public sector which could not be more efficiently made in the private sector. The comparative efficiencies of the two sectors are likewise judged in terms of cost-benefit analyses.

This recent innovation in social accounting is in part a response to rising criticism of the products of affluent societies. But even with the heightened social and ecological sensitivities of the 1970s, there is much that is dubious about the 'higher rationality' of policy-science. Are all consequences of decisions in fact foreseen? Do the money-value estimates attached to consequences which are foreseen really measure the full costs and benefits? Can ecosystem costs really be requited in a 'growth'-traumatized system? Can political decision-makers realize and accept the full costs of their actions any more than private decision-makers? Under the combined pressures of commercial and political advertising, is the public likely to be more aware of real issues today than in the past? Undoubtedly the techniques of welfare analysis will be improved. It potentially offers a more intelligent appraisal of alternative futures than either laissez faire or arbitrary decree. On the other hand, the primacy which cost-benefit analysis attaches to short-run gain, the intangible nature of so many real social values, and the ultimate weight it attaches to the most 'economical' solution, tend to narrow the range of social choices which technocratic decision-making leaves open to the future. In default of social vision, it will make not for a greater variety of urban environments but for the most uniformly 'efficient' one. To the extent that neither welfare nor growth frameworks take any account of fundamental inequalities in the distribution of wealth or income, nothing is changed very radically from the past.

Applied to the human ecosystem, cost-benefit analysis tends rather to build into policy-making much of the conventional 'growth' wisdom than to open up new patterns of life and livelihood which might embody the environmental wisdom we already possess. It is something, perhaps, that the costs of enforcing antipollution laws, many of which have rested on the statute books for more than half a century, are likely to be passed on to the consumer in the form of taxes or higher prices; this will at least reduce the amount of income available to be spent on unnecessary things. Nevertheless, to import moon dust at an admitted cost of $100,000 a pound, while reducing appropriations to prevent the unwanted production of earth dust, suggests that criteria other than welfare govern the decisions of at least one of the two largest industrial-urban polluters in the contemporary world. Cost-benefit calculations seem merely to confirm the existing order of social priorities.[74] This not only bodes ill for the actual calculations that are made regarding the relative private and social rates of return in the immediate future, but also almost ensures that the larger exploration of resource-uses will, from an ecosystem viewpoint, be suboptimal. While this situation is obviously critical in the case of non-renewable resources, it also indicates that in the longer run we are more likely to achieve the quantitative expansion of industrial-urban civilization, whose end John Stuart Mill once described as 'a stationary state,' than a planned 'steady state' of near-zero population growth, stable *per capita* energy-consumption, and effective recycling of resources called for by an ecological moderate like René Dubos.

A third possibility, that conditions for human life will rapidly become untenable and *pari passu* the chances of provoking nuclear disaster increased, cannot entirely be ruled out. The scenario has already been written in the apocalyptic message of the biologist, Paul Ehrlich.[75] In light of critical currents loose in the contemporary world, it is no less far-fetched to postulate the achievement of telesis: an intelligent direction of natural and social forces toward a benign end. As the future unfolding between these extremes becomes more random, historians of the Victorian age can only turn back to one of its prophets, Patrick Geddes, who lived long enough into this century to mark off the progress of the city along its course:[76]

> In all the great cities—especially the great capitals—you have in progress the history of Rome in its decline and fall. Beginning as Polis, the city, it developed into Metro-polis, the capital; but this into Megalo-polis, the city overgrown, whence megalomania. Next, with its ample supply of 'bread and shows' (nowadays called 'budget') it was Parasitopolis, with degeneration accordingly. Thus, all manner of diseases, bodily, mental, moral: hence Patholo-polis, and finally, in due time Necro-polis—the city of the dead, as its long-buried monuments survive to show.

By the 1970s, at a conservative estimate, the Victorian city is already a well-advanced Megalopolis. In the contemporary ambience, it is still too early to say whether it will have even a surviving monument.

Notes

1 G. Barraclough, *An Introduction to Contemporary History* (1965), pp. 1–35.
2 Data from A. F. Weber, *The Growth of Cities in the Nineteenth Century: A study in statistics* (New York, 1899), pp. 20–154.

3   E. E. Lampard, 'Historical Contours of Contemporary Urban Society: A Comparative View,' *Journal of Contemporary History*, iv (1969), 3–10.
4   W. C. Taylor, *Natural History of Society in the Barbarous and Civilized State* (1840); *Notes of a Tour in the Manufacturing Districts of Lancashire* (2nd edn, 1842). Taylor expressed the view, nevertheless, that some critics had exaggerated the dangers from the 'new element of society' hoping to destroy it 'instead of regulating its courses.'
5   A. de Tocqueville, *Democracy in America* (ed. P. Bradley, New York, 1945), I, pp. 299–300; J. S. Mackenzie, *An Introduction to Social Philosophy* (Glasgow, 1890), pp. 101–4.
6   J. B. Priestley, *English Journey* (1934), *passim*; P.E.P., *Report on the Location of Industry in Great Britain* (1939), pp. 294–8.
7   Weber, op. cit., p. 1.
8   United Nations Department of Economic and Social Affairs, *Report on the World Social Situation 1957—Including Studies of Urbanization in Underdeveloped Areas* (New York, 1957), pp. 111–12. Unless otherwise indicated, *level* of urbanization, u/p, in this study refers to the proportion of total population resident in agglomerations with 20,000 or more inhabitants. The terms 'rate of urbanization' and 'rate of concentration' are used interchangeably throughout as a measure of *absolute* structural change in the residential distribution of total population between urban agglomerations, $\geq 20,000$, and small town and rural areas, $\leq 20,000$. The 'rate' is simply $\delta(u/p)$ over a given time interval and is usually expressed as a percentage point shift. N.B. The size threshold $\geq 20,000$ tends to underestimate the level of urbanization given by most national census definitions of 'urban' population: see Table 1.5.
9   Sen Dou Chang, 'The Million City of Mainland China,' *Pacific Viewpoint*, ix (1968), 128–53, and S. R. Rein, 'The World's Great Cities: Evolution or Devolution,' *Population Bulletin*, xvi, no. 6, 109–30.
10 Demographic and economic data in the text from B. R. Mitchell and P. Deane, *Abstract of British Historical Statistics* (Cambridge, 1962). Comparisons of Table 3, columns 2 and 3, 4 and 5, 7 and 8, indicate that in certain decades (1830s, 1860s through 1890, in England and Wales; the 1860s and 1880s in the U.S.) the growth of the labor force outside of agriculture was chiefly in the service sector and not the manufacturing-mining-and-construction sector.
11 An extreme example of this dependence on migration is provided by London in the eighteenth century. The crude death-rate, 1701–50, averaged 49 per 1,000 while the rate for England and Wales averaged only 33 per 1,000. Since the crude birth-rate in London, 1701–50, averaged 38 per 1,000, the average rate of natural increase for the capital area was *negative*, −11 per 1,000, while for the country as a whole an average birth-rate of 34 per 1,000 assured a slight positive increase. Both the natural deficit and the net increment (which may have reached 125,000 over the fifty-year span) were made up by migration of people from other, mostly rural, parts of England and Wales. Even if London had not grown rapidly, net migration must have been considerable, merely to have maintained the existing population. E. A. Wrigley estimates a net in-migration of 8,000 to 10,000 a year 'to make good the burial surplus and allow the city to continue to grow';

Population and History (1969), p. 150. See also R. Mols, *Introduction à la démographie historique des villes d'Europe du xive au xviiie siècle* (Louvain, 1955), II, Bk IV.

12   Weber, op. cit., p. 236; H. A. Shannon, 'Migration and the Growth of London, 1841–91,' *Economic History Review*, v (1935), 79–86; A. K. Cairncross, 'Internal Migration in Victorian England,' *Manchester School*, xvii (1949), Table VII. The early surplus of London births cited by Weber is clearly a calculation based upon place of birth and death and not upon place of residence.

13   *Statistisches Jahrbuch der Stadt Berlin*, xix (1892), 94–5. The adjustment for Liverpool annexations is based on Mitchell and Deane, op. cit., pp. 25–7, note (e). City-born children of migrants are counted in the city's increase.

14   E. Levasseur, *La Population française* (Paris, 1889–92), II, p. 386; R. Boeckh in *Statistisches Jahrbuch Berlin*, 1892; I. B. Taeuber, 'Urbanization and Population Change in the Development of Modern Japan,' *Economic Development and Cultural Change*, ix (1960), Part 1, 4–14; Weber, op. cit., p. 240.

15   The Massachusetts census for 1895 makes this point clear. Over the preceding decade the population of the thirty-two incorporated cities of 1895 increased by 38 percent, or 38 per 1,000 for one year; but since the annual excess of urban births was averaging about 7 per 1,000 and the political process of annexation becoming more difficult by the year, the bulk of the urban increase must actually have been of foreign birth. In 1880 Boston, with 31·6 percent of its population foreign-born, ranked ninth among the nation's large cities by this measure but by 1910, with 35·9 percent, it ranked second; 74·2 percent of its population was foreign or had one or both parents born abroad.
See Massachusetts *State Census of 1895*, I, pp. 49, 220; *49th Registration Report*, 1890, pp. 156, 372, 374; F. S. Crum, 'The Birth Rate in Massachusetts, 1850–90,' *Quarterly Journal of Economics*, xi (1897), 259; *Summary of the Vital Statistics of the New England States*, 1892, p. 56; *10th U.S. Census, 1880, Population*, pp. 538–41; *13th U.S. Census, 1910, Population*, I, pp. 178, 826–8, 1007.

16   C. Booth, 'On the Occupations of the People of the United Kingdom, 1801–81,' *Journal of the Royal Statistical Society (JRSS)*, xlix (1886), 329; G. B. Longstaff, *Studies in Statistics: Social, political, and medical* (1891), p. 25. Supplement to the Swedish Census of 1890, *Befolkningsstatistik*, n.s. xxxii, no. 1, cited by Weber, op. cit., p. 237.

17   W. F. Willcox, 'The Federal Census,' American Economic Association *Publications*, no. 2 (March, 1899), 24; E. G. Ravenstein, 'The Laws of Migration,' *JRSS*, xlviii (1885), 167-235; Weber, op. cit., Table CXXI, pp. 249, 264–5.

18   The 'rural population' here is that portion given in the *Census of England and Wales* as 'rural districts and towns under 2,000.' See also, G. B. Longstaff, 'Rural Depopulation,' *JRSS*, lviii (1895), and more generally J. Guillou, *L'Émigration des campagnes vers les villes* (Paris, 1905), pp. 143–295.

19   Weber, op. cit., Table CXXI; *Historical Statistics of the U.S., Colonial Times to 1957* (U.S. Bureau of the Census, 1960), Ser. C, pp. 2–3; H. Llewellyn Smith, 'Influx of Population,' in *Life and Labour of the People in London*, C. Booth, ed. (1892–7), III, chs 2 and 3; H. Rauchberg, 'Der Zug nach der Stadt,' *Statistische Monatsschrift*, xix (1893), 125–71. W. Köllmann, 'The Process of Urbanization in Germany at the Height of the Industrialization Period,' *J. Contemp. Hist.*, iv (1969), 59–76.

20 Weber, op. cit., pp. 287–8; *11th U.S. Census, 1890, Vital Statistics of Cities*, p. 13; W. F. Willcox, 'Distribution of the Sexes in the US,' *American Journal of Sociology*, i (1896), 732.

21 Since registration data from a number of countries indicated a smaller preponderance of male births in large cities than in the countryside, the heavier infant mortality in the cities contributed to the urban female surplus as well as the in-migration differential: see data in M. A. Legoyt, *Du Progrès des agglomérations urbaines et de l'émigration rurale* (Paris, 1870), p. 69; Weber, op. cit., pp. 298–9; Massachusetts State Board of Health, *28th Annual Report, 1896*, p. 753. Also C. Walford, 'On the Number of Deaths from Accident, Negligence, Violence and Misadventure,' *JRSS*, xliv (1881), 28.

22 P. Meuriot, *Des Agglomérations urbaines dans l'Europe contemporaine* (Paris, 1897), gives an illuminating graphic representation of the age and sex structures of the populations of Paris and France respectively. *11th U.S. Census, 1890, Vital Statistics of Cities*, p. 16. R. Kuczynski, *Der Zug nach der Stadt* (Munich, 1897).

23 Booth, *Life and Labour*, III, p. 139; N. Brückner, 'Die Entwickelung der grossstädtischen Bevölkerung im Gebiete des Deutschen Reiches,' *Allgemeines Statistisches Archiv*, i (1899), 632–50. Also, A. J. Coale, 'How a Population Ages or Grows Younger,' in *Population: The Vital Revolution*, R. Freedman, ed. (Chicago, 1965), pp. 47–58.

24 Weber, op. cit., p. 324; W. Ogle, 'On Marriage-Rates and Marriage-Ages,' *JRSS*, liii (1890), 253, 267.

25 Mass. St. Bd of Health, *28th Annual Report, 1896*, p. 826; F. S. Crum, 'The Marriage Rate in Massachusetts,' American Statistical Society *Publications*, iv (1895), 338. *11th U.S. Census, 1890, Population*, I, pp. clxxxvi, 851, 858.

26 M. Rubin and H. Westergaard, *Statistik der Ehen* (Jena, 1890); Rauchberg, *Stat. Monatsschrift*, XIX, pp. 136–7; Brückner, *Allg. Stat. Archiv*, i, 640–1; Weber, op. cit., p. 320. E. A. Wrigley, *Industrial Growth and Population Change* (1961), pp. 143–5.

27 *Zeitschrift des K. Sächsischen Stat. Bureaus*, 1885, cited by Weber, op. cit., p. 333. J. E. Wappäus, *Die allgemeine Bevölkerungsstatistik* (Leipzig, 1861), II, pp. 481–4; Levasseur, op. cit., II, pp. 77–81, 390–8.

28 *11th U.S. Census, 1890, Population*, I, p. cxc; Crum, 'The Birth Rate,' p. 259. H. Bleicher, 'Die Eigenthümlichkeiten der städtischen Natalitäts und Mortalitätsverhältnisse,' 8th International Congress of Hygiene and Demography, *Proceedings*, Budapest (1894), VII, p. 468.

29 Weber, op. cit., p. 337. There is, of course, much evidence to confirm the fact that rural fertility exceeds urban fertility and even that the refined birth-rate tends to vary inversely with size of city: see W. S. Thompson, *Ratio of Children to Women 1920* (Washington, D.C., 1931) p. 142, and A. J. Jaffe, 'Urbanization and Fertility,' *Amer. J. Sociology*, xlviii (1942). Nevertheless, an analysis of U.S. data by states in B. Okun, *Trends in Birth Rates in the United States since 1870* (Baltimore, 1958), pp. 52–101, shows that neither cityward migration nor resulting size and density are *per se* explanations of lower urban fertility. Socio-economic class structure and what Okun calls 'urbanism,' the spread of middle-class 'urban' ideas and attitudes among lower classes and to segments of the rural population, are more directly related to the fall in fertility. Significantly, 'rural birth ratios are lowest in the most

urbanized and industrialized states.' Some of the remaining confusion in the literature about urbanization and fertility probably arises because some scholars employ *structural* or *behavioral* definitions of 'urbanization': on this point, see E. E. Lampard, 'Historical Aspects of Urbanization,' in *The Study of Urbanization*, P. M. Hauser and L. F. Schnore, eds (New York, 1965), pp. 519–20.

30 *55th Annual Report of the Registrar General for England and Wales*, 1893, *Supplement*, Pt I, p. xlvii, Pt II, p. cxvii. *11th U.S. Census, 1890, Report on Vital and Social Statistics*, Pt I, pp. 17–19. In 1890 the census registration states were: N.H., Vt., Mass., R.I., Conn., N.Y., N.J., Del., and D.C. The 'urban' segment is that population resident in incorporated places ≥ 5,000; the 'metropolitan district' of New York includes New York, Kings, Queens, Richmond, and Westchester counties in N.Y., Hudson and Essex counties in N.J., and the cities of Paterson and Passaic (ibid., appendix). Weber, op. cit., p. 346.

31 Data from *Encyclopaedia Britannica* (11th edn), XVI, p. 946. If changes in mortality are easier to interpret than changes in fertility they nevertheless remain very hard to measure precisely; apparent differences between town and country and among towns should be treated with due caution in light of D.V. Glass, 'Some Indicators of Differences between Urban and Rural Mortality in England and Wales and Scotland,' *Population Studies*, xvii (1964), 263–7.

32 See A. Shadwell, *The London Water Supply* (1899); G. P. Bevan, *The Statistical Atlas of England, Scotland, and Ireland* (Edinburgh, 1882), pp. 65–8; London County Council, *London Statistics*, xix, 1909, map. Also N. M. Blake, *Water for the Cities: A History of the Urban Water Supply Problem in the United States* (Syracuse, 1956).

33 E. R. Dewsnup, *The Housing Problem in England, Its Statistics, Legislation, and Policy* (Manchester, 1907); A. Newsholme, *The Elements of Vital Statistics* (3rd edn, 1899), pp. 140, 155.

34 Data from Dewsnup, op. cit.

35 A. Shadwell, 'Housing,' *Encycl. Brit.* (11th edn), XIII, pp. 819–20. Also D. Ward, 'A Comparative Historical Geography of Streetcar Suburbs in Boston, Mass. and Leeds, England, 1850–1920,' *Annals of the Association of American Geographers*, liv (1964), 477–89.

36 *51st Annual Report of the R.G., 1891*, p. lvii; *55th Annual Report of the R.G., 1893, Supplement*, Pt II, p. cxi. In addition to the seven types of milieu noted in the text, the report also covered *dockyard towns* and *London*. Also E. B. Collett, 'The Extent and Effects of Industrial Employment of Women,' *JRSS*, lxi (1898), 219–60.

37 *55th Annual Report of the R.G., 1893, Supplement*, Pt II, p. cx; T. A. Welton, 'Local Death-Rates in England,' *JRSS*, lx (1897), 65–6. Infant mortality data from Mitchell and Deane, op. cit., pp. 36–7. But note D. V. Glass, 'Some Indicators of Differences,' cited above, note 31.

38 F. von Juraschek, 'Die Sterblichkeit in den Oesterreichischen Städten,' 8th Int. Congr. of Hyg. and Dem. *Proceedings*, VII, pp. 491, 502; Bleicher, 'Die Eigenthümlichkeiten,' ibid., p. 477. On the German housing problem, see J. Faucher, 'Die Bewegung für Wohnungsreform,' *Vierteljahrschrift für Volkswirtschaft und Kulturgeschichte*, 3rd Year, iv (1865), 127–99; Verein für Socialpolitik, *Die Wohnungsnoth der ärmeren Klassen in Deutschen Grossstädten* (Leipzig, 1886), 2 vols; T. C. Horsfall, *The Improvement of the Dwellings and Surroundings of the*

People: The Example of Germany (Manchester, 1904), pp. 21–7, 138–61; A. Skalweit, 'Die Wohnungszustände in den Deutschen Grossstädten und die Möglichkeit ihrer Reform,' *Städtebauliche Vorträge aus dem Seminar für Stadtleben*, J. Brix and F. Genzmer, eds, VI, no. 6 (Berlin, 1913); Kuczynski, op. cit., p. 199.

39 W. Thompson, *Housing Up-to-Date* (London, 1907) gives details of housing conditions and public programs from correspondents in a number of countries. Weber, op. cit., p. 362.

40 New York Board of Health, *Report for the Year ending December 31, 1896* (New York, 1897), p. 1; Milwaukee Commissioner of Health, *Annual Report, 1908*, p. 18; Crum, 'The Birth Rate,' p. 259. Infant mortality in *all* Prussia, nevertheless, remained well above Massachusetts levels in prewar years.

41 C. D. Wright, *The Slums of Baltimore, Chicago, New York, and Philadelphia* (U.S. Commissioner of Labor, *Special Report No. 7*, 1894); *11th U.S. Census, 1890, Abstract*, p. 223; G. K. Holmes, 'Tenancy in the United States,' *Quart. J. Econ.*, x (1895), 37; N.Y. Tenement House Committee, *Report*, 1894, pp. 23–5; Weber, op. cit., pp. 460–2. Also R. Lubove, *The Progressives and the Slums: Tenement house reform in New York City, 1890–1917* (Pittsburgh, 1962).

42 *Historical Statistics of the U.S.*, pp. 18, 25–6. On the problem of the pure milk supply in the U.S.: E. E. Lampard, *The Rise of the Dairy Industry in Wisconsin, 1820–1920* (Madison, 1963), pp. 227–36.

43 Milwaukee Commissioner of Health, *Annual Report, 1908*, p. 18; Shadwell, *Encycl. Brit.*, XIII, p. 827. Strictly speaking, international comparisons of crude mortality are misleading owing to possible differences in sex-age composition of the populations involved, hence the greater emphasis given here to infant mortality which is not affected by age-composition; close comparisons among crude mortality-rates of individual cities, of course, are subject to the same shortcomings.

44 Weber, op. cit., p. 341n.; T. H. C. Stevenson, 'The Fertility of Various Social Classes in England and Wales from the Middle of the 19th Century to 1911,' *JRSS*, lxxiv (1911). Class differences in fertility were not a novelty of the late-Victorian era. A population's birth-rate is always an average of the fertility of different groups comprising the population. From the standpoint of interpretation, the critical question is the *basis* for differentiating people into groups; thus one can compare the fertility of different age-cohorts, communities by size, occupational groups, etc. as well as socio-economic classes. It is revealing that urban professional and business classes, as well as members of the older landed interest, deplored the relative fecundity of the lower strata of large cities once the heavy toll of their mortality was lessened. Concern was also expressed in different countries about the sectarian, 'biological,' and 'psychological' characteristics of the urban masses. Compare D. H. Wrong, 'Class Fertility Differentials Before 1850,' *Social Research*, xxv (1958), 70–86.

45 A. Smith, *An Inquiry into the Nature and Causes of the Wealth of Nations* (1776), Bk V, ch. 1.

46 J. A. Hobson, *The Evolution of Modern Capitalism: A study of machine production* (1894), pp. 340–2. Hobson removed some of these strictures on town populations from later editions of his classic study.

47 W. Z. Ripley, 'The Racial Geography of Europe,' *Popular Science Monthly*, lii

(1893–4), 479–80; A. T. Hadley, *Economics—An Account of the Relations between Private Property and Public Welfare* (New York, 1896), pp. 48–9.

48 Cited by Weber, op. cit., p. 474. C. Kingsley, *Miscellanies* (1860), II, pp. 318–45; A. J. Downing, *Cottage Residences* (New York, 1842).

49 S. J. Low, 'The Rise of Suburbs,' *Contemporary Review*, lx (1891), 545–58; 'London ... surrounds itself, suburb clinging to suburb, like onions fifty on a rope'; J. F. Murray, *The World of London* (1843), cited by D. A. Reeder, 'A Theatre of Suburbs: Some Patterns of Development in West London, 1801–1911,' in *The Study of Urban History*, H. J. Dyos, ed. (1968), pp. 253–71. H. Herzfeld, ed., *Berlin und die Provinz Brandenburg im 19. und 20. Jahrhundert* (Veröffentlichung der Historischen Kommission zu Berlin, xxv, 1968) contains an extended account of organizations and interests proposing and opposing annexation of areas around the spreading capital. In the United States political annexation of peripheral areas around the major cities had slackened by the 1890s, giving rise to 'metropolitan suburbs and satellites': American Municipal Association *Report No. 127*, 'Changes in Municipal Boundaries Through Annexation, Detachment, and Consolidation,' (Chicago, 1939), and L. F. Schnore, 'The Timing of Metropolitan Decentralization,' *Journal of American Institute of Planners*, xxv (1959), 200–6.

50 A. Marshall, *Principles of Economics* (3rd edn), pp. 263–85, 305n.; Marshall, 'The Housing of the London Poor. Where to House Them,' *Contemporary Review*, xlv (1884), 226–9; H. Solly, *Industrial Villages: A remedy for crowded towns and deserted fields* (1884).

51 *Garden Cities of Tomorrow* (ed. F. J. Osborne, 1945), introductory essay by L. Mumford, p. 35 (italics in original). For the development of the town- and country-planning movement after Howard, see W. Ashworth, *The Genesis of Modern British Town Planning* (1954), pp. 167–237. The issue underlying Howard's plan for a trust to buy and control land, as distinct from taxing the 'economic rent,' was broadly anticipated by the German movement to control land uses and structures: R. Baumeister, *Stadt-Erweiterungen in technischer, baupolizeilicher, und wirtschaftlicher Beziehung* (Berlin, 1876).

52 P. Geddes, *City Development*, pp. 221–2; *Cities in Evolution, An Introduction to the Town Planning Movement and to the Study of Civics* (1915), p. 258. P. Mairet, *Pioneer of Sociology: The Life and Letters of Patrick Geddes* (1957). Geddes' spirit infuses the remarkable M. Safdie, *Beyond Habitat* (ed. J. Kettle, Cambridge, Mass., 1970). H. W. S. Cleveland, *Landscape Architecture, as Applied to the Wants of the West, 1873*, ed. R. Lubove (Pittsburgh, 1965), is a pioneering American contribution.

53 On the metropolitan phase of the urban transformation: T. W. Freeman, *The Conurbations of Great Britain* (Manchester, 1959), pp. 3–15; A. H. Hawley, *The Changing Shape of Metropolitan America: Deconcentration since 1920* (Chicago, 1956), pp. 1–33; L. O. Stone, *Urban Development in Canada* (Ottawa, 1967), pp. 127–42; J. R. Scobie, *Argentina: A city and a nation* (New York, 1964), pp. 160–88.

54 Rates based on recent U.N. data. The rate of urban population growth is $\delta(u/u)$ as distinct from the rate of concentration, $\delta(u/p)$. Annual rates of urban population growth of from 3 to 4 percent involve doubling urban population every seventeen to twenty-four years.

55 At the outset of the urban demographic transition in late-Victorian times, fertility

was already stable or falling when mortality began to fall rapidly. In the newly urbanizing countries mortality has recently fallen without as yet a corresponding reduction in fertility. See G. J. Stolnitz, 'Comparisons Between Some Recent Mortality Trends in Underdeveloped Areas and Historical Trends in the West,' *Trends and Differentials in Mortality*, Millbank Memorial Fund (New York, 1956), p. 34. On ecological 'backfire' from predator control programs, see R. Rudd, *Pesticides and the Living Landscape* (Madison, 1964).

56 Data drawn from P. M. Hauser, ed., *Urbanization in Asia and the Far East* (UNESCO: Calcutta, 1957), pp. 123–7, Appendix Tables A–F. Consistent with the emphasis on the *infant* death-rate above, George C. Whipple terms it 'the most sensitive index of social welfare and of sanitary improvements which we possess': cited by T. L. Smith, *Fundamentals of Population Study* (New York, 1960), p. 352.

57 In addition to low vital rates and infant mortality shown in Table 1.7, the *absolute* size of rural and small-town population in Japan fell by an estimated 1·6 millions in 1950–70. These three movements were cited earlier as indicators of the urban demographic transition. The gross reproduction rate was already below unity. On the 'economic miracle' in Japan since 1945, see W. W. Lockwood, ed., *The State and Economic Enterprise in Japan* (Princeton, 1965), Pt III.

58 Data in these paragraphs principally from P. M. Hauser, ed., *Urbanization in Latin America* (UNESCO: Paris, 1961), ch. 3. In N. America the negative association between fertility and size of community was said to indicate structural and behavioral change rather than urbanization of population as such: note 29 above. More generally, T. L. Smith, 'Urbanization in Latin America,' *International Journal of Comparative Sociology*, iv (1963), 127–42. The six countries mentioned are Argentina, Brazil, Chile, Cuba, Mexico, Venezuela.

59 Hauser, ed., *Urbanization in Asia*, pp. 130–3. While rejecting the notion of 'over-urbanization,' J. Friedmann and T. Lackington, 'Hyperurbanization and National Development in Chile: Some Hypotheses,' *Urban Affairs Quarterly*, ii (1967), 3–29, argue, nevertheless, that prolonged discrepancy between rates of urbanization and *per capita* income generates a socio-political 'crisis of inclusion.' P. W. Amato, 'Elitism and Settlement Patterns in the Latin American City,' *J. Amer. Inst. of Planners*, xxxvi (1970), 96 argues that slums are developing in the core of certain Latin American cities as well as on the perimeters and that the 'upper classes' are now fleeing to the suburbs; thus the traditional settlement pattern has already collapsed.

60 E. E. Lampard and L. F. Schnore, 'Urbanization Problems,' *Research Needs for Development Assistance Programs* (Brookings Foreign Policy Studies Program: Washington, D.C., 1961), pp. 28–31; B. Harris, 'Urbanization Policy in India,' Regional Science Association *Papers and Proceedings*, v (1959), 196.

61 Data from Weber, op. cit., ch. 2; Mitchell and Deane, op. cit., pp. 25, 27. Also, R. Murphey, 'New Capitals of Asia,' *Econ. Dev. & Cult. Change*, v (1957), 216–43; and S. K. Mehta, 'Some Demographic and Economic Correlates of Primate Cities: A Case for Reevaluation,' *Demography*, i (1964), 136–47.

62 See B. J. L. Berry, 'City Size Distributions and Economic Development,' *Econ. Dev. & Cult. Change*, ix (1961), 573–88. Nevertheless, as of 1955–60 there was no country with *per capita* G.N.P. above $750 with any marked degree of primacy in its

urban structure, with possible exceptions of Denmark and France; there were many countries with *per capita* G.N.P. below $250 with both high and low degrees of primacy.

63 See, for example, H. Jacob, *German Administration since Bismarck: Central Authority versus Local Authority* (New Haven, 1963); E. Lavoie, 'La Décentralisation en France au xixe siècle: une étude bibliographique et sémantique,' *Canadian Journal of History*, v (1970), 43–70.

64 R. Morse, 'Trends and Issues in Latin American Urban Research, 1965–70,' unpublished paper, Faculty Seminar on Comparative Urban Societies, Yale University, spring 1970, 81–8. Also E. E. Lampard, 'Historical Aspects of Urbanization,' *The Study of Urbanization*, Hauser and Schnore, eds, pp. 531–42.

65 E. E. Lampard, 'The Evolving System of Cities in the U.S.: Urbanization and Economic Development,' in *Issues in Urban Economics*, H. S. Perloff and L. Wingo, eds (Baltimore, 1968), pp. 81–139. E. Juillard, 'L'Urbanisation des campagnes en Europe occidentale,' *Études Rurales*, i (1961), 18–33, provides a closely reasoned analysis of urban-rural relations during industrialization. Also, W. Christaller, 'Die Hierarchie der Städte,' I.G.U. Symposium in Urban Geography *Proceedings* (Lund Studies in Geog. Series B, Human Geography no. 24, Lund, 1962), 3–11.

66 E. E. Lampard, 'Urbanization and Social Change,' in *The Historian and the City*, O. Handlin and J. Burchard, eds (Cambridge, Mass., 1963), pp. 234–8. F. Adickes and R. Baumeister, *Die unterschiedliche Behandlung der Bauordnungen für das Innere, die Aussenbezirke, und die Umgebung von Städten* (Brunswick, 1893) is one of the first studies to advocate social controls over differential building and land uses in *and around* large cities. H. Hoyt, *The Structure and Growth of Residential Neighborhoods in American Cities* (Washington, D.C., 1939). On the relation of 'modern mentality' to the spread of the vital revolution among all classes, see R. von Ungern-Sternberg, 'Die Ursachen des Geburtenrückganges im westeuropäischen Kulturkreis während des 19. und 20. Jahrhunderts,' *Congrès International de la Population 1937*, vii (Paris, 1938), 16–34; N. E. Himes, 'Contraceptive History and Current Population Policy,' ibid., 200–10, on successive waves of alarm concerning population growth or decline. Also, E. Lewis-Faning, *Family Limitation and Its Influence on Human Fertility During the Past Fifty Years* (Papers of the Royal Commission on Population), I, (HMSO, 1948), 10, and D. H. Wrong, 'Trends in Class Fertility in Western Nations,' *Canadian Journal of Economics & Political Science*, xxiv (1958), 216–29. On the persistence of other differences in class-related behavior and urban life styles, see M. Young and P. Willmott, *Family and Kinship in East London* (1959) and B. M. Berger, *Working-class Suburb* (Berkeley, 1963).

67 T. Yaziki, *The Japanese City: A sociological analysis* (Rutland, Vt., trans., 1963), ch. 3; C. D. Harris, *Cities of the Soviet Union: Studies in their functions, size, density, and growth* (Chicago, 1970), ch. 8; A. Bose, 'Six Decades of Urbanization in India,' *Indian Economic & Social History Review*, ii (1965), 23–41; Q. Azad, 'Indian Cities— Characteristics and Correlates,' University of Chicago, *Dept. of Geography Research Paper No. 102* (Chicago, 1965); T. O. Wilkinson, 'Patterns of Korean Urban Growth,' *Rural Sociology*, xix (1954).

68 S. Kuznets, *Modern Economic Growth: Rate, structure, and spread* (New Haven, 1966), Table 5.7, Table 8.1, lines 83–110.

69 Ibid., Table 8.2. G. Ohlin, *Population Control and Economic Development* (Paris, 1967), is a model treatment of this problem. Also Lord Boyd-Orr, 'Food Enough for Everyone,' *New York Times*, 17 December 1970.
70 Statement of G. H. Brown, Director U.S. Census Bureau, reported *New York Times*, 8 October 1970. President Richard Nixon hailed the U.S. trillion dollar G.N.P. as now giving the nation the means of making social improvements 'that no other country in the world' can match. The *New York Times*, 17 December 1970, commented editorially on this silly boast, that only minute quantities of resources consumed annually in the U.S. are used for social purposes and less for maintenance of the environment: 'these resources go overwhelmingly into a luxurious private consumption such as the world has never known issuing ultimately in a waste such as the world has never imagined: 7 million cars junked a year, 20 million tons of paper, 48 billion cans and the like—all costing close to $3 billion a year just to dispose of.'
71 R. J. Dubos, *So Human an Animal* (New York, 1968), passim; G. D. Bell, ed., *The Environmental Handbook* (New York, 1970), passim.
72 See R. A. Bryson, 'Climatic Effects of Atmosphere Pollution,' paper delivered at Amer. Assn for the Advancement of Science Meeting, 1968; 'All other Factors being Constant,' *Weatherwise*, xxi (1968), 56–61; G. R. Taylor, 'Trends in Pollution,' *Futures*, ii (1970), 105–13; J. K. Page, 'Possible Developments in the Urban Environment,' *Futures*, ii (1970), 215–21. Also, S. F. Singer, ed., *Global Effects of Environmental Pollution* (Amer. Assn Adv. Science, 1970).
73 E. J. Mishan, *Costs of Economic Growth* (1967) for a general critique of the G.N.P. accounting framework. J. Breslaw, 'Economics and Eco-systems,' *Environmental Handbook*, Bell, ed., pp. 102–12; S. Tsuru, 'The Economic Significance of Cities,' *The Historian and the City*, Handlin and Burchard, eds, pp. 44–55.
74 The inherent limitations of economics as the 'science of control' are indicated by P. A. Samuelson: 'Economics cannot tell us what to believe; it can help us to sort out the costs and benefits of various arrangements, as those costs and benefits are defined by the ethical value systems that we bring to economics,' *New York Times*, 26 December 1970. Also K. E. Boulding, 'The City as an Element in the International System,' *Daedalus*, xcvii (1968), 1111–23.
75 P. Ehrlich, *The Population Bomb* (New York, 1968).
76 Cited by Mairet, op. cit., pp. 125–6.

2 Voices from Within

Paul Thompson

The daily life of the Victorian city seems superficially so remote that we can too easily assume it irrecoverable: as dead as the eighteenth century. Yet it survives among us in the minds of the old, who can often remember their nineteenth-century childhood with astonishing clarity. It is still possible to discover, through asking them, answers to some of the questions which interest us but which contemporaries did not think worth recording; and more important, since the great majority of the population do not write autobiographies, these old people's stories can convey real experiences and provide imaginative insights from points of view rarely found in documents.

If, for example, we want to define more precisely the extent and strength of the late Victorian labour aristocracy, we can still ask how they themselves drew the line —if they did—between the rough and the respectable working classes. We can at the same time discover their family occupational patterns and thus compare subjective views of social class with objective social situations. Similarly, we can supplement existing reports of the very poor and the criminal classes (which mostly come from the police or from paternalistic social workers) with their own accounts. Or if we wish to know whether urbanization in the nineteenth century produced a distinct way of life not simply explicable in terms of occupation and social class, we can examine leisure habits, relations between neighbours, eating, and religious behaviour of different social classes, and see how they vary with the move from village to small town, from large town to conurbation, from suburb to inner city. We can also compare migrants from the countryside with those born in the cities.

Inevitably this fresh evidence as often alters the social historian's question as answers it. In contrast to the changing forms of adult leisure, for example, one is more impressed by the uniformity and persistence of children's games in the face both of Victorian urbanization and the spread of cheap manufactured toys. The chief exceptions are found in one of two situations: either where children were deliberately isolated by the refusal of their parents to allow them to play outside their own house and garden, or in the intense poverty and overcrowding of inner slum districts where a child might be more concerned with feeding itself than with playing. Such inner districts were also exceptional in not being within walking distance of open fields. Since 1900, while fewer children are no doubt kept from traditional play by sheer hunger, perhaps the appearance of motor traffic in the streets, the distancing of the countryside, and the spread of lower middle-class notions of respectability have resulted in more isolated childhoods. Thus we move from the straightforward assessment of 'Urbanism as a Way of Life' to more complex—and more real—problems.

Similarly, when, in order to test the assumption that parents seventy years ago were harsh, distant, and frequently violent, one listens to descriptions of real families by their own members, it is less the change than the continuity which is striking. There is a timeless quality in many of the emotional relationships which emerge. Interviews can demonstrate conclusively that, rather than a wholesale revolution in parent–child relationships since 1900, there have been gradual changes in distribution of the various types which then, as now, may be found.

Interviews, if collected on a sample basis, can also be used to estimate the scale of this kind of change. The descriptions of city life which follow are in fact taken from a national survey of family, work, and the community before 1918, which is based upon a quota sample of occupations derived from the 1911 census.* At the time of writing it is not, however, complete. In any case, although the descriptions which follow relate entirely to the years before 1900, it is designed to be representative of the Edwardian rather than the late-Victorian period. As it happens, the seventeen interviews used here, which are those completed first, present quite a typical range. Eight are from London, two from an old cathedral city, and seven from the north. Of the heads of the families in which these childhoods were spent, two were professional, four were in trade, one was a clerk, ten were working-class; five were in serious poverty. Nevertheless, it would be misleading to treat them other than as a series of individual cases. At this stage of the work they are simply presented for their intrinsic interest.

Let us begin with two working-class families from a pre-industrial city, a cathedral and market town. City life here was still entangled with that of the countryside,

* The survey is supported by a grant from the Social Science Research Council, and is described more fully in *SSRC Newsletter*, June 1969. It was preceded by pilot interviewing, aided by the Nuffield Foundation and the University of Essex. The respondents quoted here were all born before 1892, and the oldest in 1873. I should like to acknowledge them and their interviewers by name, but in order to preserve confidentiality they must remain anonymous. Personal names and some place names have been altered, but the quotations are otherwise as far as possible literal transcriptions, except for the elimination of some repetitions and hesitations. The recordings of the interviews are to be preserved as an archive.

so that these families provide a good point of departure for our movement towards the inner districts of the great cities. They also bring us directly to the issue of respectability, subjective and objective.

Both families were relatively poor. With the *Pococks* this was because the father, although a skilled man, was old. He had been a railway guard, but was now reduced to casual work such as window cleaning and carpet beating. To make ends meet, Mrs Pocock took in tailoring work: 'mother was always at the machine . . . right into the night sometimes.' They remained nevertheless a proud family. 'You used to get the rough people as well you know, and we were told of course not to mix with those people . . . Well, my father used to say we don't owe anyone anything so therefore we can walk about with a free head. High head.' They were active churchgoers, observed the sabbath, said grace at meals. Perhaps one sign of demoralization was that Mr Pocock began to drink more as he aged, but his main response to the family's difficulty was to help more in the house, even doing all the cooking. This may be why some meals, in characteristically urban fashion, were fetched from a nearby restaurant which served meals on to the customers' own plates 'and that was put in the oven ready for the boys when they came in.' But in many ways the family's food was distinctly rural. Rabbits, crayfish and pheasants poached from a nearby ducal estate by an uncle were roasted on the open fire. There were wild mushrooms. Fresh milk came from the cattle market: 'all the beasts were brought in by road, walking in . . . They needed milking before they could go back and we were sent with a big jug and for twopence we had a big jug of milk straight from the cow. You see the farmers were glad to relieve the cows of it.' Similarly, fresh fruit was easy to get if you lived near 'a man who worked on the railway, an engine driver or something like that, and they had the run through to Evesham: they could buy what they called a pot of plums which was . . . a big basket which held 40 pounds . . . They would share them with their neighbours and sell them.' And again, in the children's play, urban and rural pleasures were juxtaposed: fishing in the river with jam-jars, the traditional autumn fair, the theatre, bicycling to country pubs for cider, and playing in a pub yard where the bargees' horses from the canal were stabled. 'They had a big rough yard with a nice big loft up above, you know. We used to think that was wonderful.'

Similar pleasures also mitigated the poverty of the *Bell* family: fishing again, playing in the fields, and sometimes on Sundays an outing to a riverside pub. But for the Bells poverty was absolute, rather than in contrast to earlier prosperity. Of fourteen children, six died in childhood. They could only afford to buy pieces of meat, the leftovers, and relied for bread on the cast-offs of a Co-operative bakery. Mr Bell was a bricklayer's labourer, quite a young man. 'He used to get 4d. an hour for carrying bricks up a ladder and keeping three bricklayers supplied, and he was only a little man too. He gloried in it too, seeing how much he could do.' Although Mrs Bell worked two days a week as a washerwoman to supplement his earnings, he would not help at all with the house, or with the children, unless to 'shut you up, with a good clout perhaps'. He was illiterate, and regarded education as 'just a waste of time . . .

"Ah well, my boy, I'm all right." ' His attitude to wealth was also defiant. 'He hated the sights of motor cars and being in the building trade, you see, he had plenty of nails, and he used to stand on the pavement when they used to come down the bridge just over here . . . and he used to chuck all these nails so they'd get punctures. He was very bitter that way.'

Although he was a rough man, this did not mean that Mr Bell did not expect strict standards of behaviour in the family. For example, at meals 'Mother used to keep a cane where she sat at the head of the table and if you started talking, you had a smack across the arms. Oh, we had no chance to talk.' They were punished for lies in the same way. Sunday was observed strictly: no playing football in the street, but instead 'three times on Sundays we used to go in the choir. And then that was a farthing a time, and if you misbehaved yourself when you was in the chancel the old vicar used to say, when we got down, "Sixpence off your money, Bell, for talking." So I didn't get much money when it came to the end of the year.' On weekday evenings they were sent to a boys' club, where there was boxing to keep them out of mischief.

Despite these precautions, less respectable forms of fighting remained favourite pastimes. Boys from another neighbourhood 'used to call us the St James's Bulldogs and we used to call them the Thorney Arabs and throw stones at one another up the lane there.' And on quarter days, when army pensions were paid, they could be sure of a good fight to watch among adults. 'You could see them come down the yard, out of the pub they'd come. Course we enjoyed it . . . We boys at school, "Come on, it's pension day today, kids, come and see some fights," out we used to come.' On the other hand, it is apparent that the district was much quieter than it had been in his father's youth. At that time the police had not dared enter it and the vicar had been thrown in the ditch. 'They were rough and ready, I mean. They'd got no discipline, no nothing. They were just ignorant.' Without protection from the police, a man had to be more self-reliant. 'My great-uncle he had . . . a well known pub, the Blue Boar . . . and great-uncle used to keep his double-barrel . . . a muzzle loader . . . loaded in the bar, because of the customers here . . . And if he had any trouble with his customers, which there was plenty, he'd get his gun down, and say, "Look, if you don't get out of here, I'll shoot your legs off." He'd ha' done it too, ooh.'

Here, as elsewhere, it is noticeable that schools did not provide a lead which might help to explain this gradual quietening of a rough district. On the contrary, they stood out as examples of institutionalized violence.

> I was a bit of a rip when I was a boy, high spirits . . . slap a boy's face in
> class for one thing . . . The master says to me, 'Bell, come out, stand on that
> form,' so I stood on the form, and the form used to go round the back,
> the windows was here . . . horses and carts would go by . . . He gets his cane
> and he slashes me all across the legs, and he never stops, so, I had a very
> powerful voice, and at the top of my voice I screamed 'Murder!'
> He stopped . . . 'None of that!' That's how he hit us.

It is a far distance to move from this childhood to that of a middle-class child in a suburb of Liverpool. 'I liked school very much . . . and I liked the teachers, and they were very fond of me.' For Katherine *Bowie* and her two sisters, in fact, school was almost her only experience of the world outside her home. Her father was a ship's officer, away most of the year; Mrs Bowie ran the house without any servants. Being Scottish, she felt 'she didn't understand the people round about her', and made no friends among them. The children were taken shopping, and sometimes in the evening to the park. There was the annual Sunday School picnic, and sometimes a summer holiday in Scotland with her father when 'oh, we just absolutely ran about wild. Go down to the shore and, you know, bathe, and . . . out in the small boats, and the fishing boats, and oh, it was lovely. And then go up to the hills to the farms . . . to help with the haymaking.' But, at home, they were not allowed into other people's houses, or to play in the street. She did not play games with other children, or wander, or have pocket-money, or go to the theatre. Only when Mr Bowie was at home were guests entertained in the house. And of all these days of isolation,

> Sunday was the quietest day on earth. We weren't allowed to do anything. Everything was done on Saturday and there wasn't a dish washed on Sunday. No. We got up in the morning and we had, as I said, we had prayers in the morning, everyone of us knelt down by the chairs. And we got our breakfast . . . And then go at 11 o'clock out to the morning service, come home, and the dinner was all ready cooked the day before, except for a few potatoes being put on. We'd have that and then go back to the Sunday School . . .

For this family, the city in which they lived remained as remote as if their mother had never left the Scottish countryside. Liverpool's cultural and social diversity was beyond their experience. The social system, for example, was conceived by them simply in terms of those who were 'nice people', and those who were not. Apart from school teachers, the music teacher and the dressmaker were 'nice people. My mother wouldn't allow us to be with, you know, people that wasn't nice. We had to speak properly.' These improper others were represented by 'people that would come round, the coal man and the bin men and people like that. You'd hear swear words and things like that.' And Katherine's world, even when she grew up, remained a tiny hedged garden, for she married her father's image, another ship's officer. 'When I was introduced to him, of course, it was the brass buttons that took my eye. My father had brass buttons.'

The second professional family was in some ways similar: suburban, religious, few visitors allowed to the house, and no living-in servants. 'Mother didn't want people to help with the children, she didn't like interference in that way.' But the *Lindsells* were artistic and this made a great difference. The family would perform plays, black minstrel shows, music on the piano, clarinet, mandolin, and banjo: 'we all played something.' Occasionally they were taken to the theatre and 'we did sometimes go to cinemas if the programme was suitable, you see, if father went to see what they were like first of all.' They would play tennis, and go out cycling with their father on Saturday afternoons. There were plenty of books to read in the house, and the girls were also very happy at school.

Nevertheless, the children had their anxieties. Not so much owing to discipline, for this chiefly took the form of strong moral pressure. For example, food 'had to be finished. We were told of all the children who hadn't enough to eat and that sort of thing—and I say one day, "Well couldn't we give it to the children who haven't enough to eat?" They told me to be quiet and get on with my meal.' But it was less easy for the child to handle her fear of divine wrath, and she suffered from constant nightmares. 'I was a mischievous child and I was terribly afraid of going to sleep in case I burnt up. I suppose, you know, I was brought up in the old way of the hell fire.'

Her parents, moreover, had definite psychological problems. 'Mother . . . would have made a marvellous actress, you see, but she just married and had a domestic life and—but she used to be very gay very often and sing, we used to sing operatics and she'd sing, be gay. But she was very frustrated and sometimes very bitter . . . She found it very difficult to show affection. Father was better.' He was, in fact, 'a very sociable man, father; he would have liked a much more sociable life.' But he was unwilling to bring his friends to the house, because they could have no drink there. 'Mother wouldn't allow it in the house, because she was afraid that if father took to it, you see he would drink too much.' For in the background Mrs Lindsell always held the haunting memory of her own parents' broken marriage. 'Grandmother was a very delightful woman but she made the wrong marriage . . . She'd got long dark ringlets —and she was so lovely.' She started to drink, perhaps because her husband's career took him away too much, but more likely because she couldn't stand him when he was at home. 'Grandfather would never drink tea or coffee—he believed in water— "God's ale," as he called it. He was a natural vegetarian. He didn't believe in killing anything.' But these high principles did not protect him from infidelity, or grandmother from drinking herself to death.

There must, of course, have been few Victorian middle-class families completely free of moral fears of Armageddon, drink, sex, or other sins. They can be discerned in the first of our lower middle-class families, who were again strongly religious supporters of a mission chapel. Mr *Barrett* was a London master chimneysweep, employing up to twenty men each night. The Barretts belonged to a Temperance Club; but at one point Mrs Barrett started to drink. There was again a memory in the family, of a great-grandmother, deserted by a ship's captain, who 'died with a broken heart'. The girl's own fears were that the 'end of the world was coming. Oh, I thought myself, I'm not going out tonight, not leave my mother and father.'

If such feelings left little mark on the Barretts, this may be because their economic situation drew them closer together. Mr Barrett expected the children to help with his chickens and garden, and he ran his business from his home. There were also advantages in living in a less isolated house. The children were allowed to play in the street, and, later on, even to get parts in a pantomime. Although guests were rare and the house door was firmly locked at ten o'clock, there was plenty to look at from the bedroom window. 'We used to watch the children over this park... There was a stall, fish stall, and then on each side was these shops, and afterwards they didn't shut till... 12 o'clock at night, and the poor little children from Kensal Town they used to have no socks or shoes on, terrible, and they used to pick all the stuff out and eat it, they were so hungry.'

For the two publicans' families, there was of course no escape from contact with poverty and drink: it was part of their business. In one the father was himself an unemployed alcoholic, drinking himself to an early death. Mr *Venables* had been a newspaper lawyer, but after persuading a capable widow who ran her own pub to marry him, he gave up work. 'He never did a thing, never did a stroke, only sit with the customers, drinking and enjoying himself. He was no use.' His only contribution to the household was to keep pigeons and bantam cocks and hens, mainly so that he could watch them fight, but sometimes for use as pigeon pie. He would invite the boy in to show him off to the customers—'put the tip of my head on one chair and the tip of my feet on the other and I would be stiff'—and then send him away so that he could tell some smutty jokes. Although their standard of living was comfortable, with good food, and the services of a maid, charwoman, and seamstress, 'there's no family life, really.' They did not eat together, and there were no family games or music. The parents kept apart although there were occasional fights—and the father was left behind when they attended church, and went on their annual holiday. The boy was not really close to his mother either: she 'was always attentive and looking after me, you know; but there was no loving or anything like that shown... She had her own troubles, I think... with the bar and the barmaids and probably they'd be pilfering... drinking the stock or something like that.'

Consequently the boy spent much of his time in the street, although the pub was in a very mixed district near the Liverpool docks. One of his playmates, for example, was a Chinese girl. Mrs Venables had 'a lot of poorer customers, but they were always treated. They could say what they wanted and buy a pint and have their dinner out of the bread and cheese basket for nothing... That was put on the counters, and in the evenings probably small plates of peas and beans or something like that. Given free.' And as so often, the roughness of the area was formalized with peculiar nastiness in the school, where the headmaster kept his personal torture instrument in his study—'a train window flap strap—slit, you know—and rap you on the knuckles.'

In the other publican's family, the *Glanvilles*, the boy also mixed with poor children in the streets, sharing their warfare with the policeman. Here too the adults kept on the right side of the police by knowing when to send over the barman with a pint of beer. But in other ways family life was a complete contrast: family meals with grace, games, magic-lanterns, and the piano, and no violence—even punishment was never severer than being sent to bed. Such gentle discipline may seem remarkable from a father who was a prize-fighter as well as a publican, and was to die as the result of a punch on the kidneys. Mr Glanville, however, seems to have been a man of unusual talents. London born, he had been brought up in Scotland, and joined a Scottish cavalry regiment. He then returned to London, first as a dairyman in Hoxton, keeping fifteen cows under his shop. He then ran an antique-furniture business, and finally a public-house. He was an amateur lawyer. 'He'd jump his own height: very fine swordsman, good runner, good jumper, good swimmer . . . Very light on his feet too. He was a Highland Fling dancer.'

Apparently he was as respected in the community as in his family. He was a churchwarden (although perhaps this was simply because the vicar was a very regular customer). There would be banquets, supervised by Mrs Glanville: 'lovely glasses she had, all colours, like a long stem and a yellow top, a blue top, a fawn top—made a nice show.' There would be elaborate flower decorations, and a menu of 'sucking pigs, H-bone of beef, shoulder of lamb, lobsters—oh, a lovely dinner.' And for the poor of the district, he organized a soup-kitchen. When the London cabmen went on strike, 'they were allowed to come to our house and have a glass of stout and bread and cheese every day they were on strike. Well, when my father died there was 95 empty cabs followed the funeral and the wreath they sent was as large as a hansom cab was . . . It was a procession. They had a Salvation Army Band, East Bloomsbury Radical Club Band . . . from City Road to Old Street, right down Old Street to Shoreditch.'

The father of the last of the four tradesmen's families was, by comparison, unenterprising, for both Mr *Timbs* and his wife came from shopkeepers' families. One grandfather kept pigs in the country and sold them in his city shop, and the other also had a shop in the city centre.

Compared with his grandparents he was not, however, a successful tradesman, and he eventually abandoned his grocer's and baker's shop to become an insurance agent. Resentment at her declining social standing may partly account for Mrs Timbs's harshness to her five children. She even burnt the girls' books to remove a distraction from household chores, and, although a charwoman was employed, the girls were made to empty bedroom slops during the school lunch hour. She was in fact a paradigm of frigidity, as cold to her husband as to her children. 'I remember once seeing my father at Christmas put his arm round my mother and she did this—she brought her hands together. You know, she almost froze. Eh dear, my mother never taught us any affection—never, never, never. It was always right and wrong and how much work you could do.' The effect of this example on the girls proved disastrous. One sister rebelled, stole, became promiscuous, and then lived with a married businessman until she had a child by him, which she handed over to her parents to bring up. The younger sister was obedient, but as unaffectionate as her mother, and when she later found a husband, she was unable to hold him.

As it happens, the father of the one clerical family had made the same move out of trade: Mr *Enfield* had been a country grocer before his appointment as a senior railway clerk. But he was a much more successful man, later becoming a station-master, and there is, compared with the Timbs, a general buoyancy in the family atmosphere. The children were allowed out in the street skipping, spinning tops, and running iron hoops until the evening meal. Life in the suburbs included pleasant walks to fetch eggs 'through the buttercup fields over Highbury Rise'. Friends were made welcome, there was family singing, reading aloud and plenty of books. The children were encouraged at school ('Dad would always help me with my sums in the evenings') and a favourite school teacher 'used to walk home with two or three of us hanging on her arm and have . . . a cup of tea and a piece of cake.'

The family were keenly religious, although far from fanatical, and on Sundays went to church three times. They regarded themselves as 'upper working-class' rather than middle-class, and had chosen a select artisan estate for their home. On Sundays their neighbours could see visible proof of their respectability: Mr Enfield in 'his frock coat and silk hat and mother with her silk cape . . . and little bonnet'. Mr Enfield was particularly proud of a fancy blue waistcoat of Chinese silk, with crocheting and braiding, which 'was looked upon as a little bit of gold dust'. Yet their main contact with the poorer working class, none of whom lived on the estate, was through the church, which had a 'sunshine committee'. The members made clothes for poor children, took them on a summer outing, bought them boots, and gave them a January tea-party with coconut ice.

It is hardly surprising, although we now cross the boundary between manual and non-manual workers, that a skilled manual worker's family can easily be found whose way of life was essentially similar to that of the 'upper working-class' but non-manual Enfields. The *Towlers*, craftsmen prosperous enough to employ a regular dressmaker, in fact thought of themselves as middle-class. Basketmakers, they made cradles, pushcarts, fancy chairs, and tables, 'and all of us used to work in it . . . We was a united family.' Mrs Towler came from a middle-class London French family, and before marriage had given public readings of Dickens's works. 'They all used to pay a penny to go and the halls were always full . . . She was a wonderful reader. And a speller too.' Mr Towler, who worked all hours, was from a humbler background, but 'aristocratic looking, with the goat's beard . . . And he was very gentle and very intellectual.' They appear to have been obliged to move a number of times because of difficulties with rent, and for about six years they gave up basketmaking and took a house with a large garden and greenhouses 'and the whole family worked in the gardens.' With their own meat, vegetables, and fruit, fresh or preserved, they ate well. At this time they 'used to employ a woman from the union to do the washing . . . and she was allowed a pint of beer and bread and cheese for dinner.'

The children clearly imbibed a strong sense of respectability. At school, for example, 'there was two girls, used to come from the fried-fish shop, and they'd hang their clothes up and we'd all rush round and take our clothes from the pegs as far as we could from them.' Another incident, however, revealed that others too might have their pride. The children had always thought of a navvy working in a gang of roadmen as 'someone different. But when I spoke to his daughter . . . she said, that nobody could do his job. It was a high skilled job to keep that hammer going on that top of that knob, you know how—five—one after the other banging. She said it was the most skilled job that was going. So they were proud of that job, see, and thought nobody was so good as them.'

Religion once more served to emphasize the special standing of their own family. The Towlers were undenominational 'Peculiar People', and at one time held services in their own house, which were attended by neighbours and children and a few other supporters. Both parents also went visiting, the mother reading to the gypsies in an encampment opposite the house, the father calling on the poor, 'cheering them up', and sometimes if they were ill staying the night. He organized a soup kitchen for the poor in his house, with the help of local butchers and corn chandlers and their own vegetables. Tickets would be sent for distribution to the Board School 'for the children to get a pint of pea soup'.

Another artisan family, although less prosperous themselves, also organized charity for the poorer working classes. The *Whitworths* lived in central Salford. Mr Whitworth was a Corporation horsekeeper, living next to the stable yard, where 'he went day and night—sat up all night sometimes with a sick horse.' The family were well fed themselves, with hens kept in the cellar, and special advantages from the father's position.

> Me father got plenty of things given to him . . . It was the farmers that supplied the Corporation with hay and corn and stuff for the horses, and they always sent . . . a hamper every Christmas with perhaps a turkey in and a goose and onions, apples, sage, parsley all round, right big hampers, you know. And then the vet that was with . . . the horses, he always sent a big round of beef.

The children, too, profited from their situation, and organized macabre tea parties.

> Me brother had two or three lads, friends, but on Saturday our Albert he used to get some bricks and cover them with bird lime to catch sparrows . . . and he'd pluck them, clean them, him and me together, I helped him.
> Me dad had a little room like he called his surgery where he kept all his horse medicines and things and he used to let us play in there. And our Albert and me, we used to clean these sparrows and then we'd get me mother's roasting tin, big lump of dripping in it. And in this surgery there was a big range . . . and our Albert'd make a fire and get the oven hot and we used to cook them . . . and me mother'd make a big jug of tea for us and we had a tea party.

The facilities of the yard also proved useful when it came to helping to feed the poor.

> I can remember one winter and it was a very long cold winter . . . and there was a lot of poverty about. And me father, he went to different shopkeepers on Charlotte Road and he got promises of scrapmeat, bacon, bones, you know, lamb bones, and all sorts of things, peas, beans, and of

> course he had carrots and turnips and . . . he got some oatmeal promised him and a baker promised him some bread . . . Me father had a very big iron boiler . . . that he used to make gruel for horses and he cleaned it all out and they made soup in this big boiler and I can remember seeing the people come. They had to come on a certain time at a certain day twice a week. They came for two or three weeks that winter. They had . . . to bring their own jugs . . . They was that hungry and . . . the children running about barefoot, even in the cold.

Hunger had its effect, at a distance, even with this well-fed family. The girl found her mother cold. 'I don't think she wanted me. I don't think she did. Well she'd had four boys. And I don't think she wanted any more because they weren't well off.' Her brother had passed on the story of an exchange with a neighbour at the time of her birth, who had asked, ' "Aren't thou pleased that thee's got a little sister?" He said, "We could have done without her. It's only another mouth to feed".'

Mr Whitworth, however, was a more affectionate man who helped a lot with the children; for example, he would bath them. It was a gentle home, in which punishment of any kind was rare. The children could enjoy the entertainments of the city, such as a round-the-world panorama shown in the Free Trade Hall, as well as outings into the nearby countryside. It was walking down a country lane one evening that the girl met her husband; he was a young Co-operative shop assistant out on a bicycling excursion, who stopped to ask her the way to the inn.

Both of them were city people, strangers in the countryside, although her father had started work on a farm. The distance which her family had travelled in moving to the city was indicated by an incident when the grandmother came to stay. 'Me mother persuaded dad to take her to the pantomime. So me dad worked to persuade her to go.' She had never been to the theatre, and as a Particular Baptist, was very suspicious of it.

> Anyway, he got a cab and he took her, and it was Bo-Peep, Little Bo-Peep. Well it started and when the ballet came on and there were dancing girls, and they'd be in tights, she says to me dad, she said, 'William, take me home.' She said, 'Take me home.' He said, 'It's only just started mother.' She said, 'Take me home. I'm not sitting here watching them dancing and kicking their legs about . . .' she says. 'And wagging their fat tails behind them,' she says, 'the brazen hussies.'

The great gulf between city and country as well as between the generations was also felt by a boy from central London who went in 1897 to stay for a holiday with his grandfather in an Essex village. Tom *Farrow* took a friend with him, also a post-office telegraph boy. 'We paid him, of course, the old man was just on his own.' They were amazed by the 'old cronies' whom they found at the village pub.

> They'd sit round smoking, and gossiping—best part of 'em couldn't read. They'd have a pot and they'd keep passing it round and—one pot—they'd keep passing it round and having a sip out of it, see. When it was empty somebody else'd pay for the next pot, and so on . . . We'd buy a bottle of lemonade for a penny . . . and we'd sit there among them and read the newspaper. Read to them out of a newspaper and you know they'd sit there going 'Ah, ah.' Well they didn't know anything much . . . We were always having little wars at some place such as little African places . . . and of course, they'd not heard about this. Some of 'em hadn't heard of the Crimea war you know—and of course they'd all be people who was alive then.

One can imagine how far this village pub must have seemed from the Victoria Palace music-hall where 'when I was young I'd take a girl friend . . . A shilling each for the seats and there was always a waiter hopping about, you know; if you wanted a drink they'd fetch you one.' Inevitably, living where he did, Tom Farrow saw something of the extremes of social life. One nearby street 'was such a rough place [the police'd] go down there four at a time. If one went down there, he'd get a good hiding.' At work he also got to know the seamier side of upper-class life. One of his workmates was an allowance man—

> a nuisance and the families used to pay 'em, make an allowance, three or four pounds a week to keep away. [This allowance man] was a gentleman born. And he used to go to church on Sundays and he went to St Peter's Eaton Square. He'd have a tall hat, frock coat and gloves and all the rest of it, see. People didn't recognize him you know because in the week he'd just be wearing the postman's uniform. And, anyway, this particular Sunday morning he went to church. As he walked into church and took his tall hat off, he—he—he'd inadvertently got a postman's hat under it . . . It was a by-word all round the office.

Tom Farrow was also introduced through work to the servants of the rich. The night boys at the post office could not go out much in the evenings, so

> in the morning in the summer time we'd go to Battersea Park to have a game of cricket. We used to find kindred souls like ourselves. There was a club; we called ourselves The Owls. They called themselves The Early Birds. And they were composed of footmen and butlers and gentlemen's servants generally. Well, then, they'd turn out and meet us perhaps at 5 o'clock in the morning.

There was also contact through courting. 'Course when I was single I used to know a lot of maids . . . Probably they had one day a week or something and take 'em for a walk in the park . . . They had to be home by nine or something like that . . . If you popped down the area steps you might get in.' You could then enjoy gossip with the other servants about their employers—'they'd all generally got some spicy little tale to tell you and—"Here, Postie," so, see . . .' He suspected some of the maids were wearing their mistress's clothes when he took them out. He also observed how the cook would enter extra milk in the account book, which the milkman would not deliver. Not that the milkman swindled only the rich. 'They always used to give you "a little drop over", so they said. But what used to happen was this. They [were] what they called "quick on the handle". . . They wouldn't pour all of it out.' Similarly, but in a more straightforward fashion, at the local brawn shop 'they never bothered to weigh it, they just cut it off.'

He had direct experience of shopkeeping, for although Mr Farrow was a postman the family kept an umbrella shop. 'It was quite a paying business . . . Mother used to do the sewing work . . . I had to do a lot of running about for the shop, you know. Go to school and take a note to be let out at four o'clock and then I had to rush off to the City, see, warehouse, for something or other. Or else I'd have to go over to the Borough for a walking stick.'

With his own family, however, it was not shop practices which induced cynicism. More relevant was his parents' attitude to religion. They never went to church, but taught him prayers and sent him to Sunday school. 'That went on for a time but I mean it was never serious . . . The local vicar came round . . . and had a chat with mother but . . . I mean, it petered out. They were never very serious on these things.' The children were also sent to the Wesleyan chapel when there was something to be had there. 'Once a year we used to have a service and a tea there and after the tea we had to spend the evening in this church . . . I know we was always glad to get out.' And although the parents were not churchgoers they insisted on Sunday suits, Sunday boots, 'and you mustn't have a ball or read a paper nor have a comic or anything like on a Sunday, no.'

Some of their other demonstrations of respectability were equally hollow, even if they may not have been atypical of artisan culture. In the parlour above the shop, which was reserved for week-end use when relatives came, a piano was kept, 'but nobody could play it, see.' One might have thought that in a house with nowhere to hang clothes, 'not an inch of back yard', the room could have been better used. Certainly 'so far as working people were concerned we were tolerably comfortable', with meat most days cooked on the gas-stove, and in the evenings perhaps fish bought from the fried-fish shop, or shrimps or winkles from a barrow. Food, indeed, had to be just right: when an aunt served suet pudding before the Sunday joint in the old fashion, 'we just looked at it and they got on. "Come on then—get on." "No, no, we'll wait till we have it all together." So we wouldn't touch it till we got the rest.'

With purchases, it was always 'cash down and we never owed.' But the Farrows let their attitude to money go to extremes. Tom was never allowed to accept money for running an errand, yet he was worked exceptionally hard at home, not merely having to clean the knives and boots and do other housework but being responsible for the younger children. 'I had very little leisure . . . because I was always in charge of the younger ones. See? I used to have one pal, and occasionally I'd be allowed to go out with him.' When his parents went out at night he had to baby-sit, 'scared stiff' they would wake and interrupt his reading in the kitchen. His parents also 'made all sorts of promises... "When you get married we'll give you a house to live in and the deeds of it", you know. But it didn't happen.' The worst deception, however, was the double trick of the money box.

> I had a money box and I was taught as a small boy not to accept anything from anybody, see. I used to [say] 'No thankyou. Father's got some', see. But of course I had to have a money box but I was never allowed to spend any and never had any pocket money... I know some children used to get a Saturday penny but I never did. No... I can see that blessed thing now, mahogany affair, see. And then one day it disappeared. And it turned up a few days after tied up with string and sealing-wax and all the rest of it. And of course, opened in front of me—'Oh, let's see what this is'—and that'd been to the baby shop to buy me a new sister. And that was that... That's where the money went.

This incident must have occurred in the 1880s. Nevertheless, it is striking to find that ten years later Dan *Shaw*, a Liverpool dock carter who had none of these pretensions, gave his daughter Annie threepence pocket money a week—'that was a penny for morning service, a penny afternoon service and penny night.' Himself no churchgoer, he was proud of his daughter's singing and if she practised at the piano with her choir book 'he'd say, "Huh, singing on Sunday, wench? I'm going to church." And he'd come to church that Sunday. Yes he'd come to the mission hall that Sunday.'

The Shaw family regarded themselves as 'just ordinary working class', and their life was indeed typical of a group which is under-represented in this set of interviews. In the district 'we were all pretty well of the same living—just carters and cotton porters and all like that... Very, very few on the rough side... all pretty well respectable.' The children played in the street, although certain companions from bad homes were forbidden, at games such as 'skipping rope, hop scotch, shuttlecocks, rounders... kick the can... blind man's buff... and Jacks and Hollies—four jacks and a holly, kneeling on the floor.' Relationships with neighbours were warm but limited. Mrs Shaw was 'very well liked. And no matter what went wrong in the street—if anybody died—"Go tell Mrs Shaw and she'll lend you the sheets".' Similarly, when she was ill a neighbour would come in to 'do a few messages or any hard jobs'. But Annie's parents did not regard neighbours as friends. 'No they just kept to themselves—just was sociable with all the neighbours and that but that was all ... Mother always taught us to keep neighbours outside.'

The family's standard of living was reasonable. Clothes were new, and there was meat most days. Nevertheless, there was not always enough to go round, and Mrs Shaw would often go short. When the father's earnings were down, 'she used to do a day's washing... at home. The neighbours used to bring it up to her. It was only neighbours she did it for.' As a dock carter Mr Shaw had a very irregular income.

He was more out of work in my early years than he was in work . . . You had to go and stand on the corner . . . You'd be glad to see the chap coming along and you'd be praying he'd say, 'Come on, come on, come on, I want you, you, you'—you'd be glad to be among the five he'd picked. And if you didn't get picked your heart'd be in your boots . . . You had to come home and go on the corner next day.

Under favourable circumstances it was possible for a family considerably poorer than this to maintain quite respectable standards of behaviour—for example, the *Ranns*, a family of nine children who lived in a London mews. Mr Rann was a navvy, illiterate but independent-minded, and 'out of work very often . . . He was a man that wouldn't be shown his work . . . He knew his work and he wouldn't be shown.' Mrs Rann couldn't read either, and they refused to allow books or newspapers in the house, no doubt regarding them as a threat to their authority. Their poverty was severe. The children were sent out to 'pick up firewood'. Apart from some chickens they were able to keep in an empty stable, they could only afford pieces of meat or pork-rind soup. Similarly, they bought fish pieces and for bread 'you used to go to the baker's and get 3d. worth of stale bread . . . if there was any there, mind you.' Often the boy, Sidney, would miss breakfast, for he had to do a milk-round before school, to earn money; and, if he was in time, there would be just bread and dripping. He also earned money in the evenings, running errands or cleaning boots or silver for the wealthy. On Saturdays he was again expected to 'find a job somewhere . . . Perhaps go and help a baker on his round.' In some ways it was best to be out of the house, which was desperately overcrowded.

Nevertheless, Mrs Rann was clearly of some local standing, for she worked as a midwife until uniforms were demanded. The children were brought up to 'raise the hat to the ladies'. They were dressed in new clothes. Bedtime was strictly imposed. Grace was said at meals 'and we weren't allowed to talk over the table . . . You had to sit right and you had to hold your knife and fork right.' Cards were not allowed in the house, and although Sidney had to fetch drinks for his parents from the pub, his father 'would never take us in the public house . . . he always used to say, "Nobody'll ever say I learned my children to drink".' In the same way, although not themselves interested in church, 'they saw that *we* went.' On some evenings the Band of Hope would hold outdoor services: 'we used to go to the bottom of the mews here and sing . . . If the cabmen were here, in the week, they used to join in . . . We was all one happy family.' Later, indeed, when he married, it was to a girl from the same mews.

In a hostile environment it was difficult to maintain such standards of behaviour, even when financial poverty was less acute. A smaller family, the *Doyles* with four children, lived in Blackgate, a central slum district in Salford. Although clearly poor, they could always afford some kind of meat every day, and on Saturdays with the week's pay in hand would often buy a chicken from a 'Chicken Ann' who toured the streets with her push cart. Mrs Doyle had worked as a weaver 'until several of the children were born'. She now was a respected figure in the district, 'a little stout old lady' in middle age. 'She was a marvel... She couldn't read or write, but she was a good woman... [It] was a pretty tough neighbourhood and the people in trouble, it didn't matter how bad they were, she'd go and try to help 'em... She didn't believe there were bad people, I don't believe.' Her particular friends, however, were undoubtedly from the elite of the district. She used to go out some evenings to an 'old-fashioned pub' to chat with the lady publican, a tobacconist, and a tripe shopkeeper. Occasionally the four ladies would go out into the countryside on a picnic together.

Mr Doyle was semi-skilled, an iron fettler at a foundry, and his income regular, for he was only unemployed when, about once in every five years, he decided to break out of the routine of his existence. 'He used to occasionally get fed up with the job and he purposely used to stop off and have a bit of a booze and get the sack. Then he'd be some time getting a job.' Even during these lapses he maintained appearances by drinking in some other part of the city. His privacy was fiercely maintained. 'A neighbour daresn't step in the house. They daresn't come past the partition. If they put their nose past the partition he'd say, "What do you want?"'

The rough standards of the neighbourhood had, however, affected him more than he would have liked to recognize. He was harsh with the children, never playing with them and reluctant even to talk. 'He expected you to obey his look, never mind his order... He was a bit of a savage.' Meals were silent, although parents were not 'dressed up over the etiquette... We had to have a knife and fork mind you.' Afterwards they would sit round the hearth. 'There wasn't much room... and he used to get in front of the fire with his chair and you try to get near and he'd say, "You shouldn't be cold".' His evening slippers were discarded iron moulds which he brought back from the foundry. 'When I wanted chastizing he used to call me over and say, "Turn round", and he was a giant and he used to hit you on the back with this shoe and it'd knock you across the kitchen... Very strict man—he used to swear a bit and you weren't to swear though.'

In the house he was no help at all. He expected the boys to fetch beer for him and read to him, because he was illiterate, but he tried to prevent them helping their mother. 'A lad hadn't to do anything. We had flag floors and we musn't clean the floor and "They're not going to make a girl of my lad"—that was his idea. Ignorance, really. But anyway I took no notice of him: I used to help my mother.'

Because of the atmosphere at home when his father was about, the boy 'didn't spend a lot of time in the house. He was too tough.' But the entertainments he could find outside were quite as savage. A particular pleasure, for example, was observing at the slaughter house—'Oh aye, I've seen them slaughter them all and cut them up.' Nor was it much more edifying to watch some of the neighbours. For although they were all poor, he felt strongly the distinctions between them.

> Our class was a respectable working man's class. But on the poor side...
> Some of them... couldn't even keep their houses decent. They used to call
> our street 'good husbands' street' because they cleaned the flags...

> Crompton Street was one standard poor. Lyme Street [the next street] was, oh it was awful, much lower than normal. They were living under a horrible condition. Broken windows. Really you couldn't believe the difference in two streets. Mixed up with Jews and thieves. Crompton Street they weren't too bad . . . They were very hard worked anyway. They didn't all work in Lyme Street . . . They lived like chip choppers . . . chopping chips and selling them in bundles [for] firewood . . . children going round with them. They were a better type of people in Crompton Street. Although as I say there was a criminal opposite to us. The lady that her husband was in jail and she had another pal. Oh, she had more than one. Smart woman too. But that was the way of life I understand with these people. My mother used to warn me against her.

Round the corner, however, such people were typical.

> A family named O'Brien had the first house in Lyme Street . . . He was a high-class tailor. They hadn't a stick in the house . . . Nice man, Mr O'Brien. Their children were no good. And the mother . . . used to take several policemen to lock her up and they always had to take her on a hand cart. That was Lyme Street. And O'Brien the poor fellow, he put up with it.

The Stacey family were still worse.

> Everybody knew them. They were downright villains. They were thieves. The women were bad women, real bad women. One eloped with a doctor. The doctor must have been mad. And they were barmaids some of them till they got a bit older and then they went to lob. Went to bed. Lob—made love. The eldest son, I think it was Paddy, he was always in and out of prison. Violence and drunkenness . . . He died from head injuries in the prison . . . Saw him one day in Chapel Street, walking, drunk, talking to himself with a brick in his hand. He used to throw bricks through pub windows . . . I saw another criminal drop dead out of a pub. Paddy Talman. He was a villain . . . He lived off Blackgate, married to a fine young woman and I was coming home from work and I saw this fellow. I'd a bit of a truckle with him once, I was glad they'd got rid of him . . . Doctor came, he was dead, shot through the heart. Now what had happened—the publican said that he'd raised a stool up to hit him and this revolver—well he might have done, I don't know. But there was one of the Stacey girls mixed up with him in the pub in that.

In a district where, to a boy, the police were 'brutes', collusion was probably the best form of protection from the criminals. One incident reveals something of the nature of this protection, and the organization of the criminal world.

> The decent people in Blackgate was safe from the roughs. Safe from the criminals . . . Well I used to come home early and go and stand on the

corner of Gravel Lane and I could see my mother come out of the Bull's Head. I was watching for her safety come home. No local'd interfere with her but a stranger might. Well a chap comes out of the pub and he offered me a handkerchief. I didn't know him, so I knew he was a stranger. Something like this—'Give me five bob for this.' And he started edging towards me. Well, the only way for me to stop him was to hit him or something. I'd no intention of giving him five shillings. And a woman came along. I'd never spoken to her in my life but she was known—as a few things. It was doubtful whether she was in touch with the police to give criminals away . . . She had a shop down some cellars, a big place. The thieves used to deal with her . . . She had a daughter that was a thief . . . and her husband they called a megs man [confidence man] . . . And she said, 'What's to do?' So I said, 'He wants five shillings for this.' And another criminal, schoolboy same time as me, came out of the pub. He'd had penal servitude—Freddy Barnes. And she said to Freddy Barnes, 'Give him a good hiding. He's trying to get five shillings off Dicky Doyle'. . . Freddy Barnes he wiped the floor with this fellow. And she said to him, 'If you're in Blackgate tomorrow morning you'll know about it. Get out tonight.'

The territorial arrangements of the adult criminal world were probably particularly well defined, but there were parallel struggles taking place in other spheres. Jews, for example, were as far as possible kept out of the district by systematic window-breaking. There were also the territorial bases for the various youth gangs of the district, known as the 'Scuttlers'. 'Gangs used to meet and fire bottles at one another, and belts with buckles on . . . There was the Savoy gang—Red Shelley was the leader of them. And Johnny Hoddy was the leader of the Blackgate mob—the King Street mob it was.' These two gangs apparently differed, the Savoy gang being more likely young criminals, the King Street mob simply a street-corner group, whose leader 'anybody could have blown down with a feather. He was only the leader because he happened to be the oldest . . . married to quite a nice girl.'

Of the two Doyle boys, one seems to have been engulfed by the way of life of this rough district, but Dicky proved more successful than his parents. He was sent to the Roman Catholic school in Blackgate, where he experienced savage caning. The school's one advantage was that he was able to make friends in a more respectable district, and go there to play, although there were penalties involved in this too. The local boys 'used to think I was stuck up . . . I had to fight my way through. Now if I got a good hiding outside [my father]'d give me another one for getting a good hiding.'

The real turning point came when he started work with a portmanteau worker in 1895. It was here, rather than at school, that he felt he learnt manners, especially from 'the man I served my time under'. Even during the first few months when he was an errand boy, buying cigars and delivering love letters for the manager, new worlds were opened to him. Once he was sent out into the country to deliver some engraved cutlery. 'Oh, she was a lovely lady. She sat in the hall. And she had a lot of money there. She give me a sovereign—"That's for you. Put it away." So I put it in my pocket—six shillings was the most I'd ever had. I was in my element. Anything to help me mother. I was happy.' After a few months his father insisted that he should be properly apprenticed (although normally only sons of portmanteau workers were accepted) so that he was trained in the skills of fancy leather work—'crocodile, seal skin, Russian skin . . . The boys used to have to sit down at a bench and stitch for a week. Oh, I can stitch real—I think I can stitch with my eyes shut . . . You gradually learnt to do other things . . . Oh, I loved it.'

Once Dicky Doyle had become a journeyman he very quickly became prominent in the craftsmen's union. The trade was already in decline, its standards undermined by competition from sweaters, and the union had an appropriately moribund air. The delegates 'used to wait with tall hats and frock coats and they hadn't tuppence . . . They used to borrow money to go and have a drink in the pub . . . [and] come drunk to the meeting.' At the age of twenty-two he found himself elected president of the union. His father at any rate was forced to recognize his new status. Dicky refused to continue 'going for free beer for him . . . I said, "You're finished." I told him, I said, "You telling me what to do"—and I told him what I had accomplished at my age—"You've not accomplished nothing, not a heap . . . And I'm not going to read for you," I said, "and Nelly's not going to read for you." But she used to when I was out . . .'

This boy's rise was matched by the descent of others into the semi-criminal world. The *Colliers* lived in a district close to Blackgate. There were seven children, and the household was in most ways exceptionally rough. Neither Mr Collier, who was a blacksmith's striker, nor the children were expected to help in the house. They spent their time 'fighting with one another' while he would belt them (if he was not at the pub) for causing damage. 'He'd knock us about terrible you know . . . We deserved it, you see.' He was entirely illiterate, so that he could not even recognize the names on a ballot paper—he would just vote blind. 'He just put [the cross] down. He didn't know what there was.' None of the family ever went to church or to Sunday school. The children bathed in a canal, warmed by discharges from the boilers of a factory, rather than at home. Their patched trousers were 'hanging out at the back' from fights. If they had any decent clothes, for most of the week they would be in pawn. They had no underwear—'just our shirts and our trousers. That's all we had . . . Went to school in winter in bare feet. We'd no clogs to put on, until about Sunday.'

Their meals were eaten without knives and forks, and the children were allowed to chatter and even to read comics while they were eating. The diet was extremely meagre, with scarcely any meat. Bread, tea, and kippers were the staple, with potato pie the Sunday special. Because there were so many children the youngest were forced to eat from a ledge underneath the table, while their parents and elder brothers sat above. 'I'll never forget—I think I've got the mark yet—my father was having his dinner and I put me hand under his knee like that and I took a piece of meat off his plate. Well, he didn't see me first time. So the next time I comes again, but he had the knife, and he chopped me fingers—he nearly chopped me three fingers off.'

The Victorian City: Past and Present

The children got into the habit of petty thieving when money was short. 'We couldn't see the old woman short . . . We two or three others we'd go together . . . outside the shops . . . We'd perhaps take a couple of taters out of that bag . . . and get onions, you know, where there was onions. We'd get a dinner that way and take it to the old woman and let her stew it up. She'd know where it came from.' It is hardly surprising that the children should think of this when they were out on the streets, continually being chased by the police for playing football, bathing in the canal, shouting in the streets at night, or trying to sell papers on Sundays. 'My mother was sick and tired of policemen coming to door.'

The children's regular pleasures included watching the police fish drowned men from the canal, and 'every morning we used to go to the police station to see if anybody was been locked up—the drunks.' Drunken women, even with children in arms, were a familiar sight. 'They used to feed 'em with the breast at one time . . . sit there in the pub with their babies.' There was also the pathetic example of Neddy Tattler who used to steal clothes 'when the washing was out in the entry', and pawn them for cash. 'The detective says, "He's the cheekiest fellow that ever we knew," he says. "He's even come back for pegs," he says. "That's how he got caught".' Even the sixpenny doctor found when he was called out that his patient was incapable, probably drunk, and the call was at his own expense. ' "I've come down for sixpence to see a fellow what's looney," he said, "and he's not got sixpence to pay, and they've stolen my dog while I've come." '

The dog, like other stolen goods, would be quickly taken to the local receiver. Also, as in Blackgate, there were gangs of Scuttlers who 'always wore belts . . . with brass buckles on . . . I knew a lot of them. One called Red Heriots. He was one of the leaders. Wore a jersey which said "Red Heriots" on it and "King of the Scuttlers". Aye, they stabbed a fellow on Oak Street Bridge . . . It was over the girls, jealousy with one another, you see.'

Another regular battle was with the landlord. If the rent was in arrears he would come round himself, and he'd say, 'Spent?'

> That's the way they'd say to him, you know—'bills'. And in question, 'Just bets and spending on horses?' And he'd get taking his coat off to have a fight, you know, for your rent. Aye, he did one person down there. He couldn't get no rent off him so he went and got a fellow . . . [to] go and put a slate on top of chimney so he try and smoke him out . . . Well them days we used to flit from one house to another . . . If there was a house empty you used to go to the woman and give her a shilling for the key . . . and keep missing rent there you see until he was going to throw them out, so they got another one, going, perhaps . . . only next street.

The short distances involved in these moves no doubt help to explain why despite the transience of many neighbours, 'in the street we knew every one of each others, you see, what they did and everything.' When the family was in trouble 'they'd come in and do any mortal thing for you.' Mrs Collier had, in fact, a rather special role in the street, for people 'used to come to her' to hear their fortunes. She knew a few primitive rituals. If a young woman had been deserted by her suitor, she would buy dragon's blood from a chemist, and put it in a frying-pan. As she turned it over above the fire, she would say:

> It is not this dragon's blood I wish to burn,
> It is my lover's heart I wish to turn.
> May he never eat or sleep until he returns to me.

After one success, her reputation spread. But the children remained cynical. 'We used to kid me mother over it.'

In spite of their shoplifting, none of the Collier children was ever caught stealing by the police. One brother even had a short rise towards respectability as a shopkeeper himself, selling greengrocery. 'He'd start with a basket and he'd go round with a basket on his head. About two or three months after that he'd have a donkey and cart. And then he'd get to a pony and cart.' Eventually he had three shops—until he ran away with Dublin Ginny, a prostitute, who robbed him.

Two of the boys started as miners. The younger, himself, went down the pit when he was thirteen and gave up at seventeen when he married a girl from the same street. 'We was young then like and we wanted a bit of pleasure.' For many years he worked on the roads with a steam-roller and they lived in a caravan, travelling the country. Before that, however, he had nearly been forced to travel in a less independent way. In 1900, when he was still working in the pit, he had been sent up the street 'for some sugar for me mother . . . And I'm going up the street, up Hesketh Street to the top and the lads are stood at the corner, you see. Well when they sees [the police] coming they flew away.' So that he was arrested instead, for loitering. He preferred going to prison for three days to paying a fine. The next time he was sent for seven days. The third time the magistrate said, 'I think you've got a lot of bad pals . . . Will you go in the Navy if I let you go?'

He joined a ship in Scotland. It seemed far tougher than the mine—'up aloft at four o'clock in the morning with weather, you know, bare feet, and you had to do knots and splices, bends and hitches, rope pulling, swimming . . . with your clothes on in the sea.' After a fortnight he was birched in front of the ship's company for smoking—'they sound the bell on ship to let them see me getting flogged.' So Bill ran away home to go back to the pit. He was caught, brought back, and flogged again, but this time he was so badly cut that he had to go to hospital where a kindly doctor arranged for his release. ' "I think you'll be best away." So he put me down for the discharge. "Unfit".'

He was lucky, for it was families like his own that fed the armed forces. We conclude with another very rough family, this time Londoners. Of the nine *Burns* children, two were to go to sea and a third into the army. Mr Burns, who was Scots, had also been a soldier, and had a good job as a foreman packer, but, since both parents drank heavily, life was only slightly worse for the children after he died. He was a rough man who did nothing for his family. 'I think if he'd ever took me out he'd have threw me in the canal.' They lived in two rooms in central London, sleeping five in a bed. They had no regular baths, no shoes, no new clothes. 'We wore one another's cast-offs. Oh, we wore practically nothing.' The clothes were ragged, not even well patched, and 'we were never clean.' There were no regular meals, and 'we was all hungry—always hungry.' They relied on bacon bones, pieces of leavings of meat, and 'pick-ups' of fruit from Covent Garden. They also made use of a nearby soup-kitchen. Only the parents sat on chairs. The children sat on the floor and ate their food with their fingers. 'I don't remember using a knife and fork as a youngster. Me Mum used to say, "Fingers before forks".'

With no chairs to sit on, guests could not be invited in. But a home which 'only had a few sticks in it' was easier to move. 'We shifted so many times . . . Some place, sometimes, we had to move quick—couldn't pay the rent.' The parents did not trouble much over school attendance. 'I was always one for running away from school. I didn't like school at all. I detested it. I . . . didn't go school any day if I got the chance.' To stop him 'hopping the wag', the headmaster would tie him to the radiator. Nor was he sent to church or Sunday school. His parents never went to church, unless something useful such as free coal was offered. Even so, his father seems to have been pleased by an occasional call from a missionary visitor who 'used to say, "Mr Burns, you'd be a fine Christian and you're a real gentleman." It was a good thing you know.'

There were no family outings or holidays, no games or sing-songs. In fact, the family did nothing together. 'We was all gone poles apart. We all used to go our own way.' For the children, this meant out into the street, dodging the policemen, cheering motor cars, picking up food from shops. 'I used to steal sweets . . . Used to be a little shop . . . I used to go in there and buy something for me mother and me hand used to creep up and steal a lump of coconut cake and she told me mother. Mother belted the living daylights out of me for it.' They would be sent out to beg, however. 'We used to go to different coffee shops and stand outside with a bag waiting for the girl who was in the serving or something to call us in and give us the leavings. Collect the leavings.' Or just to wander aimlessly, seeing what might be picked up from the gutter. 'We used to walk the streets all night at times . . . miles and miles and miles all along Tottenham Court Road, Oxford Street, all along the gutters . . . We used to walk from King's Cross to Hampstead Heath. And it was fun of the fair in happy Hampstead in those days.'

Fun the boy might watch, but could not afford to join. So we end with a child who saw the wealth and pleasure of the Victorian metropolis from a position of utter deprivation. Yet he noticed other children, whose voices are unrepresented here, in a state still worse than his own. There was a foundling hospital nearby, and 'every morning the people that was in the foundling hospital there, they used to come out and put fresh straw round the door and women used to leave their unwanted babies there.' And at night, a few hundred yards from the crowded main street with its gas-lamps and gin-palaces, a dedicated figure could be seen on his round. It was the Salvation Army man, who 'used to go down the mewses and stables and lift up the tarpaulins to see if there was any derelict children under.'

II Numbers of People

3 The Human Aggregate

Asa Briggs

To understand the nature of Victorian civilization it is necessary to understand Victorian cities—visually, through their forms and formlessness; socially, through their structures and the chronology of their processes of change, planned and unplanned; symbolically, in literature and the arts, through their features and images; together for the light they throw on the processes of urbanization; separately and comparatively in order to understand particularity and the sense of place. The world of Victorian cities was fragmented, intricate, eclectic, messy; and no single approach to their understanding provides us with all the right questions and answers or leads us to all the right available evidence.

In studying such complex questions there is a danger in following singly one or other of two different kinds of approach, which all too seldom are considered together —the approach through 'qualitative' evidence derived from a wide range of sources, documentary and non-documentary, public and private, and the approach through the accumulation and analysis of 'quantitative' evidence—the vast store of measurable data which the men of the nineteenth century produced in greater and greater quantities. It is important that the two approaches should be considered together since there already tends to be a gulf between them. On one side are the literary historians and the architects, on the other side the historical demographers and the economists. A few social and economic historians and a few general historians straddle the divide. The dangers of this situation do not lie simply in failures of interdisciplinary communication: they prevent us from understanding many problems which are key problems in Victorian studies.

© 1973 Asa Briggs

The Victorian City: Numbers of People

The fact that the quantitative approach is now becoming increasingly feasible and increasingly fashionable is of the utmost significance in this context. There is obviously much that we can find out quantitatively—in general and in detail—not least about the 'quality of urban life' that has not been found out before. By the intelligent use of social indicators, by the employment of a retrospective cost-benefit-analysis approach, by an analysis, with the help of modern statistical techniques, of factual material which the Victorians collected simply for immediate purposes, and by asking questions which for various reasons the Victorians did not themselves ask, at least in precise form, about their own society and culture, we can go some way towards 'measuring' the quality of life in nineteenth-century cities.

Before exploiting our own approaches, it is as wise to turn back to the attitudes of the Victorians themselves as it would be in studying the history of vocabularies or of tastes and styles. There is little doubt that throughout the whole period they approached the growth of their cities first and foremost in terms of *numbers*. They were aware—either with fear or with pride—that they were living through a period of change of scale—change in the size of industrial plant, change in the size of social organization, change, above all, in the size of towns and cities. They liked to collect facts about all these phenomena—sometimes for reasons of curiosity, sometimes for reasons of what we would now call 'social control'—and they developed machinery for doing so. The facts which they collected were set out in trade directories and brochures for local Chambers of Commerce as well as in the national Census reports and Blue Books: alongside successive editions of vast treatises like George Porter's *Progress of the Nation* (1836) or J. R. McCulloch's *Descriptive and Statistical Account of the British Empire* (1837) (against which Dickens reacted so sharply when he wrote *Hard Times*), there were detailed guides, heavy with statistics concerning what were often called 'the large towns and populous districts'.

Some of the facts were crude indicators of what the Victorians themselves thought of as 'material progress', aggregates derived from the success stories of economic individualism. Other facts dealt with the unplanned collective problem areas—with fertility—or mortality-rates for example, with crime, or with the supply of houses and schools. Others were far wider in scope, relating demography to social class, to economic structure, or to moral behaviour.[1] From the outset the facts were collected because many of the social statisticians were anxious not merely to present information but to propound a message, sometimes a gospel. Even when they were employed by the government as inspectors or servants of commissioners, they were seldom merely agents of the state. Moral instructors like Charles Kingsley eagerly translated the facts into the language of the sermon and the lecture. For example, in his lecture on 'Great Cities and their Influence for Good and Evil' delivered at Bristol in October 1857, Kingsley deliberately crossed the frontier between statistical and non-statistical methodology, stating that 'the moral state of a city depends—how far I know not, but frightfully, to an extent as yet uncalculated, and perhaps incalculable—on the physical state of that city; on the food, water, air, and lodging of its inhabitants.'[2]

As I have argued in *Victorian Cities*, there was no agreement on the implications, moral or social, either of the acknowledged facts or of the general laws which Victorians liked to derive from them.[3] In the most general terms, on the one side was fear—fear of a change in the pattern of social relationships associated with change in the scale of the city; fear of the emergence and of the mounting pressure of new social forces which were difficult to interpret, even more difficult to control; fear about the capacity of society to deal quickly enough with urgent urban problems before the social fabric was torn apart. On the other side was pride—pride in achievement through self-help and, through self-help, in economic growth; pride in local success through rivalry with other places, not only in the tokens of wealth and in the symbols of prestige but also in the means of control—mileage of sewers, number of water-taps or water-closets, number of school places or of policemen.

There was sometimes no give-and-take in this clash of values: the chimneys spoke for themselves. Yet there was usually ambivalence, with the ambivalence stretching to the statistical method itself. Dickens was not attacking all statistics in *Hard Times*, but rather a naïve reliance on certain kinds of statistics and on nothing more.[4] *Punch* made fun of statistics, particularly 'useless' statistical research, on many occasions, but recognized that the great utility of some kinds of research and calculation was so very obvious that it need not point out the value of the labours of contemporary statisticians. Wordsworth, of all people, wrote to H. S. Tremenheere, the 'classic' Victorian inspector, that 'we must not only have knowledge, but the means of wielding it, and that is done infinitely more through the imaginative faculty assisting both in the collection and application of facts than is generally believed.'[5]

Given such ambivalence, much of which pivots on the Victorian idea of 'fact' itself, it is necessary to explore the growth of statistics as a mode of enquiry and a means to reform, not to take the mode and the means for granted. More work has recently been carried out on 'literacy', albeit still an under-developed subject, than on 'numeracy'. It may be true that 'statistics is to industrialism what written language was to earlier civilization.'[6] Yet it was in the pre-industrial seventeenth and eighteenth centuries that the origins of statistical preoccupation are to be found. 'Great cities', J. C. Lettsom had written in 1774, 'are like painted sepulchres; their public avenues, and stately edifices, seem to preclude the very possibility of distress and poverty: but if we pass beyond this superficial veil, the scene will be reversed.'[7] Statistics pierced the veil, and throughout the last decades of the eighteenth century there were as many statistical enquiries into living-conditions in pre-industrial communities (Chester or Carlisle, for example) as there were into those in the new English industrial cities.[8] Although 'abstract' classical political economy, as developed in the early nineteenth century by Ricardo, was by its very nature (like the developing natural sciences) non-statistical, this does not mean that political economists were uninterested in statistics. Ricardo himself agreed that 'speculation' had to be submitted to the 'test of fact', and the first Fellows of the Statistical Society of London, founded in 1834, included Nassau Senior, Malthus, and McCulloch.[9]

The Victorian City: Numbers of People

They were doubtless as sceptical as Dickens concerning the proposition that everything could be reduced to 'Two and Two are Four', 'Simple Arithmetic', or 'A Mere Question of Figures', the possible titles for *Hard Times*, although by 1838 some of their number were expressing the hope that 'the study of Statistics will, ere long, rescue Political Economy from all the uncertainty in which it is now enveloped.'[10]

The purposes which the Statistical Society of London was established to serve were set forth in a prospectus as 'procuring, arranging, and publishing "Facts calculated to illustrate the Condition and Prospects of Society"'.[11] Its provincial counterpart, founded in the same year in Manchester, a city of unbridled economic individualism, certainly had from the start a distinct social leaning.[12] Businessmen might be concerned in their daily affairs with the statistics of the counting-house, but the little formative group of people, related by kinship and religion as well as by common interest, who created the Manchester Society were from the beginning preoccupied with the statistics of social relevance. They were anxious above all to show that Manchester was being misrepresented in official reports. The first complete available paper (1834) of the Society, was entitled 'An Analysis of the Evidence taken before the Factory Commissioners, as far as it Relates to the Population of Manchester and the Vicinity Engaged in the Cotton Trade.' The paper, prepared by the Greg brothers—one of whom was to clash with Mrs Gaskell on her interpretation of the texture of social relationships in Manchester[13]—was a secondary analysis of data gathered earlier by a parliamentary committee. It had five sections dealing, not with 'the facts of progress' as registered in profits and wages, but with health, fatigue, alleged cruelty towards factory children, education, morals, and poor-rates. The first completely original survey produced by the Society dealt with the provision of education,[14] and the first annual report of the Society spoke of 'a strong desire felt by its projectors to assist in promoting the progress of social improvement in the manufacturing population by which they are surrounded.'[15]

In the case of some of the other early statistical societies, all of them expressions of the great burst of provincial intellectual and social energy during the 1830s and 1840s, the purposive emphasis was even more marked.[16] The full title of the Birmingham Society was 'The Birmingham Statistical Society for the Improvement of Education'; the Glasgow Society spoke of collecting 'facts illustrative of the condition and prospects' of the community 'with a view to the improvement of mankind'. The London Society itself referred to the 'careful collection, arrangement, discussion and publication of facts bearing on and illustrating the complex relations of modern society in its social, economical, and political aspects', with the adjective 'social' being placed first. By 1838 the fourth annual report of its Council noted that 'the spirit of the present age has an evident tendency to confront the figures of speech with the figures of arithmetic; it being impossible not to observe a growing *a priori* assumption that, in the business of social science, the principles are valid for application only inasmuch as they are legitimate deductions from facts accurately observed and methodically classified'.[17]

Such a statement was at the opposite end of the scale from romantic styles of social criticism; it was at some distance, too, both from generalized Owenite social science and Benthamite reasoning. Yet the Owenites sometimes used statistics. Bentham argued powerfully that the systematic collection and annual publication of returns would furnish data for the legislator to work on, and the political economy of Edwin Chadwick, unlike that of J. S. Mill, needed to be fed on a diet of statistics. Chadwick, indeed, was an energetic member of the Statistical Society of London, even if the limitations of his statistical methods were to be emphasized later in the century.[18] He told the Political Economy Club in 1845 that of two types of economists —'the hypothesists' who reasoned deductively from 'principles' and those who believed in 'the school of facts' and worked inductively—he belonged to the second.[19] Nor was he alone. By the time of the tenth report of the Statistical Society of London in 1844, the Council was claiming that 'the pursuit of statistical enquiries has already made such progress . . . as henceforth to be a necessity of the age, and one of its most honourable characteristics.'

The statistical method was deliberately employed during the 1830s and 1840s to identify 'problems', to spread 'knowledge of social facts', and to educate 'opinion'. It was because the pioneering statisticians of this period were thought of as explorers of society—and, particularly, of urban society within which the statistical societies were created—that they were able to influence both the collective will and the individual literary imagination. Their chief merit seemed to be that they discovered 'facts' at first hand, and Mill deferred to Chadwick on the grounds that Chadwick got his information direct while he (Mill) 'could only get it second hand or from books.'[20] The notion of 'exploration' which recurs time and time again in the imagery of early urban studies carried with it a sense of adventure. It also carried with it the sense that nineteenth-century cities, in particular, were 'mysterious places' where one section of the community knew very little directly about the rest. 'Why is it, my friends,' the great American preacher W. E. Channing asked in Boston in 1841—in a sermon in which he anticipated Disraeli in speaking of 'two nations coexisting side by side in the same community'—'that we are brought so near to one another in cities? It is, that nearness should awaken sympathy; that multiplying wants should knit us more closely together; that we should understand one another's perils and sufferings; that we should act perpetually on one another for good.'[21] In reality, however, the nineteenth-century city, with all its varieties of experience, did not permit this intuitive understanding. There was a striking contrast between *ought* and *is* which the statistician could expose. At his best, indeed, he could be a mediator as well as an explorer, not dwelling on the 'mystery' of the city—or exploiting it, as G. M. W. Reynolds and some of the purveyors of romantic fiction did—but rather on the dissipating of it.

Beyond a certain point it was clearly impossible to talk of 'numbers', the starting point of urban exegesis, without also talking about 'relationships', social and geographical. Most pre-industrial cities had been places which were small enough and for all their complexities, simple enough to generalize about or about which to moralize as 'wholes' or to satirize in terms of galleries of urban 'types'. It was far more difficult, in socially and geographically segregated cities with unequal spatial densities and with manifestly unequal conditions of 'class', segregated district by district, to grasp the idea of the city as a whole and to identify its 'problems' except through the collection and deployment of statistics. Doctors, who were particularly prominent in the deliberations of the Manchester Statistical Society, might be in a more favourable position than most to move easily from one area to another, relying on knowing directly rather than knowing at second-hand. So also might at least an active minority of ministers of religion. Yet with the separation of work-place and home, the growth of single-class living areas and the decline in what later urban sociologists were to call primary or 'face-to-face' relationships, statistics as a mode of enquiry easily came into its own.[22]

It was because 'numbers' could be used not simply for the purpose of rhetoric but for purposes of disclosure—exposure in many cases is not too strong a word—that journalists as well as 'men of good will' were deeply concerned with statistics—the famous articles in the *Morning Chronicle* in 1849 and 1850, written in the aftermath of Chartism and in the alarming presence of cholera, provide the outstanding example[23]—or that novelists also used statistics when they presented social comment in fictional form during the 1840s. As Arnold Kettle has pointed out, they were often addressing 'the downright factual ignorance of the middle class'.[24] 'The facts—the facts are all in all; for they are facts', wrote the reviewer of Mrs Gaskell's *Mary Barton* in *Fraser's Magazine* in 1849.[25]

There were obvious difficulties when 'disclosure' was extended to 'analysis' and 'analysis' was related to 'action'. First, despite Mill's comment on Chadwick, there was a problem of knowledge. The statisticians were observers who often started their enquiries not only in ignorance of what facts they would find but also with different values from those of the people they were observing. At best, they were able to display the need for sympathy which Channing emphasized. At worst, they were handicapped by their lack of the kind of personal knowledge that Mrs Gaskell, for instance, possessed, whatever the social limitations which restricted her approach, of how working-class families thought and felt.

Henry Mayhew, who throughout his work deliberately interposed individual vignettes and statistics (many of his statistics related to people whom he was not directly describing—employers, customers, and clients), sometimes seemed to be trying to move one or two steps beyond the position of an observer. He took the trouble, for example, to provide written answers to queries he received from individuals mentioned in his surveys, and he was willing to hold public meetings. He also appeared freer from many of the stifling inhibitions and restraining 'value frames' which limited the social comment of most of his contemporaries. Yet even in Mayhew's case, as E. P. Thompson has remarked, it would be 'ludicrous' to suggest that he 'discovered Victorian poverty'. 'The poor had long before discovered themselves, and the *Northern Star* contained a part of their own testament.'[26] Their testament, moreover, necessarily had policy implications written into it which diverged from the policy implications of social statisticians like Chadwick.[27] The work of the statisticians was important, in Mayhew's opinion, in that it encouraged a questioning of untested middle-class assumptions and prejudices, and he complained bitterly, if a little too comprehensively, that 'economists, from Adam Smith down, have shown the same aversion to collect facts as mad dogs for the touch of water.'[28]

There was more, however, to the question of 'values' than this. Not every middle-class critic of society was impressed by statistical exposure or by the conclusions which statisticians were prone to reach about the viability of the 'social system' and its inherent possibilities for 'improvement'. The kind of people who turned for guidance to Thomas Carlyle, the prophet of his age, spurned statistics, as Ruskin was to spurn them. What could be more 'general' than the opening sentences of Carlyle's *Past and Present* (1843)—'England is full of wealth, of multifarious produce, of supply for human want in every kind' or Ruskin's remark in *Unto this Last* (1862) that 'our cities are a wilderness of spinning wheels ... yet the people have not clothes ... Our harbours are a forest of merchant ships, and the people die of hunger'? 'Where men formerly expended their energy on scholastic quibbles,' a writer in the *Edinburgh Review* complained during the late 1850s, 'they now compile statistics, evincing a mental disease, which may be termed the colliquative diarrhoea of the intellect, indicating a strong appetite and a weak digestion.'[29]

The preference for 'generality' was reinforced by the feeling, first that statistics was something of a 'fad', second that there could be more evasion and misrepresentation than disclosure in the work of statisticians, and third that individuals counted for more than 'averages'. 'It is astonishing', *Punch* wrote in 1848,[30] having regularly satirized such pursuits,

> what Statistics may be made to do by a judicious and artist-like grouping of the figures; for though they appear to begin with a limited application to one subject, there is no end to the mass of topics that may be dragged in collaterally on all sides. A few facts on mendicancy, introduced by one of the members [of the British Association for the Advancement of Science] became the cue for an elaborate calculation of how many meals had been given to Irish beggars in the last twenty years; and this was very near leading to a division of the meals into mouthfuls, with a table showing the number of teeth, subtracting the molars and taking out the canine, employed in the mastication of these twenty years' returns of meals.

There are innumerable satirical references to statistics in *Punch*, where the quality of life in London was a subject of frequent pictorial comment, but it was the novelist, in particular—and given the educational and social patterning of the time, perhaps the woman novelist even more than the rest—who made the most of the inadequacies of 'averages'. There appeared to be something misleading as well as arid in constructing systems of classification and statistical tables. One man's death was more 'real' than a statistic in a bill of mortality or what Mr Gradgrind called 'the laws which govern lives in the aggregate'. The preacher shared this same preoccupation with the individual (real-life even more than fictional) and the plot.

The Victorian City: Numbers of People

'We may choose to look at the masses in the gross, as subjects for statistics,' wrote Kingsley in 1849, 'and, of course, where possible [he went on characteristically], of profits.' Yet there was 'One above who knows every thirst and ache, and sorrow, and temptation of each slattern, and gin-drinker, and street boy. The day will come when He will require an account of these neglects of ours not in the gross.'[31] For Kingsley and those who thought like him, 'the Sanitary Idea' might depend for its ammunition on statistics relating to the unhealthy environment, but its force as a 'gospel' rested on an appeal to deeper forces within the individual, through the operations of the conscience, and, possibly unconsciously, through psychological pressures which the men of the nineteenth century did not understand. It is interesting to compare Elizabeth Barrett Browning with Kingsley when she wrote in her verse-novel, *Aurora Leigh* (1856),[32] the lines

> A red-haired child
> Sick in a fever, if you touch him once,
> Though but so little as with a finger-tip
> Will set you weeping; but a million sick . . .
> You could as soon weep for the rule of three
> Or compound fractions.

Ruskin went far beyond 'the Sanitary Idea' in his writings on political economy, drawing an explicit general distinction between questions of quality and quantity: it was not the sum of products but the quality of a people's happiness which constituted the wealth of the community, he argued. 'But taken as a whole,' he wrote, throwing statistics to the winds, 'I perceive that Manchester can produce no good art, and no good literature; it is falling off even in the quality of its cotton.'[33]

In any discussion of the approach of statisticians and the response of contemporaries to their methods and conclusions it is important to bear in mind that there were changes in statistical preoccupations from one part of the century to another. The earliest statistical surveys of city populations, some of which were based on questionnaires or what Kay-Shuttleworth called 'tabular queries', were already giving way in the late 1840s to less controversial analyses of more narrowly defined specific questions, with at least one precocious pre-Le Play essay on family budgets, pointing the way forward to new modes of enquiry.[34] By the 1850s and 1860s, substantial city surveys based on original exploration had been almost completely replaced by social-science essays concerned with secondary material: they were being debated by the members of the National Association for the Promotion of Social Science, founded in 1857,[35] with the aged Brougham, a link with Bentham, as first president, but they received far less widespread public attention from contemporaries than the surveys of ten to fifteen years before. The title of the Statistical Section of the British Association, which had been created in 1833, was changed to 'Statistical Science' in 1857 and in 1863 to 'Economic and Statistical Science', and the proportion of papers devoted to economics increased significantly. During the late nineteenth century, the social survey was to come into its own again, but by then some of the older provincial statistical societies had disappeared, the Royal Statistical Society had acquired its prestigious adjective (in 1875), and statisticians had developed an embryonic professional sense. 'We have learned', William Newmarch was once quoted as saying, 'that in all questions relating to human society . . . the only sound basis on which we can found doctrines . . . is not hypothetical deductions, however ingenious and subtle, but conclusions and reasoning, supported by the largest and most careful investigation of facts.'[36]

The middle years of the century belonged to the National Association for the Promotion of Social Science which, on its visit to Manchester in 1866, inspired the *Manchester Examiner and Times* to comment that 'the mass of miseries which afflict, disturb or torment mankind have their origin in preventable causes. They can be classified just as drugs are classified and they may be employed with almost the same certainty of operation. It is to social science that we are indebted for a knowledge of their character, and it is to its progress we must look for the amelioration of our home miseries.'[37] On the same occasion the *Manchester Guardian* noted how social science thrived in a city atmosphere just because of the multiplicity of urban problems. 'Nowhere are the social changes which are now in progress and which are viewed with hope or fear according to the temper of the observer, more manifest than among the teeming population of which this city is the centre and metropolis.'[38]

Characteristic of the new style of 'learned' paper in 'social science' was Dr William Ogle's fascinating piece on 'Marriage-Rates and Marriage-Ages with special reference to the Growth of Population', read before the Royal Statistical Society in 1890.[39] This paper, which neatly sets out a problem—why had the nineteenth-century marriage-rate fluctuated in 'a very irregular manner'?—first disposed of a fallacy propounded by, amongst others, J. S. Mill and Henry Fawcett, who had not bothered to give the actual figures on which their statements were based, that the marriage-rate varied inversely with the price of wheat. Second, it went on with the aid of graphs to correlate the marriage-rate with the value of exports per head of the population. Third, it explored the relationship between export data and the very patchy trade-union statistics relating to employment and unemployment. Fourth, it developed a fascinating theory on the basis of the statistical evidence, explaining why there were marked variations of marriage-rates in different registration counties (arguing that marriages were more numerous in those counties where women earned independent wages). Fifth, it looked at age and occupation structure. Sixth, it related historical evidence to anticipations of the future. The limitations of statistical method were noted, even if they were not fully or critically investigated. This kind of article needed ingenuity and skills of the highest order, far removed from Dickens's 'Simple Arithmetic', and Dr Ogle certainly could not be placed in the camp of those who saw 'figures and averages and nothing else'.

There is still much to sort out in relation to the detailed history of the use of statistics. As late as 1887 Wynnard Hooper, the writer on Statistics in the ninth edition of the *Encyclopaedia Britannica*,[40] had to devote much of his limited space to the dispute between those who believed that there was a science of statistics—with its own specific content—and those who believed that there was only a statistical method, a convenient aid to investigation in the majority of sciences. The former group was still of strategic significance—not surprisingly so, perhaps, when thinking in the natural sciences remained for the most part non-statistical. It had moved from political to social arithmetic, with some of its members following Maurice Block and anticipating Louis Chevalier in giving a new name to their branch of study—'demography'. The President of the Royal Statistical Society argued boldly and passionately at the jubilee meeting in 1885 that statistics was superior in method to social science or sociology and that it amounted to 'the science of human society in all its relations'. Statistics *was* sociology.[41]

Hooper was more cautious, as he had every right to be when the map of both natural and social sciences was changing as significantly as it was at that time. After noting—and how common a note it was becoming in so many areas of English life—that there had unfortunately so far been 'no attempt in England to deal with the subject . . . in a systematic way', though 'the practice of statistical inquiry of scope and method has been carried on in England with a high degree of success'—he refused to identify all sociology with statistics.

> The statistical method is essentially a mathematical procedure, attempting to give a quantitative expression to certain facts; and the resolution of differences of quality into differences of quantity has not yet been effected even in chemical science. In sociological science the importance of differences of quality is enormous, and the effect of these differences on the conclusions to be drawn from figures is sometimes neglected, or insufficiently recognized, even by men of unquestionable ability and good faith.

The term 'values' was not used in this context. 'Society is an aggregate', wrote Hooper, 'or rather a congeries of aggregates.' And he went on to draw an 'expert' conclusion, concerned solely with statistical techniques—a conclusion which was sharply different from that current in the statistical societies during the pioneer amateur phase of the 1830s and 40s before the rise of professionalism: 'the majority of politicians, social "reformers" and amateur hoarders of statistics generally were in the habit of drawing the conclusions that seem good to them from such figures as they may obtain, merely by treating as homogeneous and comparable qualities which are not comparable. Even to the conscientious and intelligent inquirer the difficulty of avoiding mistakes in using statistics prepared by other persons is very great.'[42]

This cautious conclusion, set out two years after Charles Booth had started his huge statistical enquiries by challenging Hyndman's simple, unsophisticated, and politically orientated figures of London poverty, should be set alongside Booth's ambition, far-reaching in scope, to prepare a detached and impartial presentation of the social situation through the use of statistics. 'A framework can be built out of a big theory and facts and statistics run in to fit it, but what I want to see instead is a large statistical framework which is built to receive accumulations of facts.'[43] Seebohm Rowntree likewise turned to statistics because they provided a less 'sentimental' foundation for policy recommendations than straight appeals to human feeling or to political prejudice.[44] It was Mrs Webb who wrote of Booth's study that 'prior to this enquiry, neither the individualist nor the Socialist could state with any approach to accuracy what exactly was the condition of the people of England. Hence the unreality of their controversy.'[45]

The problem of relating quantity to quality, which Booth never tried to baulk in the same way as he baulked most questions of theory, could be tackled, he said, 'given sociological imagination'. 'The statistical method was needed to give bearings to the results of personal observation and personal observation to give life to statistics ... It is this relative character, or the proportion of facts to each other, to us, to society at large, and to possible remedies, that must be introduced if they are to be of any value at all in social diagnosis. Both single facts, and strings of statistics *may* be true, and demonstrably true, and yet entirely misleading in the way they are used.'[46] Booth added to the list of questions for full quantitative examination those relating to poverty, seeking 'to connect poverty and well-being with conditions of employment'. And, in this attempt to connect, he was to query (like Rowntree later) and with the blessing of the great neo-classical political economist of late-Victorian England, Alfred Marshall, the non-quantitative basis of much of earlier nineteenth-century political economy. It was not only sociology that had to be statistical. Economics had to be statistical, too, as it had been for Porter earlier in the century and as it was at the time for Robert Giffen.[47]

However great the changes in mood and context, in one important respect Booth followed directly in tradition from writers of the 1830s and 1840s. He never concerned himself very much with the state. The city was still *the* place to study if you wished to understand society. 'It is not in country', he wrote in a famous passage,[48]

> but in town that 'terra incognita' needs to be written on our social map. In the country the machinery of human life is plainly to be seen and easily recognized: personal relations bind the whole together. The equipoise on which existing order rests, whether satisfactory or not, is palpable and evident. It is far otherwise with cities, where as to these questions we live in darkness, with doubting hearts and ignorant unnecessary fears, or place our trust with rather dangerous consequences in the teachings of empiric economic law.

Booth's own preferences—subjected to searching self-criticism and criticism—quickened the enquiry. London was a stage, not a laboratory: it had its drama, and the drama was perpetually interesting, as it was for Henry James who, while complaining of London's 'horrible numerosity', none the less concluded that it offered 'on the whole the most possible form of life'.[49] H. L. Smith, one of Booth's assistants, emphasized that among the attractions of London was 'the contagion of numbers' and that, taking into account all the problems of the city, it was 'the sense of something going on . . . the difference between the Mile End fair on a Saturday night, and a dark and muddy country lane' which drew the young in particular into the 'vortex'.[50] Number and quality were being related in a more sophisticated way. 'New York's the place for me,' Booth himself once wrote to his wife. 'There seems something subtle, an essence, pervading great metropolitan cities and altering everything so that life seems more lively, busier, larger, the individual less, the community more. I like it. It does me good. But I know it has another aspect and I am not surprised when people feel crushed by the wickedness of it, the ruthlessness, heartlessness of its grinding mill, as you did in Paris.'[51]

By the end of Queen Victoria's reign, very much in English life had been measured. The great official enquiries into environment at the beginning of the reign —in the name of 'the Sanitary Idea'—had their counterpart at the end of the reign in the great unofficial enquiries into poverty. In between, while the cities grew, as did the proportions of the population who were city-dwellers, the collection of many of the relevant statistics ceased to be a major exercise in difficult and uncharted social investigation and became, like so much else in Victorian life, an institutionalized routine with decennial census reports, annual medical officer of health reports, financial returns, and so on. There were still big gaps in relation to both economic statistics (including the statistics of employment) and cultural statistics, but there were now statistical experts who were taking over or seeking to take over previously debatable areas of policy and administration. From its inception in 1889 the London County Council appointed a full-time statistical officer to collect such data for the use of its various committees.[52] As for the dedicated non-experts, they were moving again from urban detail to more general schemes of social regeneration, to the nature of what Sir John Simon had called earlier in the century the 'underframework of society',[53] from issues related to municipal action within the particular city to national welfare policy which would iron out some of the differences between cities.

It is interesting in the light of this story that the writer on the history of statistical method in *The International Encyclopaedia of the Social Sciences* states that 'if we have to choose a date at which the modern theory of statistics began, we may put it, somewhat arbitrarily, at 1890.' Pointing to the work of F. Y. Edgeworth, Karl Pearson, Walter Weldon, and G. U. Yule, he notes also that this was the birth year of R. A. Fisher. 'Life was as mysterious as ever,' he goes on, 'but it was found to obey laws. Human society was seen as subject to statistical enquiry, as an evolutionary entity under human control.'[54] This realization, as we have seen, came earlier in studies of the city, but the refinements of statistical analysis came later. Full sophistication was a twentieth-century achievement.

Before turning briefly from what the Victorians themselves did or failed to do with their limited techniques and the quantitative evidence at their disposal to what we in the twentieth century can and should do with statistics in our interpretation of the past, it is important to make two basic points. The first concerns the rates of measurable change in relation to qualitative evidence offered by or available to the Victorians themselves. The second concerns the relationship between facts and theories centred specifically on the city, and facts and theories relating to the constitution and development of society as a whole.

On the first point, it is obvious from the history, particularly, of urban public health that the noisiest and most exciting periods of debate did not necessarily coincide with the periods of greatest demographic and social change. Because rates of measurable progress did not reflect the power of language or of argument, qualitative evidence by itself may often be misleading. The fierce debates in the 'age of Chadwick' did little to force down crude death-rates which did not fall substantially until the 1870s when, on the whole, questions of public health stirred contemporaries less. What William Farr, the statistical genius behind the mid-Victorian censuses, described as 'one of the most important series of facts relating to the life of a nation ever published' was the revelation that the annual mortality for all ages in the population had scarcely altered throughout the period 1838–71.[55] The infant mortality-rate, rightly considered in the twentieth century to be one of the critical indices of social control, remained more or less constant around 150 per thousand live births until the twentieth century.

Earlier historians of public health concentrated on the study of qualitative materials relating to 'the age of Chadwick'. Confronted with the mid-Victorian inability to bring down death-rates and infant mortality-rates, the social historian with quantitative interests may now seek to explain the statistics in one or all of four ways: (1) qualitatively by pointing to the fact that it was not until the 1870s that advances in bacteriology began to produce results and that the limitations of 'environmentalism' were overcome in an age of increasing concern for personal health and housing; (2) qualitatively by examining those changes in the economy, in the 'administrative system', and in society which distinguished late-Victorian from early-Victorian England and permitted effective 'amelioration'; (3) quantitatively by breaking down national aggregate-rates within short periods and examining, as best he can, the pattern of death-rates in particular places, for particular groups or for particular diseases; (4) counter-factually, as is now beginning to be common in the new quantitative economic history,[56] by seeking to analyse what would have happened if there had been no health legislation, local or national, during the 1840s and 1850s.

There is little doubt that in this field quantitatively based urban history enables us to know far more and to understand what we do know far more clearly and in greater depth than if we rely, as did whole generations of historians, on the colourful and rhetorical Victorian discussion of sanitary matters, from the propaganda of the Health of Towns Association to an article by the Bishop of Bedford on 'Urban Populations' in the *Fortnightly Review* of 1893 which did not give one single precise statistic.[57] It forces attention to local variations in the period from the 1830s to the 1870s, gives new interest to the late-Victorian years when rates actually did fall, and provides the basis for a not unfavourable assessment of the mid-Victorian achievement. 'Stable death rates', it has been argued, 'conceal a considerable victory; for by holding in check the powerful forces against health which swiftly growing population and rapid urban agglomeration naturally generated, the sanitary pioneers could congratulate themselves on a valuable, if negative achievement.'[58] Finally it leads to a new evaluation (but not to a dismissal) of the qualitative evidence itself. The early language of 'civic economics', the kind of language Chadwick talked naturally—and some of it was couched almost in terms of cost-benefit analysis[59]—gave way to the specialist language of doctors, engineers, and housing economists. At the same time, the fervour of the first novelists and poets when they took up 'the Sanitary Idea' disappears, and with it the sense of drama.

The second important point is related to the first. The local element in initiative for health improvement during the mid-Victorian years, which sometimes led to fierce battles between 'clean' and 'dirty' parties at the city level, changed in relative significance from 1870 onwards as the demand for a 'national' health policy gained in strength. Statistical information was a necessary, and, as time went by, an uncontroversial instrument in the reforms achieved through legislation and administration between 1869 and the foundation of the Ministry of Health at the end of the First World War. Indeed, it was through a detailed consideration of health statistics (along with statistics of education) that the balance between local and national action began to change. The significance of what was happening can be well illustrated also from a brilliant review of Seebohm Rowntree's *Poverty* by C. F. G. Masterman, who was to make his reputation as author of the Edwardian classic, *The Condition of England*. Masterman, who was a master of qualitative argument and a brilliant coiner of original phrases which still stick, was forced by Rowntree's bleak and ungarnished statistics on poverty in York to this conclusion. 'The social reformer, oppressed with the sense of the ... poverty of London,' Masterman pointed out, 'is apt to turn with envy towards the ideal of some flourishing provincial town.' What better town than York, he might ask, an ancient community revitalized by the railway in recent Victorian history, a community contrasting in every way with the 'homogeneous matrix' and 'sheer immensity of the aggregation' of London? Would it not be desirable, he might go on, to break up the giant city of six millions into sixty cities of 100,000 each, 'not too large to cause congestion nor too small to prevent the intertwining of varied industries necessary for permanent stability'?

Such an approach, Masterman concluded, was quite wrong. Rowntree's demonstration that the proportion of primary poverty in York was the same as in Booth's London proportion was a devastating answer—'a thunder-clap', Masterman called it—to such social reformers. A national policy was needed to tackle poverty, not a series of local urban expedients, however enlightened. From this angle Masterman found Rowntree's detailed statistical work far more persuasive than Booth's, for at the end of reading Booth's 'nine bulky volumes, mazes of statistics ordered and classified, maps of picturesque bewilderment of colour, infinite detail of streets and houses and family lives' what was left was 'a general impression ... of something monstrous, grotesque, inane: something beyond the power of individual synthesis; a chaos resisting all attempts to reduce it to orderly law.'[60] Rowntree's 'definite and limited material' was manageable enough and comprehensible enough to guide policy —and, we can add in retrospect, to guide it towards the peculiarly English twentieth-century balance between the urban and the national.

There are more than administrative implications to this second important point. It is easy for writers on the Victorian city or for that matter on Victorian urbanization to concentrate so much on the city or on what happened in all cities that they forget that urbanization is a 'societal' process which not only precedes the formation of particular cities but also shapes their role as agencies in a developing society.[61] Quantitative analysts have rightly pointed to other implications of this, relating changes in cities to changes in the countryside—this was beginning to be a favourite late-Victorian theme, very dear, it may be noted, to Rowntree—or to the map of world trade. They have also qualified conceptions of urban causality by looking more closely at the components of class behaviour and at the logic of industrialization.[62] In this context, it is not strange that studies which begin with the Victorian city end with the twentieth-century state. Quantitative investigation—concerning rates of change and differentials between cities, about the balance between urban and national finance, and about shifts in politics and communications—is a necessary element in the approach to a fuller understanding of this range of issues which cannot easily be grasped in terms of qualitative evidence or argument taken by itself.

Any account of the Victorians' approach to statistics and the way in which they handled quantitative evidence must be qualified and supplemented in the light of recent historical scholarship. Yet the wisest of the Victorians themselves knew that their evidence was in places incomplete and in other places defective. Some of them even anticipated recent social scientists who have argued that 'both quality and quantity are misconceived when they are taken to be antithetical or even alternative. Quantities are *of* qualities and a measured quality has just the magnitude expressed in its measure.'[63] In Britain, as in France, current research, in Louis Chevalier's words, is 'a continuation of earlier interpretative efforts which have gone on as long as urbanization itself'.

Different kinds of quantitative enquiry are being carried out in Britain at the present time, though perhaps less ambitiously than in several other countries. First, historical demographers are relating data about birthplace and migration to occupational data, family size, and varieties of environment within the same urban community, with W. A. Armstrong, for example, using enumerators' books for nineteenth-century censuses to explore urban social structure through 'still glimpses of a moving picture'. Following in the footsteps of Rowntree in exploring York, he is explicitly concerned not with the social pathology of the city—the subject which has always captured most attention, qualitative and quantitative—but with 'normal' communities and classes. He is also interested in the relationship at the family base between experience in Victorian urban society and what had gone before.⁶⁴

Second, H. J. Dyos, with the assistance of computers, is seeking to define and explain the social changes which occurred in the making of Camberwell, a suburb of Victorian London. His interest is not in 'still glimpses' but in processes—'how particular neighbourhoods came to be occupied, held, or vacated by different social classes, and just what kinds of communities these were in terms of the structure of families and households'—and he suggests cautiously that 'quantitative techniques can give precision to history, but they do not seem capable of formulating new hypotheses, much less of bringing to it objectivity.'⁶⁵

Third, J. R. Kellett, who has made the most of urban railway statistics, has, in fact, advanced new hypotheses as a result of his detailed studies, pointing out on the way how the techniques of enquiry into the social effects of railway building on the urban way of life became far more sophisticated by the beginning of the twentieth century than they had been during the 1840s. His study is useful also in underlining that even in twentieth-century cost-benefit studies 'the most adequate and detailed statistical analysis tends to become only one of several factors moulding final decisions.'⁶⁶

Fourth, in relation to political decision-making in nineteenth-century cities, where studies of city 'personalities' and attempts at 'city biography' have hitherto dominated a largely underdeveloped field, J. R. Vincent has directed attention to the analysis of poll-book material, making generalizations on the way about political statistics and the political motivation which lies behind them. A number of other writers have collected statistics about the social composition of town councils—here, as in occupational studies, there are very real difficulties in categorization—in an effort to relate leadership to class and occupational structure.⁶⁷ Other studies have focused on minority groups and sub-cultures. So far, however, it is obvious that the note of caution sounded by H. J. Dyos is particularly necessary in any consideration of 'the behavioural aspects of urban communities', and L. F. Schnore has even argued that quantitative studies are 'practically impossible'. 'We shall have to continue depending', he concludes, 'upon impressionistic accounts concerning the attitudes and values of our urban forebears.'⁶⁸

Finally, on one subject where quantitative approaches seem possible and inevitable, very little statistical work has been carried out on local finance, where there are complicated issues centring on rateable values and annual rates in the pound (there were for long, of course, whole clusters of urban rates) and the relation of rating statistics to land values, building, and economic activity. What work has been started points to the difficulty of making even simple comparisons.

How far is it feasible to go beyond all these individual studies and produce a Victorian counterpart for the useful study of C. A. Moser and W. Scott, *British Towns: A statistical study of their social and economic differences* (1961)? In *Victorian Cities* I suggested that 'a whole Victorian urban typology could be constructed on these lines.' In pursuing the matter a little further—and some historians should pursue it a lot further—it is useful to start with Moser and Scott's conclusion about 'the striking diversity' of the 157 English towns with a population of more than 50,000 in the middle of the twentieth century. Some are more or less self-contained towns with rural districts at their limits: others are grouped together, with no obvious boundaries, in the tangle of conurbations. 'Between the extremes ... lie every variety of urban species both simple and complex' so that 'no single formula can describe them all.' The statistical diversity even after decades of national social policy-making was extremely striking: one household in five in Gateshead lived in overcrowded conditions, for example, as against one in sixty in Coulsdon: infant mortality-rates in Rochdale were three times as high as in Merton and Morden.

Moser and Scott followed eight main lines of statistical enquiry—into population size and structure; demographic change over twenty years; households and housing; economic characteristics; social class; voting; health; and education. What they left out was just as interesting as what they put in—local government finance, on the grounds that it 'needed more detailed treatment' than they 'had time to give'; employment statistics, on the grounds that these were collected from different units; all cultural data, because the statistics were too patchy; all information on crime, religion, and 'the physical characteristics of the town and its amenities'.[69] Given the superiority of twentieth-century statistical sources over those of the nineteenth century, the omissions look discouraging to the historian.

Indeed, the historian of Victorian cities, fascinated as he would be by the diversity of his communities—there were twenty-one of them (plus London) in 1901—would not be happy on historical grounds about most of these exclusions or with the authors' statement that while 'the change of population between 1931 and 1951 is, on the whole, an inadequate index of the age of a town' they did not feel that they could delve further back into history. As a minimum he would wish to introduce chronological tables setting out the timetables of change in different cities as far, at least, as the creation of new institutions or activities was concerned, and more ambitiously he would want to go on to relate classificatory systems to urban 'profiles' or 'images'.[70] He could, none the less, experiment happily with schemes of urban classification both in terms of Moser and Scott's list of primary variables, and, if he had the skill, with component analysis.

It is apparent at the outset that there are fascinating contrasts of experience between the nineteenth and twentieth centuries, with no Victorian town ever containing, as Worthing did in the period covered by Moser and Scott, 58·8 percent of its population over the age of sixty-five. Bradford, for instance, never had more than 4 percent at any census between 1841 and 1901, while the town with the lowest proportion in Moser and Scott's account—Dagenham—had 4·9 percent in 1951. Middlesbrough had 36 percent of its population below the age of fifteen in 1841 and 1901, a significantly higher figure than the English town with the highest 1951 figure —31·1 percent—Huyton-with-Ruby. Bradford had as high a figure as 45 percent in 1841. The male/female ratio is also particularly interesting to compare. Middlesbrough, with more males than females both in 1851 and 1901, stood out in this connection as much as in its remarkable statistics of growth, a matter of civic rhetoric and one of the topics covered qualitatively in my *Victorian Cities*. It was, in fact, described in 1885 by the statistician E. G. Ravenstein as a town which by 'its rapid growth, the heterogeneous composition of its population, and the preponderance of the male sex, recalls features generally credited only to the towns of the American west.'[71] Statistical series of many of Moser and Scott's social indicators are missing for the nineteenth century, but far more can be done with the nineteenth-century 'mix' of the population and the relationship between 'native' and 'extraneous' elements than they have tried to do. There would be value, too, through econometric and other network studies in relating recent urban experience to that of the nineteenth century. Of Moser and Scott's urban groups 1, 2, and 3 in 1951—mainly resorts, administrative, and commercial towns—five out of thirty-seven were already towns of over 50,000 in 1851 and thirteen in 1901; whereas of their groups 4, 5, 6, 7 and 8—mainly industrial towns—sixteen out of sixty-seven were in the 1851 list and forty-seven in that of 1901. Of the fifty-one towns in their groups 9, 10, 11, 12, 13 and 14—suburbs and suburban-type towns—there were no towns in the 1851 list and only seven in 1901.

Much of the history of Victorian urbanization can be studied in such terms, although full explanations of historical change depend on relating statistics to evidence of a qualitative kind derived from impressions of particular places as seen by inhabitants, visitors and 'experts'. Such evidence must be accumulated from a wide variety of sources, local and national. No urban history can afford to neglect the 'sense of place' which must be a main theme of all studies of Victorian cities, or fail to consider the distinctive and the 'unusual' as well as the general and the commonplace. It was Henry James, writing of London, who commented aptly that 'when a social product is so vast and various, it may be approached on a thousand different sides, and liked and disliked for a thousand different reasons.'[72] All such visions illuminate the city as it actually is or was. In the famous third chapter of Macaulay's *History of England* (1848), the historian drew an unforgettable picture of seventeenth-century London which relied heavily, through contrast, on the character of mid-Victorian London. 'The town did not, as now, fade by imperceptible degrees into the country. No long avenues of villas, embowered in lilacs and laburnums, extended from the great centre of wealth and civilization almost to the boundaries of Middlesex and far into the heart of Kent and Surrey.' By the middle of the nineteenth century 'the fireside, the nursery, the social table' were no longer in the City of London: 'the chiefs of the mercantile interest are no longer citizens.' The use of space had changed. 'He who then rambled to what is now the gayest and most crowded part of Regent Street found himself in a solitude, and was sometimes so fortunate as to have shot at a woodcock.' Relations had changed, too, along with the ecology. Unimproved London was less socially segregated: 'In Covent Garden a filthy and noisy market was held close to the dwellings of the great.' At the same time, there was a more absolute difference between the metropolitan and the Londoner. 'A Cockney in a rural village was stared at as much as if he had intruded into a Kraal of the Hottentots. On the other hand when the Lord of a Lincolnshire or Shropshire manor appeared in Fleet

Street, he was as easily distinguished from the resident population as a Turk or a Lascar.'

Macaulay had little to say about statistics—far less, for instance, than Mayhew, writing soon afterwards, or Henry Buckle—yet his picture holds and illuminates. He set out less to explain or to interpret than to describe and to evoke. Quantitative approaches to social history do not destroy the value of this kind of writing. What they do, indeed, is to sharpen the appeal of the artist's vision, enabling us to relate special experience or special ways of viewing common experience to common experience itself.

Notes

1 See, for example, R. W. Rawson, 'An Enquiry into the Statistics of Crime in England and Wales', *Journal of the Statistical Society of London (JSS)*, ii (1839–40), 316–45, and J. Fletcher, 'Moral and Educational Statistics of England and Wales', ibid., xii (1847), 151–76.
2 The address is printed in Charles Kingsley, *Sanitary and Social Lectures and Essays* (1880), pp. 187–222.
3 *Victorian Cities* (1963), pp. 57–8.
4 'My satire is against those who see figures and averages, and nothing else ... the addled heads who would ... comfort the labourer in travelling twelve miles a day to and from his work, by telling him that the average distance of one inhabited place from another in the whole area of England, is not more than four miles.'
From a letter to Charles Knight, 30 January 1855, quoted in the Norton Critical Edition of *Hard Times*, ed. George Ford and Sylvère Monod (1966), p. 277.
Dickens admired H. T. Buckle's *History of Civilization in England* (2 vols, 1857–61) with its statistical preoccupations.
5 His letter of 16 December 1845 is quoted in H. S. Tremenheere, *I Was There: Memoirs*, ed. E. L. and O. P. Edmonds (1965), p. 54.
6 Harold Perkin, *The Origins of Modern English Society 1780–1880* (1969), p. 326.
7 J. C. Lettsom, *Medical Memoirs of the General Dispensary in London* (1774), p. x.
8 See M. W. Flinn's admirable introduction to his recent edition of *Report on The Sanitary Condition of the Labouring Population of Great Britain, 1842*, by Edwin Chadwick (1965), pp. 21 ff.
9 M. Blaug, *Ricardian Economics* (1958), pp. 182–3. See also J. A. Schumpeter, *History of Economic Analysis* (1954), p. 524.
10 Review of *An Address Explanatory of the Objects and Advantages of Statistical Enquiries* by J. E. Portlock, F.R.S., *JSS*, i (1838–9), 317.
11 See [James Bonar and Henry W. Macrosty], *Annals of the Royal Statistical Society, 1834–1934* (1934), p. 22. The first meeting, in the convening of which the Rev. T. R. Malthus played a leading part, took place on 15 March 1834.
12 Its story is well told in T. S. Ashton, *Economic and Social Investigations in Manchester, 1833–1933* (1934).
13 W. R. Greg (1809–81) critically reviewed *Mary Barton* in the *Edinburgh Review* in April 1849 (lxxxix, 402–35) as being unfair to the mill-owners. Mrs Gaskell in

her preface had stated that she was concerned to show how things *seemed* to the poor. See A. Pollard, *Mrs Gaskell, Novelist and Biographer* (1965), pp. 59–60, and E. Wright, *Mrs Gaskell* (1965), pp. 231–2.

14 *Report of Committee of the Manchester Statistical Society on the State of Education in the Borough of Manchester, in 1834* (1835). The Committee comprised seventeen men, and the main object of the report was to show that Parliament had underestimated the number of schools in Manchester. A second edition containing 'a more minute classification' appeared in the same year.
15 Quoted in Ashton, op. cit., p. 13.
16 See 'Provincial Statistical Societies in the United Kingdom', *JSS*, i (1838–9), 48–50, 115–17; ii (1839–40), 132–3.
17 See also B. Kirkman Gray, *Philanthropy and the State, or Social Politics* (1908), Appendix to ch. 2: 'The Origin of the Royal Statistical Society'.
18 See A. Newsholme, *The Elements of Vital Statistics* (1889), pp. 111–12.
19 MS. draft, June 1845, quoted in R. A. Lewis, *Edwin Chadwick and the Public Health Movement 1832–54* (1952), p. 12.
20 Quoted in Lewis, op. cit., p. 15.
21 William E. Channing, *A Discourse on the Life and Character of the Rev. Joseph Tuckerman, D.D.* (Boston, 1841), p. 4. For Dickens's view of Channing, see *American Notes* (1842), ch. 3.
22 For the classic statement, widely reprinted, of urban sociology in terms of numbers and relationships, see Louis Wirth, 'Urbanism as a Way of Life', in *American Journal of Sociology*, xliv (1938), 1–24, and its critique by R. N. Morris, *Urban Sociology* (1968). See also R. E. Park, E. W. Burgess and R. D. McKenzie, *The City* (1925).
23 The first *Morning Chronicle* article by Mayhew on what was later incorporated in his *London Labour and the London Poor* (1861–2) appeared in October 1849.
24 Arnold Kettle, 'The Early Victorian Social Problem Novel', in Boris Ford, ed., *The Pelican Guide to English Literature*, vi. *The Nineteenth Century* (1958), p. 171. For the novel as a favourite form of social commentary during the 1840s, see Kathleen Tillotson, *Novels of the Eighteen-Forties* (1954).
25 *Fraser's Magazine*, xxxix (1849), 430.
26 E. P. Thompson, 'The Political Education of Henry Mayhew', *Victorian Studies*, xi, no. 1 (September 1967), 43–62.
27 For Chartist attacks on Chadwick, see, for example, *Charter*, 28 April and 23 June 1839. Yet in 1848 Chadwick wrote to the Bishop of London that he did not see 'how any one could get up in the Commons and contend that where there was a heavy infantile slaughter, or where the working classes are ravaged by epidemics, there shall be no intervention except on the initiation of the middle classes.' (Quoted in Lewis, op. cit., p. 170.)
28 Henry Mayhew, *Answers*, July 1851, quoted in Thompson, loc. cit., p. 56.
29 Quoted in Ashton, op. cit., p. 51.
30 *Punch*, xv (1848), 92: it added, more succinctly, a misquotation of Pope's line from his 'Epistle to Dr Arbuthnot': 'They lisped in numbers, for the numbers came.'
31 *Charles Kingsley. His letters and memories of his life*, edited by his wife (1892 edn), p. 88. Compare the speech by Dickens with that of Lord Carlisle at a Festival of the Metropolitan Sanitary Association in 1851 as described in the *Illustrated London News*, 7 May 1851: K. J. Fielding, ed., *The Speeches of Charles Dickens* (1960), pp. 127–32;

the occasion is briefly described in Humphry House, *The Dickens World* (1942 edn), pp. 195–6. See also the account of the implications of Tom-all-Alone's 'filth and slime', with its retribution theme, in *Bleak House* (1853).

32 Book II.
33 John Ruskin, *Fors Clavigera*, Library edn, xxix (1905), p. 224.
34 Henry Ashworth, 'Statistics of the Present Depression of Trade at Bolton', *JSS*, iv-v (1842-3), 74–81. See also T. C. Barker, D. J. Oddy and J. Yudkin, *The Dietary Surveys of Dr. Edward Smith, 1862–3* (1970).
35 See Brian Rodgers, 'The Social Science Association, 1857–1886', *Manchester School of Economics & Social Studies*, xx, no. 3 (September 1952), 283–310.
36 Frederic J. Mouat, 'History of the Statistical Society of London', *JRSS*, Jubilee vol. (1885), 50.
37 *Manchester Examiner and Times*, 5 October 1866.
38 *Manchester Guardian*, 1 October 1866.
39 *JRSS*, liii (1890), 253–80.
40 Vol. XXII, pp. 461–6.
41 Sir Rawson W. Rawson, *JRSS*, Jubilee vol. (1885), 8–10.
42 *Encyclopaedia Britannica*, loc. cit.
43 Quoted in T. S. Simey and M. B. Simey, *Charles Booth, Social Scientist* (1960), p. 77.
44 B. Seebohm Rowntree, *Poverty, A Study of Town Life* (1901), pp. 133–4. For the method and its implications, see Asa Briggs, *Social Thought and Social Action: A study of the work of Seebohm Rowntree* (1961), ch. 2.
45 Beatrice Webb, *My Apprenticeship* (1926), p. 216.
46 T. S. Simey and M. B. Simey, op. cit., p. 78.
47 T. W. Hutchison, *Review of Economic Doctrines, 1870–1929* (1953), p. 426. See also Alfred Marshall, *Principles of Economics* (1920 edn), p. 492: 'No doubt statistics can be easily misinterpreted, and are often very misleading when first applied to new problems. But many of the worst fallacies involved in the misapplications of statistics are definite and can be definitely exposed, till at last no-one ventures to repeat them even when addressing an uninterested audience.'
48 Charles Booth, *Life and Labour of the People in London*, 2nd series: *Industry* (1903), I, p. 18.
49 *The Notebooks of Henry James*, ed. F. O. Matthiessen and Kenneth B. Murdock (New York, 1961), pp. 27–8. He wrote this in a Boston hotel in 1881.
50 H. Llewellyn Smith, 'Influx of Population', in Charles Booth, ed., *Life and Labour of the People in London* (1892), III, p. 75. There is something of the same sense in Ruskin's *Praeterita* (1886).
51 T. S. Simey and M. B. Simey, op. cit., p. 80.
52 See *London Statistics*, 1889 onwards.
53 *City Reports* (1849), pp. 44–57, quoted in Royston Lambert, *Sir John Simon* (1963), p. 150.
54 M. G. Kendall, 'The History of Statistical Method', in *The International Encyclopaedia of the Social Sciences* (1968 edn), XV, p. 227.
55 That for males was 23·3 per 1,000 for 1838–54 and precisely the same for the whole period 1838–71, and for females 21·6 and 21·5, respectively. See William Farr, *Vital Statistics* (1885), p. 183.
56 See R. W. Fogel, 'The New Economic History, its Findings and Methods',

The Victorian City: Numbers of People

Economic History Review, 2nd series, xix (1966), 642–56. See also D. K. Rowney and J. Q. Graham, eds, *Quantitative History* (1969).

57 *Fortnightly Review*, liii, new series (1893), 388–93.

58 Lambert, op. cit., p. 602.

59 See, for example, Morpeth's speech in the debate on the Public Health Act of 1848 in Hansard, xcvi, cols 385–428, and Asa Briggs, *Public Opinion and Public Health in the Age of Chadwick* (Chadwick Trust, 1946).

60 C. F. G. Masterman, 'The Social Abyss', *Contemporary Review*, lxxi (January 1902), 23–35.

61 See Eric E. Lampard, 'Historical Aspects of Urbanization', in Philip M. Hauser and Leo F. Schnore, eds, *The Study of Urbanization* (1965), pp. 519–54.

62 A. Kaplan, 'Measurement in Behavioral Science' from *The Conduct of Inquiry* (1964), reprinted in M. Brodbeck, ed., *Readings in the Philosophy of the Social Sciences* (1968), pp. 601–8.

63 Louis Chevalier, *Classes laborieuses et classes dangereuses à Paris pendant la première moitié du xixe siècle* (1958). See also his article 'A Reactionary View of Urban History' in *The Times Literary Supplement*, 8 September 1966.

64 See W. A. Armstrong, 'Social Structure from the Early Census Returns', in E. A. Wrigley, ed., *An Introduction to English Historical Demography* (1966), pp. 209–38, and 'The Interpretation of the Census Enumerators' Books for Victorian Towns', in H. J. Dyos, ed., *The Study of Urban History* (1968), pp. 67–85.

65 H. J. Dyos and A. B. M. Baker, 'The Possibilities of Computerising Census Data', in ibid., pp. 87–112. See also H. J. Dyos, *Victorian Suburb: A study of the growth of Camberwell* (1961).

66 J. R. Kellett, *The Impact of Railways on Victorian Cities* (1969). See also A. R. Prest and R. Turvey, 'Cost Benefit Analysis, A Survey', *Economic Journal*, lxxv (1965), 683–735.

67 J. R. Vincent, *Poll Books: how Victorians voted* (1967); E. P. Hennock, 'The Social Composition of Borough Councils in Two Large Cities,' in H. J. Dyos, ed., *The Study of Urban History*, pp. 315–37.

68 Leo F. Schnore, 'Problems of Quantitative Study', in ibid., pp. 189–208.

69 C. A. Moser and W. Scott, *British Towns: A statistical study of their social and economic differences* (1961), p. 8.

70 See A. L. Strauss, *The American City: A source book of urban imagery* (1968); Kevin Lynch, *The Image of the City* (1960).

71 See Asa Briggs, 'The Sense of Place' in Smithsonian Annual, II, *The Fitness of Man's Environment* (1968), pp. 77–99.

72 Henry James, *English Hours* (1963 edn), p. 10. The article first appeared in 1888.

4 The Contagion of Numbers

J. A. Banks

Between 1841 and 1901 the population of England and Wales more than doubled, rising from 15,914,148 to 32,527,843 persons.[1] Inevitably this growth was not distributed evenly throughout the country. By 1841 the agricultural and industrial changes of the previous century had already made their impact on the social life of the people in the form of a great expansion of town dwelling, yet over half the population still lived in what for convenience may be loosely termed 'rural' as compared with 'urban' districts. By 1901 this proportion had fallen to about one-fifth, and it has remained more or less at this level ever since.[2] The last fifty years or so of the nineteenth century, that is to say, saw the consolidation of a process of urbanization in what had been at the beginning of the century an agrarian society. Towns of over 100,000 inhabitants increased from six in 1841 to thirty in 1901—only London had been so large in 1801. Towns of 50,000 to 100,000 inhabitants, of which there had been five in 1801, increased from twenty-two in 1841 to forty-nine in 1901.[3] Many rural areas, whole counties even, became depopulated in the sense that by 1901 their populations were smaller than they had been in 1851.[4] A simple graph or histogram showing the expansion of urbanism should, therefore, be interpreted as indicating rather more than a mere change in the proportions living in urban as opposed to rural districts.[5] Behind the steady advance of the town lies a history of movement into and out of the countryside, although with the balance always in favour of urban growth.[6]

To the Victorians this pattern of rural-urban migration and the accompanying growth of town life was probably more obvious in England and Wales than might have been the case had the island been geographically much larger. A journey across

© 1973 J. A. Banks

the north of England from Liverpool to Leeds, for example, could pass through two other major cities, Manchester and Sheffield, and not one of these is more than forty miles from the next. Sheffield, moreover, is at the northern end of another such journey, less than forty miles from Nottingham, which is less than fifty miles from Birmingham. Of course, all these towns were large for their time in 1801, having more than 20,000 inhabitants, but by 1901 even the laggard amongst them was over six times as big. In such a relatively small total area the growth of population rapidly made the English a nation of townspeople. The Victorians, indeed, created a new civilization 'so thoroughly of the town' that it has been said to be the first of its kind in human history.[7] The task of this chapter is to attempt an analysis of what this notion of such a civilization entails, especially as a sociologist sees it in its demographic aspects, with particular emphasis on the patterns of migration.

The differences in the quality of town life as contrasted with that of the countryside have often been attributed to population growth and the increase in density. The movement into the towns in the nineteenth century inevitably meant that more people lived in the same space. Thus London, always in the vanguard, already housed 20·9 persons to the acre in 1801, but even London doubled this number over the hundred years. 112 other towns with populations of 4,000 persons and over in 1801 had increased their densities from 12·5 to 22·9 to the acre by 1891. 170 towns of from 2,000 to 3,999 inhabitants in 1801 increased from 5·8 to 13·5 over the same period, while 224 towns of from 1,000 to 1,999 inhabitants increased from 3·0 to 10·6.[8] A growth in size was accompanied by an increase in density—and such figures have regularly been interpreted by social historians as the background cause to much of the misery, squalor, and vice which were found in the Victorian cities at this time. Thus Kitson Clark has written of them that they were 'singularly ill prepared' to receive the millions who went into them.[9]

> Suitable housing did not exist and the additional numbers were crammed into every nook and cranny from attic to cellar of old decaying property, or into cottages run up hastily in confined spaces with little or no access to light and air ... Water and sanitation were often not provided at all, and where they were provided there was often a judicious mingling of cesspools and wells with an occasional overstocked graveyard or active slaughter house to add to the richness of the mixture ... Since many industrial processes now needed coal furnaces, and by this time probably most domestic fires burned coal, from many towns, particularly in winter, a heavy sulphurous smoke cloud was emitted to combine with other atmospheric conditions to make the fogs which were such a feature of Victorian England, and which probably slew their thousands.
> Such conditions were not new, nor probably were they inherently worse than what had existed before ... But as numbers increased so these

evils increased in the area they affected, and probably certain factors in them, as for instance, the problems of the provision of water and the disposal of sewage, came to be less manageable and more pregnant with danger.

Or again, Ashworth has put the beginning of town-planning in Britain into the context of growing public recognition, or at least middle-class recognition, that the towns of the 1840s constituted a social problem of considerable magnitude.[10]

> Even if he were not his brother's keeper every man of property was affected by the multiplication of thieves; everyone who valued his life felt it desirable not to have a mass of carriers of virulent diseases too close at hand ... It was morality (or, more exactly, criminality) and disease that were causing concern. Overcrowding and congestion, poverty, crime, ill health and heavy mortality were shown to be conditions commonly found together ... there was nothing new in the existence of congested criminal quarters. In this as in other matters it was the changed scale of things that gave to an old problem the appearance of something new.

The growing density of the population in the cities produced acute over-crowding which in its turn was the cause of social evils, growing awareness of which led to the rudiments of a Welfare State.[11]

The implication of such quotations and of the argument underlying them is that the quality of mid-nineteenth-century urban living was as a matter of fact inferior to that of earlier periods, and even, perhaps, inferior to that of the areas from which the migrants to the towns had come. The impact of urbanization on these people, indeed, has been summed up[12] as a

> grave social damage. A family dwelling in an industrial town found itself not only divorced from nature and from the particular place of its origin, but cut off from other families. To isolation were added serious stresses within the family. Not all its members were engaged, as formerly, as a group, doing essential farming tasks, and sharing the benefits on the basis of a family communism administered by the senior members. Instead, the members of the family might well be employed in different occupations, with a premium placed by the wages system upon youth and vigour, so that the rights and authority by which the older generation traditionally maintained their status and self-respect into old age were impaired or destroyed. The new mobility made possible by the railway operated to disperse families as their more enterprising members sought new opportunities. Many of these became lodgers in the houses of others, or servants living in dingy conditions in the lesser middle-class families. So there developed an element of the population with a kind of sub-status, neither heads of families nor true members. Others of the unassimilated were the growing

hordes of homeless children, the street-arabs who became a kind of nomad class in the heart of the great cities.

Thus, although it may be legitimately concluded, with Hobsbawm, that in recent years the 'pessimistic' and the 'optimistic' schools have become overshadowed by 'agnostics' in the long-continued debate over the impact of the Industrial Revolution on the level of living of the working class at this time,[13] a residue of this controversy has been the assumption that the 'deprivations', if not definitely economic, were harmful in their influence on the 'personality'.[14] This assumption will not be considered in quite this form here, largely because of the difficulty of deciding how to deal with such an elusive concept when the reference is to large numbers of people over a generation or so. What will be challenged, nevertheless, is the view that the people concerned *felt* the hardships of town life so acutely, as compared with life in the countryside. The sociologist is bound to ask why it was that people continued to move into towns if conditions there were so manifestly worse than elsewhere. Undoubtedly it was the case that some of the worst urban areas were inhabited by the Irish, and it was also true that Irish vagrants constituted a large proportion of those figures of the migrant unemployed which Hobsbawm has used with such telling effect in his presentation of the anti-optimistic point of view;[15] but the further interesting fact about the Irish was their preference for an urban rather than for a rural alternative to the life they had left behind them in their country of origin. In the United States, for example, where throughout the nineteenth century and especially after the Homestead Act of 1862 a policy was deliberately fostered to make it easy for immigrants to settle on the land,[16] the Irish were notorious for remaining in the commercial and manufacturing centres, and it has been estimated that although some 80 percent of the Irish immigrants were of rural origin, no more than about 6 percent in all settled permanently on the land in that country. No doubt, as MacDonagh has pointed out, most emigrants from Ireland, during the period of the famines at least, were cottiers or squatters accustomed to living in close communities rather than in rural isolation.[17] Nevertheless it would be hazardous to claim that the concentration of Irish in the urban areas of England and Wales was a simple consequence of their failure to find work on the land at some distance from the cities, although it is true that the decline in agricultural employment could have made settlement on the land extremely difficult for those Irishmen who would have preferred to live in the countryside.

The alternative, therefore, must be to explore the possibility that the movement into the towns was a direct response to some feature of town life which was attractive to the rural population of the nineteenth century. Economic opportunities, clearly, there must have been, since the Industrial Revolution was an urban rather than a rural phenomenon; but such opportunities should not be interpreted as implying that the economic state of the rural areas was necessarily consistently worse than that of the urban and hence drove people into the towns in search of work. One assessment of migration patterns in Victorian England has concluded that 'both in

North and South the rise and fall in the movement of population from the rural areas had little to do with agricultural prosperity and depression',[18] although there is other evidence that emigration abroad was related to long-term economic decline in the rural areas of Cornwall and Gloucestershire.[19] A sociologist might thereupon suggest that some attempt should be made to examine the quality of town life to determine whether it was superior to that in the countryside, not so much perhaps in terms of material levels as in terms of those social relations, which in this context the eminent French sociologist, Émile Durkheim, called in 1893, a 'dynamic or moral density' (*densité dynamique ou morale*). Indeed, one implication of this position is that it would tend to make the series of changes which we call the Industrial Revolution a result of urbanization, in the sense that the rate of growth of industry, as well as its initial establishment in the towns, would seem to be dependent upon the social attractiveness of town life. Stripped of its other nuances, this was essentially Durkheim's claim. Eager to refute the assumption of his day that the division of labour was the driving force behind social change, he put forward a sociological plea for considering 'moral' density to be intimately connected with physical density (*densité matérielle*), each influencing the other and both responsible for the growth of specialization and the division of labour in society. In a series of deductions from the assumption that individuals possess a need for as intimate a social contact with one another as they can find, he demonstrated that a denser population increases the possibility of greater intimacy and also makes possible a diversification of interests and functions in social life. The outcome is that such diversification, as a matter of fact, occurs in the form of economic, political, and social specialization.[20]

Empirical justification for this point of view, is, to be sure, very difficult to obtain except in a very broad sense. Thus, although the Registrar-General's 'dictionary' of occupations is not altogether a reliable guide, because of variations in linguistic usage in different parts of the country at different times, it is the best guide we have, and it identified about 15,000 distinct occupations in 1901 as compared with only 7,000 in 1851.[21] The population had doubled and the number of occupations had more than doubled in fifty years. The overwhelming number of new occupations, it can be safely assumed, were urban occupations, and urban densities had grown much faster than rural densities,[22] some of which had certainly declined. Such figures, taken together, suggest that there is at least a tenuous relationship between specialization and density, although of themselves they do not indicate which is the cause and which is the effect. In any case there is no hint here of any measure of that elusive concept, '*densité morale*'. Durkheim, indeed, spent the rest of his professional life wrestling unsuccessfully with the attempt to get a clearer understanding of what is implied in the idea and others like it,[23] and, it must be admitted, sociologists and social anthropologists studying urban-rural phenomena have added little to what he achieved. But, for all that, there appears to exist a general consensus amongst them that the urban and the rural differ in important respects in the nature of the social relationships which they display.[24]

Durkheim's *Division of Labour in Society* is perhaps best understood in the present context as a contribution to a discussion of social relationships in terms of what has been called 'unquestionably the most distinctive development in nineteenth-century social thought, a development that extends well beyond sociological theory to such areas as philosophy, history and theology to become indeed one of the major themes of imaginative writing in the century'. The characteristic feature of this development was the 'rediscovery of community', that is, the rediscovery of 'forms of relationship which are characterized by a high degree of personal intimacy, emotional depth, moral commitment, social cohesion, and continuity in time'.[25] Among sociologists Auguste Comte, Frederic Le Play, and even Karl Marx had some part to play in the discussion, but the man whose name is still most closely associated with the notion of a marked contrast between rural and urban ways of life in this respect was Ferdinand Tönnies, whose book, *Gemeinschaft und Gesellschaft*, published in 1887, gave to sociology the two words of its title as distinctive terms by which to refer to the contrast. Tönnies, it is true, did not draw the line of distinction between the village and the town as such, but rather between the town and the city; yet his description of the latter as 'essentially a commercial town and, in so far as commerce dominates its productive labour, a factory town' makes plain that the kind of town which emerged in Britain in the nineteenth century was a city in his sense and hence 'typical of *Gesellschaft* in general'.[26]

The word *Gesellschaft* in this context largely eludes translation, for Tönnies regarded it essentially as a negation of *Gemeinschaft* which can easily be replaced by 'community' as described above. What Tönnies had in mind was what he conceived to be an

> artificial construction of an aggregate of human beings which superficially resembles the *Gemeinschaft* in so far as the individuals live and dwell together peacefully. However, in the *Gemeinschaft* they remain essentially united in spite of all separating factors, whereas in the *Gesellschaft* they are essentially separated in spite of all uniting factors . . . everybody is by himself and isolated and there exists a condition of tension against all others . . . nobody wants to grant and produce anything for another individual, nor will he be inclined to give ungrudgingly to another individual, if it be not in exchange for a gift or labour equivalent that he considers at least equal to what he has given.

Thus the sharp contrast is between 'all intimate, private, and exclusive living together' as in the family, and the 'mere co-existence of people independent of each other' as in the joint-stock company. City life for Tönnies was dominated by the latter spirit, although of course, because people lived in families even in the cities, *gemeinschaftlich* relationships were not unknown there. Nevertheless,[27]

> in the city . . . and especially in the metropolis, family life is decaying . . . In the village the household is independent and strong, also in the town the household is preserved and has a certain beauty; only in the city does

the household become sterile, narrow, empty, and debased to fit the conception of a mere living place which can be obtained everywhere in equal form for money. As such it is nothing but shelter for those on a journey through the world.

From such passages and others it is quite clear that Tönnies saw nineteenth-century city life as dominated largely by relationships that were impersonal, cold, morally indifferent, socially alienated, and transient, and the whole tenor of his book indicates that he associated himself with those who praised rural life with the claim that 'the *Gemeinschaft* among people is stronger there and more alive: it is the lasting and genuine form of living together.'[28] The lack of authenticity in town life which is central to this theme was directly challenged by Durkheim, although without specific reference to Tönnies. 'Cities', he wrote, 'always result from the need of individuals to put themselves in very intimate contact with others. They are so many points where the social mass is contracted more strongly than elsewhere. They can multiply and extend only if the moral density is raised.'[29] Of course, Durkheim did not attempt to argue that the high moral density of cities was characterized by that kind of community which Tönnies attributed to the village. Rather did he point to the stifling effect which *Gemeinschaft* could have on individual initiative and to the contrast between this and 'the free expression of individual variations' which obtained in large urban areas.[30]

> To assure ourselves of this it is sufficient to compare great cities with small. In the latter, whoever seeks to free himself from accepted customs meets with resistance which is sometimes very acute. Every attempt at independence is an object of public scandal, and the general reprobation attached is of such a nature as to discourage all imitators. On the contrary, in large cities, the individual is a great deal freer of collective bonds.

Tönnies had called *Gemeinschaft* 'a living organism' and *Gesellschaft* 'a mechanical aggregate and artefact'.[31] Durkheim referred to the same kind of evolution from the former to the latter as a movement from mechanical to organic solidarity; that is, from relationships which constrained people to behave in similar ways to those which united them in different but complementary roles. In this sense each type of solidarity was as genuine and authentic a form of co-operation as the other. Thus, if Tönnies was inclined to romanticize the past in his critique of the present,[32] Durkheim underplayed the element of repression in modern society in order to redress the balance,[33] concluding that[34]

> great cities are the uncontested homes of progress; it is in them that ideas, fashions, customs, new needs are elaborated and then spread over the rest of the country. When society changes, it is generally after them and in imitation. Temperaments are so mobile that everything that comes from the past is somewhat suspect. On the contrary, innovations, whatever they may be, enjoy a prestige there almost equal to the one the customs of

ancestors formerly enjoyed. Minds naturally are there oriented to the future. Consequently, life is there transformed with extraordinary rapidity; beliefs, tastes, passions, are in perpetual evolution. No ground is more favourable to evolution of all sorts. That is because the collective life cannot have continuity there, where different layers of social units, summoned to replace one another, are discontinuous.

Durkheim, it should be understood, saw the growth of cities as dependent materially on migration. 'Far from owing their existence and progress to the normal preponderance of births over deaths, they present, from this point of view, a general deficiency. It is, then, from without that they receive the elements to which they owe their daily increase.' But why do people move from the countryside to the town? Because 'it is inevitable that the greatest centres, those where life is most intense, exercise an attraction for the others proportionate to their importance.' This is hardly illuminating as it stands, although it becomes clearer if seen against Durkheim's view that the traditions of societies dominated by relationships of 'mechanical' solidarity were perpetuated by old people who could exercise authority over their children. Migrants to cities, he argued, were mainly people 'who, on becoming adult, have left their homes and been freed from the action of the old'; and this is because 'the prime of youth' is 'the time when men are most impatient of all restraint and most eager for change'.[35] Durkheim was relying here on what Weber referred to as a matter of 'common observation' in the nineteenth century, namely that the migrants to the cities were 'chiefly young people'.[36] The statistical evidence in support of this observation is, as a matter of fact, not altogether so strong as Weber (and Durkheim) seem to have thought, since there were cases of marked losses from rural areas of women aged over thirty-five in the period 1881–1901.[37]

Nevertheless, it does seem to have been generally true that girls of from fifteen to twenty and men of from twenty to thirty-five formed the great bulk of migrants from the countryside at this time, and probably earlier. Economic explanations were usually put forward to account for this movement, although the emphasis therein was not necessarily placed on destitution on the part of the migrants. Thus, H. L. Smith's discussion for Charles Booth of the influx into London asserted that 'the countrymen drawn in are mainly the cream of the youth of the villages, travelling not so often vaguely in search of work as definitely to seek a known economic advantage.' Moreover, even here there was an emphasis on an attraction in London beyond what he called the 'gigantic lottery of prizes'. What brought them to live in the city was 'the contagion of numbers, the sense of something going on, the theatres and the music halls, the brightly lighted streets and busy crowds—all, in short, that makes the difference between the Mile End fair on a Saturday night, and a dark and muddy country lane, with no glimmer of gas and with nothing to do. Who could wonder that men are drawn into such a vortex, even were the penalty heavier than it is?'[38] Possibly such a 'contagion of numbers' and the lure of 'busy crowds' were what Durkheim had in mind when he wrote of the intensity of city life in the context of 'moral density'.

Of course, as the Claphams pointed out, 'until the forties, neither the man of business nor the clerks or operatives who worked for him had much leisure in the factory towns';[39] or rather, until the Ten Hours Act was amended in 1850 to create the 'normal' working day with much of Saturday afternoon free,[40] working people had few opportunities to take advantage of the public pleasures of town life when they were fully employed, and when they were unemployed they could hardly afford to. Nevertheless, in these respects they were no worse off than the country-dwellers who, indeed, along with shop assistants still at the end of the century had no weekly half-holiday.[41] On the other hand, so long as the towns were physically small it was possible for their inhabitants to continue the inexpensive leisure-time pursuits to which they had become accustomed in their villages of origin. Thus, the Leicester Domestic Mission Society reported in 1846 that 'groups of ten or twelve men of the very poorest classes would get out of their alleys and back streets to roam the fields near the town and perhaps poach a little.'[42] For those who could read, twenty-five towns had opened *free* public libraries by 1860[43] and Co-operative Societies, following the practice of the Rochdale Pioneers, often put aside some of their funds for educational purposes, opening reading-rooms and lending-libraries for their members.[44] Undoubtedly, only a very small number of persons took advantage of such opportunities, or of the excursions arranged by Mechanics' Institutes, which gave Thomas Cook the idea of the excursion train in the 1840s;[45] and it is likely that most of the new, urban activities of the second half of the century—rowing, sailing, cricket, musical concerts, and drama—in the industrial towns of the Midlands and the North, were patronized by the middle classes rather than by the rest of society.[46] However, it is clear that the opportunity for a richer, fuller life lay in the future development of the town, rather than in the countryside. Although as the towns grew in size and outdoor pursuits had perforce to be confined to suburbia, the impression grew that these were poor substitutes for the changing pleasures of Arcadia, denied to those 'whom cruel fate prevents from living in the real country'.[47]

For women, one of the attractions may have been marriage, and indeed the fact that women were more likely than men to be short-distance migrants was explained categorically in terms of 'the marriages which take women into a neighbouring town'. Presumably this is meant to imply that some countrywomen married town-dwellers, possibly following men who had moved earlier from their village; but there may have been a marriage attraction in the town from the possibility that, in spite of the more unfavourable sex-ratio there,[48] a woman's chances of marriage may have been better, because both men and women married at younger ages in the town than in the country and a larger proportion of them altogether married. Again, an economic explanation has been put forward for this, namely 'the degree to which women find industrial occupations' which was claimed to be 'one very powerful factor in determining the marriage-rate'.[49] It is open to question, however, whether economic opportunity in the case of women, especially when linked with migration from home, might better be interpreted as a means of independence from the often severe restraints on behaviour inherent in rural family life, dominated by the Victorian paterfamilias. Young women copied the example of the young men, becoming lodgers in the factory towns nearby, or fending wholly for themselves, where they did not marry early.

As early as the 1830s, it should be noted, observers had been commenting on the growing economic independence of children in urban, working-class families,[50] and Engels in 1844 referred to the practice on their part of paying their parents a fixed sum for board and lodging, keeping the rest of the wages for themselves. Occasionally they moved out from their parental home to set up houses for themselves, either alone or with a friend.[51] Charles Booth, at the end of the century, wrote similarly of the 'weakening of family ties between parents and children' which he thought had occurred during the twenty years previous to his survey. 'Nowadays the home tie is broken early ... The growing independence on the part of the children is frequently spoken of.'[52] In all, this economic opportunity *may* be seen as a determining factor but an alternative possibility is to regard it as influential mainly in the sense that it made possible the children's independence from parents. That some sections of the population migrated in search of the new jobs has been taken here to indicate that they were attracted as much by the freedom from traditional family ties as by the opportunities of employment.

The demographic consequences of this development are also of some significance. With a relatively younger population marrying earlier, and with a larger proportion of the population marrying, the Victorian towns might have been expected to show a higher birth-rate and a larger completed family than prevailed in the countryside. On the whole, however, the opposite was the case,[53] and in certain comparisons the difference was marked.[54] This is particularly noticeable when the analysis is conducted in occupational, rather than in urban-rural terms. Amongst the working class the highest fertility, as demonstrated by the 1911 Family Census data, occurred in mining families, with those engaged in agriculture second. Both groups were engaged in rural pursuits and both had a higher fertility than the average for unskilled workers taken as a whole. Textile families, on the other hand, were low in fertility, below the average for skilled workers, taken as a whole. A movement of women from the countryside into the textile towns might hence be seen as involving a transition to a type of family life which was very different from that of their family of origin, and this was all the more emphasized once people began to restrict the size of their families; for in the 1860s the practice began first in the towns, or at least amongst urban occupations, and proceeded much further there.

Of course, as is well known, this vital revolution was started initially and proceeded in the nineteenth century most rapidly amongst the upper and middle classes. It was only later that it was adopted by the working class. However, even amongst the middle classes it was an urban rather than a rural phenomenon, the 1911 census showing that farmers and graziers had larger families throughout the period 1861–86 than any other section of the middle class. Indeed, their families were larger than those of the working-class textile workers.[55] Thus, the distinction existed independent of social class, and, however much later developments in the countryside may have invalidated many more recent urban-rural comparisons,[56] it can hardly be denied that the formative years of British urban civilization witnessed a change in the character of the evaluation of children in the family, which has been revolutionary in its impact on the position of married women in society and hence on society generally.

For present purposes it is not necessary to go into the factors responsible for the inception and spread of family limitation. What is rather more to the point is that one facilitating means appears to have been the comparisons which members of the middle class were able to make of their own levels of living at different times with those of their immediate neighbours and acquaintances.[57] Hence what Himes called 'the democratization' of birth-control might possibly be explained, not so much by the publicity which was given to the Bradlaugh–Besant trial and which he thought to be important,[58] as by the comparisons which members of the urban working class were able to make of their situation with that of their wealthier neighbours. This may also have been related to the pattern of migration into the town, to the degree that many of the young girls from the country originally went into domestic service, and it is possible that the very great growth in the demand for such servants on the part of the middle class between 1851 and 1871[59] provided an economic opportunity for migration at the time. Later, it is true, the supply of domestic servants declined, especially amongst girls under twenty.[60] However, by that time the lesson had been learned, at least in the sense that some of the middle-class standards and aspirations had been adopted by the working class. It is not suggested here that they learned to use birth-control from being employed in the middle-class families. So far as it is possible to tell, working-class birth-control is a twentieth-century practice, on the whole; and there seems to have been a 'conspiracy of silence' on the issue on their part long after the Victorian age had come to an end. Nevertheless, there can hardly be any doubt that by the end of the century working-class women were expressing a need for relief from the burden of too frequent childbearing, as their letters to the Women's Co-operative Guild showed,[61] and in this respect the demand for emancipation on the part of working-class women followed a generation later than that made by the wives and mothers in the middle class.

Thus, Hobsbawm's argument for the earlier period—'on a gloomy interpretation, the popular discontent of the early nineteenth century makes sense; and on an optimistic interpretation it is almost inexplicable'[62]—is a clear *non sequitur*. The discontent of women with their traditional lot of childbearing and child-rearing, in families of an average size of at least six children, arose not because they had now become worse off than their mothers and grandmothers had been, but because they saw themselves worse off than they *might* be. It is, of course, true that such discontent did not manifest itself in the kind of collective organization intent on reform that Hobsbawm may be presumed to have had in mind, but it seems equally true that the feminist movement as such was organized to claim equal rights with men in the educational, occupational, and political fields, in response to social changes which made manifest that there was a large section of the population for whom such rights were essential. There is no question that they were claiming rights which had once been theirs and had latterly been lost. The issue, indeed, centred round a demographic fact, namely that throughout the nineteenth century there was a great growth of 'surplus' women. Already by 1841 there were 358,976 more women than men in England and Wales. By 1901 this number had increased to 1,070,015, and the sex-ratio had risen over sixty years from 1·046 to 1·068 women to each man.[63] Although about one-third of these were below the customary age for marrying, the number who could never marry while monogamy lasted was large and grew disproportionately throughout the century.

A traditional agricultural society could possibly cope with such a problem in that there is always productive work in and around the homestead for an unmarried woman to do and so to augment the family's worldly goods; but the kind of pecuniary calculations which is typical of urban industrialization effectively separates home and work for the middle and working class alike, making home activities overwhelmingly a form of consumption and work a form of family-income earning. The problem of the surplus woman, accordingly, was that of finding some place for her in such a civilization; and the feminist movement obtained its greatest impetus from the need to find for spinsters and widows, especially of middle-class origins, outlets for their energies which would also be remunerative.[64] For working-class women the factories already existed, and where there were no factories they could find employment as dressmakers, admittedly at very low wages. The danger here was that they might easily drift into prostitution, the great social evil of the Victorian era, and there was always a section of the feminist movement which saw its task, ideologically speaking, as that of a struggle of woman versus man. Hence, although organized feminism was led by middle-class women it was never a class movement, and some of the issues it adopted, such as the abolition of the Contagious Diseases Acts and the ending of the double standard of sexual morality, were clearly not of direct advantage to middle-class women.[65] It is, that is to say, an error to regard social protest as indicative of deterioration in the situation of those who mount it. Urban life, by bringing ever larger numbers of people in close proximity to one another, increased the possibility of invidious comparisons, and it was this which produced social movements aimed at an even better future for all.

Again, it is to be expected that towns might differ in such respects according to the degree to which the sex-ratio, for example, and the numbers of surplus women varied markedly from the national average. Thus, Middlesbrough was selected for special mention by Ravenstein in 1885 on the ground that 'its rapid growth, the heterogeneous composition of its population, and the preponderance of the male sex, recall features generally credited only to the towns of the American West.'[66] Does this mean that, in the face of aggressive masculinity, feminism was a movement which found it difficult to take roots there? Historians have apparently found no reason to comment on this point,[67] and clearly a much more intensive study of such questions than has been presented here would be necessary to decide the issue. In so far as a town like Middlesbrough could be shown to display characteristics of 'moral density' distinct from what obtained generally, and more like what was typical of the American frontier, added emphasis would be given to the analysis of urban life in Durkheimian terms. Thus, although Durkheim's notions cannot claim to have been rigorously tested here, sufficient illustrative material has been presented to support the general argument that the analysis of the process of urbanization in terms of growing social solidarity makes good historical and sociological sense.

The thesis that the quality of social life in cities attracted migrants to them from the countryside, in spite of the greater morbidity and mortality associated with overcrowding, should not be interpreted as implying that there were no disadvantages to be discerned in so far as social solidarity was concerned. Durkheim, indeed, proffered an explanation of the fact that 'suicide, like insanity, is commoner in cities than in the country', in terms of a direct relationship between 'voluntary deaths' and the *intensity* of social life. Individuals who, for one reason or another, find themselves isolated from strongly integrated groups are more prone to suicide,[68] so that the stranger in the town can be especially vulnerable. The basis of this situation, as another sociologist put it about this time, is 'reserve'.[69]

> If so many inner reactions were responses to the continuous external contacts with innumerable people as are those in the small town, where one knows almost everybody one meets and where one has a positive relation to almost everyone, one would be completely atomized internally and come to an unimaginable psychic state. Partly this psychological fact, partly the right to distrust which men have in the face of the touch-and-go elements of metropolitan life, necessitates our reserve.

Hence the warmth of social relationships with friends in the city is matched by what appears to be coldness in relationships with mere acquaintances, and a man who cannot find or make friends is in some peril.

This point should not be laboured unduly. After all, most migration was over short distances. Ravenstein, who was probably the first to emphasize this fact, asserted that even long-distance migration was undertaken in stages. Among the Irish, for example, he said, 'some of them landed at Liverpool, and gradually worked their way through Cheshire, Stafford, Warwick, Northampton, and Birmingham, whilst another stream, and perhaps the more voluminous one, passed through Plymouth, Hampshire and Surrey.'[70] The concept of short distance is, of course, meant to be comparative. Some migrants went a long way, even further when railway traffic became commonplace, although the general effect of the railways on the pattern of migration was negligible in this respect, since their greatest effect was that many more people moved a short way. Thus, most migrants were never very far from their villages of origin and could maintain some contact with their families and perhaps with friends, even where economic circumstances made it difficult for them to return home for long or permanently. The element of strangeness would in any case be small for those who had not moved very far. Local dialects would probably not have presented the kind of communication problem that existed for long-distance migrants and this was probably also true of cultural differences generally. It seems reasonable to assume, therefore, that the rural hinterland for most towns provided a population more easily assimilated into urban life than likely to remain aloof and apart. When they moved, moreover, they probably went to those parts of the town where people from their villages already lived, as a recent study of Preston has shown.[71] In this sense the adjustment to the new way of life could be made gradually.

Occasionally this clustering resulted in some part of the town perpetuating distinguishing features and even influencing the life of the rest, although this seems to have been rare. For example, Liverpool in 1881 contained only 81 percent English-born inhabitants, as compared with West Bromwich, Norwich, Ipswich, Leicester, and Nottingham with over 98 percent English-born; and 12·8 percent of the Liverpudlians were Irish.[72] The existence of a relatively large Roman Catholic 'enclave'[73] in a predominantly Protestant area gave Liverpool a characteristic ethos which has set it off as different even to the present day. At a time when working-class attendance at church was minimal, the fact that the Irish continued the habit of going to church regularly gave it an influence in the life of the city. Possibly 'worship with their fellows, led perhaps by an Irish priest, was one of the few familiar and comforting things available to them in ugly industrial England';[74] but the effect was to maintain a sense of community, of having interests in common which were different from those of their neighbours, and which influenced not merely their religious but also their political life. The only Irish Nationalist candidate ever elected to Parliament from an English constituency, T. P. O'Connor, was returned for the Scotland Division of Liverpool in 1885 and held the seat until 1929.[75] What the example of Liverpool emphasizes, however, is that it was exceptional for migrants to carry their way of life into a town, in the sense of it remaining an abiding source of strength and separateness. Rather was it that the town's way of life became theirs. Thus, the kind of social solidarity which Durkheim had in mind as characteristic of rural life was to some extent perpetuated in local pockets within the new urban areas; but the overall character of urban life in this new civilization invented in nineteenth-century England was one of integration through the co-operation of people of diverse sorts—a truly new way of life graphically labelled by him as organic solidarity.

Notes

1 United Kingdom Registrar-General, Census 1961, *England and Wales Preliminary Report*, HMSO (1961), Table 6, p. 75.
2 John Saville, *Rural Depopulation in England and Wales, 1851–1951* (1957), Table VI, p. 61.
3 Details obtained from B. R. Mitchell, 'Population and Vital Statistics 8,

Population of the Principal Towns of the United Kingdom, 1801–1951', *Abstract of British Historical Statistics* (Cambridge, 1962), pp. 24–7.

4 Details obtained from Mitchell, 'Population and Vital Statistics 7, Population of the Counties of the British Isles 1801–1951', pp. 20, 22; cases of county decline in England were Cornwall, Huntingdonshire, Rutland, Somerset; in Wales, Anglesey, Breconshire, Cardiganshire, Montgomeryshire, and Radnorshire. See also Saville, op. cit., Table V, pp. 56–7. For details of rural decline in two counties of growing population, see Richard Lawton, 'Population Trends in Lancashire and Cheshire from 1801', *Trans. Historic Society of Lancashire and Cheshire*, cxiv (1962), 197–201, and Figure 20. A detailed analysis of areas smaller than the county is to be found in Richard Lawton, 'Population Changes in England and Wales in the later Nineteenth Century: an Analysis of Trends by Registration Districts', *Trans. Institute of British Geographers*, no. 44 (1968), Figure 1, 59.

5 See, for example, Mary P. Newton and James R. Jeffery, *Internal Migration: Some aspects of population movements within England and Wales*, General Register Office Studies in Medical and Population Subjects, no. 5, HMSO (1951), histogram on p. 9.

6 Richard Lawton, 'Rural Depopulation in Nineteenth-Century England', in Robert W. Steel and R. Lawton, eds, *Liverpool Essays in Geography* (1967), pp. 233–47.

7 Thomas W. Freeman, *The Conurbations of Great Britain* (Manchester, 1959), p. 1.

8 Thomas A. Welton, 'On the Distribution of Population in England and Wales, and its Progress in the Period of Ninety Years from 1801 to 1891', *JRSS*, lxiii (1900), 529, 533.

9 G. Kitson Clark, *The Making of Victorian England* (1965), p. 79. The implication is that this covers the period 1821–51. For a sample of contemporary documents of the 1840s see 'A Gazateer of Disgusting Places' in E. Royston Pike, *Human Documents of the Industrial Revolution in Britain* (1966), pp. 305–35.

10 William Ashworth, *The Genesis of Modern British Town Planning* (1954), pp. 47–8.

11 David Roberts, *Victorian Origins of the British Welfare State* (New Haven, 1960), ch. 1. Notice that Asa Briggs does not mention the population issue in his 'The Welfare State in Historical Perspective', *European Journal of Sociology*, ii (1961), 221–258.

12 Sydney G. Checkland, *The Rise of Industrial Society in England, 1815–1885* (1964), pp. 263–4. See also John L. and Barbara Hammond, *The Town Labourer, 1760–1832. The new civilization* (1917), ch. 3.

13 Eric Hobsbawm, *Labouring Men: Studies in the history of labour* (1964), pp. 64, 121. Although he concludes that the 'negative case is now pretty generally accepted', and although he himself has contributed to it in the tradition of the pessimistic school, chapters 5, 6, and 7 of this book comprise an excellent and fair survey of the debate.

14 Checkland, op. cit., ch. 7. Section 9 of this chapter is entitled: 'The Impact on Personality' and begins with the sentence: 'Even more fundamental to happiness than monetary rewards and living conditions was the state of mind of men, women and children who needed to be able to find comfort and even pride in their relations with one another and with the productive tasks upon which their livelihood depended' (p. 263).

15 Hobsbawm, op. cit., pp. 78–9.

16 Stanley C. Johnson, *A History of Emigration from the United Kingdom to North America, 1763–1912* (1913), pp. 203–7.
17 Oliver MacDonagh, 'Irish Emigration to the United States of America and the British Colonies During the Famine', in R. Dudley Edwards and T. Desmond Williams, eds, *The Great Famine: Studies in Irish history 1845–52* (New York, 1957), pp. 383–4. See also Johnson, op. cit., p. 183, n. 2, for details of the numbers of Irish in the farming and industrial states in 1880.
18 A. K. Cairncross, *Home and Foreign Investment 1870–1913* (1953), p. 75.
19 Ross Duncan, 'Case Studies in Emigration: Cornwall, Gloucestershire, and New South Wales, 1877–1886', *Economic History Review* (second series), xvi (1963–4), 1.
20 Émile Durkheim, *De la division du travail social*, 3rd edn (Paris, 1911), p. 238.
21 Interdepartmental Committee on Social and Economic Research, *Guide to Official Sources, No. 2: Census Reports of Great Britain, 1801–1931*, HMSO (1951), pp. 31, 34.
22 Welton, op. cit., Summaries 1 and 2, pp. 529, 533.
23 Talcott Parsons, *The Structure of Social Action*, 2nd edn (Chicago, 1949), chs 8–11, and ch. 18, pp. 708–14. A very careful and accurate survey of Durkheim's position on moral density is given in Leo F. Schnore, 'Social Morphology and Human Ecology', *American Journal of Sociology*, lxiii (1957–8), 620–34. See also Harry Alpert, *Émile Durkheim and his Sociology* (New York, 1961 edn), pp. 91–3.
24 See the useful summary in Ronald Frankenberg, *Communities in Britain: Social life in town and country* (Penguin, 1966), pp. 286–92.
25 Ronald A. Nisbet, *The Sociological Tradition* (New York, 1966), p. 47.
26 Ferdinand Tönnies, *Community and Association*, trans. and ed. Charles P. Loomis (London and New York, 1955 and 1963), p. 227 (page references are to the New York edition).
27 Ibid., pp. 64–5, 33–4, 229, 162.
28 Ibid., p. 35: 'Gemeinschaft ist das dauernde und echte Zusammenleben, Gesellschaft nur ein vorübergehendes und scheinbares.'
29 Durkheim, op. cit., p. 239. Cf. English translation by George Simpson with the title, *The Division of Labor in Society* (New York, 1933), p. 258.
30 Ibid., English version, pp. 297–8.
31 Tönnies, op. cit., p. 35.
32 Rudolf Heberle, 'The Sociological System of Ferdinand Tönnies: "Community" and "Society"', in Harry E. Barnes, ed., *An Introduction to the History of Sociology* (Chicago, 1948), pp. 242–3.
33 Alpert, op. cit., p. 185.
34 Durkheim, op. cit., p. 296.
35 Ibid., pp. 342 and 295.
36 Adna F. Weber, *The Growth of Cities in the Nineteenth Century: A study in statistics* (New York, 1899), p. 280.
37 See the summary conclusion in Thomas A. Welton, *England's Recent Progress. An investigation of the statistics of Migrations, Mortality, &c.* (1911), p. 9.
38 H. Llewellyn Smith, 'Influx of Population', in Charles Booth, ed., *Life and Labour of the People in London* (1892), III, pp. 75, 120.
39 J. H. Clapham and M. H. Clapham, 'Life in the New Towns', in G. M. Young, ed., *Early Victorian England, 1830–1865* (1934), I, pp. 230–1.

40 Frank Tillyard, *The Worker and the State* (1922), pp. 111–12; Barbara L. Hutchins and A. Harrison, *A History of Factory Legislation*, 3rd edn (1926), pp. 105–7.
41 J. A. R. Pimlott, *The Englishman's Holiday* (1947), p. 142.
42 Page 12 of the Report quoted in A. Temple Patterson, *Radical Leicester: A history of Leicester, 1780–1850* (Leicester, 1954), p. 378, n. 8.
43 E. L. Woodward, *The Age of Reform, 1815–1870* (Oxford, 1949 edn), p. 475, n. 1.
44 G. D. H. Cole, *A Century of Co-operation* (Manchester, 1944), pp. 227–32. For a description of the social aspects of a Co-operative Store on a Saturday evening see George J. Holyoake, *The History of the Rochdale Pioneers* (1900 edn), p. 39.
45 John Pudney, *The Thomas Cook Story* (1953), pp. 53–9.
46 Clapham and Clapham, op. cit., pp. 237–44.
47 Mrs C. S. Peel, *The New Home* (1898), quoted in H. J. Dyos, *Victorian Suburb* (Leicester, 1961), p. 26.
48 Weber, op. cit., pp. 278, 286.
49 William Ogle, 'On Marriage-Rates and Marriage-Ages, with special reference to the growth of population', *JRSS*, liii (1890), 268.
50 See the evidence cited in Ivy Pinchbeck, *Women Workers and the Industrial Revolution, 1750–1850* (1930), p. 313.
51 Friedrich Engels, *The Condition of the Working Class in England*, trans. and ed. W. O. Henderson and W. H. Chaloner (Oxford, 1958), pp. 164–5. Notice that Harriet in Disraeli's *Sybil* (1845) left her parents to set up lodgings with a friend because she was tired of supporting her family.
52 Charles Booth, 'Notes on Social Influences and Conclusion', *Life and Labour of the People in London* (1903), final volume, p. 43.
53 David V. Glass, 'Changes in Fertility in England and Wales, 1851 to 1931', in Lancelot Hogben, ed., *Political Arithmetic, a Symposium of Population Studies* (1938), pp. 161–212.
54 Francis Galton, 'The Relative Supplies from Town and Country Families to the Population of Future Generations', *JSS*, xxxvi (1873), 19–26.
55 John W. Innes, *Class Fertility Trends in England and Wales, 1876–1934* (Cambridge, 1938), Table XIII, p. 42.
56 See the useful discussion in Peter H. Mann, *An Approach to Urban Sociology* (1965), pp. 96–105.
57 J. A. Banks, *Prosperity and Parenthood: A study of family planning among the Victorian middle classes* (1954), ch. 9.
58 Norman E. Himes, *Medical History of Contraception* (Baltimore, 1936), pp. 239–45. For an alternative assessment of the effect of the trial see J. A. and Olive Banks, 'The Bradlaugh–Besant Trial and the English Newspapers', *Population Studies*, viii (1954), 22–34.
59 Banks, *Prosperity and Parenthood*, ch. 9 and pp. 83–4.
60 United Kingdom Registrar-General, *Census of England and Wales, 1901: General Report with Appendices*, HMSO (1904), pp. 95–6.
61 *Maternity: Letters from working-women collected by the Women's Co-operative Guild* (1915), pp. 18–190.
62 Hobsbawm, op. cit., p. 241.
63 United Kingdom Registrar-General, Census 1961, p. 75.

64 J. A. and Olive Banks, *Feminism and Family Planning in Victorian England* (Liverpool, 1964), ch. 3.
65 J. A. and Olive Banks, 'Feminism and Social Change—A Case Study of a Social Movement', in George K. Zollschan and Walter Hirsch, eds, *Explorations in Social Change* (Boston, 1964), pp. 552–4.
66 E. G. Ravenstein, 'The Laws of Migration', *JRSS*, xlviii (1885), 215.
67 Cf. Asa Briggs, *Victorian Cities* (1963), ch. 6.
68 Émile Durkheim, *Suicide: A study in sociology*, trans. J. A. Spaulding and G. Simpson (1952), pp. 70, 121–2, 208–16. *Suicide* was first published in French in 1897.
69 Georg Simmel, 'Die Grossstädte und das Geistesleben', in Karl Bücher *et al.*, *Die Grossstadt* (Dresden, 1903), trans. Kurt H. Wolff, *The Sociology of Georg Simmel* (Chicago, 1950), p. 415.
70 Ravenstein, op. cit., p. 183. See also Weber, op. cit., and Arthur Redford, *Labour Migration in England, 1800–1850* (Manchester, 1962).
71 Michael Anderson, *Family Structure in Nineteenth Century Lancashire* (Cambridge, 1971), pp. 152–8.
72 Ravenstein, op. cit., pp. 173, 176.
73 Twenty-three percent of Roman Catholic places of worship in England and Wales in 1851 and 1861 were found to be in Lancashire. See William G. Lumley, 'The Statistics of the Roman Catholics in England and Wales', *JSS*, xxvii (1864), 183.
74 K. S. Inglis, *Churches and the Working Classes in Victorian England* (1963), p. 121.
75 John A. Jackson, *The Irish in Britain* (1963), p. 122.

5 Comers and Goers

Raphael Samuel

The 'migrating classes', as they were called by the promoters of the Leicester Square Soup Kitchen in 1850,[1] have left remarkably few traces of their existence, even in the places which once served them as town lairs. No railway hotels mark the terminal point of their journeyings. No imposing clubs stand as memorials to the seasons they spent in town. The back-street lodging houses where many of them put up have long since been swept away, though not their forbidding institutional successors—the 'Free Dormitories' and 'Night Shelters' promoted and endowed by late Victorian philanthropy. The no-man's-land where the travelling showmen drew up their caravans and the gypsies encamped—the wasteland edge of the nineteenth-century town—has been built over by houses and streets. It is not easy to imagine a time when men slept rough in the shadows of the gas-works, the warmth of the brick-kilns, and the dark recesses of places like London Bridge; or lined up in their hundreds with tin cans or basins at Ham Yard, Soho, or the midnight soup-kitchens of Whitechapel and Drury Lane.[2] Perhaps the most complete change has been on the waterfront, which has been robbed of all its life. There is no Scandinavian settlement in Rotherhithe, no Tiger Bay in Cardiff. Rambling down Ratcliffe Highway, the curious observer is likely to find himself alone, flanked by derelict land and a low brick wall instead of the crowded dancing-saloons and rifle galleries, photographers' booths and boarding-houses of a hundred years ago. Then it was a great night-time pleasure strip, drawing sailors of all nations—red-shirted Americans, 'chewing indefatigably', Chinamen and Lascars, 'smoking Trichinopoly cheroots', 'shivering' Italians, 'piratical-looking' Greeks.[3]

© 1973 Raphael Samuel

Travellers in nineteenth-century England played a much greater part in industrial and social life than they do today. Among them were to be found some of the country's major occupational groups, as well as many hundreds of miscellaneous callings and trades. On the canals there were the boatmen and their families, who remained an important element in the carrying trade long after the coming of the railways—a half-gypsy population who owed their very existence to the Industrial Revolution. Tens of thousands of navvies followed in the track of Victorian 'improvement' and the great public works, a class of men 'very fond of change', and forever on the move, especially the more skilled among them.[4] The building trades were chronically migratory, with men moving constantly from job to job, sometimes covering very great distances in the tramp for work. (Some hundreds of stonemasons emigrated seasonally to the United States, leaving in the early spring and returning in the fall.)[5] The summer harvests depended to a considerable degree upon travelling labourers, most of them recruited from the countryside, but some coming out from the towns. James Greenwood met one of these on the road to Hitchin:[6]

> He seemed to be a decent sort of man, and, for a wonder, was not an Irishman. He lived and worked, all the winter, at the Potteries at Shepherd's Bush, he told me, and every June set out on tramp, working his way at any kind of field labour, and winding up with the Northern late corn harvests, when he returned home with a pound or so in his pocket, besides what he was able to send, from time to time, to keep his old woman.

At the heart of the wayfaring constituency were those whom Mayhew called the 'wandering tribes'; people who had been either born and bred to a roving life (like the gypsies and travelling showmen) or forced into it when settled occupations failed them, like the travelling Irish, who came over each year for the harvest, or the old soldiers and army reserve men, who lived as trampers on the road, and were said to constitute 'probably not less than one-fourth or one-fifth of the whole class of destitute homeless persons.'[7] It is possible to distinguish four different classes among them. First there were the habitual wanderers, who flitted about from place to place, with no regular settlement at all. Second there were those who spent the greater part of the year in the country, but kept regular winter quarters in the town. Third there were the 'fair weather' travellers, who went out on 'summer tours', but for the rest of the year stayed in one place. Finally there were those who made frequent short turns into the country, but never moved far from their home base. Examples of these types could be found in each of the wandering tribes. Amongst the packmen and dealers, for instance, there were some who followed a weekly itinerary, 'pitching it' in town on Saturday-night markets, and going on country rounds during the week;[8] there were others who turned out only for special occasions, like the army of free-lance hucksters who made an annual appearance on Derby Day at Epsom Downs; and yet others who left home for months at a time, like the 'muggers' and 'potters' who dealt in cheap crockery and earthenware,[9] and the cheap-jacks, whose yearly round began in February and often did not end until November.[10] 'I used to go round the country—to Margate, Brighton, Portsmouth—I mostly travelled by the coast, calling at all the sea-port towns', a cutlery seller told Mayhew; 'I went away every Spring time, and came to London again at the fall of the year.'[11] Similarly among the gypsies there were comers and goers within every migration range, including the most limited (the Battersea gypsies kept very close to London, and a few of them remained encamped all the year round).[12] Race-meetings and fairs attracted both short- and long-distance migrants, some following them round as part of a regular circuit, others going out on individual expeditions, with a basket, or a show, or with a tray suspended from the neck.

Wayfaring life had a definite place in the moral topography of the nineteenth-century town. Railway stations always attracted a floating population to the vicinity; so did the wholesale markets, especially those frequented by the drovers. Brickfields, on the outskirts of town, were regular dossing-places, along with gasworks, railway bridges, and viaducts. There were Irish quarters in every centre of industry and trade, 'Little Irelands', whose inhabitants remained notoriously migratory in their habits, 'exceedingly fluctuating and unsettled'.[13] The Irish were among the first to settle in the boom towns of mid-Victorian England (Middlesbrough, Barrow-in-Furness, West Ham). They were among the first to leave when there was a depression.[14] 'Our population is rather a floating one ... following the up and down ... of industry', wrote a Catholic priest of his Irish flock at Sunderland. 'St. Patrick's mission is not like many other ones.'[15] In the summer many of the Irish went off to 'country' work. A few of the York Irish were still doing this in Rowntree's time,[16] while in London the Cockney Irish constituted a lasting element in the annual exodus of hop-pickers.[17] Waterfront districts, too, always bore a peculiarly migratory character, not only in the big sea-ports, where there were fully-fledged sailortowns, but also inland, wherever there were wharves or docks. 'The quarters of every town that lie near the wharves and banks always seem to deteriorate', wrote Lady Bell, in her book about Middlesbrough. 'There is something in the intercourse of sailors from other ports who come and go, nomadic, unvouched for, who appear and disappear, with no responsibility for their words or their deeds, that seems to bring to the whole world a kinship of lawlessness and disorder.'[18] Quite apart from the native sailors—perennial comers and goers—there were the sailors from other ports, an even more nomadic element. At Shields, for instance, it was computed that besides the 15,000 sailors belonging to the port there were no less than 40,000 seamen from other places who annually visited the port.[19] On the Middlesbrough waterfront there were 'many ... foreigners'. 'Almost daily we come in contact with Germans, Greeks, Swedes, Danes, Norwegians, Arabians, Chinese, Lascars and Spaniards', remarked a chaplain to the seamen in 1892.[20] Sailors were not only more numerous and varied than their counterparts today; they were also more visible, spending far more time ashore, especially in winter.

Regular 'trampers'—'comers and goers who ... prefer darkness to light'—had their own peculiar lairs. An example in Nottingham is Narrow Marsh ('this provincial Whitechapel') where J. Flanagan, the Bermondsey evangelist, once spent eighteen months in rescue work 'chiefly among the crimps, outcasts and tramps, who were ever turning up'.[21] Another provincial example is Angel Meadow ('the lowest, most filthy, most unhealthy and most wicked locality in Manchester') whose inhabitants were accounted for by the *Morning Chronicle* special correspondent as 'prostitutes... bullies, thieves, cadgers, vagrants, tramps and... those unhappy wretches, the "low Irish"'.[22] In Merthyr Tydfil there was 'China', a maze of courts and tortuous lanes which the Education Commissioners of 1847 described as 'a sort of Welsh Alsatia'. It was, they said, 'a mere sink of thieves and prostitutes such as unhappily constitutes an appendage to every large town'.[23] There was a little enclave of this kind even in a sleepy place like Hitchin. James Greenwood came upon it in 1872, a long, narrow street which he describes as 'evidently the headquarters of the tramping fraternity':[24]

> It is like a slice from the backslums of Whitechapel, or Kent Street in the Borough. As, in the delectable localities named, at least one house in a half-dozen throughout its length is a lodging house for travellers—travellers, however, who are not so worn-out and leg-weary but they prefer, on a sultry night in June, to sprawl in the house passages and on the steps...

London had a number of little enclaves of this kind. Some of them were ancient haunts of the travelling fraternities (St Giles; Tothill Fields, Westminster; the Mint, Southwark); others were the accompaniment of the city's nineteenth-century growth. One of the newer tramp quarters was in Mill Lane, Deptford and its environs, 'known to tramps and low-class prostitutes throughout London', according to Booth.[25] Another was in the Arpley Road area of Anerley and Penge, 'a resting place for tramps entering London from the South'.[26] In the north there was Campbell Road, Islington, the occasion of much police and clerical disquiet.[27] To the south-east there was the 'Dust Hole', Woolwich, the subject of Canon Horsley's anti-vice campaign in the 1890s.[28] Booth wrote that it was a 'house of call' for tramps passing in and out of London on the high road to Kent, and a regular junction for the outer London tramp circuit. 'Policemen from Notting Dale', it was said, 'find old friends in Ropeyard Rails. The casual loafer floats between the two.'[29] Notting Dale, 'the resting place for tramps entering London from the North or West',[30] was the largest of these new 'tramp' districts. It was first settled by some refugee pig-keepers from Marble Arch, in flight from the sanitary authorities, and they were soon followed by gypsies, brickmakers, and (in later years) lodging-house keepers. Bangor Street, the most frequently noticed of its streets, was occupied by lodging-houses from end to end, and we are fortunate to have a good account of how a tramping family made its way there in the 1890s. It comes from the autobiography of Sam Shaw, a little boy of ten at the time. The family had set out from Birmingham 'without any fixed stopping places or definite... sleeping places':[31]

Father told us that we were going on the road to a big place where matches and newspapers sold better than in Birmingham. So 'on the road' we went . . . For weeks we tramped . . . each stage was from workhouse to workhouse where we were provided with bed and breakfast. If we arrived too late for admittance then we begged a shelter in a barn . . . or slept under the stars . . . Day after day, begging our way . . . We reached Edgware . . . and . . . spent the night in a workhouse . . . reaching London . . . next day . . . While we children played in one of London's parks our parents searched for a room . . . Bangor Street, Notting Hill Gate, eventually provided us with a home . . . It consisted of only one room and was furnished with two beds and two rickety chairs which were all the family exchequer could afford. Those who hadn't chairs to sit on sat on the bed.

Within such quarters, and scattered on all the tramp routes of the country, were the common lodging-houses, the night-time haven of the wandering tribes, or at least of the better off portion among them, hawkers and travelling labourers especially. Travelling people used them as regular staging posts, and laid up in them for the winter when the season's journeyings came to an end. There were few towns without a street or two largely given over to them. Around mid-century (a time when they were the subject of much anxious investigation) Gloucester's lodging-houses were concentrated in Leather-Bottle Lane;[32] Banbury's in Rag Row (a back street of the proletarian suburb of Neithrop);[33] Huddersfield's in the narrow courts off Kirkgate and Castlegate (the Irish ones in Windsor Court);[34] Doncaster's in Skinner's Yard, Far St Sepulchre Gate;[35] Brighton's in 'those bad streets' Egremont Street and Nottingham Street.[36] In Derby they were in Walker Lane, 'the St Giles of Derby'.[37] At Ashton-under-Lyne the twopenny lodging-houses were in Crab Street and the twopenny-halfpenny ones in Duncan Street.[38] London, of course, was full of them—no fewer than 988 were registered in 1889, with accommodation for over 33,964 people, quite apart from the coffee-shops which offered twopenny and three-penny beds, and the rooming-houses.[39] There were great numbers of them, too, in Liverpool, not only the sailors' boarding-houses of the waterfront, but also the 'emigrant houses' where travellers put up while waiting to make the transatlantic passage.

Common lodging-houses were condemned by sanitary reformers on account of the 'promiscuous' mixing of the sexes, and the crowded, impromptu conditions in which the inmates ate and slept. But to the footsore wanderer they offered warmth and a cheerful shelter. The kitchen was the hub of lodging-house life; it was usually to be found in the basement and served as a drying-room, a workplace (for those with a basket of merchandise to prepare), an eating room, and sometimes (though illegally) as a place to sleep. In the evening, when travellers returned from their rounds, it was turned into a common-room, thick with the fumes of coke cooking and tobacco, and warmed by huge, blazing fires. Even hostile observers had sometimes to admit that the atmosphere was companionable. 'The night being wet, enormous fires blazed in both rooms', wrote Dr Coulthart, investigating one of the low lodging-houses of Ashton-under-Lyne:[40]

> Groups of evidently abandoned creatures of both sexes, many of them dripping with rain, were drying themselves before the fires; while others, more jovial than the rest, were doing ample justice to the merits of ale, porter, porridge, beefsteaks, cow-paunch &c., nearer the door. Among the numbers was a woolly-haired negro who, the police officer with me said, had been driven from the streets a few hours before for ballad-singing. In vacant corners were hawkers' baskets, pedlars' boxes, musical instruments, and beggars' crutches.

Lodging-houses offered cheap overnight accommodation for prices ranging at mid-century from a penny to threepence, and in later years from fourpence to sixpence. Even so, they were expensive for travellers of the poorer class, who were able to use them only when they were 'flush'. There was a good deal of interchange between the lodging-houses, on the one hand, where accommodation had to be paid for, and the Casual Wards and Refuges, on the other, where it was free.

Men who slept rough (by no means only tramps) had an easy time of it in the country, when they made their summer rounds, and could pick and choose their places to sleep al fresco. Things were more difficult when winter drove them to the towns, and they were subject to a good deal of police harassment. But they were never without refuge. Police supervision was slight on the ragged peripheries of town, where many of them made a halt. The kilns on the Brent Brickfields at Willesden were an overnight stopping-place for men who tramped up to London. 'The cavities . . . afford warmth and shelter,' says a police report, 'hence the . . . numbers'.[41] Outside Manchester, too, 'the numerous brickfields on the outskirts of the city' were frequently resorted to by tramps.[42] In town itself there were certain places where men could escape the glare of the bull's-eye lantern: empty or ruinous buildings, shop doorways, omnibus depots, railway stables, coal-holes and boiler-rooms, cul-de-sacs and covered passageways, impromptu shelters arising, as it were, within the crevices of city life. In London, the wholesale markets were always a draw to tramps, partly because of the all-night coffee-shops and stalls, partly because they provided odd jobs, but also because of the shelter: any corner might be used for a kip—even the water-closets at Covent Garden.[43] There were certain other places, too, where tramps could assemble together more or less undisturbed. The Thames bridges served them for many years as 'Dry Arch' hotels, 'open houses for the houseless wayfarer'.[44] As late as 1869 Daniel Joseph Kirwan came upon a dozen people who had made their home in the underground recesses of London Bridge, and were burning driftwood for their fire: he called it a 'perfect gypsy encampment'.[45] The most sensational of these shelters in inner London were the Adelphi Arches off the Strand, a series of underground chambers and vaults 'running here and there like the intricacies of catacombs'. Thomas Miller described them as a 'little subterranean city'[46] and in the 1860s (according to one excited account) 'no sane person would have ventured to explore them without an armed escort'.[47]

Victorian 'improvement' swept away many of these nooks and crannies, but it created others in their stead. Street lamps yielded the homeless wanderer a certain starveling warmth; so did the boiler-rooms of the factories, the newspaper offices, and (in later years) the big hotels. Model dwellings, with their open landings, stairways and passages gave a rudimentary shelter to sleepers-out.[48] The colonnade of St George's Hall ('our great municipal building') sheltered so many of Liverpool's homeless that it was described in 1890 as the 'lodging-house' of the destitute.[49] The Thames Embankment, the most spectacular of mid-Victorian 'improvements' in inner London, very soon became a by-word for the number of its tramps, some of whom filled the seats beneath the plane trees ('No. 2 bench' was recommended to Duckworth by a former *habitué*),[50] and others of whom used it as an all-night promenade. Its character was reinforced by the Shelters built at either end (the Salvation Army's 'Penny Sit-Up' at Blackfriars, and the *Morning Post*'s 'Embankment Home' on Millbank); by charitable distributions of food (such as the Eustace Miles Food Barrow at Cleopatra's Needle); and by the nightly distribution of relief tickets underneath the Craven Street arches (2,000 men were assembling there nightly in October 1908).[51] The 'lynx eyed metropolitan police' (as they were described by General Booth),[52] after attempting to drive the tramps away, were by 1910 treating the Embankment as a 'kind of corral' where large numbers of tramps were conveniently assembled under the direct observation of law and order.[53]

Railway arches—both brick viaducts and iron bridges—were by far the most frequented of these new al fresco shelters, and they were resorted to by the homeless from their earliest days.[54] The Craven Street Arches at Charing Cross were a nightly assembly point for trampdom; so was Byker Bridge at Newcastle and (in later years, at least) the 'Highlanders' Meet' along Argyle Street in Glasgow. In Croydon the Windmill Bridge of the London & Brighton railway was a night-time resort for sleepers-out;[55] in Spitalfields the Wheler Street arches of the Great Eastern railway (a long, low, tunnel-like bridge) were used by some forty to sixty people a night, according to a police report; at Rotherhithe there were the 500 arches of the South Eastern and the London & Brighton railways, whose night-time *habitués* were not deterred by occasional prosecutions before the local magistrates.[56] The building of the Overhead Railway along the docks was a late addition to tramp facilities in Liverpool, skirting the seven-mile length of the waterfront and open to all comers—it became locally known as 'the Docker's Umbrella'.[57]

Gypsies made their encampments on the outer edge of town, where streets and houses gave way to waste. The Everton gypsies pitched their tents on a piece of waste ground near Walton Breck; in 1879 they were summoned before the magistrates for having failed to supply themselves with water 'as required by the Public Health Act', but the camp was still flourishing seven years later.[58] The Smethwick gypsies occupied a piece of waste ground near to the Navigation Inn.[59] At Charlton the local gypsies occupied an open space near Riverside—with six travelling-vans and a number of tents.⁶⁰ At Plumstead they camped out in the marshes—a stretch of land drained by deep ditches 'like the fen country' and used as a shooting range; their tents were made of old skirts 'stretched over hurdles'.⁶¹ In west London one of their camps was by Latimer Road, 'the ugliest place . . . in the neighbourhood of London . . . half torn up for brickfield clay, half consisting of fields laid waste in expectation of the house-builder';⁶² another was in West Kensington (in the days before it was overrun by the bourgeoisie) close to the market-gardens and the brickfields, between Gloucester Road and Earls Court.⁶³ In north-east London the chief gypsy settlement was at Hackney Wick, 'where the marsh-meadows of the River Lea, unsuitable for building land, seems to forbid the extension of town streets and blocks of brick and stuccoed terraces';⁶⁴ there was a smaller settlement near Finsbury Park, in the dust-heaps at the bottom of Hermitage Road.⁶⁵ The Battersea gypsies camped in Donovan's Yard, a plot of ground near the South-Western railway, 'commanding an unpicturesque prospect of palings, walls, and arches'. The encampment, in 1900, was occupied for about six months every year, 'from October till the flat-racing season'. It was made up of two long lines of wagons 'broken here and there by a firewood-dealer's hut'. The horses had been sold off 'to save the cost of keeping them in idleness during the cold months'. T. W. Wilkinson, writing about the camp,⁶⁶ remarked on the

> curious air of domesticity: . . . women, most of them stamped with their tribal characteristics, sit on the steps of the waggons, some at needlework, some merely gossiping. Other housewives are engaged on the family wash. Bent over tubs and buckets in close proximity to the fire, on which clothes are boiling briskly, they are rubbing and rinsing with a will, now and again going off for more water to a tap at one end of the ground.

This 'domesticity' was less extraordinary than Wilkinson implies. However far they travelled, the gypsies, like others of the wandering tribes, usually retained a base to which they regularly returned. The larger gypsy-colonies were well established. At Notting Dale a nucleus of families (most of them Hearnes) remained in occupation all the year round. Thomas Hearne, the father of the community, and a chair-bottomer, had converted his van into a makeshift cottage, with an old tin pail serving for a chimney, and a signboard announcing his trade.⁶⁷ The Wandsworth gypsies were also well entrenched. 'The houses are many of them owned by the richer members of the clan; and room is found for vans, with wheels or without, in which the poorer members crowd. The gypsies regard their quarter as their castle.'⁶⁸ They had been subject to some harassment in the 1870s, when they were driven from the Common by the Metropolitan Board of Works, and later on were hauled up before the local magistrates on account of the 'intolerable nuisance' of their tents, hovels, and vans. But a gypsy capitalist called Penfold had bought up cottages and lands in the vicinity, and twenty years later the colony was flourishing undisturbed, '3 families to a house'.⁶⁹ The Battersea colony, too, had acquired the status of semi-permanent residents when Booth enquired about them: 'These people, living in their vans, come and go, travelling in the country part of every year . . . They move about a good deal within the London area as well as outside, but are usually anchored fast all winter, and throughout the summer one or another always occupies the pitch.'[70]

The ebb and flow of wayfaring life in nineteenth-century England was strongly influenced by the weather. The months from March to October were the time when travelling people were to be found on the roads, and when they were joined by every class of occasional itinerant. With the approach of cold weather, in October and November, the season of journeyings came to an end and the wandering tribes returned to town. Spring was the time when long-distance travellers left town. Country labourers, who wintered in the metropolitan Night Refuges, were said to 'fly off' about March.[71] So did the men who sailed in the Greenland whalers, as one of their songs reminds us:[72]

> Twas eighteen hundred and twenty four,
> On March the eighteenth day,
> We hoist our colours to the top of the mast,
> And to Greenland bore away, brave boys,
> And to Greenland bore away.

The trampers' London season ended when the Night Refuges closed down:[73]

> The winter is the homeless man's London season; and most of the refuges, especially the larger ones, are open only four or five months in the year, from November or December till about April. In mid-winter they are most crowded, by April they are usually comparatively empty. Still, even to the last there is a substantial number at the large refuges. What becomes of them when turned out is not very clear. Probably the majority go . . . into the country.

Travelling showmen began to leave town quite early in the spring. The peep-show caravans (Mayhew tells us) generally left London between March and April 'because some fairs begin at that time', and were seldom seen again until October, 'after the fairs is over'.[74] Some showmen had already travelled considerable distances by the time of the Easter Fairs. Manchester showmen turned up regularly at Blackburn for the annual fair which opened on Easter Monday: on one occasion there was even a troupe of strolling players from London.[75] Travelling circuses began their tenting tours in March.[76] Old Joe Baker's Circus, which wintered in the Bristol slums, had 'taken to the road for the Easter Fairs', and had already reached Worcester when the young Ben Tillett caught up with them.[77] David Prince Miller, 'weather-bound' one winter in Carlisle, began the next year's season with a February fair at Dumfries, then doubled back to Carlisle, and worked his way over to Newcastle upon Tyne for the Easter Monday 'hopping'. Another year he had an indoor engagement in Birmingham, which kept him for the winter, and then proceeded to Manchester in the spring, to take part in Knott Mill Fair, 'the great Manchester carnival'.[78] Broadly speaking one may say that the showman's season began with the Easter fairs, but the starting-point was different on certain circuits. In East Anglia the season opened with the great Charter Fair at King's Lynn, which began on 14 February and lasted six days. It was here that Batty, the circus proprietor, began the 'outdoor business' of the season, after putting on a Christmas show indoors; here too, according to one of Mayhew's informants, cheap-jacks started their rounds.[79] In the Thames valley, when 'Lord' George Sanger was a boy, the season opened with the May-Day Fair at Reading. 'Showmen of all descriptions moved out of their winter quarters to attend it', Sanger recalls. Among them was his own father, who travelled the fairs with a peep-show during the summer months, and in the winter returned to his carrying-business in Newbury.[80]

All through the spring men were beginning to move from indoor to outdoor jobs, sometimes taking the step of exchanging a fixed occupation for a roving one. Gas-stokers ('regular winter men')[81] were giving in their notices—or getting the sack—as early as February and March, when the retort-houses began to close down. Only about a third of them were employed all the year round. 'What becomes of the extra men who are employed in the winter?' an official of the Gas, Light & Coke Company was asked by the Labour Commissioners in 1893, and he replied, 'They are only too ready to leave us in the spring.'[82] At the South Metropolitan Gas Works the bulk of them went bricking. 'Our best stokers, at any rate, do', a Charity Organisation Society enquirer was told. 'They give us notice early in the year . . . and then they go to the brickfields in the summer.'[83] The chimney-sweeper's town season ended in May, and both masters and journeymen were thrown out of employment. 'Some turn coster-mongers, others tinkers, knife-grinders &c., and others migrate to the country and get a job at hay making or any other kind of unskilled labour', wrote Mayhew.[84] Those who still worked in the trade went on country rounds, travelling from job to job, and making up their money by the country sale of soot.[85] The maltings season in the breweries ended about the same time, and some thousands of country labourers, who had come up to town for the winter, and spent seven or eight months at the work, returned to their native villages. 'We used to finish at the maltings towards the end of May,' a Suffolk labourer told George Ewart Evans, 'but before you went home you had to have a new suit. You dussn't come home from Burton wearing the suit you went up in . . .'[86]

The spring migration out of London occurred in stages; it began with an outward drift from the city to the suburbs rather than with a single clean break. Regular trampers hung about town as long as they could after the Night Refuges closed down, 'tiring out' the London and suburban workhouses (as one of Mayhew's tramp informants put it) before finally cutting adrift.[87]

The outward movement of the London gypsies only became general in May, and those who gained their living as itinerant agriculturalists moved out even more slowly: 'Christians who wish for opportunities of doing good to the Gipsies in and about London will find many of them in the suburbs in . . . April, May, and June, when they generally find work in the market gardens. In . . . July and August they move into Sussex and Kent, and are engaged in the harvest.'[88] Showmen also frequently spent a month or two in the suburbs before 'pitching it' out in the shires. Stepney Fair ('then the biggest gathering of the kind in England') was the first place where the Sangers pitched, when they wintered in Mile End, and they followed it up with King's Cross Fair, one of the new impromptu fairs which had grown up on the wasteland edge of the city.[89] In later years (when the city extended outwards), the Easter Fair on Wanstead Flats took its place: it was known as the 'Gypsy Fair' from the number of them who made it the first 'gathering' of their season.[90]

By the beginning of June a fresh series of migrations was under way. Gangs of mowers moved about the country, and haymakers followed in their wake. Later on in the summer the corn harvest set up a demand for extra helpers on all sides. At Kenilworth the extra men who came to do the work were from 'Coventry, Ireland, Buckingham and Berkshire'.[91] At Holbeach, Lincolnshire, they included numbers of Irishmen from the big towns of the Midlands—'English Irish' as they were called locally; at Godstone in Surrey there was an immigration from London, Croydon and elsewhere—'frequently travellers on their way to the hop-picking in Kent.'[92] Some of this movement took place within a short migration-range, the surplus labour of the towns being absorbed in the fields of the immediately surrounding countryside. But some of it occurred over longer distances. A town missionary who boarded a steamer at London Bridge in August 1860 found that a 'considerable number' of the 300 passengers were labourers from the southern counties 'going to Yorkshire for the harvest' (the steamer was bound for Hull).[93]

During the summer months there was big money to be earned in the fields, for the man who was prepared to rough it, and to try his luck on tramp. Early in July 1872, for instance, wages were said to have risen as high as eight to ten shillings a day in some of the suburban hayfields of Middlesex.[94] Harvest rates could be nearly as high, even in a low-wage county like Oxfordshire: in August 1872 two builders' labourers at Woodstock were demanding seven-and-six a day to dig the site of a new gasholder 'because the harvest was about to begin'. When the rate was refused as 'exorbitant' they went off to find work as harvesters instead.[95] For a few brief months industrial and agricultural employments faced one another as direct competitors, and the worker who was disgruntled, or ill-paid, or out on strike, was not slow to take advantage of the situation. Those who worked in the 'dangerous' trades, for example—the lowest class of Victorian town labour—had a chance to escape from the wretched conditions in which they worked. 'It is surprising to me how persons can breathe here', a factory inspector wrote of a London enameller. 'One man is sensibly affected; he goes away each year hop picking, for the purpose (as he says) of cleaning himself from chalk.'[96] Amongst London fur-pullers the escape into hopping was general. 'The work, disagreeable at best, is unendurable in hot weather, and when hop or fruit picking in the country offers as an alternative, it is gladly accepted.'[97] There was a similar efflux of labour from some of the white-lead factories, with their dust-laden, poisonous atmosphere. At H. & G. Grace's, Bethnal Green, for example, as many as a third of the employees were said to go off hopping, even though there was no summer shortage of work.[98] Similarly, at the Millwall Lead Company, 'All of them go hopping in the autumn for a month or two, that is one of their chief occupations, but of course that lasts only a short time; and they hawk fruit about if they can; but when they have nothing of that kind to do, then they come back to the white lead works.'[99] In the Brough Lead Works, Sheffield, haymaking rather than hopping provided a seasonal escape. 'Have you any difficulty in getting men to come here?' the manager was asked by the Labour Commission in 1893; 'only just in the summer time', he replied, 'in the middle of the summer, when there is hay harvest and other jobs.'[100]

The summer harvests prompted a whole series of different itineraries. Some labourers followed the harvests round—haymaking in June, turnip-hoeing or pea-picking in July, corn-harvesting in August, hop-picking or fruit-picking in September. Others went out only for a single crop or relied upon the odd jobs in the countryside to support them on the road. Those who went for one harvest sometimes stayed for two, either because the work was available or because they had given up their lodgings in town. Others alternated between town and country. Even those who followed the 'long' harvest were by no means all of one type, and there seems to have been a broad division between those who visited a variety of *places*, travelling from county to county, and those who did a variety of *jobs*, all of them within a single district. In Lincolnshire and the East Riding, for instance, migrant labourers were divided between those who worked the country in regular rounds, 'beginning further south and working northwards as the harvests successively ripen', and others, 'less migratory than this' who came into a district at hay harvest 'and manage to find sufficient work at odd jobs in the same district to keep them till corn harvest commences.' Both, it may be added, hailed from the Midland manufacturing towns.[101]

Haymaking was the starting point for many of these summer rounds. Cobbett called it the 'first haul' of the 'perambulating' labourer,[102] and the London tramp regarded it as 'just the proper season' for leaving town. 'Down I strolled into Sussex, towards the border of Hampshire, and soon got a job', one of Thor Fredur's informants told him, 'half-a-crown a day, my food, a corner of a barn with clean straw to lie at night, as much beer as I would drink while at work ... I ... always begin with a spell at haymaking.'[103] The Irish came over for it *en masse*, leaving the mowing for native-born labourers, but following closely behind them. Haymaking in the nearby countryside was one of the resources which enabled the Padiham weavers to support themselves during the long strike of 1859.[104] It was still a standby for out-of-work labourers in York when Rowntree and Lasker made their study for *Unemployment*, along with 'carting', 'droving', 'farm work', 'snow sweeping'; the weekly itinerary of an unemployed grocer's assistant shows the way in which it was dovetailed into the local network of odd jobs:[105]

Monday—Called on ... [grocer] in answer to an advertisement. Was told I had been too long out of the trade. Then searched advertisement in the Library.

Tuesday—Got job of digging up sand at boat landing-place (having drawn owner's attention to need for the job).

Wednesday—Same.

Thursday—Worked for boat owner. Earned 3s. 6d.

Friday—Library for advertisement.

Saturday—Applied for work at stores and for horse clipping at . . . without success. Spent afternoon outside the town looking for a job haymaking.

Monday—Started at 4 a.m. seeking haymaking job at three villages [named]: got work at 9 a.m. Came home early owing to rain. Earned 3s.

Market-gardens, too, drew upon migrant labour from the towns. The London Irish were prominent amongst them, and according to 'A Wandering Celt', writing in the *Labour News* for 1874, did most of the pea- and fruit-picking for the London market. 'Young Irish women, who in London during the winter are fruit-sellers or working in dust-yards, pickle-factories, sack-making, etc., go into the market-gardens for the summer.'[106] In Essex there was a regular influx of Londoners in June and July—most of them rough women from the East End, who took on summer employment as market-garden hands, and then crossed over to Kent, in time for the start of the hopping. James Greenwood came upon a little colony of them at Rainham in 1881, 'browner by many shades than gipsies'. 'There ain't no men among 'em,' a local told him, 'only women, and girls, and a few lads.' The work (chiefly pea-picking) lasted from eight to ten weeks, and was paid at a flat rate of eighteenpence a day. But much of the work could be done piece-work, which suited a woman with children to work alongside her: 'in fine weather and at certain work—onion-pulling for instance—some families earned as much as four shillings a day.'[107]

Summer was the height of the travelling season. There were more navvies on the road, moving from job to job,[108] more tramps, more travelling hawkers and dealers. Street-arabs left London in shoals, traversing the country in every direction, and trading (or thieving) as they tramped: 'they sometimes sleep in low lodging-houses . . . frequently "skipper it" in the open air . . . and occasionally in barns or outhouses.'[109] Vagrants, too, preferred to spend the summer out of town, and 'seasoned it' al fresco, or staged their way across the country with the aid of the Casual Wards. In the workshop trades the tramping artisan, who trudged along under his oil-skin knapsack, was a familiar figure at this time, as Thomas Wright, the journeyman engineer, reminds us. According to him, it was a 'frequent practice' for men on tramp to do their travelling by night (when the weather was hot) and to have their sleep by day, in an orchard or a field 'conveniently near the roadside'.[110]

The wayfaring constituency was further enlarged by summer newcomers. There was a seasonal influx of Italians, for example. Some seven or eight hundred of them crossed over to England every year for the ice-cream trade. 'Each spring brings a contingent', wrote Ernest Aves, in his notes on the Italian colony in Clerkenwell.

'They come in small parties from all parts of Italy, travel slowly, take their food with them and when autumn comes go back to their wives and their vineyards.'[111] The organ-grinders arrived a little later. 'June July Aug. Sept, are the busiest months, when a great number migrate and travel the country through visiting Birmingham Bradford, even Scotland, Wales & most seaside resorts . . . It is then that a great number of fresh arrivals are seen, but as winter arrives they gradually depart.'[112] On a larger scale there was the annual influx of harvesters from Ireland who began to arrive in large numbers at the end of May and followed the harvests round. The coming of really warm weather, in May and June, also tempted a weaker and more occasional class of traveller to venture out of town. Some hundreds of sandwich-board men, for instance, took to the roads at this time, most of them army pensioners and 'beyond middle life'; 'in the summer . . . numbers of them go into the country and by pea, hop or fruit-picking, or in some other way, obtain a livelihood until September or October, when they return to their old haunts.'[113] Even in the workhouse there was a class of inmate who had their summer 'tour'—old men and women, who laid up in the 'house' for most of the year, but enjoyed a brief spell of freedom when conditions outside allowed it. Booth came upon some of them in Stepney Union workhouse.[114]

Male	Married	69	Carpenter	Wife left him and went to her son's. Man goes on tramp during summer months.
Male	Married	56	General labourer	Man was doing casual work when relief was first given . . . admitted in 1886 . . . Only out for short periods in the summer since.
Female	Married	63		Woman had medicines in 1882. She goes out nursing or fruit picking in summer, and winters in workhouse.
Female	Widow	60	Washing	Husband died (1849). Woman had out-relief, but it was stopped. Goes hopping with daughter.

The largest movement out of town took place in August and September. It was the season both for feasts and fairs ('Wakes' weeks in the Potteries and Lancashire), and for holidays on the sands. All kinds of opportunities opened up for those who followed the track of holiday-spending, and the regular showmen were joined by a whole army of itinerant hucksters ministering to the pleasure-goers' needs. It was also, in a different branch of summer activity, the height of the harvest season. The corn harvests which began in August and which lasted (until the coming of the mechanical reaper and binder) for up to six weeks, set up an enormous demand for extra hands and, as in the case of haymaking, some of this was supplied from the towns. Fruit-picking and hopping, the September harvests, offered a choice of less strenuous opportunities.

The late summer exodus corresponded to a general slackness in town employment. August and September were the 'dead months' of the year in many of the indoor trades (especially those which depended upon the world of rank and fashion), in heavy industry (on account of the heat), and in certain branches of factory work. During the 'Long Vacation' even the occasional law-writer was said to go off hopping.[115] In the London docks 'very many men' found work elsewhere in the months from July to September, 'the time of the harvest, and... the militia... in training'.[116] The same was true in the coalfields, an important source of harvest labour in the Midlands and South Wales. In the Black Country many smithies were closed up when the inhabitants took themselves off *en masse* to the fruit-picking in the Vale of Evesham, and the hopping in Herefordshire. Women and girls found work particularly hard to come by at this time. A 'little army'[117] of ironers and laundresses was thrown out of work when the middle class went off on holiday (some followed them to the sea-side);[118] the tailoring trade was invariably dull at the end of the London season, and the number of women and girls thrown out of employment was said to be 'incredible'.[119] Factory work was also scarce. At Allen & Hanbury's, Bethnal Green, one of the biggest factories in the East End, something like a quarter of the girls were sacked every summer;[120] the same was true of the match-factory girls (some found alternative employment in the jam factories, others went hop-picking or fruit-picking in Kent),[121] and in the bundlewood yards on the Surrey side of the Thames: 'Nothing is made up for stock... the women go fruit and hop picking, and the men find casual employment as best they can.'[122]

This is no doubt one of the reasons why women and girls figure so largely in the late-summer movements out of town. Another was that the work was of a kind which gave them a positive advantage over men—work for the nimble-fingered. Hopping was often undertaken by family groups with a woman in command and the husband (if he existed) elsewhere. It was piece-work, and children could make as great a contribution to earnings as grown-ups. Fruit-picking seems to have been largely in the hands of itinerant girl labourers. London work girls who went 'fruiting' in the orchards of West Middlesex were joined by others who had come south from Staffordshire and Lancashire: there is a good description of them in Pask's *Eyes of the Thames* (1889):[123]

> the North Country girls... look forward to this fruit harvest ten miles from London to find them the means to form a little nest-egg to help them through the coming winter. Their work, when they choose to take overtime, which they generally do, is, in the early summer, from half-past three in the morning until eight at night. By this custom, they can always, if they choose, earn over eighteen shillings a week, doing piece-work, or, as it is termed in market-garden parlance, 'great' work... In their short print dresses and with their red cotton handkerchiefs tied over their heads, the girls look well enough to form a pretty study for any follower of the Fred Walker school. Still, despite rosy cheeks, blue eyes, and agile forms,

the romance is soon broken when they open their mouths. If ever a rival could be found for Billingsgate, it would be some London market-orchard. Even the Irish girls, who can boast a far higher standard of morality, are as foul-tongued as a lighterman working on the Pool.

The September hop-picking was the jamboree of the wandering tribes. Mayhew called it the 'grand rendezvous for the vagrancy of England and Ireland.'[124] Gypsies came to it from every part of southern England—'nearly all the gypsies in England', according to one inflationary account.[125] In London the common lodging-houses were said to be 'almost deserted' on account of it, 'the Bohemian inmates having betaken themselves *en masse* to the pleasant fields of Kent'.[126] Even the workhouse population was notably affected.[127] The Irish poor had a 'positive mania' for hopping, and the 'wild unrestrained kind of life' which it allowed;[128] for those who travelled the harvests it was the climax of the summer's round; for many more it was the one departure of the year from town: 'It is no uncommon thing for the houses of rooms to be shut up and for whole families to go off together', wrote Denvir. 'In the season of 1891 as many as eight hundred, chiefly Irish, went from Poplar alone, and it is the same among our poorer fellow-countrymen in other parts of London.'[129] Trampers were moving off to the hop-fields fully a month before the picking season began. 'They are gone for about two months and then we have another rush', Mr Duffus, the Superintendent of St Giles workhouse, told an enquirer in 1891.[130] In September they 'infested'[131] the hopping counties of Hereford and Kent, a ragged army of followers, some of them quite indigent: when an inspection was made of the casual ward at Hollingbourne in 1868, only two of the 289 inmates ('all hoppers') had a sum of money amounting to twopence on them. 'The great majority had nothing, and were partly without clothing.'[132]

The social composition of the migrant picking-force gradually changed in the course of the nineteenth century, partly because of the greatly increased demand for hopping-labour, partly because of improved methods of cultivation (which shortened the season to as little as three weeks) and partly because of the cheap trains promoted by the railway companies which put the hop-fields within reach of the proletarian family group, instead of only its more able-bodied members. In earlier times the French, the Welsh, and later on the Irish migrant labourers had been prominent: in the second half of the nineteenth century the hopping became increasingly an affair of women and children from the towns. Long-distance migration did not cease (a man from Warrington who went hopping in Kent in 1893 found himself working alongside two Yorkshire colliers who were out on strike)[133] but most recruitment was from nearer localities. At Martley in Worcestershire the extra hands were reported as coming from 'Stourbridge, Dudley and the mining districts',[134] those at Bromyard in Herefordshire were from Cradley Heath.[135] In East Kent many of the hoppers came from the Sussex villages and the seaport towns.[136] In mid-Kent and west Kent the immigration was more Cockney, being drawn from the Medway Towns (2,000 hoppers came from Gravesend in 1876),[137] from Croydon, and, above all from Inner

London (according to one account, Poplar sent to East Farleigh, Bermondsey to Wateringbury, Shadwell to Paddock Wood).[138] The Farnham hoppers in Surrey (said to number about 5,000 in 1887) were attributed by Sturt to 'the slums of Reading and West London',[139] in earlier years they had come from Portsmouth and the south-coast towns.[140] There were slum districts in which something like a general turn-out took place as hop-picking approached. Ellen Chase, one of Octavia Hill's property managers, says that 'exciting rumours' of the size of the hop harvest filled her Deptford street for weeks beforehand;[141] and George Meek, who went hopping in Mayfield, Sussex, in 1883, remarks on the 'rough lot from the purlieus of Edward Street Brighton' whom he found there.[142] Many of the hoppers travelled down together in family and neighbourhood groups, and worked together for the same farmer in their own companies. The frequency of hop-pickers' strikes, one of the more affecting if least noticed features of this seasonal migration, may be partly accounted for by the fact that many of them were already closely knit together.

After the harvests the movement from town to country was reversed. Country occupations began to grow scarce, while in the towns, on the other hand, there was a general revival of trade. The season of journeyings came to an end, and with the approach of cold weather the wandering tribes returned to town 'with the instinct which sends some birds of passage southwards at the same season'.[143] 'I like the tramping life well enough in summer', a girl tramp told Mayhew, ''cause there's plenty of victuals to be had then . . . it's the winter . . . we can't stand. Then we generally come to London.' Her sentiments were echoed by another of Mayhew's informants—a girl who passed the winter in the Metropolitan Asylum for the Houseless Poor. 'I do like to be in the country in the summer-time', she told him. 'I like hay making and hopping, because that's a good bit of fun . . . It's the winter that sickens me.'[144] Travelling in winter was 'an unusual thing for the gipsies':[145] it was very little practised by vagrants,[146] or by tramping artisans.[147] Even the regular tramper, who moved about for the sake of keeping on the move, rather than with any particular destination in mind, deserted the roads.

Travelling people began to drift back to town in October, and by November the movement was general. 'All over England', wrote an observer in 1861, 'a characteristic migration sets in . . . tens and hundreds of thousands . . . driven by necessity . . . swarm into the towns.'[148] The largest movement, and certainly the most frequently commented on, was in the direction of London, the winter Mecca of the wandering tribes. But it had its counterpart in a whole series of local migrations from country to town, and in a general change-about, which continued right through the autumn, from outdoor to indoor jobs, and from summer to winter trades. General labourers took their navvying skills into the gas-works. Travelling sawyers exchanged the saw-pit and the woodland clearing for a workshop bench ('towards winter time . . . a roof over their heads became desirable').[149] Migratory thieves, who conducted their summer business al fresco, at the race-grounds and the fairs, turned to a spell of safe-cracking or burglary in town aided by the long dark nights. (November, according to Manby Smith, was the month when many of them came back to London.[150]) Cheap-jacks rented shops and conducted mock auctions from their own premises instead of from temporary pitches in the open air.[151] Not all the exchanges were as regular or predictable as this: Booth gives us a glimpse of one or two in the street notes which he collected from London School Board visitors: 'Punch and Judy show in summer and makes iron clamps in winter'; 'Works at watercress beds in season, and sweeps chimneys in winter.'[152]

A first wave of travellers returned to London immediately after the hop-picking. 'Within a few days' (according to one account) some of the gypsies were back in town: 'hopping over, they go, almost *en masse* . . . to buy French and German baskets . . . in Houndsditch.'[153] In the Kentish suburbs (Gravesend, Woolwich, Greenwich) there was 'always' an influx of unskilled labourers at this time, 'men who . . . resort to the casual labour afforded by the revolution of the seasons.'[154] Some of them found local employment as rubbish-carters and scavengers, others drifted on into town. In Notting Dale, the common lodging-houses, 'comparatively empty' during hay-making, hop-picking, and fruit-picking, filled up with returning travellers:[155] Bangor Street ('one of the most dangerous streets in London') was said to be inhabited 'almost entirely by' them. St Giles, '*le quartier général des vagabonds*', was very soon packed: tenements, closed up for the summer, were once again re-occupied, as the inhabitants returned with their summer earnings from Kent; the lodging-houses took on their winter complement of sandwich-men, loafers, and touts; and the casual ward of the workhouse was crammed with travellers (many of them country labourers) making their way back home.[156] The Holborn Irish returned home in mid-October, much to the dismay of the local Medical Officer of Health, who complained of 'crowds of squalid Irish people . . . returning from the country to their winter haunts in the Courts and Alleys.'[157] Even on the London waterfront the end of hopping made itself felt: 'I never saw so many callers', wrote a correspondent of the *Labour News* in October 1874, 'the dock labourers having returned from the hop gardens . . . seem to have grown in number.'[158]

Not all the travelling harvesters returned at this time. Some of them jobbed about the country until the frosts came, or worked their way home in stages. Potato-lifting kept some of them out of town until November—the West Ham Irish[159] in the early nineteenth century, for example, and the travelling Irish of York in later years.[160] Trade tramps, too, sometimes delayed their arrival in town. Travelling coopers had a mid-autumn season at Lowestoft and Yarmouth, where they were employed in the herring trade.[161] In the Northampton shoemaking trade there was a 'very large influx' of travellers a little before Christmas, 'when the better sort of work is more brisk, and when there are generally more orders in the "bespoke" department'.[162] Cheap-jacks and showmen were among the latecomers. Many of them followed the autumn fairs, such as the great autumn cattle fairs, and the Michaelmas and Martinmas 'Hirings', or wound up their season by 'pitching' at the late town fairs. Some showmen delayed their wintering almost till Christmas, when the 'World's

Fair' ('the great event in the showmen's year') opened at Islington and van-dwellers from all parts of the country made a winter camp in the yard of the Agricultural Hall.[163]

'Wintering in town' was a regular part of the showmen's round, and the difficulty of making it pay was one which cost them much ingenuity. Some, it seems, let out their shows, and lived upon the proceeds, like the little 'half gypsy' colony whom Hollingshead came upon in Owen's Yard, Lambeth, settled in the midst of dust-heaps and factories.[164] The more enterprising adapted themselves to the conditions of town life by hiring temporary premises and putting on their shows indoors (the penny sideshows and 'gaffs' which figure so largely in descriptions of nineteenth-century street-life were often promoted in this way). 'Lord' George Sanger, who seems to have been very successful in making wintering pay, has left an excellent account of his repertory. One winter he went in for *poses plastiques* and conjuring, and hired a warehouse in Bethnal Green Road where the crowd sometimes was so great 'that we had to square the policeman not to interfere'. Another year he took on an empty chapel in Clare Market, and fitted it up as an impromptu theatre, playing a round of pieces, 'gaff fashion', and for Christmas put on a pantomime. Wintering in Liverpool, during the Crimean winter of 1854–5, he took on a large piece of ground near 'Paddy's Market' ('the lowest part of Liverpool'), and built a board and canvas theatre, with admission charges of a penny:[165]

> Here we had a semi-dramatic-cum-circus sort of entertainment that exactly suited the neighbourhood ... what we mostly did was acting on the gaff principle, and there was nothing we were afraid to tackle in the dramatic line, from Shakespeare downwards ... One of our best and most popular actors ... was Bill Matthews. He ... made a big hit ... by his impersonation of Paddy Kelly, an Irishman who had distinguished himself as a soldier at the Alma, news of which battle, fought on September 20th, had thrilled the nation. Well, Matthews did a riding act, 'Paddy Kelly, the hero of the Russian war', and in his uniform, slashing at the enemy with a sword and plentiful dabs from a sponge of rose-pink, excited the audience to frenzy.

Trampers 'led by an instinct somewhat analogous to that of ... animals who lurk in holes from the inclemency of the season',[166] came up to town for the shelter and the warmth. In London, according to Mayhew, they turned up each year 'as regularly as noblemen' to season it in town.[167] 'In the winter season', the Chief Constable of Manchester complained,[168]

> tramps flock to large towns such as Manchester, where they can obtain warm sleeping-quarters in the various brick-yards, boiler-houses, and different buildings connected with factories and workshops; also they can generally obtain free meals, which are provided by the various philanthropic societies in Manchester during the cold season. In the summer

> months they migrate to the country . . . sleeping out in the open when the weather is good.

Night Refuges ('strawyards' as they were known to *habitués*)[169] brought many of them to town. In London, critics alleged, there was an immediate increase in vagrancy when they opened up for the winter (usually in November, but the precise date depended on the state of the weather).[170] 'A great number of persons come to London in November, when the refuges are generally opened', an officer of the Mendicity Society complained. 'It is not an unfrequent answer, when they are asked, "How is it you have come to London again? You were here last year". "Oh, I thought the Houseless was open".'[171] Night Refuges (which were financed by private subscription) were much more popular with travellers than the workhouse, and, while they were open, slept many more people than the metropolitan Casual Wards.[172] The regime, though spartan, was comparatively kindly. The stranger was offered warmth and shelter without any of the humiliations and restraints associated with the Poor Law. Inmates could stay for as long as a month at a time (in the Casual Wards the rule was two nights only), and they were allowed to go out when they liked (in the Casual Wards shelter had to be paid for by hard labour and forcible detention for a day). At the Playhouse Yard Asylum, Cripplegate, the oldest and largest of the metropolitan asylums (it later removed to Banner Street) the dormitories were kept 'always . . . heated', and there was a gigantic communal fire. 'As these are lighted some time before the hour of opening, the place has a warmth and cosiness which must be very grateful to those who have encountered the cold air all the day, and perhaps the night before.'[173] Night Refuges existed in a number of the larger towns: Manchester, Birmingham, and Edinburgh each had one; in London there were seven (more, when the Salvation Army embarked on its social work), quite apart from such specialized institutions as the Destitute Sailors' Asylum, in Well Street, Ratcliffe Highway, and critics may well have been right to hold them responsible for so many homeless men making London their winter retreat.[174]

The autumn migrations, like those of the spring, took place in stages, and once again it was the outskirts and environs of the town which felt them first. Vagrants, it seems, made their way into London sideways, circling the outer ring of the metropolis, and testing the hospitality of the brickfields—or the suburban Casual Wards—before making their way into town. Travelling prostitutes seem also to have arrived back in this way, with preliminary comings and goings. 'They travel round the country in summer and come into London in November', an officer of the Mendicity Society told an enquiry in 1846. 'If they come before the refuge is open they go to Peckham, then they go to St Olave's, and then to Greenwich, and other unions.'[175] Gypsies ended their autumn journeys on the outer peripheries of town ('as close as you please to the skirts of civilization'),[176] but as the weather grew more severe some of them moved further in, and went to live in rooms. 'They leave the country, and suburban districts of London', wrote a City Missionary in 1860, '. . . and make their dwelling in some low court . . . 2 and 3 families . . . in one small room.'[177] George

Smith came upon a little colony of this kind when he visited Canning Town in the winter of 1879–80, seventeen families crowding together in two small cottages, where they had crept 'for... the winter'.[178] Families like this occasionally visited Deptford. Ellen Chase recalls that they would 'tide over the rough weather' by renting temporary accommodation. 'The walls would remain as bare as they found them... young and old sat upon upturned boxes about the small grate, as contented as if it were a camp fire.'[179]

As well as the travellers, returning to base, the towns received a large winter influx of refugees. Some came for the charities and shelter. Many were winter out-of-works, who came up to town (sometimes unwillingly) when every other resource had failed them. 'It was when the snow set in... I thought I would come to London', an inmate at the Houseless Poor Asylum told Mayhew:[180]

> The last job I had was six weeks before Christmas, at Boston, in Lincolnshire. I couldn't make 1s. 6d. a day on account of the weather. I had 13s., however, to start with, and I went on the road... going where I heard there was a chance of a job, up or down anywhere, here or there, but there was always the same answer, 'Nobody wanted—no work for their own constant men'. I was so beat out as soon as my money was done—it lasted ten days—that I parted with my things one by one. First my waistcoat, then my stockings (three pair of them), then three shirts... After I left Boston, I got into Leicestershire, and was at Cambridge and Wisbeach, and Lynn, and Norwich; and I heard of a job among brickmakers at Low Easthrop, in Suffolk, but it was no go. The weather was against it, too. It was when the snow set in. And then I thought I would come to London, as God in his goodness might send me something to do.

London was full of such winter refugees, caricatured, and yet in some sense truly represented, by the 'froze out gardeners', who regularly appeared in winter as beggars, or buskers, about the city streets. Night Refuges ('the outcast's haven') catered largely for men of this class. 'Travelling tradesmen' were said to compose the bulk of the inmates of the Ham Yard Hospice—trade tramps who had been reduced to a state of complete destitution, and could not afford even the price of a lodging-house bed; country labourers figured largely at Banner Street, 'a very rough class of men, who will work if they can get it... digging among fruit trees and market-garden work—in the fields'.[181]

A certain number of farm labourers, turned off after harvest, drifted into the towns, and they were joined by others as country employment grew scarcer and the weather more severe. In Norfolk and Suffolk there was a class of freelance labourers, known as 'joskins', who went off after harvest to Lowestoft and Yarmouth and got work in the autumn fishings (October and November were the height of the East Anglian herring season); another class migrated to the breweries of Burton-on-Trent.[182] Farm labourers in Monmouthshire and Nottinghamshire took up winter employment in the pits;[183] in Carmarthenshire, at mid-century, some of them went off to the ironworks ('in the summer they return home or go to England for the harvest').[184] In Sussex, according to a report of 1895, 'the more helpless class' made for Brighton and Hove, 'large towns . . . where there are many charities'.[185] Most of these migrants disappeared, for the season, into obscurity—mere 'birds of passage' in the town—and the historian is fortunate when he comes upon an individual case, like the one recorded by Steel-Maitland and Miss Squire in Jenner's Row, Birmingham:[186]

> Mr. J. and his wife, in third floor front, were occupied in making straw-baskets, which they sell to some of the family shops. Mr. J. about fifty, seemed a superior type of man. He said he should not take to regular home-life now. For the last twenty years he had tramped from town to town. During the winter he and his wife took a furnished room, and in the summer they walked into Herefordshire, where he did apple-pulling, and other odd jobs at the cider harvest, for a farmer. They were allowed to lock up a few pots and pans and a bed in a shed on the farm, and to this they returned every summer. Mr. J. said he had wintered in Plymouth, Manchester, Bristol and in London. They generally reckoned to be in the country for about five months of the year. If times were good, 'The missus' might perhaps go by train, but he always walked . . . He thought he should go to Cardiff next winter, if all was well . . . He had found Birmingham an expensive place, and did not think he would return. (This was his first visit.)

One of the most enduring of these autumn migrations was that which brought many hundreds of country labourers to work in the breweries. It has been vividly documented from oral tradition by George Ewart Evans, and his account can be supplemented by documentary references from earlier years. Maltsters were usually taken on at the beginning of October and continued till about the latter end of May, 'being about seven months of the year'.[187] The great majority of them were drawn from the country—'big-framed men, strong enough to handle the comb-sacks (sixteen stones each) of barley'. At Hertford and Ware, according to a *Morning Chronicle* account in 1850, 'nearly the whole of them are employed as agricultural labourers when not engaged in malting.'[188] At Newark, where 460 maltsters were employed in the 1890s, only about a third of the men were engaged all the year round, 'the rest go into the kilns in September, and remain till May or June.'[189] Farm labourers in Derbyshire went up to Burton-on-Trent. 'The winter employment in Burton helps to keep wages up', the Agricultural Employment Commissioners were told in 1868, 'especially for the hired single men.' 'A good many men go to Burton in the winter, where they get 13s a week and beer . . . many of them would be out of work in winter if they didn't.'[190] The catchment area for Burton was very wide indeed. In the later nineteenth century it extended as far as East Anglia, and George Ewart Evans has collected some remarkably detailed testimonies from labourers who made the last of these autumn journeys in the years 1900–30. He has also recovered a

Burton labour list for 1890–1 which shows the extent of the East Anglian hirings. It is taken from the records of Messrs Bass, Ratcliff & Gretton. Here are the first twelve entries.[191]

Name of worker	Home village	Nearest railway station
ADDISON, George	Melton	Woodbridge
ASHEN, Henry	Flempton	Bury St. Edmunds
ASHEN, William	Flempton	Bury St. Edmunds
BALDWIN, William	Aldburgh	Harleston
BARBER, Walter	Martlesham	Woodbridge
BEAUMONT, Peter	Baylham	Ipswich
BETTS, Arthur	St. Cross	Harleston
BACKHOUSE, Jesse	Sutton	Woodbridge
BLOOMFIELD, Richard	St. Lawrence	Harleston
BRAGG, John	Bardwell	Bury St. Edmunds
BRETT, Charles	Martlesham	Woodbridge
BROOKS, Alfred	Pakenham	Bury St. Edmunds

Building-workers flocked up to town when the frosts put a stop to the country trade. In London they figured very largely among the winter refugees—pick and shovel men, who tramped the metropolitan building-sites looking for work ('*bona fide* navvies, up to "Lunnun" in search of a job'); country craftsmen like the 'strong and handy carpenter' whom the roving correspondent of the *Labour News* met in Greek Street, a 'most desponding' man who had left Watford just after Christmas, and tramped it inside and outside the town, without finding himself a place; painters and decorators 'calling at every job, and offering, in many instances, to work at half-starvation wages'. 'Outside the heavy jobs on hand', he writes in January 1873 '... may be seen building hands and labourers ... asking the foreman to put them on, and the reply is that they have already too many men.'[192] Painters were particularly badly placed, and are often singled out for attention in distress reports. At the Newport Market Refuge for instance, they were by far the most numerous group of inmates—57 of the 644 men admitted in 1889.[193]

The months from November to February were always a bad time in the building trade, but it was only in the country and suburban branches of the trade that the stand-still was complete.[194] In the towns there were big contract jobs where work continued in all but the worst of weather. Stonemasons, 'though apt to be severely hit by a really hard winter', stood a fair chance of getting work; so did navvies. The ordinary builders' labourers were less well placed, but some got employment with the vestries, some went into the gas-works, and there was always a chance of employment wherever there was a heavy job in hand. In the mid-Victorian years 'no end of work' was provided (even in winter) by town improvement schemes, by the building of Board Schools and Model Dwellings, churches and chapels, town halls and commercial offices, by extension work on the railway terminals and the docks; by road-widening work, tramways and the laying out of drainage works and sewers.[195] In

London such great undertakings as the building of the Metropolitan Railway (where the writer of the *Reminiscences of a Stonemason* was taken on for tunnelling work in January 1866),[196] the Thames Embankment, and the Law Courts, were a continuing source of winter employment.[197] Cubitt's, one of the largest London builders, seem actually to have put on extra men in winter: 'when the summer comes they go brickmaking', a branch manager told an enquiry in the 1890s, 'the Brick men are very good workers, and we always give them a job. They come and go.'[198] It was the same in some of the big provincial towns, to judge from a trade report which appeared in January 1882.[199]

> LIVERPOOL ... the works in hand are of almost unexampled importance and magnitude. The City Corporation is proceeding actively with its immense operations for supplementing the present water supply with water from the Vyrnwy reservoir. Each of the railway companies having a terminus at Liverpool is making extensions. The Mersey Docks and Harbour Board is engaged in improving at various points its vast system of docks. The scheme for tunnelling the Mersey is being pushed forward actively. A new university is being erected. New commercial and trading edifices are springing up in all directions.
>
> MANCHESTER. In this district the building trade is fairly active so far as heavy work for public and business purposes is concerned but in house-building, either of the cottage, or in the better class of dwellings, there is comparatively little doing ... The Manchester Corporation have several important works in hand. A new free library and reading-room with the basement occupied by shops ... recently erected on the old Knott Mill Fair ground, are on the point of completion ... New baths are ... being erected by the Corporation for the Rochdale Road district ... the contractor ... Mr. James Hind ... has also in hand for the Corporation the erection of women's swimming baths, as an addition to the present Leaf Street baths ... Amongst other important work at present in hand ... is a new General Post Office, a new railway station for the London and North Western Railway Company, and new business premises for the proprietors of the *Manchester Guardian* ...

Perhaps sailors might be classed among the winter refugees. There were many more of them on shore in winter than in summer, weather-bound or without a berth. Some found shore-going occupations for the season—for instance as dock labourers or as shipwrights—but most of them swelled the ranks of the unemployed: at the Destitute Sailors' Asylum, Well Street, London, established in 1827 to supply shelter, food and clothing to distressed seamen, 'and to keep them until they can obtain employment', winter admissions were more than double those of the summer months.[200] The months of December, January and February 'usually' found the shipping trade at its lowest point in the West Coast ports—at Fleetwood in Lancashire, for example,[201] at Liverpool (where the emigrant traffic seasonally collapsed), and at Milford Haven, where a great number of vessels, 'large and small', and manned by men of many nations, sheltered for the winter.[202] Many thousands of sailors were shored up in the East Coast ports, when the Sound froze over and the Baltic trade came to a stop. The period of this winter standstill varied from year to year, depending on the state of the weather: in 1895 it was causing unemployment among the seamen of North Shields and Grimsby as late as May;[203] in mild winters it could be quite brief. Bagshawe has left a vivid description of the winter scene at Whitby when the Baltic traders laid up:[204]

> The old quays were thronged with . . . lads . . . home from long cruises . . . Grave old skippers stood in knots at the Bridge-end and fought their battles with gales and bad holding-ground over again, and discussed the chances of good freights in the coming spring, when the ice should loosen its hold on the northern waters, and each of them would strive to be the first of the year to break into the silent fiords and gulfs.

For the man who was looking for work, October was a good time for coming up to town. Vestries were beginning to recruit sweepers and street orderlies, in anticipation of the late autumnal muds; extra men were taken on at the public parks for end-of-season gravelling and repairs.[205] The gas-works were making up their winter labour force; the breweries were beginning their 'regular busy season' (in London the biggest brewings took place in October, immediately following the arrival of the hops).[206] In the building industry there was a late burst of employment—'an early covering-in process seems to be the one thing aimed at', wrote the *Labour News* in October 1876.[207] Some of the workshop trades enjoyed a 'second season' in mid-autumn (hatters and brushmakers, for example);[208] in others it was the peak period of the year, notably amongst the journeymen coopers,[209] the bookbinders, and in the printing trades, where the production of Christmas numbers, almanacks, and the 'great variety of . . . literary productions that usually crop up about this season of the year', kept 'grass hands' in full work.[210]

A certain amount of extra employment became seasonally available on the waterfront. In the Liverpool docks some 2,500 extra porters were taken on when the cotton season began: 'A certain number of these are men who systematically follow another trade in one season or go to sea and come back for the busy period between October and March.'[211] In the London docks there were more men employed in December than at any other time of the year. Waterfront industries too were seasonally brisk, notably the oil mills at Hull, 'the largest seed-crushing centre in the United Kingdom',[212] and cotton-picking at Liverpool. There was a steadily increasing volume of work at the riggers, the sail-makers, and the shipwrights as the weather grew more severe. Winter was the height of the repairing season, with many ships laid up for the purpose of dry docking, or put on the slip to have their bottoms caulked, coal-tarred, and blackleaded. At Whitby, when the Baltic traders laid up, the town presented a 'stirring scene', with the ships moored in tiers across the upper harbour, and many men at work. 'The caulking mallets rang merrily in half-a-dozen shipyards; rope-walks and sail-lofts worked overtime, and the air was redolent of pitch-kettles and new timber.'[213] Deep-sea ships continued to be treated in this way in the days of steam: the *Great Eastern* was put on the gridiron after its first transatlantic crossing, and laid up at Milford Haven for the winter.[214] The winter harvest of wrecks brought more work to the repairing yards as well as providing salvage men (such as the Yarmouth beachmen) with a full-time occupation. 'Seamen seem more abundant than berths, but not more so than is usual at this season', wrote the West of England correspondent of the *Labour News* in January 1873, 'the many shipping casualties having found temporary employment for many of them, and abundant work for the shipwrights and sail makers.'[215]

Industrial employment was less open to the wayfarer, but a limited amount of it became seasonally available. For example, oil mills ('very warm work') recruited their winter labour force from those who followed the summer trades.[216] So did some of the coal-mines. Both steel-smelting, which was slackest in June and July, and iron-puddling, which was at full stretch in November, were to a certain extent winter trades,[217] with extra jobs at times for the rough class of general labourer. Gas-works were by far the most frequent employers of this class of labour. They began putting on extra men 'about the latter end of August'[218] and took in many more in mid-winter, when the 'dark . . . days of fog and cold' drew in. The work (stoking and firing) was intensely laborious (there was said to be no other trade in England where a man lost weight and size more quickly),[219] and the hours were incredibly long (a seventy-eight hour week was quite normal in 1882).[220] But it was very well paid. Winter 'so much dreaded by others', was hailed by gas-stokers as an old friend (wrote an observer) 'the harbinger of . . . plenty'.[221] Winter hands at the gas-works were largely recruited from migrant labourers who spent the summer out of town, brickmakers especially, but also a 'good proportion' of builders' labourers, navvies, and 'many . . . who . . . go into the country for farming work in the summer'.[222] Will Thorne, who went navvying and brick-loading in the spring and summer, has left a very good account of the way men tramped up to town for the work:[223]

> I had always wanted to go to London, and my desire . . . was stimulated by letters from an old workmate . . . who was now working at the Old Kent Road Gas Works . . . I finally decided to go . . . in November, 1881. With two friends I started out to walk the journey, filled with the hope that we would be able to obtain employment, when we got there, with the kind assistance of my friend . . . We had little money when we started, not enough to pay for our food and lodgings each night until we arrived in London. Some days we walked as much as twenty miles, and other days less. Our money was gone at the end of the third day . . . For two nights we slept out—once under a haystack, and once in an old farm shed . . .
> On arrival in London we tried to find . . . my friend . . . but . . . were unsuccessful. Our money was all gone, so there was nothing for us to do but to walk around until late at night, and then try to find some place to sleep.

We found an old building and slept in it that night. The next day, Sunday, late in the afternoon, we got to the Old Kent Road Gas Works, and applied for work. To my great surprise, the man we had been looking for was working at the time. He spoke to the foreman and I was given a job.

Quite apart from the regular winter trades, such as gas-stoking and cotton-portering, there was a multitude of chance occupations and residuary employments which the migrant classes were well-placed to take up. Street-trading was a major resource to which many of them turned during their winter stay in town. Gypsies hawked their clothes-pegs and basketry about the suburbs, canvassing from door to door; and they turned up in force on Fridays for market day at the Caledonian Road. Travelling Italians took up position, with tin cans and braziers, outside the pubs, selling roast potatoes or hot chestnuts; organ-grinders perambulated the streets. The migrant Irish often turned trader for the season when winter drove them back to town. The street trade in cutlery, which was particularly brisk in winter, seems to have been largely in the hands of those who went on summer rounds, like 'Showman George', the man who makes a brief appearance in *The Life and Adventures of a Cheap Jack*, 'a big, stout, free-spoken, and rather jolly fellow, who kept a large drinking-booth at the fairs and races during the summer months, and in the winter hawked butchers' cutlery.'[224] So was the winter sale of nuts and oranges, which in London was the special province of the Irish. 'When we got to London', one of Mayhew's informants told him,'... we got to work at peas-picking, my wife and me, in the gardens about. That is for the summer. In the winter we sold oranges in the street, while she lived, and we had nothing from the parishes.'[225] The development of Christmas as a great spending holiday increased the possibility of impromptu sales, and produced its own fugitive callings, such as the kerbstone trade in novelties and toys (especially penny toys), the crying of almanacks and Christmas numbers, and the street-sale of holly and mistletoe. Christmas also helped to loosen the purse-strings of the rich, and made life temporarily easier for the 'griddlers' and 'chaunters'. As children were taught in the nursery:

> Christmas is coming
> The goose is getting fat
> Please put a penny in the old man's hat.

London in the weeks before Christmas was a paradise of odd jobs. Extra hands were taken on at the Post Office, to cope with Christmas deliveries; at Covent Garden, where there was a 'second season' in fresh fruit and flowers; at the railway stations; and at the docks. The Christmas pantomime season gave temporary employment to a whole army of 'extras'—scene-shifters, stage-hands, ballet-girls and 'supers'.[226] Christmas was the height of the advertising season, and men of the lodging-house class were widely employed in delivery work on tradesmen's hand-bills (a winter refugee from Stockton, lodging in St Giles, was earning 2s. 6d. a day for this in November and December 1877);[227] also hand-bill distribution at the street corners, and board-carrying.

In London something like six or seven thousand men were employed as board-carriers at this time, more than twice as many as at other times of the year, 'the extra contingent being provided by those who have spent their summer months in agricultural pursuits.'[228] The work was paid for at rates varying from one shilling to one-and-eightpence, the 'highflyers' (who carried over-head boards) being paid at a somewhat higher rate than the others ('except for theatrical & publisher's work which is always the worse paid').[229] Some of this work was done on a casual basis, but Nagle's, one of the leading London contractors, employed men for as long as a month at a time,[230] and according to Booth's investigators, a man, if known to the contractors, might reckon on employment 'throughout the season'. The chief employment office for the West End boardmen was at Ham Yard, Leicester Square, where an enterprising contractor had established himself next door to the soup-kitchen. In the early morning it served as the sandwich-board equivalent of the dock-gate call. A forest of grimy hands shot up for each of the jobs, and little knots of the chosen came forward from the throng. At night a 'Doré-like' group of figures were to be seen, 'camping .. as near as they can to the office which doles them out their jobs'.[231] On 29 January 1904, when the L.C.C. enumerators of the homeless counted forty-nine of them, they were sleeping out (or making themselves comfortable for the night) by the air vents of the Palace Theatre 'on account of the heat coming from the boilers through the grating'.[232]

Christmas was the winter harvest of the wandering tribes. When it was over things changed for the worse. January and February were bad months for the working man, and especially for the poor and insecure. Almost every trade experienced a lull after the Christmas rush of work, and unemployment became widespread as the weather grew more severe. Even so, the balance of advantage, from the point of view of the homeless, remained overwhelmingly on the side of the town—in fact it was after Christmas that the last of the migrations to town took place. There was warmth and shelter in the town, even if work was impossible to come by, soup to be obtained at public kitchens, open to all comers, Night Refuges and Asylums in place of the workhouse, public works (always started up by the vestries when the season was particularly severe) in place of the humiliations of the parish stoneyard.[233]

Bad weather itself, in the conditions of town life, was a prolific source of occasional opportunities. Wintry weather added urgency to the street beggar's cry. Rainy days were a godsend to the cab touts who loitered about the railway stations and the theatres, and to the crossing-sweeper, who levied a small tribute on the wealthier passer-by. After a big snow thousands of hard-up opportunists took to the streets. Augustus Mayhew noticed how 'The whole town seems to swarm with . . . sweepers, who go about from house to house, knocking at the doors, and offering to clear the pavement before the dwelling, according to the Act of Parliament, for twopence.'[234] Frosts, too, had their collateral advantages for the wide awake. In London the 'ice harvest' on the northern heights brought some hundreds of men foraging, with ice-carts and shallows, to Finchley Common and Hampstead Heath,[235] while at the frost fairs which sprang up in the public parks, the hard-up could earn

1 *above* Sailors' Home in the East End, from *Illustrated London News*, lxiii (1873), 600.

2 *below* The Strangers' Home, Limehouse, from *Illustrated London News*, lvi (1870), 253.

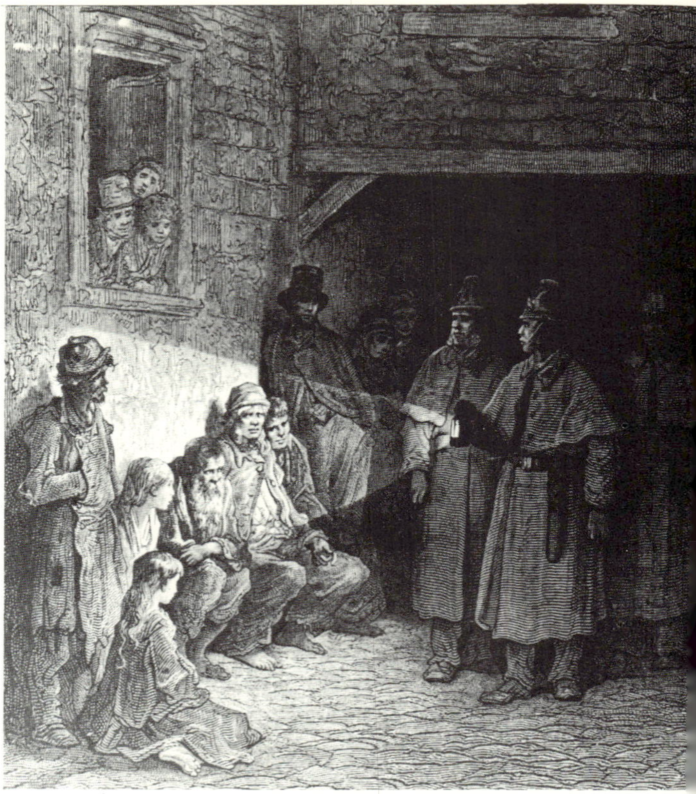

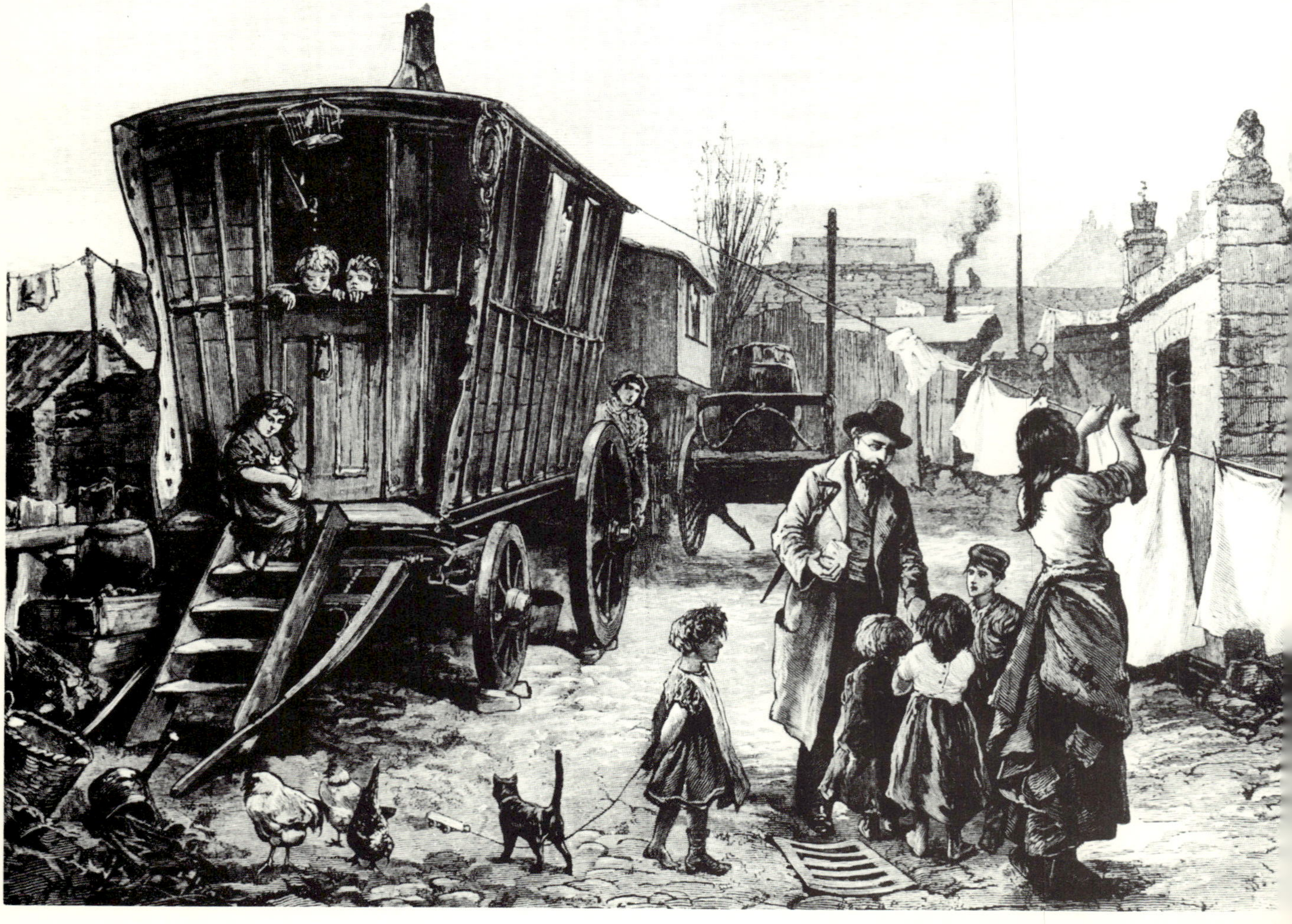

3 *above left* The Bull's-Eye, from Gustave Doré and Blanchard Jerrold, *London. A Pilgrimage* (1872), facing p. 144.

4 *left* Gypsy encampment in Notting Dale, from *Illustrated London News*, lxxv (1879), 504.

5 *above* A gypsy bivouac on Epsom Downs before the Derby. Not less than 1,500 'trampers, gypsies, and one sort or another' camped here overnight but many more arrived by morning. From *Illustrated London News*, xxxii (1858), 513.

'OUT OF THE PARISH'

Sir Giles Overreach. 'NOW, THEN, MY MAN! YOUR WORK'S DONE, SO BE OFF OUT OF THIS PARISH.' Agricultural Labourer. 'AH! SIR GILES! IT BE BETTER NOR FOUR MILE TO T'TOWN.' Sir Giles Overreach. 'CAN'T HELP THAT! NO "UNION CHARGEABILITY" FOR ME.'

6 *above* From *Punch*, xlviii (1865), 213.

7 *above right* 'All the way from Manchester, and got no work to do-o-o.' London out-door music, from *Illustrated London News*, xxxiv (1859), 13.

8 *right* Sale of spring herrings at Yarmouth, from *Illustrated London News*, xxviii (1856), 373.

'ALL THE WAY FROM MANCHESTER, AND GOT NO WORK TO DO–O–O.'

9 *above left* New ward for casual poor at Marylebone workhouse, from *Illustrated London News*, li (1867), 353.

10 *below left* Labour-yard of the Bethnal Green Employment Association, from *Illustrated London News*, lii (1868), 156.

11 *above right* Frost fair in St James's Park, from *Illustrated London News*, xxxviii (1861), 63.

12 *below right* A thaw in the streets of London, from *Illustrated London News*, xlvi (1865), 184.

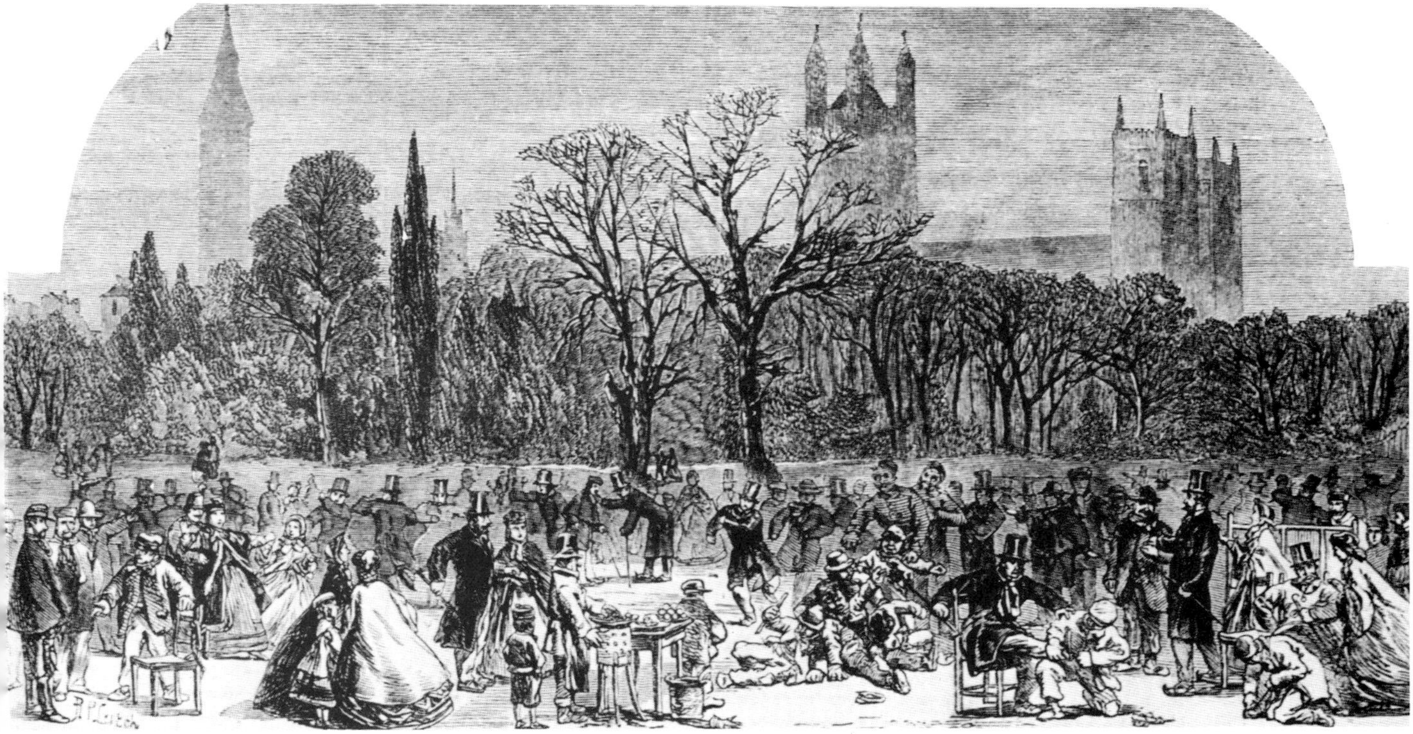

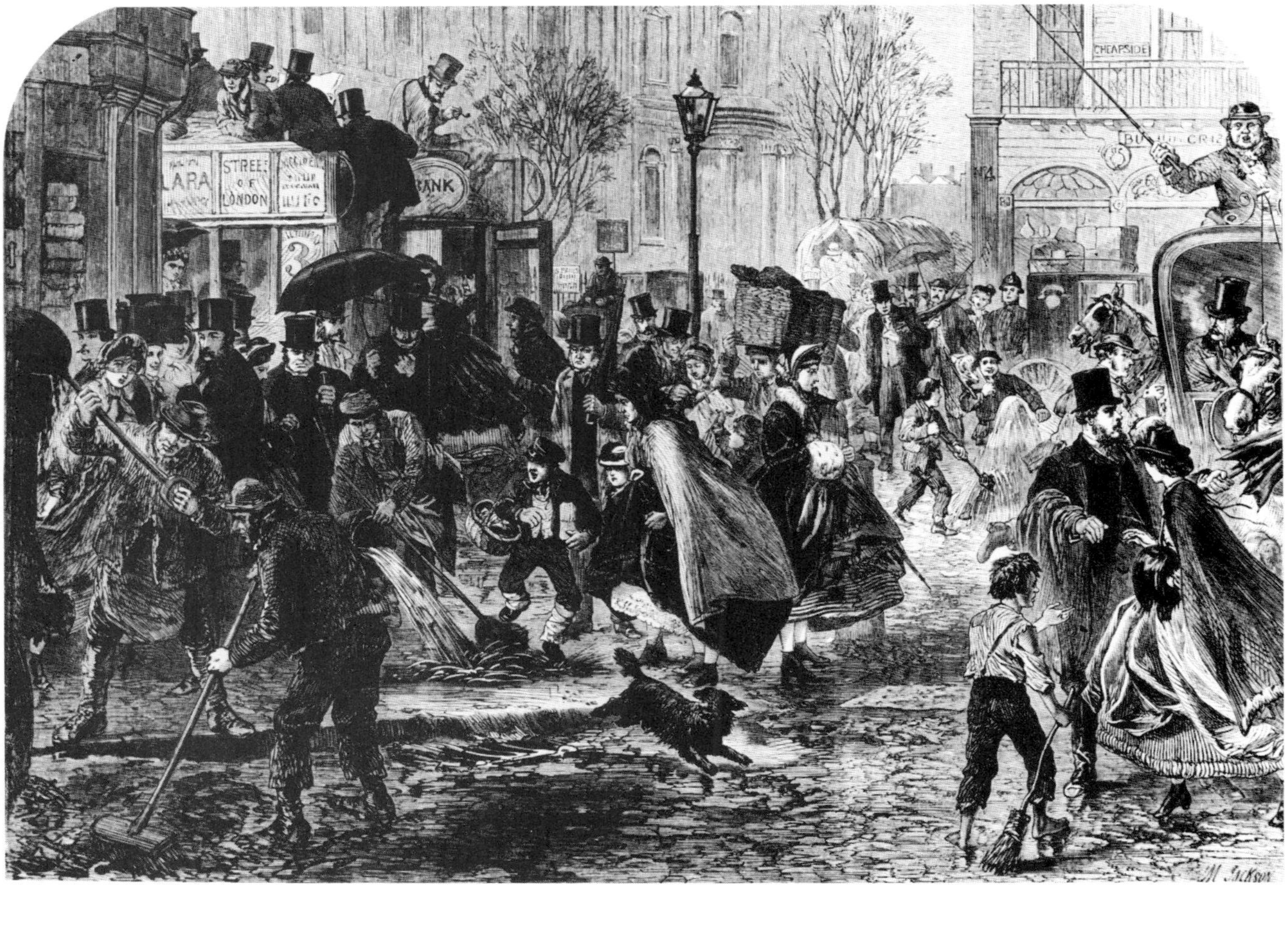
LARA
LONDON
BANK
CHEAPSIDE

13 *above right* London gypsies, from [John Thomson and Adolphe Smith], *Street Life in London* (1877), facing p. 1. *Guildhall Library*

14 *right* Moved on by the police: St James's Park, December 1873, from *Illustrated London News*, lxiii (1873), 601.

15 *above far right* On the tramp, from J. H. Crawford, *The Autobiography of a Tramp* (1900), facing p. 104.

16 *below far right* Van-dwellers at the Agricultural Hall, London, from George R. Sims, ed., *Living London* (1901), III, p. 321.

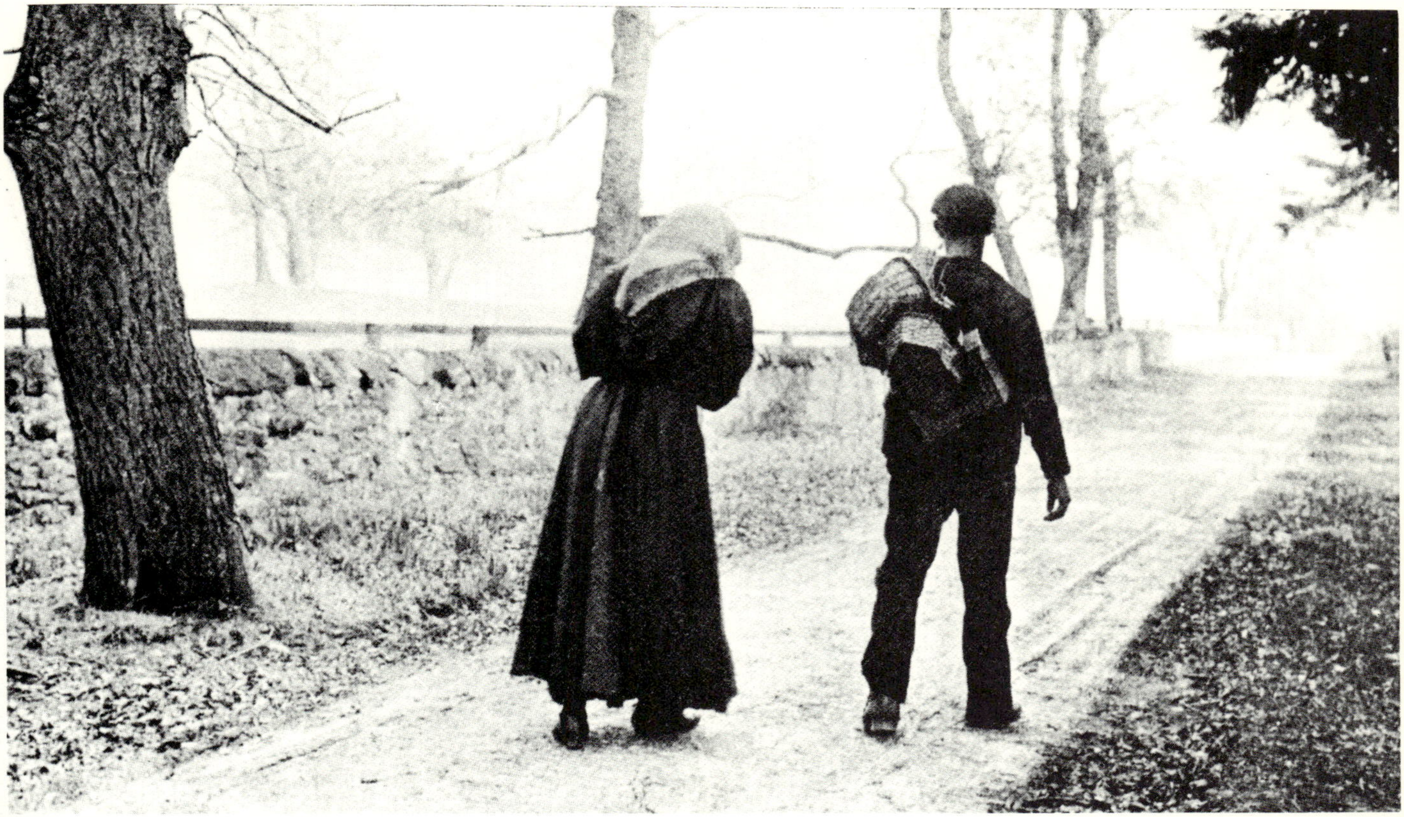

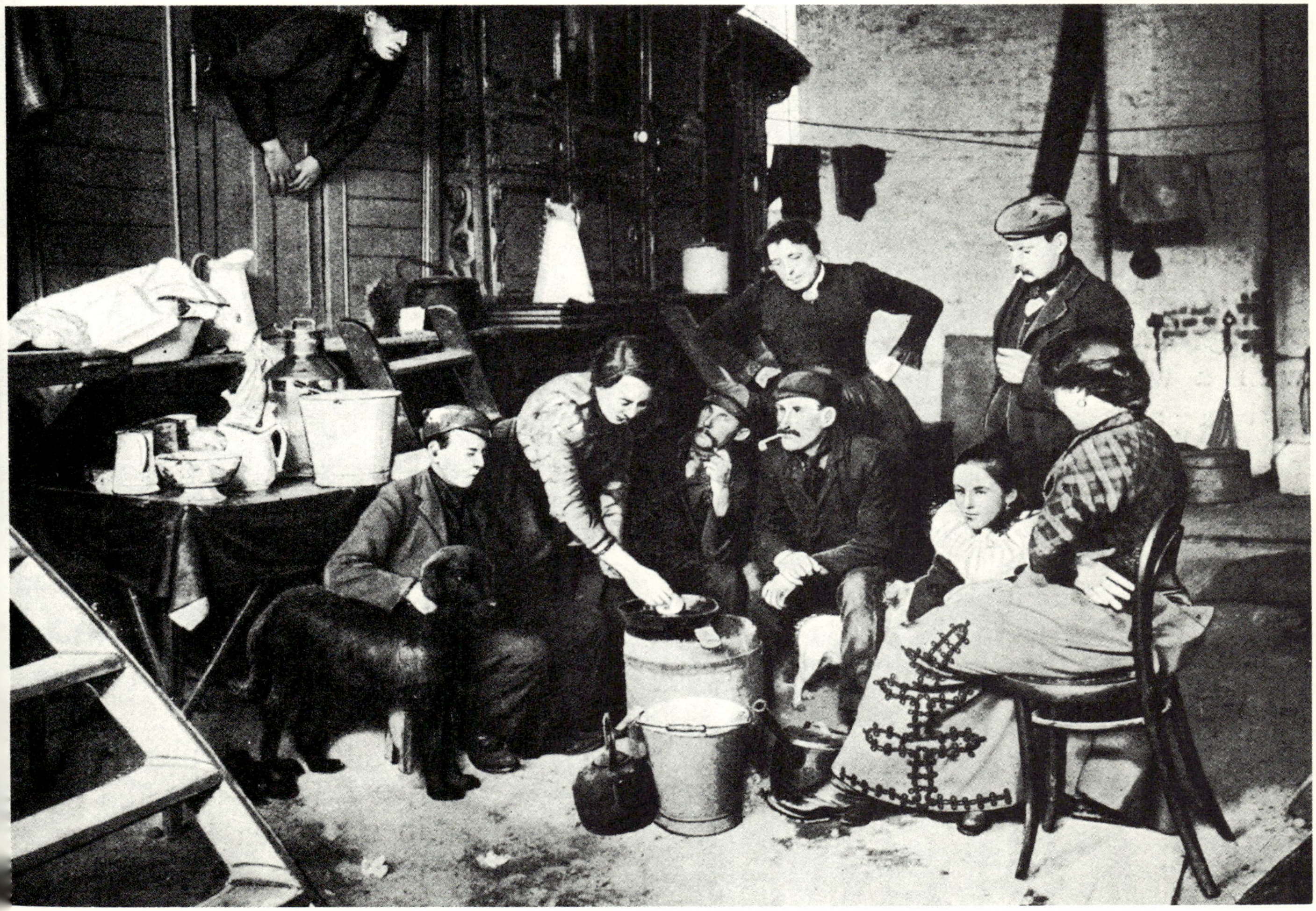

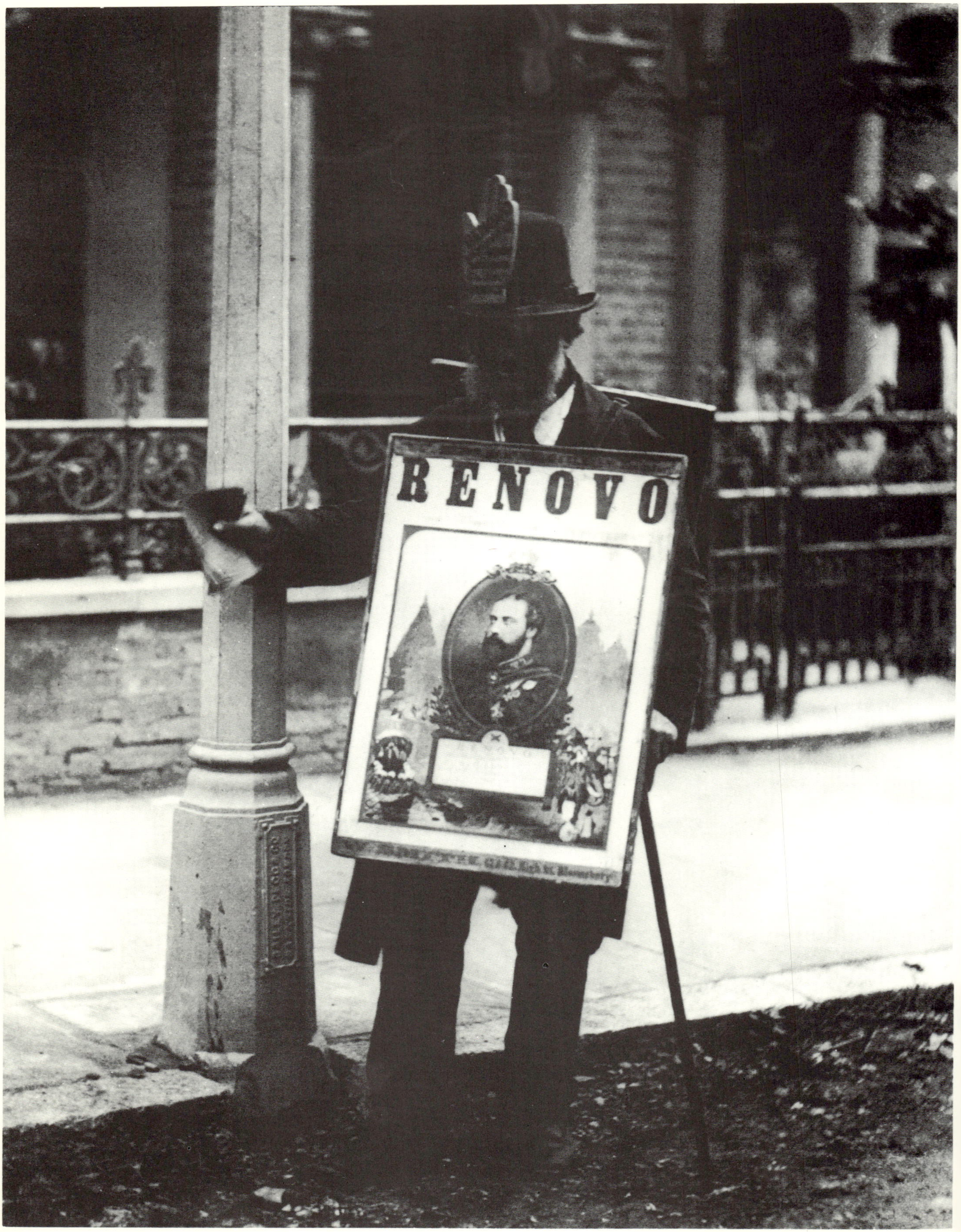
RENOVO

17 *left* A sandwich-board man, from [John Thomson and Adolphe Smith], *Street Incidents* (1881), p. 21.
Guildhall Library

18 *above right* Italian street musicians in London, from [John Thomson and Adolphe Smith], *Street Life in London* (1877), opposite p. 85.
Guildhall Library

19 *right* A mobile circus in a London street, from a lantern slide, *c*.1890.
David Francis

20 *below* Suffolk maltsters at Burton-on-Trent, *c.*1906, from George Ewart Evans, *Where Beards Wag All* (1970), facing p.145.

21 *right* Navvies at the Crystal Palace, Sydenham, 1854.
Mansell Collection

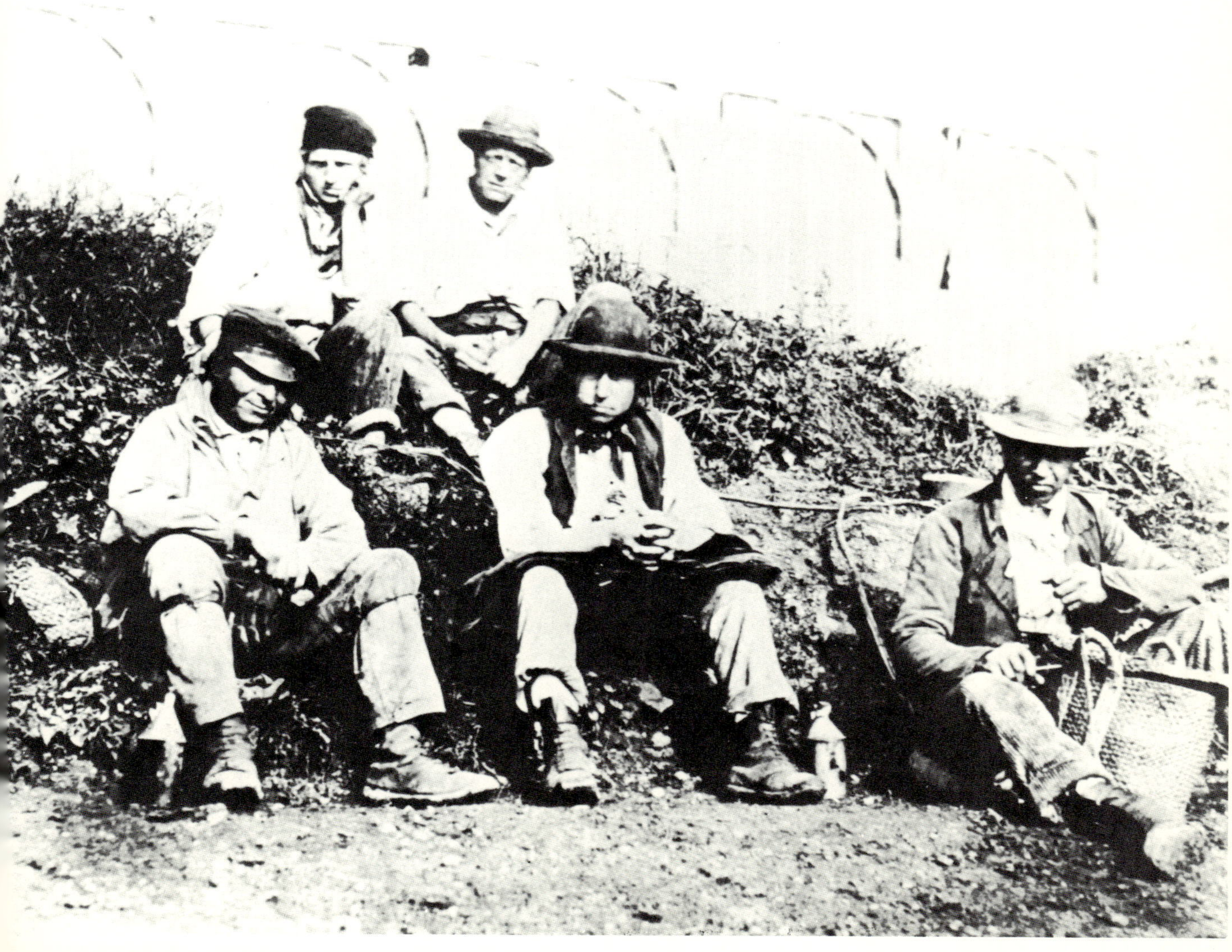

22 Hop-pickers, September 1875.
Mansell Collection

23 Fishermen in Whitby harbour, from a lantern slide.
Photo: Frank Sutcliffe
David Francis

24 A Dundee whaler putting to sea about 1900, from Basil Lubbock, *The Arctic Whalers* (1937), facing p. 406.

odd pennies by sweeping the ice for the skaters, putting on and hiring out skates, or by trading in comforters and sweets.[236] 'Lord' George Sanger, wintering in London one year during the 1840s, found this a profitable line:[237]

> A terrible winter it was, with an unusually hard, long spell of hard frost. Our funds in hand were not very heavy, and seeing all our cash going out and none coming in made me very unhappy. At last, however, I struck a new line with considerable success. Wandering on to Bow Common and Hackney Marshes I found numbers of people sliding and skating on the large ponds there. They were trying to keep warm in the bitter weather, and I noticed that, despite the crowds gathered there, nothing was being sold or hawked. That gave me an idea.
>
> I knew how to make rock and toffee, such as was sold at the fairs, for I had assisted in the process many times. Here was my chance. I went and bought about ten pounds of coarse moist sugar, at that time sevenpence a pound, and some oil of peppermint, borrowed some pans to boil it in, and very soon had a nice little stock of strong, good-looking peppermint rock. Then I took it to Hackney Marshes near the biggest piece of ice, and at a penny a lump it sold like wildfire. I was cleaned out in an hour, and had made several shillings profit.
>
> I could see I had hit on a good thing, and at once went to work on a bigger scale. I borrowed what little money my brothers William and John had saved, added my stock to it, and then went and purchased a big parcel of sugar from a grocer in the Whitechapel Road and more oil of peppermint. This I boiled into rock, which was cut into penny lumps, and having pressed my brothers William and John into the service we started out. The rock sale proved as brisk as ever, and we came home with our pockets loaded with coppers and silver, having made over two pounds profit.
>
> The problem of how to live through the winter in London without trenching on the savings from the summer show business, savings that were always needed to give a good start to the caravans when the time came for the road again, was solved.

With the return of spring the wandering tribes began to stir. Sailors, no longer weather-bound, signed on at the Registry Offices. Showmen put their caravans in harness, and set off for the early fairs. Cheap-jacks and packmen resumed their country rounds. By March the emigrant traffic, almost at a standstill in the three winter months, was moving to the first of its seasonal peaks. In the workshop trades the 'regular roadster'—the congenital nomad—showed signs of restlessness as soon as the sun began to rise higher in the sky. Such a man was Dominic Macarthy, the travelling compositor affectionately recalled in W. E. Adams' *Memoirs of a Social Atom*.[238] Every winter he came to London and supported himself as best he could by getting an occasional job as a 'grass hand'. In the spring his wandering life was resumed. He was a good workman, Adams tells us, but incurably nomadic, 'whenever the proper season came round'. George Acorn, recalling his childhood in Bethnal Green, describes a similar type, a tramp shoemaker called 'Old Bill', who turned up in the neighbourhood every winter, 'bronzed and tattered', and left again in the spring. He was employed at 'The Little Wonder', a back-street cobbler's shop, and while winter lasted, and the nights were long, he would sit contentedly at his bench, heel-balling or sewing 'as patiently as anybody':[239]

> But as soon as the sap began to rise, and the buds to burst in the trees, he would get fidgety, would rise from his stool, and, going to the door, would look at the sky, with his hand shading his eyes.
> 'Weather breaking, eh?' Jordan commented.
> 'Yes', the old cobbler would reply, as if a new spirit had entered into him.
> 'Want to be off?' His employer took a delight in putting these leading questions to him.
> 'Not just yet', 'Old Bill' replied, 'but very soon, very soon.'
> As the days lengthened his eyes fairly glowed with anticipation, his restiveness increased.
> One evening I called in at Jordan's to find a vacant chair.
> 'Where's Old Bill?' I inquired.
> 'God knows', was the reply. 'Somewhere in the country by now, getting fresh air, and seeing things.'
> 'Does he go away every year?' I asked.
> 'He has, ever since I've known him, George. He's got the wandering spirit, and when he sees the green leaves a-coming on the trees he has to go out and taste the country air; it would kill him to stop here all the year round . . .'

The wandering tribes found their place in the underlife of the nineteenth-century town, and it is not easy to log their comings and goings with precision. Their circuits were innumerable, their settlements obscure, and their interconnections with more settled lives can often only be conjectured. Numbers are difficult, perhaps impossible, to arrive at, since they varied with the changing of the seasons and the ups and downs of trade. Nor is it easy to define the boundaries of each individual group. The wayfaring constituency was in a constant state of flux. The tramp, the navvy, and the pedlar might be one and the same person at different stages of life, or even at different seasons of the year; the 'gaff' proprietor might spend his summer on the roads; the free-lance labourer turn to busking, or board-carrying, or gas-stoking, when winter drove him into town. The distinction between the nomadic life and the settled one was by no means hard and fast. Tramping was not the prerogative of the social outcast, as it is today; it was a normal phase in the life of entirely respectable classes of working men; it was a frequent resort of the out-of-works; and it was a very principle of existence for those who followed the itinerant callings and trades. Within the wandering tribes themselves the nomadic phase and the settled were often intertwined, with men and women exchanging a fixed occupation for a roving one whenever conditions were favourable.

One thing at least is clear. The wandering tribes (like other nomadic peoples) followed well-established circuits, and journeyed according to a definite plan. There were comparatively few who moved about the country simply for the sake of keeping on the move, or who travelled hither and thither, as the spirit moved them, without a springboard, a haven, or regular ports-of-call. Some kept a foothold in town all the year round; many of them wintered there, and turned up again 'as regularly as noblemen' when the long nights drew in. Their comings and goings were closely bound up with the social economy of the town, and the openness (or otherwise) of its employment and its trades. The wandering tribes were often the object of hostile legislation, whether to bring their lodging-houses under inspection and control, to bar them from using city wastes, or to harass them from pursuing their callings about the city streets. Their children, after 1870, were subject to the eager ministrations of the School Board Visitors; the camping sites of those who lived in moveable dwellings fell one by one to the enterprise of the speculative builder, or the railinged enclosure of the public parks. But it was economic change, in the later Victorian years, which really undermined them—the growth of more regular employment, especially for the unskilled, and the decline of the 'reserve army of labour' in both the country and the towns; the mechanization of harvest work, and the displacement of travelling labourers by regular farm servants;[240] the rise of the fixed holiday resort in place of the perambulating round of wakes and feasts and fairs; the extension of shops to branches of trade which previously had been in the hands of itinerant packmen and dealers. Towards the end of the century the towns began more thoroughly to absorb their extra population, and to wall them in all the year round.

Notes

1 *A Plan for Preventing Destitution and Mendicancy in the British Metropolis* (1850), p. 6.
2 Public Record Office (P.R.O.), Home Office Papers (H.O.) 45/10499/117669/10.
3 *Household Words*, 6 December 1851; Richard Rowe, *Jack Afloat and Ashore* (1875), p. 74.
4 On the Settle–Carlisle railway, which was building in the early 1870s, more than 33,000 men found employment on a single section of the line, although the greatest number of men employed at any one time was never more than 2,000: F. S. Williams, *The Midland Railway* (1876), p. 522.
5 R. T. Berthoff, *British Immigration in Industrial America* (Cambridge, Mass., 1953), pp. 82–3. For a parallel migration of stonecutters see 50 Cong. 1, Misc. Doc. 572, part 11, Dip. and Consular Reps. on Immigration, p. 10; 51 Cong. 2, Rep. No. 3472, Select Committee (S.C.) on Immigration, pp. 301, 305, 352, 870.
6 James Greenwood, *On the Tramp* (1872), p. 26.
7 Charity Organisation Society (C.O.S.), *Report on the Homeless Poor* (1891), p. xx.

8 For an example, see *Sir James Sexton, Agitator, An autobiography* (1936), pp. 21–2.
9 F. Groome, *In Gypsy Tents* (Edinburgh, 1880), p. 286.
10 Charles Hindley, ed., *The Life and Adventures of a Cheap Jack* (1881 edn), passim.
11 Henry Mayhew, *London Labour and the London Poor* (1861), I, p. 339.
12 Charles Booth, *Life and Labour of the People in London* (1902–4), 3rd series, V, p. 157.
13 Lancashire Record Office, RCLv, Visitation Records, 1865.
14 John Denvir, *The Irish in Britain* (2nd edn, 1894), p. 411.
15 'St. Patrick's Church Sunderland', notes in the possession of the Rev. Vincent Smith; for a similar situation at Warrington, 'a large thoroughfare for Irish people', see P.R.O., H.O. 129/466, St Alban's, Warrington.
16 B. S. Rowntree, *Poverty* (2nd edn, n.d.), pp. 31–2; for earlier references to the travels of the York Irish, see *Parliamentary Papers* (*P.P.*), 1867–8, XVII, Royal Commission (R.C.) on Employment of Children, Young Persons, and Women in Agriculture, 1st Report: Appendix Pt II (4068–I), pp. 255, 258.
17 Denvir, op. cit., p. 400.
18 Lady Bell, *At the Works* (1907), p. 8.
19 *The Word on the Waters*, i (December 1858), 258.
20 Ibid. (April 1892), 332.
21 J. Flanagan, *Scenes from My Life* (1907), p. 36.
22 *Morning Chronicle*, 4 January 1850. Sixty years later, Angel Meadow still bore a 'peculiar reputation' on account of its common lodging-house. See *P.P.*, 1909, XLIII, R.C. on the Poor Laws, Reports on the Relation of Industrial and Sanitary Conditions to Pauperism (Cd. 4653), App. XVI, p. 102.
23 *Morning Chronicle*, 29 April 1850; *P.P.*, 1847, XXVII-Pt 1, R.C. on Education in Wales (870), p. 304.
24 Greenwood, op. cit., pp. 27–8.
25 Booth, op. cit., 3rd series, V, p. 75; cf. London Mendicity Society, 76th Annual Report (1894), p. xiii; Mayhew, op. cit., I, p. 337.
26 Booth, op. cit., 3rd series, VI, p. 136.
27 London School of Economics, Booth MSS., B. 267, pp. 145–9; Booth, op. cit., 3rd series, I, p. 138.
28 Booth MSS., B.281, pp. 83–5; B.371, pp. 117–29, 143–61; John W. Horsley, *I Remember* (1911), pp. 125–31; Booth, op. cit., 3rd series, V, pp. 90–1.
29 Booth, op. cit., 3rd series, V, pp. 90–1.
30 Ibid., III, pp. 151–2.
31 Sam Shaw, *Guttersnipe* (1946), p. 29.
32 *P.P.*, 1842, XXVI, Report of the Poor Law Commissioners on an Inquiry into the Sanitary Condition of the Labouring Population of Great Britain: Appendix.
33 T. W. Rammell, *Report to the Board of Health: . . . Banbury* (1854), p. 11. There is an excellent description of Rag Row in Barrie S. Trinder, *Banbury's Poor in 1850* (Banbury, 1966).
34 P.R.O., H.O. 107, Census Returns; Huddersfield Ref. Lib., Lodging House Committee M.B., 1 May 1854 and *passim*.
35 W. C. E. Ranger, *Report to the Board of Health: . . . Doncaster* (1850), p. 38.
36 Sanitary Condition of the Labouring Population, Local Reports, pp. 63–5, 78.

37 Ibid., p. 174; E. Cresy, *Report to the Board of Health ... Derby* (1849), pp. 13–15.
38 J. R. Coulthart, *Report on Ashton-under-Lyne*, in *P.P.*, 1844, XVII, R.C. on the State of Large Towns and Populous Districts: 1st Report (572), Appendix, p. 84.
39 C.O.S., op. cit., p. xvi.
40 Coulthart, op. cit., pp. 36–7.
41 P.R.O., Metropolitan Police (Mepol.), 2/1490.
42 *P.P.*, 1906, CIII, Departmental Committee on Vagrancy (Cd. 2891), II, QQ. 7768, 7947; ibid. (Cd. 2892), Report on Vagrancy, III, App. XXXII; Mary Higgs, *Glimpses into the Abyss* (1906), p. 51; S. and B. Webb, *The Public Organisation of the Labour Market* (1909), pp. 81, 83.
43 Bishopsgate Library, Mansion House Committee on the Unemployed in London (1885), transcript of proceedings, fol. 10[r].
44 John Fisher Murray, *The World of London* (1844), I, p. 247.
45 Daniel Joseph Kirwan, *Palace and Hovel* (1963 edn), pp. 64–70.
46 Thomas Miller, *Picturesque Sketches of London* (1852), p. 207.
47 *London in the Sixties, by One of the Old Brigade* (1914 edn), pp. 61–2.
48 P.R.O., H.O. 45/10499/117669; Booth, op. cit., 1st series, I, p. 68; 3rd series, II, p. 244; *The Times*, 28 August 1894.
49 *Liverpool Review*, 19 April 1890.
50 Booth MSS., B. 152, p. 104.
51 P.R.O., Mepol. 2/645, Mepol. 2/1068, Mepol. 2/1425, Mepol. 2/1490; H.O. 14571/20236; T. W. Wilkinson, 'London's Homes for the Homeless', in G. R. Sims, ed., *Living London* (1903), I, p. 337.
52 General Booth, *In Darkest England and The Way Out* (1891), p. 25.
53 P.R.O., Mepol. 2/1068.
54 Terry Coleman, *The Railway Navvies* (1965), p. 49; John R. Kellett, *The Impact of Railways on Victorian Cities* (1969), p. 346.
55 *Surrey Gazette*, 10 November 1863.
56 P.R.O., H.O. 45/10499/117669; Mepol. 2/1490.
57 Stan Hugill, *Sailortown* (1967), p. 112; Rowland Kenney, *Westering* (1938), p. 82.
58 *Porcupine*, xxi (1879), 409; George Smith, *Incidents in a Gipsy's Life* (Liverpool, 1886), p. 11.
59 *Labour Press and Miners' and Workmen's Examiner*, 15 August 1874.
60 Booth MSS., B. 371, p. 55.
61 Ibid., p. 239.
62 *Illustrated London News*, lxxv (1879), 503.
63 Reginald Blunt, *Red Anchor Pieces* (1928), p. 110.
64 *Illustrated London News*, lxxvi (1880), 11; Booth MSS., B. 346, pp. 165, 231; George Smith, *Our Gipsies and their Children* (1880), p. 267.
65 Booth MSS., B. 348, p. 97.
66 T. W. Wilkinson, 'Van Dwelling London', *Living London*, III, pp. 321–2.
67 Henry Woodcock, *The Gipsies* (1865), pp. 144–7; V. Morwood, *Our Gipsies in City, Tent, Van* (1885), pp. 338–40; George Sims, *Off the Track in London* (1911), pp. 36–7.
68 Booth, op. cit., 3rd series, V, p. 206.

69 J. J. Sexby, *Municipal Parks* (1898), pp. 237–8; British Museum, Leland Collection, newspaper cutting, 2 January 1879; Booth MSS., B. 298, pp. 91–7.
70 Booth, op. cit., 3rd series, V, p. 157; cf. Booth MSS., B. 366, pp. 183–5, 190; *Building Trade News*, December 1894.
71 C.O.S., op. cit., QQ. 1892–4.
72 'The Greenland Whale Fishery', in R. Vaughan Williams and A. L. Lloyd, eds, *The Penguin Book of English Folk Songs* (1968), pp. 50–1; in 'The Whale-catchers', ibid., p. 100, the date is 23 March. For the spring departure of the whalers at Dundee, see G. N. Barnes, *From Workshop to War Cabinet* (1924), pp. 14–15; for Hull, see *Autobiography of Thomas Wilkinson Wallis* (Louth, 1899), pp. 14–15.
73 C.O.S., op. cit., pp. xv–xvi.
74 Mayhew, op. cit., III, p. 88.
75 P. A. Whittle, *Blackburn as It Is* (Preston, 1852), pp. 31–2; Hindley, op. cit., pp. 280–1.
76 James Lloyd, *My Circus Life* (1925).
77 Ben Tillett, *Memories and Reflections* (1931), pp. 33, 37, 38.
78 David Prince Miller, *The Life of a Showman* (1849), pp. 65, 67, 83–4, 86; *Free Lance*, viii (1873), 125; Hindley, op. cit., p. 149.
79 [C. Thomson], *The Autobiography of an Artisan* (1847), pp. 240–1; Mayhew, op. cit., I, p. 329.
80 'Lord' George Sanger, *Seventy Years a Showman* (1952 edn), pp. 42–4, 66.
81 C.O.S., *Report on Unskilled Labour* (1890), Q. 885.
82 *P.P.*, 1893–4, XXXIV, R.C. on Labour: Minutes of Evidence (Group 'C') (C. 6894–IX), Q.26,435.
83 C.O.S., *Unskilled Labour*, Q.282; cf. Frank Popplewell, 'The Gas Industry', in A. Freeman and S. Webb, eds, *The Seasonal Trades* (1912), pp. 168–71; Booth MSS., A.3 fol. 209; Will Thorne, *My Life's Battles* (1925), p. 36.
84 Mayhew, op. cit., II, p. 375.
85 G. Elson, *The Last of the Climbing Boys* (1900), p. 199.
86 George Ewart Evans, *Where Beards Wag All* (1970), p. 266.
87 Mayhew, op. cit., III, p. 399.
88 B.M., Leland Coll., cutting, 16 May 1872; J. Crabb, *The Gipsies' Advocate* (1831), pp. 136–7.
89 Sanger, op. cit., pp. 141, 147, 195.
90 George Smith, *I've Been a-Gipsying* (1883), pp. 39–59.
91 *P.P.*, 1868–9, XIII, R.C. on Children in Agriculture, 2nd Report (4202–I), App. II, A.i, p. 61.
92 *P.P.*, 1893–4, XXXV, R.C. on Labour: The Agricultural Labourer, Assistant Commissioners' District Reports (C. 6894–VI), B.–VI [Holbeach], para. 12; ibid. (C. 6894–V), B.–VII [Godstone], para. 10.
93 *Seaman's and Fisherman's Friendly Visitor*, iii (1860), pp. 111, 126.
94 *Labour News*, 13 July 1872.
95 *Jackson's Oxford Journal*, 3 August 1872. (I am grateful to David Morgan for this excellent reference.)
96 *P.P.*, 1880, XIV, Factory Inspectors' Report for 1879 (C. 2489), p. 54.
97 Booth, op. cit., 2nd series, II, p. 140.

98 Booth MSS., B. 93, fol. 88.
99 *P.P.*, 1893–4, XVII, Departmental Committee on Conditions of Labour in Lead Industries (C. 7239–I), p. 181. (I am grateful to Anna Davin for this excellent reference, as also for the refugee London enameller.)
100 Ibid., Q.5216.
101 R.C. on Labour: Agricultural Labourer, Summary Report (C. 6894–VI), para. 20.
102 William Cobbett, *Rural Rides* (Everyman edn) I, p. 84.
103 Thor Fredur, *Sketches from Shady Places* (1879), p. 24.
104 William A. Jevons, 'The Weavers' Strike at Padiham in 1859', National Association for the Promotion of Social Science, *Trade Societies and Strikes* (1860), pp. 468–9.
105 B. S. Rowntree and B. Lasker, *Unemployment* (1911), p. 63; cf. pp. 30, 99, 152, 160.
106 'The Irish in England, III', *Labour News*, 14 March 1874.
107 'Whitechapel Villagers' in *Toilers in London by One of the Crowd* (1883), pp. 99–101; for some Essex evidence, see G. A. Cuttle, *The Legacy of the Rural Guardians* (Cambridge, 1934), pp. 267, 273; for Irish pea-pickers at Stoke Poges, *P.P.* 1867–8, XVII, R.C. on Children in Agriculture, 1st Report: App. Pt II, p. 539; for a Bloomsbury pea-picker, see P.R.O., N.H. 13/268, paper dated 2.4.1851.
108 Thomas Fayers, *Labour among the Navvies* (Kendal, 1862), p. 12.
109 Mayhew, op. cit., I, p. 478.
110 Thomas Wright, *The Great Unwashed* (1868), pp. 261–2.
111 Booth MSS., B.210, pp. 1–3.
112 Booth MSS., A.28.
113 Booth, op. cit., 2nd series, II, pp. 277–8.
114 Ibid., 2nd series, IV, App. B.
115 Booth MSS., B.152, p. 104.
116 Booth, op. cit., 2nd series, III, p. 411.
117 *Labour News*, 20 September 1873.
118 A. M. Anderson, *Women in the Factory* (1922), p. 38. (I am grateful to Anna Davin for this reference.)
119 *Labour News*, 20 September 1873.
120 Booth MSS., B.93, fol. 69.
121 Booth, op. cit., 1st series, IV, pp. 286, 313–14, 324.
122 Ibid., 2nd series, I, p. 220.
123 A. T. Pask, *Eyes of the Thames* (1889), pp. 148–9.
124 Mayhew, op. cit., II, p. 299.
125 'The Irish in England, II', *Labour News*, 28 February 1874.
126 Denvir, op. cit., p. 400; *P.P.*, 1906, CIII, Report on Vagrancy (2891) II, QQ. 5820–2; J. Ewing Ritchie, *Crying for the Light* (1895), p. 123.
127 *Indoor Pauper, by One of Them* (1885), pp. 54–5.
128 'The Irish in England', *Dublin Review* (1856), 508.
129 Denvir, op. cit., p. 401.
130 C.O.S., *Homeless Poor*, Q. 398; P.R.O., Mepol. 2/1490, 1 August 1913.
131 Report on Vagrancy, II, Q. 4180.
132 R.C. on Children in Agriculture, 2nd Report, App. Pt. II, p. 139.

133 *Runcorn Examiner*, 30 September 1893.
134 R.C. on Children in Agriculture, 2nd Report, App. Pt II, Ai, p. 135.
135 R.C. on Labour: Agricultural Labourer, District Reports (C. 6894–IV), B.–V [Bromyard], para. 9; Cf. Stourbridge Observer, 10 September 1881.
136 R.C. on Children in Agriculture, 2nd Report, App. Pt II, G, p. 43.
137 *Labour News*, September 1876.
138 'Three Weeks with the Hop-Pickers', *Fraser's Magazine*, n.s., xvi (1877), 635.
139 George Sturt, *A Small Boy in the Sixties* (Cambridge, 1923), p. 76; R.C. on Children in Agriculture, 2nd Report, App. Pt II, C, p. 4.
140 William Marshall, *The Rural Economy of the Southern Counties* (1798), II, p. 69.
141 Ellen Chase, *Tenant Friends in Deptford* (1929), pp. 102–3.
142 George Meek, *Bath Chair Man* (1910), p. 55.
143 'Shelter for the Homeless', *Leisure Hour* (1865), p. 11.
144 Mayhew, op. cit., III, pp. 405, 406.
145 Alexander Somerville, *The Autobiography of a Working Man* (1848), p. 56.
146 *P.P.*, 1895, VIII, S.C. on Distress from Want of Employment, 2nd Report (253): App., p. 33.
147 'Trade Tramps', *Leisure Hour* (1868), 358.
148 *London City Mission Magazine*, 2 December 1861.
149 George Sturt, *The Wheelwright's Shop* (Cambridge, 1963 edn), p. 29.
150 Charles Manby Smith, *The Little World of London* (1857), p. 143.
151 Mayhew, op. cit., I, p. 329; Hindley, op. cit., p. 209.
152 Booth MSS., B.50, p. 58; B.82, p. 14.
153 B.M., Leland Collection, undated cutting [1879?].
154 Mayhew, op. cit., II, p. 335.
155 C.O.S., *Homeless Poor*, Q.1,498; Sir Henry Smith, *From Constable to Commissioner* (1910), p. 165.
156 L.N.R. [Mrs Ellen Ranyard], *The Missing Link* (1859), pp. 29–30, 51; *P.P.*, 1833. XVI, S.C. on Irish Vagrants (394), pp. 175, 193; Mayhew, op. cit., IV, p. 297.
157 Holborn Board of Works M.B., Report of Septimus Gibbon, 6 October 1856 (Holborn Reference Library).
158 *Labour News*, 10 October 1874.
159 Edward G. Howarth and Mona Wilson, *West Ham* (1907), pp. 306–7.
160 R.C. on Children in Agriculture, 2nd Report, Appendix.
161 Verbal communication from Charles Connor.
162 *Morning Chronicle*, 23 January 1851.
163 T. W. Wilkinson, 'Van Dwelling London', in Sims, op. cit., III, pp. 319–20; according to J. Howard Swinstead, *A Parish on Wheels* (1897), p. 194, the Drill Hall, Portsmouth, seems to have served a similar winter function.
164 John Hollingshead, *Ragged London in 1861* (1861), pp. 181–2.
165 Sanger, op. cit., pp. 135–6, 175, 204–5.
166 *London City Mission Magazine*, 2 December 1861.
167 Mayhew, op. cit., III, p. 407.
168 Report on Vagrancy, App. XXXII.
169 Mayhew, op. cit., III, p. 381.
170 26th Report of the Mendicity Society (1844), pp. 13–14.
171 *P.P.*, 1846, VII, Report on District Asylums (368), Q.1903.

172 P.P., 1914–16, XXXII, Report of the Metropolitan Poor Law Inspectors' Advisory Committee on the Homeless Poor (Cd. 7840), p. 7.
173 C.O.S., *Homeless Poor*, p. xiv; Report on District Asylums, Q.1812; Mayhew, op. cit., III, p. 410.
174 For some venomous attacks, see C. E. Trevelyan, *Three Letters . . . on London Pauperism* (1870); C.O.S., *Conference . . . on Night Refuges* (1870).
175 Report on District Asylums, Q.1909.
176 James Greenwood, *Low Life Deeps* (1876), p. 212.
177 *London City Mission Magazine*, 2 January 1860.
178 George Smith, *Our Canal, Gipsy Van and other Travelling Children* (1883), pp. 17–18.
179 Chase, op. cit., p. 96.
180 Mayhew, op. cit., III, p. 412.
181 C.O.S., *Homeless Poor*, QQ.2, 141–2.
182 Evans, op. cit., pp. 235–6.
183 R.C. on Labour: Agricultural Labourer, District Reports (C. 6894–IV), B.–V [Bromyard], paras 8, 9.
184 A. H. John, *The Industrial Revolution in South Wales* (Cardiff, 1950), p. 66.
185 S.C. on Distress from Want of Employment, 2nd Report, App., p. 53.
186 R.C. on the Poor Laws, Reports on Relation of Industrial and Sanitary Conditions to Pauperism p. 364.
187 Evans, op. cit., p. 243; cf. Leone Levi, *Wages and Earnings* (1885), pp. 112–13.
188 *Morning Chronicle*, 8 May 1850.
189 R.C. on Labour, Agricultural Labourer, District Reports (C. 6894–VI), B–V, 62.
190 R.C. on Children in Agriculture, 2nd Report, Aj2, 33, 36[b], 37[a], 38.
191 Evans, op. cit., Appendix I.
192 *Labour News*, 19 October and 30 November 1872; 18 January 1873.
193 C.O.S., *Homeless Poor*, p. xxii.
194 N. B. Dearle, *Unemployment in the London Building Trades* (1908), p. 80.
195 *Labour News*, 5 December 1874 and *passim*.
196 *Reminiscences of a Stonemason*, by A Working Man (1908), p. 75.
197 51st Report of the Mendicity Society (1869), p. 11; *Labour News*, passim.
198 C.O.S., *Unskilled Labour*, QQ. 1722–3.
199 *Building and Engineering Times*, 7 January 1882.
200 *Fisherman's Friendly Visitor and Mariner's Companion*, March 1844, p. 27; cf. *Morning Chronicle*, 19 April 1850.
201 S.C. on Distress from Want of Employment, 2nd Report, App., p. 159.
202 G. Holden Pike, *Among the Sailors* (1897), p. 127; *The Word on the Waters*, iii (1860), 270.
203 *Labour Gazette*, 11 May 1894.
204 J. R. Bagshawe, *The Wooden Ships of Whitby* (Whitby, 1933), pp. 84–5.
205 *Labour News*, 14 October 1876.
206 Booth MSS., B. 122, fol. 63.
207 *Labour News*, 21 October 1876.
208 Webb, op. cit., p. 256.
209 *Labour News*, 1 November, 1873; Booth MSS., B. 84, fols 2, 9, 18, 25, 65.
210 Ibid., *passim*; Booth MSS., B. 104, fols 12, 26, B.101, fol. 75; P.P., 1876, XXX, Factories and Workshops, Q. 4604.

211 R. Williams, *The Liverpool Docks Problem* (Liverpool, 1912), p. 42; and 'The First Year's Working of the Liverpool Docks Scheme', *Trans. Liverpool Economic and Statistical Society* (1913–14), p. 99.
212 R.C. on Labour: Minutes of Evidence (C. 6894–IX), QQ. 31,598; 31,616; 31,501; *The Port of Hull* (Hull, 1907), p. 189.
213 Bagshawe, op. cit., pp. 84–5.
214 *The Diaries of Sir Daniel Gooch* (1892), p. 78.
215 *Labour News*, 25 January 1873.
216 *Paddy the Cope, My Story* (1935), p. 61; R.C. on Labour, QQ. 31,501; 31,598; 31,616; Booth MSS., B. 94, fols 5,14; B.117, fols 1,18.
217 Webb, op. cit., pp. 256, 257.
218 R.C. on Labour, Q. 26,439.
219 'The Irish in England', *Labour News*, 28 March 1874.
220 *Labour Standard*, 15 April 1882.
221 *Labour News*, 4 November 1876.
222 'The Irish in England', *Labour News*, 28 March 1874.
223 Thorne, op. cit., pp. 49–50.
224 Hindley, op. cit., pp. 188–9; Mayhew, op. cit., I, p. 338.
225 Mayhew, op. cit., I, p. 104, 105; III, p. 413.
226 Ibid., III, p. 94; *Labour News*, 8 January 1876; Booth, op. cit., 1st series, I, p. 211; 2nd series, IV, p. 130; Webb, op. cit., p. 257.
227 C.O.S., *Report on Soup Kitchens* (1877), p. 13.
228 T. Camden Pratt, *Unknown London* (1897), p. 32.
229 Booth MSS., A. 17 part A, fol. 82.
230 C.O.S., *Homeless Poor*, Q. 1,871.
231 Alsager Hay Hill, *The Unemployed in Great Cities with suggestions for the better organisation of labourers* ... (1877), p. 15; Wilkinson, in Sims, op. cit., I, p. 332; J. B. Booth, *London Town* (1929), pp. 302–3.
232 P.R.O., H.O. 45/10499/117669; Mepol. 2/1490.
233 C.O.S., *Exceptional Distress* (1886) and, for a hostile scrutiny, *Winter out-of-Work, 1892–3* (1893). Booth wrote that the vestries, were 'the principal extra source of casual employment during the winter' (op. cit., 2nd series, IV, p. 40).
234 Mayhew, op. cit., p. 6.
235 Ibid., p. 6; Greenwood, op. cit., pp. 152–6.
236 Walter Besant and James Rice, *The Seamy Side* (1880), p. 10; *Labour News*, 8 November 1873; Booth, op. cit., 2nd series, I, pp. 94, 131.
237 Sanger, op. cit., pp. 128–9.
238 W. E. Adams, *Memoirs of a Social Atom* (1893), I, pp. 304–7.
239 George Acorn, *One of the Multitude* (1911), pp. 92–4.
240 David Morgan, 'The place of harvesters in nineteenth-century village life', in Raphael Samuel, ed., *Work: Industrial work groups and workers' control in nineteenth-century England* (forthcoming).

6 Pubs

Brian Harrison

Comparison in urban history is best conducted at the level of particular institutions within the town, rather than between towns as a whole.[1] The pub and the temperance society, which can be found in most Victorian towns, demand such an approach. Vigorously competing for the attention of the new urban masses, they symbolized alternative styles of urban life. They had not always been rivals. In a predominantly rural society, the pub complemented the church; vestry meetings and Sunday schools were often held in the pub, parsons promoted church-ales, and church-goers from a distance took a dram before they prayed. Vicars of Wakefield had long been exhorting 'the married men to temperance, and the bachelors to matrimony', but only after the 1820s did they begin to attack pubs as such. Teetotallers before the 1830s were isolated eccentrics, and it was only with the launching of the anti-spirits and teetotal movements in 1829–34 that formal associations of abstainers appeared.

Throughout the nineteenth century, the temperance movement remained a predominantly urban movement, and in its early years attracted those urban and nonconformist personalities who supported the Anti-Corn Law League, the Liberation Society, and the Chartist movement. Total abstinence gradually became a passport to social respectability. In a society relatively starved of recreation, working men had to choose the life of the pub and the music-hall or the life of the temperance society, mutual improvement society, and chapel: there was nowhere else to go. The steady extension of temperance ideas to the countryside throughout the century helped to advance the Liberal Party's rural frontier; and, given the striking success of the Church of England Temperance Society in the 1870s, there was nothing to

© 1973 Brian Harrison

prevent even rural communities from polarizing round pub and temperance society. Whereas in the early nineteenth century the Protestant Dissenting Deputies' watchdog committee met regularly at the King's Head Tavern in the Poultry, such a venue would have been inconceivable by the 1870s; and whereas temperance in the 1830s was a politically neutral issue, by the 1890s it lay at the heart of party politics.

The survival of two drink maps, for 1887 and 1899, enables us to study London's pub geography in some detail; similar analyses could in principle be conducted for other Victorian towns, and London is prominent in the subsequent discussion only because information about its pubs is so readily accessible. Temperance societies will receive rather less attention, because London temperance geography was never mapped out, though maps could perhaps be created from temperance society subscription-lists. In 1896 London had 393 persons per pub—a lower pub density than that of any other major English city: Leeds had 345 persons per pub, Liverpool 279, Birmingham 215, Sheffield 176, and Manchester 168. Pub density was lower than in London in only nine of the 235 boroughs outside London. But London's pub geography clearly reveals the pub's three major roles in nineteenth-century society: transport centre, recreation centre, and meeting place.[2]

As transport centre, the pub's role strikingly diminished during the nineteenth century, because railways greatly reduced the quantity of long-distance road travel. Until the 1830s, the great London coaching-inns were the equivalents of railway termini: one of the most famous, the Bull and Mouth in St Martin's-le-Grand, had underground stables for 400 horses. Passengers loaded up in the great backyards, for in stage-coach days the whole inn was oriented round a galleried rear-courtyard. Coaches entered and left through a large archway into the main road—the most impressive of these being the Bull and Mouth's classical arch rebuilt in 1823. Only after the pub had lost its special connection with horse-drawn transport did the street façade become overwhelmingly important. Regular timetables were issued, and, at public-house stops between the termini, passengers waited in the large glazed front rooms for the coach to arrive (Plates 25 and 26). The great transport-inns clustered together—in Borough High Street, Bishopsgate, Piccadilly, and the inner suburbs of the City—and competed for custom. It is not surprising that pubs are so prominent in a book like *Pickwick Papers*: as B. W. Matz put it, 'the book . . . opens in an hotel and ends in one'.[3]

The railway completely upset this pattern, for by speeding up travel and increasing its comfort, the railway reduced the need for refreshment on the journey. Railway companies did not at first perceive these implications, and provided generous drinking facilities at their stations. One of the earliest London termini, the London & Croydon Railway's Bricklayers' Arms, actually took the name of a pub, and two of the company's earliest stations—the Dartmouth Arms and the Jolly Sailor—were not renamed Forest Hill and Norwood Junction till 1845. Temperance reformers, who eagerly invested in the early railways and saw them as instruments of progress, fought to alter this policy; in 1881 they prevented even Gladstone from allowing the sale of drink on the train to railway passengers.[4] The pub's billeting function declined for the same reason: railways enabled soldiers to travel faster and to lodge at their own specialized version of the hotel—the barracks.

In towns, pubs suffered partly from the rerouting of transport (which lent attractions to station-hotels as against the old coaching-inns), but also from the sheer volume of railway travel. Hotel-keeping grew up as a trade in its own right, where drink accounted for a smaller proportion of the profits. The old coaching-inns had to adapt; the Bull and Mouth converted its great archway into a front entrance and became a hotel (Plate 27), and a London travel guide in 1871 claimed that the coaching-inns had 'all become comfortable middle-class hotels, with railway booking-offices attached'.[5] The less enterprising among them decayed until they were pulled down: the Green Dragon, last of the great coaching-inns in the Gracechurch Street/Bishopsgate complex, was torn down in 1877. Nevertheless, the 1899 map shows the persistence of the link between pubs and railway travel; there were three pubs actually on the station premises at St Pancras, and two each at King's Cross, Euston, and London Bridge. Several Waterloo Road pubs faced the passenger as he left Waterloo station, and of the thirty-five railway stations on Booth's map, twenty-six had pubs on or immediately outside the premises.

Publicans are resilient creatures: they may suffer by some social changes, but they take care to profit by others. Long-distance railway travel, by greatly increasing the amount of short-distance road travel (particularly in London, where through-journeys by rail could not be made), enabled the urban publican to recoup some of his losses. Furthermore, the railways were slow to cater for suburban commuters: roadside pubs therefore gained from the consequent growth of bus and tramway systems which linked suburbs to city-centre. All five of the pioneer tram-routes out of Leeds city-centre launched under the Leeds Tramways Order of 1871 terminated at suburban pubs; London Transport's maps still advertise the Royal Oak, the Angel, Swiss Cottage, and Elephant and Castle. But although the increased speed and utilization of urban road transport benefited some urban roadside pubs, it reduced pedestrian travel and thereby reduced passengers' opportunities for drinking while travelling. Hence John Dunlop's comment in 1839: 'we have known the establishment of a coach or omnibus on some road, abolish public-houses along its line'.[6]

The mark of the commuter lies heavily upon late-Victorian London's pub geography. The 1887 map shows pubs concentrated far more closely in the city-centre than in the suburbs; in 1896 the number of persons per pub ranged all the way from 116 in St James's to 727 in Edmonton. Unfortunately we have no figures to reveal contrasts in pub density *within* provincial towns; it is therefore impossible to see whether St James's high concentration of 0·7 acres per pub was a record for the country as a whole. But the figure certainly represents a pub concentration far greater than in any other complete town in 1896; the City (1·0 acres per pub) and the Strand (2·0) followed close behind. With fifty persons per pub in 1896, the City of London had a pub density greater than any other of the 235 boroughs. Yet the City

The Victorian City: Numbers of People

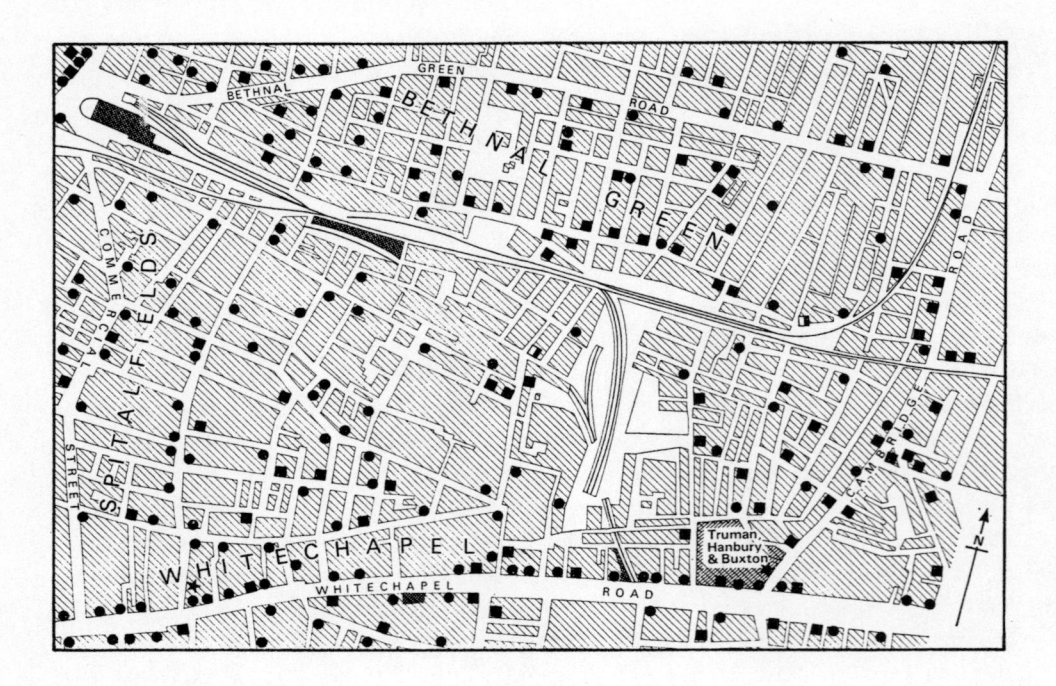

- Brewery
- Fully licensed house
- Beer-house with 'on' and 'off' licence
- Beer-house with 'off' licence
- Grocer with licence to retail wine, beer or spirits in bottles
- Restaurant with wine, beer or spirit licence, but without a 'bar'
- Railway
- Railway station

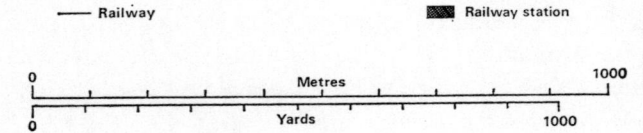

I Licensed premises, Bethnal Green and Spitalfields, 1899. This is an area of small-scale manufactures and of working-class housing. Whitechapel Road, as a major east–west thoroughfare, attracted many pubs. There were only two grocers' licences in the whole area, but many beer-shops and a famous brewery. Note the siting of the pubs on street-corners. Source (as for II–IV): Charles Booth, *Life and Labour of the People in London*, final volume (1903), 'London, 1899–1900'.

pubs were really catering for ten times as many people, resident in the suburbs; if these are allowed for, its pub density was low—479 persons per pub.[7]

Commuter travel is also reflected in the long lines of pubs which straddle the major east–west routes and, to a lesser extent, the north–south routes. Until the eighteenth century, London's east–west transport axis had been waterborne as well as landborne, but by the Victorian period river transport had markedly declined, and relatively few riverside pubs survived. In the nineteenth century, imported goods, after being unloaded, were moved west across London by road to be sold or consumed: much of the labour force—particularly after the late-Victorian slum clearance at St Giles—lived in the east, but moved westwards for most of its working day. Before the large popular restaurants appeared, strategically-placed pubs provided the working man with breakfast on his journey to work, and with refreshment on his journey home. In the one mile of Whitechapel Road from Commercial Street to Stepney Green, there were in 1899 no less than forty-eight drinking-places: in the three-quarter mile stretch of the Strand from Trafalgar Square to St Clement's there were forty-six (Maps I and II). Drinking-places also clustered along the major north–south axes: Bishopsgate/Shoreditch/Borough High Street and Tottenham Court Road/Waterloo Road. At both ends of the London river bridges, pubs closed in on

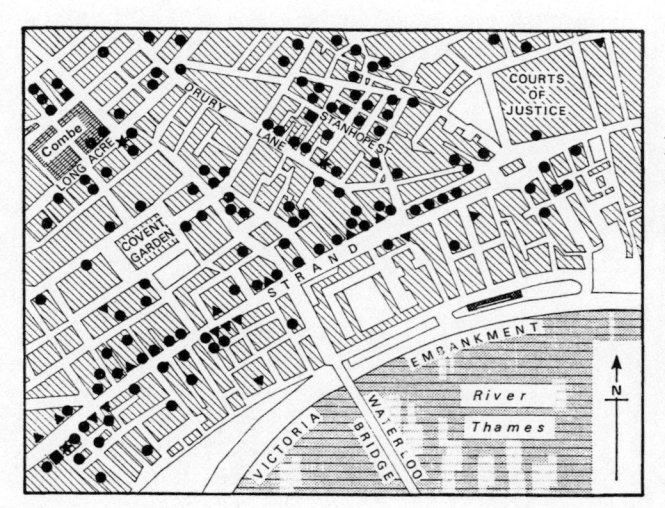

II Licensed premises in the Strand, 1899. Drinking facilities clustered along the Strand's major east–west route, but they have been excluded from the Victoria Embankment. Gladstone's refreshment-house licences accommodated the many middle-class commuters who staffed the shops and offices in the area. But poverty lurked in the back-streets, and in the Stanhope Street area Booth found 55·5 percent of the population in poverty; the 'large proportion of Irish, many rough characters' there ensured that pubs clustered in the locality. But for the Duke of Bedford's policy, the Covent Garden porters would have ensured a more impressive cluster further south.

Key and scale as for Map I

the passer-by, as they also did at the major road-junctions—there were six at Elephant and Castle, five at Cambridge Circus, four at St George's Circus.

Many of the West London pubs were therefore catering, not for the privileged classes, but for working people resident in East London, or for the personal servants of the well-to-do. Areas of working-class employment in the west—the Wellington and Knightsbridge Barracks and Covent Garden (Map II), for example—were well supplied with pubs. The 1899 map also shows a remarkable concentration outside the Mint, and in dockside areas like Shadwell High Street and West India Dock Road (Map III). Pubs occupied all four corners of the central square in the cattle market opened at Copenhagen Fields in 1855, for at this time bargains were sealed over a drink. Pubs were the focus for the market-town's servant-hirings and fairs, and temperance reformers felt obliged to attack both. Some nineteenth-century employers—notably the Richardsons and Titus Salt, in their temperance utopias of Bessbrook and Saltaire respectively—tried to promote work discipline by banning pubs from the factory area.

The commuter's needs help to conceal a fundamental class-contrast in Victorian drinking habits; for, whereas at the beginning of the century different classes patronized the same pubs, by the 1860s the respectable classes were drinking at home, or not drinking at all. If free trade had been allowed to operate, if working men had laboured where they lived, and if recreational and business commuters had not existed, pubs would have been far less common in West than in East London. These complicating factors do not in fact completely conceal this underlying contrast. The 1887 map shows that the best London streets—Whitehall, Pall Mall, Piccadilly, and the Embankment—had very few pubs. Over a wide area to the north and west of the West End, pubs were again very sparse—in Bloomsbury (Map IV), Belgravia, and Kensington, all fashionable areas whose landlords (notably the Dukes of Bedford and Westminster) pursued a temperance policy. William Hoyle the temperance reformer

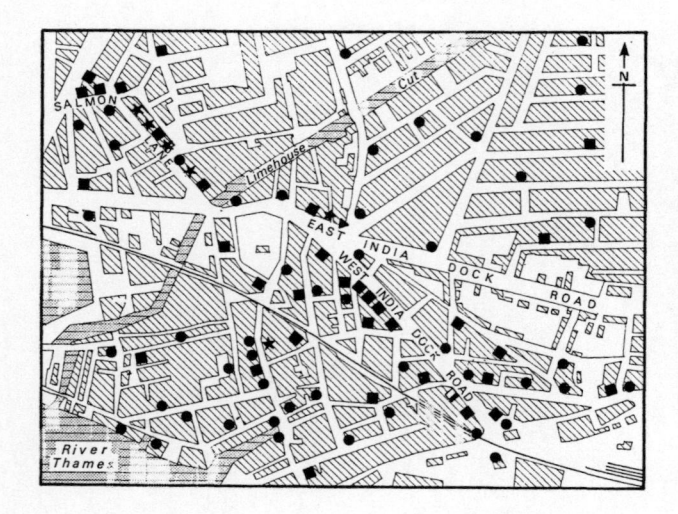

III Licensed premises in dockland, 1899. The many pubs on the West India Dock Road were a measure of the thirst of dockers working locally. There were many beer-shops, and the grocers' licences in Salmon Lane probably catered for the many Italians living nearby. Pubs clustered on the main thoroughfares, and in the area with the most poverty (41·4 per cent) to the south-west.

Key and scale as for Map I

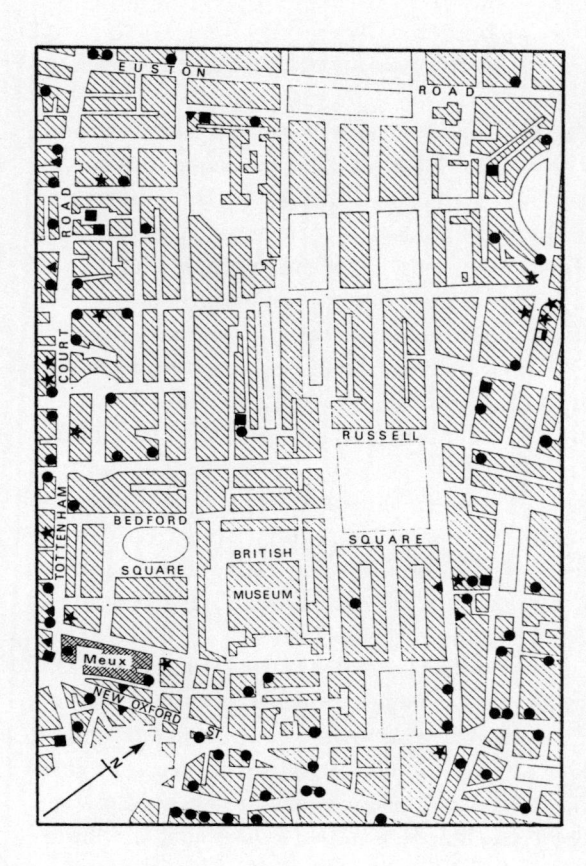

IV Licensed premises in Bloomsbury, 1899. The Dukes of Bedford enhanced the value of their estates by virtually excluding pubs, but drinking facilities (especially Gladstone's grocers' licences) abounded on the fringes because aristocratic cellars had to be supplied.

Key and scale as for Map I

noted that in Lancashire's industrial towns, more socially homogeneous than London, pubs were distributed far more evenly: 'this morning I have been for about an hour riding about the West End', he told a select committee in 1877, 'where I did not see a public-house at all'.[8]

If we turn from the quantity to the category of licences, the contrast becomes very clear. By the mid-Victorian period the well-to-do were much more likely to patronize some types of off-licence than to enter the pub. The off-licence was a Victorian invention, and a peculiarly urban phenomenon: the more scattered the population, the less prevalent it became. The nineteenth century saw a marked shift from on-consumption to off-purchase of drink. The change stems partly from changing class relationships and religious attitudes, and from the accompanying growth of the suburb, but also from technological developments in the drink industry, which expanded the trade in bottled beer. Evangelical families with a conscience, like the Fremlins of Maidstone, felt fewer qualms at promoting the relatively respectable 'family trade'. By an Act of 1834, beer-house licences were divided into two categories: for on- and off-sales. In late-Victorian England and Wales there were over twelve thousand of the latter, as against about thirty thousand of the former; by

167

1896 off-licences of all types accounted for a sixth of all drink-retailing licences in England and Wales.

There were relatively few beer-house off-licences in late-Victorian London, and these were not located in any particular area. But the relative prevalence in well-to-do areas of the off-licence in general stands out clearly from the L.C.C.'s licensing statistics for 1903–4. Among the eighteen London boroughs with fewer than average on-licences per head, there are eleven which have more than the average proportion of off-licences per head. There seems to be an almost inverse relation between the number of on- and off-licences per head in any London area. Furthermore, if one lists the eleven areas relatively well-endowed with off-licences, one finds oneself reading out a roll-call of what were then London's well-to-do suburbs: Westminster, Stoke Newington, Paddington, Lewisham, Kensington, Hampstead, Hammersmith, Hackney, Fulham, Chelsea, and Battersea.[9]

It is Gladstone's refreshment-house and grocers' licences for the sale of wine which explain the relative prevalence of the off-licence in wealthy areas. Gladstone claimed in 1860 that his licensing scheme would provide the poor with drinking facilities which were then being enjoyed only by the rich. But although his reforms did greatly open up the wine trade, and virtually created the firm of Gilbey, the 1899 map shows his wine-shops clustering only on the fringes of Clubland and the Bloomsbury estate. If we analyse thirty-three of Charles Booth's London districts comprising all levels of poverty, there is a negative association (significant at the 10 percent level) between the percentage of the population living in poverty in a London district and the proportion of its total drink licences accounted for by wine licences.[10] Booth's map shows only three refreshment-house licences east of Shoreditch, whereas in the Strand alone there are thirteen. Gladstone's grocers' licences, which in 1880–1900 accounted for nine-tenths of all his wine licences, had merely enabled the rich to replenish their cellars: his refreshment-house licences had merely enabled the businessmen to enliven their midday meals. If the temperance reformers are to be believed, wine licences also encouraged drinking among women—a view which gains support from Seebohm Rowntree's figures for York in 1899. He found that the customers of a local grocer's off-licence divided between 37 percent children, 35 percent women, and only 27 percent men. In the three pubs whose customers he surveyed on three days of the week, the percentage of men never fell below 47 percent and in one case rose as high as 81 percent.[11]

The distinction between on- and off-licences, and between work-time and recreational drinking, helps to explain why the association between pub density and the incidence of poverty is so weak. When one adds to these factors the campaign by late-Victorian magistrates and the L.C.C. to reduce the number of licences in slum areas, the fact that there is any association at all testifies to the immense importance of the pub in the residential areas of the late-Victorian working class. It also emphasizes the importance of considering the second major function of pubs in Victorian towns—the recreational function.

Late-Victorian London's breweries can almost all be found in working-class areas: in that inner ring of decayed suburbs which encircled the City (Map I). This forcibly impressed the young F. N. Charrington when, walking one evening to a local ragged-school from the family brewery in Mile End, he saw a working man come out of a pub and knock down into the gutter his wife, who was pleading for money; Charrington looked up to see his own name in huge gilt letters on the façade of the pub.[12] Particularly prominent in slum areas was the beer-house. In Booth's thirty-three London districts, there is a positive association (significant at almost the 10 percent level) between the percentage of the population in poverty and the number of beer-houses, expressed as a percentage of all local drinking-places. Created by the 1830 Beer Act, beer-houses were forbidden to sell spirits and were at first relatively free from magistrates' control. Beer-selling was popular with the prosperous working man who could employ his wife behind the bar. Pubs occupy strategic positions in the Victorian slum. Of the 160 pubs in the area bounded by Bethnal Green Road/Commercial Street/Whitechapel Road/Cambridge Road (Map I), 131 were situated on corners or opposite road junctions; two pubs stood guard over the entrance to many a side-street. Pubs with more than one entrance simultaneously attracted pedestrians from more than one thoroughfare. Statistics and maps cannot tell us what these pubs were doing: the historian must try to supplement the techniques of the social scientist with the insights of the novelist.

The Victorian slum pub must be seen in the context of street-life. All but the busiest streets at that time united rather than divided the community: in working-class areas 'the emphasis is not so much on the individual home, prized as this is, as on the informal collective life outside it in the extended family, the street, the pub and the open-air market'. Furthermore, economic, technological, and recreational factors ensured that, by modern standards, the pavements were alive with pedestrians many of whom felt obliged to subscribe to drinking customs on the way. Many people earned their living on the pavement—the beggars, stallholders, acrobats, organ-grinders, pedlars, whom Mayhew interviewed so brilliantly. In 1877 William Hoyle explained why pub density was so much greater in Manchester than in Bolton: 'there are a great many men in Manchester who are common porters, who loiter about the streets, and there is comparatively little of that in Bolton'. Nineteenth-century shops were more akin to stalls than they now are, less shut off from the street; glass seldom intruded between goods and purchaser. The pub was different, in that its goods and customers were shut off from public view by doors and frosted glass; but as its role was more recreational than that of the shop, this was a positive asset. Besides, the barrier between pub and street was easily crossed, for the two worlds were connected by great areas of glass, by first-floor balconies, by pavement seats and tables, by potmen carrying cans of ale in wooden frames to customers in nearby premises, and by a multitude of entrances; the 250 central London pubs investigated in 1897 had an average of three entrances apiece (Plates 28 and 30). Much to the temperance reformer's disgust, the street-corner was a social centre: the epitome of everything he disliked.[13]

The pub was marked off from nearby houses by large and colourful signboards; it became even more distinctive during the nineteenth century owing to at least two waves of expenditure on its façade and interior. It may not be coincidental that both these waves of expenditure occurred during decades when the temperance movement was booming. In the early 1830s the publican had every reason to follow the fashion for enriching shop façades which had gripped London in the 1820s, for the Beer Act of 1830 had created a new rival for his custom. A new phenomenon, the 'gin-palace'—with plate-glass windows, richly ornamented façade, gilded lettering and brilliant lamps—began to arouse comment. Its style soon became almost uniform in urban pubs, and still influences their design today. Its splendour accentuated the contrast between the pub and the squalor of its surroundings.

In towns very much darker at night than they now are, the huge and elaborate wrought-iron gas-lamps (Plate 30) which hung over the pub entrance extended the brilliance of the interior into the street, just as the pavement extended its floor. Temperance reformers were half-intoxicated by the brilliance of the scene: at 8.30 one Sunday evening in 1836, a temperance reformer observed the gin-palaces in the Ratcliffe Highway:

> at one place I saw a revolving light with many burners playing most beautifully over the door of the painted charnel-house: at another, about fifty or sixty jets, in one lantern, were throwing out their capricious and fitful, but brilliant gleams, as if from the branches of a shrub. And over the doors of a third house were no less than THREE enormous lamps, with corresponding lights, illuminating the whole street to a considerable distance. They were in full glare on this Sunday evening; and through the doors of these infernal dens of drunkenness and mischief, crowds of miserable wretches were pouring in, that they might drink and die.

Here, as elsewhere, publicans were profiting from the parsimony of the public authorities, and were introducing new upper-class comforts which working people could enjoy only communally.[14]

This external change accompanied an even more significant interior transformation. Simmel claims that 'cities are . . . seats of the highest economic division of labor': the urban predominance of the nineteenth-century gin-palace certainly illustrates his theory, for it was specially designed for the casual urban drinker. Whereas the pub had originally been nothing more than an enlarged home, city life demanded a large shop with specialized equipment and a bar. The cluttered interior which suited a small and relatively slow-moving community did not suit the London of the 1830s. What was needed was a long bar, enclosing several assistants; easy access to large quantities of alcoholic drinks, racked in attractively-painted casks on the back wall; a large area before the bar, where many customers could move freely under the manager's close supervision; and separate entrances for wholesale and family transactions.[15]

A second wave of expenditure took these developments further in the 1860s when, according to the slum missionary, the Rev. J. M. Weylland, 'there was a sudden enlargement of public-houses, and in the attractiveness of them'. By the early 1870s Ruskin was complaining that 'there is scarcely a public-house near the Crystal Palace but sells its gin and bitters under pseudo-Venetian capitals copied from the Church of the Madonna of Health or of Miracles'. The interior transformation also continued: 'many public-houses which had tap-rooms and parlors [sic] now have those rooms thrown into large bars', said Weylland in 1877; 'the people merely go in and drink their drams, stand there but for a short time and pass out again, hence it is that there are not so many people found seated and drinking themselves drunk as there used to be years ago.' Such expenditure must often have bewildered the seasoned drinker, especially when (as later in the century) elaborate carved mirrors and screens were introduced: like Tinker Taylor, he would have found such places 'too stylish . . . now for him to feel at home in'. But the sit-down drinker was never forgotten: there were always pubs whose many internal divisions promoted privacy, reduced opportunities for disorder, and preserved class distinctions.[16]

These changes are important, not so much for the luxury which so impressed contemporaries, as for the increase in the *size* of the pub. Looking back over the century, Rowntree and Sherwell in 1899 found that 'while Temperance workers have been steadily endeavouring to reduce the *number* of public-houses, the publicans and brewers have been busily intent upon increasing their *size*.' This probably helps to explain the fact that in the boroughs of 1896, holding size of household constant, there is a strong negative association (significant at the 1 percent level) between a town's population and its pub density; a doubling in the population is associated with a 20 percent rise in the ratio of persons per pub. The large size of urban pubs, together with licensing policy in the late-Victorian slum, may help to explain what is at first sight the surprising fact that, holding size of town constant, the more crowded the housing conditions (persons per house) in any town, the *lower* the pub density (significant at the 1 percent level). An increase of one in the average size of household is associated with an increase of thirty in the ratio of persons per pub. The fact that pub density rises the more scattered the population, and that the relatively small rural pubs were assessed to relatively low rates, reminds us that the steady reduction in pub density in England and Wales since the 1860s owes something to increased urbanization as well as to increased sobriety.[17]

The extravagant and crowded atmosphere of the pub helped Victorian publicans to pursue their important recreational role. Imagine the dram-shop's impact on a tired and bored working man, fleeing from his drab home, nagging wife or landlady, and crying children: 'there is light enough for the transformation scene of the pantomime, and noise enough for a fair', wrote 'Saunterer' of the *Bradford Observer*, who toured local dram-shops in 1871. 'We have some difficulty in obtaining a standing place beside the bright, pewter-topped counter, the crowd of drinkers being so great' (cf. Plate 29). The pub never became a mere shop, whose assistants were indifferent to its customers: it remained a social centre whose hold on working people was the greater because of their migratory lives. If Paris is any guide, in the rapidly-growing nineteenth-century towns which absorbed so many rural immigrants, there would probably be a temporarily low proportion of women to men, and a high proportion of adults to children: all the more need, then, for a 'masculine republic' in every street.[18] If working men were unemployed, in lodgings, 'flitting' from one house to another, or trapped in the bachelor trades, their dependence on the publican was complete. Pub density in some types of town increased, in 1896, with the likelihood of non-resident visitors. Holding size of town and size of household constant, pub density is greater in garrison towns than in market towns (significant at the 10 percent level); the same applies to resorts, though the association is not significant. In industrial towns and seaports, however, pub density is lower than in market towns, though not significantly so.

The close concentration of pubs outside the Knightsbridge Barracks on the 1887 map, and outside the Wellington Barracks on the 1899 map, reflects the heavy dependence of servicemen upon the pub. 'Have you ever thought of collecting some facts showing the demoralizing influence of the barracks in our large towns...', wrote that shrewd campaigner Richard Cobden to the pacifist Henry Richard in 1850; 'you might, through your own friends and members of the [Peace] Society, collect some startling information upon these points'. Publicans helped to recruit, to billet and to entertain the armed forces. Government policy, by discouraging servicemen from marrying, inevitably encouraged publicans in garrison towns and seaports to provide prostitutes, and to devour the savings which servicemen accumulated while overseas.

Pubs also lay at the heart of London's underworld. Although by the early-Victorian period the salacious public-house 'cock and hen' clubs may have been declining, the famous Cyder Cellars (Plate 32), Coal Hole, and Evans's Cave of Harmony still catered for the rabelaisian young men who kept late hours in the West End. Here one could find apprentices, clerks, guardsmen, undergraduates, clubmen, lawyers, and commercial men up from the country; 'it was no double meaning, but plain out', said the temperance missionary Jabez Balfour, of the songs he heard there. Renton Nicholson, a major figure in London's underworld, continued to preside at the Coal Hole's indecent Judge and Jury Society right into the 1860s; mock matrimonial cases were tried with the aid of 'female' witnesses (men dressed up in women's clothes), and in the view of one observer 'everything was done that could be to pander to the lowest propensities of depraved humanity.'[19] Robert Hartwell, standing as working-men's candidate at Stoke in 1868, felt obliged vigorously to rebut local accusations that he had been connected with the Cyder Cellars as a mock barrister: 'he had never been in the Cyder Cellars, and knew no more of that place than what he had heard by report.' Without such a denial, Hartwell had no hope of winning local dissenters and respectable working men to his side.

Here the two extremes of society met, for the pub's internal divisions could not prevent young bloods and aristocrats from sharing the pleasures of their humblest inferiors. 'When we dip down below the bourgeois and the regular working-classes', wrote J. A. Hobson, '... we find a lower leisure class whose valuations and ways of living form a most instructive parody of the upper leisure class.' *Household Words* in 1857 spoke of the Haymarket's 'sparring snobs, and flashing satins, and sporting gents, and painted cheeks, and brandy-sparkling eyes, and bad tobacco, and hoarse horse-laughs, and loud indecency'; a cruder sort of Haymarket could be seen every Saturday night in the East End's Ratcliffe Highway. Sir George Grey's Public-House Closing Act of 1864, which closed London's pubs between 1 a.m. and 4 a.m., was directed as much against prostitutes as against publicans. As seen from below in *My Secret Life*, it ensured that 'many nice, quiet accommodation houses were closed, and several nice gay women whom I frequented disappeared'; by the 1860s the puritans had made serious inroads on London's underworld.[20]

Sexual adventures were not the only pleasures which united the two social extremes: for generations, aristocrats and working men had centred their sporting activities on the pub. For whereas in the countryside one retreated there from the harshness of the elements, in the town one retreated into the pub from the harshness of urban and industrial life. Sport was a major consolation; the urban pub has always, even in its architectural styles, embodied the urban Englishman's desire to flee into the country. We have rightly been reminded that 'most of the new industrial towns did not so much displace the countryside as grow *over* it': the traditions, the ceremonial, the sociability, and even the sports of rural life were all preserved by urban publicans. Richard Nyren, innkeeper at Hambledon's Bat and Ball, is prominent in the history of cricket, and hanging over the entrance to Lord's first cricket-ground was an advertisement for wines and spirits, Lord's supplementary source of income. When Mayhew in the late 1840s attended the Graham Arms ratting session, all classes were present, and 'a young gentleman, whom the waiters called "Cap'an"' featured prominently in the proceedings. This was the world over which the Marquis of Hastings presided in the 1860s: 'his advent at a ratting match or a badger drawing was a signal to every loafer that the hour of his thirst was ended, and that henceforth "the Markis was in the chair".' When the R.S.P.C.A.'s early-Victorian inspectors interrupted pub cock-fights, they often found army officers presiding.[21]

The connections between drinking and racing were equally strong: the retired prize-fighters and sporting men who kept London sporting-pubs followed their customers to Epsom with tents and booths, and one day in the 1860s Taine was reminded of a recent racing event by the drunken men to be seen all along the road from Epsom to Hyde Park. 'Tom Spring's Parlour' (Plate 31), *alias* the Castle Tavern, Holborn, was conveniently placed midway between the East and the West End; it was a major centre of sporting gossip, attended by early-Victorians of all classes. Tom Spring, an impressive figure, had succeeded another prize-ring hero, Tom Belcher, as landlord in 1828, and died in 1851. On the night before a great (and illegal) prize-fight, his house was crowded by customers eager to learn the *venue*. Spring was deeply implicated, for instance, in the great Caunt–Bendigo contest of 1845. He was manager and chief supporter of Caunt, himself a London publican, and on the preceding night Caunt distributed coloured handkerchiefs as favours at a dinner held at the Castle. Frederick Gale claimed that Tom Spring, like other sporting men, had 'acquired that natural good-breeding which is engendered by associating with people much above them in society'.[22] Similar gatherings assembled at the Green Dragon, Fleet Street, in the 1860s, to meet Marwood the public executioner, who held court there at noon on execution days 'in the "select" section of the pub'. Here, then, are neglected haunts of popular Toryism in Victorian England: sources of that 'unsystematic, unintellectual support of familiar standards and habits of life' which has been attributed to the Tory party. Missionaries were not the only well-bred Victorians to bridge the social extremes by touring London slums. In the 1860s fashionable young aristocrats took a delight in exploring low taverns incognito at night. 'The aristocracy and the working class', Lord Randolph Churchill once declared, 'are united in the indissoluble bonds of a common immorality'.[23]

To a large extent the pub was a centre for male recreation, and the temperance movement liked to see itself as defending helpless women and children against male selfishness (Plates 33 and 34). But this was not the whole story. The 1899 map shows pubs strategically placed at the main entrances to the East End's Victoria Park. Suburban London publicans catered for Londoners taking a week-end walk, and at a pub like the Rosemary Branch, Islington, the entertainments provided (Plate 35) could be quite elaborate. During London's Easter recreations of 1825, balloons went up from the Eagle, and from the Star and Garter near Kew Bridge, while Greenwich Fair on the evening of Easter Monday ensured that 'every room in every public-house is fully occupied by drinkers, smokers, singers and dancers'.

The reputations of particular suburban pubs rose and fell according to their distance from the built-up area. Against this background, one can understand the passion aroused during the early 1850s in the debate on whether the Crystal Palace should sell drink and open on Sundays, for this was the first major opportunity for providing Londoners with a large drink-free suburban week-end pleasure-ground. If sabbatarians kept the place closed, or if temperance reformers kept it dry, the alternative was drunken sabbath-breaking at London's many suburban public-house tea-gardens—not pious and domestic respectability.[24]

Publicans catered for indoor recreational needs with the music-hall. The 1887 map, with its high pub density in the West End, clearly reveals the close link between pubs and recreation, for pub geography was governed by the evening recreational commuter as well as by the daytime business commuter. The early-Victorian music-hall evolved from three distinct institutions—the supper room, the variety saloon, and the tavern concert-room. The first of the music-halls proper evolved naturally out of the free-and-easies held at Charles Morton's Old Canterbury Arms at Lambeth, an important locality for popular recreation. In 1852 Morton opened his Canterbury Hall, which admitted ladies to every performance; despite puritanical criticism, its entertainment improved markedly on what had gone before. This was so successful that in 1854 Morton erected a larger building, and in 1856 added a picture-gallery. Preceded by Edward Weston in 1857, Morton crossed the river in 1861 and by 1866 London boasted twenty-three music-halls. Outside London there were at least three hundred in 1868—including nine in Birmingham, ten in Sheffield, and eight each in Leeds and Manchester. By 1908 there were as many as fifty-seven music-halls in London, attracting groups well below the theatre-goer in social grade: as a printer informed a parliamentary enquiry in 1892, 'it is to the music halls that the vast body of working people look for recreation and entertainment.'[25]

The links between the pub and the stage persisted for generations: in 1943 'Mass Observation' found that in Bolton 'the touring company is still on the side of the publican, constantly boosting beer, and often being sarcastic about teetotallers, temperance and parsons.' The temperance movement evolved an independent set of entertainers—notably the Shapcott and Edwards families, and the Poland Street handbell ringers—to provide music-hall recreations for teetotal audiences. But temperance reformers like F. N. Charrington and Cardinal Manning who tried to deprive the music-hall of its drink were merely accelerating an inevitable differentiation: for music-hall proprietors soon found that mass entertainment was profitable in its own right (Plate 36).[26] Music-halls had from the first been as Matthew Hanly described them in 1892: 'temperance halls at all times'.[27] Here, as elsewhere, publicans had the initiative to provide the community with a new service, then specialized in it, then lost control of it altogether.

The pub's third major social function—the provision of opportunities for public meetings of all kinds—flowed naturally from its prominence in the transport system and from the lack of alternative accommodation. *Laissez-faire* attitudes meant fewer public buildings; religious control over education meant that schools were seldom open to working men for political or trade purposes. 'Large rooms' rarely existed in the working man's home, whereas they could be taken free of charge from publicans confident of drink profits. Inevitably the pub became a centre for all kinds of working-class activity. Trade societies made pubs their 'houses of call', and customer and publican co-operated in building up the reputation of a house. A fashionable pub like the Cheshire Cheese stored up the autographs of its well-to-do customers, just as Charlie Brown's Railway Tavern in Limehouse stored up the valuable curios contributed by its sailor clientele. The Mile End Road's Bell and Mackerel displayed 20,000 specimens of animals in cases, originally collected by the East London Entomological Society.[28]

Many popular reforming movements originated in pubs. The London Corresponding Society was founded at a meeting in October 1791 between Thomas Hardy and three friends at the Bell, Exeter Street; the Hampden Club originated in 1812, at the Thatched House Tavern; the National Union of the Working Classes originated in some carpenters' meetings in the Argyle Arms, Argyle Street; and many of the factory movement's short-time committees in the North of England met in pubs. The government monitored such meetings through the licensing system, and London publicans in 1839 were threatened with the loss of their licence if they let their rooms to Chartists. There was an economic argument for instituting free trade in drink in the early nineteenth century, but it was the political argument which lent this radical campaign its fire. By the 1830s moral pressures against 'pothouse politicians' had become almost as powerful against radical meetings as threats to withdraw the licence of the publican who accommodated them. A committee of the National Union of the Working Classes in 1833 recommended that meetings be held outside pubs whenever possible because 'nothing will more effectually contribute to the success of our cause ... than sobriety and good conduct'; but the committee admitted that 'this may, in the first instance, be attended with a little difficulty, and probably a diminution of numbers'. And when the Reform League's executive committee considered meeting outside pubs in 1865, the former Chartist J. B. Leno insisted that the League was 'compelled to use the public houses at present as other places were not open to us. It was a matter of necessity not choice.'[29]

Working men were not alone in holding meetings at pubs in the early-Victorian period. In 1856 Robert Lowery recalled that in Newcastle pubs during the 1830s

> all classes met ... to compare notes and to hear individual remarks and criticisms on what occupied public attention ... Every branch of knowledge had its public-house where its disciples met. Each party in politics had their house of meeting—there was a house where the singers and musicians met—a house where the speculative and free thinking met—a house where the literate met—a house where the artists and painters met—also one where those who were men of science met.

In mid-Victorian London, there were pubs for everybody's taste—for medical students, prostitutes, servicemen, sportsmen, actors, foreigners, and lawyers—in which it was often possible to reserve one's own seat. In the Strand at the Crown and Anchor, where the London Working Men's Association held its first public meeting in February 1837, John Bright made his London debut as an Anti-Corn Law League orator in February 1842. And according to Robert Owen, 'bigotry, superstition, and all false religions received their death blow' in the large room of the City of London Tavern, at his famous public meeting on 21 August 1817.[30]

London pubs feature prominently in the 1831–2 reform crisis. The Crown and Anchor, for instance, saw the origins of the National Political Union, accommodated the office of the Parliamentary Candidates' Society, and housed the finance committee of the Loyal and Patriotic Fund. Indeed, it was only in the mid-Victorian period that pubs ceased to be at the centre of party political organization. The Whig party could hardly have survived its long period of opposition before 1830 without the aid of publicans prepared to distribute its propaganda and accommodate the party's monthly dinners in London. A report of the London Tavern's dinner for General Sir Charles J. Napier (Plate 37) shows how lavish these entertainments could sometimes be. By 1866 the London Tavern was providing annual banquets for the officers of twenty-eight different regiments during May, for twenty-four of the City companies, and for many London charities.[31]

The pub's extensive social functions are perhaps best summarized with a brief glance at the pub geography of two provincial towns, Oxford (Map V) and York, in

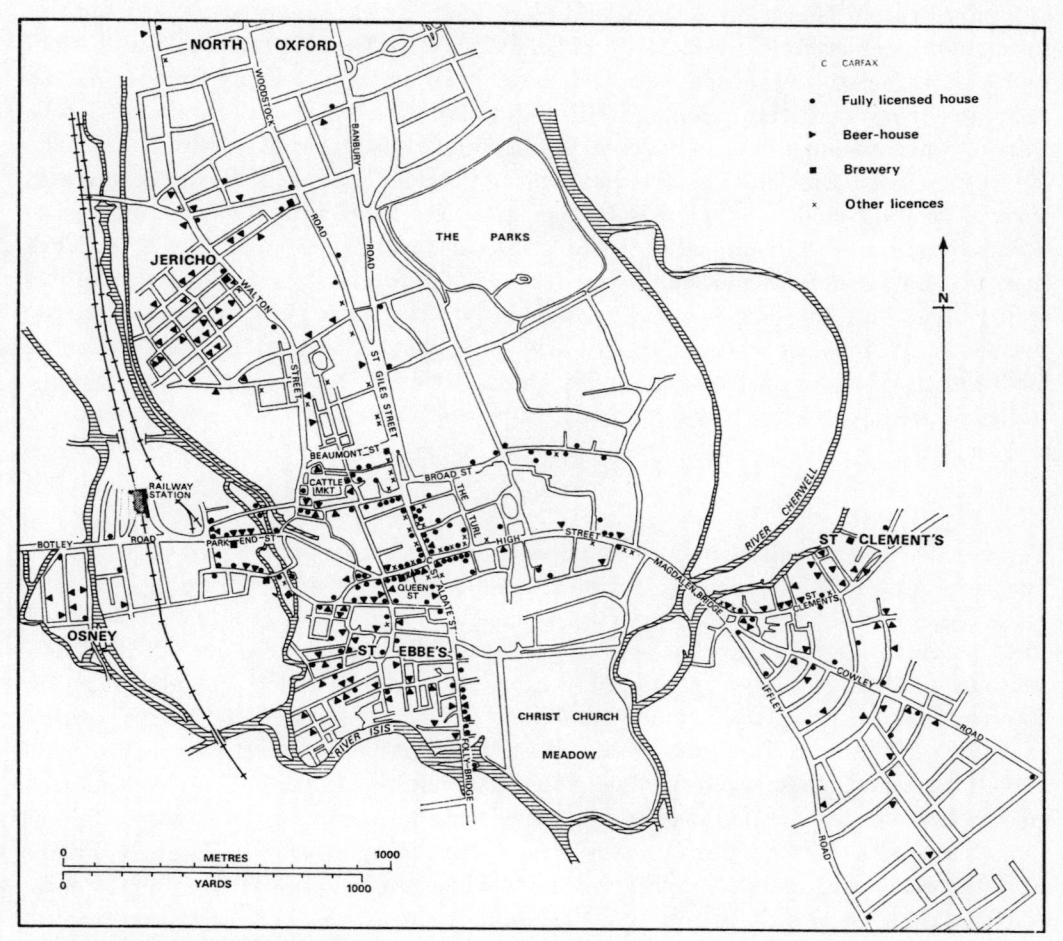

V Drink map of Oxford, 1883
Source: Bodleian Library, Ref. No. C 17. 70 Oxford (7)

1883 and 1902, respectively. In both cities we can see the same concentration of breweries and pubs in the poorer central areas of the town—at Oxford in St Ebbe's and at York in the central areas coloured grey on Seebohm Rowntree's drink map; beer-houses cluster in the working-class areas of Oxford—in Jericho, St Clement's, and St Ebbe's. In both cities there is a concentration of pubs in the central trading areas to cater for the suburban visitor: in Oxford near Carfax and the cattle market, and in York's main Fossgate-Walmgate thoroughfare. But in both cities pubs are almost absent from the *best* central streets—from Oxford's Beaumont Street, and from York's Bootham, Monkgate, and Clifton. In Oxford, of course, the colleges had their own drink facilities, and there was no need for many pubs east of the Turl. In both towns, pub geography closely reflects major transport routes. Pubs cluster at the main entry-points to the city created by Oxford's peculiar water geography—

at Folly Bridge, Magdalen Bridge, and Park End Street. Rowntree reports that of the 236 on-licences at York in 1899, 111 had two entrances, 11 had three, and 1 had four.[32] Significantly, the one approach to Oxford notably lacking in pubs is the approach from the north, along St Giles and the Woodstock and Banbury Roads. Pubs are almost entirely absent from the fashionable suburb of North Oxford. The only type of drink required by the dons and professional people of North Oxford was wine to be consumed at home or in college, and wine-shops probably account for the licences marked 'other' on the west side of St Giles in the 1883 drink map. At York there are 136 licences of all kinds within a circle of a quarter-mile radius in the centre of the city; but doubling the radius adds only 121. With pubs as with churches, twentieth-century rationalization involved substituting suburban going concerns for central decayed institutions: with pubs, through compensation and licence transfer: with churches, through reunion of warring sects and concentration of resources.

These extensive transport, recreational, and political roles of the pub were for the first time powerfully challenged during the nineteenth century—partly by economic influences like the railway, and partly by religious and social groups who believed that social stability was best preserved through fostering sobriety and not drunkenness. In 1873 T. H. Green passionately rebuked W. V. Harcourt, Liberal M.P. for Oxford, over his libertarian attack on temperance legislation: 'I can scarcely think that, if you had seen much of the life of the working classes at close quarters, you c[oul]d have had the heart to speak as you did. Even here in Oxford . . . any one who goes below the respectable classes finds the degradation & hopeless waste wh[ich] this vice produces meet him at every turn.' Hear also the cries of Beatrice Potter from the East End in 1886: 'there are times when one loses all faith in *laisser-faire* and would suppress this poison at all hazards, for it eats the life of the nation.' It became common at temperance meetings to rebut the free traders and libertarians by quoting the working man's wife who said 'I can get my husband, sir, past two public-houses, but I cannot get him past twenty.'[33] When 'Saunterer' toured Bradford dram-shops in 1871, his response epitomized hostile reactions. 'Everywhere, the sights we see are saddening in the extreme', he wrote; 'poverty, misery, and wickedness go hand in hand.'

Still more serious, it was widely believed that pubs fostered revolutionary activity. Pubs assembled working people *in crowds*, and encouraged them to consume an article which might subvert their customary rational appreciation of the Government's power: wine-shops had indeed helped to transmit revolutionary ideas in France after 1789, and temperance reformers were eager to substitute a safe domesticity for the ominous public life of the working class. Many plots were hatched in pubs, whether it be for a poaching raid, for the Luddites' attack on Cartwright's Mill in 1812, or for the German Democratic Society's Continental conspiracies at the Red Lion, Soho, in the 1840s. Some pubs were notorious shrines for the worship of criminals like Jack Sheppard or Dick Turpin; several were well-known 'flash houses', where the police bargained regularly with the criminal underworld. In 1884, J. M. Weylland described the slums which were destroyed when New Oxford Street was built:[34]

> a criminal once reaching this 'city of refuge', considered himself safe from arrest. Within its precincts were many low beer-houses. The chief of these was known as 'Rat Castle'. Forty years ago, just before its removal, we penetrated to it, and found its low ceilinged, wretched rooms filled with desperadoes and youthful thieves. Bull-dogs, and others trained for ratcatching, mingled with the people, all of whom were more brutalized than the animals.

To the outsider, public-house debating clubs seemed merely to foment popular discontent. There were, of course, a few middle-class discussion groups in early-Victorian London pubs (Plate 38), but respectable debaters soon moved off to their clubs, temperance halls, literary and philosophical societies, and mechanics' institutes, leaving the publican to preside over the growth of working-class articulateness and self-confidence. During the 1839 Chartist convention Peter Bussey sent back reports to his Bradford beer-house, which was 'like a theatre; there was a rush for early places, and all paid for admission.' London's debating clubs were regularly reported by Ernest Jones in his *People's Paper* during autumn 1858, and in the Cogers' Hall at the Barley Mow off Fleet Street, many famous mid-Victorian radicals could be heard. In this 'long low room, like the saloon of a large steamer', dim with tobacco smoke, about a hundred debaters sat at the three long narrow tables which ran the length of the room; they were surveyed by departed distinguished members from the dingy portraits on the walls, and by the chairman from his elaborate seat on the dais. This was only the most famous of several such London debating clubs.[35]

At Birmingham's Hope and Anchor Inn, Navigation Street, a Sunday Evening Debating Society met continuously from the mid-1850s to 1886, and greatly fortified local radical opinion. The survival of its minute-books enables us to discuss its activities quite fully. At the meeting on 17 January 1886 George Bill, an old and regular speaker, described the Society as 'the best Sunday School for politics that had ever been established in our Town', and thanked Robert Edmonds, the landlord, who had 'thrown open his rooms for more than 30 years to enable us to educate each other'. From 8.30 to 11 every Sunday evening, almost without a break, twenty to forty members and (if the year 1864 was typical) over 150 spectators gathered to hear the topics introduced by regular speakers and then debated. The range and quality of the debates are remarkable: between 19 July 1863 and 6 November 1864, thirteen debates went on domestic politics, fourteen on foreign affairs, six on literary and cultural topics, five on religion, four each on labour questions, social problems and crime, two on science, and one on local affairs.[36]

The debaters' stamina is also remarkable. On 15 August 1858, so vigorous was the discussion on the motion 'Does the Mind of Man eminate [*sic*] from the Brain or from a higher Source', that an adjourned debate lasting two-and-a-half hours had to be arranged. From 7 August 1870 to 5 March 1871, thirty-two consecutive weekly debates were held on the Franco-Prussian War; about a hundred voted in the concluding debate, which condemned the French almost unanimously. In 1874 there were eight consecutive debates on miracles and seven nights were later devoted to Gladstone's pamphlet on Vaticanism. So excited was the Society in 1876 that it discussed the Eastern Question continuously once a week for over six months. The debates were permeated by a deep historical awareness, and could be ferocious in their moral judgments. When the Society debated the motion 'Was the Motive of Henry the 8th. at the Reformation of a Religious, Political, or Libidinous Character' on 9 June 1861, four votes were cast for the first, two for the second, and twelve for the third. In debating the motion 'What would be the best Mode of ensuring the Moral Character of the Priesthood?' on 3 August 1862, Mr Moreten, a regular speaker, won thirteen votes for castration—as against eleven for improving the education of women in morality, and only two for selecting priests more skilfully.

There was much to frighten the outside observer: the motion on 23 January 1859 for working-class enfranchisement was unopposed, and the monarchy lost by thirty-nine votes to sixteen in a debate on republicanism on 22 March 1863. On two occasions, 26 August 1866 and 21 April 1867, the Society was visited by prominent London Reform Leaguers: these included George Howell, who joined in the debates. The Society formally sent its condolences to the family of Ernest Jones on 1 February 1869: 'his Memory will long be cherished as one of the most able and honest advocates of the rights and liberties of the people that has ever lived to adorn our history with matchless eloquence.' And by twenty-three votes to fifteen, on 11 June 1871, the Society supported the Paris Communists. Nor was its radicalism exclusively political. On 21 August 1859, the strikers in the London building trade were supported by twenty-two votes to two: on 16 October 1864 the Midlands miners, on strike at the time, were supported by forty-five votes to eighteen: and on 12 March 1865, thirty-three voters branded the ironmasters' lock-out as 'unjust and tyranical [sic]', whereas only fifteen supported a milder motion hoping for arbitration. On 6 February 1859 the motion 'Has the Establishment and extension of Machinery tended to decrease the Wages of the employed' received unanimous support.

The cultural and political threat from pub culture lurked at the back of many a Victorian politician's mind: it was occasionally discussed in public when uncovered by the indiscretion of a Robert Lowe or by the wrath of a Carlyle. When the slums had been cleared from St Giles, the East End became identified with everything feared by the well-to-do, and the drinking habits of the poor were regarded with a mixture of fear and fascination. Several temperance reformers specialized in retailing to the rich the barbarities of the poor. In his prurient and unpleasantly moralizing *Night Side of London* (1857), the prohibitionist J. Ewing Ritchie exposed slum-life to upper-class eyes; and by the 1870s G. W. McCree, prominent missionary and temperance reformer, was profiting from lectures on topics like 'Lights and Shades of Life in London'.

In late-Victorian London, as in early nineteenth-century Paris, the worlds of crime and of labour were merging in the middle-class mind; in the melodramatic 'contrasts' often described in temperance works, Ratcliffe Highway's drunkenness symbolized the East End's threat to the West End's prosperity: 'there is not a sin which the imagination of man can conceive which is not rife in that north Bank of the river Thames', Archbishop Manning told a prohibitionist meeting in 1871.[37]

Yet it would be quite wrong to imply that a united working class, entrenched behind the pub, faced a united and exclusively middle-class temperance public. Such an interpretation would make nonsense of nineteenth-century temperance history. In reality the temperance movement was one of several Victorian reforming crusades which transcended the gulf between employer and employee, and emphasized instead the gulf between 'rough' and 'respectable' working men. During the nineteenth century, at least three ways of containing pub culture presented themselves: all three attracted considerable support from working men. Briefly, these three temperance strategies were restriction, prohibition, and insulation.

The first was embodied in the steadily extending Victorian restraints on drinking hours and on children in pubs. Although the Sunday-trading riots of 1855 held up restriction of opening hours for many years,[38] working-class opinion was divided both during 1855 and during the better-known licensing crisis of 1871–2. Some libertarian working men pointed to the aristocratic clubs, and branded licensing legislation as class legislation; but many others (Lovett and several ex-Chartists not least among them) were more apprehensive about a drunken and ignorant populace. Such men attacked the pub, not because they shared middle-class fears that it fomented radicalism, but because they believed that only a *sober* radicalism could ever be really effective. Working-class temperance enthusiasts are therefore sometimes found outside the formal temperance movement—operating either as individuals or as members of the secularist and self-improving working men's temperance groups which congregated on the fringe of official temperance organizations. But many respectable working men lent impetus to the second temperance strategy: to prohibitionism, as championed by the United Kingdom Alliance.

Founded at Manchester in 1853, the Alliance argued that ratepayers should enjoy the landlord's power to ban the drink trade from any locality. Several London landlords, including the Marquis of Northampton and the Dukes of Westminster and Bedford, excluded pubs from their property. Whereas there had been seventy-four pubs on the Bedford estate in 1854, by 1893 the owner's policy had reduced their number to thirty-four. On the 1887 map, the Duke of Bedford's Bloomsbury and Fig's Mead estates stand out from their surroundings by their lack of red spots. It was even argued that leasehold enfranchisement would deprive society of the great landlords' benevolent policies. But to prohibitionists, who were always democrats above all else, temperance by fiat was a very bad second-best. Besides, the objective of these landlords was not so much temperance as increased prosperity for their estates.

Prohibitionists were fired by zeal for local self-government and for the moral progress of the masses. Their cry was the Permissive Bill, or local option—that is, for prohibition enacted by a two-thirds ratepayer majority in a locality. The Alliance temptingly invited its supporters to mount simultaneous attacks on aristocratic privilege and pauper idleness. Prohibitionism attracted the middle levels of society, cemented the radical alliance between Nonconformists and respectable working men and, by making local option a major late-Victorian political issue, extended the Liberal Party's popular base. It was a policy so reliant upon the moral idealism of slum-dwellers that the Fabians later regarded it as utopian: 'legislation on the principle of asking the blind to lead the blind', they pronounced in 1898, 'is not up to the standard of modern political science': only a more broadly-based representative assembly could be expected to curb the drink trade.[39] Not till the end of the century did the Alliance begin to lose its democratic image and seem to be promoting a measure which would merely enable wealthy property-owners to exclude vulgarity from their midst; in the mid-Victorian period, Alliance leaders were in the vanguard of progressive thought.

'There are more than 600 Members sitting in this House', declared Samuel Smith in a local option debate of 1883, 'and I will venture to say that there are very few of them who have a public-house within sight of their residence.' The prohibitionists' parliamentary agent, J. H. Raper, regularly held up a sovereign at temperance meetings and offered it to anyone who could name a magistrate who had licensed a pub next door to his own house. With pubs, as with brothels, the rich were better placed than the poor to defend their respectability and property. Lady Henry Somerset, a prominent late-Victorian temperance reformer, produced maps before the Royal Commission on Liquor Licensing in 1897, and asked why a thoroughfare was held to justify pub licences in Whitechapel, but not in Belsize Park. Pubs, she concluded, 'exist absolutely regardless of the needs of the neighbourhood, and . . . the only places which are comparatively free are those places which are rich enough to maintain an effective opposition to their establishment'.[40]

Prohibitionist speakers often advertised the plight of the respectable working man, beleaguered by the pubs he never patronized. At a temperance meeting in 1871, Manning[41] claimed that the rich banned pubs from upper-class areas, yet if a pub were established

> in a part of the town where honest working men and their families live, where peace and quiet have hitherto reigned . . . between the dwellings of two honest labouring men . . . their wives and children . . . have to live all the day long and all the week round, with this moral pestilence on their threshold; and . . . the walls of the dwellings are so thin that the noise of revelry and the words, it may be, of impurity and blasphemy are heard in the chambers where they rest. But this atmosphere of pestilence will hang about the dwelling of the poor man, and he has not power to abolish it.
> He has no means whatever to purify the street where he lives, and to protect his family from infection.

25 A typical inn yard—the King's Head, Southwark, in 1900. The inn's major transport functions had long since been taken over by the railway, but such yards sometimes became loading bays for warehousemen. There are two such firms operating from this one, and the galleries themselves no longer serve the inn alone.
Greater London Council

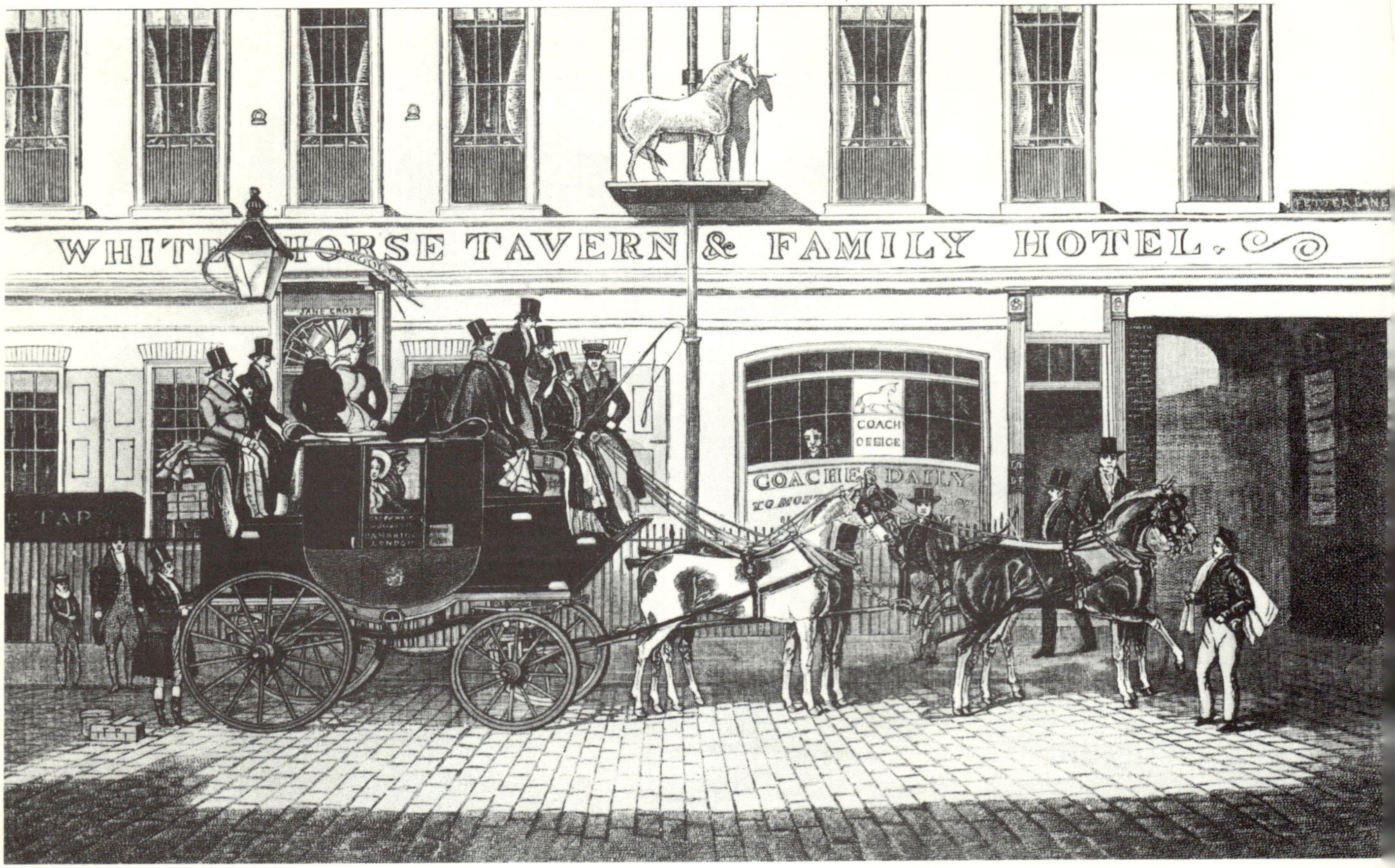
WHIT
ORSE TAVERN & FAMILY HOTEL.
COACH
COACHES DAILY

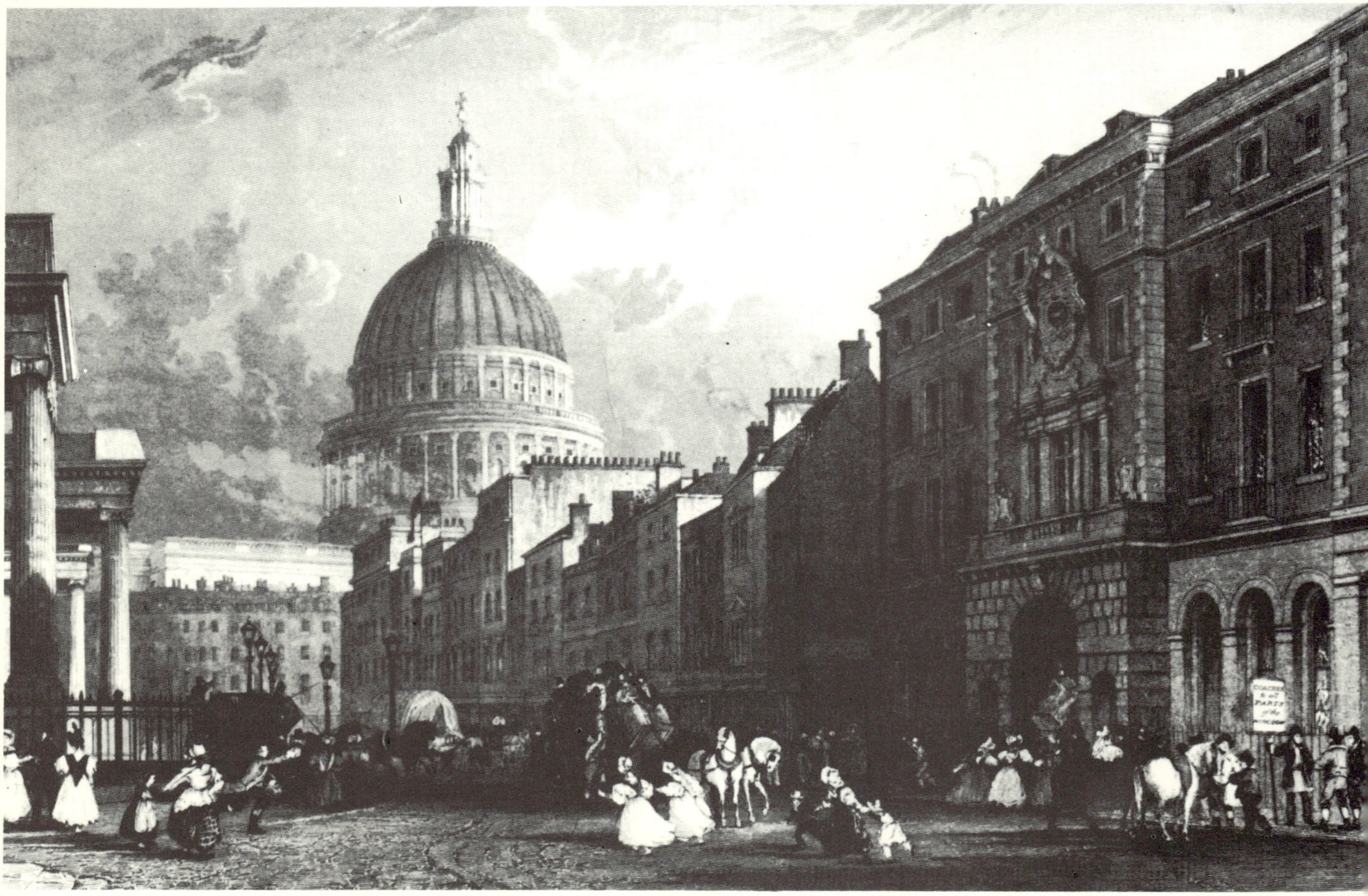

26 *above left* The pub serving road transport in its heyday. The *Cambridge Telegraph* setting off from the White Horse in Fetter Lane. After the print by J. Pollard in the John Johnson Collection.
Bodleian Library

27 *left* A coaching inn nearing the end of its coaching days. The Bull and Mouth about 1830, shortly before becoming the Queen's Hotel, St Martin's-le-Grand. The placard in the bottom right-hand corner advertises coaches to all parts of the kingdom. After a steel engraving of a drawing by T. Allom.
Guildhall Library

28 *above* Idleness, extravagance, drunkenness, squalor: a London street-corner scene on a Sunday morning, from *Illustrated London News*, xxix (1856), 578.

29 A London ginshop interior in 1852. Crowded, smoky, noisy, this was the kind of atmosphere which struck 'Saunterer' so forcibly when he visited the Bradford ginshops in 1871. The combination of sporting raffishness with humiliating squalor was attacked at all points by the temperance movement. This engraving by T. B. Smithies comes from *Working Man's Friend*, i, new series (1852), 56.

30 The Old Oak, Mansfield Road, Hampstead. A typical street-corner brewers' pub of the late nineteenth century, complete with discreet frosted glass, prominent signboards, several entrances, the faded relics of the gin-palace's classical style, and a magnificent set of wrought-iron gas-lamps. *R. B. Sawrey-Cookson*

31 *above* Tom Spring's Parlour, *c.*1840. A centre of sporting gossip for all classes, run by a retired prize-fighter at the Castle in Holborn. *Guildhall Library*

The Harmonic every Evening.
New and Splendid Rooms.
CYDER CELLARS,
MAIDEN LANE.   COVENT GARDEN.

JOHN REGAN

Having at an immence expense completed his improvement trusts his exertions will merit from a discerning Public, a continuance of that support and general Patronage which this Old Established and Favourite Place of Resort has for so many Years experienced.

Wines, Chops, Spirits Dinners, &c., of the best Quality,   **Moderate Charges.**

32 *below left* The Cyder Cellars, Maiden Lane, a centre of London's underworld, and one of the ancestors of the music-hall, from a handbill in the Norman Collection. *Guildhall Library*

33 *above left* The bright window of the pub contrasts strongly with the drab surroundings, as wives wait for their husbands to leave off drinking on pay night. Temperance reformers sided firmly with wives and children in their claim on working men's wages. From S. C. Hall, *The Trial of Sir Jasper* (1872), p. 3

34 *above right* Temperance reformers claimed, sometimes with excessive sentimentality, to be defending women and children against male selfishness. This picture records an alleged intervention at an Aylesbury temperance meeting by a railway employee. He said that on this occasion his five-year-old daughter, whom he carried every night to the beer-shop, begged him 'Father, don't go.' He turned back (and became a teetotaller) when he felt a tear fall from her as he was about to enter. Temperance tracts often emphasized the theme of the virtuous child reforming the drunken parent. From an engraving in *Band of Hope Review*, 1 March 1867, p. 297.

35 *above* Suburban entertainments at the Rosemary Branch, Islington, one of several suburban tea-gardens catering for weekend recreation in the early nineteenth century. Their prosperity was short-lived because the demand for building land took away their gardens and 'views'. From the Crace Collection.
British Museum

36 *below* The Surrey Music-Hall, Lambeth, where the audience was free to move about during performances; in other music-halls the audience remained seated while waiters took the orders. From an undated picture in *Paul Pry* no. 12, unpublished collection 'Pleasure Gardens of South London', II, p. 96. By courtesy of the Trustees of the London Museum.
London Museum

37 Dinner at the London Tavern for Sir Charles Napier in 1849. The leading pubs of London could still cater lavishly for the best company. For humbler gatherings in slum areas, the publican's 'Large Room' was indispensable. From a print in the Norman Collection.
Guildhall Library

38 A middle-class discussion group in session at the Belvedere, a London pub in Pentonville in the 1850s, at ten o'clock in the evening. From G. A. Sala, *Twice Round the Clock* [1859], p. 288.

39 Signing the pledge of total abstinence in Sadler's Wells Theatre, 1854. Temperance meetings were sometimes held in theatres during the 1850s, partly because large numbers were interested in the subject, but also because such things could also be entertaining. This drawing was made by George Cruikshank, a temperance zealot himself, and comes from *Illustrated London News*, xxiv (1854), 465.

40 The Band of Hope, Littlemore, Oxon. Towards the end of the nineteenth century, the temperance movement, having failed to win over the adults, sought to protect the children. Children could hardly refuse to abjure temptations they had scarcely experienced, especially when asked to do so in the schoolroom, as in this watercolour by Miss E. D. Herschel.
Bodleian Library

Prohibitionists were among the first free-trading Liberals to seek social progress through state intervention, for a respectable working man surrounded by drink-shops could hardly be expected to self-help his way to prosperity.

Prohibitionists spiced their diagnosis with hints of aristocratic conspiracy: they claimed that pubs were 'imposed' on respectable working people by an aristocracy anxious to keep them socially, educationally, and politically subordinate. 'I hold that no more efficient means for corrupting a people can be found', wrote William Lovett, 'than that of blending their amusements with the means of intoxication.' He was particularly bitter on this score because his National Hall school at Holborn was denied the music licence which the magistrates granted to the publican who succeeded him there in 1857: 'publicans can always have such licences', he wrote tartly, 'but not so those who would have music apart from the means of intoxication.' Pub geography in nineteenth-century London did of course protect the sobriety of the rich and obstruct the respectability of the poor; but to go further, and to ascribe this to deliberate aristocratic design, was to lose a sense of proportion. As Baron Bramwell pointed out, existing pub geography stemmed only from the fact 'that a public-house in a square in which the rich live would not pay.'[42]

The third temperance strategy was to insulate oneself from temptation. Insulation could be sought in two ways. The easy way was for the abstainer to husband his savings and move away from the slum: 'the home that had satisfied my wants as a drinker was not in harmony with my self-respect as a teetotaller', wrote Thomas Whittaker of his conduct after signing the pledge in the 1830s, 'and I soon put myself in possession of a house rented at twelve pounds a year.' In the 1820s, licensing restriction had been opposed by speculative builders eager to add value to their suburban estates by providing drink facilities. But by the 1880s, the climate had so changed that it was profitable to *exclude* pubs from one's estate. The Artisans, Labourers and General Dwellings Company specialized in building drink-free suburban utopias—at Queen's Park, Shaftesbury Park, and Noel Park. Like so many utopias, the nineteenth-century temperance estate suffered by the fact that it was embedded in a corrupt society: pubs clustered round its fringes. Yet Shaftesbury Park had charms for at least one of its residents: during the past year, he wrote in 1875, 'I have never heard the song of a drunkard, but during the summer evenings I have often listened to the songs, &c., of the people from the open windows.'[43]

This residential separation of teetotallers from drinkers in late-Victorian London probably accentuated the slumland weakness of a Liberal Party which sought to attract all classes. It was certainly more courageous to face the evil directly from inside the slum, for this involved braving ridicule and even violence from one's own class. Slum-dwellers disliked working men who 'gave themselves airs', and teetotallers were often insulted. Hence arose a second form of insulation. By taking the pledge, the sober slum-dweller could protect himself against temptation and insult by joining a group of the like-minded (Plates 39 and 40). If successful, such groups could purchase a temperance 'hall', or attach themselves to a Nonconformist chapel or mission.

The Victorian City: Numbers of People

But it was always a struggle, particularly in London, where dissenting chapels were weaker than in the North of England, and where temperance effort within the working class was fragmented by the gulf between Nonconformity and the relatively powerful secularist movement. In Booth's thirty-three districts, religious institutions exceed pubs in number only in the freak district of Holborn. The 1899 map shows that religious institutions, unlike pubs, were poorly sited for influencing the passer-by. Anglican churches (until the 1860s, likely to be anti-temperance) were sometimes well placed, because the community had often grown up around them. But Nonconformist chapels and missions usually huddled in backstreets and backyards; unlike pubs, they seldom clustered at the crossroads. In Booth's districts, an increased percentage in poverty is not accompanied by an increase in the ratio of religious institutions to drinking places: nor is there any significant relation, when population is held constant, between the number of religious institutions and the number of pubs. Yet the drinker lacked the temperance reformer's skill in making his political presence forcefully and continuously felt. From a host of obscure contests between drinkers and teetotallers in Victorian towns, the stuff of Victorian politics was made. Here religion impinged upon social class, recreation upon party politics.

These contests were at times very bitter: yet the bitterness conceals a fundamental similarity between the roles of pub and temperance society. Both provided individual working men with the chance of rising in society; both gave working men collective experience which could easily be applied to politics and wage-bargaining. Both recognized the social evils rife in Victorian towns: the publican palliated them, the temperance reformer tried (however ineffectually) to remove them. Teetotallers and heavy drinkers may even have tended to appear together in the same communities. We know from recent social surveys that teetotalism is now most common in areas of heavy drinking, and we have several individual nineteenth-century assertions that drunkenness was more prevalent in early-Victorian Glasgow and Lancashire, where the temperance movement was strong, than in London and the southern rural counties where the temperance movement was weak. But we lack statistical proof of these assertions.[44]

There are many other similarities between pub and temperance society. Both struggled to protect the newcomer from the loneliness and strangeness of town life. Urban sociologists tell us, somewhat pretentiously, that 'the city is characterized by secondary rather than primary contacts'—contacts which are relatively anonymous, segmentalized, utilitarian. Social anthropologists tell us that modern Africans moving from country to town often form new types of non-tribal voluntary association; these small private societies unite the like-minded, replace the community ties which lent structure to rural life, integrate the newcomer into urban society by initiating him into business habits, and assign prestige through elections to petty offices.[45] Organizations of this sort existed in nineteenth-century England, yet serious historians have so far ignored them. Instead, they tend to use the term 'urbanization' rather clinically, and conceal from us the shock of transition experienced by so many rural immigrants at the time. We need to know *how* the immigrant was socialized, what organizations catered for him, and how these were related to the contemporary social and political structure.

The temperance society, like the pub, could ensure that the newcomer to Victorian towns received recognition, education, employment, friends, advice, and support; it could encourage him to resist the temptations which the anonymity of urban life made so seductive; and it could provide a new framework for daily conduct. Like the urban chapels to which they were so often attached, temperance societies were 'havens of refuge in an utterly strange and alien landscape. Here, and here only, could the old links of identity with home be maintained, the familiar forms of worship be recovered.'[46] Their very names, not to mention their roles, closely resemble those of the voluntary association formed by newcomers to the modern African city.

Nor did pub and temperance society differ so markedly in their political tendency. Pub society was never so subversive as it seemed to the respectable outsider. The Conservative Party realized this when it promoted working men's drinking-clubs from the 1860s; indeed, in some ways the temperance society was far more radical. Pub society's traditionalism and chauvinism made drinkers unreceptive to Continental doctrines of revolution. 'Saunterer''s dram-shop criminals were less dangerous than he thought; they did not threaten the successful entrepreneur—they were his mirror-image. The Birmingham Sunday Evening Debating Society was really quite a conservative body. Mr Nuttall on 12 September 1864 raised twenty-two votes for his motion that the Chartists failed from propagating 'wild and impract[ic]able ideas and endeavouring to carry them out by physical force [,] thus setting all other classes against them'. The Society usually ensured that both sides of any question were presented, and like many other pub associations it encouraged respect for House of Commons procedure. By the 1880s and perhaps before, it was operating decidedly from inside the existing political party system. There was much presenting of testimonials, mutual congratulation, and Christmas punch-drinking at the landlord's expense: in short, much consciousness of respectability. The Society was, in fact, like the temperance society, a 'small-scale success system': it did not threaten the social structure.[47]

It is clear, then, that to a large extent pub and temperance society fought so fiercely because their roles were so similar. Indeed, as their rivalry grew more intense, temperance society and pub appropriated each other's attractions. Pubs acquired organs, hymn singing, and soft drinks; temperance societies acquired professional entertainers, sent their members into pubs to spread the new gospel, and even purchased pubs (though only for the purpose of turning them dry). In this context, then, as in the very different context described by Gladstone, 'there is often, in the courses of this wayward and bewildered life, exterior opposition, and sincere and even violent condemnation, between persons and bodies who are nevertheless profoundly associated by ties and relations that they know not of.'[48]

The Victorian City: Numbers of People

Notes

1   See J. R. Kellett, *The Impact of Railways on Victorian Cities* (1969), and the essays by John Foster and E. P. Hennock, in H. J. Dyos, ed., *The Study of Urban History* (1968). I acknowledge gratefully here the most generous help I have received in interpreting the statistics in this article from my colleague Mr A. J. Glyn, Corpus Christi College, Oxford; Miss H. Karayannis helped me with the computations. Professor Peter Mathias, All Souls College, Oxford, was kind enough to provide valuable bibliographical help at an early stage.

2   The two maps are the National Temperance Publication Depot's *The Modern Plague of London* (c. 1887), and 'London 1899–1900', a map in the final volume of Charles Booth's *Life and Labour of the People in London* (1903 edn).
Miss Joan Pollard of the London Museum kindly guided me to the former, which is kept in the London Museum, Kensington Palace; she also gave me much other painstaking help. For ideas on the study of local temperance societies, see my 'Temperance Societies', *Local Historian*, viii, nos 4 & 5 (1968–9). The social and other roles of drink and drinking-places in the 1820s and 1870s are fully discussed in my *Drink and the Victorians* (1971), chs 2, 14, 15. There is a mass of antiquarian literature on pubs, most of it worthless; but for an excellent discussion, see W. B. Johnson, 'The Inn as a Community Centre', *Amateur Historian*, ii, no. 5 (1955).
The term 'pub' is used in this article to denote any type of public drinking-place, and not synonymously with the specialized term 'public-house'; there were several categories of 'pub' in nineteenth-century England, and these have been distinguished where necessary. The expression 'pub density', as used here, refers to density in relation to population, not in relation to area.

3   B. W. Matz, *The Inns and Taverns of 'Pickwick'* (1921), p. 9; I am most grateful to Professor Philip Collins of Leicester University for generous help on Dickens's attitude to pubs.

4   For early railways, see George Stephenson, in *Temperance Monthly Visitor* (Norwich), October 1860, 79; J. A. R. Pimlott, *The Englishman's Holiday* (1947), p. 88.
For Gladstone, see National Temperance League, *Annual Report 1881–2*, 26.

5   *Collins Illustrated Guide to London & Neighbourhood 1871* (1871), p. 118;
see also N. C. Selway, *The Regency Road: The coaching prints of James Pollard* (1957), pp. 21, 29; Charles Knight, *London*, IV (1843), pp. 310ff.; VI (1844), p. 313.
For canals, see L. T. C. Rolt, *Narrow Boat* (1944), p. 59—a reference I owe to Christopher Harvie of the Open University.

6   J. Dunlop, *The Philosophy of Artificial and Compulsory Drinking Usage* (1839), p. 306; cf. *Parliamentary Papers* (*P.P.*), 1898, XXXVI, Royal Commission (R.C.) on the Liquor Licensing Laws, Third Report: Minutes of Evidence (C. 8694), Q. 26,312.
For Leeds, see G. C. Dickinson, 'The Development of Suburban Road Passenger Transport in Leeds, 1840–1895', *Journal of Transport History*, iv (November 1960), 215; see also J. R. Kellett, op. cit., pp. 87ff., 139–40, 278, 281; and E. A. Pratt, *The Policy of Licensing Justices* (1909), ch. 8.

7   *P.P.*, 1898, XXXVII, R.C. on the Liquor Licensing Laws: Statistics (8696), p. 44 et seq.

8   *P.P.*, 1877, XI, Select Committee (S.C.) (H. of L.) on Intemperance, Third Report: Minutes of Evidence (418), Q. 8404.

9 Figures in G. B. Wilson, *Alcohol and the Nation* (1940), pp. 394–6; R.C. on Liquor Licensing Laws, Statistics, pp. 30–1; *London Statistics 1903–1904*, xiv (1904), 369; cf. H. J. Dyos, *Victorian Suburb* (Leicester, 1961), p. 154, and the interesting table in Charles Booth, *Life and Labour*, final volume (1903), p. 221.

10 The boundaries and poverty classifications of Booth's districts are in *Life and Labour*, 1st series, II, Appendix, pp. 1–60; drinking facilities have been calculated from Booth's 1899 map. The following districts have been analysed: 44C, 2B, 48E, 48C, 2D, 22E, 40A, 75B, 79B, 66A, 40B, 60A, 78A, 48D, 43A, 23A, 59B, 42A, 74C, 43F, 43G, 41E, 2A, 63B, 80C, 23B, 41D, 48F, 92A, 43E, 86E, 41A, 88F. Two difficulties arise: the districts were classified ten years before the map was compiled, and during that decade the L.C.C. was pursuing a vigorous temperance policy. Second, Booth gives no details of the *size* of the drinking places.

11 B. Seebohm Rowntree, *Poverty: A study of town life* (Nelson edn, n.d.), pp. 371–83, 390.

12 For brewery locations, see *Handbook to London As It Is* (John Murray, new edn, 1879), p. 79; G. Thorne, *The Great Acceptance* (1913), pp. 20–1.

13 Quotations from M. Young & P. Willmott, *Family and Class in a London Suburb* (1960), p. 130; S.C. on Intemperance, Q. 8408. For street recreations, see A. R. Bennett, *London and Londoners in the Eighteen-Fifties and Sixties* (1924), pp. 47, 60. For drinking customs, see J. Dunlop, op. cit., p. 306; for pub entrances, see J. Rowntree and A. Sherwell, *Temperance Problem and Social Reform* (5th edn, 1899), p. 84.

14 Quotation from *Temperance Penny Magazine*, January 1836, 6. On gin-palaces, see *Sketches by Boz* (2nd edn, 1836), I, pp. 280, 282; S. and B. Webb, *The History of Liquor Licensing in England, principally from 1700 to 1830* (1903), pp. 120–1; John Hogg, *London As It Is* (1837), p. 287.

15 Simmel quoted in P. K. Hatt and A. J. Reiss, eds, *Cities and Society* (Chicago, 1957), p. 643; for an excellent account of the evolution of pub interiors, see M. Gorham and H. McG. Dunnett, *Inside the Pub* (1950), pp. 64–5, 94–113.

16 Weylland, S. C. on Intemperance, QQ. 9141, 9124. Ruskin is quoted in E. T. Cook, *Life of John Ruskin* (1911), I, p. 308; cf. Ruskin's preface to the 3rd edn of his *Stones of Venice*, in *Works*, ed. Cook and Wedderburn (1903), IX, p. 12. Tinker Taylor is in Thomas Hardy, *Jude the Obscure* (1957 paperback edn), p. 187; I owe this reference to Mr John Stothard, formerly of Corpus Christi College, Oxford. See also A. Shadwell, *Drink, Temperance and Legislation* (1902), pp. 204–5.

17 Quoted from J. Rowntree and A. Sherwell, op. cit., p. 83. For rating, see G. B. Wilson, op. cit., pp. 291–2; he prints figures for persons per pub on p. 236.

18 'Saunterer' in *Bradford Observer*, 19 January 1871: I owe this reference to Miss Peggy Rastrick, Killinghall Road, Bradford, who is studying the history of local temperance activity. The phrase 'masculine republic' was coined by H. W. J. Edwards, *The Good Patch* (1938), p. 158; I owe this reference to Dr W. R. Lambert, formerly of the Department of History, University College, Swansea. See also L. Chevalier, *Classes laborieuses et classes dangereuses à Paris pendant la première moitié du xixe siècle* (Paris, 1958), pp. 225, 299.

19 Cobden quoted by J. A. Hobson, *Richard Cobden: The international man* (1918), p. 63; for Balfour, see *P.P.*, 1854, XIV, S. C. (H. of C.) on Public Houses:

Minutes of Evidence (367), Q. 1265; the observer was J. E. Ritchie, *The Night Side of London* (1857), p. 79. For Hartwell, see *Staffordshire Advertiser*, 14 November 1868, p. 6. I am most grateful to Paul Anderton, of the Historical Association's North Staffordshire branch, for this reference. For 'cock and hen' clubs, see *P.P.* 1849, XVII, S.C. (H. of C.) on Public Libraries: Minutes of Evidence (548), Q. 2783 [Lovett]. For London's underworld, see Guildhall Library, *Norman Collection*, G.R.1.1.5 (Inns and Taverns, II), 'Coal Hole' and 'Cyder Cellars'. For generous help in guiding me through this excellent collection, and for much other assistance, I must thank Mr Ralph Hyde, the Guildhall Library. For Judge and Jury clubs see James Greenwood, *The Wilds of London* (1874), pp. 99ff., and the illustration in T. McD. Rendle, *Swings and Roundabouts: A yokel in London* (1919), p. 196.

20 Quotations from J. A. Hobson, *Work and Wealth* (1914), pp. 155–6; William Acton, *Prostitution* (2nd edn, 1870), p. 21; *My Secret Life* (Amsterdam, n.d., British Museum copy), VI, p. 127. For Sir George Grey, see *P.P.*, 1871, XIX, R.C. on the Contagious Diseases Act: Minutes of Evidence (C. 408), Q. 18,152.

21 Quotations from E. P. Thompson, *The Making of the English Working Class* (2nd edn, 1968), p. 445; H. Mayhew, *London Labour and the London Poor* (1861), III, p. 6; Anon., *London in the Sixties (With a Few Digressions)* by One of the Old Brigade (1908), p. 35. For cockfights, see my 'Religion and Recreation in Nineteenth-Century England', *Past and Present*, no. 38 (1967), 118.

22 F. Gale, *Sports and Recreations in Town and Country* (1888), p. 4. See also Taine's *Notes on England*, ed. E. Hyams (1957), p. 36; Guildhall Library, *Norman Collection*, G.R.1.1.5 (Inns and Taverns, II), 'Thomas Spring's Parlour'; John Timbs, *Club Life of London* (1866), II, p. 235; K. Chesney, *The Victorian Underworld* (1970), pp. 268–9.

23 For Marwood, see *London in the Sixties*, p. 161; cf. p. 91ff. Other quotations from E. L. Woodward, *The Age of Reform* (1938), p. 120; Sir Oswald Mosley, *My Life* (1968), p. 18.

24 For 1825, see W. Hone, *Every-Day Book*, I, part 1 (1825), cc. 438, 442. For the Crystal Palace, see Percy Cruikshank, *Sunday Scenes in London and its Suburbs* (1854), p. 12; S. Couling, *History of the Temperance Movement* (1862), p. 221; P. T. Winskill, *The Temperance Movement and its Workers* (1892), III, p. 54.

25 Quotation from *P.P.*, 1892, XVIII, S.C. (H. of C.) on Theatres and Places of Entertainment: Minutes of Evidence (240), Q. 5177 [Matthew Hanly], Q. 864. Music-hall statistics from R. Mander and J. Mitchenson, *British Music Hall: A story in pictures* (1965), p. 19; *P.P.*, 1866, XVI, S.C. (H. of C.) on Theatrical Licences: Report (373), p. 307; *Collins Illustrated Guide to London & Neighbourhood* (1871), p. 124; *Era Almanack*, 1908, pp. 102–3. The *Almanack* does not confront the serious problems involved in defining the term 'music-hall'.

26 'Mass Observation', *The Pub and the People: A worktown study* (1943), p. 160. See also Mander and Mitchenson, op. cit., p. 21.

27 *P.P.*, 1892, XVIII, S.C. (H. of C.) on Theatres and Places of Entertainment: Minutes of Evidence (240), Q. 5171; cf. P. Snowden, *Socialism and the Drink Question* (1908), p. 82.

28 C. E. Lawrence, 'Public-House Museums', *Ludgate* (n.d., *Norman Collection* C. 23.1 T. 1895); *Graphic*, 8 December 1928, 410.

29 For the N.U.W.C. see *Poor Man's Guardian*, 6 July 1833, 217; I owe this reference to Dr Patricia Hollis, the University of East Anglia. Leno is in Bishopsgate Institute, *Reform League Executive Committee Minutes*, 25 August 1865; see also W. Lovett and J. Collins, *Chartism* (1840), p. 48.

30 Lowery, *Weekly Record of the Temperance Movement*, 21 June 1856, 107; Owen, quoted in J. F. C. Harrison, *Robert Owen and the Owenites in Britain and America* (1969), p. 92.

31 Guildhall Library, *Norman Collection*, Box C.23.1 (London Tavern); see also John Timbs, op. cit., II, p. 278.

32 B. S. Rowntree, op. cit., p. 363. The York map is in the 2nd edn, 1902. *The Drink Map of Oxford, 1883* is in Bodleian Library, ref. no. C.17.70 (Oxford), 7.

33 Quotations from Balliol College, Oxford: T. H. Green MSS., Black Box: T. H. Green to W. V. Harcourt, draft reply to Harcourt's letter of 7 January 1873; Beatrice Webb, *My Apprenticeship* (2nd edn, 1946), p. 238. For the working man's wife, see Rolleston, in *Alliance News* (Manchester), 19 October 1872, 739.

34 Quoted from J. M. Weyland, *These Fifty Years* (1884), p. 8. For plots, see F. Peel, *The Risings of the Luddites, Chartists and Plugdrawers* (1968 edn), p. 53ff.; A. R. Schoyen, *The Chartist Challenge: A portrait of George Julian Harvey* (1958), p. 135. For wine-shops in the French Revolution, see G. Rudé, *The Crowd in the French Revolution* (Oxford, 1959), p. 217.

35 For Bussey, see A. J. Peacock, *Bradford Chartism 1838–1840* (York, 1969), p. 20. Cogers's quotation from Guildhall Library, *Norman Collection*, G.R.1.1.5 (Inns and Taverns, II).

36 See the Society's MS. Minutes (2 vols) in Birmingham Central Library, ref. nos 103138–9. Mrs Corke of the Leamington Historical Association kindly referred me to this source.

37 *Alliance News*, 21 October 1871, 674; see also A. E. Dingle and B. H. Harrison, 'Cardinal Manning as Temperance Reformer', *Historical Journal*, xii (1969), 504ff.

38 For a full account of this incident, see my 'Sunday Trading Riots of 1855', *Historical Journal*, viii (1965), especially 238–9. See also *Daily Telegraph*, 20 April 1871, 5; 10 May 1871, 4; 19 June 1871, 5.

39 Fabian Society, *Municipal Drink Traffic*, Tract no. 86 (1898 edn), p. 5. For the Bedford Estate, see D. J. Olsen, *Town Planning in London: The eighteenth and nineteenth centuries* (New Haven, 1964), p. 165.

40 3 Hansard 278, c. 1324 (27 April 1883); R.C. on Liquor Licensing Laws, Q. 31, 377.

41 Manning, *Alliance News*, 5 October 1872, 706–7.

42 Quotations from W. Lovett, *Life and Struggles* (1876), p. 374; National Library of Scotland, Edinburgh, Combe MSS. 7365 f.89: Lovett to Combe, 25 November 1857; C. Fairfield, *Some Account of George William Wilshere, Baron Bramwell* (1898), p. 272.

43 Quotations from T. Whittaker, *Life's Battles in Temperance Armour* (1884), p. 66 and *Alliance News*, 2 October 1875, 636. See also P. T. Winskill, *Temperance Movement and its Workers*, III, p. 113; J. N. Tarn, 'Some Pioneer Suburban Housing Estates', *Architectural Review*, cxliii (1968), 367; Rowntree and Sherwell, op. cit., pp. 212–15. I am most grateful to Dr D. A. Reeder, Garnett College, London, for help on this topic.

44 For Glasgow and Lancashire, see W. Reid, *Temperance Memorials of the Late*

Robert Kettle Esq. (Glasgow, 1853), p. 13; British Museum Add. MSS. 44428 (Gladstone Papers), f.178: Bishop Fraser to Gladstone, 5 November 1870; Joseph Livesey, *Preston Temperance Advocate* (Preston), July 1834, p. 53. For recent surveys, see F. Zweig, *The Worker in an Affluent Society* (1961), pp. 130–1; D. E. Allen, *British Tastes* (Panther edn, 1969), pp. 89, 143–4.

45 Wirth is quoted in Hatt and Reiss, op. cit., p. 54, but see also pp. 61, 631. For social anthropologists, see W. B. Schwab, 'Oshogbo—an Urban Community', in H. Kuper, ed., *Urbanization and Migration in West Africa* (University of California Press, 1965), pp. 102–6; M. Banton, 'Social Alignment and Identity in a West African City', ibid., pp. 143–4; P. Marris, *Family & Social Change in an African City: A study of rehousing in Lagos* (1961), pp. 39, 42; G. Breese, *Urbanization in Newly Developing Countries* (New Jersey, 1966), pp. 87–8, 98. Dr Alan Macfarlane kindly directed me towards much of this literature.

46 I. G. Jones, in G. Williams, ed., *Merthyr Politics* (Cardiff, 1966), p. 52; cf. the revealing quotation from B. G. Orchard, in R. B. Walker, 'Religious Changes in Liverpool in the Nineteenth Century', *Journal of Ecclesiastical History*, xix (1968), 204.

47 J. Foster, in H. J. Dyos, ed., *The Study of Urban History*, p. 294.

48 W. E. Gladstone, 'The Evangelical Movement; its Parentage, Progress, and Issue', *Gleanings of Past Years, 1843–79* (1879), VII, p. 224.

7 The Literature of the Streets

Victor E. Neuburg

The massing of people in cities in the nineteenth century made a vast difference to the scale of many things. It meant not only higher densities on the ground but also new opportunities for communication on many different levels. These concentrations of people rapidly became the means and the ends of mass production, but they also provided the basis for new ways of life that depended at certain levels less on the household than on the street. The urban mass provided, too, a demand for various kinds of mass journalism, of which perhaps the most characteristically urban was that known as 'street literature'. It is a vague category and ought for our present purposes to be fairly narrowly defined. At its widest it can comprise cheap newspapers; sporting journals; song books; almanacks; broadsides; advertisement hoardings; political, religious, freethinking, or commercial leaflets or circulars; printed wrapping material; 'penny dreadfuls' like those issued by Lloyd, Purkess, and others; and even —though this is perhaps more arguable—the dumpy little books, both fiction and non-fiction, of which Milner and Sowerby's 'Cottage Library' was the most comprehensive series, although by no means the only one. More narrowly we can say that the most characteristic and typical kind of street literature was the broadside or street ballad.

  The value of this narrower definition is based upon two factors: first, the Victorian broadside was produced almost entirely for the poor; second, it bore a superficial resemblance to a traditional form of popular literature. From the earliest days of printing, ephemeral sheets of this kind had circulated throughout Europe.[1] The broadside was the most substantial element in the literature of the poor, and there seems no reason to doubt that people living in the Victorian city accepted it as an unquestioned part of their background.

The broadside consisted of popular ballads or prose printed in 'broadside' form and hawked around the streets and in public houses. The sheets were essentially simple to produce. They were printed upon one side only of paper which was nearly always of the flimsiest; and sometimes they carried woodcut illustrations which were not seldom lively, and on occasion lurid. They varied in size, and the use of *ad hoc* typography, combined with the illustrations, gave many of them considerable visual appeal. The cries of the sellers—'Three yards a penny!' or 'Two under fifty for a fardy!'—were one kind of sales promotion; there were other sellers who sang the words of the ballads they had for sale; and there were the 'pinners up' who attached samples of their wares to a convenient wall or hoarding, or who exhibited their stock by affixing it to a pole.[2]

Henry Mayhew moved amongst this fraternity, and what little we know of them is due to his investigations.[3] 'Do I yarn a pound a week?' asked one street-seller of ballads rhetorically in response to Mayhew's query. 'Lor' bless you, no. Nor 15s., nor 12s. I don't yarn one week with another, not 10s. sometimes not 5s. . . . I am at my stall at nine in the morning, and sometimes I have walked five or six miles to buy my "pubs" before that. I stop till ten at night oft enough. The wet days is the ruin of us; and I think wet days increases.' This man sold his sheets at a halfpenny a time, a penny if he could get it; and this seems to have been the usual price, although it must be said that the economics of the street trade are obscure. Depending upon the quality of the paper, songs were bought usually at twopence or even twopence-halfpenny a dozen. Manifestly weather was a crucial factor—on summer days trade was likely to be a great deal brisker than in winter, and it is clear that the seller of street literature depended upon the kindness of the weather for his livelihood quite as much as the mediaeval peasant had done centuries earlier.

From Mayhew's pages we are able to gather something at least of the life of the street-seller. It was precarious, and except upon the few occasions when a sensational murder took the public fancy, little money was to be made out of it. But in general the world of the itinerant vendor remains—Mayhew notwithstanding—obscure. There are one or two books which shed a flickering and uncertain light upon it. *The Life and Adventures of a Cheap Jack* is one, and 'Lord' George Sanger's *Seventy Years a Showman* is another: both show something of the shifting, insecure background of wandering sellers and showmen. From a somewhat earlier period there is David Love's autobiography, and William Cameron, a Scottish pedlar, wrote his autobiography.[4]

The most artless writing of this kind can often provide an imaginative key which sets more orthodox material into a sharper focus. Often this will be in the form of the memories of an old person who recalls the streets of his childhood more than seventy years ago. A description of Hoxton by Albert Jacobs, who was born in Nile Street in 1889, has this precise quality of simplicity and conveys a vivid sense of what it was like to be there:[5]

Life in the streets seemed quite interesting, first thing in the morning the lamplighter came round with his long pole turning out the street lamps and in the evening lighting them. There were many hawkers singing their wares, gypsies to sell you something or offer to tell your fortune, the cats meat man with his one wheel barrow followed by hungry cats hoping he would drop a piece while he cut slices for a customer, the milkman ladling out milk from a churn on a barrow into an oval metal container with a hinged lid or into your own jug. Sunday was the day of the Muffin Man, he came round balancing a large tray on his head packed with Muffins and Crumpets and ringing a bell while calling out 'Muffins'. He often had to go to the bakery for fresh supplies and there would be people there handing in tins containing their Sunday dinners to be cooked. In the afternoon there would be the winkle man calling out from his barrow with winkles, shrimps and cockles . . . There was the Ballad Singer who sung a popular song and sold copies of it for a penny and again there was always the Hurdy-gurdy man, an Italian with a small organ on a short pole and a monkey trained to turn the organ handle . . .

When, however, we come to ask what was the nature of the stock-in-trade of the street-sellers, then we are upon very much firmer ground, for much of it has survived; and this adds a significant historical dimension to the understanding of an important element in Victorian popular culture. In the British Museum there are the Baring-Gould and the Crampton collections, and, in a number of the larger provincial libraries, locally printed sheets are preserved. Not only, indeed, is there no shortage of original material, but also the rambling, untidy books—compilations, perhaps, is a better word—by Charles Hindley offer examples of street literature which but for him would have disappeared.[6] Nevertheless, this mention of the debt we owe to Charles Hindley does highlight the fact that there has been no real attempt to classify or quantify the output of broadsides. The two collections in the British Museum offer no classification either within periods or around episodes, presenting only selections of these sheets mounted in guard books; and it must be borne in mind that, because of the very nature of this literature, its survival poses an immediate problem. It may be that the most popular of its items were those most readily available for preservation; or is it in fact more likely that the less popular items, where copies remained unread and undiscarded, had the better chance of survival?

The material which we have is certainly sufficient to make possible some valid generalizations about its nature. There were four broad categories of street ballad: street drolleries; ballads about the Royal Family or politics; 'ballads on a subject'; and ballads concerning crime.[7]

The first of these covered general themes, in prose as well as in verse, some of which were 'cocks' or 'catchpennies', fictitious narratives offered to a gullible public as though they were true and topical. The keywords in the sales patter for such sheets were 'Horrible', 'Dreadful', 'Murder', 'Love', 'Seduction', 'Blood', and so on.

Mayhew has described the way in which these were sold: 'Few of the residents in London—but chiefly those in the quieter streets have not been aroused, and most frequently in the evening, by a hurly-burly on each side of the street. An attentive listening will not lead any one to an accurate knowledge of what the clamour is about. It is from a "mob" or "school" of running patterers, and consists of two, three, or four men.'[8] They shouted their wares, and the noisier they were the better they deemed their chance of good sales. Imaginary murders, usually with a lurid love interest, seem to have been the most popular, and could by a practised patterer be made to sound extremely plausible. Typical of these—and there were very many of this type—was one entitled 'Shocking rape and murder of two lovers'.[9] This was an account in prose and verse, embellished with a horrifying woodcut, of how one John Hodges, a farmer's son, raped Jane Williams and afterwards murdered her and her lover, William Edwards, in a field near Paxton. 'This', declares the anonymous author, 'is a most revolting murder. It appears Jane Williams was keeping company, and was shortly to be married to William Edwards, who was in the employment of Farmer Hodges.' John, his son, made approaches to the young lady 'who although of poor parents was strictly virtuous.' After the rape, Edwards came upon Hodges, who immediately turned upon him with a bill-hook, which he afterwards used to kill the girl. He was then apprehended and 'committed to take his trial at the next Assizes'. The last two of the 'Copy of Verses' which follows the prose accounts are typical:

> Now in one grave they both do lie,
> These lovers firm and true,
> Who by a cruel man were slain,
> Who'll soon receive his due.
>
> In prison now he is confined,
> To answer for the crime.
> Two lovers that he murdered,
> Cut off when in their prime.

In every way this 'cock' or 'catchpenny' is representative of its type of street literature. Less repetitive, and certainly more interesting, is the large sub-class of 'street drolleries' proper, though it is of course very much more difficult to select any title which typifies this category. Examples are 'The full particulars of "taking off" Prince Albert's inexpressibles'; 'The perpetual almanack'; 'How to cook a wife'; 'The Dunmow Flitch'; 'Secrets, for ladies during courtship'. Some of these are illustrated with woodcuts of which by no means all were relevant to the text—a characteristic of street literature, whose producers appeared to use woodblocks with inconsequential abandon. 'How to cook a wife', for example, has a dramatic woodcut showing a soldier with a knife, menacingly approaching two distressed females—one swooning—in a sylvan glade.

One of the most interesting titles is 'Railroad to hell, from dissipation to poverty, and from poverty to desperation. This line begins in the brewery, and runs through all public-houses, dram-shops, and jerry-shops, in a zigzag direction, until it lands in the Kingdom of hell.' Although undated, the imprint 'T. Such, Union street, Boro'' suggests that this broadside was issued, almost certainly not for the first time, in the 1870s.[10] Such was a commercial printer who carried a large stock of sheets, and the fact that he thought it worthwhile to exploit a temperance theme throws an interesting light upon the saleability of street ballads which took a strong line over the demon drink. Not many did, but the subject was not entirely unusual. More immediately striking, perhaps, is the railway imagery which is used in these leaden-footed verses:

>   Such Taverns as these are Railroads to Hell,
>   Their barrels are engines which make men rebel;
>   Their jugs and their glasses which furnish their Trains,
>   Will empty their pockets and muddle their brains.
>   And thus drunkards ride to Hell in their pride,
>   With nothing but steam from the barrels inside.

A companion piece from the same printer is 'The railway to heaven. This line runs from Calvary through this vain world and the Valley of the Shadow of Death, until it lands in the Kingdom of Heaven.' In this ballad the railway imagery is more pronounced:

>   The Railway mania does extend,
>   From John o' Groats to the Land's End;
>   Where'er you ride, where'er you walk
>   The Railway is the general talk.
>
>   Allow me, as an old divine,
>   To point to you another line,
>   Which does from earth to heaven extend,
>   Where real pleasures never end.
>
>   Of truth Divine the rails are made,
>   And on the Rock of Ages laid;
>   The rails are fixed in chains of love,
>   Firm as the throne of God above.

The final couplet, too, is worth quoting:

>   'My son', says God, 'give me thy heart,
>   Make haste, or else the train will start.'

Examples of technology impinging upon popular literature are unusual. The railway symbolism had its origins in what Americans knew as 'The Great Awakening'— a shrill evangelical revival which found expression in the camp meetings held mostly in upper New York State during the 1830s. The evangelists later brought their message and their sons to England, where 'The railway to heaven' became popular at revival meetings in the North. In fact, Americans absorbed machinery into the imagery of their popular literature much more rapidly than did the English.

The tone is broadly evangelical, and mention is made—always with approval—of the Church of England, the Quakers, the Baptists, the Independents, and the Methodists 'both old and new'. Such an approach to religious matters is characteristic of street literature, which rarely if ever reflects the religious conflicts of the nineteenth century, and a simple explanation of this is the desire to achieve the widest possible sales. The fact that religion formed the subject-matter of broadsides at all appears interesting, for recent research has indicated that the working classes had little or no interest in formal religious practices. Why then should commercial printers have concerned themselves with the subject? The answer is probably that such sheets in most cases had a seasonal sale. A number of Christmas ballad sheets has survived, and it is possible that at this time of the year, and perhaps at Easter, ballads could be sold which appealed to a residual religious emotion which had nothing to do with church-going.

Many street ballads took as their subject the Royal Family, while others of course had political themes. The former often exhibited an adulatory, unsophisticated attitude to Royalty which remains a characteristic of much journalism today. Fairly typical of this group is 'A new song on the birth of the Prince of Wales', with its chorus:

> So let us be contented and sing with mirth and joy
> Some things must be got ready for the pretty little boy.

Where critical attitudes are to be found, they fall far short of anything approaching a full-blooded republicanism.

Political street ballads are in every way more interesting, and the striking feature of most of them is their moderation. There is seldom a strident note, and little rabble-rousing. 'The temper of the ballads on such questions as strikes and lock-outs', wrote a contemporary,[11] 'has struck us as singularly fair and moderate... In the middle of the bitter struggle of the last three years in the building trades, we find nothing really violent or objectionable.' A verse, quoted in the *National Review* of October 1861, from 'The glorious strike of the builders' (i.e. the strike of 1859), catches the mood of many political broadsides:

> They locked us out without a cause—
> Our rights was our desires,—
> We'll work for Trollope, Peto, Lucas,
> For all the world, and Myers.
> If we can only have our rights,
> We will go to work much stronger:
> Nine hours a day, that's what we say,
> And not a moment longer.

Strikes apart, there were other issues which provided material for the political broadside. Reform was one, political personalities another. Then there were political litanies of various kinds, similar in form to those for which William Hone had been tried in 1817. Crimean ballads, too, fall into this category.

The political tone of street literature, then, was muted. 'The great battle for freedom and reform' is clearly reformist, but by no stretch of imagination extremist:

> With Gladstone, Russell, Beales and Bright
> We shall weather through the storm,
> To give the working man his rights,
> And gain the Bill—Reform.
>
> We want no Tory government
> The poor man to oppress,
> They never try to do you good,
> The truth you will confess.
>
> The Liberals are the poor man's friend,
> To forward all they try,
> They'll beat their foes you may depend,
> And never will say, die.

A sense of bitterness with their lot—and this is very different from revolutionary fervour—is to be found in those ballads which deal with wages and the cost of living. 'Fifteen shillings a week' provides a complete budget for a working man who was in receipt of this wage. Husband, wife and seven children! Rent cost one shilling and ninepence, tobacco eightpence, tea eightpence, fuel one shilling and tenpence halfpenny. Clothes were bought second-hand—a jacket for sixpence, threepence for socks. This particular ballad is of considerable interest, purporting as it does to outline working-class expenditure in the 1870s (a date which can be inferred from Such's imprint, though earlier versions may exist).

Similarly, 'How five and twenty shillings were expended in a week', gives in detail the weekly expenditure of a tradesman and his wife. This sheet is probably earlier than 1870, but there is no way of verifying this. In this case no children are mentioned. Rent costs three shillings and twopence, meat four shillings, tobacco sixpence, and so on. Most interesting, however, are the incidental comments:

> Last Monday night you got so drunk, amongst your dirty crew,
> It cost twopence next morning for a basin of hot stew.

Breakages, for example:

> There's a penny goes for this thing, and twopence that and t'other,
> Last week you broke a water jug, and I had to buy another.

Hygiene:

> A three farthing rushlight every night, to catch the bugs and fleas.

So far, then, as the cost of living is a political issue, these ballads were in essence political. More obviously so was 'Dizzy's Lament':

> O dear! Oh dear! What shall I do
> They call me Saucy Ben the Jew
> The leader of the Tory crew
>   Poor old Benjamin Dizzy.
> I'd a great big house in Buckinghamshire
> My wages was Five Thousand a Year;
> But now they have turned me out of place,
> With a ticket for soup, in great disgrace.
> I had a challenge last Monday night
> Billy Gladstone wanted me to fight;
> The challenge was brought by Jackey Bright
> To poor old Benjamin Dizzy.

Disraeli, Gladstone, Bright—clearly these were names which entered the consciousness of the humblest reader. It would of course be possible to offer further extracts from ballads of this kind. Almost always they were anti-Tory, but their keynote was reform rather than revolution. Ernest Jones's 'The Song of the Lower Classes' was circulated as a broadside round about 1848, and was later reprinted, once, together with an English translation of 'The Marseillaise', as a Socialist League broadside; there were titles like 'The Chartists are coming'; a number of Corn Law ballads were produced; but certainly in all the street ballads I have examined it remains true that anything more than a fairly mild reformism, and occasional support for Chartism, is entirely absent.

Those pieces which fall under the heading of 'ballads on a subject' show the kind of themes which could take the fancy of the public, and illustrate too the ingenuity of the unknown hack-writers who produced these effusions. Here we find them seizing upon every event which the contemporary scene provided; and, in earning commercial reward for the printers and street-sellers, they now provide for us aids towards re-creating that scene. As we might expect, the Tichbourne claimant, 'Bloomers', 'Wonderful Mr Spurgeon', the Volunteer forces, the opening of Holborn Viaduct, were all celebrated in street literature. So, too, were the death of the Duke of Wellington, the popularity of the polka when it reached this country about 1844, and the Great Exhibition of 1851, in ballads to which reference will be made later. The category of 'ballads on a subject' might be extended to include popular songs, some of the older ballads, and even music-hall songs, although the relationship between these and the street ballad is imprecise and as yet unexplored.[12]

The most popular subject of all, however, was crime. Sheets concerned with the execution of a criminal were the most numerous of all, and were bought with 'singular eagerness'. Almost all of them were illustrated—it was usual to see the criminal or criminals dangling from the gallows—and there was an account of the execution in

41 *above* James Catnach's shop in Great St Andrew's Street, Seven Dials, from *Punch*, ii (1842), 183.

42 *right* A ballad-seller, as drawn by Doré. From *London. A Pilgrimage* (1872), p. 84.

MARRIAGE OF THE QUEEN.

TUNE.—"Billy O'Rooke."

Come grave or sad, and dull or mad,
  I'll not detain you long, sirs,
While I relate some odds and ends,
  I've worked into a song, sirs.
The other day the Park I cross'd,
  And in a crowd did mingle,
A snob sung out "God Save the Queen,"
  She no longer will live single.
    CHORUS.
Prince Albert's come from Germany,
  To change his situation;
So the house of Hanover now may last
  Another generation!

Spoken.—Why do you know Mrs. Tomkins, as I came by Buckingham Palace the other day, I was told the Queen had summon'd all her councillors into her Privy, to let them know she wanted a husband. Oh! stop there, Mrs. Knowall, that is wrong; you mean her Privy Councillors, but I'll not interrupt you, so go on, well; I heard there was the Count of Strasbug, the Duke of Humbug, the Prince of Wirtenbug, and Albert of Cobug, and a whole host of other bugs; but out of them she fixed her eyes on Albert of Cobug, and to the astonishment of all present, bawled out aloud, "that's the man for me!"

In the Council hall the Queen did bawl,
  Prince Albert I will marry,
Don't fret your kidneys, gentlemen,
  For him I mean to tarry;
The day I'm married I'll present,
  Aye, in my Royal Passage.
All single maids, from Albert's store,
  A polony or a sausage.

Spoken.—Oh, Crikey, Mrs. Skinflint, won't St. Giles's be alive the day the Queen is married, for the Prince will send a waggon load of herrings, a ship load of murphys, and a quartern of the cratur to every mother's son, that we may toast his most Gracious Majesty's Royal Highness, that he may live long, and be surrounded by a host of little bugs, to the honour of the Queen, long life to hers good luck to the pair of them say I, Mother Flaherty—Dan O'Connell, Prince Albert, & Erin-go-Braug.

Through Ireland they will rejoice,
  The day our Queen is married,
And many a wench will cry aloud
  Sure, I too long have tarried;
Victoria has a husband got,
  May nothing ere distress her,
And Albert gain the people's love,
  Then toast "the Queen, God bless her."

Spoken.—Oh! Mrs. Jenkins, won't we have a flare—the day the Queen is married,—there is to be a bullock roasted in Clare Market another in Smithfield, lots of pigs and other poultry, a large German sausage will be hung up to every lamp post in St. James's Park, solely for the feminine gender; a man is not to have even a smell of them, and should any man be found daring enough to cut a slice off the forbidden meat, he will be placed in the same state and along side of Achiles in Hyde Park, for 48 hours. Oh! dear me, Mrs Bustle, O dear me, I would not look at it for life, and perhaps without even a fig leaf to hide—oh horrid, &c.

A bullock will be roasted whole,
  The day our Queen is married,
And German sausages on poles,
  Will through the streets be carried.
Such glorious sprees will be that day,
  Old Sal will treat her crony,
And Dublin Set of Saffron Hill,
  Will sport a large polony.

Spoken.—Lord bless you, Mother Thingamy, there will be such fun, nothing was ever like it, nobody must be gloomy that day, under pain of Royal displeasure. Now I should like to know who this Albert is? Why, that is his name, Oh, nonsense! I know that already; but where does he come from? That I will tell you directly; he is the second son to the Grand Duke of Germany, Master Manufacturer of German Sausages to the British Empire, and worth 600 millions of screwbles, and he will give the Queen all he has got.

May Albert with her happy live,
  In gay and sweet content, sirs,
And their wedding day through England be,
  A day of merriment, sirs,
Oh! may she have a happy reign,
  On her subjects never frown, sirs,
But quickly have a son and heir,
  To wear Old England's Crown, sirs.

BIRT, Printer, 39, Great St. Andrew Street; Seven Dials

43 A characteristic broadside of 1840. The vignette came from stock. Birt, the printer, was an important supplier to the street trade.
Leslie Shepard

JOHN BULL & HIS PARTY
Or, Do it Again.

Tune—Do it Again.

As the shamrock, the rose, and the thistle were meeting
Together one morning, so jovial and gay,
First in popped a Welchman and then came a Frenchman,
And loudly to old Farmer Bull thus did say ;
I think this great nation wants some alteration,
All over the land and far over the main,
You once was victorious, done deeds bright and glorious,
But now do you think you can do it again.
Do it again, do it again, but now do you think you can do it again.

I met with young Albert, so buxom and all pert,
He was down in the nursery telling some tales,
In the royal palace to Addy and Alice,
The young Duke of York, and the great Prince of Wales ;
Then he went in the passage and took up a sausage
And strove for to banish all sorrow and pain,
He played up so charming, oh ! it was alarming,
Buy a broom, ax my eye, and we'll do it again.

Then up stept Victoria, all England adores her,
Russians, Prussians, and Frenchmen she loves to invite,
She likes you know what then, to travel to Scotland
And gaze on the Highlands with joy and delight
Towns, counties, & cities, with sweet little kiddies
By steam she'll supply, & send them by the train
If old Bull don't like it, why then he may pipe it,
She has done it before and she'll do it again.

Then up stept old Nosey so blythe and so cosey,
With his cocked hat and feather, believe it's true
His cannons rattled as if 'twas in battle,
When he faced poor Boney at famed Waterloo,
Like a soldier afloat with his sword, sash, & coat,
He had travelled through Houndsditch and Petticoat Lane,
Singing, buy old clothes, hey day, elts, powder, and pipe clay,
I am getting too old for to do it again.

Then ratcatching Bobby jumped up in the lobby,
And to old Neddy h— these words he did say,
You know all I will state they have broke down the toll gates,
While you was a hopping in Kent t'other day ;
Jemmy G— with letters of hymen and fetters,
Came creeping and weeping in sorrow and pain,
When his nose it went right slap into Bobby's rat trap,
And old Bull holloa'd out will you do it again.

Up came Gladstone a blinking with old Mr. Lincoln,
And as in the Enclosure together they stood,
Said Gouldbourn to Lincoln by jingo I'm thinking
They will hang us all up in the forests & woods ;
Then old Harry Hardinge with a bag of half-farthings,
Got a good situation right over the main.
So says he off I'm jogging, I was once fond of flogging,
And I shan't come to England to do it again.

Like winking did up run Mr. Tommy Duncombe,
And in Finsbury Square he began for to dance,
He was followed so quickly by Alderman Wakley,
Who out of his pocket did pull a great lance,
Saying Duncombe so clever we'll struggle together
And throw our enemies into great pain,
We will strive for to righten, we must not be frightened,
I have bled them before and I'll do it again.

Up came Mr. Daniel saying now don't I stand well,
Against all oppression and shocking bad law,
Repeal now and glory and no whig and tory,
The Lakes of Killarney and Erin go bragh ;
I will not be frighted till Erin is righted,
For the repeal of the union my nerves I will strain
You know well Hibernia, repeal does concern her,
I have beat them before and I'll do it again,
Do it again, do it again, I have beat them before and I'll do it again.

BIRT, Printer, 39, Great St. Andrew Street, Seven Dials.

44   An unillustrated sheet on sale about 1850, also supplied by Birt.
Leslie Shepard

What do you think of Mister Billy Roupell.

It is now all over Lambeth, there's a grand flare-up we see
Abnot a well-known gentleman, who once was their M.P.;
A worthless, cunning fellow, everybody knew him well,
And many a jolly row has been about Billy Roupell.

Funny things we see, he's a man you know full well,
In every street in Lambeth they knew Billy Roupell,
And Billy was resolved to have the whole lot or none.

Though Billy was a natural son, he a gentleman would be,
It was Bill Roupell, Esquire, and afterwards M.P..
When Bill Roupell possession of the property had got,
He didn't care who went without so he had got the lot.

He got Roupell Park, at Brixton and every other place,
By roguery and forgery now he is in disgrace;
And when his poor old father died, was not he a naughty chap,
He even prigged his father's breeches, and his poor old mother's cap.

Was'nt he a mem'er of Parliament, a wicked naughty man,
Was'nt he, good folks of Lambeth, a disgrace unto the

Twas once, Roupell for ever, sounding through every street,
And now I say, 'tis Bill Roupell, how is your poor feet?

Bill carried on a stunning game, he wanted the whole lot,
He even sold the frying-pan, the scrubbing brush and mop;
Bill was a wicked brother, and a good-for-nothing son
Not satisfied without he had the whole lot or none.

Billy was a gentleman who'd have thought it, I declare?
Billy was an officer in the Rifle Volunteers;
He was a member of Parliament, a rogue to the back-bone
Determined he was for to have the whole lot or none.

Billy made the money fly, how he would sport and play,
And when the game was finished then Billy ran away.
They had him down at Guildford, for the naughty deeds he'd done,
When he told them to their face he'd have the whole lot or none.

It is no joke, you Lambeth folk, for one avoricious brother,
To guilty be of forgery, just for to rob another;
But I'll tell you what the end will be for the wicked deed he's done,
In a little time he'll have to go the whole hog or none.

Well, now you Roupell, M.P. in sorrow may bewail,
For forgery and perjury he lies within a gaol;
Reflecting on the wicked deeds which he for lucre done,
So Bill Roupell will have to go the whole hog or none.

This gentleman round Lambeth is
By every one known well.
He was, you see, the great M.P.,
The wonderful Roupell.

TAYLOR,
Printer, 92, Brick Lane, Spitalfields.

45 A topical broadside of 1862, telling a story of belated justice. William Roupell forged a will in 1856 but was not brought to book for six years, when he was given penal servitude for life. *Leslie Shepard*

THE STRIKE
Of the London Cabmen!

Ryle and Co., Printers, Monmouth Court,
Seven Dials.

OH! here's a great and glorious row,
  All over London, sirs, I vow,
A few words of which I'll tell you now,
  The strike of the London Cabmen.
Some days ago they pass'd an act,
About the cabs—it is a fact—
Which made old swells and codgers smile,
That they should ride at a tanner a mile;
And if not pleased, then all around,
The cabman had to measure the ground,
And that's the reason I'll be bound,
  Has caused the strike of the cabmen.

Cut him! slash him! here's a go,
All over town come up, gee wo,
The law was hard you all do know,
  So strike did the London Cabmen

On Tuesday last, oh! what a sight,
The thing was done and all was right,
Just at the hour of twelve at night,
  All strike did the London Cabmen.
There was not a cab upon the ground,
And never a Cabman to be found,
Swells and cripples on did steer,
Singing out, oh! law, oh! dear;
They did not know there was a strike,
They bawl'd and squall'd with all their might,
They hunted up and down all night,
  But could not find a Cabman.

Soon after twelve the sky look'd dark,
A City Barber, and his clerk
On London Bridge made this remark,
  'There's a strike among the Cabmen.'

Now I did this expect awhile,
When whip along, a tanner a mile,
We through London streets could go,
In wet and dry, come up! gee wo.
Well I must really now confess,
Indeed I did think nothing less,
This tanner job did so oppress,
  That strike would the London Cabmen.

As I was going down the Strand,
I met a poor old feeble man,
Upon his legs he scarce could stand,
  And he roar'd aloud for a Cabman
But never a one was to be had,
He bawl'd and holloa'd cab, cab, cab,
I am tired out and nearly mad,
Now is not my condition sad?
He was unable on to roam,
When down he squatted on the stones,
Saying, I shall never more get home,
  All through the London Cabmen.

While coming home from Drury Lane,
I tumbled over Bet and Jane,
Who said the Act was much to blame,
  To cause the strike of the Cabmen.
Indeed! indeed! said lovely Kate,
I know they gave in all their plates,
And told their masters they should go,
To dwell in South Australia, O.
Behold how dismal is the streets,
No cab or carriage can we meet,
And we must travel on our feet,
  All through the London Cabmen.

Now if comes a thunder shower
If it rains, or if it pours,
You may be about the streets for hours,
  And never find a Cabman.
And won't it too make many fret,
To see the ladies muslin wet,
And see them thro' the streets to scud,
With their backs behind all covered in mud
And won't it cause them to be ail,
To wash their petticoats and tails,
Cause thro' the dirt they have to sail,
  All through the London Cabmen.

46   Catnach retired in 1838 and left his business to his sister, Annie Ryle, who continued it till 1841 as Ryle and Co. This ready commentary on a passing event appeared during this period. *Leslie Shepard*

SHOCKING MURDER OF A WIFE And Six Children.

Attend you feeling parents dear,
While I relate a sad affair;
Which has filled all around with grief and pain,
It did occur in Hosier Lane.

On Monday, June the 28th,
These crimes were done as I now state,
How horrible it is to tell,
Eight human persons by poison fell.

In London city it does appear,
Walter James Duggin lived we hear,
And seemed to live most happily,
With his dear wife and family.

They happy lived, until of late,
He appear'd in a sad desponding state,
At something he seem'd much annoy'd,
At his master's, where he was employ'd

He was discharged, and that we find,
It prayed upon his anxious mind,
Lest they should want—that fatal day
His wife and children he did slay.

Last Sunday evening as we hear,
To the Wheatsheaf he did repair,
Then homewards went as we may read,
For to commit this horrid deed,

To the police he did a letter send,
That he was about this life to end,
And that he had poisoned, he did declare
His wife, and his six children dear.

To Hosier Lane in haste they flew,
And found it was alas, too true,
They found him stretched upon the bed
His troubles o'er—was cold and dead.

They searched the premises around,
And they the deadly poison found;
And the shocking sight, as you may hear
Caused in many an eye a tear.

They found upon another bed,
The ill-fated mother, she was dead,
While two pretty children we are told,
In her outstretched arms she did enfold

It is supposed this wretched pair,
First poisoned their six children dear,
Then took the fatal draught themselves,
Their state of mind no tongue can tell.

Of such an heartrending affair,
I trust we never more may hear,
Such deeds they make the blood run cold
May God forgive their sinful souls

This wholesale poisoning has caused much pain,
It did take place in Hosier Lane.

Disley, Printer, 57, High Street, St, Giles

47 This tissue of fact and fiction belonging to the 1860s comes from a man who once worked for Catnach and specialized in crime sheets.
Leslie Shepard

prose, followed by a 'copy of verses' often alleged to have been written by the condemned felon in his cell on the eve of execution.

Occasionally such a broadsheet was memorable. The 'Confession of the Murderess', which was published on a sheet entitled 'The Esher tragedy. Six children murdered by their mother', has a quality of horror which is both credible and moving:

> 'On Friday last I was bad all day; I wanted to see Mr Izod, and waited all day. I wanted him to give me some medicine. In the evening I walked about, and afterwards put the children to bed, and wanted to go to sleep in a chair. —About nine o'clock, Georgy (meaning Georgianna) kept calling me to bed. I came up to bed, and they kept calling me to bring them some barley water, and they kept calling me till nearly 12 o'clock. I had one candle lit on the chair—I went and got another, but could not see, there was something like a cloud, and I thought I would go down and get a knife and cut my throat, but could not see. I groped about in master's room for a razor. I went up to Georgy, and cut her first; I did not look at her. I then came to Carry, and cut her. Then to Harry—he said "don't mother". I said, "I must" and did cut him. Then I went to Bill. He was fast asleep. I turned him over. He never woke, and I served him the same. I nearly tumbled into this room. The two children here, Harriet and George were awake. They made no resistance at all. I then lay down myself.' This statement was signed by the miserable woman.

Few other broadsides rival this in style; the mounting tension and understatement give one the sense of being present at a nightmare. For the most part these murder sheets possess for us all the dullness of sensational news that has passed into obscurity. The prose and verse are largely stereotyped, while the illustrations are scarcely credible. This, however, is what the nineteenth-century public wanted—the continuing popularity of broadsides dealing with crime bears this out. As a street-seller said to Henry Mayhew, 'There's nothing beats a stunning good murder after all.'

The producers of street literature were printers in London and in the larger provincial towns and cities.[13] The Worrall family in Liverpool, Bebbington in Manchester, Harkness in Preston, were amongst the leading provincial printers in this field, and the list could be considerably extended. A great deal of plagiarism went on —successful items were shamelessly copied—and local themes were of course exploited by local printers. Production of street ballads could be readily and easily undertaken by any printer when there was material about to catch the public fancy, and it was easy to combine this kind of work with the usual jobbing tasks—labels, notices, letterheads, catalogues, and so on—which formed the greater part of his work.[14]

The most noted specialist in producing street literature was James Catnach,[15] who supplied many hawkers in London and elsewhere with the wares which they cried in the street, sold at fairs and races, or offered for sale by 'pinning up'. His business was founded in London in the second decade of the nineteenth century, and was flourishing in the hands of his successors for well over fifty years. Catnach himself died in 1841 and was buried in Highgate Cemetery, having made a great deal of money out of street literature (and incidentally out of publishing cheap books for children). It is said that his sheets were sold to their sellers for coppers, his employees were paid in pennies and halfpennies, and he would hire a hackney coach each week to convey the coins to the Bank of England. Even so, he was the recipient of many bad pence, and legend has it that these were embedded in plaster of Paris in the kitchen behind his printing office. Catnach was the doyen of street printers and a specialist in this kind of publication, besides being sufficiently astute a businessman to pick up job lots of printers' old stock, including blocks a century and more old which he used to excellent effect.

It is almost entirely due to the industry of Charles Hindley, Catnach's biographer, that we know something of the life and work of this printer. Catnach's neighbour and rival printer, John Pitts, has been the subject of a recent book; and in it, rescued from the limbo of the nineteenth-century periodical press, there is a vivid account of Pitts's manager, Bat Corcoran, selling ballad sheets to hawkers and others:[16]

> But let us see Bat amidst his customers—see him riding the whirlwind—let us take him in the shock, the crisis of the night when he is despatching the claims of a series of applicants. 'I say, Blind Maggie, you're down for a dozen "Jolly Waterman", thirteen to the dozen.—Pay up your score, Tom with the wooden leg. I see you are booked for a lot of "Arethusas".—Master Flowers, do you think that "Cans of Grog" can be got for nothing, that you leave a stiff account behind you.—Sally Sallop, you must either give back "The Gentlemen of England" or tip for them at once.—Friday my man, there are ever so many "Black-eyed Susans" against you.—Jimmy, get rid of the "Tars of Old England" if you can; I think "Crazy Janes" are more in vogue. What say you to an exchange for "Hosier's Ghost"?'

Although this account dates from 1825, the method of selling ballads wholesale cannot have changed significantly throughout the century. It is, moreover, the only description of this aspect of the trade which appears to have been written, or at any rate to have survived.

Little was said anywhere about the writers of street literature. Who were they? Unknown hacks in the truest sense. Henry Mayhew tracked one of them down; so too did Charles Hindley in 1870.[17] This was John Morgan, who had worked for Catnach, and recalled rather bitterly how difficult it had been to secure adequate payment from him. The printer, of course, was in a difficult position, for if a ballad sheet was successful it was immediately pirated, and much of the profit was dissipated. How then could he afford to pay more than one shilling, or more rarely half-a-crown, for a work which would cease to be his as soon as it was published? A persistent legend has it that Catnach himself was the author of many of his own broadsheets, and this may have been true. What is more certain is that of the approximate sum of £12,000 which Henry Mayhew estimated was spent upon broadside ballads and the like in the late 1840s and early 1850s,[18] only a very tiny part could have found its way into the pockets of the unknown authors of so much of it.

All these factors must be borne in mind in assessing street literature and its place in working-class life. Attitudes and tendencies within a society do not, of course, change their emphasis or direction at points which can be committed to even a verbal chart; they develop gradually, and these ballad sheets reflect such a development. The authors received a trifling recompense for a product which was short-lived in the extreme, and there was the pressing commercial need for its sellers to reach as wide a public as possible in order to earn a meagre living. The printers were the only ones who did reasonably well out of it, and even then Catnach's success was unusual, and due partly to the fact that he exploited other markets with skill. There is no evidence to suggest that any of those concerned in the street trade ever set himself the task of forming attitudes. Events were followed, seized upon, and exploited.

The method of exploitation was simple and direct, and a striking example of this is provided by the death of the Duke of Wellington. As victor of Waterloo he had been a popular hero: as a Conservative Prime Minister this was clearly not the case. His death on 14 September 1852 was followed by a lying-in-state at Chelsea Hospital from 10 to 17 November, and he was buried at St Paul's on 18 November. The demise of a man so widely known, and the splendour of his funeral, presented street printers with an opportunity for unusually large sales, and it was seized in a characteristic way. Wellington is presented simply as a popular hero, with no mention of his political activities:

DEATH OF WELLINGTON

On the 14th of September, near to the town of Deal,
As you may well remember who have a heart to feel,
Died Wellington, a general bold, of glorious renown,
Who beat the great Napoleon near unto Brussels town.

Chorus
So don't forget brave Wellington, who won at Waterloo,
He beat the great Napoleon and all his generals too.

He led the British army on through Portugal and Spain,
And every battle there he won the Frenchmen to restrain,
He ever was victorious in every battle field,
He gained a fame most glorious because he'd never yield.

He drove Napoleon from home, in exile for to dwell,
Far o'er the sea, and from his home, and all he loved so well.
He stripped him quite of all his power, and banished him away,
To St Helena's rocks and towers the rest of his life to stay.

> Then on the throne of France he placed Louis the king by right,
> In after years he was displaced all by the people's might,
> But should the young Napoleon threaten our land and laws,
> We'll find another Wellington should ever we have cause.
>
> He's dead, our hero's gone to rest, and o'er his corpse we'll mourn,
> With sadness and with grief oppress'd, for he will not return,
> But we his deeds will not forget, and should we ere again,
> Follow the example that he set, his glory we'll not stain.
>
> So don't forget brave Wellington, who won at Waterloo,
> He beat the great Napoleon and all his generals too.

The language is simple, and the emphasis is upon Wellington's martial glory. It requires no great leap of the imagination to picture this sheet selling in its thousands to the crowds who watched the funeral procession or who filed past the catafalque. Not all its purchasers would have been working-class readers, but in both style and form it is a typical product of street literature. One of the surviving copies of this ballad, together with a similar one entitled 'Lamentation on the death of the Duke of Wellington', was printed by John Harkness of Preston, Lancashire, and this illustrates the way in which topical themes, usually originating in the metropolis, were pirated immediately by provincial printers. The converse is less true, and examples of ballads on anything like national themes originating in provincial towns and being subsequently printed in London are virtually non-existent.

A topical theme of a rather different kind is represented in a ballad called 'Jullien's Grand Polka'. Louis Antoine Jullien, born in France in 1812, was one of the many foreign musicians who performed in England during the nineteenth century. He was the first to take advantage of the popularity which the polka was achieving since its introduction into this country. The street ballad on the subject is satirical about the craze, and illustrates the way in which a theme which took the public fancy could be exploited right across the social spectrum: at the fashionable end the dance itself at routes and parties, the vogue for 'polka' jackets and bonnets; at the other a street ballad comment tilting against the pretensions of the upper and middle classes. Much earlier in the century, when Pierce Egan's *Life in London* had achieved an immense success both in book form and on the legitimate stage, Catnach had brought out a twopenny edition which, together with the versions which were staged at the penny theatres of Lambeth and elsewhere, provides an almost exact parallel—the adventures of Corinthian Tom and Jerry Hawthorne being enjoyed simultaneously at opposite ends of the social scale.

> JULLIEN'S GRAND POLKA
> Oh! sure the world is all run mad,
> The lean, the fat, the gay, the sad,—
> All swear such pleasure they never had,
> Till they did learn the Polka.

Chorus
First cock up your right leg so,
Balance on your left great toe,
Stamp your heels and off you go,
To the original Polka. Oh!

There's Mrs Tibbs the tailor's wife,
With Mother Briggs is sore at strife,
As if the first and last of life,
Was but to learn the Polka.

Quadrilles and Waltzes all give way,
For Jullien's Polkas bear the sway,
The chimney sweeps, on the first of May,
Do in London dance the Polka.

If a pretty girl you chance to meet,
With sparkling eyes and rosy cheek,
She'll say, young man we'll have a treat,
If you can dance the Polka.

A lady who lives in this town,
Went and bought a Polka gown,
And for the same she gave five pound
All for to dance the Polka.

But going to the ball one night,
On the way she got a dreadful fright,
She tumbled down, and ruined quite,
The gown to dance the Polka.

A Frenchman he has arrived from France
To teach the English how to dance,
And fill his pocket,—'what a chance'—
By gammoning the Polka.

Professors swarm in every street,
'Tis ground on barrel organs sweet,
And every friend you chance to meet,
Asks if you dance the Polka.

Then over Fanny Ellsler came,
Brilliant with trans-Atlantic fame,
Says she I'm German by my name,
So best I know the Polka.

And the row de dow she danced,
And in short clothes and red heels pranced,
And, as she skipped, her red heels glanced
In the Bohemian Polka.

> But now my song is near its close,
> A secret, now, I will disclose,
> Don't tell, for it's beneath the rose,
> A humbug is the Polka.
>
> Then heigh for humbug France or Spain,
> Who brings back our old steps again,
> Which John Bull will applaud amain
> Just as he does the Polka.

As will be seen, 'Jullien's Grand Polka' did convey a hint of class antagonism—but it could justify no stronger description than that. At the same time two of the lines in the ballad drew upon a rich vein of folklore:

> The chimney sweeps on the first of May
> Do in London dance the Polka.

The survival of this custom in Victorian cities demonstrates not only its persistence, but also a range of shared, or perhaps remembered, experience which was rarely referred to in nineteenth-century popular literature. The Victorian street ballad throws very little light upon urban folklore, and the extent to which the popular customs of a rural past survived in an urban setting remains obscure.

Another ballad for which there would have been a ready public deals with the Great Exhibition of 1851, and is called 'Crystal Palace':[19]

> Britannia's sons an attentive ear
> One moment lend to me,
> Whether tillers of our fruitful soil,
> Or lords of high degree.
> Mechanic too, and artizan,
> Old England's pride and boast,
> Whose wondrous skill has spread around,
> Far, far from Britain's coast.
>
> Chorus
> For the World's great Exhibition,
> Let's shout with loud huzza,
> All nations never can forget,
> The glorious first of May.
>
> From every quarter of the Globe,
> They come across the sea,
> And to the Chrystal Palace
> The wonders for to see;
> Raised by the handiwork of men
> Born on British ground,
> A challenge to the Universe
> It's equal to be found.

Each friendly nation in the world,
Have their assistance lent,
And to this Exhibition
Have their productions sent.
And with honest zeal and ardour,
With pleasure do repair,
With hands outstretch'd, and gait erect,
To the World's Great National Fair.

The Sons of England and France
And America likewise,
With other nations to contend,
To bear away the prize.
With pride depicted in their eyes,
View the offspring of their hand,
O, surely England's greatest wealth,
Is an honest working man.

It is a glorious sight to see
So many thousands meet,
Not heeding creed or country,
Each other friendly greet.
Like children of one mighty sire,
May that sacred tie ne'er cease,
May the blood stain'd sword of War give way
To the Olive branch of Peace.

But hark! the trumpets flourish,
Victoria does approach,
That she may long be spared to us
Shall be our reigning toast.
I trust each heart, it will respond,
To what I now propose—
Good will and plenty to her friends,
And confusion to her foes.

Great praise is due to Albert,
For the good that he has done,
May others follow in his steps
The work he has begun;
Then let us all, with one accord,
His name give with three cheers,
Shout huzza for the Chrystal Palace,
And the World's great National Fair!!

The tone here is unmistakably one of triumph and self-congratulation. Chauvinism and fervour for the Royal Family are the keynotes. This fact in itself is hardly remarkable, and the lines—

> O surely England's greatest wealth
> Is an honest working man

are entirely consistent with the tone of the ballad. There is, however, some evidence to suggest that working men did not see the matter in this light. In 1851 G. J. Holyoake produced a pamphlet entitled 'The Workmen and the International Exhibition', and although priced at a halfpenny it was in fact designed to be given away to visitors to the exhibition. Its aim was to show the conditions in which many working men lived, and the misery which existed in workshops and houses where so many of the striking exhibits had been prepared. The cost in human terms was thus stressed by Holyoake.

Now Holyoake's pamphlet was not in any sense street literature, and the comparison does seem to show that the tendency of the latter was to romanticize reality—to offer a kind of cultural jingoism—in order to achieve as wide a sale as possible where vast concourses of people were gathered together on specific occasions.

The three street ballads quoted are typical products of their time. Quite apart from the verse, which is often slipshod, they exhibit, like most street literature, a superficiality which provides heroes whose less attractive qualities are conveniently ignored. If themes are exploited with any directness and clarity there is little evidence in the sheets of anything more than a vaguely sketched class-consciousness—class and economic antagonisms, indeed, are played down despite the general mildly reformist nature of the political ballads. Of evocative phrases there are few.

Street literature is, however, worth a closer scrutiny than it has yet received. A contemporary, writing in the *National Review* for October 1861 (to which reference was made earlier), said of these ballads that 'they are almost all written by persons of the class to whom they are addressed', and urged that they were worthy of study because they provided 'one of those windows through which we may get a glimpse at that very large body of our fellow-citizens of whom we know so little'. This was the literature of the urban working class, and with all its defects it provides one of the few insights we have into their popular culture. Occasionally there are ballads which look back to a rural past, but these are few. In its format, even in its size, the Victorian street ballad perhaps bears a resemblance to the tabloid newspaper which in many ways—concern with sex, crime, Royalty—it anticipated in a somewhat crude way. Only sport is missing.

This sense of looking forward to the tabloid newspaper of the twentieth century is further strengthened when the Victorian street ballad is contrasted with the eighteenth-century chapbook. The ballad sheet attempted, often with some success, to deal with topical themes and to exploit news and events of the day. The chapbook, on the other hand, preserved the fragments of an older tradition petrified in print, with 'Guy of Warwick', 'The Seven Wise Masters', 'Old Mother Shipton', and others together with a good deal of folklore, and many of the jests and rhymes which had earlier formed an important element in the oral lore of the English peasantry.[20]

Clearly these differences in the popular literature of the eighteenth and nineteenth centuries[21] are evidences of a fundamental change in mass reading habits, and in attitudes to life. But there is another reason for looking more searchingly at this kind of publication. Although we cannot be absolutely certain who bought and read such sheets, contemporaries suggested strongly that they circulated almost entirely amongst the poor,[22] and such circulation figures as we have are impressive. 'The execution of F. G. Manning and Maria, his wife'—based on a celebrated case in 1849—sold two and a half million copies, and 'The trial and sentence of Constance Kent' one hundred and fifty thousand. These are considerable figures, and not by any means unique for the street trade. They seem even more striking when we compare them with the sales figures for Henry Milner's cheap editions of Burns and Byron, which sold 183,333 and 126,514 copies respectively over a much longer period.[23]

Figures of this kind for the sale of broadsides, taken in conjunction with the enormous stocks carried by Catnach, are a strong indication that such ephemeral literature was an important element in the development of the mass reading-public. Like the chapbook in the eighteenth century, the street ballad provided the means by which those who could read—an ability which was not always gained too easily by the nineteenth-century poor—could exercise their skill. Ballad sheets were readily available, and they were short. This last point is an important one, for as a London vicar pointed out, 'it needs more than an average love for reading to be able to turn with interest or profit to books, when mind and body are wearied with a long day's toil'.[24]

Through street literature we are able to penetrate, however vicariously, the world of feeling of the poor in Victorian cities. The ephemeral nature of the street ballad symbolizes the precarious quality of their lives. It was, of course, the cheap newspaper towards the close of the nineteenth century which killed the street ballad, and the singer in Hoxton whom we noticed earlier, who sang a popular song and sold copies of it at a penny a time about seventy years ago, must have been one of the last of his kind.

What was the quality of life in the Victorian city? How did it seem to those who lived in the slums and rookeries of Victorian London, Manchester, Liverpool? These are large questions. The street ballad offers us one view, from below as it were, of elements in the culture of the urban poor.

Notes

1. In Germany, they were called 'fliegende Blätter'; in Russia, the 'lubok'; in England, 'broadsides'. See, for example: R. Lemon, *Catalogue of a Collection of Printed Broadsides in the Possession of the Society of Antiquaries of London* (1866); William A. Coupe, *The German Illustrated Broadsheet in the Seventeenth Century:*

Historical and iconographical studies (2 vols, Baden-Baden, 1966, 1967); Y. Ovsyannikov and A. Shkarovsky-Raffé, *The Lubok* (Moscow, 1969), in which substantial English abstracts are given throughout.

2   See Henry Mayhew, *London Labour and the London Poor* [4 vols, 1861–2], I, p. 222, for a daguerreotype of a vendor of street literature. (Facsimile reprints of the four volumes published in 1967 by Cass and by Dover, New York.) There is a picture of two female ballad singers in 'P. Parley' [George Mogridge], *Tales about Christmas* (1838), p. 65. William B. Jerrold, *The Life of George Cruikshank* (1883 edn), p. 90, shows Cruikshank's delineation of a mendicant ballad seller.

3   Mayhew, op. cit., I, pp. 213–51, 272–85.

4   Charles Hindley, ed., *The Life and Adventures of a Cheap Jack* (1881 edn); 'Lord' George Sanger, *Seventy Years a Showman* (1910); *The Life, Adventures and Experiences of David Love, written by Himself* (3rd edn, Nottingham, 1823); William Cameron, *Hawkie: The autobiography of a Gargrel*, ed. John Strathesk (Glasgow, 1888).

5   'A. Jaye' [Albert Jacobs], 'Looking Back', *Profile* (Hackney Library Services), II, no. 11 (August 1969), 2; no. 12 (September 1969), 2.

6   C. Hindley, *The Catnach Press* (1869); (ed.) *Curiosities of Street Literature* (1871); *The Life and Times of James Catnach, late of Seven Dials, Ballad Monger* (1878); *The History of the Catnach Press* (1886). See also John Ashton, *Modern Street Ballads* (1888). Several of these titles have been reprinted in facsimile editions.

7   I have followed Hindley's arrangement of street ballads. His *Curiosities of Street Literature* (1871) is divided into four sections, and this seems the most suitable broad classification of such material.

8   Quoted by Hindley, *The Catnach Press* [p. 15].

9   Unless otherwise stated, the ballads and prose quoted are from Hindley's *Curiosities of Street Literature* and Ashton's *Modern Street Ballads*. Not only are their contents representative of the genre as a whole, but they are also the range of material which is most readily available for study. Patrick Scott, *Index to Charles Hindley's Curiosities of Street Literature* (Victorian Studies Handlist 2, University of Leicester Victorian Studies Centre, 1970) greatly facilitated the use of this anthology.

10 In 1871, when ballad printing had passed its heyday, there were still four printers in London who specialized in this kind of work:
W. S. Fortey (Catnach's successor), Monmouth Court, Seven Dials.
Henry Disley, 57 High Street, St Giles.
Taylor, 92 & 93 Brick Lane, Spitalfields.
H. Such, 177 Union Street, Borough. (Such's business had been established in 1846, and in 1950 some of his woodcuts were said to survive in the premises of S. Burgess in Of Alley, off Villiers Street, in London. The present writer attempted to see them, but had no success.) There is a copy of Such's *Catalogue of Songs Constantly kept in Stock* (undated), listing over 800 titles, in the Mitchell Library, Glasgow.

11 Anon., 'Street Ballads', *National Review*, xiii (1861), 412.

12 Even the extent to which favourite music-hall songs were published as street ballads is a matter for conjecture.

13 No satisfactory account of London street ballad printers exists. Catnach and his successors, Ryle, then Fortey, dominated the trade in the metropolis, but a large

number of other printers were at one time or another involved in it. V. E. Neuburg, *Chapbooks: A Bibliography of References to English and American Chapbook Literature of the Eighteenth and Nineteenth Centuries* (1964), pp. 16–27, lists, with dates, a number of provincial printers.

14 For a valuable account of such a local printer see [C. R. Cheney *et al.*], *John Cheney and his Descendants, printers in Banbury since 1767* (Banbury, for private circulation, 1936).

15 Little or no recent research has been done upon Catnach. Various magazine articles have appeared, and in 1955 the Book Club of California published in a limited edition *Catnachery* by P. H. Muir, which is delightfully produced but nothing more than a newly arranged account of what is already known, with no attempt at critical judgments of any kind.

16 Leslie Shepard, *John Pitts: Ballad printer of Seven Dials, London, 1765–1844* (Private Libraries Association, 1969), p. 71.

17 Mayhew, op. cit., I, pp. 301–2; *History of the Catnach Press*, pp. xiv–xxx.

18 Quoted by Shepard, op. cit., p. 81.

19 This is the version given in Ashton, *Modern Street Ballads*, pp. 284 ff.

20 For chapbooks see John Ashton, *Chap-books of the Eighteenth Century* (1882). (Facsimile reprints published in 1970 by the Seven Dials Press, Hatfield, and also in 1967 by Benjamin Blom, New York.)

21 See Neuburg, op. cit., for a guide to sources in chapbook and street literature.

22 C. J. Montague, *Sixty Years in Waifdom; or The ragged school movement in English history* (1904), pp. 17–20, provides an example of this. (Reprint published in 1969 by Woburn Press.)

23 Figures for the sale of broadsides are those quoted by Hindley, *Curiosities of Street Literature*, p. 159. Those for Milner's books are from Milner and Co. stock book, quoted by H. E. Wroot, 'A pioneer in cheap literature, William Milner of Halifax', *Bookman*, xi (1897), 174.

24 Robert Gregory, *Sermons on the Poorer Classes of London* (1869), p. 109.

8 The Metropolis on Stage

Michael R. Booth

The theater was quick to recognize in the growing numbers and multifarious activities of urban life in the nineteenth century a really dramatic opportunity. It was not only that these human concentrations provided more promising openings for theatrical managers or emotional reactions of a peculiar kind and intensity among the crowds themselves. The real achievement of the theater in this age of cities was to make theaters of the cities themselves. A deliberate artistic and thematic use of the city as a moral symbol and an image of existence, as well as a strikingly visual and human presentation of the realities of its daily living, originates in the theater with the Victorian stage rather than with any earlier period in the development of English drama. It is true that scenes are laid in London and that characters are drawn from the London milieu in the drama of the seventeenth and eighteenth centuries, but in such drama the urban setting is a background for plots and intrigues rather than a foreground that frequently dominates both characters and narrative. The conflict between courtier and citizen is an interesting aspect of seventeenth-century comedy, but attention here focuses on intrigues and tensions between the two parties and not on the locale of their conflict. The wits and gallants of Restoration comedy inhabit a world of chocolate- and coffee-houses, the Mall, the Park, and all the resorts of fashion; in comical inferiority, aldermen and rich cits leave their counting-houses and lock up their wives and daughters, usually in vain. Once again the scenery provides illusional backing to the action on the forestage, and, given English staging techniques in the late seventeenth century, it could hardly do anything else. In the eighteenth century the counting-house and the merchant became more respectable,

Copyright 1973 Michael R. Booth

the class structure of both the drama and its audiences broader, and yet there was no attempt (again there were technical difficulties) to put the real life of the city on stage. In the nineteenth century, however, the attempt virtually to make the city a character in the drama was made and became common practice. The urbanized drama was born.

For the Victorians the city of drama was London. Despite the growing network of provincial circuits and the expansion of theater in the new manufacturing towns, the English drama and theater remained very much London-centered. First the London star and then the London touring company went out to the provinces; conversely, the desire of all provincial actors was to secure a hearing and a reputation in the West End. A dramatist like H. J. Byron (1834–84) might write melodramas specially for provincial audiences and a star like Barry Sullivan (1821–91) might be more popular in the provinces than in London, but nevertheless London was the heart of the theatrical world. Importing young actors and dramatists, London exported stars and the drama to all parts of the British Isles. As communications improved and provincial audiences grew, London companies found touring appearances profitable, and their tours slowly strangled the old provincial stock companies. These new audiences wanted to see London successes, not local plays, and by the end of the nineteenth century the metropolis, like a great leech, was sucking the native theatrical life of the provinces dry. Indeed, London and the life of London became one of the dominant themes in the drama of the day. *The Heart of London* (1830), *The Scamps of London, or The Crossroads of Life* (1843), *London by Night* (1845), *London Vice and London Virtue* (1861), *The Work Girls of London* (1864), *The Streets of London* (1864), *The Poor of London* (1864), *Lost in London* (1867), *The Great City* (1867), *The Great Metropolis* (1874), *The Lights o' London* (1881), *The Great World of London* (1898)—almost all these, and many more with similar titles, are Victorian in origin and character.

Although in the first half of the nineteenth century the relative growth in the population of the larger manufacturing towns was even greater than that of London, the number of theaters catering to the new urban proletariat was much higher in London than anywhere else. Moreover, the drama produced for this proletariat (preserved, when it was preserved at all, in cheap acting editions), though it was diffused all over the country and adapted to local situations where necessary, remained recognizably a London export. During the first third of the century the drama and theater ceased to be the mainly middle-class territory they had been in the eighteenth century. Then, if the lower classes went to the theater at all, they inhabited the upper galleries of the patent theaters—Drury Lane, Covent Garden, and the Haymarket—and watched a drama largely aimed at the tastes and concerns of their betters. Now, however, for the first time in the history of the English theater, playhouses were built and a drama was written exclusively for the working and lower-middle class. By 1837 most of these playhouses were already open: the Surrey, the Victoria, the Bower, and Astley's on the South Bank of the Thames; the City of London, the Pavilion, and the Standard in the East End. The Britannia in Hoxton was to follow in 1841, the Effingham in Whitechapel and the Grecian in Hoxton in 1843. At these theaters the physical and moral simplicities of melodrama, farce, and pantomime were paramount; their audiences were mostly illiterate and cared nothing for the tragedy and refined comedy of an earlier age. What they wanted was the colour, action, excitement, illusion, poetic justice, and moral satisfaction that constituted an escape from the dreary monotony and daily discomfort of lives spent mostly in the business of survival. Such an escape the dramatists and managers of the popular theater set out to provide for them, and they succeeded so well that the types of drama that they produced froze quickly into a set of readily identifiable conventions which entertained millions of spectators for the rest of the century. Dickens concisely summed up this new popular taste when he described the ordinary patron of the Victoria Theatre in 1850:[1]

> Joe Whelks, of the New Cut, Lambeth, is not much of a reader, has no great store of books, no very commodious room to read in, no very decided inclination to read, and no power at all of presenting vividly before his mind's eye what he reads about. But put Joe in the gallery of the Victoria Theatre; show him doors and windows in the scene that will open and shut, and that people can get in and out of; tell him a story with these aids, and by the help of live men and women dressed up, confiding to him their inmost secrets, in voices audible a mile off; and Joe will unravel a story through all its entanglements and sit there as long after midnight as you have anything left to show him. Accordingly, the Theatres to which Mr Whelks resorts are always full; and whatever changes of fashion the drama knows elsewhere, it is always fashionable in the New Cut.

Quite possibly Dickens's Mr Whelks had grown up in a small village and come to London either by himself or with his parents. In any event, emigration had its effect on urban theater as well as on urban economics, and two of the particular characteristics of Victorian melodrama were its idealization of the village home and its denigration of London. The former was symbolically equated with the state of man before the Fall and represented virtue, innocence, beauty, peace, and a known place in God's world, while the latter stood for vice, the loss of innocence, squalor, degradation, bitter suffering, and helpless anonymity. The song 'Home, Sweet Home,' written for an operatic melodrama of 1823, *Clari, or the Maid of Milan,* by John Howard Payne, referred to a village home. The heroine of Edward Lancaster's *Ruth* (1841), after asking 'Who would dwell in cities, where our days are passed in obscurity?' sings 'My Own Village Home.' Such songs were common in village melodrama. The most convenient device for carrying this weight of moral dogma was the heroine, at first happy and tranquil in her village home, then miserable and imperiled in London. Once in the second state she invariably recalled the first. T. P. Taylor's temperance melodrama, *The Bottle* (1847), set in London, has the agonized Ruth, wife to the drunken Thornley, beg the bailiff to allow her to keep one precious possession:

The Victorian City: Numbers of People

I must beg you not to take that; it is the picture of the village church where I worshipped as a girl, that saw me wedded in my womanhood; there are a thousand dear recollections connected with it, humble though they be. There was a meadow close by, over whose green turf I have often wandered, and spent many happy hours, when a laughing merry child; and dearer far it is to me, for beneath a rude mound in that sad resting-place poor mother and father lie.

In a later scene Ruth, ill, starving, and separated from her drunken husband, is looking for work; she is assisted by her friend Esther, who is trying to keep alive by

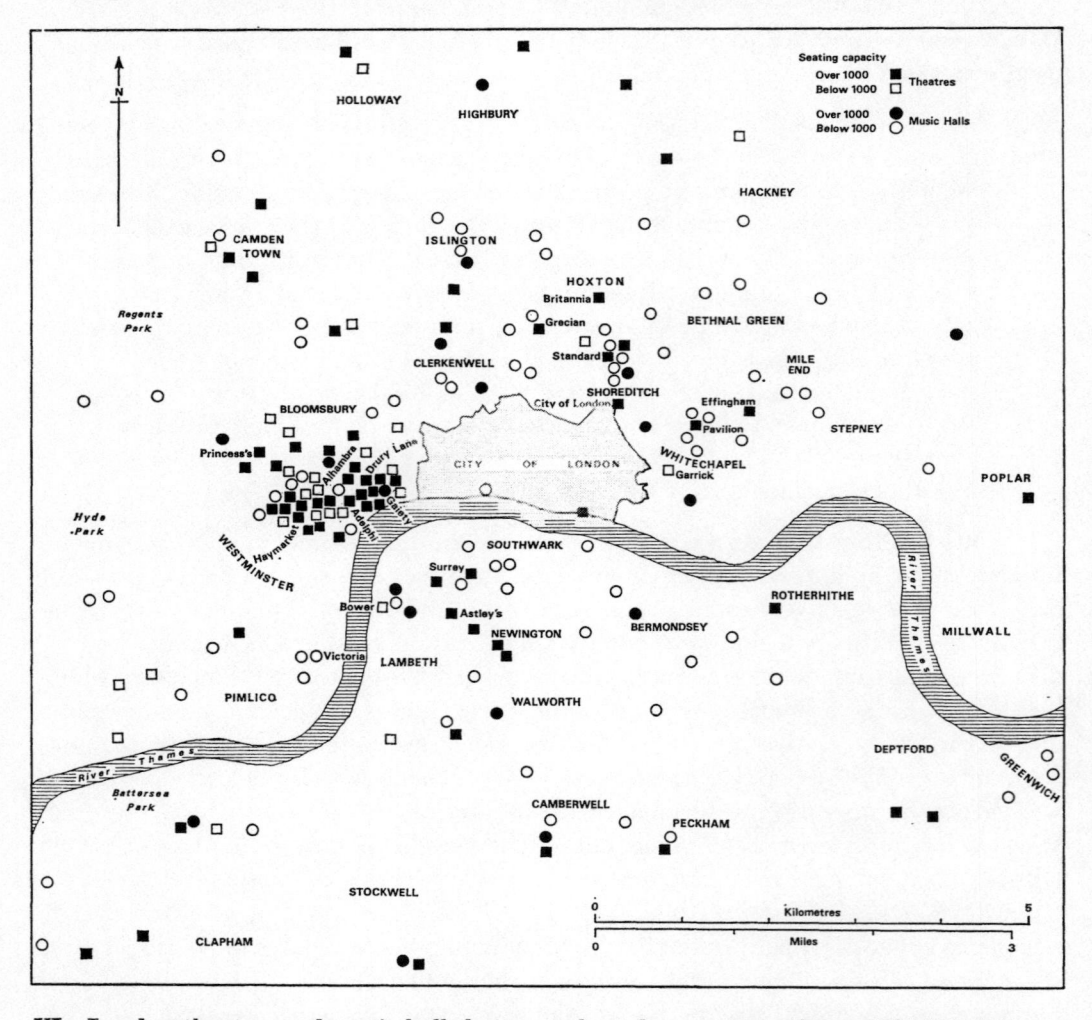

VI London theaters and music-halls known to have been open in the period 1875–1901. Places mentioned in the text are named. Source: Diana Howard, *London Theatres and Music Halls 1850–1950* (1970), which contains an extensive gazetteer and bibliography, and some illustrations.

SADLER's WELLS.

Licensed by the Lord High Chamberlain, pursuant to the Act 6 & 7 Vic. Cap. 68
Under the Management of THOMAS LONGDEN GREENWOOD,
Melbourne Cottage, White Hart Lane, Tottenham.

☞ **First Night of a New National Local Drama.**
Last Six Nights OF THE ENGAGEMENT OF Mr. R. W. PELHAM.

MONDAY, Nov. 13th, 1843, and DURING THE WEEK.
The Performances will commence with (FIRST TIME) an entirely new National Local, Characteristic, Metropolitan, Melodramatic Drama of the day, in 3 Acts, correctly exhibiting Life and Manners in innumerable novel and interesting Phases, called The

CROSS ROADS
OF LIFE!
OR THE
SCAMPS
OF LONDON.

The Groundwork of the Drama founded on the celebrated Play, "LES BOHEMIENS," now attracting the attention of all France, and applied to the circumstances and realities of the present moment, by the Author of TOM & JERRY, &c

The New Scenery, (from Actual Authorities) by **Mr. F. FENTON** and Assistants. The New Flash Medley Overture, and Slang Dramatic Music, by Mr. W. MONTGOMERY. The Action of the Piece arranged by Mr. C. J. SMITH. The Dresses by Mr. HAMPTON & Miss BAILEY. The whole produced under the direction of Mr. H. MARSTON.

Deverex, alias ▬ ▬ (a Swell Cove out of Luck, King of all the Scamps & Greeks in London) Mr H. MARSTON
Mr Dorrington (a wealthy Liverpool Merchant, on a visit to London) MR. ROMER
Frank Danvers .... (a British Naval Officer, just arrived from the Indies) .... Mr BIRD
Herbert Danvers .... (His Younger Brother, a ruined Roué, pigeoned by the Greeks) .... Mr MELVIN
Mr Hawksworth Shabner, { Principal Proprietor of a Silver Hell at the West End, Director of a Company, Capital One Million, Bill Discounter & Anythingarian where there's anything to be got } Mr P. WILLIAMS
Bob Yorkney .... (the Duffer, tired of the Lay) ... Mr W. H. WILLIAMS
Tom Fogg, alias Old Deady, alias The Animal, { a Gin-drinking Vagabond suffering under delirium tremens } .... Mr C. J. SMITH
Ned Brindle .... (the Magsman—a Half-and-Half Cove) .... Mr CORENO
Joe Onion (the Crocodile—an Out-and-Out Cove, Cadger and Creature of Deverex's) Mr C. FENTON
Dickey Smith (the Wakeful Bird a young Gentleman in no ways particklar to a shade, picking up a living how he can) Mast. G. MASKELL
Ikey Bates { Landlord of Rats' Castle, otherwise the Dyot Street Hotel, Proprietor of twenty two-penny dabs and a most respectable bagatelle Board, having cut bumblepuppy as too low } Mr LAMB
Waiter at the Cat and Bagpipes Tavern, Mr SMITHSON. Inspector of the XYZ Division, Mr FRANKS.
Louisa, .. (the Victim of an ill-requited attachment) Miss CAROLINE RANKLEY
Miss Charlotte Willers, { a young Lady with her Cat, &c.. from the Country, betrothed to Mr Yorkney amusing her leisure hours in shoe-binding, waistcoat and gaiter making } Mrs R. BARNETT

After which, the Interesting Drama of

GWYNNETH VAUGHAN

Evan Pritchard, Mr C. J. SMITH. Owen Williams, Mr BIRD. Morgan Morgan, Mr W. H. WILLIAMS.
Hugh Morgan, Mr ROMER. David, Mr LAMB. Thomas Johns, Mr SMITH. Pryce Mr GRAMMAR.
Gwynneth Vaughan, Miss C. RANKLEY. Lyddy Pryce Mrs R. BARNETT.
Betsy Thomas, Miss STEPHENS. Taffline, Miss MORELLI. Phœbe, Mrs ANDREWS.

Mr. R. W. PELHAM will, for the First Time at this Theatre, give his
NIGGER LECTURE ON LOCOMOTIVE.
And Description of his Courtship wid de Black Girls down in Louisiana.
To conclude his Black Art of de Evenin, he will, for de benefit of all dose who hab de Blues, dance his new Solo Reel
HOP, SKIP, AND A JUMP, and dat will sure to cure dem.

To conclude with a Melo-Drama, in Two Acts, by T. E. WILKS, Esq, entitled The

ROLL OF THE DRUM.

Ernest Viscount d'Obernay, Mr BIRD, Captain Charles Aubri, Mr LAMB, Lucius Junius Brutus, Mr ROMER,
Oscar, Mr HENRY MARSTON, Valentine, Mr CORENO, Peter Peaflower, Mr W. H. WILLIAMS,
Bailie, Countess de Renville, Miss CAROLINE RANKLEY, Martha, Miss COOKE, Rosalie, Mrs R. BARNETT

48 An early metropolitan drama of low life, better known by its sub-title, *The Scamps of London*. It was adapted in 1843 by W. T. Moncrieff from a novel by Eugène Sue. The full supporting programme set forth on this poster was a well-established practice.
Victoria & Albert Museum: Enthoven Collection

PRINCESS'S THEATRE.

LICENSED BY THE LORD CHAMBERLAIN TO

MR. VINING

(ACTUAL & RESPONSIBLE MANAGER), UPPER MONTAGUE STREET, RUSSELL SQUARE.

**Doors Open at Half-past Six o'clock,
To commence at Seven.**

Saturday Next, August 8th,

The Performances will commence at SEVEN with the Popular FARCE, by J. M. MORTON, Esq., of

POOR PILLICODDY.

Mr. Pillicoddy,	Mr. DOMINICK MURRAY.
Captain O'Scuttle,	Mr. W. D. GRESHAM.
Mrs. Pillicoddy,	Miss EMMA BARNETT.
Mrs. O'Scuttle,	Mrs. ADDIE.
Sarah Blunt,	Miss POLLY MARSHALL.

After which, **AT EIGHT O'CLOCK,**

A NEW DRAMA, IN FOUR ACTS, entitled

AFTER DARK

A TALE OF LONDON LIFE.
BY
DION BOUCICAULT.

The subject of this Work is derived from a Melodrama by Messrs. D'ENNERY & GRANGÉ, with their permission.

Gordon Chumley,	(Light Dragoons)	Mr. J. G. SHORE.
Sir George Medhurst,	(under the assumed name of Hayward)	Mr. H. J. MONTAGUE.
Chandos Bellingham,		Mr. WALTER LACY.
Old Tom,	(a Bearbman)	Mr. VINING.
Dicey Morris,	Keeper of a low Gaming House, near Leicester Square, and Proprietor of the Elysium Music Hall, Broadway, Westminster	Mr. D. LEESON.
Pointer,	(A Division)	Mr. W. D. GRESHAM.
The Bargee,		Mr. R. CATHCART.
Crumpett,		Mr. MACLEAN.
Area Jack,	(a Night Bird)	Mr. HOLSTON.
Jem and Josey,	Karkstone Vocalists by day, and Elysium Artists by night	Messrs. H. & J. MARSHALL.
The Coleman, Mr. TAPPING.		Nick, Mr. CHAPMAN.
Marker,		Mr. TRESIDDER.
Eliza,	(once a Barmaid at the Elysium, Sir George's Wife)	Miss ROSE LECLERCQ.
Rose Egerton,	(an Heiress, Sir George's Cousin)	Miss TRESSY MARSTON.

THE PERIOD, THE SUMMER OF 1868.

ENTIRELY NEW SCENERY by Mr. F. LLOYDS,
Mr. W. MANN, and numerous Assistants.
Music Composed and Arranged by Mons. E. AUDIBERT.

ACT I.

Scene 1. VICTORIA STATION AND GROSVENOR HOTEL.

Returning from the Derby—The Sensation Column in the Times—Fifty Pounds Reward—Dicey Morris finds a Lost Heir—The Hansom Cabman No. 8941—Old Tom the Boardman—How the Medhurst Estate was Left to the Lost Heir.

Scene 2. No. 5½, LITTLE COMPTON MEWS.

Eliza, the Cabman's Wife, refuses to be accommodating—The Compact.

Scene 3. SILVER HELL.

Chicken Hazard—Dicey gives wrong change for a five pound note—Sir George arrives, and he renews his acquaintance with Champagne—The Proposition—The Barmaid and the Baronet.

Scene 4. No. — RUPERT STREET, HAYMARKET.

The Gambler's Wife—How Eliza proposes to Liberate her Husband.

Scene 5. THE STRAND, NEAR TEMPLE BAR.

Old Tom's History—Fanny Dalton and Richard Knatchbull—Father and Daughter.

Scene 6. Blackfriars Bridge ON CRUTCHES, AND THE THAMES BY NIGHT.

The Nest of the Night Birds—Bankside Hotel (Limited)—Airy Rooms—Water always laid on—The Kerbstone Drama—The Suicide.

ACT II.

Scene 1. DRY ARCHES Under Victoria Street.

Old Tom sends for his Friend, and takes his Farewell to Liquor—After Dark, Light will come—How Eliza got a Situation.

Scene 2. THE LILACS.

How the Confederates worked—Rose and Sir George—The Confession—How Eliza finds an unexpected Friend in her New Mistress.

Scene 3. GARDEN GATE.

How Bellingham meets with his match, and Old Tom finds a clue.

Scene 4. GREEN CHAMBER.

How Rose showed Eliza the Bracelet—The Bridegroom makes a Confession and a Happy Mistake.

ACT III.

Scene 1. ELYSIUM MUSIC HALL, IN BROADWAY, WESTMINSTER.

The Provident Mudlarks' Benefit—How Bellingham dealt with Chumley—Old Tom objects, and how he is suppressed—A visit of the Police—All serene.

Scene 2. WINE CELLARS.

Old Tom's Adventures in Subterranean London—A Murder in the Dark.

Scene 3. UNDERGROUND RAILWAY.

Something on the Track—Old Tom destroys wilfully the Property of the Company—The Night Express.

ACT IV.

THE LILACS.

Bellingham's Triumph—Eliza in search of a Father—Dicey Morris throws over an old friend.

To conclude with the FARCE of

NO. 1, ROUND THE CORNER.

Flipper, - - - Mr. J. G. SHORE.
Hobbler, - - - Mr. R. CATHCART.

Stage Manager, - - Mr. J. G. SHORE.
Musical Director, - Mons. E. AUDIBERT.
Secretary, Mr. T. ROBERTS. Treasurer, Mr. T. E. SMALE.

N.B.—Box-office Open Daily from Ten o'clock until Five, under the Direction of Mr. WADES.

Any person wishing to secure places, can do so by paying One Shilling for every party not exceeding Six, which places will be retained the whole Evening.

Dress Circle, 5s. Boxes, 4s. Pit, 2s. Gallery, 1s.
Orchestra Stalls, 6s.
Private Boxes, £2 12s. 6d., £2 2s. and £1 11s. 6d.

49 *After Dark* was a 'sensation' drama by Dion Boucicault, derived from a melodrama by D'Ennery & Grangé but set in the entirely contemporary London scene of 1868. The cheapest seats were a shilling. *Victoria & Albert Museum: Enthoven Collection*

THEATRE ROYAL, DRURY LANE.

Sole Lessee and Manager — Mr F. B. CHATTERTON.

ATTRACTION FOR THE HOLIDAYS.

EASTER-MONDAY, April 22nd, & EVERY EVENING DURING THE WEEK

At SEVEN o'clock, HER MAJESTY'S SERVANTS will perform a New Comedy-Drama, in Four Acts, entitled—THE

GREAT CITY

(Written by ANDREW HALLIDAY.)

THE NEW AND MAGNIFICENT SCENERY BY

MR. WILLIAM BEVERLEY.

The Music and Original Overture composed and arranged by Mr. J. H. TULLY.
The Character Dances arranged by Mr. J. CORMACK.
The Dresses by Mrs. LAWLER and Mr SWAN. Machinery by Mr. J. TUCKER. Properties by Mr. NEEDHAM.
The whole produced under the direction of .. .. Mr. EDWARD STIRLING.

Lord Churchmouse	Mr C. WARNER
The Hon. Mr Dawlish	Mr F. MORTON
Major O'Gab (Half-Pay)	Mr FITZJAMES
Arthur Carrington (a Young Gentlemen)	Mr C. HARCOURT
Mendez (a Jew)	Mr F. VILLIERS (His First Appearance at this Theatre.)
Jacob Blount, M.P.	Mr J. C. COWPER (His First Appearance at this Theatre.)
Mogg (a Convict)	Mr W. McINTYRE
Jenkinson (a Footman and a Man of Business)	Mr J. ROUSE (His First Appearance at this theatre.)
Ragged Jack (a Street Arab)	Mr J IRVING (His First Appearance at this Theatre.)
The Bos (Steward of the Beggars' Club)	Mr J. B. JOHNSTONE
Doctor	Mr NAYLOR
Railway Porter	Mr CULLEN (His First Appearance at this Theatre.)
First Policeman	Mr J. MORRIS
Inspector Brown	Mr WEAVER
Edith	Miss MADGE ROBERTSON [later Mrs Kendal] (Her First Appearance at this Theatre.)
Mrs Mauvray (a Young widow)	Miss R. G. LE THIERE
Aunt Judith	Mrs WARLOW
Fanny	Miss C. THOMPSON (her First Appearance at this Theatre)

Street Passengers, Beggars, Police, Paupers, Swells-Extremes of St. James's and Giles's.

SCENERY AND INCIDENTS:

CHARING CROSS HOTEL.

Alone in the Great City—The Faithless Lover—A Friend in Need—Home Again.

A STREET NEAR ST. PAUL'S.

The Plot—The Meeting—The Recognition—The Compact.

50 A programme of a performance of Andrew Halliday's *The Great City* in April 1867. This was another 'sensation' drama of considerable topographical ingenuity, culminating

WATERLOO BRIDGE.

A Discovery—Despair—The Rescue.

BELGRAVIA!

DRAWING ROOM IN EDITH'S HOUSE.

Fortune—Grand Society—Pride—A Proposal and a Rejection—A Mysterious Visitor—Pride Humbled—The Awakening of Love and Gratitude—A Midnight Mission.

GATES OF THE WORKHOUSE.

The Theory and Practice of the Poor Law—A Jew's Commentary—Meeting of the Lovers—Suspicion.

THE JOLLY BEGGARS' CLUB.

The Beggars at Home—The Supper—JOLLY BEGGARS' DANCE—A Strange Story—Tracked—The Escape.

THE BOARD-ROOM.

The Renewal of Love—Jenkinson's Grand Scheme—How to get up a Company (Limited)—Startling Information—A Jew's Friendship—A Jew's Vengeance.

ROOM IN EDITH'S HOUSE.

A Daughter's Devotion—The Lover's Trial.—A GARRET—The Hole in the Wall—The Escape and Pursuit.

LONDON by NIGHT

THE HOUSETOPS.—Desperation—The Last Chance.

CHAMBER IN EDITH'S HOUSE. The Accusation—Counter Accusation—The Last Wish—Happiness.

STREET NEAR ST. PAUL'S.

THE RAILWAY STATION.

REALISATION OF FRITH'S CELEBRATED PICTURE.

THE ARREST.   THE END.

To conclude with the Farce, in One Act, by T. L. GREENWOOD, Esq., entitled

THAT RASCAL JACK!

Mr Maddleton	... ...	Mr J. NEVILLE
Mr George Granby	... ...	Mr C. WARNER
Rascal Jack	... ...	Mr JOHN ROUSE
Waiter	... ...	Mr BEDFORD
Amelia	... ...	Miss BESSIE ALLEYN
Lucy	... ...	Miss C. THOMPSON

Stage Manager	... ...	Mr EDWARD STIRLING.
Musical Director	... ...	Mr J. H. TULLY.
Treasurer	... ...	Mr JAMES GUIVER.

Private Boxes, One, Two, Three, Four, and Five Guineas.   Stalls, 7s.   Dress Circle, 5s.   First Circle, 4s.   Balcony Seats, 3s.
Pit, 2s.   Lower Gallery, 1s.   Upper Gallery, 6d.   **NO HALF PRICE.**

Box Office open from Ten till Five daily.   Doors open at Half-past Six, the Performances to commence at Seven o'clock.

E. W. MORRIS & COMPY.'S STEAM PRINTING WORKS, WILDERNESS LANE, WHITEFRIARS.

in a reconstruction of W. P. Frith's painting of Paddington station. See nos 63 and 342.
Victoria & Albert Museum: Enthoven Collection

THEATRE ROYAL, NEW ADELPHI

SOLE PROPRIETOR AND MANAGER, MR. BENJAMIN WEBSTER.

Licensed by the Lord Chamberlain to Mr. BENJAMIN WEBSTER, Actual and Responsible Manager, Kennington Park.

LOST IN LONDON.

Mr. BENJAMIN WEBSTER, finding recovery from his present serious Illness to be impossible without perfect rest and relief from professional anxieties, has yielded to the advice of his Medical Adviser, EDWIN CANTON, Esq., and his Friends, and prevailed upon Mr. HENRY NEVILLE to undertake the Character originally written for Mr. WEBSTER.

FIRST NIGHT
OF A
New and Original Drama, in Three Acts,
By **WATTS PHILLIPS, Esq.**,
Author of "THE DEAD HEART," &c., entitled

LOST IN LONDON,
WITH
ENTIRELY NEW SCENERY AND ADELPHI EFFECTS,

Mr. HENRY NEVILLE, (His First Appearance at this Theatre.)
Mr. J. L. TOOLE,
Mr. PAUL BEDFORD.   Mr. ASHLEY,
Miss NEILSON, (who has been expressly engaged for this Drama.)
Mrs. ALFRED MELLON, (Late Miss Woolgar.)

The MOUNTAIN DHU!
EVERY NIGHT.

SATURDAY, MARCH 16th, MONDAY, MARCH 18th, 1867,
AND DURING THE WEEK.

To commence at SEVEN with (FIRST TIME) a New and Original DRAMA, in Three Acts, by WATTS PHILLIPS, Esq., Author of the "Dead Heart," &c., entitled

LOST IN LONDON.

THE NEW SCENERY BY MESSRS. DANSON.
The Machinery by Mr. CHARKER.   The Appointments by Mr. T. IRELAND.
The Music Arranged by Mr. EDWIN ELLIS.
The whole Produced under the Direction of Mr. R. PHILLIPS.

Gilbert Featherstone,	(Owner of the Bleakmore Mine)	Mr. ASHLEY,
Sir Richard Loader,	(his Friend)	Mr. BRANSCOMBE,
Job Armroyd,		Mr. H. NEVILLE, (His First Appearance here).
Jack Longbones,	(Miners)	Mr. PAUL BEDFORD,
Dick Raine,		Mr. ALDRIDGE.
Noah Morehead,		Mr. TOMLIN,
Benjamin Blinker,	(a London Tiger)	Mr. J. L. TOOLE,
Thomas,	(a Footman)	Mr. W. H. EBURNE,

Tops,	(a Post Boy)	Mr. C. J. SMITH,	
	Guests, Miners, &c. ;		
Nelly,			Miss NEILSON,
		(Who has been expressly engaged for this Drama.)	
Tiddy Dragglethorpe,		Mrs. ALFRED MELLON, (late Miss Woolgar.)	
Florence,		Miss A. SEAMAN.	

LANCASHIRE.
ACT I.
Scene 1. **JOB ARMROYD'S COTTAGE.**
Scene 2.—**BLEAKMOOR.**
Scene 3. **BLEAKMOOR MINE!**

LONDON.
ACT II.
Scene 1.—**INTERIOR OF THE FERNS' VILLA.**
Scene 2.—**ANTE ROOM AT THE FERNS' VILLA.**
Scene 3.—Exterior of the Ferns' Villa in Regent's Park (BY NIGHT.)
Scene 4.—**ANTE ROOM AT THE FERNS' VILLA.**
Scene 5—**BALL ROOM.**

ACT III.
INTERIOR OF A COTTAGE IN THE NEIGHBOURHOOD OF LONDON.

PROGRAMME OF MUSIC.

Overture,	"IL BARBIERE DI SIVIGLIA,"	Rossini.
Valse,	"DEBATEN,"	J. Gung'l.
Quadrille,	"LINDA,"	D'Albert.
Galop,	"WEDDING,"	C. Coote, Junr.
Valse,	"SHADOWS OF DESTINY,"	Captain Colomb, R.A.
Overture,	"MOUNTAIN DHU."	Edwin Ellis.

Musical Director, Mr. EDWIN ELLIS. Solo Cornet, Mr. W. H. HAWKES.

To conclude with the New and Original GRAND EXTRAVAGANZA, by the Author of the Celebrated Burlesque of "KENILWORTH," entitled The

MOUNTAIN DHU!
OR, THE KNIGHT, THE LADY AND THE LAKE.

Fitzjames (an errant knight [wi' Burns] who, pursuing his little *game* among the mountains, discovers a little *dear*, and being engaged with the *dear's talking*, finds himself on the horns of a dilemma, by which he is naturally *stag*-gered) Mrs. A. MELLON

Malcolm Græme (the original "bonnie laddie, highland laddie," who will give himself all sorts of Scotch airs [with v-air-iations] it his hoped to the satisfaction of the public) Miss HUGHES

Roderick (the genuine Mountain Dhu, a fiery spirit considerbly O. P. [which being interpreted means over proof], the wool-gatherings of Toole-ittle John—also a Mountain Do, whose crest, being a Caledonian Pine [*Fur Scotticus*] sufficiently indicates his character) Mr. J. L. TOOLE

The Douglas (the definite article of his clan, a Scotch heavy *f*[e]*ather*, very much *down on his luck*) Mr. PAUL BEDFORD

Malise (a young [*electric*] spark of the period, who flashes about the Highlands and kindles the flame of freebooting patriotism) Miss EMILY PITT

Red Murdoch (Rough and *Red*-dy, Roderick's *Mac*totum, descended [very *low*] from an ancient line of [*Highway*] Robbers) Mr. ASHLEY

Norman (a Scotch Norseman, or *Sandy Knavian*) Mr. R. ROMER

Ellen (the Blue Belle Ellen of Scotland, lately taken from the Greek by Paris, now taken from Paris by the Scotch, also very much taken with Fitzjames, and no doubt the public will be very much taken with her) Miss FURTADO

Allan Bane (a transparent medium, a spirit wrapper and a spirit tapper, gifted with second sight, or the art, when in liquor, of seeing double) Mr. C. J. SMITH

Box-Office Open Daily from 10 till 5, where Private and Family Boxes and Places can be taken without Fee a Fortnight in advance.

N.B.—No gratuities to Box-keepers, or for Bills of Performance allowed.

Acting Manager, Mr. J. KINLOCH. Treasurer, Mr. J. W. ANSON. Stage Manager, Mr. R. PHILLIPS.
Nassau Steam Press—W. S. JOHNSON, 60, St. Martin's Lane, W.C.

51 The first night of *Lost in London*, 16 March 1867. This was a drama by Watts Phillips and dealt with the seduction of a coalminer's wife by an aristocratic mine owner who takes her to live at a villa in Regent's Park. See no. 64.
Victoria & Albert Museum: Enthoven Collection

52 *right* The Surrey Theatre in Blackfriars Road, Lambeth. It was opened in 1782 as the Royal Circus but was converted into a theater in 1810. It later became a variety theater and eventually a cinema. It was demolished in 1934. From an engraving in R. Wilkinson, *Londina Illustrata* (1819–25), II, facing p. 190.

53 *right* The Adelphi Theatre Royal in the Strand, one of the most famous of the minor theaters. Opened as the Sans Pareil in 1806, it took its new name in 1819. Moncrieff's *Tom and Jerry; or, Life in London* had a long run here in the 1820s. From a watercolor.
Photo: R. B. Fleming & Co. Ltd Guildhall Library

54 *far right* The Adelphi in 1840, suitably enriched so as to give 'the spectator a correct idea of the entrance to a theatre.' The frontage was enlarged still more in the rebuilding of 1858, when the seating capacity reached 1,500. From *Mirror*, xxxvi (1840), 271.

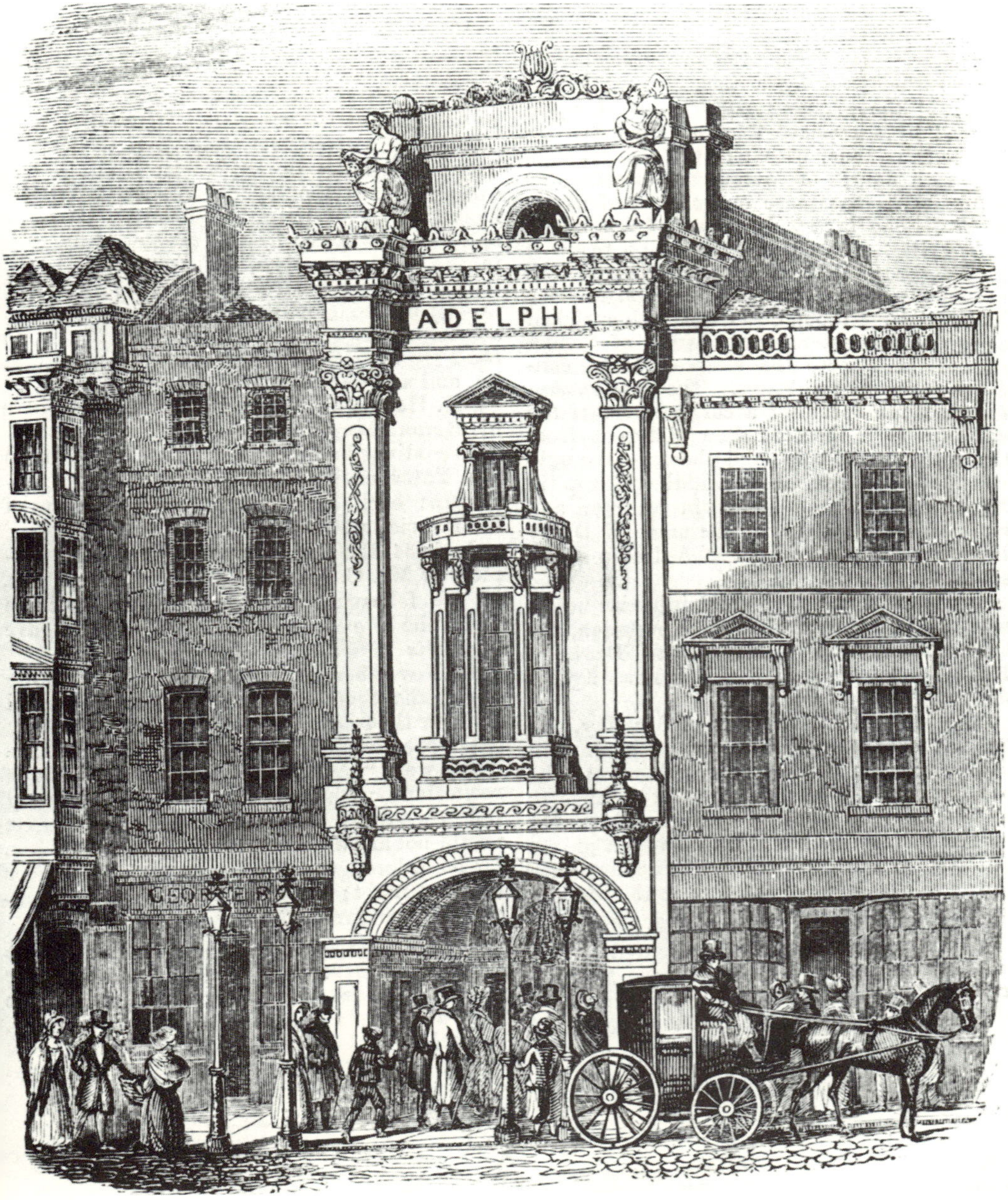
ADELPHI

55 *right* The London Pavilion, Tichborne Street, *c.*1870. This theater grew out of a hall erected on the yard of the Black Horse inn on the left, first used as a waxworks exhibition. It became a sing-song saloon in 1859 and a music-hall with a capacity of 2,000 two years later. It was rebuilt on a new site in 1885 following street improvements.
Guildhall Library

56 *below* The Alhambra Theatre of Varieties, Leicester Square. This building replaced the original structure of 1854, built as an exhibition hall known as the Panopticon but soon converted into a theater and destroyed by fire in 1882. The second building had a maximum capacity of over 3,000.
Photo: G. W. Wilson
Aberdeen University Library

57 *left* The second Theatre Royal, Haymarket, on its opening night in 1821. From an engraving in R. Wilkinson, *Londina Illustrata* (1819–25), II, facing p. 160.

58 *below* The Garrick Theatre in Whitechapel, a low working-class theater, presenting *Starving Poor of Whitechapel*, while a policeman gets the worst of it in the pit. The original theater was opened in 1831 and burnt down in 1846. The auditorium shown here was raised in tone in 1873. It was closed finally about 1880. From Gustave Doré and Blanchard Jerrold, *London. A Pilgrimage* (1872), p. 164.

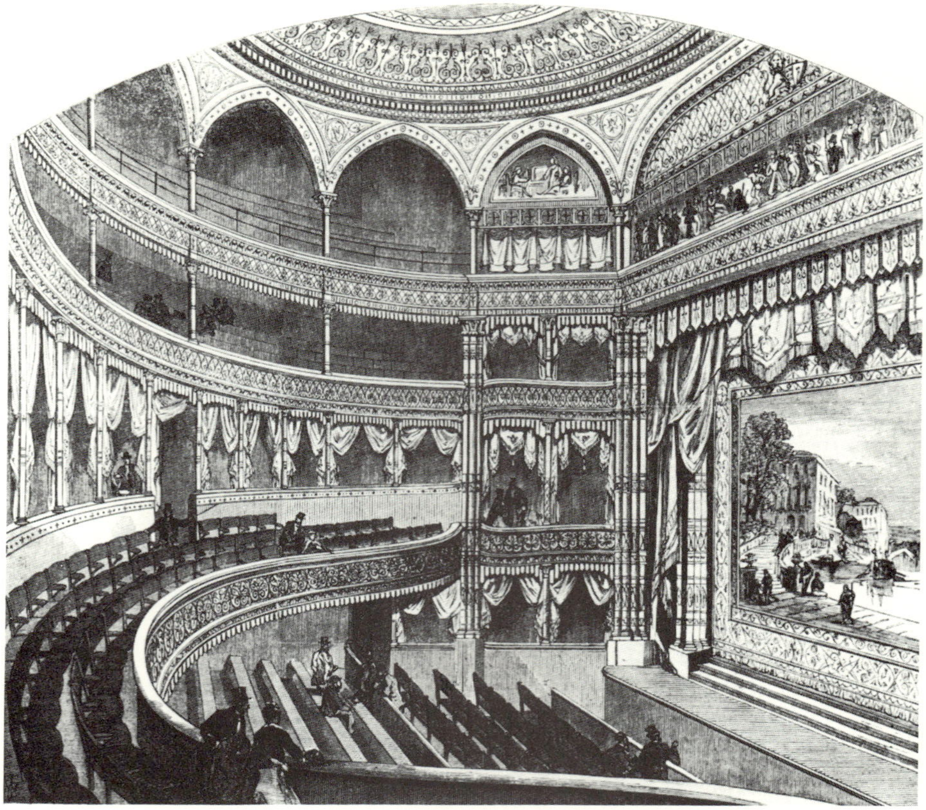

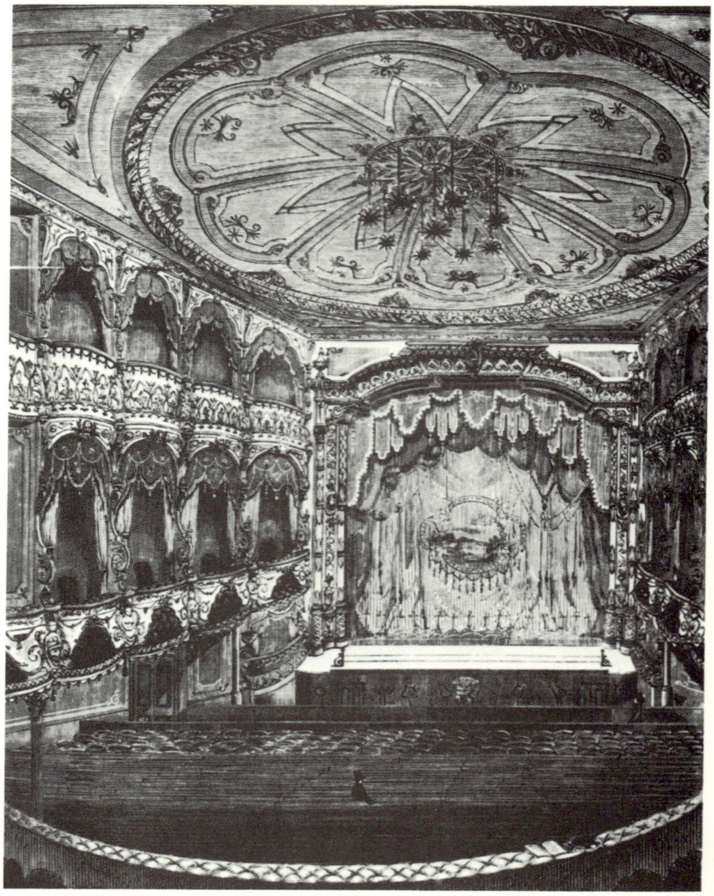

59 *above left* The Gaiety Theatre, Strand, opened by John Hollingshead in 1868. An earlier building on this site was known as the Strand Musick Hall. George Edwardes's famous productions began here in 1886. The building was demolished in 1902. From *Illustrated London News*, liv (1869), 20.

60 *left* The Grecian Theatre, Shoreditch, built on ground once occupied by pleasure gardens, and managed by the Conquests from 1851 to 1879. It had by then become the largest theater in the East End, eventually holding 3,400. It was sold to the Salvation Army in in 1881 and staged no more productions. From *Builder*, xxxv (1877), 1105.

61 *above* The Britannia Theatre, Hoxton, the most famous of East End theaters, opened in 1841 at the back of a tavern and remained under the management of the Lane family till 1899. This auditorium dates from 1858. From *Builder*, xvi (1858), 763.

62 *above* *The Streets of London*, Princess's Theatre, August 1864. This was another Boucicault play. One reviewer wrote of the 'union of an exact picture with all the movement and mechanical aids which are appropriate to the scene . . . of Charing-cross at midnight, with its lighted lamps, its Nelson column, its gleaming windows of Northumberland House, its groups of rich and poor wending their way to club or garret . . . perhaps the most real scene ever witnessed on the stage in London.' From *Illustrated Times*, v (1864), 129, 131.

63 *above right* *The Great City*, Drury Lane Theatre, May 1867. The scenes were set 'amidst the maze of the metropolis' and contained some improbable developments, including a house-tops chase of a convict who escaped by hanging from telegraph wires. From *Illustrated Times*, x (1867), 276.

64 *right* *Lost in London*, Adelphi Theatre, April 1867. The final scene. From *Illustrated London News*, l (1867), 341.

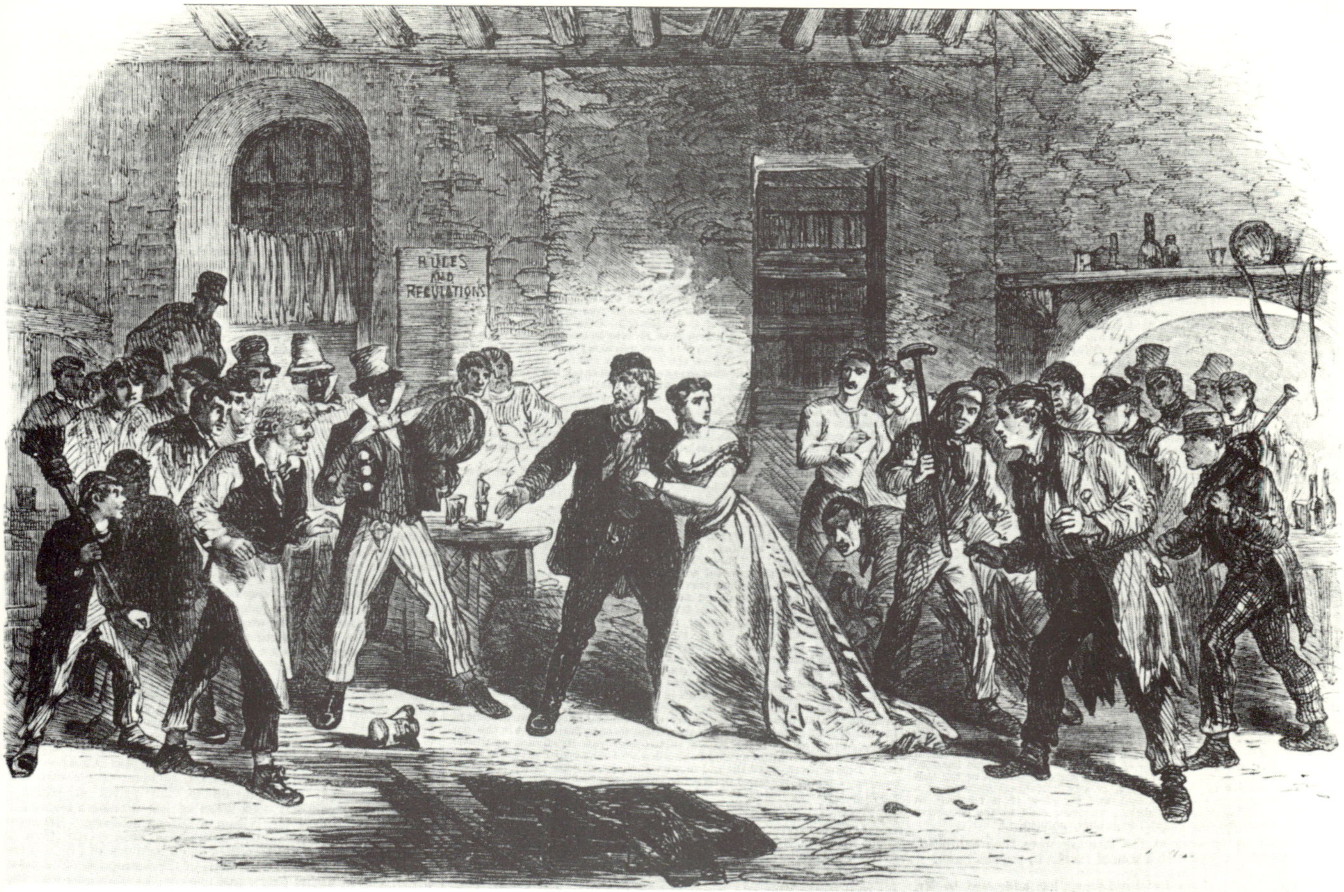
RULES
AND
REGULATIONS

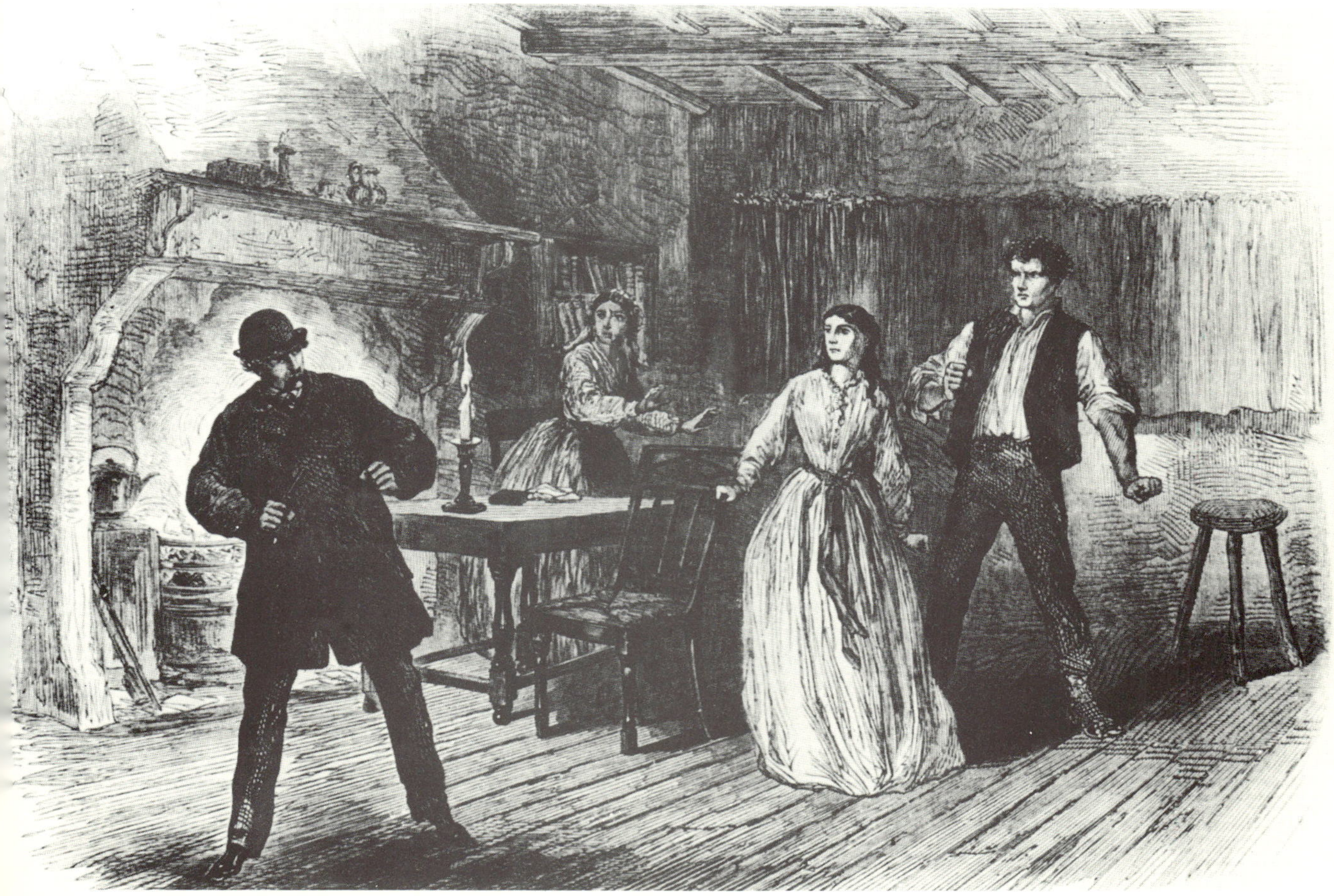

65 *The Long Strike*, Lyceum Theatre, September 1866. A play by Boucicault adapted from *Mary Barton*, which 'endeavours, after the example of the late Mrs Gaskell in her novels, to place before the audience the actual life of the world —a narrow world, too—the world of Manchester, with its discontented workmen and its manufacturing despots.' A metropolitan viewpoint, from *Illustrated London News*, xlix (1866), 290, 309.

sewing. Fearful of the landlady's demands for rent, which she cannot pay, Esther says, 'Work, work, work, and yet of no avail; it will not clear away the poverty by which I am surrounded. The dreadful threat of the few things I have got together being taken from me, the fear of being thrust forth homeless, checks every zealous intention, defies all industrious efforts. Well, well, I must try—still struggle on.' On Esther's wall is the same picture of the village church, which she has rescued from a sale; Ruth sees it and cries, 'Why does it hang there, as if to remind me of the past—to tell me what I might have been?'

London was indeed the City of Dreadful Night for melodramatists, and since all melodrama dealt in moral and material extremes, the village innocent often became, in the course of the same play, the urban sufferer caught in a vicious circle of poverty and sin. J. T. Haines's *The Life of a Woman, or The Curate's Daughter* (1840) is based upon 'The Harlot's Progress' and 'realizes' Hogarth's pictures on the stage. Fanny, the daughter in question, is seduced in London; her aged uncle suffers greatly; her honest village lover seeks her; and Fanny herself is consumed with guilt and agony of mind. Similarly, Susan, the village heroine of John Stafford's *Love's Frailties, or Passion and Repentance* (1835), is lured to London by her seducer. Maddened by guilt and horror, she throws herself into the Serpentine, but is rescued by her brother, who has come to London to look for her. (Carefully included in this scene is *'the view of the new Bridge across the Serpentine'*.) The virtuous Agnes of W. T. Moncrieff's *The Lear of Private Life* (1820) is betrayed by her city lover into leaving beloved father and village home and eloping with him to London. Living there as a kept woman she realizes that he means never to marry her, and she flees to her father and her childhood home. In Douglas Jerrold's *Martha Willis, the Servant Maid* (1831), the heroine, 'a good girl, the darling of our village,' goes into domestic service in London, is falsely accused of theft, and suffers under sentence of death in Newgate until the real criminal is discovered. The preface points directly to the moral-urban theme: 'It is the object of the present drama to display, in the most forcible and striking point of view, the temptations which in this metropolis assail the young and inexperienced on their first outset in life.' The blissful innocence of Martha's village life, in contrast to the corruption that later overwhelms her, is evident from her speech when she arrives in London and unpacks her box, which contains

> my best stuff gown, my four cotton ones, and my white aprons for Sundays, and here are the ballads, 'Crazy Jane,' 'The Dusty Miller,' 'Sheep Sheering,' and 'Blue-eyed Mary,' and there's the Charm for the Tooth Ache, and there's 'The Babes in the Wood,' and 'Lady Godiva,' and my grandmother's wedding-ring, and the needle case Ralph Thomas would give me, and there's the dream book, and Doctor Watts, and my sampler when I was a little girl. Oh, those were happy days! And here's the picture of our church and village, that Mr. Carmine painted for me, and told me always to keep by me, and—but where are the ribbons?

Nelly, the heroine of Watts Phillips's *Lost in London* (1867), carried off from her

Lancashire mining village to a life of kept luxury in London, awakens after repentance and severe illness and looks out of the window:

> London! (*She gazes for some moments steadfastly at the distant city, the red light of the setting sun falling full upon her face.*) The shining city of my dreams—my dreams! Its spires are bathed in light. (*As she gazes the light fades from her face and her voice changes from one of exultation to one of deep sadness.*) But the darkness is creeping down, and a shadow rises between me and the fading light.

At the end of the play Nelly dies and her husband Job tears aside the curtains, revealing '*the distant city, now brought out in strong light by the rising moon.*' Job declares that 'though lost in London (*he indicates by a gesture the city now bright with moonbeams*) I shall foind her theer,' pointing to Heaven. The polarity of London and Heaven is an obvious one, and to make the identity of London as a symbolic character in the drama even plainer, Phillips adds a note requiring a great width of window at the back of the scene, because 'I wish the great city to appear *most distinctly*, as a background to the last act of the drama.' This symbolic view of London is by no means at variance with the life-like treatment of the London streets and the people to be found on them; symbol and verisimilitude go hand in hand in nineteenth-century melodrama.

The men of Victorian drama go as badly astray in London as the women, although their suffering is caused mainly by gambling and drinking. Drink is a separate problem whose place in urban drama will be briefly examined below, but the village innocent in the capital of vice takes just as quickly to gambling. Young Wildflower, up from a Yorkshire village to buy an estate for his father with £5,000, loses it to gamblers in A. V. Campbell's *The Gambler's Life in London* (1829). Stephen Lockwood, a good farmer in W. B. Bernard's *The Farmer's Story* (1836), draws a lottery ticket worth £20,000 and sets up fashionably in London, where, ruined by drink and gambling, he sinks into poverty. The first two acts are entitled 'The Village—Labour and its Lesson' and 'The Metropolis—Wealth and its Temptations.' The honest Lancashire hero of Tom Taylor's *The Ticket-of-Leave Man* (1863), befuddled by the unaccustomed pleasures of London and duped by a gang of counterfeiters, complains, 'I used to sleep like a top down at Glossop. But in this great big place, since I've been enjoying myself, seeing life—I don't know (*passing his hand across his eyes*)—I don't know how it is—I get no rest—and when I do, it's worse than none—there's great black crawling things about me.' Job Armroyd, the betrayed miner husband of *Lost in London*, sums it up simply: 'It be a dreadful and a dreary place, this Lunnon, for them as are weak an' wi' no hand to guide 'em.'

Praise of country life at the expense of the city has been a theme common in Western literature since the days of Rome, but its expression in nineteenth-century drama is rather different and peculiarly of the age of great industrial cities—peculiarly Victorian. There is nothing like it in previous English drama and it can hardly be found before the 1820s. What is praised is specifically the village, not rural life in general, a village that is no longer a real place but a dream, a lost Eden in the sharpest contrast to the dirty streets and the wretched dwellings which the rural poor now inhabited. That life in the village was so hard did not, apparently, take long to forget, and the village of Victorian melodrama is like no other village that ever existed or could exist; it was a village of shimmering sweetness, sentiment, nostalgia, and beauty. It appears again and again in plays seen and enjoyed by metropolitan audiences in their tens of thousands, and the great London in which these audiences lived, in these same plays, vanished before the Ideal Village from which thousands of them came (or thought they came), many of them long ago. This is what countless Londoners saw as the stage version of their own past—illusion, it is true, but with all the complex significance of illusion. After 1860 drama of this sort was not so prevalent. Perhaps it was because the villager had stopped seeking his economic salvation in London or other cities, and because urban audiences had ceased comforting or amusing themselves with illusions of a simpler and purer rural life.

The living conditions of rural migrants to the city had become a theme of the Victorian popular theater. The only form of Victorian drama that dealt with social issues, clumsily and emotionally oversimplified in presentation though they were, was melodrama, which was also the form in closest touch with the needs and dreams of the urban masses. It is therefore to the melodrama that we must turn to see some reflection of their lives. By the 1840s melodrama had already become a crude theater of social protest and had taken up such matters as slavery, industrialization, class conflicts, game-laws, and other problems of contemporary society. As for London, melodrama was full of its homeless poor, and of their crime, drunkenness, and nostalgia for the lost life of the village. To give only one example of its social range, melodrama was a powerful weapon in the armory of temperance reform. *Fifteen Years of a Drunkard's Life* (1828), *The Drunkard's Doom* (1832), *The Drunkard's Glass* (1845), *The Drunkard's Fate* (1847), *The Drunkard's Progress* (1847), *Drink, Poverty, and Crime* (1859), *Drink* (1879), *Destroyed by Drink* (1879), *Intemperance* (1879), *Gin* (1880)—these are titles of some of the English temperance melodramas. Most of them end in the redemption of the drunkard and his family by the influence and direct intervention of the temperance spokesman, but in a few drink pursues its course unchecked, causing poverty, crime, madness, and death.

Two such plays can be examined as metropolitan examples of their school: T. P. Taylor's *The Bottle* (1847) and the same author's *The Drunkard's Children* (1848), both based on Cruikshank's temperance engravings, each one being 'realized' in a stage picture. The opening scene of *The Bottle*, entitled 'The Happy Home,' shows Richard Thornley, a mechanic, with his wife and two children, celebrating a tenth wedding anniversary in a '*neatly-furnished room*' in Finsbury. The bottle is brought out for the first time despite the wife's admonitions and fears. Soon there is no stopping Thornley on his slippery downward path. Ignoring the pleas of his loving wife and his best friend (who refuses to touch a drop), he continues to drink, both at home and in the 'High-Mettled Racer,' and is fired from his job for continued absences from work; at the end of Act I the Thornley furniture is removed for non-payment of debts, all having been spent on drink. In Act II, three years later, Thornley's son is taken into a gang of thieves to save him from starvation; his daughter becomes a prostitute for the same reason; his baby dies; he murders his wife in a fit of drunken rage ('Ha, Ha! The bottle has done its work!') and dies raving in Bethlehem madhouse, which he describes in a sane moment as 'warmer here, and better than the cold and muddy streets.' *The Drunkard's Children* is a sequel to *The Bottle* and traces the fortunes of Thornley's two surviving children. The son, now a dissipated thief, frequents beer-shop and gin-palace alike, gambles, is convicted of robbery, transported, and dies in the hulks. The daughter falls into a life of drink and prostitution, and, homeless and starving, ends her life in the dark, cold waters beneath Waterloo Bridge. The sins of the father have been visited on the children, and the Demon Gin (that favorite figure of temperance literature) has done his work.

The saloon or public-house figures prominently in *The Bottle* and *The Drunkard's Children* and most temperance dramas. It was almost always depicted as a place of evil and corruption, although in J. P. Hart's *Jane, the Licensed Victualler's Daughter* (1840), one character speaks in favor of the public house: 'Next to my own house I consider this my resting place from the toils of business, and I contend that a respectably-conducted tavern is one of the greatest blessings a nation can enjoy—it is in the social parlour that rational pleasure sweetens the care of trade.' Because of the high moral tone of Victorian melodrama, sentiments like this were rarely heard on the stage, although one hardly needs to draw attention to the ubiquity and popularity of the public house. Not only do the heroes of temperance melodrama fall further into physical and moral degradation in the public house, but it is also the scene of unseemly brawls and sometimes murders. In some plays the hero progresses downward from his pleasant rural home, through the horrors of a vile saloon, to poverty and despair on the streets. In an American temperance play with a happy ending, W. H. Smith's *The Drunkard, or The Fallen Saved* (1844), the reformed drunkard and his wife are happily reunited, not in New York, where in his unregenerate state the husband had been fighting in saloons and sleeping in the street, but in the same country cottage in which they lived happily in the first act, with '*vines entwined, roses &.c. The extreme of rural tranquil beauty.*' To this cottage, away from the foul metropolis, they return at the end of the play, '*everything denoting domestic peace and tranquil happiness. The sun is setting over the hills.*' To the melody of a flute the welcoming villagers quietly sing 'Home, Sweet Home.'

In order for the dramatization of urban social problems such as poverty, homelessness, and drink, to hold the attention of London working- and lower-middle-class audiences—who knew these things well from first-hand experience of them—at least a surface realism had to be created. By the early years of Victoria's reign the theater was able, through a series of technical advances such as the development of the box set, the controlled use of gas lighting, and the perfection of stage mechanisms, to present a more realistic and detailed picture of contemporary life than was possible before. At the same time the tendency of the arts, including the theater, was to move toward a greater fidelity to the surface of life, a tendency that faithfully reflected the ever-increasing materiality and emphasis on the business of daily living. A stage art that concerned itself primarily with reproducing the surface details of life began reconstructing the immediate physical environment of the lives of London audiences, as well as exterior views of the main sights of the city. In this way the drama was, in a sense, true to life, and in this way its presentation of character and situation could carry sufficient conviction for the occasion. The fact that the basic content of such drama was in many respects notably unreal, the dream world of the popular melodrama or the middle-class 'drama,' was an added reason for enjoyment rather than the reverse; a taste for the real and an indulgence in the illusory could be satisfied simultaneously. Such a duality lies at the heart of Victorian drama.

The stage realism of the 1840s was greatly stimulated by *Les Bohémiens de Paris* (1843), a dramatic version of Eugène Sue's novel *Les Mystères de Paris* (1842–3), which contains many scenes of Parisian low life. In London there were several adaptations of *Les Bohémiens de Paris*, and they all concentrated on the realistic depiction of London scenes. In Moncrieff's *The Scamps of London* (1843), the opening stage direction sets the scene with some care:

> *London Terminus of the Birmingham Railway. Curtain rises to bustling music.*
> DICK SMITH (*with Congreves*), *Cabmen, Baked Taters, Fried Fish, Lucifer Matches, and other Vendors and Hawkers, with Miscellaneous Vagabonds, discovered.* TOM FOGG *seen lying on the ground, leaning against a kerb stone, on one side, in a half-stupefied state, taking no notice of any one. Various cries of* 'Baked Taters, all hot,' 'Fried Fish, a penny a slice,' 'Lucifer Matches,' *&.c. heard confusedly mingling together.*

Later in the scene '*eight o'clock strikes, and bell rings, whistle of engine is heard, and train arrives*'; the train being, at this juncture in theater technology, offstage. Another scene displays Waterloo Bridge by moonlight, with the homeless settling down for the night under the dry arches; in a few moments the heroine throws herself off the bridge (a popular activity in Victorian melodrama) and is saved by a boat putting off below. Other scenes include the pleasure gardens of the suburban Cat and Bagpipes Tavern and Terpsichorean Saloon, and the '*Interior of miserable room in Rat's Castle, the Rookery, Dyot Street, St. Giles's. At the back of bagatelle board,* NED BRINDLE, JOE ONION, DICK SMITH, CADGERS, *and* VAGABONDS *of every description divided into groups, sitting, smoking, drinking, &.c.*' Charles Selby's *London by Night* (1845) adds the brickfields of Battersea and a low public house; in Dion Boucicault's *After Dark* (1868), the climactic scene occurs in the Underground, with semaphore signals working, signal lights changing, tunnels converging in perspective, and a train (onstage this time) bearing down on the drugged body of the hero. *The Drunkard's Children* contains scenes of a beer-shop, a gin-palace, and the sleeping-room of a cheap lodging-house. Henry Leslie's *Time and Tide, A Tale of the Thames* (1867) depicts among other things the Embankment Works, Waterloo Bridge, the Houses of Parliament, and the departure of an emigrant ship from the London docks. In

The Victorian City: Numbers of People

Andrew Halliday's *The Great City* (1867) audiences were much excited by a real hansom cab driving through the Waterloo Bridge toll-gate.

Scenes of high life could be presented as thoroughly as scenes of low life or the common sights of London. In *Lost in London* the third scene of Act II is an exterior view of a villa in Regent's Park with '*a picturesque view of other villas in varied perspective:*'

> *This set should partake of those characteristics which form what is called a realistic and sensational scene. A great snow effect. Scene brightens gradually, as the various windows and distant gas lamps are lighted up. As scene progresses, broughams, &c. can be driven on if necessary, and all the minor out-door details which accompany the giving of a grand evening party.* FEATHERSTONE'S *villa has handsome portico, with large practical doors . . . Snow falling in scattered flakes at first, afterwards more thickly.*

Two scenes later the interior of the house is revealed, and the effect is equally elaborate and equally detailed:

> *Handsome suite of rooms in* FEATHERSTONE'S *house. Decorations blue and white, profusely relieved by gilt work. Furniture rich, and elegant mirrors adorn the walls, so as to multiply the reflection of the vases and statuettes placed about; chandelier hangs from ceiling of inner room. The two rooms open into each other by a broad arch, surmounted by a handsome cornice, from which fall velvet curtains, drawn up at sides so as to show table spread with refreshments, wines, fruits, &c. &c., the whole giving idea of elegant but prodigal luxury. Music and laughter as* GUESTS *(all in full toilet) come crowding from inner room.*

Time and Tide was performed at the Surrey, one of the working-class theaters on the South Bank in the same year (1867) as *Lost in London* was put on at the Adelphi, a West End playhouse catering predominantly for the middle class. Whatever their class, London audiences wanted to see the city scene recreated faithfully before their eyes, and demands for this kind of realism not only affected the development of Victorian dramas, but also enabled playwrights, managers, stage carpenters, and gasmen to present a composite stage-picture of London that was indeed metropolitan in its scope and variety.

This realism was not, however, merely a realism of display, of bridges, buildings, squares, and railway stations. Such a realism was only half of it; the other half had to do with people. Londoners, especially the poorer classes, daily rubbed shoulders with countless other people; their pleasures were communal, their apartments were overflowing, their streets were crowded. The stage could not reproduce the street-thronging thousands, but it could offer both serious and comic characters pursuing a variety of occupations, characters entirely recognizable to audiences who encountered them every day in the course of ordinary business. The preface to *Martha Willis* points out that 'this great metropolis teems with persons and events, which, considered with reference to their dramatic experience, beggar invention. . . It is these scenes of everyday experience—it is these characters which are met with in our hourly paths that will be found in the present drama.' *Martha Willis* contains only a coach-office porter, a pawnbroker, a shopman, a bawd, a gang of thieves, and prison officers, but no one could see (or even read) melodramas like *The Bottle* or *The Drunkard's Children,* both essentially plays of the streets, without being conscious of the drama's wide selection from contemporary London character-types and occupations. The *dramatis personae* of *The Bottle* include, besides the hero (a mechanic) and his family, a recruiting sergeant, a police constable, a bailiff, a thief, a needlewoman, a master engineer (the hero's employer), and a lodging-house keeper. The comic woman is a shoe-binder, the comic man, Coddles, a potboy and then a pieman. Scenes are laid in apartments, streets, garrets, and public houses in the areas of Finsbury, the Bank, Chick Lane, and Moorfields. The geography and the characters would have been perfectly well known to the audience, for *The Bottle* was first given at the City of London Theatre in Norton Folgate. Many aspects of the play's life and sentiments would have been known to them, too—drink, crime, the rent, poverty, hunger, the police, and the old village home. A character like Coddles, trying to sell an unwanted stock of meat pies, with his livelihood depending on their sale, but keeping a sense of humor about his failure, would have been dear to their hearts. *The Drunkard's Children* is even more crammed with life: safecracker, thimble-rigger, costermongers, dancers, publicans, casino manager, prison chaplain and jailor, barmaid, threepenny lodging-house keeper, dog fancier and stealer, prizefighter, dustman, lascar, and others.

Over thirty years later the staging of London scenes had become elaborately realistic and the trouble taken to establish local atmosphere more painstaking. A rather shocked Clement Scott described the last-act setting of George Sims's *The Lights o' London* (1881), the Borough market on a Saturday night:[2]

> It is a marvellous example of stage realism, complete in every possible detail ... If anything, it is all too real, too painful, too smeared with the dirt and degradation of London life, where drunkenness, debauchery, and depravity are shown in all their naked hideousness. Amidst buying and selling, the hoarse roar of costermongers, the jingle of the piano-organ, the screams of the dissolute, fathers teach their children to cheat and lie, drabs swarm in and out of the public house, and the hunted Harold, with his devoted wife, await the inevitable capture in an upper garret of a house which is surrounded by police.

These examples of the realistic portayal of London street life and the surface of London character have been drawn mainly from the lower-class popular melodrama, but enough evidence from the whole range of Victorian drama could be produced to show that the eye of playwright and stage manager for the details of urban character extended to all classes of drama. From whatever social class they came, London audiences liked their stage London to look like real London and their stage Londoners to act like people they knew; in this wish they were satisfied.

The Victorian City: Numbers of People

Because Victorian melodrama and the 'drama' possess a certain sweep of external events and often chose to depict a cross-section of metropolitan life, they were better suited to a broad portrayal of the London environment and the handling of urban themes than was the more circumscribed comedy or the domesticated farce. The farce was often citified in settings and characters, but since it was commonly a one- or two-act afterpiece with a necessarily inexpensive interior set, it lacked any sort of urban scope and contented itself largely with the comical and extravagant treatment of domestic trivia and misunderstanding. Comedy, the more serious parts of which were almost indistinguishable from melodrama, had a wider range, and from comedy, as well as from melodrama and 'drama,' we can select a Victorian theme that had not properly appeared in drama before and was intimately connected with the City of London: the theme of commercial life, high finance, and speculation. Such a theme had previously only touched drama, and that incidentally, as in Steele's *The Conscious Lovers* (1722), Lillo's *The London Merchant* (1731), and Cumberland's *The West Indian* (1771). Hovering in the background of these plays and others like them are beneficent merchants and noble men of commerce; since Addison's Sir Andrew Freeport, the merchant (and later the substantial tradesman), ceased to be the automatic figure of fun and scorn that he had been in seventeenth-century comedy. However, on the Victorian stage the world of business does not merely supply the plot and intrigue mechanism of an earlier drama of settlements, contracts, wills, and property dealings; it now becomes the thematic center of plays that occupy themselves with this world. Here, of course, we have moved out of a working-class and into a middle-class theater; to be 'something in the City' was taken to be the sole ambition of generations of middle-class youths and their parents.

The general theme of business and its accompanying ambitions appears in many plays of the period, some of them comedies, but most of them serious. A dramatist who turned more than once to it was Tom Taylor, in his vigorous career at one time or another tutor, barrister, journalist, Professor of English, and editor of *Punch*. His comedy *Still Waters Run Deep* (1855) treats of the final and rightful assertion of authority in the home by an apparently down-trodden husband, Mildmay. One of the means by which Mildmay commends himself to the audience is his refusal to be taken in by the speculating villain, Hawksley, and the shares the latter is selling in his fraudulent Galvanic Navigation Company. Hawksley declares proudly that a man may 'chance his hundred thousand on the up or down of the Three percents, every month of the twelve, and he may cultivate domestic felicity at his box at Brompton in the respectable character of a man of business.' He then lectures Mildmay on the stock market and how huge financial advantages are to be gained from investing in Inexplosible Galvanic Boats; the lecture is a long one and must have been calculated to hold the audience's interest as well as Mildmay's. Mildmay replies with a long speech tracing Hawksley's connection with a forged bill presented at a share-discounting house in the City. The plot of the play hangs on the exposure of Hawksley as a roué and a swindler, but the business theme is more than a plot device: Mildmay, in a sense, proves himself a man by a cool and competent demonstration of business knowledge and ability, for he too was 'something in the City.' The second act of another play by Taylor, *Payable on Demand* (1859), is almost entirely taken up with the frenzied speculations of Reuben Goldsched on the eve of Napoleon's abdication, and his attempts to pour money into a falling market so that he can be sure of an immense profit when he sells on the rising market which the news of the abdication (brought to him in advance by fast carrier-pigeon from Paris) will be sure to create. The tension of the act is purely financial, mounting steadily as messengers arrive with reports of the market and Goldsched issues orders to buy and sell. The substance of *Settling Day* (1865), whose second act is laid in a carefully described City broker's office, is financial speculation and financial crisis as the hero struggles to save his bank from ruin. Again, the dialogue is full of stock-market phraseology. *The Ticket-of-Leave Man* is concerned with the attempt of ex-convict Brierly to rehabilitate himself socially. He is taken into Mr Gibson's City office, turned away when his past is revealed, and finally places Mr Gibson in his debt by preventing a robbery. (As in *Settling Day*, the act laid in the broker's office is commercially solid and authentic in setting.) The certain implications of the happy ending are that Brierly can resume his rise to City prominence and City wealth, thereby taking the respected place among the seats of Commerce to which he so deeply aspires. A revealing passage of dialogue in Act III points directly to Brierly's prospects beyond the happy ending:

> MR. GIBSON. Go on as you've begun—keep a bright eye and an inquiring tongue in your head—learn how business is done—watch the market—and from what I've seen of you the six months you've been here, I shouldn't wonder if I found a better berth than messenger for you one of these days.
> BRIERLY. Mr. Gibson—sir—I can't thank you—but a lookout like that—it takes a man's breath away.
> MR. GIBSON. In the City there's no gap between the first round of the ladder and the top of the tree.

Other dramatists also developed the motifs of success in the City, speculation, and financial disaster. George Henry Lewes's *The Game of Speculation* (1851), adapted from Balzac's *Mercadet*, presents the manipulative wizardry of Affable Hawk, who twists his creditors round his little finger as he bluffs his way towards a Stock Exchange hoax of vast proportions that will save his tottering fortunes. While there is nothing of the London scene in the play, its dialogue is heavy with investments, bonds, mortgages, loans, and stock, and the character of Affable Hawk is that of the over-ambitious speculator whose swindles on the London Stock Exchange were exposed from time to time in a blaze of publicity. 'All our morals,' he says, 'lie in dividends... In these days credit is everything—credit is the wealth of commerce, the foundation of the state.' Holding up half-a-crown, Hawk declares 'Here lies modern honour. Chivalry has shrivelled into that'—and he acts accordingly. Hawk is properly redeemed by a stroke of coincidental good fortune just as he begins to carry out his swindle. A more ruthless speculator is Bloodgood, the villain of Boucicault's *The Poor of New York* (1857), which was successfully adapted as *The Streets of London* in 1864. Two financial crises, in the 1830s and 1850s, drive the hero's family into poverty (in *The Streets of London* they beg in a snow-covered Trafalgar Square), since Bloodgood has absconded with their fortune, which was deposited in his bank. All comes right in the end, but in the meantime much dramatic interest is attached to the business transactions carried on in Bloodgood's bank in Act I, and to his reputation as a speculator on the stock market.

The London of commercial speculation, financial panic, and the promise of riches and status was, as we have noted, not a working- or lower-middle-class London at all. The most significant fact of English theatrical history in the second half of the nineteenth century is the slow but sure upper-middle-class takeover of both the theater and the drama, and the steady rise of theater into that middle-class respectability it is even now trying so hard to shake off. Consequently, the lower-class dramatic vision of London retreated to the dwindling number of neighborhood theaters still open, which by the end of the century meant the Britannia, Standard, Pavilion, Surrey, Elephant and Castle, and the West London. The West End playhouses were given over entirely to the middle class, except the galleries in theaters like Drury Lane, the Adelphi, and the Princess's, which were mainly filled by a lower class of patron. In the East End, urban themes were dramatized in much the same way as in the 1840s, and low-life scenes and characters abounded; in the West End the increasing taste for visual realism brought forth more and more ponderous reconstructions of the London landscape. The annual autumn drama at Drury Lane, an institution that lasted from the 1880s until after the First World War, offered the most elaborate effects of all, especially in portraying the pleasures, occupations, and public resorts of the upper classes: Derby Day, Ascot, Goodwood, the Grand National, Rotten Row, Hurlingham, the Stock Exchange, the House of Commons, Westminster Abbey, the promenade at the Empire, the Military Tournament at Earls Court—all staged with the full scenic and mechanical resources of Drury Lane, which were considerable.

As a spectacle of life, therefore, or as a commercial jungle, a moral lesson, a monster of poverty and darkness, or a reminder of a lost heritage, London fascinated Victorian audiences, even those outside London, and was staged for them in different types of drama, different kinds of theaters, and before different classes of people, again and again. In the theater it was London above all, and neither before nor since has the metropolis exercised such a hold on the imaginations and dreams of English theater audiences.

Notes

1   'The Amusements of the People,' *Household Words*, i (1850), 13–15.
2   *Theatre*, October 1881, 239–40.

III Shapes on the Ground

9 The Camera's Eye

G. H. Martin and David Francis

Of all the cities of man, the Victorian city was the first to come under the camera's eye, the first to transmit the lights and shadows of its pattern directly to posterity. It lives for us, therefore, in its own image, as well as in the streets and buildings that we still inhabit. Just as the experiences and expedients of the Victorian age inform our own, so we share with that time a unique self-consciousness: a twofold vision. Study it as we may, the past must fall for us, as for Victorians, into the world before photography and the world that the camera has seen.

The nineteenth century's two great contributions to historical enquiry were a rigorous technique and the photograph. Both were products of a sophisticated industrial society, and both are with us still, though they are not as often found in conjunction as they might be. It was appropriate, and perhaps natural, that the century in which, above all others, the principles of historical criticism were refined, should also have passed on to posterity the first exact representations of the contemporary scene. We have drawings and paintings from the earliest ages of mankind, and the artist's vision still extends and intensifies our knowledge of the world. The early fear that the photographer would supplant the artist was never more than a misapprehension. The graphic arts and photography have come amicably to terms, but as a historical record the photograph stands alone. Only the camera enables us to see the past as though it were the present, to dwell upon an instant that is fixed for all time.

The camera is not a faultless instrument: its lenses distort, exaggerating some horizontal lines and depressing verticals, and it can be made to utter untruths. What matters most, however, is that its eye observes, but does not select. It has rather less

© 1973 G. H. Martin and David Francis

capacity for deceit than we have ourselves, and each of its pictures offers us a unique record. At the same time, they share one important quality with the graphic artist's work, for every one is the consequence of a deliberate act. A photograph is a document, and the historian's first business is to ask of it, as he would of any other record, who made it, to whom it was addressed, and what it was meant to convey. Those questions imply others, and entail some reference to photographic techniques. A man may write what he will, within the bounds of his own skill and apprehension. He may photograph what he will, within the capacity of the camera. The first context of any photograph is the contemporary state of photography; the second, the intentions and assumptions of the photographer, so far as we can recover them. Those preliminaries established, there opens beyond them a world as wide as our curiosity.

The origin and the name of the camera lie in the *camera obscura*, an optical device which, in Leonardo da Vinci's time, consisted of a darkened room with a small hole in one wall through which a pencil of light passed to form on the opposite wall an inverted image of whatever lay outside.[1] In the seventeenth and eighteenth centuries a portable version fitted with a lens which transmitted an upright image on to a piece of translucent paper, by means of an inclined mirror, was used by painters to compose and draw landscapes, and it has a close connection with the iconography of the city. The brilliant paintings of Venice and London made by Antonio Canale (1697–1768), perhaps the most precisely observed records of the urban scene before the age of photography, rest on drawings made with the aid of the *camera obscura*.[2]

The illusory sense that those living images gave of commanding Nature, by setting a moment of reality aside for study and appraisal, excited artists and laymen alike. The *camera obscura* became a piece of picnic equipment, like the Claude-Lorraine glass, as well as, in its larger versions, a form of public entertainment which still survives in some towns. It was used as a surveyor's aid, and as the object of philosophical disquisitions upon art. The excitement of fixing its pictures by the power of their own light appealed to many of its users, but had to wait upon the ready availability of photo-sensitive chemicals. Science and art were beginning to separate, though learning was still nominally one, and the investigations upon which photography ultimately depended were the work of *savants* and industrialists rather than artists. The first report of successful photography came, appropriately enough, from the Wedgwoods, who could turn their talents to anything. Thomas Wedgwood (1771–1805), who took up scientific experiments when his health forced him to leave his father's firm, published an account of simple contact printing in 1802.[3] It was a modest triumph, because the prints, which were produced by the action of light on silver nitrate, could not be made permanent, but the salt eventually used as a fixative, sodium thiosulphate, had already been discovered by Chaussier in 1799.

It was a long step from these first principles to the photography of focused images, and the problems in the way were eventually overcome, as often happens, by several experimenters working independently. A French physicist, Joseph Nicéphore Niepce (1765–1833), already interested in the use of polished metal plates for lithography, began to copy etchings by allowing light to pass through an original made transparent with wax or oil, on to a plate sensitized with a preparation of bitumen. The idea occurred to him between 1813 and 1816; it was only a step beyond Wedgwood's technique, but by the 1820s he had devised ways of enhancing and fixing his images. In June or July 1827 he recorded a view from a bedroom window of his house at Gras with the aid of his camera. The faint lines and hazy shadows of that scene of roof-tops, *basse-cour*, and trees are our first authentic impression of time past.[4] News of Niepce's work reached Louis Jacques Mandé Daguerre (1789–1851), a scene-painter and the inventor of the diorama, whose own photographic experiments had followed the same course, though with less success. Daguerre proposed that they should work together, and after some hesitation Niepce agreed, with the result that the photographic image fixed upon a metal plate, which was substantially his work, is known to the world as the Daguerreotype.[5] We shall never know what neologism would have given Niepce his due.

Niepce and Daguerre had originally each worked with a solution of asphalt in oil, on both metal and glass plates. Their joint and perfected process used a polished silver plate made light-sensitive by iodine, and the image chemically developed and fixed, all by processes originally suggested by Niepce. Daguerre's principal contribution seems to have been the development of the latent image. In the meantime William Henry Fox Talbot (1800–77), Twelfth Wrangler in 1821, and a learned amateur of optical physics, had also begun to experiment with the images of the *camera obscura*, having pondered the idea of fixing its images by chemical means during an Italian tour in 1833. He also made contact prints of leaves and small objects. His work began with Wedgwood's process, but he worked out an elaborate technique for sensitizing a self-developing paper negative, on which the images were fixed with sodium hyposulphite. Fox Talbot eventually published his work in January 1839, a few weeks after Daguerre and Isidore, Niepce's son, had announced their own, but before they published details, and he patented his Calotype, or Talbotype, as a rival to the older process.[6] His method was overshadowed by the Daguerreotype, and his photographic printing-works at Reading were closed in 1847 as a result of the losses he suffered after agreeing to supply photographs to illustrate the magazine *Art Union*, but the calotype pointed to the future development of photography. Whereas Niepce and Daguerre envisaged and perfected a method that fixed a single image, Fox Talbot invented the photographic negative, from which any number of positive prints could be made. From its earliest days, therefore, the new technique held out the promise of unlimited exact reproductions of an exemplar: the very foundation of mechanical industry and the exacting society that it has to serve.

Although its inventors were all aware of its larger implications, especially for the graphic arts, the earliest achievements of photography were necessarily domestic in scale and relatively trifling in character. Their interest lay in their existence: cumbrously secured and limited in scope as these first reflections of life's patterns are, their appearance is a prodigy. If observation can change the character of the thing observed, then photography changed the world that it recorded. We scan early photographs for the authentic detail of the past, yet their archaic features—costume, settings, ambience—are themselves a challenge to the imagination: a paradox. Though we prize those things because they are captured for us from the past, they partake of our own world, of modernity and even of contemporaneity, because they have been photographed. It was entirely proper that our earliest photographic views should include a French and an English country house. Niepce, the retired lieutenant of the engineers, *savant* and *fonctionnaire*, Fox Talbot at Lacock Abbey, the country gentleman with an aptitude for higher mathematics and a taste for physics and drawing, each photographed what was at hand and familiar, and in doing so presented what was traditionally characteristic of the societies in which they lived.[7] Yet at the same time they transformed what was traditional, and unified what was diverse: the camera, like the steam locomotive, makes all places one.

Of all the ingredients of the new world that mechanical industry was creating, and that now came under the camera's eye, none was more important than the city. Man could reach out further than ever before, but he was enabled to do so only by concentrating his resources and efforts in an unparalleled fashion. The forces so gathered bore equally hard upon the conventions of the pre-industrial town and the elegancies of the country house, and new patterns were inexorably imposed upon both. Everything had now to relate or respond to the needs and demands of industrial society: the capacity to do so was a condition of survival. The rule applied to the newcomer, photography, as it did to farming.

In the first decades of its existence, however, photography's debt to graphic art was more evident, and probably more anxiously discussed, than its debt to science and industry. The camera and its plate were still mere contrivances, and, though they called for ingredients unknown to Roger Bacon's pharmacopoeia, the ensuing recipes would not have been beyond his ingenuity. The product was another matter: here was the tantalizing promise of the *camera obscura* fulfilled, and Nature held captive. An early photograph could be compared only to a painting, and their familiarity with the principles of the camera gave artists the less reason to feel apprehensive about a new application of them. Hippolyte Delaroche (1797–1856), a robust executant of historical set-pieces, saw in Daguerre's pictures 'an easy means of collecting studies' which might otherwise have to be gathered 'over a long period of time, laboriously, and in a much less perfect way'.[8] It was an informed assessment, but the full effects of a world-wide repertory of styles upon artistic practice were not registered until the present century. In the meantime there were other interactions. Artists discovered that they had more than the techniques of other artists to learn from photography, but were driven ever further from representational work by photography's success. They also discovered that photography could satisfy demands which the graphic arts could never have met.

Photographers for their part were at first content with a role that implied subordination. Fox Talbot's first published collections of pictures were entitled *The Pencil of Nature* (1844–6)[9] and *Sun Pictures in Scotland* (1845), and he expounded his discoveries in a work called *The Process of Calotype. Photogenic Drawing*. It was not until the second half of the century that battle was joined between photographers themselves over the artistic and documentary attributes of their work, or that the public began seriously—and even then not always accurately—to distinguish the work and aims of the photographer and the graphic artist.

One reason for that slow differentiation lay in the practical difficulties of taking satisfactory photographs, and the consequent limitations of their scope. Early studies are either landscapes or portraits, and their majestically slow techniques greatly favoured portraiture. It took hours to expose a plate out of doors, and the finished scene was devoid of life. The intrinsic interest of the operation encouraged some photographers to persist with landscapes, and there are a few early pictures of towns, including a celebrated view of Trafalgar Square in *The Pencil of Nature*, but for the moment the results had only the appeal of novelty. The prospects for portrait photography were rather different. The daguerreotype technique was literally a painful one, and entailed the subject's sitting with unnatural stillness, restrained by iron clamps, but the result was a portrait-miniature in the most respectable tradition: personal, unique, and, by the standards of competent artistry, cheap. The number of miniatures exhibited at the Royal Academy fell dramatically, from some three hundred to sixty-four, between 1830 and 1860, and commentators took a despairing view of the future of the art, although established painters and their clients were glad to use photography to reduce the tedium of long sittings. One hard-wearing convention of portraiture, in which the sitter's chin rests on his hand, or one finger extends up the cheek while the others are doubled under the chin, seems to have spread from daguerreotype to painted portraits, and to have originated as an ingenious device to keep the face immobile before the camera.

Photography could afford to pay some dividends to painting, for the daguerreotype multiplied, not only amongst those who could have afforded to have their likenesses painted, but also, and more significantly, amongst those who could not. In the welter of domestic objects that distinguished the nineteenth-century house from its predecessors, the daguerreotype and its protective leather case or metal binding took a proud place. It was a marvel, and one of a solidly reasuring kind. In early years there were some disappointments. In 1847 *Punch* had a sketch called 'Photographic failures', with verses to point the moral that portraits 'taken by the cheap process' faded and vanished, like the affections that they were meant to commemorate.[10] Those properly prepared and developed, especially those intensified by a coating of gold, survived to grace the chimney-shelf or the occasional table. Soon daguerreotypes could be coloured or retouched to repair little blemishes in Nature: their advance was triumphant, and it prepared that mass-market for photographs of all kinds which grew up in the second half of the century.

One striking use of portrait photography in the 1840s was in a record of the 474 ministers who attended the signing of the Act of Separation by which the Free Church of Scotland seceded in 1843. The photographs were calotypes prepared by Robert Adamson, an Edinburgh chemist with a studio on Calton Hill, and David

The Victorian City: Shapes on the Ground

Octavius Hill, who had been commissioned to paint a commemorative picture of the proceedings. The finished painting proved Hill to have been overwhelmed by his material, but the constituent photographs, both of individuals and of small groups, were of the highest order.[11] The series included some views of Edinburgh and St Andrews which are among the earliest photographs of Scottish towns.

Although each process was used for either purpose, the daguerreotype excelled in portraiture and in depicting the details of urban scenes—the shining surface of wet paving-stones was singled out as an impressive feature—whilst the calotype's softer tones lent themselves best to a classic portrayal of landscapes. Both could be finely handled by experts, but both processes were very slow. The deliberate ritual strengthened the analogy with painting, while its restrictions emphasized the photographer's subordination to the artist. Though photographs might be made to look like paintings, it was still possible to believe that their chief use was to collect material for the painter. The photographer's liberation depended upon improvements in his equipment.

The next major advance in technique came in 1851, displacing the calotype immediately, and the daguerreotype after a short rivalry. This was the wet-plate process, in which a glass negative coated with collodion, a solution of gun-cotton in ether, was charged with a solution of bromide or other photo-sensitive salts.[12] The method was not a simple one, but it cut the time of exposure from minutes to a matter of ten or fifteen seconds. The plate was not only more sensitive than anything used before, but also allowed substantial numbers of good positive prints to be taken. Its success established the photograph, as opposed to the photographic portrait, as a marketable commodity. Its disadvantages were a cumbrous apparatus, and a developing technique of which C. L. Dodgson made restrained fun in *Hiawatha's Photographing:*

> Secondly my Hiawatha
> Made with cunning hand a mixture
> Of the acid pyro-gallic
> And of glacial acetic
> And of alcohol and water:
> This developed all the picture.

There was no exaggeration in the satire: the plate had to be developed before the coating dried, and a rapid sequence of photographs demanded the presence of an assistant to deal with the exposed plates. The entire apparatus filled a trek-cart, and on occasion, a caravan: the camera, tripod, plate-holder, dark-tent, baths, and chemicals weighed eighty pounds or more. That was the paraphernalia that Roger Fenton took to the Crimea for his brilliant photographs of the camps and the detritus of the battles, and that Mathew Brady used even more adventurously in the American Civil War.[13]

The wet-plate process greatly broadened the scope of photography. It was still not possible to capture movement, but a great many kinds of activity could be temporarily stilled to allow the photographer to work with a full plate, and he could shorten the time of exposure even further by exposing only a part. The earliest photographs show outdoor scenes unpeopled, except when engravers were retained to copy and enliven daguerreotypes with a figure or two. Now the bystanders appear even on full plates, stiffly at first, not quite at ease for several decades: men in stovepipe hats, ghostly figures of the impatient or the ill-at-ease, philosophical dogs. The camera could record formal occasions, like the opening of the Crystal Palace at Sydenham in 1854, when the event itself imposed constraints.[14] It could also depict ordinary scenes with a modicum of life. The wet plate did not make an immediate difference to portrait photography, although it removed much of the tedium in the sitting. Its most important applications were to the stereoscopic view-card and the *carte de visite*.

The principles of stereoscopic imagery were already well-understood; Sir David Brewster (1781–1868) exhibited a stereoscope at the Great Exhibition in 1851, and Negretti and Zambra produced for the occasion a fine series of stereoscopic daguerreotypes which were particularly admired by the Queen and the Prince Consort. Popular taste endorsed the commendation: the hand-stereoscope, invented by Oliver Wendell Holmes, was sold in large numbers in the 1850s and 1860s, and after a brief intermission enjoyed even greater esteem in the last two decades of the century. It was a device well suited to domestic use. The pleasure it afforded was private, but not exclusively so, as the viewer could be easily passed from hand to hand. The fascination of the illusion, so simply created and destroyed by moving the picture, was not even limited by the number of cards to hand.

Stereoscopic cards were published in large sets, many of which, like Francis Bedford's series on English counties and regions or G. Washington Wilson's *Scottish Scenery*, include striking views of town life.[15] Wilson's collection of negatives, a most valuable storehouse of topographical material drawn from all over Britain in the middle decades of the century, is now in Aberdeen University Library. Topography, however, though a usefully broad subject, was by no means the only theme for the stereoscope. The Manchester Art Treasures Exhibition of 1857 and the International Exhibition of 1862 were the subjects of elaborate sets. The former had a particular interest as the first public exhibition of works of art gathered nationally from private collections; as an art exhibition that included photographs; and as a token of Manchester's civic vitality and cultural eminence in the period. Other and less exalted matter can be found among the many series of comic cards, which range from the mildly lubricious—lovers surprised, girls washing clothes with their arms or legs bare—to the enigmatically dull. Even those pictures, however, have their own historical interest, not only as exemplars of taste, but also for the studio-contrived interior scenes that many of them portray. At a time when ordinary indoor photography was impracticable, a stage-set that was meant to carry some degree of conviction with contemporaries can also tell posterity something. The same is true of the photographic slides that multiplied from this time onward, and of much other superficially unpromising material.

The *carte de visite* was a trifling device, patented by André Disdéri in 1854. As its name implies, it was a small paste-down print, some 2¼ by 3½ inches, that was originally designed as a cheap substitute for the conventional engraved card. Distributed as a keepsake with or without more elaborate mountings, it rapidly extended its range from personal portraits to views, and then to celebrities, to curios, and to any other theme acceptable to popular taste.[16] Like the stereoscopic card, the small print implied a small negative and therefore a short time of exposure. Photographs of that kind could capture some details of commonplace scenes more satisfactorily than full-plate studies, and it was from stereoscopic studies that painters now discovered that the human mode of walking, especially in the swing of the leg and foot, was markedly different from the established artistic convention. The motion of a galloping horse could not be analysed in the same way until Eadweard Muybridge began his ingenious experiments with a battery of cameras in 1872,[17] but in the meantime the cumulative experience of the camera's eye told upon the artist's practice and assumptions. The photographer's reactions were different, but the historical importance of these small glazed prints overrides their artistic and technical significance, even though many of them were of the highest quality. To the historian they speak of the mass-market that they commanded; their consequence lies in their numbers and diffusion. They are most literally the city-dweller's view of the world. They familiarized an urban public, avid for instruction, amusement, and discreet reassurance, with the wonders of the great exhibitions, with exotic scenery, with the triumphs of civil engineering, with the architecture and the living patterns of British and foreign towns. They showed a world complex and diverse, now subdued and even imprisoned, captured on pasteboard by the powerful devices of man. The photograph was a means of education, not just by virtue of what it depicted, but also in itself: it was part of the process by which men could persuade themselves of their mastery of material things.

One powerful extension of that process was waiting upon a further technical change: photographs could still not be reproduced with letterpress. The potential of photographic illustrations in books, periodicals, and newspapers was readily perceived, but only slowly realized. The magazine *Art Union* which ruined Fox Talbot's photographic firm was illustrated with calotypes; these were also used in the presentation copies of the catalogue of the Great Exhibition in 1851; the pasting-in was slow and expensive, but it was continued in works of limited editions to the end of the century. Some early daguerreotypes were copied by engraving, a process which, like colouring, allowed modifications to please the purchaser. As copying techniques improved, photographs were increasingly used as the basis of engraved illustrations. One notable example is Henry Mayhew's *London Life and the London Poor* (1851–62), a vivid and influential evocation of the city's street life, and an appropriate register of the camera's observations.

Work of that kind promoted in turn a general improvement in the accuracy of popular illustrations. The *Illustrated London News*, founded in 1842, had set new patterns in journalism and enjoyed a revolutionary success before any question of photogravure arose. It was itself an education in the uses of pictorial material, but its early methods were impressionistic and sometimes perfunctory. When Krook's fiery death in *Bleak House* brought men of science and philosophy to Cook's Court, Cursitor Street, it also brought, besides the ordinary journalist, 'the artist of a picture newspaper, with a foreground and figures ready drawn for anything, from a wreck on the Cornish coast to a review in Hyde Park, or a meeting in Manchester'. Afforded a seat in a neighbour's room, he proceeded to fill in his block with 'Mr Krook's house, as large as life; in fact, considerably larger', and for good measure he added the fatal chamber itself, showing it 'as three-quarters of a mile long, by fifty yards high', to the powerful gratification of his audience from the court. Historians value such evidence as they have, from any period, and they can only be grateful for early picture-newspapers and their versatile artists, even when they have to review them critically. What is remarkable is not so much that such works occasionally betrayed imaginative aberrations, as that the principles of photographic accuracy were established as early and as thoroughly as they were. Easy reference to an abundant pictorial journalism informed and improved public taste and its expectations. The fact is that artistic licence is something that can only be measured when there is a photographic standard to set against it; the historian of the Victorian city has to take a good deal less on trust than have some of his fellows.

The inevitably uneasy relationship of photography and graphic art had important effects upon both as the century advanced: an obvious one was to sharpen the professional pretensions of photographers. The notion that the photograph was a work of art as much as, and even more than, a mere record persisted, and caused much anguish both to those who entertained it and those who rejected it. The issue was complicated by the presence of painters who took up photography for its own sake, like the Swede, Oscar Rejlander (1813–75). Rejlander produced the first composite photographs, arranging and rephotographing more than thirty pictures for his moralistic study 'The Two Ways of Life' in 1857. Although Queen Victoria bought a copy of that photograph, it was subjected to some criticism because it depicted dissipation, and did so with painful realism. There was a similar reaction to a composite study of a deathbed scene by Henry Peach Robinson (1830–1901)— 'Fading Away'. It would have made an acceptable painting, but photography made verisimilitude intolerable.[18] The implication was that the public could not accept photography as an art; no matter what liberties the photographer might take with his material, its factual parts put an uncomfortable gloss of truth upon even the most imaginative whole. Some photographers were unconvinced, and the battle between the pictorialists—those who wished to create works of art with the camera— and the naturalists—who wished to record and define, though not necessarily without forethought and art—was joined noisily and repeatedly, dividing camera clubs and former friends all over the country.

Two results of the debate—at times a dispute—are particularly important to the historian. One is the emergence of a small but significant number of skilled photographic reporters: men such as Frank Sutcliffe,[19] whose studies of Whitby are among the finest we have of a nineteenth-century town, or the enthusiasts who first directed the Manchester Photographic Survey, a project of the 1880s still incomplete, but remarkable for the enlightenment and cogency of its original planning.[20] The other is more general, but not less considerable. Whichever side a dedicated photographer might choose to support, his professionalism was sharpened by the acrimony of the debate, as well as challenged by the difficulties of the medium. The wet-plate process was the salvation of photography: its elaboration helped the professional to establish himself, and paid a splendid dividend in the quality of his photographs. From the 1850s until the late 1870s amateur photographers were either deterred, or forced to take professional pains, by the patient skill and expensive care that a successful picture demanded. In the meantime the professionals were encouraged by the growing market for pictures of all kinds. The principles of dry-plate photography were discussed from the 1850s, but the sensitive emulsions and strong developers required took some years to perfect, and ready-coated plates were not generally available until the late 1870s. The dry plate needed only one-fifteenth of a second's exposure, and immediately made possible smaller cameras and rapid, informal photography. A new and splendid prospect opened before the amateur, but in the meantime a powerful tradition of deliberated and carefully executed pictures had been established.

The best town- and country-landscapes of the wet-plate camera are in many ways the most interesting of our graphic records. They did for the daily scene, or for some chosen aspect of it, what the portrait daguerreotype had done for the individual sitter, depicting its image as faithfully as the preceding formalities allowed. Their considered composition and subtle tones look to the long-established techniques of landscape and topographical painting: the calotype in its day had been esteemed in England because its soft outlines were reminiscent of brushwork. Their precision and searching detail, on the other hand, were rather material for the scientific observation in which the nineteenth century excelled. Except for the long-ranging shots exposing only a fraction of the plate, which were used for some stereoscopic pictures, they are all necessarily formal, but their formality comes not so much from the time that they took to expose as from the fact that the photographer was a conspicuous figure. His apparatus and ceremonial inspired respect, if not awe. With the notice that his preparations gave, the apprehensive and the naturally retiring could retire. Those who remained chose to be there, and their watchful gaze is not entirely a matter of muscular tension.

Professionalism assures its own satisfactions, but it also implies a sale for the skills entailed. The best prospects for the photographer of landscapes, in the widest sense, were the stereoscopic card, the lantern slide, and views of the picturesque—studies like the handsome sepia prints that the Abrahams of Keswick produced in the 1880s.[21] Although hand-drawn pictorial postcards appeared in the late 1860s, photographic printing on a large scale had to wait upon the half-tone process, which did not appear for another twenty years. Those considerations restricted commissioned and saleable work—even today photographs other than portraits are rarely bought for their own sake—and put a premium upon the picturesque, the celebrated, and the readily-assimilable among the subjects with which the camera could deal. Although the photographs of the third quarter of the century are an archive of high quality and interest, therefore, they still offer us only a selective view of the contemporary world. Rapid movement, indeed any movement, defeated the full-plate camera, and the small photographs from which stereographic cards were made, to which we owe the liveliest street scenes of that period, could not be effectively enlarged. Indoor photography was barely possible, and naturalistic work indoors out of the question. A crowd gathered and waiting for an event could be fairly taken, but anyone with business of his own, except of the most contemplative kind, was commonly beyond reach. The world of trade and manufacture, mines, factories, and moving trains, the whole strength of industrial society, were impracticable subjects in most ordinary conditions. The city itself could be captured only in moments of repose or formal excitement: in the camera's eye, the patterns of life were still or slow-moving.

Those conditions changed from the late 1870s onward, with the introduction of the dry plate and the miniature camera. The Kodak roll-film camera then appeared, in 1888, and was quickly refined, by the use first of celluloid film, and then of light-proof spools that allowed simple and rapid loading in daylight. With cameras that could lie in, and even be operated from, a coat pocket, and that could take scores of pictures in the time needed to expose and process a single wet plate, the whole concept of photography changed. The photograph as an instantaneous record came into its own.[22] Bustling streets, moving vehicles, men at work could all be taken at close range, unobtrusively and at the photographer's whim. The amateur and the casual photograph flourished, and popular interest and demand created a large photographic industry. Instantaneous photography and the flexible film made possible the invention of the cinematograph. Professional interest in the photographic analysis of rapid motion went back to the stereoscopic pictures of the mid-century, and beyond, to the optical toys of earlier decades. Muybridge's ingenious silhouettes of a galloping horse showed the most careful painters that they had entirely mistaken the sequence of movement entailed, and that in particular the rocking-horse posture, front and back legs simultaneously extended, in the best tradition of racing prints, was a crude illusion. Dry-plate photography allowed the physiologist Jules Étienne Marey to improve Muybridge's technique, and apply it to the flight of birds.[23] In 1896 the first paying audiences in London were regaled with pictures of people just like themselves, walking in the street, and other wonders of animation. The same year saw the *Illustrated London News* adopt half-tone blocks for its illustrations. The camera and its ancillaries were now the masters of all things.

Half-tone blocks had been used in the United States since 1880, but there were difficulties in adapting them to fast-running newspaper presses.[24] By the late 1890s, however, news-photographs were extensively used by the press, at least by popular and experimental papers, and they have survived in great numbers. Early cinematographic films are more rare; the nitrate stock of the oldest prints was extremely

The Victorian City: Shapes on the Ground

unstable, and little thought was given to their historical interest until 1935, by which time much valuable material had been lost. Commercial films taken before the First World War are more common than those taken later in the century, because in the early days the copies were sold outright. After hiring was introduced, the distributors destroyed all copies when the film's commercial life was over. The National Film Archive's catalogue of news films from 1895 to 1933 naturally contains many records of formal events, and particularly of royal functions, but there are also a number of more casual and, therefore, more rewarding occasions.[25] Examples from the period 1896–9 range from pictures of rush-hour traffic on London Bridge or at the Mansion House—such popular subjects with photographers—to the turn-out of a fire-brigade, a bicycling rally, and the arrival at Henley railway station of visitors to the Regatta. The life that the illusion of motion lends to these pictures enhances their chief interest, which lies in the mere extension of photography to a new range of commonplace subjects. The camera had not long been able to freeze motion; now it was able to record it. The future of the cinematographic film as an art lay in the skilful use of the illusion: its historical interest lies largely, though not exclusively, in its individual frames.

Just as the cinematographic film as a work of art was long bound by the conventions of the theatre and still-photography, and television in its turn by cinematography, so the old uses of the camera bore upon those who experimented with its new devices at the end of the century. That was not a disadvantage. The snapshot—an expression borrowed from sporting slang and transformed by its association with the hand-camera—and the photographic analysis of rapid or confused motion could not entirely lose, in the nineteenth or even the early twentieth century, the formal touch that belonged to more deliberate studies. Amateur work, both good and bad, followed professional examples. The family photograph album, though it came to contain more photographs and less matrix than its predecessors—a change that has gone on more sluggishly in records of weddings—was informed by a continuing sense of ceremony. The restrictions upon indoor photography until the very end of the century diminish the social as opposed to the personal interest of such portraits, and in any case the accidents of time have reduced the number of albums available to us, but the rapid growth of amateur photography after the introduction of the flexible film greatly enlarged the repertory. It did so, however, without much immediate loss of formality. Even today the number of people who face a camera without some perfunctory gesture to decorum is very small. An age which hardly knew how to question the psychic value of ceremony (and which would certainly have accomplished much less if it had) was not likely to underplay formality in its most solemn, because most exact, records. The consequence was that the technical standards of most late-Victorian and Edwardian photography remained high; that the camera became ubiquitous without becoming entirely commonplace; and that those uses of photography which were aimed deliberately at enlarging its scope, by photographers who revelled in the techniques denied to earlier generations, are often most enviably good. The face of the Victorian city in its latest years is presented to us in admirable detail.

That face is there for us to study, making what allowance we can for the limitations imposed by occasion, circumstances, and above all, technique. The photographer's mind, its intentions and preoccupations, lie behind the photograph, removed from us, but not beyond our reach. The applications of photography are often more valuable objects of study, in that sense, than the prints themselves. The stereoscope provides one instructive example, and the lantern slide another. The stereoscope was a private amusement, though not necessarily a solitary one. The magic-lantern was widely used as a domestic toy, but was also an engine of public instruction. Before 1870 the ordinary choice was between small single-wick lanterns of low power, and oxy-hydrogen burners which demanded skilled handling. Then Walter Woodbury introduced the Sciopticon, an American-invented lantern with a brilliant oil-light, and with it a wide range of slides made from negatives of his own, exposed in the 1850s and 1860s. Woodburytypes, which were printed in sepia, are still recognized as outstandingly good slides, and they helped to promote a large domestic market for slides and magic-lanterns, besides encouraging amateurs to make slides, and showmen of all kinds to use the lantern as a popular attraction.[26]

The Victorians esteemed rational amusement highly, and were right to do so, because the elaborate structure of their society was threatened by what passed as the alternative. Public lectures combined instruction and diversion in an unexceptionable way, and lantern slides enlivened them, giving an attractive gloss to even the most unpromising subjects. Lantern shows were older than photography, and the technical improvements made in the last quarter of the century carried them through to the age of the cinematograph. Evangelists and temperance lecturers used lantern slides as intensively as anyone, although the catalogues of commercial distributors like Newton & Co. of Covent Garden reveal a public hungry for straightforward information of any kind, on subjects ranging from the evolution of the ironclad to the manufacture of looking-glasses or of chocolate. One of the most interesting sets of privately-commissioned slides was made for, and on occasions with the assistance of, Charles Spurgeon, the son of the celebrated evangelist, Charles Haddon Spurgeon, and himself the minister of a Baptist chapel in South Street, Greenwich. Between 1884 and 1887 Spurgeon secured a striking series of pictures of street life in the borough, to illustrate his evangelical lectures.[27] The collection includes some excellent individual studies of street-traders and bystanders, and offers us a valuable conspectus of outdoor life in the borough in the mid-1880s, but Spurgeon evidently directed the photographer carefully, and on occasion he took part in front of the camera himself. He was therefore dealing in symbols rather than exclusively in facts, and what we know of his pictures we may suspect of others.

Commercial undertakings like the Bamforth Company at Holmfirth, near Leeds, produced homiletic series of slides for the general market. The firm was founded in 1870 by James Bamforth, a local man, and specialized from its beginning in slides, although it turned to picture postcards, and especially comic cards, as soon as British postal regulations allowed the blank side of a card to be divided between address and correspondence.[28] The comic cards later bred their own ribald conventions, but in its early years the firm offered many temperance lessons, and its products include various photographic scenes of low life. Slides presented as 'Life Models', or 'Taken from Life' were universally popular, and a successful series called 'slum life in our great cities', though not exactly that, did include, like Spurgeon's pictures, some striking vignettes of life in the streets. Bamforth's slides, however, portrayed interior scenes photographed in studio sets. There was presumably some measure of verisimilitude in those productions, but for the most part it is not provided for a keenly critical audience. The unheated and ill-furnished garret in which the story called 'Dot and Her Treasures' is set will serve as an example of an unheated and ill-furnished garret, but even in the absence of authentic indoor scenes, it does not advance our knowledge of the poor and their houses to any degree. There is probably more to be gained from the representation of more prosperous households in other series, for there we have furnishings and bric-à-brac surviving against which the reconstructions can be checked. On the other hand, pictures of poverty, real and simulated, can tell us something about the interests, the experience, and the acumen of those classes of society that were instructed and admonished, perhaps, on occasion gratified or reassured, by such lantern slides in private shows and public lectures. There were, after all, physical dangers as well as technical difficulties in photographing low life, and the expedients that served instead are as informative on their own terms as the truest photographic reportage could be. So is the phenomenon of a respectable audience in a gas-lit Temperance Hall, or Railway Mission, moved to compassion, unease, or censure by the plight of the heathen abroad and the unfortunate or improvident at home.

There was even more to the lantern slides, however, than foreign travel, evangelism, and social reform. The ubiquity of the entertainment, and on particular occasions its imposing scale, tell their own story. Like Daguerre's dioramas in their day, it drew upon technical ingenuities to cater for a large audience, collectively a much larger audience than Daguerre was able to reach, and one which made very uneven demands upon the material that was presented. The captive audience is not a product of the twentieth century alone, and the lantern slide entrapped its admirers long before all images were expected to move and speak. Not that all showmen were content to accept uncritical tribute: D. W. Noakes of Greenwich, who was official lanternist at the Albert Hall, took his own slides for 'dioramic' projection, and built his own steam-launch for a photographic voyage through the Midlands from London. His illustrated lecture, 'England Bisected by a Steam Launch', was delivered to an audience of 1,500 at the Crystal Palace in 1891, and his set of 300 slides under that title constitutes one of the finest topographical records of the time. H. W. Taunt of Oxford, who also produced many slides of high quality, took an equally interesting series of pictures to illustrate a guide to the Thames Valley, preparing his prints in a floating photographic laboratory. The guide was illustrated with pasted-in photographs, an effective but expensive technique used for works in limited editions until the triumph of half-tone blocks and a commercially viable collotype.[29] Taunt's photographic collections include some excellent studies of the Thames and of shipping in its lower reaches, and a most valuable record of Oxford in the last decades of the century.

Photographers like Noakes and Taunt were able to experiment and perfect their work because the market for photographs of all kinds was expanding, and the growing amateur enthusiasm encouraged rather than hindered the professionals. The census returns show an increase in the number of professional photographers in Britain from a mere 51 in 1851 to 17,268 by 1901. By 1905 it was estimated that 4,000,000 amateurs, one-tenth of the whole population, were taking photographs, although the number of those willing and able to experiment effectively and to develop and print their own pictures, was comparatively small. In 1880 there were only fourteen photographic clubs in Britain, but their numbers increased sharply from that time, and by 1900 there were 256. Their membership was often riven by the debate between the champions of photography as a pictorial art and the photographic reporter, a debate re-opened and intensified by Peter Henry Emerson in 1886,[30] but the best work on either side showed the hollowness of the dispute. What matters to the student of the age is that technical improvement and high professional standards were maintained together. The great extension of photography as a popular pastime was bound in the long run to debase the quality of much that was done, but for some time the material available to us increases without any serious loss of interest. With the improvement of electric lighting and of flash photography, besides the constant refinement of lenses, by the beginning of the twentieth century the scope of photographic reportage had become almost as wide as it is today. The whole pattern of life, and especially of life in the city, which absorbed and displayed more and more of the sum of human energy, was depicted for the wonder of posterity.

The camera's eye, therefore, enables us to observe the Victorian city from almost its earliest years, and to catch all its nuances in its prime. What is more, what we can learn of it from its photographs goes back far beyond the mere details of the scenes that they display. The development of the camera itself is a comment upon the process by which the nineteenth century, like our own, exploited and refined ideas by popularizing devices. Photography both reflected and educated taste; its cumulative effect upon the popular mind was not less than its effect upon the graphic arts, or the part that it played in the elucidation of the natural sciences. The city was the natural setting for photography, and the photographer in quest of rustic idylls took the city with him in his uniform camera-box, machined-glass plates, and nicely-calculated solutions of salts. The progress of the city, and the formidably elaborate society that it housed and sustained, informs our assessment of every photograph, whatever appears in the print.

Some distinguished individual portraiture aside, our best photographic records are architectural; the camera has played on nothing with better effect or more consistently than on the forms, the details, and the setting of buildings. The value of such records is obviously great when rapid change is effacing traces of both the recent and the remote past in our towns, but even where buildings survive it is only the searching detail of photography that enables us to take a proper measure of change and continuity in the scene. What is less well represented in the towns is, in the first period, people, and for most of the time their work. We can deduce the processes by which wealth was created and applied, but for as long as photography was confined to the out-of-doors and the static scene it could take only a superficial account of human industry. Even when rapid and indoor photography became a commonplace, the complexity of industrial processes imposed their own barriers to analytical reporting. It has taken almost another century to establish an effective symbolism in cinematography, and to unlock some of the mysteries of social organization: the process is still far from complete.

A human figure gives scale and a sympathetic interest to a picture, and photographers have generally welcomed a bystander or two. Even the early daguerreotypes had figures added to them when they were copied by engravers. In the first decades of photography such human interest had to be contrived by one means or another; it is not until the small stereoscopic plates captured distant movement that the scene could be naturally peopled; the full tide of human existence could not be recorded until the coming of instantaneous photography. Individual figures stand out at all periods, many of them with their single claim to immortality in their presence, accidental or solicited, before the camera. For the rest, clothes and the manner of wearing them, stance, individual and collective style, all have their own story to tell. Only the record of activity is defective: what is represented is apt to be either withdrawn contemplation, or motion to some undisclosed end. Except when the celebrated are frozen in formality, those who emerge are usually, and not inappropriately, the middling sort. If they have no declared purpose, then their business is to people the scene. They do it well enough, and yield us an occasional bonus as well.

The scene itself has always two striking characteristics: it is by turns familiar and unfamiliar to us, and the familiarity is in some part the work of the camera itself. The photograph is part of our own world, and though we have apparently shed the primitive fear that what is captured in that way is lost to us, we have a lingering sense of community with what the camera's eye has seen. It is not so easy as it might seem to be, therefore, to take stock of the nineteenth century on its own terms; we are likely to be distracted by the emergence of our own context, and to delineate the wrong age. The temptation is the greater because one feature that the nineteenth century had in common with the twentieth was precisely the imposition of the town and its values upon the countryside: the substitution of road metal for mud and gravel, paint for vegetation, sewers for ditches, houses and streets for fields and farmyards. It is apparently a condition of man's social organization that he should replace what grows with what he has made; in the city the process is carried to its logical extreme. The change in the physical world that the camera observes is like the technical progress of the photographic print itself, from its shadowy beginnings to the instantaneous film; it is a progression from natural uncertainty to an unsettling precision, a progression from what is instinctive and self-renewing to what is contrived and temporary. Life is a product of change, but the pace and nature of social and physical change was radically altered in the nineteenth century. It was something more than the multiplication of people that we can observe coincidentally but significantly upon the photographic plates, and more, even, than the remorseless spread of the city itself. It affected every manifestation of organized life, including our own.

The Victorians were themselves aware of the peculiar qualities of the photograph. They grasped the significance of it as an impartially searching record and they appreciated its peculiar aesthetic very keenly. They saw it, quite properly, as one of their most remarkable achievements, distinctive and powerful, and they enriched their language with photographic metaphors in tribute to it. The connection between photography and other recording arts was obvious enough, and both authors and their critics invoked the comparison. Dickens, whose descriptive writing has the sharpness of personal experience, compared himself to a camera, and commentators upon his work both approved and reprobated what they took to be its photographic techniques.[31] For many purposes, and especially for works of social comment and criticism, the analogy had a useful uncertainty about it; to liken oneself to a camera might imply either an innocent impartiality or a superhuman power of comprehension. E. C. Grenville Murray's *Spendthrifts*, published in 1887, is subtitled 'and other social photographs'. As in other matters, popular allusion kept pace with technical change. W. D. Howells published a volume of essays in 1905 as an American observer of English life; he called it *London Films*, and began the first with the words, 'Whoever carries a mental Kodak with him (as I suspect I was in the habit of doing long before I knew it) must be aware of the uncertain value of different exposures.' He might reasonably count upon a sympathetic response from four million readers.

One passing but significant reference occurs in a work by Frederic William Maitland, the most felicitously gifted of English historians. Maitland's chief interest lay in the history of law, and especially the law of medieval England. In turning to those studies, he committed himself to one of the most rigorous techniques in historical science, and he showed himself always most perspicaciously aware of its strengths and limitations. In 1884, in the introduction to an edition of the plea roll for the county of Gloucester for 1221, he said of that document 'it is a picture, or rather, since little of imaginative art went into its making, a photograph of English life as it was in the early thirteenth century, and a photograph taken from a point of view at which chroniclers too seldom place themselves.' The reference to imaginative art cuts across the debate amongst photographers, but the metaphor was entirely a just one. Its appositeness lies not in the fact that judges and cameras might alike be supposed, or even hoped, to deal impartially with what is presented before them, but because written and photographic records alike are irrefutable upon their own terms. Both are documents, both are constructed with a purpose and an audience in mind. We use judicial and administrative records to explore the distant past because they are the most significant and informative evidence that we have from the society that made them. In its own age, and subject to like analyses, the photograph may have its rivals as a source of historical evidence, but it has no superiors. It offers us in its highlights and shadows the very pattern of the Victorian age.[32]

The Victorian City: Shapes on the Ground

Notes

1. Helmut and Alison Gernsheim, *History of Photography, from the Camera Obscura to the Beginning of the Modern Era* (hereafter referred to as Gernsheim) (2nd edn, 1969), p. 19.
2. Aaron Scharf, *Art and Photography* (1968), p. 1.
3. His experiments were reported in a paper by Sir Humphry Davy in the *Journal of the Royal Institution*: Beaumont Newhall, *History of Photography from 1839 to the the Present Day* (2nd edn, New York, 1964), p. 12.
4. Gernsheim, plate 21.
5. Gernsheim, p. 62; Scharf, op. cit., pp. 5–6; Newhall, op. cit., pp. 13–16.
6. Gernsheim, pp. 76–7.
7. For one of Fox Talbot's 'photogenic drawings' of Lacock, see Newhall, op. cit., p. 33.
8. Scharf, op. cit., p. 17.
9. Scharf, op. cit., p. 9.
10. *Punch*, xii (1847), 143.
11. Scharf, op. cit., pp. 29–30.
12. Gernsheim, pp. 197–200.
13. Helmut and Alison Gernsheim, *Roger Fenton, Photographer of the Crimean War* (1945); Ray Meredith, *Mr. Lincoln's Cameraman: Mathew Brady* (New York, 1946).
14. Gernsheim, p. 266.
15. Gernsheim, p. 257. For a valuable account of the whole subject, and for dating stereo-cards, see William Culp Darrah, *Stereo Views: A history of stereographs in America and their collection* (Gettysburg, 1964).
16. On the *carte de visite* and photographic taste in the mid-century, see Helmut Gernsheim, *Julia Margaret Cameron, Her Life and Photographic Work* (1938), p. 39.
17. Gernsheim, pp. 435–7.
18. Gernsheim, pp. 246–9.
19. F. C. Lambert, *The Pictorial Work of Frank M. Sutcliffe* (1904).
20. See Harry Milligan, 'The Manchester Photographic Survey Record', *Manchester Review*, viii (1958), 193–204.
21. A selection of Abraham's work appears in A. P. Abraham, *Beautiful Lakeland* (Keswick, 1912), and in G. D. Abraham's books on rock-climbing, such as *British Mountain Climbs* (1909), but the firm is probably best known for its photographic prints, which provided images of wild nature for drawing-room walls.
22. One of the most talented photographic reporters to use a miniature camera was Paul Martin, whose vivid pictures of street life in London and elsewhere in the 1890s were published in 1939 under the title *Victorian Snapshots*.
23. Gernsheim, p. 440.
24. Gernsheim, p. 452.
25. British Film Institute, *Catalogue of the National Film Archive*, 2 parts (2nd edn, 1965). See also note 32.
26. Josef Maria Eder, *Ausführliches Handbuch der Photographie* (4th edn, Halle-an-der-Saale, 1926), *sub* 'Woodburydrück'.
27. See O. J. Morris, *Grandfather's London* (1956).

28 For a general account of Bamforth's work see Frederick Alderson, *The Comic Postcard in English Life* (Newton Abbot, 1970).
29 See, for example, H. W. Taunt, *Goring, Streatley, and the Neighbourhood* (Oxford, 1894).
30 Gernsheim, p. 457.
31 See, for example, Philip Collins, ed., *Dickens: The Critical Heritage* (1970), pp. 6, 190, 255.
32 The best bibliography of photography is Albert Boni, ed., *Photographic Literature: An international bibliographical guide* (New York, 1962). Its coverage of periodicals is particularly good, but as it is arranged in a continuous alphabetic sequence of authors and subjects the student in search of illustrative material needs either to know for what or for whom he is looking, or to use the work in conjunction with a comprehensive history of photography. There is a number of such studies which are concerned with the social and historical consequence of the subject, besides the development of techniques, and therefore valuable preliminary guides to the rich but ill-ordered fund of evidence that the camera has provided. Among the best examples are Josef Maria Eder's *History of Photography* (New York, 1945), Beaumont Newhall's *History of Photography from 1839 to the Present Day* (2nd edn, New York, 1964), and Helmut and Alison Gernsheim's *History of Photography, from the Camera Obscura to the Beginning of the Modern Era* (2nd edn, 1969). Aaron Scharf's *Art and Photography* (1968) is a perceptive study of the interplay of photography and the graphic arts, and essential reading for the historian who wishes to understand and use nineteenth-century photographs. Techniques and their bearing upon historical interpretations are also effectively discussed in Erich Stenger's *History of Photography: Its relation to civilization and practice* (Boston, Mass., 1939), and William Culp Darrah's *Stereo Views: A history of stereographs in America and their collection* (Gettysburg, 1964).

There is now a substantial literature of historical photographs, seeking either to evoke a particular period, or to illustrate a theme. Helmut and Alison Gernsheim's *Historic Events, 1839–1939* (1960), and Agnes Rogers, *The American Procession: American life since 1860 in photographs* (New York, 1933), are both of general application. Recent works on Britain include O. J. Morris's *Grandfather's London* (1956), Sir John Betjeman's *Victorian and Edwardian London from Old Photographs* (1969), and C. S. Minto's *Victorian and Edwardian Scotland from Old Photographs* (1970). The work of two distinguished photographers of urban life is recorded in Eugène Atget's *Photographe de Paris* (Paris and New York, 1930), and *Victorian Snapshots* (1939) by Paul Martin. There is a danger, even with work like that of Martin and Atget, that old photographs will be admired for their quaintness, and studied only superficially for their content. The best correctives are probably the writings of photographers themselves, particularly in the nineteenth century during the debate upon the documentary and artistic qualities of photography, and further study of the very extensive photographic archives that we possess, and upon which we have as yet drawn only lightly. Two guides to national collections, the British Film Institute's *Catalogue of the National Film Archive* (2nd edn, 1965) and D. B. Thomas's *The Science Museum Photography Collection* (1969), can stand here as exemplars, but there are still some notable collections, like that of the National Monuments Record, or the Wilson negatives in Aberdeen University Library, and the rather later collections of Henry Minn, in the Bodleian Library, which have been only inadequately discussed, or have passed unnoticed, in print. Local enquiries could well start with *The Camera as Historian: A handbook to photographic record works* (1916) by H. D. Gower, L. S. Fast, and W. W. Topley, which contains a useful list of county and other record collections. Thereafter only local enquiries will serve, until an effective national census is undertaken. There is no study of a British town to set against H. M. Mayer and R. C. Wade, eds, *Chicago: Study of a metropolis* (Chicago, 1969).

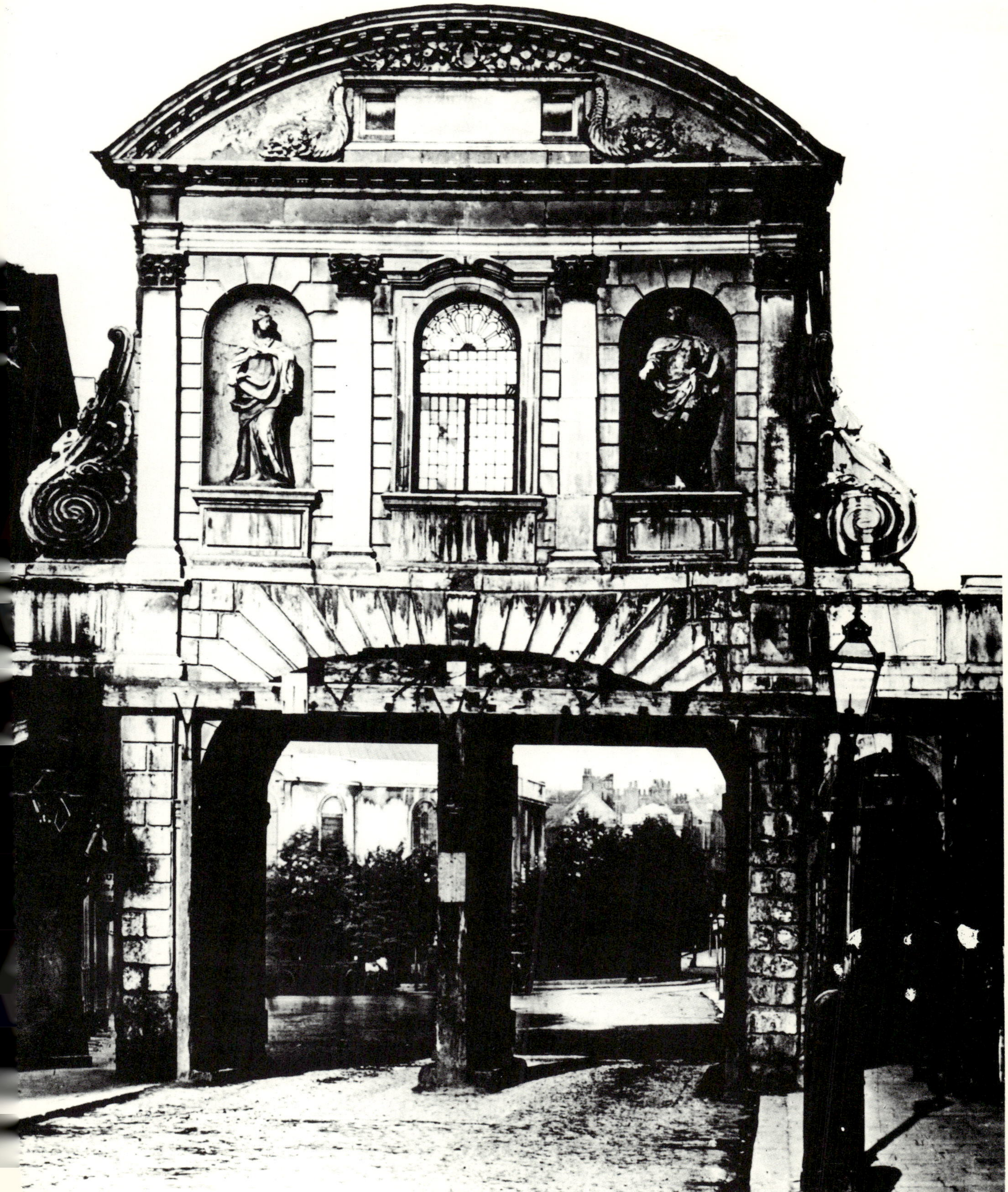

66 *preceding page* Temple Bar, a sagging relic of the old city, rescued and sentimentally mounted as a suburban conversation-piece, in a glade of Theobald's Park. The gateway, which marked the limits of the city's jurisdiction on the Strand, was the true embodiment of early-modern London: it was designed by Sir Christopher Wren, and the room above the arch was rented for many years by Child's Bank, as a store for the ledgers.
*Photo: G. W. Wilson
Aberdeen University Library*

67 *left* St John's Gate, Clerkenwell, *c.*1890. The last relic of the Hospitallers' Priory of St John of Jerusalem, and a rare survival from the suburbs of medieval London. The gateway is also celebrated for having housed the publisher's office of the *Gentleman's Magazine*, and was preserved and restored by public subscription in 1845–6, at a time when the development of Clerkenwell might have swept it away.
Greater London Council

68 *below left* The Banqueting Hall, Whitehall, from a stereo card, *c.*1865. An early picture of the historic city in repose, with only kerbstones and gaslight intruded by the nineteenth century.
David Francis

69 *right* Butcher's Row, Aldgate, 1883: buildings in the tradition of the pre-industrial city tricked out with eclectic ornament and overshadowed by the cliff-like blocks of the nineteenth century. Note the pavement canopies, one with skylights, and the open counters of the shops. The City Bicycle School's advertisement offers a prospect of country lanes from a penny-farthing; the ghostly team of the dray stands for the restless movement in the streets that at this range the camera was still unable to arrest.
Greater London Council

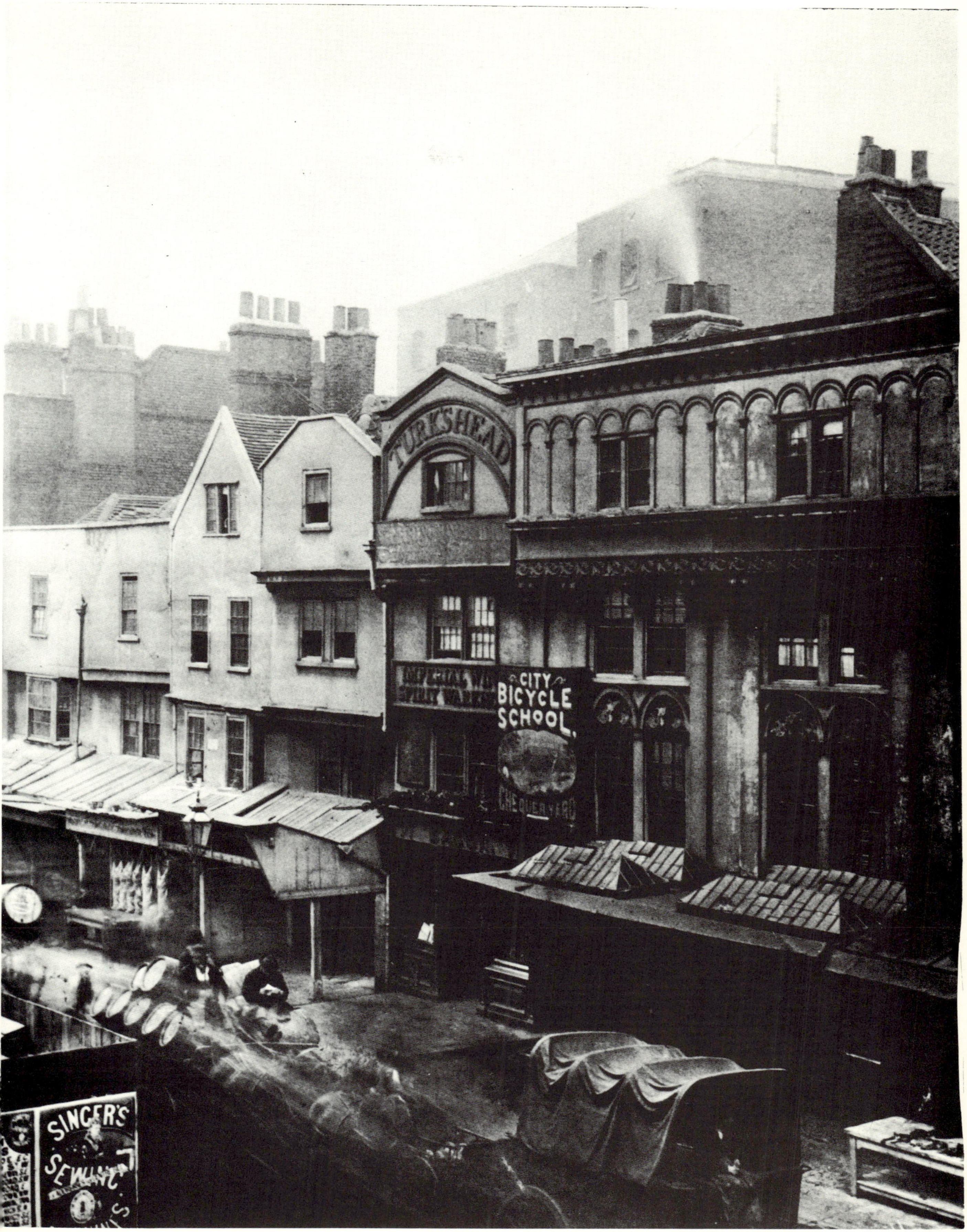
TURKSHEAD
CITY
BICYCLE
SCHOOL.
CHEQUER YARD
SINGER'S

70 *above* Butcher Row, Coventry. An undated but timeless scene: the stock of the second-hand furniture shop, fourth from the camera, suggests a date in the later decades of the century, but most of the detail belongs to the ages that escaped the camera's eye.
Photo: G. W. Wilson
Aberdeen University Library

71 *above right* A latterday inn yard: the Hotel Cecil decorated for the French president's visit to London, July 1903.
Photo: Bedford Lemere
National Monuments Record

72 *right* The Oxford Arms, Warwick Lane, Newgate, in 1875. The Oxford Arms was the chief Oxford carriers' inn in the seventeenth century, but was abandoned and turned into tenements in the nineteenth. The photograph is an effective study of a classic process of urban decay.
Greater London Council

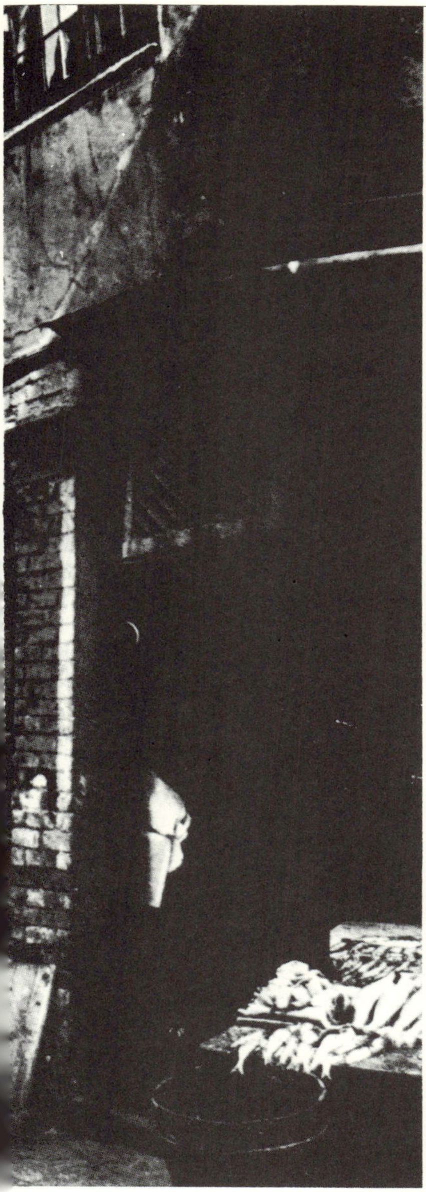

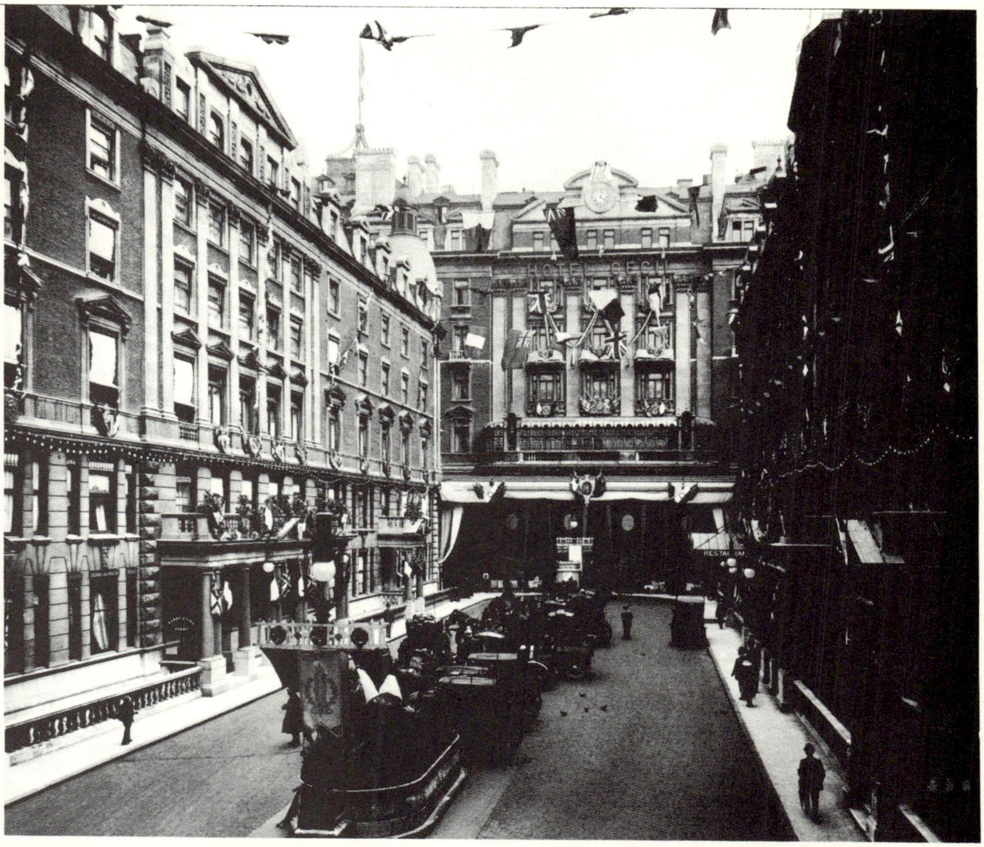

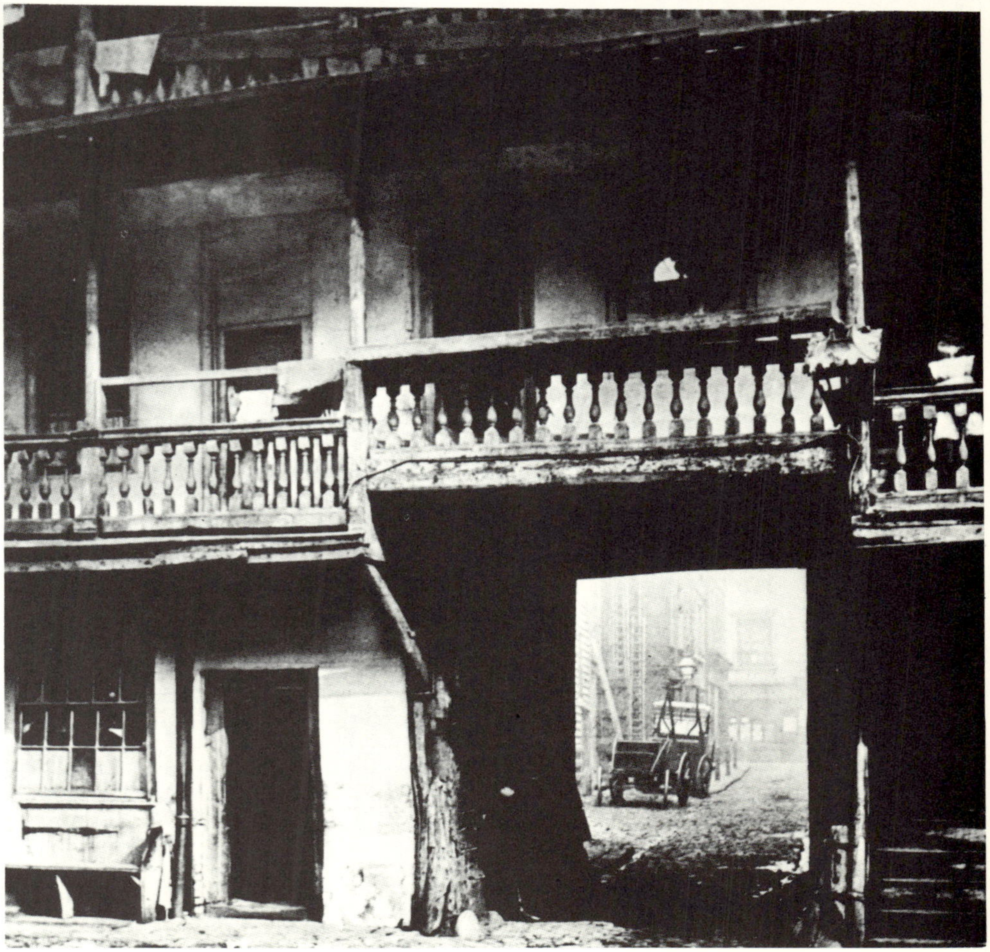

73 *above* Whitehorse Close, Edinburgh, *c.*1890. The closes were a marked characteristic of old Edinburgh; this is a tolerably good one, with low ranges of buildings, archaic features in the outside staircases, and some marks of improvement. The cambered court covers what would once have been an open drain.
Photo: G. W. Wilson
Aberdeen University Library

74 *right* Sackville Street, now O'Connell Street, Dublin, *c.*1865: the Post Office and Nelson's column, a cab and an abstracted bystander. The condition of the roadway explains the persistence of the private crossing-sweeper.
David Francis

75 *above* The Old Town, Edinburgh, from the Scott Monument. The eighteenth century drained the Nor' Loch as a step towards developing the New Town. The nineteenth century planted it, and then filled it with the railway. This picture, taken with a long enough exposure to blur the figures on the bridge, emphasizes the harshness of the railway's intrusion, both in the design of the metal bridge and its booking office pavilions, and in the uncompromising ranks of the rolling stock.
Photo: G. W. Wilson
Aberdeen University Library

76 *left* The Victoria Embankment under construction, *c.*1865. The Embankment, built between 1864 and 1870, was a comparatively late addition to London's public works, and really marks the end of the river above the bridges as an important highway, and its acceptance as a nuisance that had to be tamed into passivity. Large constructions were a good subject for stereoscopic photography, especially when work could be suspended to people the scene.
Greater London Council

77 *above far left* Leyton, the first Town Hall and Free Library, c.1865. Leyton was an Essex forest village, long favoured as a retreat by London merchants, which became a town in the middle of the nineteenth century. The urban sanitary authority and a free library, an unusual amenity in a half-formed community, was housed in this Italianate building set down in the empty spaces of Leyton Road.
Howarth Loomes

78 *above left* The Sailors' Home, Great Yarmouth, from a stereo card, c.1860. A raw piece of the new urbanity, fronted by an unmade road, and with a windmill decaying behind.
Howarth Loomes

79 *above right* Bristol General Hospital, from a stereo card, c.1865. The hospital, raised in a characteristic mixture of the utilitarian and the ornamental, rises clean and new over the clutter of the quayside. Bristol's Royal Infirmary, opened in 1735, was the first provincial hospital, in the modern sense of the word, and a measure of Bristol's early development. The General Hospital was built in 1862, after another century of urban growth.
Howarth Loomes

80 *above far right* The fifth Duke of Rutland, variously attended in Leicester market place, from a stereo card, c.1858. Not all the company stayed the course.
Howarth Loomes

81 *right* The Cobden Hotel, Corporation Street, Birmingham, c.1890: a bold marriage of Venetian Gothic with a Renaissance style, and an eloquent name. Corporation Street was vigorously promoted by Joseph Chamberlain: its objects were to improve and dignify the face of Birmingham, and to demonstrate the potentialities of municipal enterprise. It did both, though with more assurance than grace.
Photo: G. W. Wilson
Aberdeen University Library

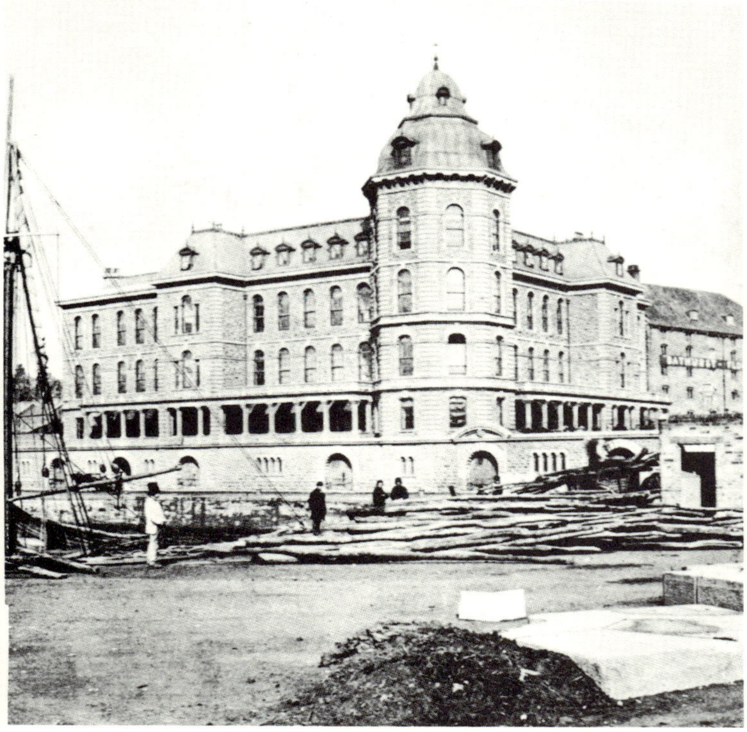

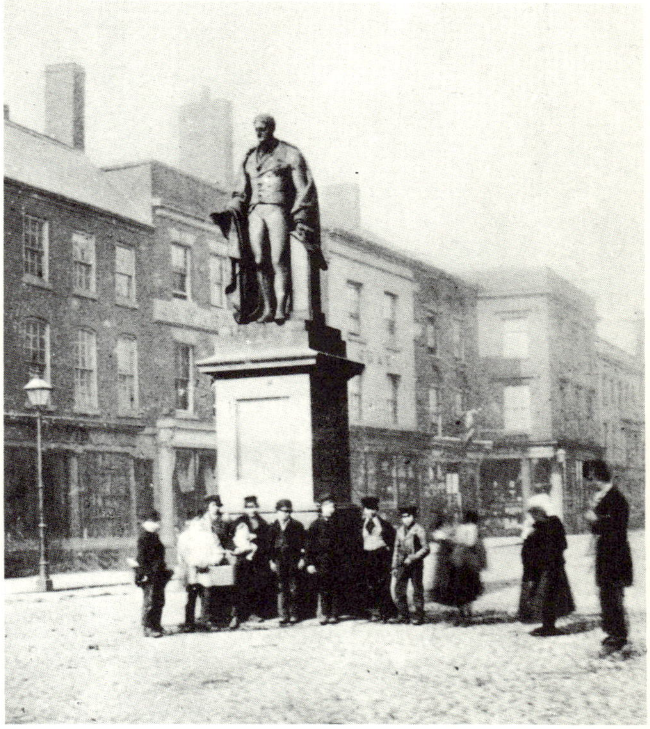

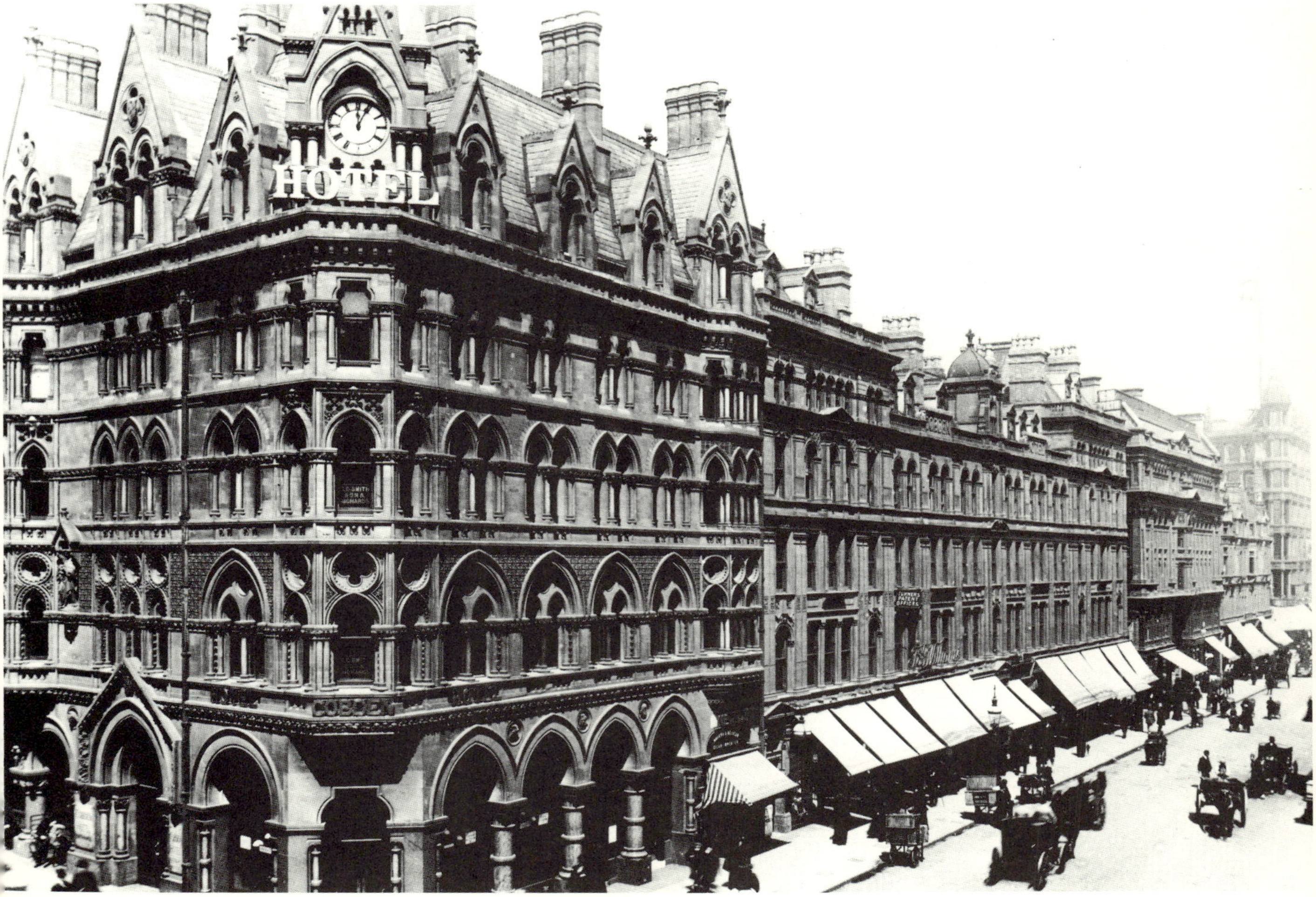
HOTEL

82 *left* Queen Victoria and Prince Albert at the reopening of the Crystal Palace, 10 June 1854. An engraving from a photograph taken by P. H. Delamotte, generally accepted as the first news photograph. Engravings from photographs remained the mainstay of the *Illustrated London News* until the adaptation of the half-tone block to rotary printing.

83 *centre* A stereo card, showing a wet-plate photographer drying plates in his garden, in the late 1850s.
Howarth Loomes

84 *below* The Free Trade Hall, Manchester, *c.*1855, from a calotype negative by James Mudd. An appropriate subject for the early urban photographer, and massively apprehensible. The image has been retouched to strengthen detail, and it is now impossible to tell whether the figures once had a ghostly substance, or whether they are entirely the retoucher's work.
Photo: James Mudd
Kodak Museum

85 *right* The Chain Pier, Brighton, from a stereo card, *c.*1865. The chain pier at Brighton has a special interest as the first seaside jetty which was formally used as a promenade. The card is an early one, and elaborately mounted.
Howarth Loomes

86 *centre* An illustration from Mayhew's *London Labour and London Poor* (1851). The illustrations in the original edition were based upon daguerreotypes.

87 *far right* An advertisement published in the *Illustrated London News* in 1856, depicting family life enhanced and beautified by the use of the stereoscope.

88 *below* George Augustus Sala, from a *carte de visite*, *c.*1865. The photographic *carte de visite* was a vulgar device, but its real popularity was as a collector's novelty. It served as the mainstay of the popular photographer's business until the arrival of simple amateur photography. Sala was a London essayist, and a very appropriate subject for an urban fad. Members of the royal family, and celebrities such as Florence Nightingale, were other esteemed figures.
David Francis

89 *left* A stereo card workshop, *c*.1860, with blocks of cards drying, and a manual roller press. The picture is elaborately posed, but photographs of any manufacturing process are rare at that time.
Kodak Museum

90 *below* A late-nineteenth-century stereoscope, made in the 1890s, but essentially the instrument invented by Oliver Wendell Holmes in 1860. The card on this model is a transparency, to be viewed by transmitted light.
Howarth Loomes

91 *right Slum life in our great cities*, the title-slide of the series, 1890. The presentation of the title, to say nothing of the title itself, is a promise of sentimentally uncritical nonsense. The interest of the slides lies partly in their existence, and the implication of a market for them, and partly for what they offer in the way of direct observation. The subjects presumably earned silver for posing, but none of them seems to have been got up as a Respectable Pieman. *Life in the back streets* would have been a better name for most of the scenes depicted: the slums themselves are more secret places.
David Francis

SLUM LIFE
IN OUR
GREAT CITIES.
PHOTOGRAPHED DIRECT FROM LIFE.

92 *below* A group of children wearing straw bottle-wrappers, from *Slum life in our great cities.*
David Francis

93 *right* Street life in Seven Dials, *c.*1890
Guildhall Library

94 *below right* A municipal set-piece: a Liverpool fire engine, *c.*1890, with the brass burnished and the crew freshly brushed to face the camera. Even instantaneous photography cannot take the formality out of an occasion like this. For a snapshot on the street, see no. 110.
Photo: Thomas Burke
Liverpool Public Libraries

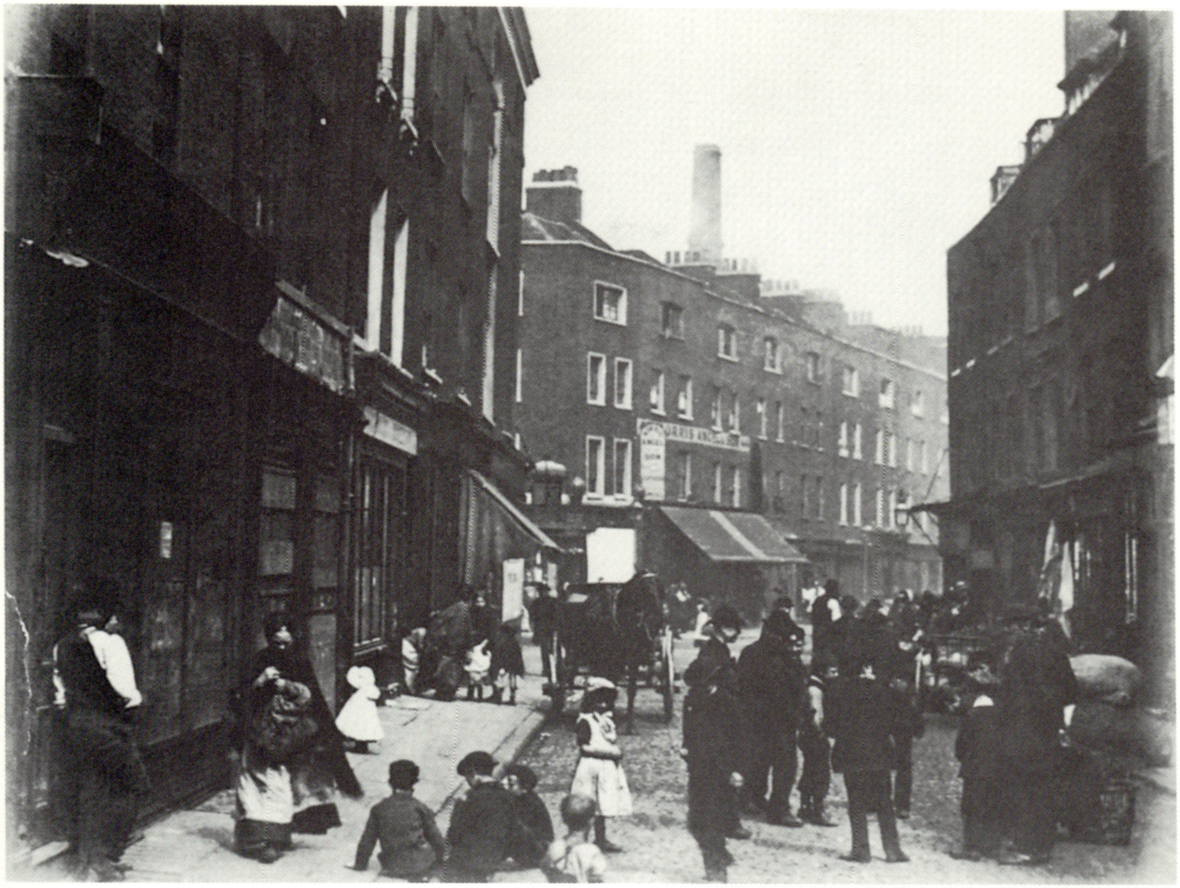

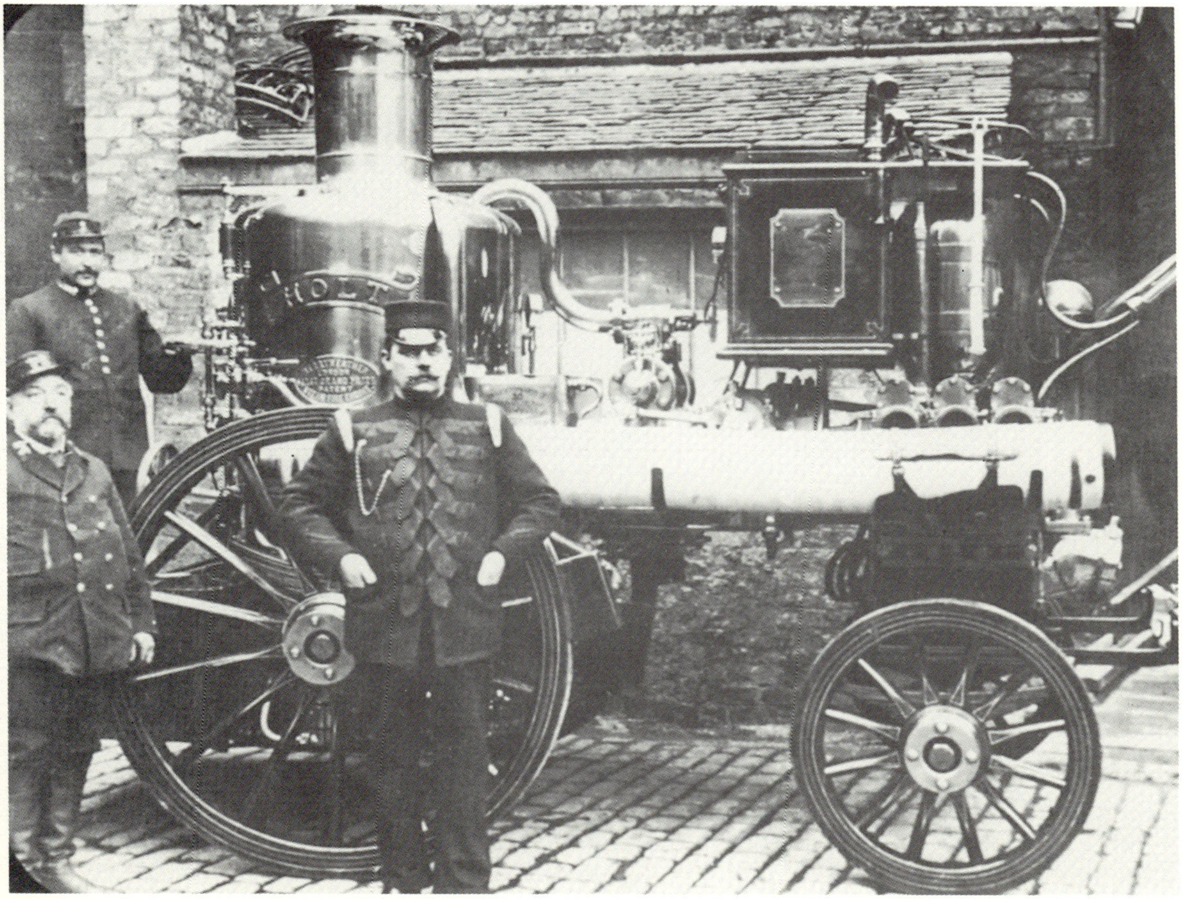
HOLT

95 *below* An unidentified street trader, posed for *Slum life in our great cities*. The basket is a swill, made in Westmorland or Furness, and the scene is probably set in Manchester or Liverpool.
David Francis

96 *right* A patriarchal figure in a Jewish street in Liverpool, from *Slum life in our great cities*.
David Francis

97 *below right* A round game, c.1890. The children are being directed and have an adult audience, but there is no general awareness of the camera.
David Francis

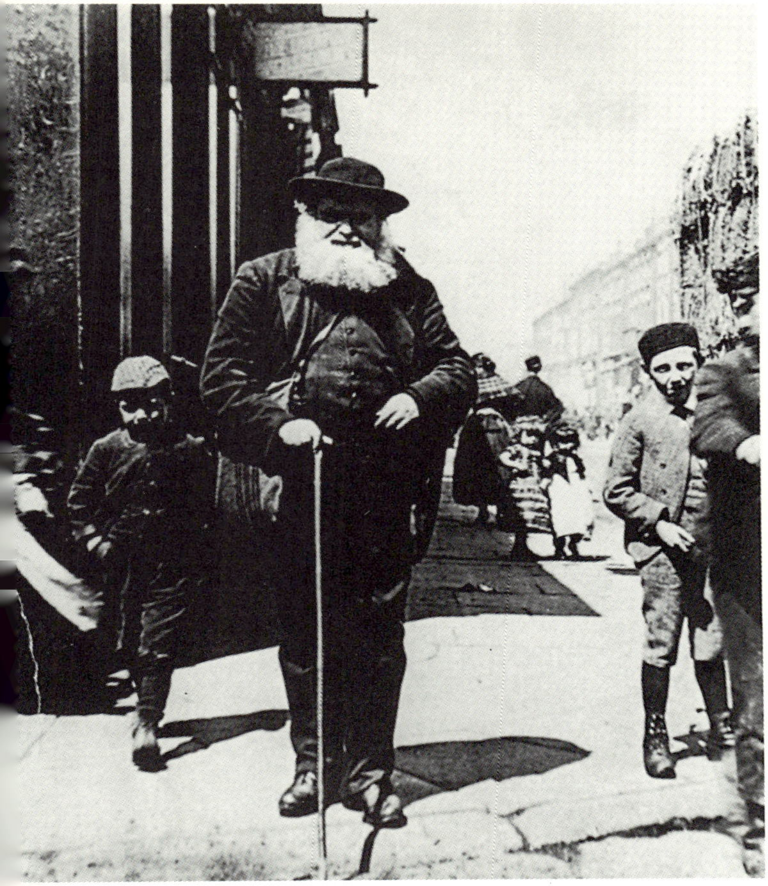

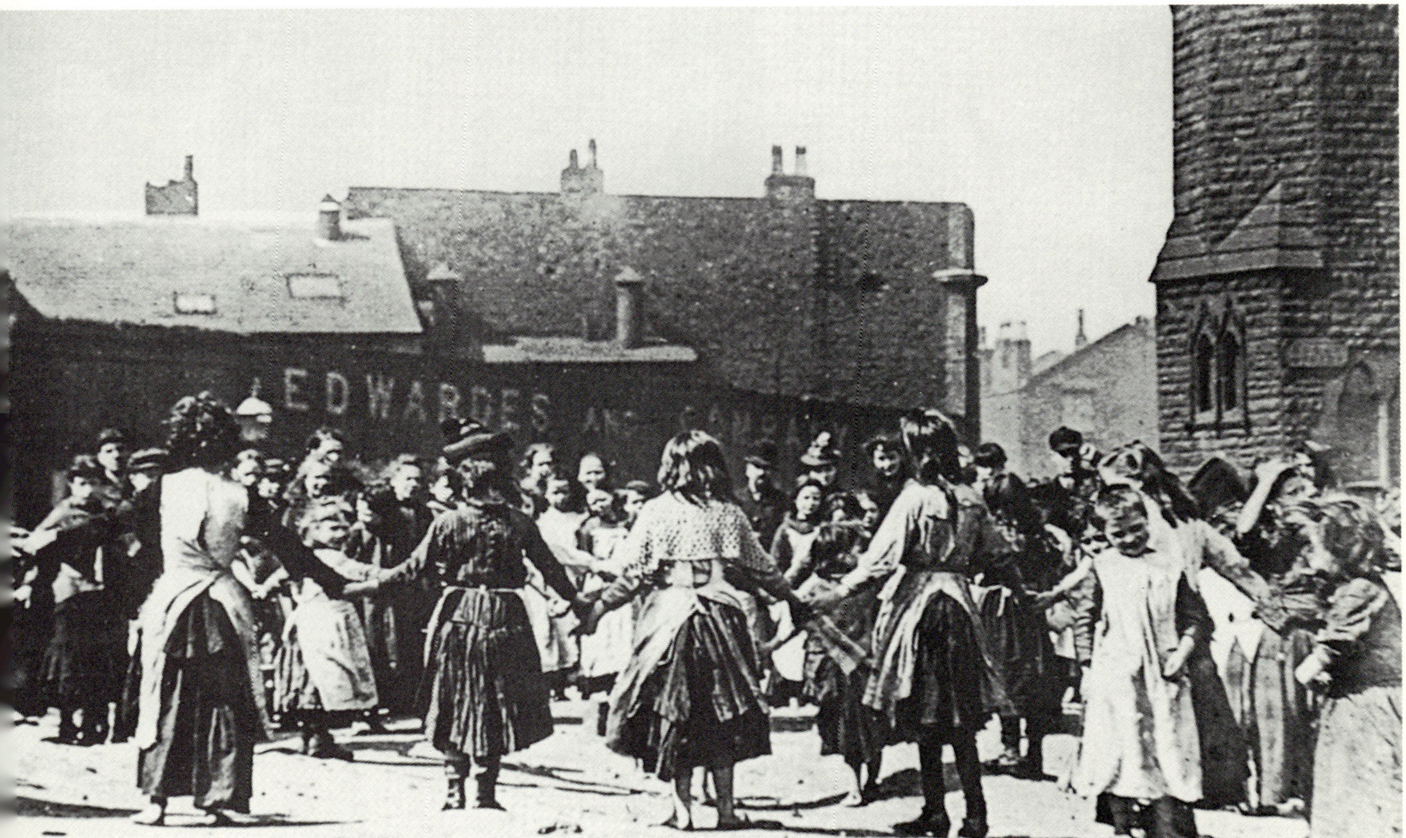
EDWARDES

98 *right* Jacks outside the barber's, Richmond Row, Liverpool, *c*.1890.
*Photo: Thomas Burke
Liverpool Public Libraries*

99 *below* Sweet vendor, from a lantern slide, *c*.1890. A scene from suburban life; the gravelled road, paved sidewalk, forbidding iron railing, and densely planted gardens imply a comfortable patronage for the Italian's sugar figures, if this was his ordinary beat.
David Francis

100 *left* A hot chestnut man, from the series *Street life, or the people we meet*, 1890. The string of advertisements, the boys' collars, and the unswept roadway are so many comments upon an industrial society. The carnation in the vendor's buttonhole is a matter of taste.
David Francis

101 *below* A second-hand shoe dealer, from *Slum life in our great cities*.
David Francis

102 *right* A bootblack at work, c.1900.
Kodak Museum

103 *below* A hokey-pokey ice-cream man, posed by his painted cart, London, c.1890. In Britain the baldacchino would more often serve to shelter than to shade the wares.
Kodak Museum

104 *left* Ice cream, the native product, from *Slum life in our great cities*. The original shows some signs of careless retouching, but the vignette of the vendor and his customer, the detail of the buildings, and the stridency of the written signs make it a telling picture.
David Francis

105 *below* A vendor of hot food, perhaps baked potatoes, with his donkey cart.
David Francis

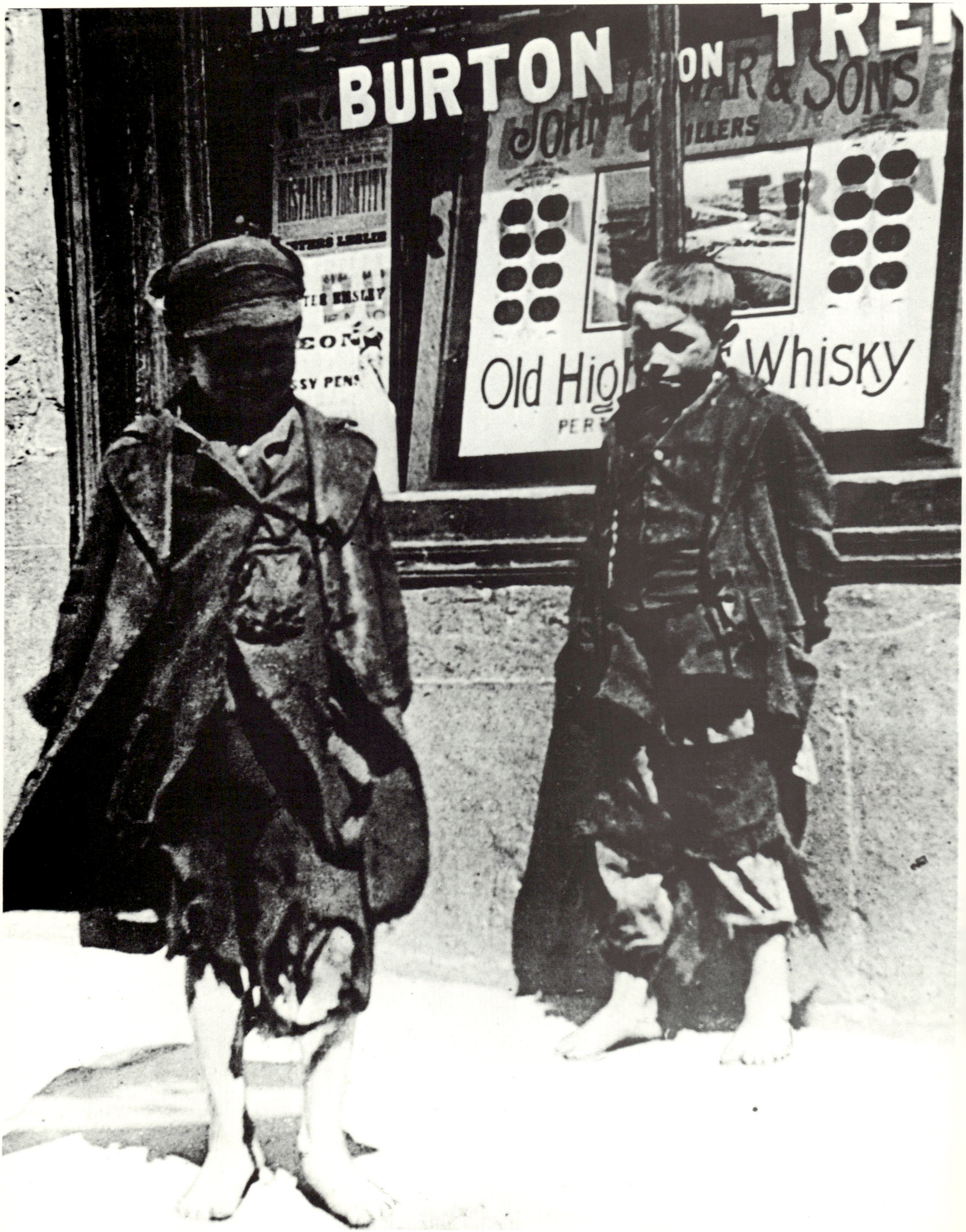
BURTON ON
& SONS
Old Hig
Whisky

106 *far left* Two boys from *Slum life in our great cities*. The authentic face of ragged poverty, although the photograph shows some signs of retouching to sharpen the picture.
David Francis

107 *left* A street-cleaner and his audience in Clerkenwell, *c*.1900. The drinking fountain, a fine specimen of its kind, was raised in 1868 by the Good Samaritan Temperance Society.
Kodak Museum

108 *below* Bricklayers posing, *c*.1890. The smallness of the fireplace suggests a bedroom; the walls are not cavity walls and are only casually bonded. Billycocks were entirely practical headgear, their comic dignity aside.
Manchester Public Libraries

109 *above left* A London carthorse at his oats, c.1900.
Kodak Museum

110 *below left* A London & North-Western Railway Company's horse collapsed in South John Street, Liverpool, c.1890. The availability of the box camera and flexible film made drama of this kind an obvious subject for the roving photographer.
*Photo: Charles Inston
Liverpool Public Libraries*

111 *above middle* Selling a horse in the street, c.1900.
Manchester Public Libraries

112 *below middle* Street life in Kensington Park Gardens: the Household Cavalry at exercise, 1889.
Guildhall Library

113 *above right* Laying Denver blocks, 1900–2. Tar-coated wooden blocks were first used in Chicago in the 1870s. They offered a reasonably quiet and resilient surface to iron-tyred wheels, but proved less satisfactory for heavy motor traffic.
Kodak Museum

114 *below right* Liverpool: *Herbert Campbell* pressing asphalt over cobbles, a street scene of 1897.
Liverpool Public Libraries

115 *left* The Rows, Chester, *c.*1860. A rare early view of a provincial town from a stereoscopic card. The Wild-West quality of such street scenes in the mid-century, which is a product of the superficially similar industrial development of England and the United States, is enhanced here by the wooden promenades of the rows.
David Francis

116 *right* Borough High Street, Southwark: a splendid display of individualism in the shopfronts and roof-lines, of street furniture of all kinds, and of heavy commercial traffic.
Photo: G. W. Wilson
Aberdeen University Library

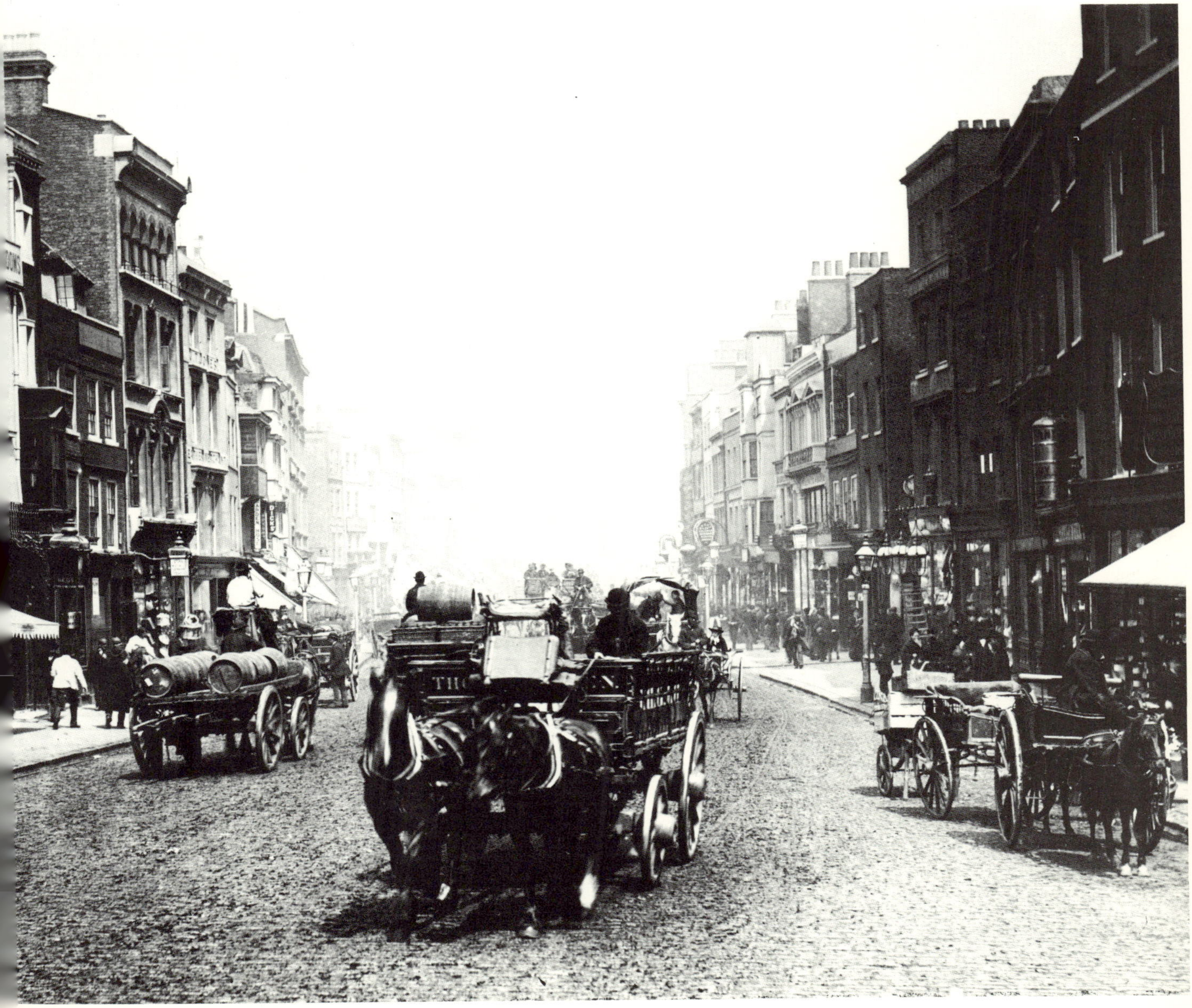

117 *below* Lime Street, Liverpool, *c.*1890.
A thinly-populated but vigorous scene.
Photo: G. W. Wilson
Aberdeen University Library

118 *right* Derby-Day traffic in the Clapham
Road, *c.*1895. The shop, a second-hand furniture
and piano dealer's built in the front garden of
the end house on the terrace, is a measure of the
road's decline into seedy commerciality.
Photo: G. W. Wilson
Aberdeen University Library

ELEPHANT & CAS
HORSE & CARRIAGE REPO
NEW KENT Rd
CLAPHAM, St GEORGE'S Ch & LORD

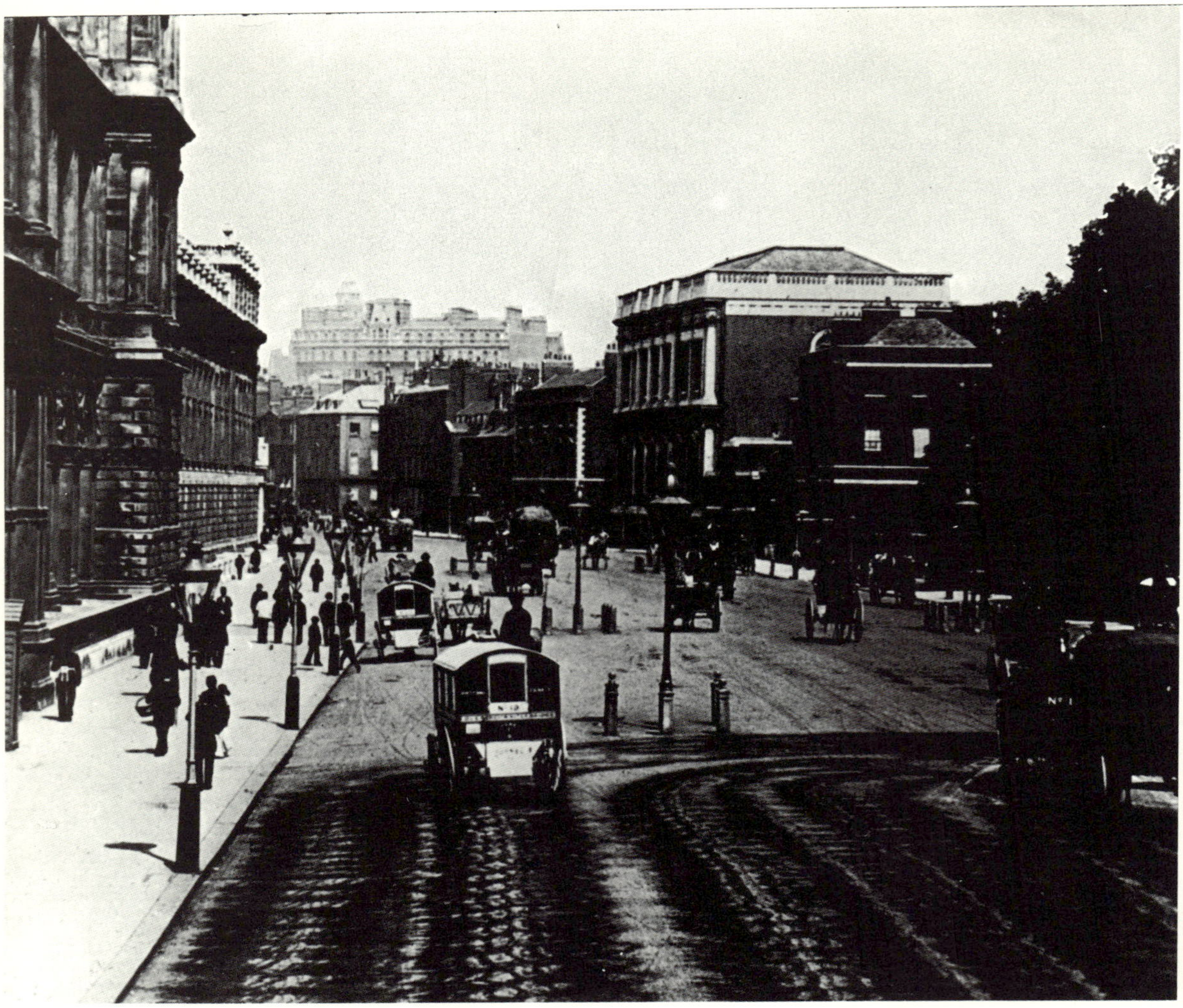

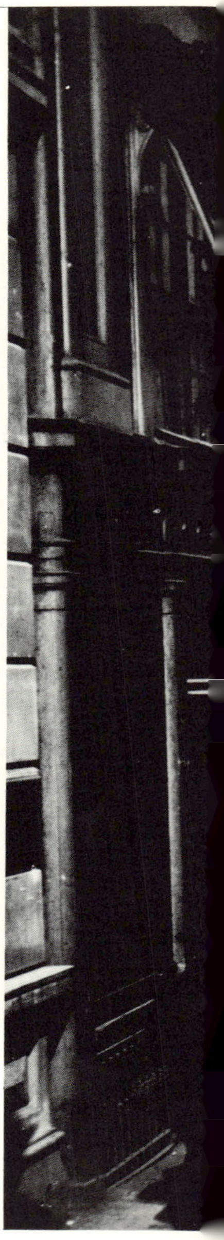

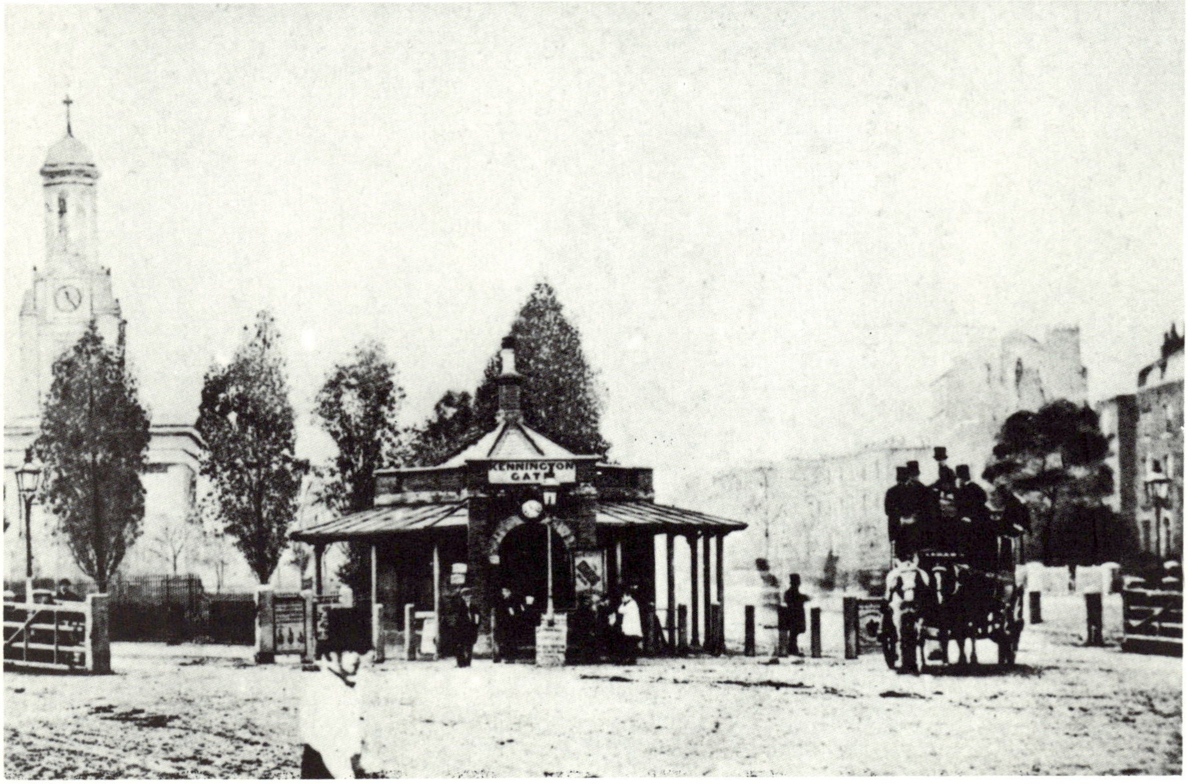
KENNINGTON

119 *above left* Whitehall in 1885, from a lantern slide. The patterns on the road are made by the water-cart at the right.
David Francis

120 *left* Kennington turnpike gate, 1865. The gate, already an archaism among the stove-pipe hats, was swept away shortly after this photograph was taken.
Guildhall Library

121 *above* Fleet Street and Ludgate Hill. A subtle and interesting contrast with the Borough High Street (no. 116), and a scene that betrays some awareness of the camera in the left foreground.
Photo: G. W. Wilson
Aberdeen University Library

LOCKHART'S
COCOA
ROOMS.
TO LET
PROSSER'S
OYSTER AND SUPPER ROOMS
WHISKY
50, HOLBORN.
PUBLIC SUPPLY STORES

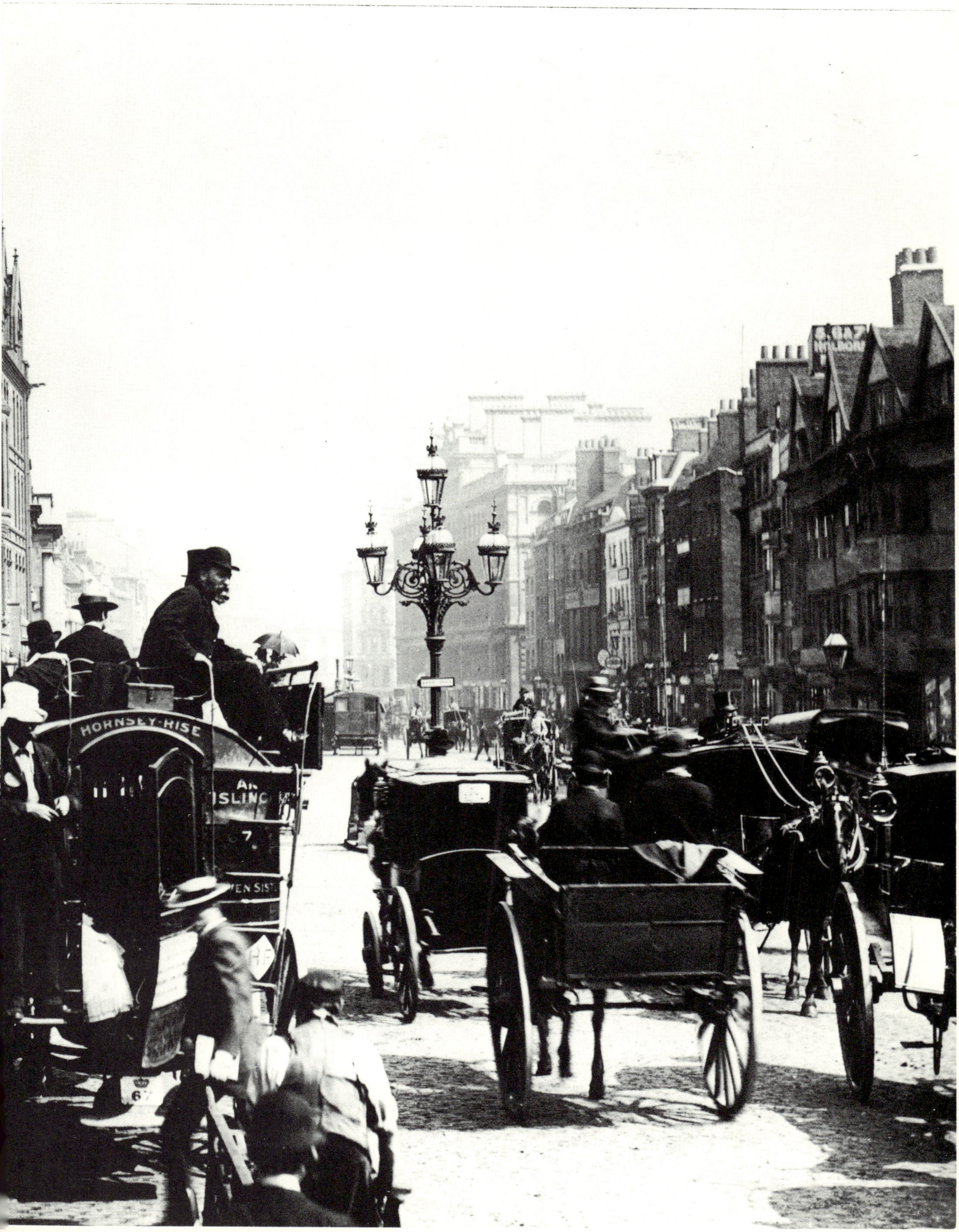
HORNSEY-RISE

122 *preceding page* Holborn in the 1890s. The open door of the omnibus shows a woman's skirts, but central London away from the shopping streets is a male preserve. The omnibus, with its spindly hand-rails and outside passengers perched like uneasy birds, makes no concessions to its passengers' dignity, but the photograph, which was probably taken with a camera concealed in a van, is a fine example of a snapshot, and has caught the whole scene off-guard.
Photo: G. W. Wilson
Aberdeen University Library

123 *above* The Trongate, Glasgow, c.1895, with horse-trams. As in Edinburgh and elsewhere in Scotland, the old buildings are taller than in English towns. The hand of the nineteenth century shows chiefly, the people apart, in the hard lines of the façades and in the sign-writing.
Photo: G. W. Wilson
Aberdeen University Library

124 A policeman on point-duty in the Euston Road, *c.*1900.
Kodak Museum

NESTLE'S MILK
NESTLE'S
MILK
NESTLE'S MILK
COOMBS'
AERATED
SWISS
MILK
THROUGH
2029

125 *left* Passengers boarding a horse-bus, probably in the Edgware Road, *c*.1900. The intensive advertising is noteworthy.
Kodak Museum

126 *above right* A comic stereo card of the 1860s, satirizing competition between omnibus companies.
David Francis

127 *below right* A horse-tram of the North Metropolitan line in Graham Road, Dalston, 1895. The horse-tram, which provided a reasonably efficient service between the inner suburbs and the centre of the city, was an important factor in urban growth before the development of mechanical road transport.
Kodak Museum

128 Bristol, the swing-bridge, *c.*1890. A fine photograph of the quays at the historic centre of the town. Bristol was displaced in the course of the nineteenth century by the growth of the industrial north, from the first to the second rank of cities. It had to change comparatively little to maintain its size and essential functions, and the difference between this and a view in the eighteenth century is a matter only of superficialities, the steamer and horse-tram among them.
Photo: G. W. Wilson
Aberdeen University Library

129 West Hartlepool, the docks, *c.*1880.
Photo: G. W. Wilson
Aberdeen University Library

130 *above* Southampton, the steamer pier, *c.*1895. An atmospheric study, showing the metabolism of the great railway port at a low ebb.
David Francis

131 *above right* Passengers for an emigrant ship, Liverpool, *c.*1895
Photo: Thomas Burke
Liverpool Public Libraries

132 *right* London, the Embankment in use, *c.*1895. The water-trough, here refreshing two cab-horses, is one of those raised by the Metropolitan Drinking Fountain and Cattle Trough Association. It is the equivalent of the motoring accessories that now dot the streets, but unlike them is the product of philanthropic enterprise.
Kodak Museum

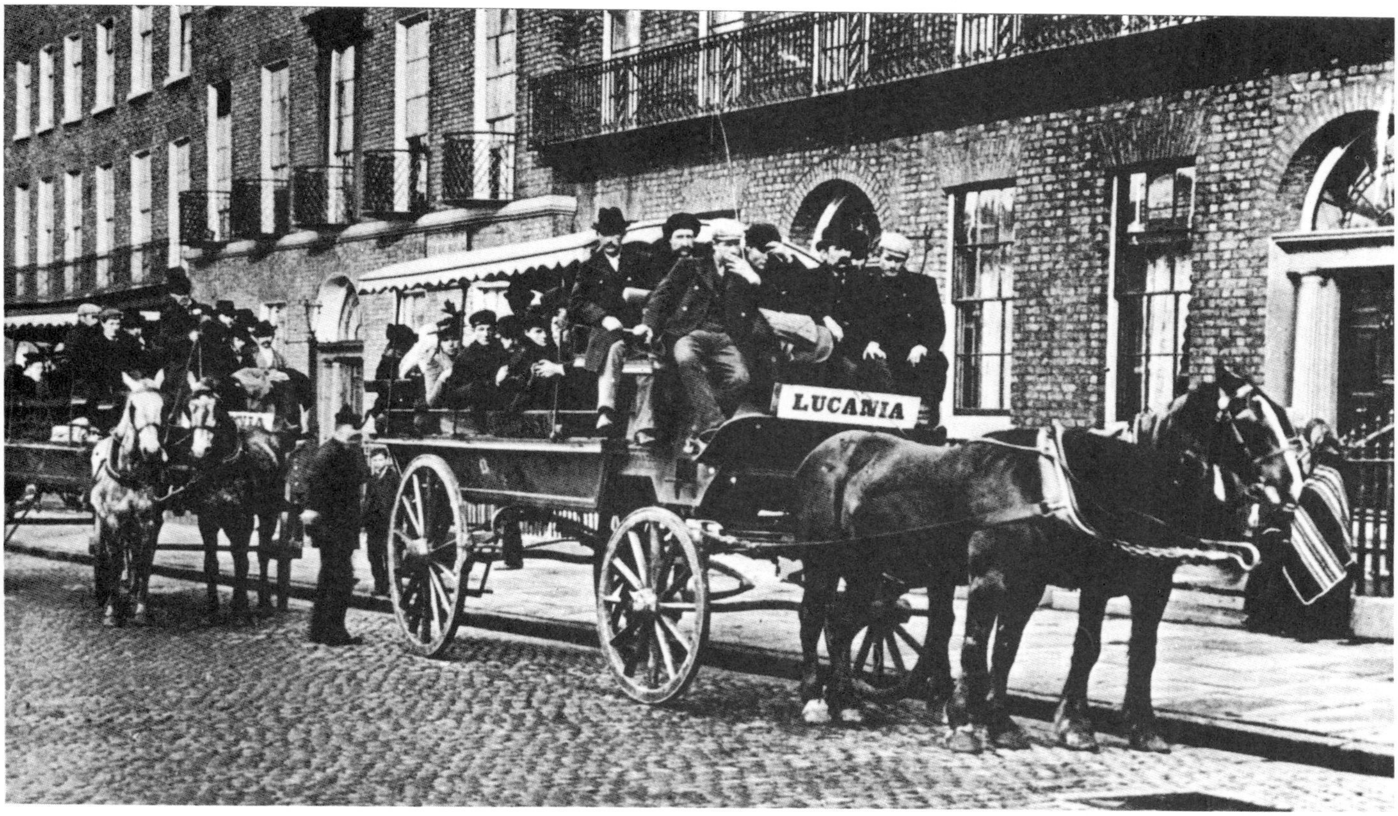
LUCANIA

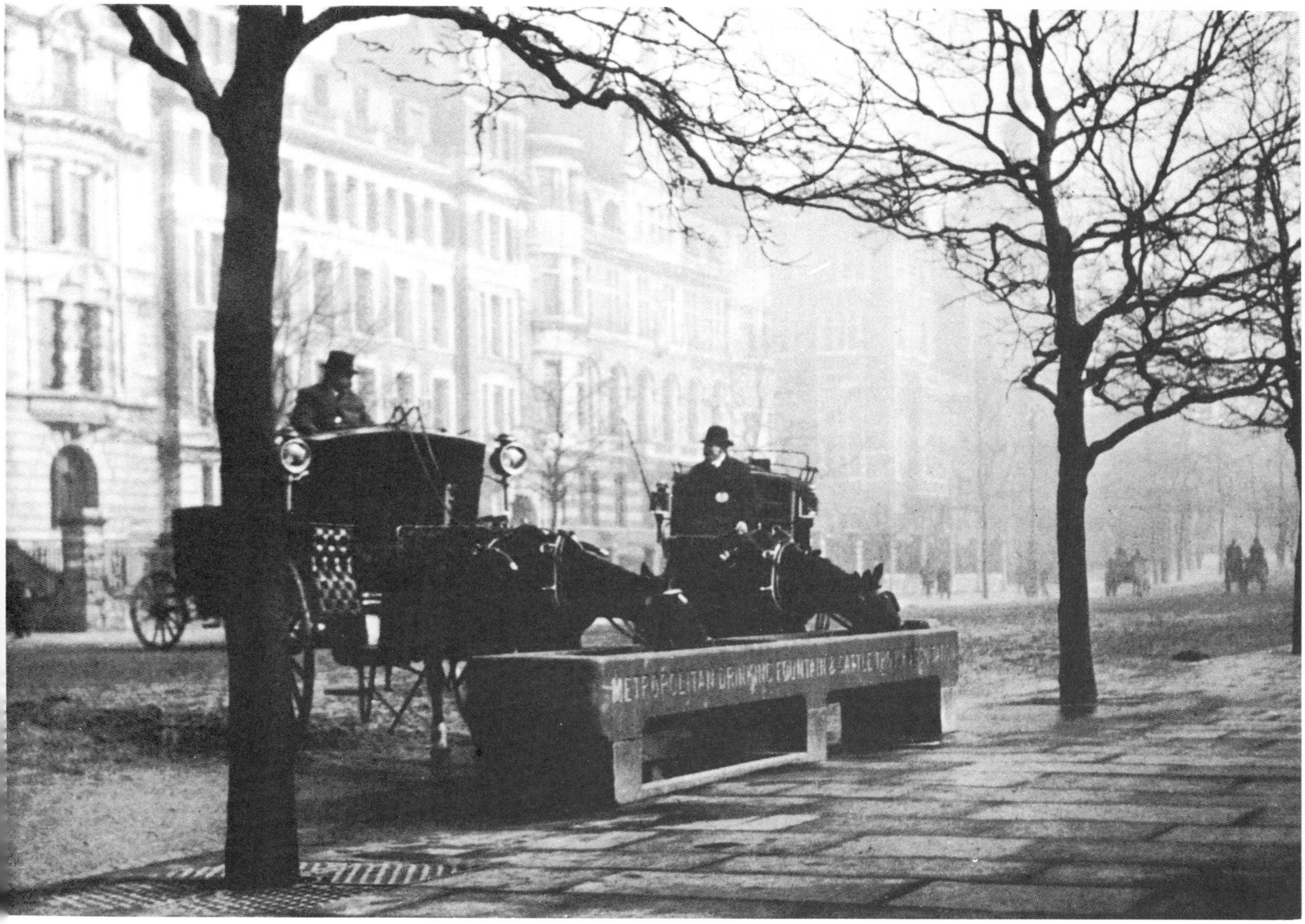
METROPOLITAN DRINKING FOUNTAIN & CATTLE

133 *above* Kingsbury & Neasden station, Metropolitan Railway, c.1894. A snapshot of the journey to work, taken with a Kodak no. 2 camera.
Kodak Museum

134 *above right* A uniformed carrier and his son, from *Slum life in our great cities*. The association of this small, but carefully-caparisoned horse and his drivers and their painted cart with the slums is a useful reminder of the hazards that beset earnest but uncritical publicists.
David Francis

135 *right* The full tide of human existence: traffic on and below London Bridge in 1895–6.
Greater London Council

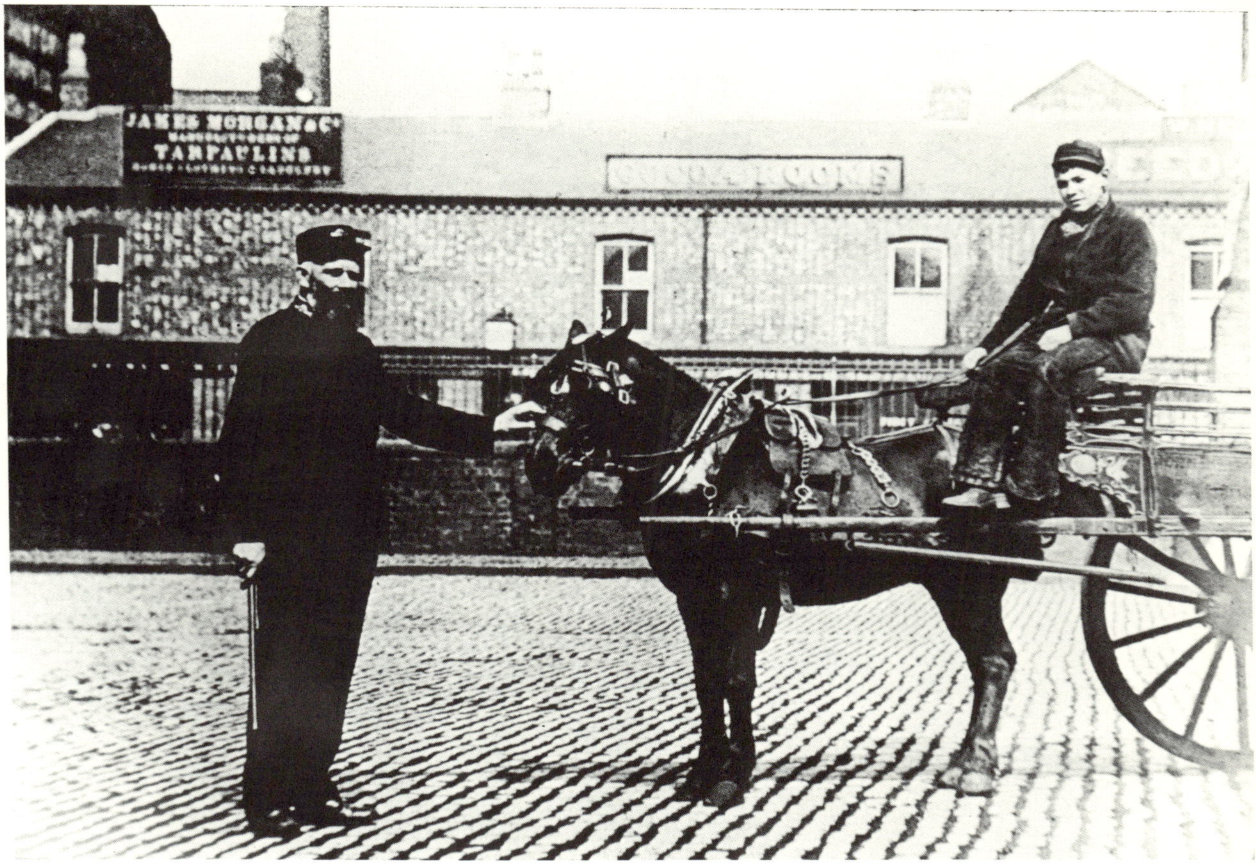
JAMES MORGAN & Co
TARPAULINS

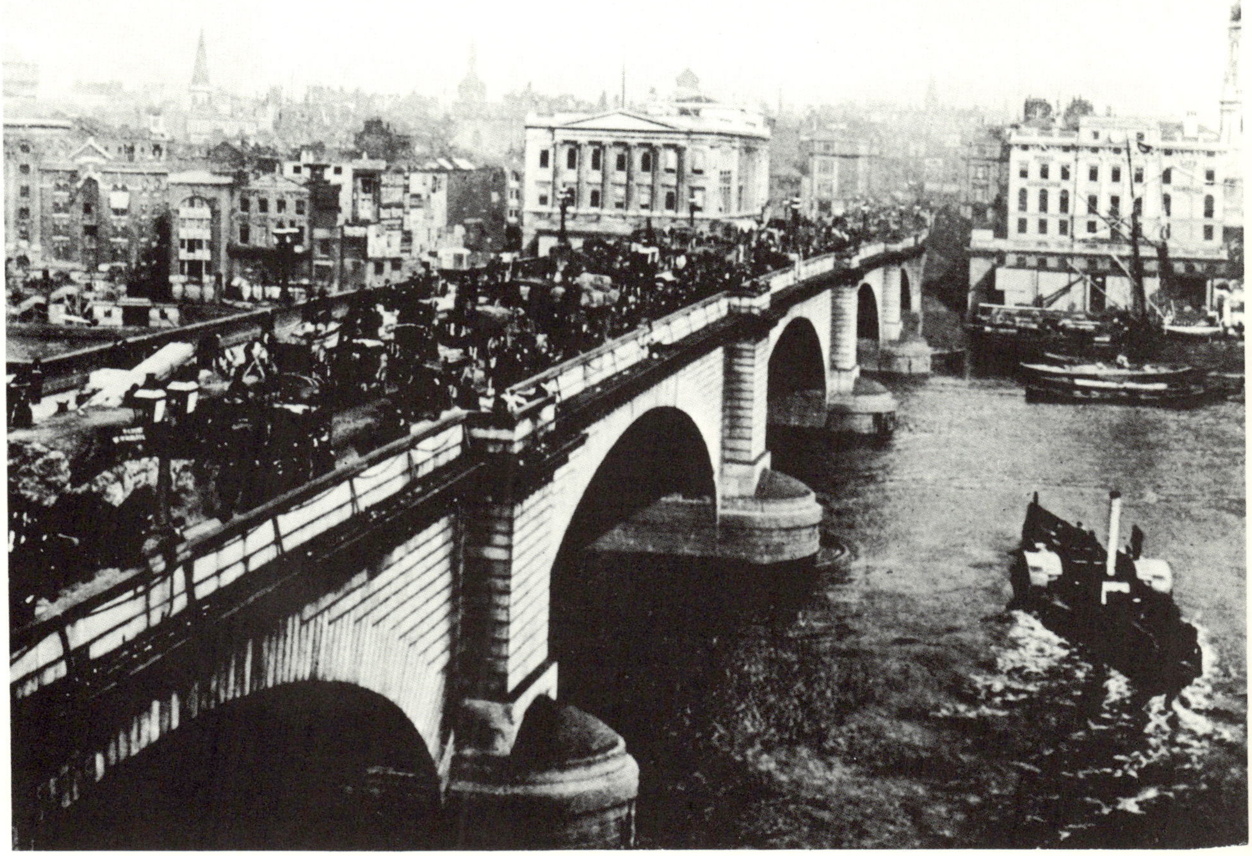

136 *above* A picturesque study of St Paul's and the river from Wren's house, by a masterly photographer.
*Photo: George Davison
Kodak Museum*

137 *above right* Barges and a cargo of hay below Tower Bridge, c.1900. Victualling the Victorian city was a matter of providing fodder for horses as well as food for men.
*Photo: George Davison
Kodak Museum*

138 *right* The Thames, below Tower Bridge, c.1900. Another painterly photograph with its own value as an historical record: the successive changes in the current and temperature of the river effected by the removal of old London Bridge, the building of the Embankment, and the general growth of the city, have made its freezing an exceptionally rare occurrence in the present century.
*Photo: George Davison
Kodak Museum*

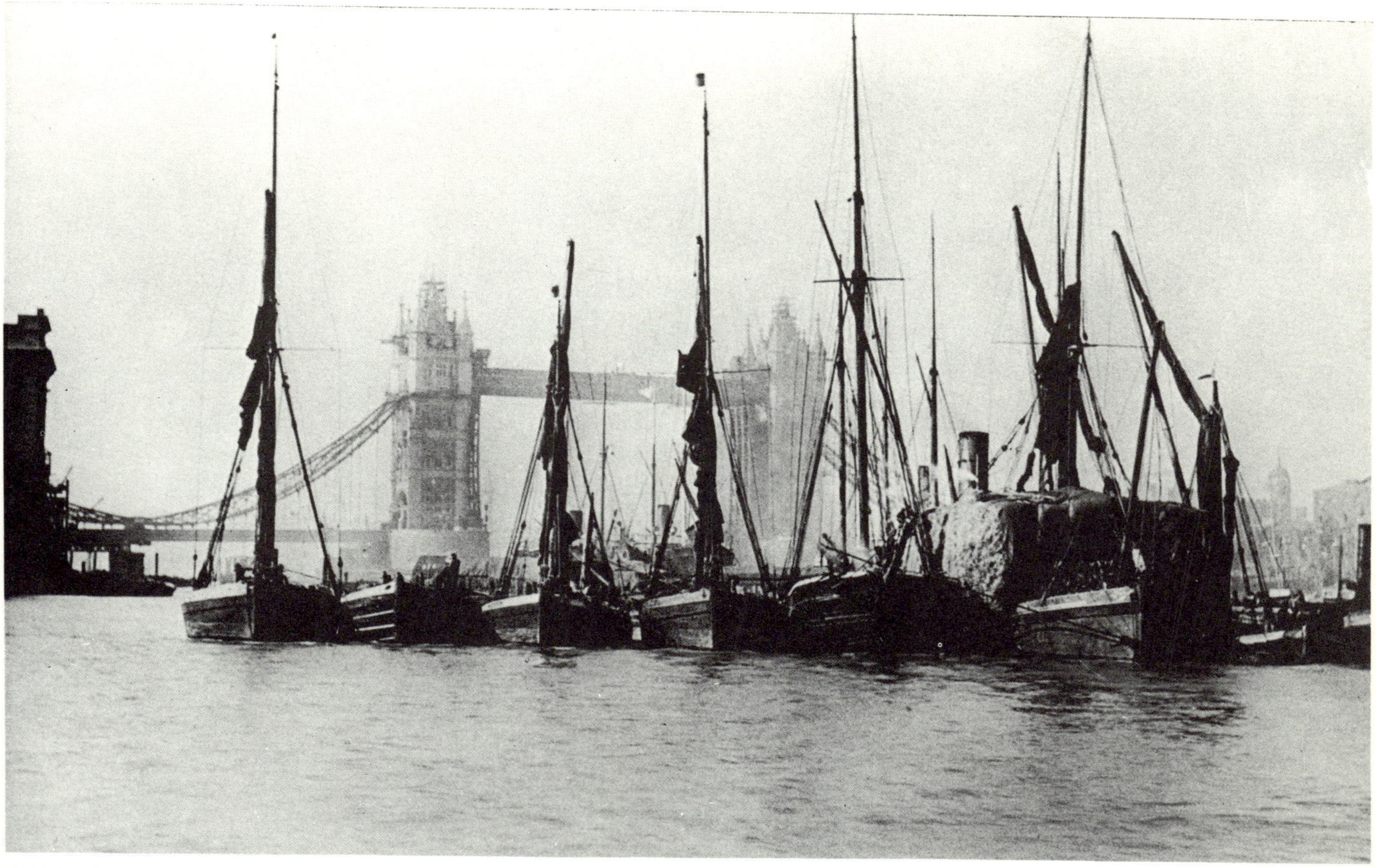

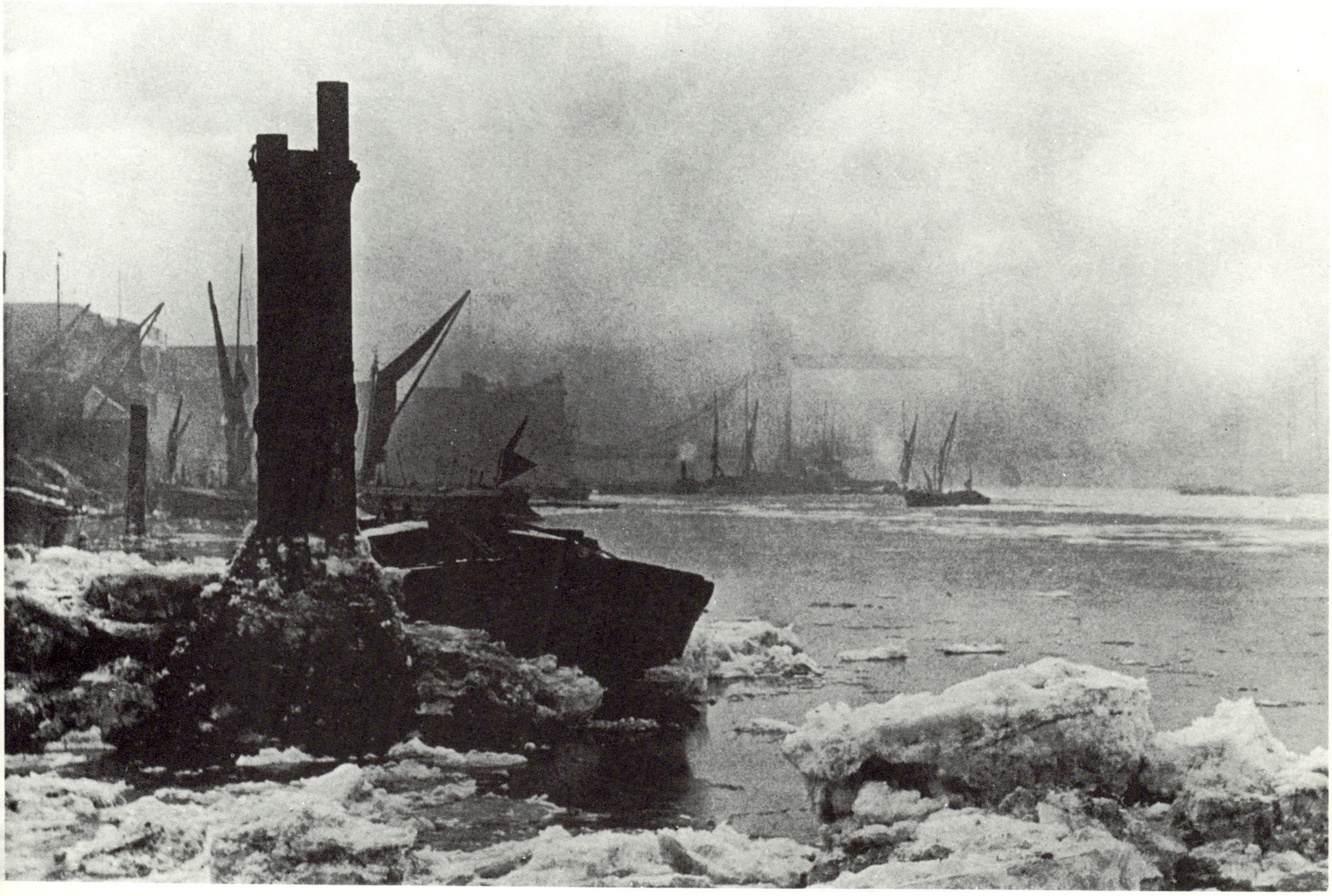

139 *above* Aberdeen at the close of the century. The tolbooth steeple and the mercat cross stand for continuity: the nineteenth century's peculiar contribution here is not so much the trams, which are eloquent enough, but in the numbers of people. Markets are traditionally thronged, but by this time densely-packed and disciplined crowds are ordinary and not extraordinary features of life.
Photo: G. W. Wilson
Aberdeen University Library

140 *above right* Provisioning the city: Covent Garden, with carts, baskets, and the Opera, in the late 1890s.
Kodak Museum

141 *right* Chapel Street, Salford: the edge of the Flat-Iron Market, 1894.
Photo: S. L. Coulthurst
Manchester Public Libraries

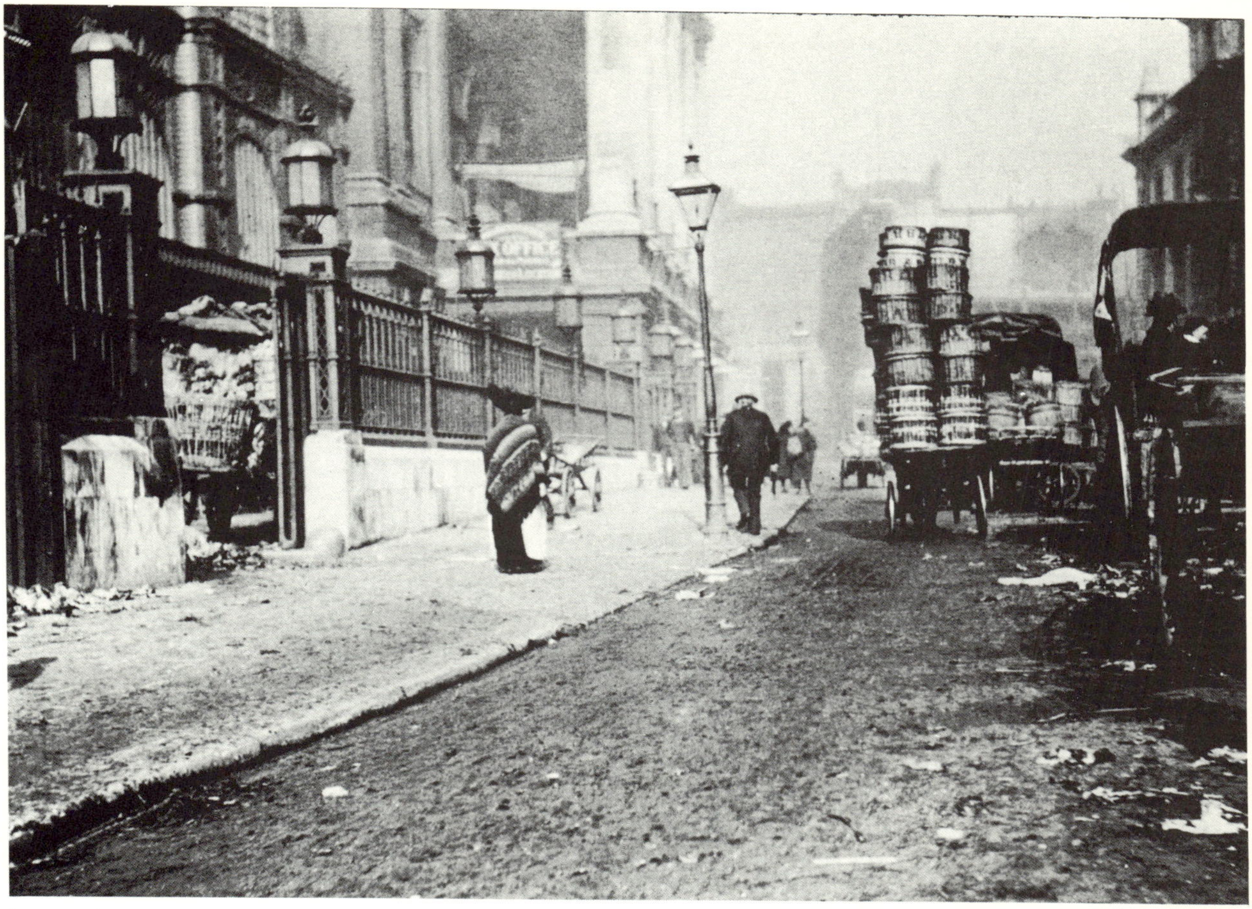

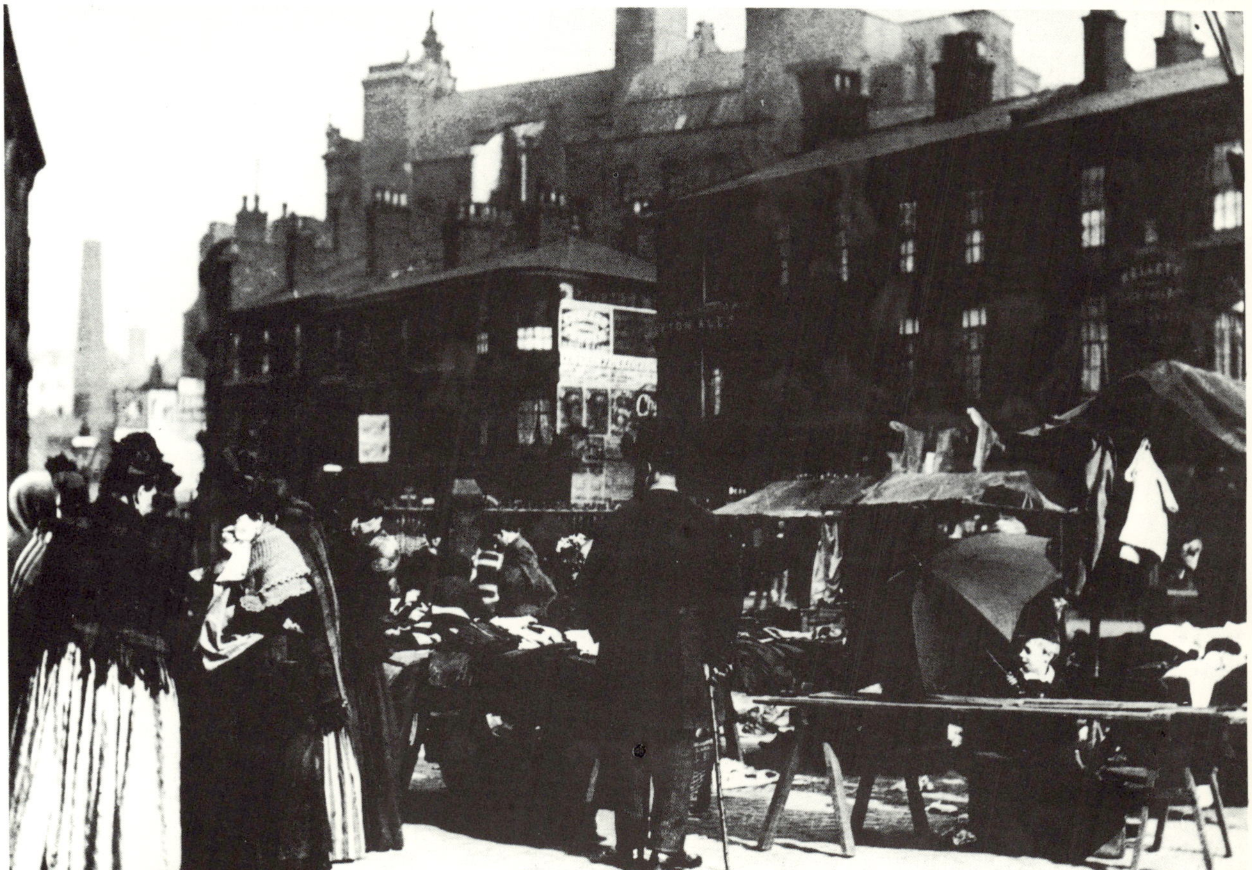

112 FARMER BATES 112
CASH BUTCHER.
WHOLESALE AND RETAIL.
FAMILIES SUPPLIED

142 *left* The enticements of abundance: a butcher's shop, c.1900.
Kodak Museum

143 *above right* No. 1 Atley Road and No. 700 Old Ford Road, in 1899. A pawnbroker's, on hard times, and an off-licence conveniently to hand. The pawnshop was an urban institution ranking with the public-house, though in this instance its general resemblance to one is a matter of the fashion of shop design.
Greater London Council

144 *right* Children outside a small provision shop: a slide from *Slum life in our great cities*. The battered panelling of the shop front, the product of much aimless scratching and kicking, is a reminder that our present scruffy habits have a long history.
David Francis

145 *above* An ironmongers shop in Athol Street, Liverpool, c.1890.
Photo: Charles Inston
Liverpool Public Libraries

146 *left* Eggs, butter carefully shaded, and fruit: a dairy shop off Great Homer Street, Liverpool, c.1890.
Photo: Thomas Burke
Liverpool Public Libraries

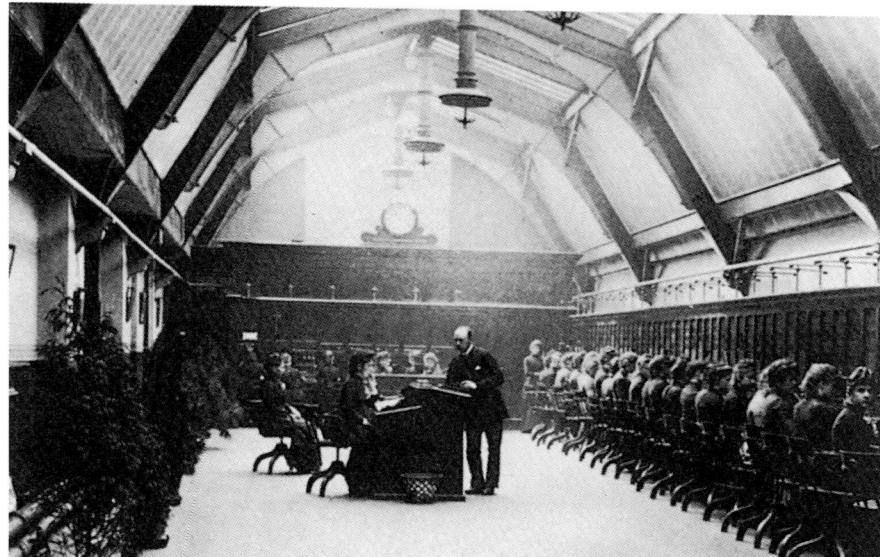

147 *above right* The Kodak Company's counting house, Clerkenwell Road, *c*.1902. The office of a vigorous and progressive company, with female clerks, typewriters, electric lighting, and pictures, albeit photographs, on the walls. Kodak's also appreciated the value of photography for advertising and recording; there would have been many obdurately old-fashioned offices at this time that escaped the camera's eye.
Kodak Museum

148 *middle right* A telephone exchange, *c*.1900. The room is an adaptation of a standard public hall, including institutional photographs and an ornamental dado. The potted shrubs are a concession to femininity, as indeed is the presence of a woman supervisor; the telephone service offered women one of the very few opportunities for advancement outside hospital nursing. The presence of the male manager marks the limits of enlightenment.
Photo: J. W. Wade
Manchester Public Libraries

149 *right* Exchange Buildings, Liverpool, 30 September 1881. A forest of silk hats at Cotton Corner.
Liverpool Public Libraries

150 *above left* The Kodak Company's head office in Clerkenwell Road, *c.*1902, unobscured by false modesty. See also no. 147.
Kodak Museum

151 *middle left* An advertising float, from *Street life, or the people we meet*, *c.*1890. Most societies have some addiction to the grotesque; the nineteenth century's came out strongly in its architecture, and in its advertisements.
David Francis

152 *below left* A sandwich-board man in Bradford, from *Street life, or the people we meet*. With casual labour cheap, a sandwich-board offered an economical way of advertising widely, and Hepworth's sandwich-board, nothing if not eye-catching, was larger and therefore more economical than other people's.
David Francis

153 *above right* The passing show: Hyde Park in the 1890s.
David Francis

154 *right* Fun of the fair: a contortionist on Hampstead Heath, *c.*1900. The Heath was a popular Bank Holiday resort; the crowd shown here is socially mixed, but by that time casual gatherings of people away from their work are better and more individualistically dressed. The great revolution of the twentieth century is still to be accomplished, but it can be said to have begun by the last years of the nineteenth.
Kodak Museum

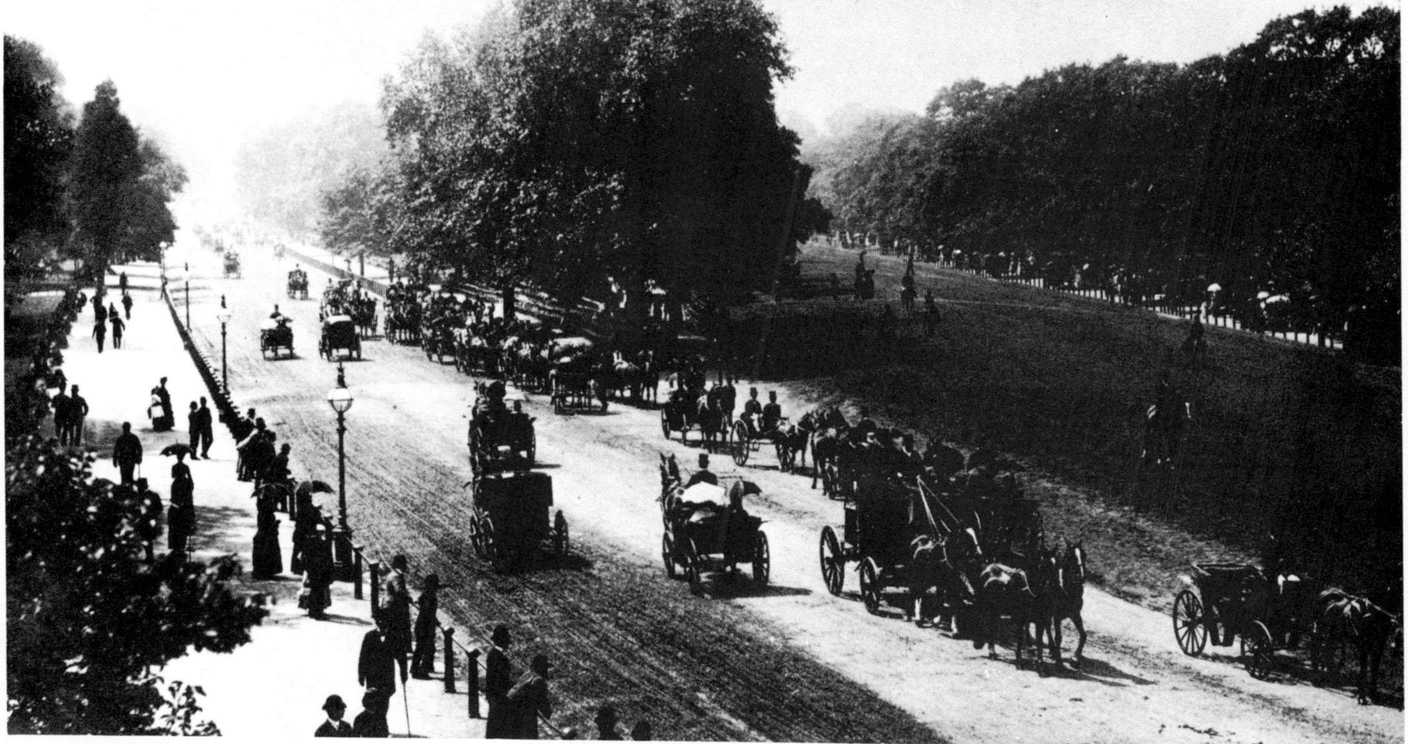

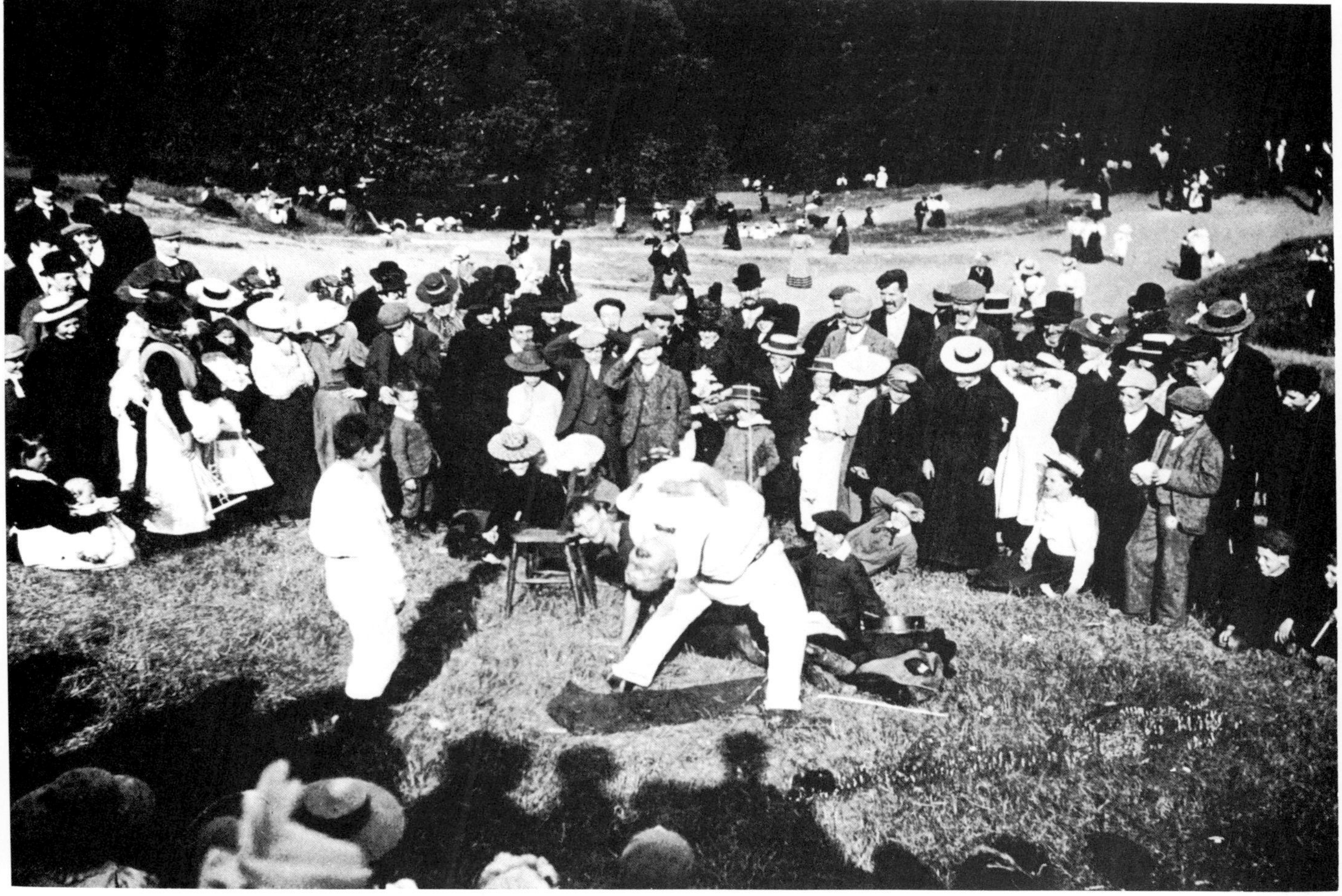

155 *above left* Surrey Gardens in the 1890s. A peep-show, becomingly brought up to date, and a cock-shy.
Greater London Council

156 *left* A child reciting in a public-house, 1895. A photograph that at first glance is straightforwardly informative, and on closer inspection unsatisfactory in various ways. The group is obviously contrived, and so wooden as even to raise doubts about the assertively-explicit background. If the picture has any general purpose it fails to make it clear. The child cannot be said either to relish her situation or be distressed by it, and her audience is cataleptic. The object of the photograph may have been to immortalize a real Infant Phenomenon, but if so, it again fails for its lack of a title.
Photo: Thomas Burke
Liverpool Public Libraries

157 *above* Members of a bicycling club, *c*.1900. Group portraiture is a good deal older than photography, but a photograph is an ideal record of an association or meeting. This picture is reproduced from a lantern slide, an additional refinement of the original plate. The bicycle was an urban product that offered agreeable exercise and a sense of freedom; it has gradually been overborne by the motor car, but it has exercised a considerable social influence in its time.
Kodak Museum

158 *right* A tipster at Aintree, 1901, decked in the rags of Empire.
Photo: R. Eastham
Liverpool Public Libraries

159 *above left* *Urbs in rure*: Epsom on Derby Day, 1898.
Guildhall Library

160 *left* The pier, Greenwich, 1865. An early and rare picture of a Citizen Steamship Company's steamer at the pier, with a hulk among the river traffic in the background. The river was used intensively by Londoners, and the Margate wherries carried a heavy popular traffic to the coastal resorts of Kent before there were railways or steamers.
Howarth Loomes

161 *above* Brighton, launching the *Skylark*, c.1890. Perhaps the original *Skylark*.
Photo: G. W. Wilson
Aberdeen University Library

162 *above left* Hastings, the beach, *c.*1890. Pleasures taken soberly; the absence of anything that might be called beachwear is notable.
Photo: G. W. Wilson
Aberdeen University Library

163 *left* The pier and the Grand Hotel, Hastings, *c.*1895. The promenade is an aristocratic institution, a place on which to stroll and be noticed. The pier is an extension of the promenade, which also serves as a landing place: pavilions and distractions are extraneous to its real purpose. This photograph, which includes one bathing machine boldly labelled 'For Gentlemen' shows how thin the veneer of urbanity can be stretched on the sea-front, where the visitor's and therefore the builder's first instinct is to stay within sight of the sea. There are haystacks and sheep on the open ground beyond.
Photo: G. W. Wilson
Aberdeen University Library

164 *above* The Grand Hotel and South Sands, Scarborough, *c.*1900. A fine photograph of the oldest of all seaside resorts in the high season. The seaside, as an institution, developed as it did because the mineral spring of Scarborough spa happened to flow out of the cliffs into the sea a little to the left of this picture. The nineteenth century's peculiar contribution was to make the seaside holiday a popular institution and one that mitigated some of the ills of an industrial society. Scarborough remained a fashionable resort (it is evident that Mr Walshaw's bathing machines bear no advertising of anything but his own business), but the orderly confusion of the beach here is at one with more simply popular places: it is informed by a gregarious individualism.
Photo: G. W. Wilson
Aberdeen University Library

165 *left* Blackpool beach, from a stereo card, 1875. The huge legs of the Tower, an extraordinary construction, dwarf the Royal Hotel and contrast sharply with the casual contrivances of the promenade at this time. Blackpool has a long history as a resort, but its present face is a creation of the late-nineteenth and early-twentieth centuries.
David Francis

166 *below* A band on a beach and a flotilla of bathing machines, from a lantern slide, c.1895. The bathing machine was concerned with bathing for health and with preserving modesty, not at all with pleasure. A study of this kind underlines the oddity of the institution.
David Francis

167 *right*   The opening of the Manchester Ship Canal, 1894. The triumphal arch is an ancient institution; an arch of firemen and their equipment over tramlines is the very essence of municipality.
Manchester Public Libraries

168 *below*   The hustings: the declaration of the poll for the coronership of Central Middlesex in Lincoln's Inn Fields, 1884. The hustings was an institution of high antiquity that survived long enough to be photographed, although the early difficulties of photographing crowds makes pictures of it a comparative rarity. The attendance on this occasion is predominantly Respectable.
Greater London Council

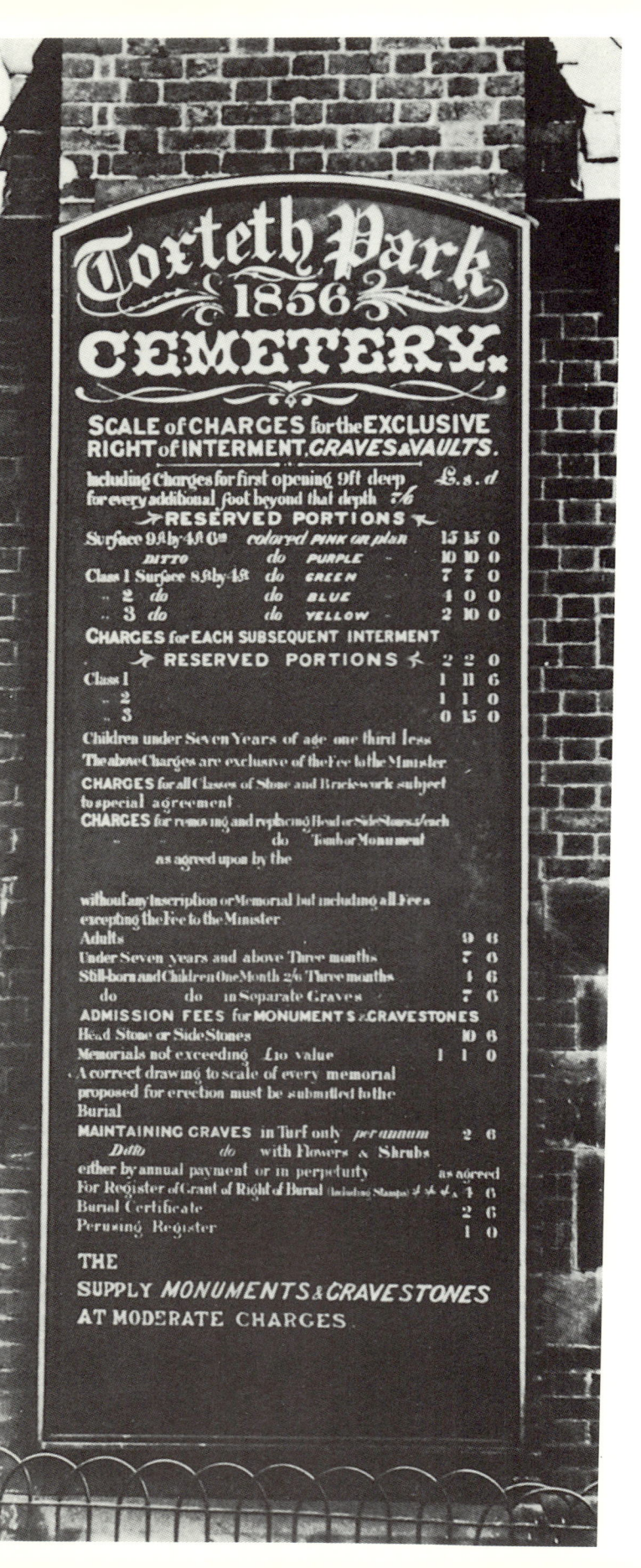

169 *left* Board of charges at Toxteth Park Cemetery, Liverpool, photographed in 1907. The teeming city implied a populous cemetery, and the cemetery was a place to which people resorted not only to tend the family graves, but to walk soberly and reflectively among the evergreens. Heavy infant mortality, and a strong family tradition that observed the anniversaries of deaths for several generations, charged the cemetery with emotion. It was also a place in which proprietorial instincts were gratified, although as the board here shows, its business-like regulations were tempered by mercy.
Liverpool Public Libraries

170 *right* Oakum-picking, *c*.1900, from a lantern slide. The setting is probably a workhouse. The parti-coloured brick walls have been familiar to later generations in other institutions.
David Francis

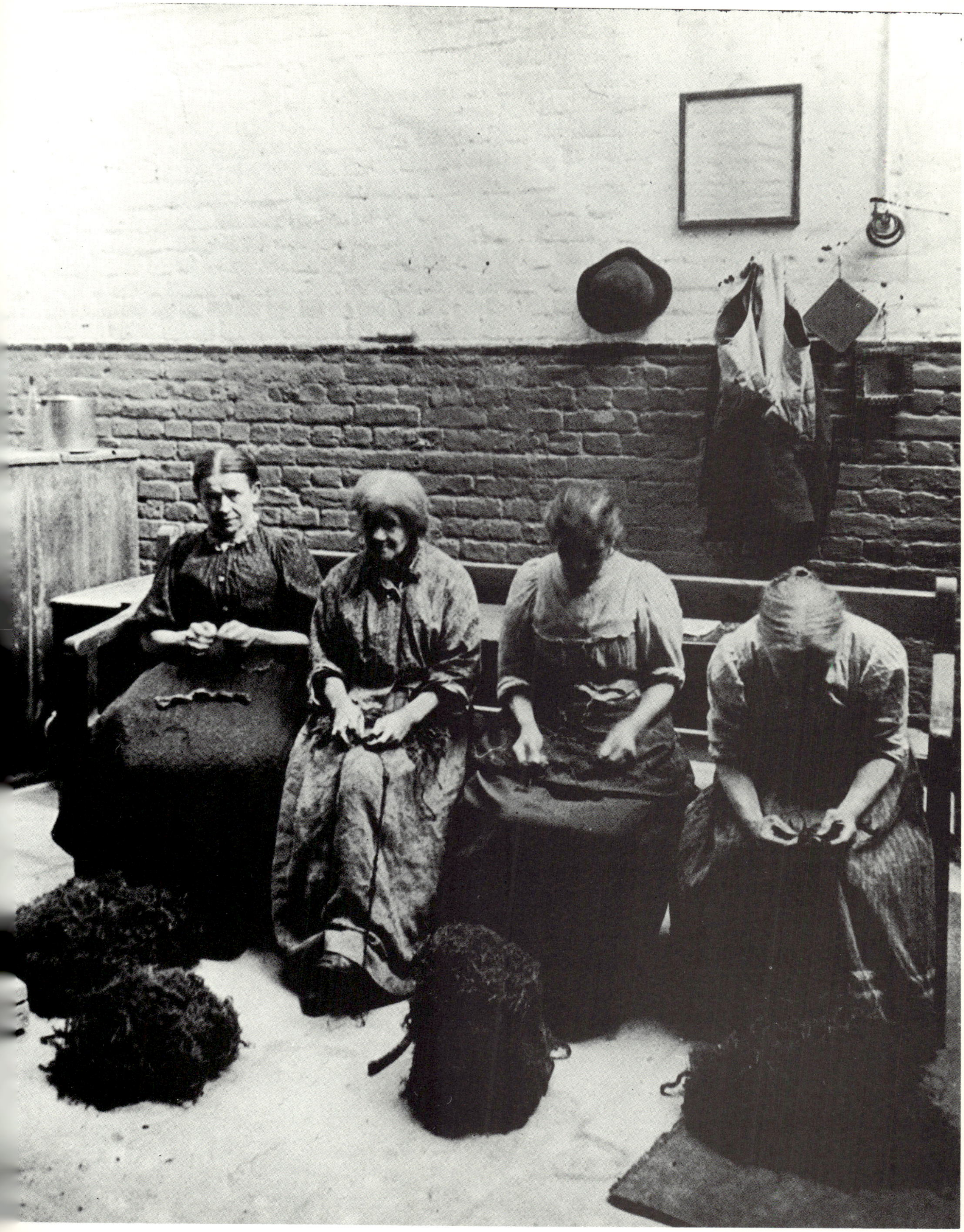

171 *above* Rational amusement: a scene in Sefton Park, c.1885.
Liverpool Public Libraries

172 *above right* Brighton, municipal gardening, c.1865, from a *carte de visite*.
David Francis

173 *right* Public benefaction: laying the foundation stone of the Brown Library, Liverpool, 15 April 1857. An early and striking photograph of a kind that becomes more accomplished and less interesting as time passes and municipal buildings multiply. The figures in front (from the left) are the Bishop of Chester (the Rt Revd Dr John Graham), Mr Brown himself, Dr Thomas Raffles (the leading Nonconformist minister in the town), the redoubtable Henry Picton, and Lord Stanley.
Liverpool Public Libraries

174 *above left* The Brown Library, *c*.1890.
Photo: Frith's
Liverpool Public Libraries

175 *below* Time to stand and stare: demolition on the Strand at Aldwych, when Kingsway was cut, photographed with a Kodak panoramic camera in 1901. The new street was to be called Queensway, but on Edward VII's accession its name was changed.
Kodak Museum

176 *right* The high pitch of civic dignity: Sheffield Town Hall, *c*.1900. Sheffield was an untidy town with no particular architectural tradition, an important manufacturing centre, but not incorporated until 1838. By the end of the century it was a rich and powerful city, and felt that it owed itself a monumental town hall, which it raised in the years 1890–7. The architect was E. W. Mountford.
Photo: G. W. Wilson
Aberdeen University Library

Willing
ROYAL PRINCESS'S
ROSE'S
CROWNED
BIRD'S
BRITISH
CAPITAL & LABOUR
CAPSTAN
CIGARETTES
FOR 3D

177 The twentieth century inaugurated in Manchester: the departure of the first electric trams from Albert Square, 6 June 1901.
Manchester Public Libraries

10 The Face of the Industrial City
Two looks at Manchester

G. F. Chadwick

To look again on the face of the industrial city of Victorian England as its own inhabitants saw it is an impossibility. The elements that compose the scene are too complex for that: complex around the edges where town becomes country, and rural villages and industrial colonies cohere with the larger mass; complex in the growing number of new paths being made through the city by new modes of transport; complex in the additions of new to traditional scales of buildings and spaces; complex in the changing scale of the city as a whole and of its related parts. These complexities were partly a function of new technologies and the economic motives from which they sprang, partly the product of historical inertia and the sheer durability of inherited structures and disposed space.

Until the Industrial Revolution the major structures of most towns were their churches. The medieval town consisted principally of low-built houses—diverse in detail but consistent in height—accented visually by numerous period churches, but above all by the central church, the abbey or minster. This pattern can still be seen from the city walls of York, save where the monstrous bulk of the North Eastern Railway offices and more recent office blocks intrude: a huddle of rectangular red-brown roof planes meeting at many angles, but remarkably consistent in their general level, punctuated by pale white-grey limestone towers and spires of the minor churches, and presided over by the vast bulk of York Minster. In the Victorian industrial town, new economic pursuits reduced such dominants, replacing duomo and campanile by factory and chimney. The process was most pervasive in the textile towns of East Lancashire and the West Riding, where mill chimneys overshot church

© 1973 G. F. Chadwick

spires as the cotton and woollen industries grew and prospered. Multi-storeyed mills were compacted to conserve the energy generated by steam plants in vertical as well as horizontal transmission, to concentrate materials, skills, work people, and production in large volumes of space, either cheek by jowl with rows of workers' houses, or on new, level sites by watercourses or railways. Even today, when so many mill chimneys have been felled, those which are left dominate the skyline, as for example on that classic train ride from Manchester Victoria to Oldham Mumps: scores of cotton-mills riding at anchor on the plain near Failsworth, like the Review of the Grand Fleet of 1914, lighted at night: 'fabulous argosies, as romantic as if they were Venetian, and in the history of the world more important'.[1]

Other places, other shapes: as the metal trades grew, so primary extractive industry grew and its technology improved: pit-head gear, blast-furnace, rolling-mill, stamping-shop, gas-works, railway yards, specialized structures of every kind, even their indestructible debris and discarded apparatus, formed indelible outlines on the ground, with every technical improvement. The Welsh valleys, the Black Country of South Staffordshire, the shipyards of Tyneside or Merseyside, all became marked with characteristic local technologies, with characteristic structures, of a size and scale never seen before. The existence, from now on, of two scales within the industrial city is perhaps the most significant visual change of all: the human, in the continuing tradition of urban housing and life on the streets; and the super-human, not in the tradition of York, the daily reminder of God's omnipotence, but in terms of a temporal power, bursting out of industrial technology from hand- and water-power, to steam-power.

If we are to have any perception of these shapes on the ground we must visit specific spots and get the best focus we can through contemporary recordings of every kind, however fragmentary. In doing so, we should remember that the synthesis will inevitably be ours. We cannot resolve through our own imagination, for example, the nagging question whether the Victorian city was actually easier to grasp perceptually—whether one could get the feel of it more readily—than today's cities. We cannot know, in the terms of these volumes, the full reality of its image. Industrial steam and domestic smoke, factory hooters and horses' hooves, smelly people and stinking waterways, a wholly different catalogue of urban incident, preclude that. But we can for one place at a time get a *visual* focus.

Much of the early-Victorian city went unrecorded in photographs, and lithographs of street scenes were not common, being often related to specific occasions like royal visits. Manchester is no exception, and few photographs earlier than 1860 survive; however, there is the beautifully drawn and exceptionally complete Ordnance Survey of 1850 (surveyed in 1849), from which a very good picture of this city can be gleaned. The maps, to a scale of 5 feet to 1 statute mile (i.e., 1:1056), give a great deal of detailed information about the use of buildings and land, and the ground-floor plans of all public buildings are shown very completely. This survey and the prints and photographs from a range of dates make it possible to reconstruct the landscape as it might be experienced on a journey through the city.

If we begin our journey to the south of the city, in the Higher Ardwick area,* we find a mixture of the still-rural and the suburban. Fields are being tended, but small colonies of villas are being settled and the large grounds of Ardwick Hall are being invaded by new terraces of houses. To the east, mounds of spoil and hollows indicate the active presence of brick crofts, found all around the growing city, which can thus be said literally to rise out of the ground on which it stands. Here too is the junction of the Manchester, Sheffield & Lincolnshire Railway and the London & North-Western Railway, both marked by approach viaducts of dull red brick. Older routes are marked by toll-bars on the main approaches, and here is one guarding the turnpike leading over the Pennines to Sheffield. At Higher Ardwick, near the Green, are houses of different sizes, all with well laid-out gardens. South of it, the villas, in small groups or dispersed in their separate grounds, are mostly stuccoed and in the faintly classical styles which have been popular for some time; occupied by relatively wealthy merchants and professional men, they are in a green and pleasant land of trees and gardens whose arabesque beds would have been approved, no doubt, by the late Mr Loudon. Off the London Road, the group called the Polygon stands out, an arc of nine large houses set back from the long, lamp-lit approach from Stockport Road, behind a series of sweeping ovals of separate drives, and overlooking open country. The private gardens are furnished with summer houses, sundials, and ample water from private pumps and wells to nurture the products of their conservatories. The Polygon is dignified, solid, respectable: the outside world looks up the flights of its front steps to porticoes and solid front doors.

There are cemeteries near by at Ardwick and Rusholme Road, many of the headstones bearing pathetic witness to the high mortality of working-class children, and to the great epidemics of twenty years ago. The entrance to Ardwick Cemetery is flanked by a chapel on one side and the block of the registrar's and sexton's houses on the other, and its simple rectilinear pattern of paths and graves is already bearing testimony to changing taste—exotic stones for memorials, less restraint in feeling, bolder decoration and lettering.

In the developing parish of Chorlton-upon-Medlock the speculative builder is active everywhere. All classes are catered for, from those inhabiting the new, substantial villas along the wide, straight Oxford Road to the dwellers in mean rows of back-to-back houses a few yards behind them. But, as we move inwards towards the city centre, even the larger houses lack front gardens and few have rear ones. In regular terraces, the chequerboard pattern of new building estates south of the River Medlock is filling in: although small, skimpily built, and mean, these monotonous, red-brick rows are an improvement upon the courts and cellar dwellings in the centre of the city. For one thing, they are on the edge of the country, within sight, still, of green fields. Nevertheless, there are shops on corners and on the main thoroughfares,

* All place-names may be identified by reference to the sections of contemporary ordnance maps which have been included among the accompanying plates.

public-houses at frequent intervals, churches less often, and the apparatus of local government. Chapels and churches, indeed, form one of the main visual features of such suburbs, relieving the two-storeyed level of building by imposing stone façades, Classical or Gothic, porticoed, or towered: even the more nondescript Nonconformist chapels make some change in bulk and skyline, if not in materials. The city is still small enough for most of it to be reached on foot in fifteen or twenty minutes from any other part. The main streets, therefore, have their daily processions of people walking to work, with the middle classes patronizing the horse omnibuses or using their own traps or carriages: horses are a common sight, both at work on the streets, in fields or paddocks towards the edges of the city, and standing tethered in front of villas and coach-houses, or on cab-ranks.

If we take Oxford Street towards the centre of the city, we soon find ourselves, on crossing the Medlock, in an area almost completely given over to manufacturing, though pockets of small back-to-back houses and courts exist. This is industrial Manchester, by and large: the Atlas Ironworks, the Caledonia Foundry, cotton-mills, saw-mills. These manufactories are located near the wharves of the Rochdale Canal, which bisects the centre from north-east to south-west and joins the Bridgewater Canal at Castle Field. Eagle Quay and other wharves and basins penetrate deep on either side of Oxford Street, almost to St Peter's Square, once the edge of the urban area, offering large sites with water frontage. Now the water is largely hidden from the street, but there are numerous works entrances through which narrow boats and horse-drays, cranes and hoists, weighing-machines and chimneys, can be seen. Despite the new railways, the trade still passes over these quays—a canal port for the sea-going port of Liverpool, the coalfields of South Lancashire, the salt-fields of Cheshire.

The railways are already beginning to parcel out the city, to put a girdle of brick, stone, and iron around its vitals. By 1836 the Liverpool & Manchester and the Bolton & Manchester Railways have penetrated to within half-a-mile of the Royal Exchange; by 1849 Victoria station and London Road station are established, the former within a quarter of a mile of the Exchange, the latter half-a-mile away at the end of Piccadilly, whilst radial lines spread outwards from the beginnings of extensive yards and terminal facilities. The new railways, being almost exclusively on viaducts, are immediately visible and intrusive, and many houses have been knocked down to make way for them, causing great distress, with housing in short supply. The price Manchester paid for its progress fell heavily on the worst-paid and worst-housed, some of whom made homes of the undercrofts formed by the new viaducts. The new stations are now providing the city with a new class of public buildings, although (with the single exception later of the Central station train-shed) they are uninteresting structures. Around London Road station, to west and south-west, are large areas of crowded, back-to-back houses and narrow streets: minimal labouring-class housing on the slope down to the River Medlock's foul waters. In London Road itself is one of the many street-markets which characterized the inner parts of the city.

Between London Road station and Piccadilly, at this time, is another area of quays and basins: another part of the inland port. Here again is a great focus of industry and commerce, with bustling horse-wagons and boats arriving and leaving. A portent for the future, though, is the Manchester, Sheffield & Lincolnshire Railway's General Stores which have already taken up occupation and will eventually swallow up the entire Stockport basin and fill it in as a site for railway warehouses.

Piccadilly itself, in 1849, is a peripheral-centre area, not the major focus that it becomes much later. The Manchester Royal Infirmary and Lunatic Asylum, a solid, porticoed stone edifice, occupies what was later to be Piccadilly Gardens, flanked by the Infirmary Baths in George Street to the west. The pond in front of the Infirmary (a remnant, like the pond on Ardwick Green, of brick-pits?) is not incorporated into a formal promenade at this time, although for the Queen's visit in 1851 'The Manchester Fountains' were to be arranged there by Freeman Roe, Hydraulic Engineer. North of Piccadilly at this time is a great mixture: industry (including the very large Newton Street Cotton Mills near Stevenson Square), livery stables, chapels, public-houses, horse and carriage repositories, and private dwellings. Most buildings are rather domestic in scale, save for the cotton-mills, and many hotels line Piccadilly itself. South of here, towards Oxford Street, is another multiplicity of buildings in what is later to become the predominantly warehouse area of Portland Street—great Renaissance palaces to replace the mixture of back-to-back and terraced housing, courtyards, public-houses, schools, timber-yards, and mills.

The changes in this area of Manchester, north and south of Portland Street, are to be of great significance for the city: functionally, in the displacement of mixed uses, including manufacture, by warehousing and commerce; visually, in that many of the city's finest nineteenth-century buildings (including those shown in Ernst's 1857 map) are to arise here. Over a century later, however, some domestic structures still remain (now used as offices), and the basic street-pattern is unaltered.

The 'Main Street' of Manchester in 1849, as now, is Market Street, running from the River Irwell to Piccadilly, before falling again south-eastwards to the Medlock. The chief landmark in the city, by virtue of function and position rather than bulk, is the Exchange, one link in a sequence of buildings and open places which makes up the historic core of the city. Here are the remnants of a medieval street and building pattern: Cathedral, Market Place, Exchange, St Ann's Square, St Ann's Church, King Street, (Old) Town Hall. In 1849, of course, these replicate Renaissance ideas, but Long Millgate, Hanging Ditch, Withy Grove, and other streets around the Cathedral, with their many timber-framed houses, are indissolubly medieval. Thus around the Chetham Hospital and the Cathedral are courts and yards, long, narrow plots running back from street frontages, short vistas along curving streets, low 'black and white' buildings. There are many markets: the covered Victoria Market for meat; the Fish Market in the Old Shambles; the Flower, Fruit, Vegetable and Poultry Market in the Market Place; and Smithy Door Market for fruit and vegetables in Victoria Street. Crowded stalls stand on cobbled or sett-paved streets; numerous lamps on posts and stalls light the scene at night.

The Victorian City: Shapes on the Ground

There are all kinds of juxtapositions in the city centre. We might well expect the Electric Telegraph Company's office to adjoin the Exchange, but be surprised that the nearby Cotton Finishing Works should so obviously point the moral of Manchester's main concern. The view along the Irwell, from one of its bridges, shows the Cathedral beset by at least half-a-dozen mill chimneys. The Old Town Hall and Cross Street Chapel are also cheek by jowl, with the Bank of England opposite, for good measure. The new commercial and public architecture is solid, even scholarly, in contrast to the sordid gimcrack aspect of the housing sandwiched in everywhere, between commerce and industry, inside the ring of railway viaducts. Little of what is to happen later is yet in evidence. Albert Square has not been contemplated, nor Waterhouse's Flemish-Gothic Town Hall, which is to rise twenty-five years later on the site of the Town's Yard, off Princess Street (the Yard's boundaries survive in the triangular site plan of the present building); and the Cross Street–Corporation Street route has not yet been pushed through. The unequal contest between living space and working space, which underlies the continuous process of rebuilding cities in a state of growth, is not yet evident. Apart from houses in the path of the railway, very little is being torn down to make room for something else. Manchester, it might almost be said, is still a coral island, growing largely by accretion, cell upon cell, haphazard, irregular, but terribly fast.

We now move forward fifty years. Manchester in the 1890s was photographed mainly in its central parts. Once again the Ordnance Survey is a great help, the 1:500 series of 1891 being most detailed and complete, though the very cartography speaks of a changing attitude, it would seem, since the series of 1849: a certain hardness and boldness of line, less elegance of lettering—but a profusion of accurate-looking detail.

The city now is much bigger, of course: it has spread inexorably outwards, especially along the railway lines. The diameter of the city has doubled, and so in the nineties it is fourfold the city of the fifties in spatial extent, whilst the population has increased by more than 300,000 in the same period. Manchester is no longer a city which can be covered on foot: transport has become more organized as the horse bus has given way to the street tramway, and suburban railway stations have begun to come into their own. The built-up area has yet to reach the flood plain of the Mersey in the south, for example, and many hamlets that are later to lose their geographical distinctness and become mere names in an urban mass are still identifiable, protected by open fields.

On the south, the city extends to Alexandra Park, newly laid out from farm land in 1867. To the south-east, it now engulfs Victoria Park, long isolated as a rural suburb; and there is an urban outlier at Birch, near Platt Hall. Eastwards, beyond the railway web, housing now creeps along the Hyde and Ashton Roads, to Beswick, and northwards towards Philips Park. To the north-east, Miles Platting and Harpurhey are unattached no longer, and a straggle of development goes northwards to Cheetham Hill. On this side of the city, buildings have now also sprung up along the Bury New Road to Broughton, and Salford, too, reaches into the great loop of the Irwell, and on its south bank north-westwards to Pendleton. West Salford reaches to Cross Lane, but Weaste and Eccles are still rural. To the south-west, Old Trafford is barely reached, whilst Trafford Park still has its deer and awaits the coming of the Manchester Ship Canal, and its later transformation into an industrial estate; Chorlton-cum-Hardy remains relatively remote.

Following again our journey of half a century earlier, we find Ardwick Green now laid out as a formal garden, with fountains in basins at either end, a ceremonial flagstaff, pavilions, and a urinal. There are tramlines now along the main roads to Hyde and Stockport, and a little more intensiveness in building. Ardwick Green is now more clearly a focus of local community services: Osborne House Institute (Young Ladies' Association) graces the north side of the Green next to Ardwick Town Hall. On its other side is the Industrial School, whilst nearby are the Post Office, Provident Dispensary, Nurses' Home, and St Thomas's Church. Such groups, such foci, are now to be found all around the parent city, representing an advance upon the formlessness of earlier dispositions of broadly similar facilities. At an equal distance north of Manchester, we find a similar situation at Cheetham Hill, with its fine Assembly Rooms, or—even more notably, because of its layout—the Chorlton complex around Grosvenor Square and All Saints' Church: Classical portico, Italianate twin towers and Corinthian columns, Gothic spire, Victorian-Georgian domestic façades. Even at Ardwick Green, though, there are still odd plots of land left vacant in the process of urbanization. The small, cramped courts of the 1830s are still there, the spectrum running through small back-to-back, tunnel-back, courtyard, terrace, substantial terrace, or semi-detached houses, in a haphazard and spasmodic arrangement.

The Polygon, to the south of Ardwick Green, is still large, spacious, and open— but now the Union Carriage Works seems perilously near, and the Baths behind Ardwick House have promoted a chimney which flagrantly overtops the suburban trees of this villa enclave. North and west of the Polygon are the very different ranks of terraced houses. These streets are exceedingly regular, with narrow back alleyways, tiny backyards, outdoor privies, and ashpits. Occasionally we meet a wider street—Brunswick Street, for example, running across the main south and southwest radial streets—but all are now paved, drained, lighted, and populated uniformly. The street-corners tend to be taken by churches and schools, counterpointing the red terraces and lopped forests of pale earthenware chimney-pots: St Paul's Church, with its school and segregated playgrounds and rectory; a Baptist Chapel (Particular); or the Balfour Club. (An alternative corner-marker, of course, is the public-house: the Salutation Inn—still standing in 1970—or the Wellington, both south of Cavendish Street.) In this area, too, is a wider spectrum of entertainment, industry, or out-of-worktime activity: a drill hall, a saw-mill, the Bridgewater Music Hall, the Ormond Hall (City Mission)—a further tier of community provision, mainly private rather than public, ranged behind the main grouping of Grosvenor

Square at a little distance. This area, developed well before 1835 in the main, has small, often short, terraces of tiny houses without gardens, whilst court-housing is quite common. Steps often lead up to front doors, so that there are basements, and coal-holes in the pavement, but local flavour may appear here and there in such things as street names, which appear occasionally carved in Pennine gritstone, like pathetic little Mahogany Street, off Oxford Road, only demolished in 1969 after about 140 years of mean life. Oxford Road itself, in the nineties, is a main artery, sett-paved, broad, and tramwayed. (It is interesting to speculate how much, in fact, of a late-Victorian city like Manchester was covered over by hard pavings or buildings or railway lines and other structures: little of the original ground surface can have been left uncovered, and streams like the Cornbrook were mercilessly culverted.) The shopping function of Oxford Road must have been considerable, complementing the many corner shops inside each housing area. Southwards along Oxford Road, the Gothic bulk of Waterhouse's Owens College is dominant on the right, its Flemish-roofed tower answering the calm, majestic statement of Holy Name Church a little further along to the left. The university atmosphere is limited as yet: there is no Whitworth Hall, though there is a Gymnasium, whilst a cab-rank and cabmen's shelter occupy the roadside. A suburban feeling is uppermost here: an increasing number of larger houses, like Arlington Place opposite Owens; broad approaches to substantial, stuccoed Classical-feeling villas; some of these are detached, some semi-detached, some large houses even in terraces, around Holy Name and Lime Grove. There is a High School on Dover Street, destined to survive as a university faculty building. The pattern continues southwards to Whitworth Park, and beyond to Victoria Park; more gardens to the houses, but a very mixed arrangement for most classes except the poorest, though the content of each area changes perceptibly as one moves away from the main street into quieter byways. There is, of course, still Victoria Park itself, its landscape maturing, its villas still protected by gate and lodge from the world without, but its atmosphere now sullied by the pall of urban smoke that cannot be dispersed.

Retracing our steps and moving north, rather than south, towards the city centre, we find that Oxford Street still has a partly industrial face, though commerce and other activities intrude increasingly. The Atlas Ironworks and the Eagle Foundry are still there, and the Eagle Mill (Smallwares), fronting or visible from Oxford Street; the canal wharves are still in use and teams of horses still draw ironshod carts across the granite carriageways; mills and timber-yards still flourish by the Rochdale Canal. Interspersed with the wharves and their travelling cranes are clusters of small houses: even the centre of the city still boasts a large resident population. But change is in the air already: northwards, major monuments like the Free Trade Hall in Peter Street have risen, flanked by the Theatre Royal, Y.M.C.A., Comedy Theatre, and Café Royal (all still identifiable in 1970). The Central station's great arched roof has been built (though not its attendant offices, which entirely failed to materialize); and the Cheshire Lines Railway goods station, of great bulk, raises its sombre brick walls behind the Florentine splendour of the Free Trade Hall.

The Midland Hotel has not yet arrived, but the St Peter's Square area, already distinguished in the image of the city by its intersection of streets at the front of the Church—a *tour de force* of civic design of long standing—is reinforced by these new additions, and, of course, by Waterhouse's masterpiece, the Town Hall, rising behind, its tower a majestic assertion of civic pride and financial irresponsibility. Albert Square is not fully laid out even in 1891, its ornaments then being the Albert Memorial, Bishop Fraser's statue, a urinal, and diagonal tramtracks.

The streets leading down to Deansgate from Albert Square are as mixed in their buildings and uses at this time as anywhere in the city: courts of small houses, pubs, clubs, and all manner of industry and public endeavour. For industry, there are the City Printing Works, the Albert Packing House, and the Gas Meter Testing Office; whilst the St Anne's Schools (Girls and Infants), the Fire Station, the Grand Circus, Manchester Technical School, St Mary's Chapel (Roman Catholic), and the High Court of Justice (Probate Division) represent different forms of public endeavour. But the subdued screen of warehouses and offices lining Albert Square itself must have produced an impression of greater uniformity, quite at odds with the circumstances behind them.

Passing from St Peter's Square again, Piccadilly-wards, we see many changes in the area south-eastwards, between Mosley and Portland Streets. Here are many of the newer, purpose-built warehouses for King Cotton: substantial, regular in plan, of masonry outside and framed structure inside, many of them with glazed internal courts and light-wells. Such structures are new, but some of the earlier buildings, like Watts's warehouse in Portland Street, remain the grandest of them all. Despite this, many of these façades will appear in the *Builder*'s pages, and be approved of generally, adding their little touches to an area distinguished already by Sir Charles Barry's Art Gallery, the Portico Library, and others portrayed by Mr Mennie's pencil in the frieze to Ernst's plan of the city in 1857.

In Piccadilly itself, the Hospital has gone, only the Dispensary for Outpatients remaining in its grounds. The formal layout of today has been completed along the north frontage, with the Queen's statue flanked by Dalton and Watt and Peel, and underground conveniences, but much of Parker Street and Piccadilly is given over to the tramcar. There are many hotels here and along Market Street or towards London Road: Mosley, White Bear, Albion, Royal; the Electric Hotel on Swan Court, off Market Street, particularly arouses the curiosity. Towards London Road station, the major hotel blocks complement the bulk and richness of the warehouses further west: the Grand, the Queen's, with lesser but loyally-named fellows nearer the station: Brunswick, Clarence, Waterloo, Imperial. Behind these, the Venetian-Gothic tower of the Police Courts in Minshull Street provides an unexpectedly rich vertical thrust to the dominantly horizontal canyon-like quality of most streets in the area (and a landmark from the railway). There are chapels, Sunday schools, and cotton-mills on the way across the Rochdale Canal back to London Road and its station, still jointly owned, but now rebuilt in a totally undistinguished fashion, and flanked to its north by the London & North-Western Railway's goods shed and the vast

The Victorian City: Shapes on the Ground

Manchester, Sheffield & Lincolnshire Railway's Goods station near the Dale Street wharves of the Rochdale Canal.

Retracing our steps to Piccadilly, and down Market Street we might now let the photographs take over the story in greater detail—here in Market Street is the focus for the shopper newly arrived by tramcar or train from any point in south-east Lancashire; for the businessman at the Exchange lower down, or the warehouse near Piccadilly; for the roisterer in hotel, music-hall, or public-house near by. Here is the High Victorian City *par excellence*, hard and gritty, even proud, externally, but warm and human within.

Note

1   Keith Dewhurst, 'When Cotton was King', *Guardian*, 9 December 1967.

178 River Irwell, 1859: the Victoria Bridge, as seen from Blackfriars Bridge. The Cathedral is the only non-industrial building in sight. Industry hereabouts turns its back to the river and to neighbouring Salford. The stream itself is thickly polluted but is a traffic artery none the less. A small steam packet is moored between the wooden jetties on the right.
Photo: George Grundy
Manchester Public Libraries

179 *above* Victoria Fish Market in Victoria Street, 1860. The classical style at its most elemental for a public building meeting a basic need. The cast-iron grilles with their daisy-chain pattern admit the necessary breath of air; skinned rabbits hang outside; a restless knot of people appear as ghosts. The haberdasher's shop adjoining is a mongrel: Norman, Gothic, Byzantine, and Classical influences.
*Photo: George Wardley
Manchester Public Libraries*

180 *left* The Wellington Inn, Old Shambles, 1865. A sixteenth-century building made grotesque by Victorian commercial zeal and the Florentine grace of the new watchmakers' premises next door. The giant spectacles enable us to read the date when the inn gave up its top floor—1802. A man is sitting on the pavement. This ancient place is a hub of impending change. It survives still.
Photo: George Wardley
Manchester Public Libraries

181 *above* The Vintners' Arms, Smithy Door, 1865. A survival from the middle ages. The shuttered shops and the Three Tuns pub on the right are soon to disappear though the wine and brandy stores will hang on.
Photo: George Wardley
Manchester Public Libraries

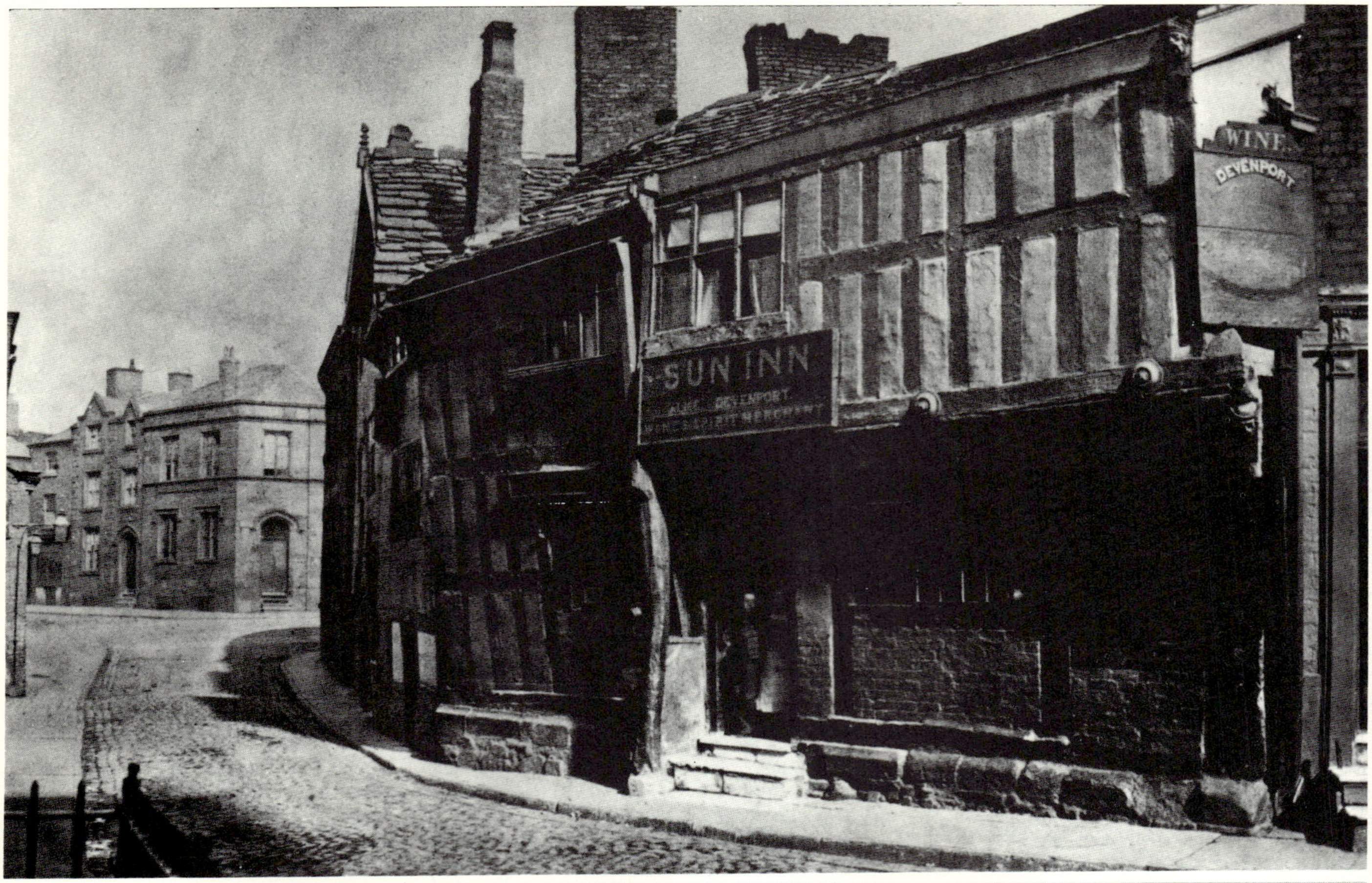
WINE
DEVENPORT
SUN INN

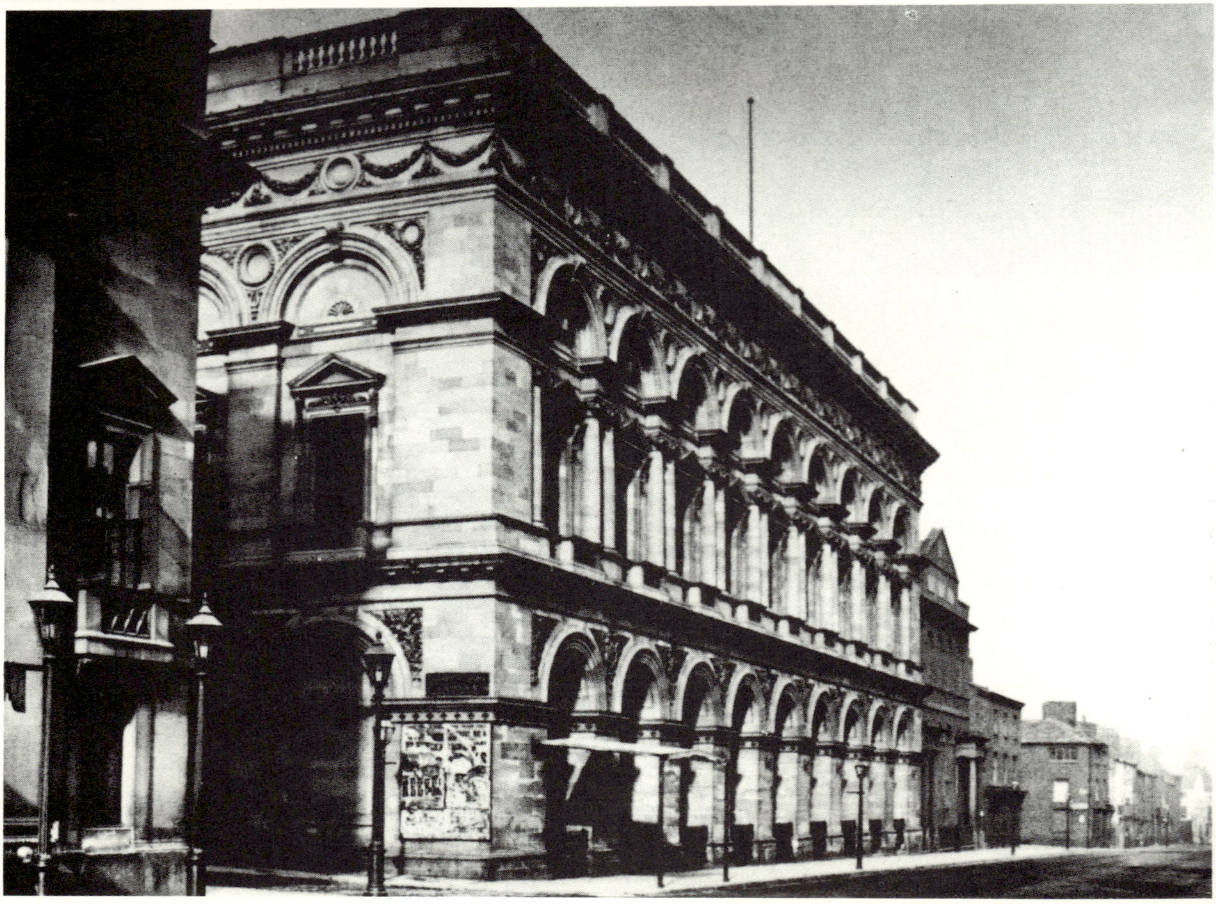

182 *above left* The Sun Inn, Long Millgate, 1866. Known locally as 'Poets Corner', this was the meeting-place of the group described in Chapter 31.
Photo: George Wardley
Manchester Public Libraries

183 *left* Free Trade Hall, Peter Street, 1861. A handsome symbol of Manchester's commercial ethos that overwhelms the lesser property downhill. The new Free Trade Hall, erected on the site of the original building of 1843, was opened in 1856. It was the forum for political agitation, temperance demonstrations, and religious revivalist meetings, rather than musical festivals—the great meeting-place of Manchester designed by Manchester's leading architect of the day, Edward Walters. It contained a pair of public halls and a foyer. In the spandrels between the nine arches are the arms of neighbouring towns—Liverpool, Rochdale, Oldham, Bolton, Stockport, Ashton-under-Lyne, and Wigan, grouped around Manchester and Salford in the centre. Free trade, the four continents, industry, commerce, agriculture, and the arts are garlanded above them. See also no. 84.
Photo: James Mudd
Manchester Public Libraries

184 *above* Newall's Buildings, Cross Street and Market Street, 1867. Boarded up and awaiting demolition for the new Royal Exchange.
Photo: James Mudd
Manchester Public Libraries

185 *above* Cotton warehouses, Portland Street, 1858. Such buildings had an almost swaggering confidence and completely dwarfed any people on the street. The more princely of this pair is Watts's warehouse, perhaps the finest building in Manchester at this time. Each of the four main storeys is in a different style, topped off by a prodigious attic. See also no. 327.
Howarth Loomes

186 *above* King Street, 1866. A good-class shopping street barricaded for the night.
Manchester Public Libraries

187 *right* Bank of England branch, King Street, 1866. A doorless neo-Classical façade with a daunting array of ironwork and blinkered windows, designed by C. R. Cockerell in 1845. The entrance is at the side.
Manchester Public Libraries

188 *above* Royal Exchange, just before being demolished in 1868–9. It was founded in 1806 and replaced an earlier building demolished in 1790. A building of compact power by Thomas Harrison.
Manchester Public Libraries

189 *above right* Queen's Hotel, *c*.1866. A recently completed stucco building by Edward Walters, a symptom of commercial coming and going on a growing scale.
Manchester Public Libraries

190 *below right* Albert Memorial, 1866, by Thomas Worthington. A rather lonely outrider of the Gothic revival, placed in 1862 facing the site of the new town hall. The very similar design of the London monument by Scott was executed fifteen months later.
Manchester Public Libraries

191 *below far right* Manchester & Salford (later Williams Deacon's) Bank, on the corner of Marble Street and Mosley Street, 1866. The last of Walters's buildings, built six years earlier, and a rich celebration of Manchester's natural identification with the Renaissance, full of subtlety and power.

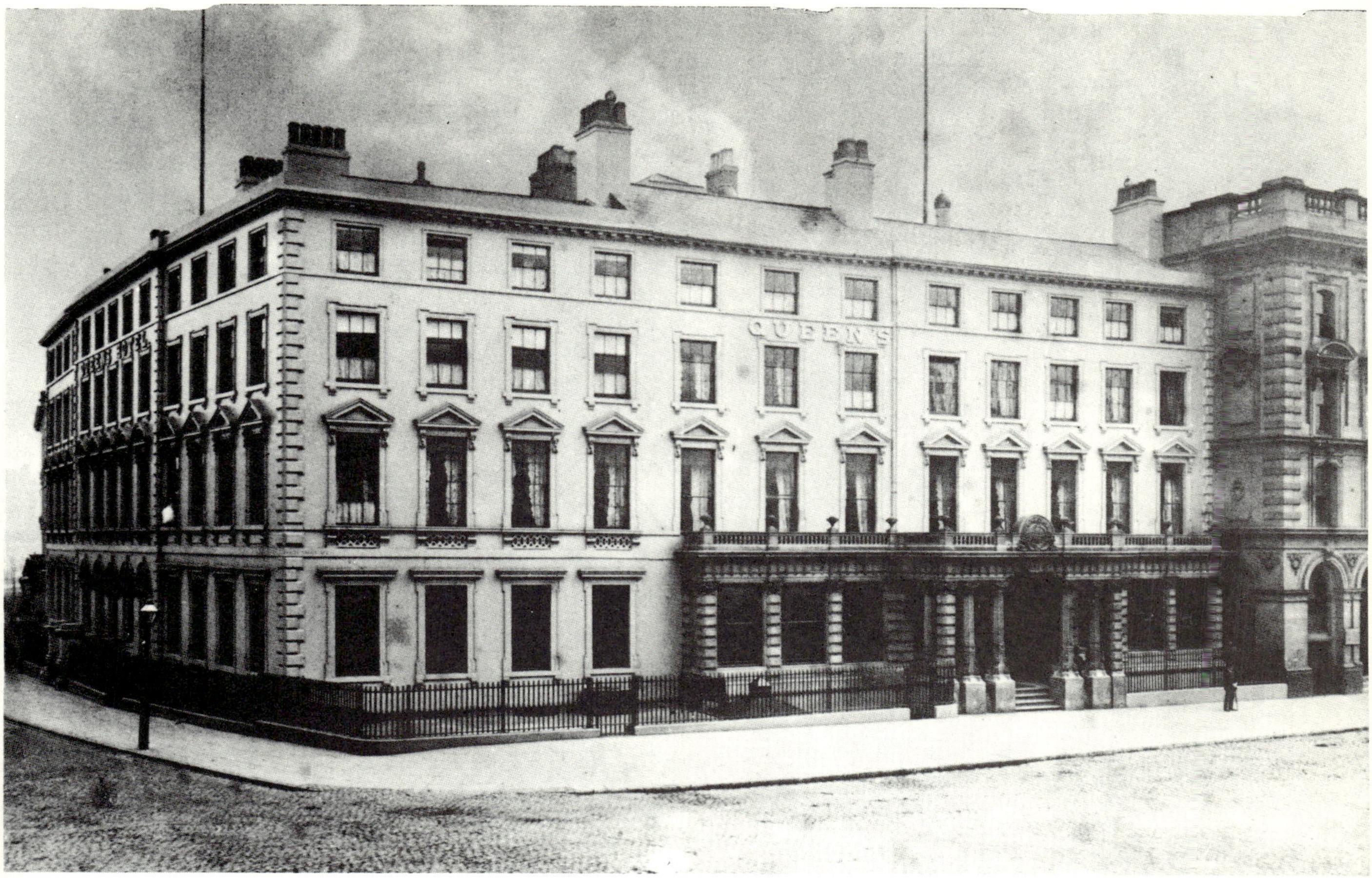
QUEEN'S

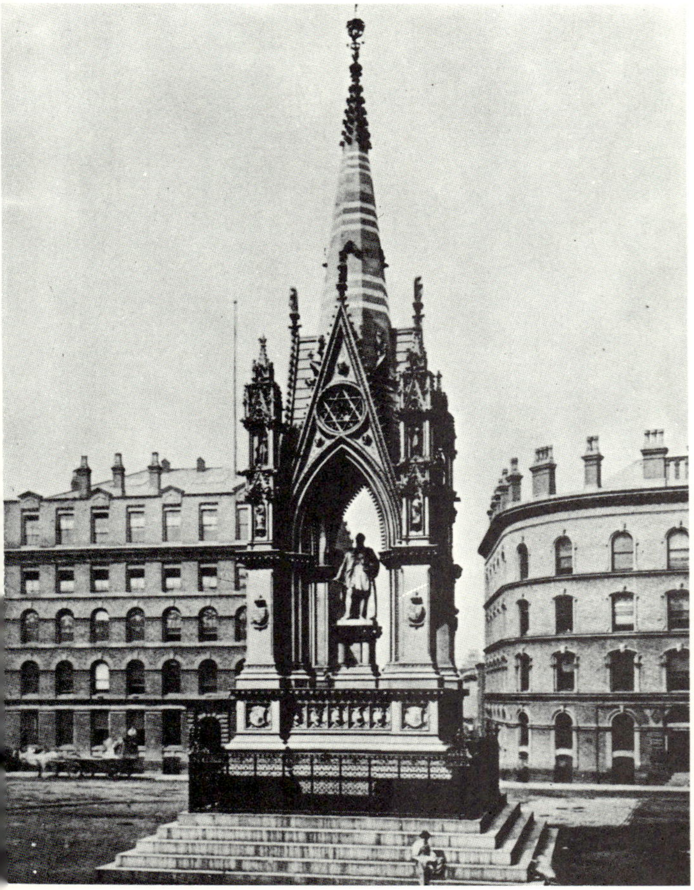

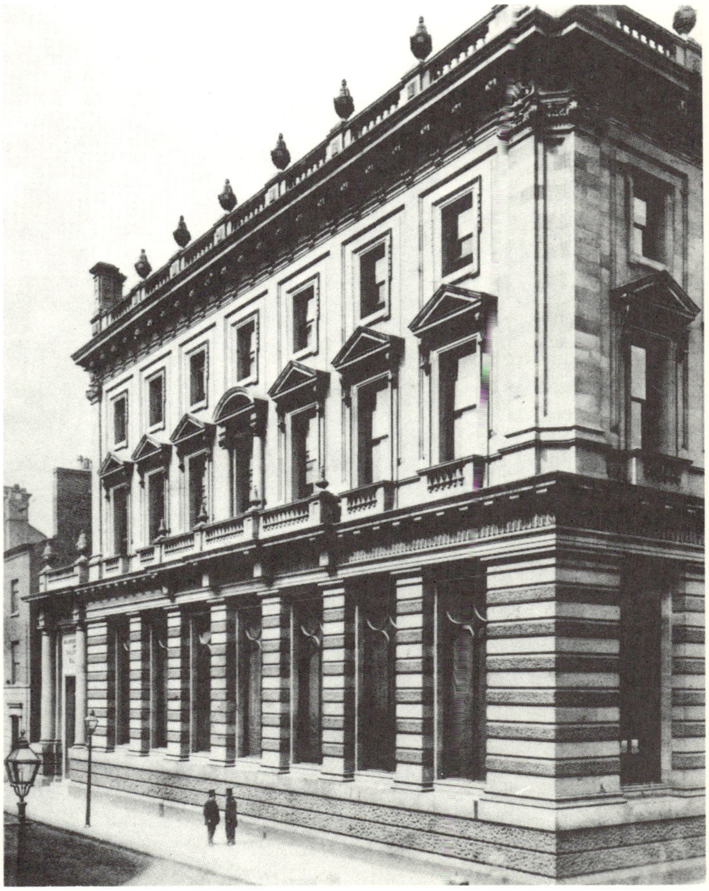

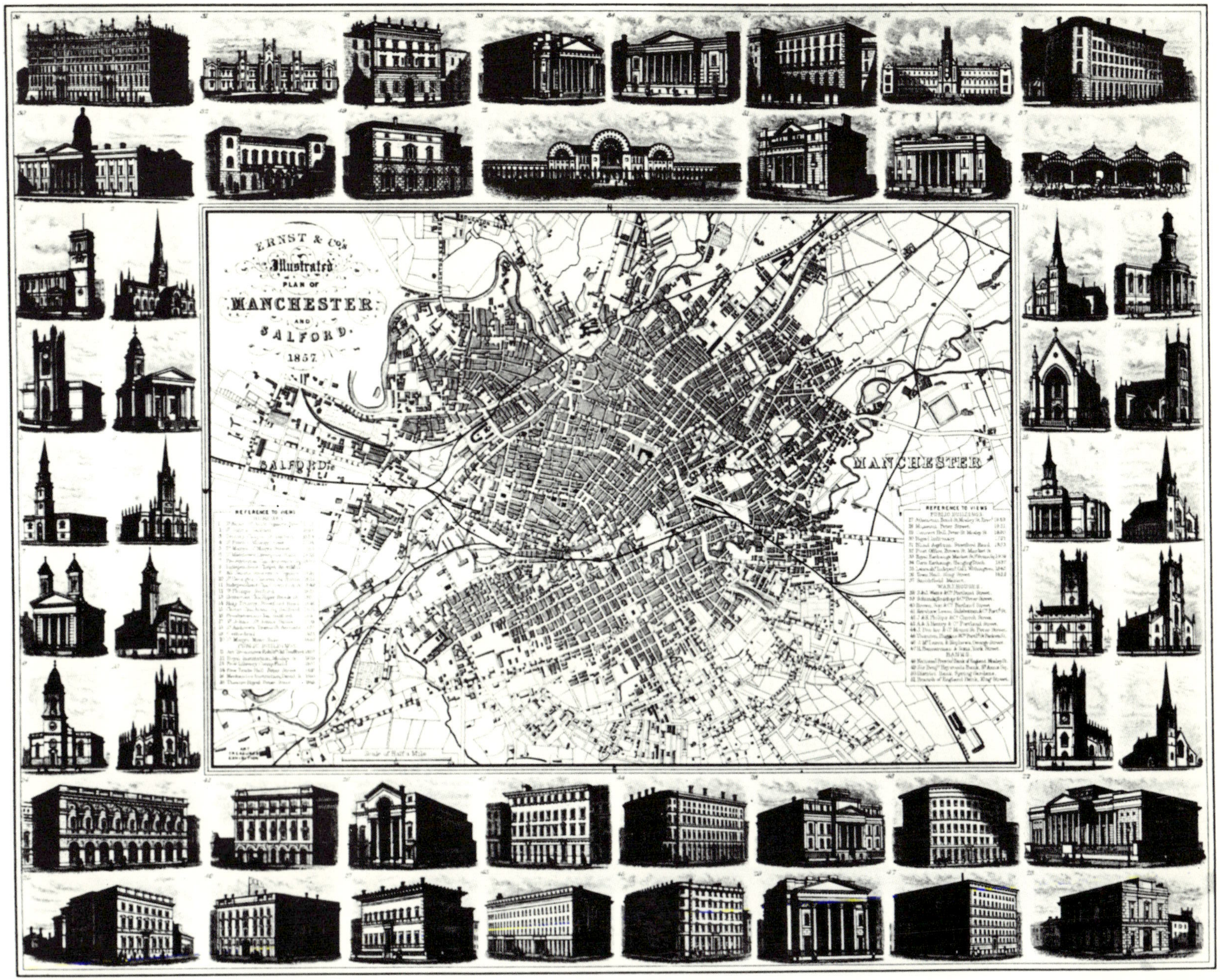
ERNST & Co's
Illustrated
PLAN OF
MANCHESTER
AND
SALFORD
1857
SALFORD
MANCHESTER
REFERENCE TO VIEWS

192 *above* Ernst & Co.'s *Illustrated Plan of Manchester and Salford*, 1857. The changing scale of the city centre is suggested by the frieze to this plan. The buildings differ stylistically but there is a feeling of solid uniformity about the way they are presented.

193 *below* The street plan of the High Victorian city: the central area around 1900. Seven railways converge on the city centre but do not pierce it; three rivers and four canals surround it; four major passenger stations and depots lie within a radius of just over two miles. The ancient centre of the city lay around the cathedral and the main axis of growth ran south and south-east. The more exclusive residential districts tended to be on this side, while those upwind of the gasworks along the Rochdale Road remained in possession of the working classes. For more detailed maps of parts of this area see nos 229–34.

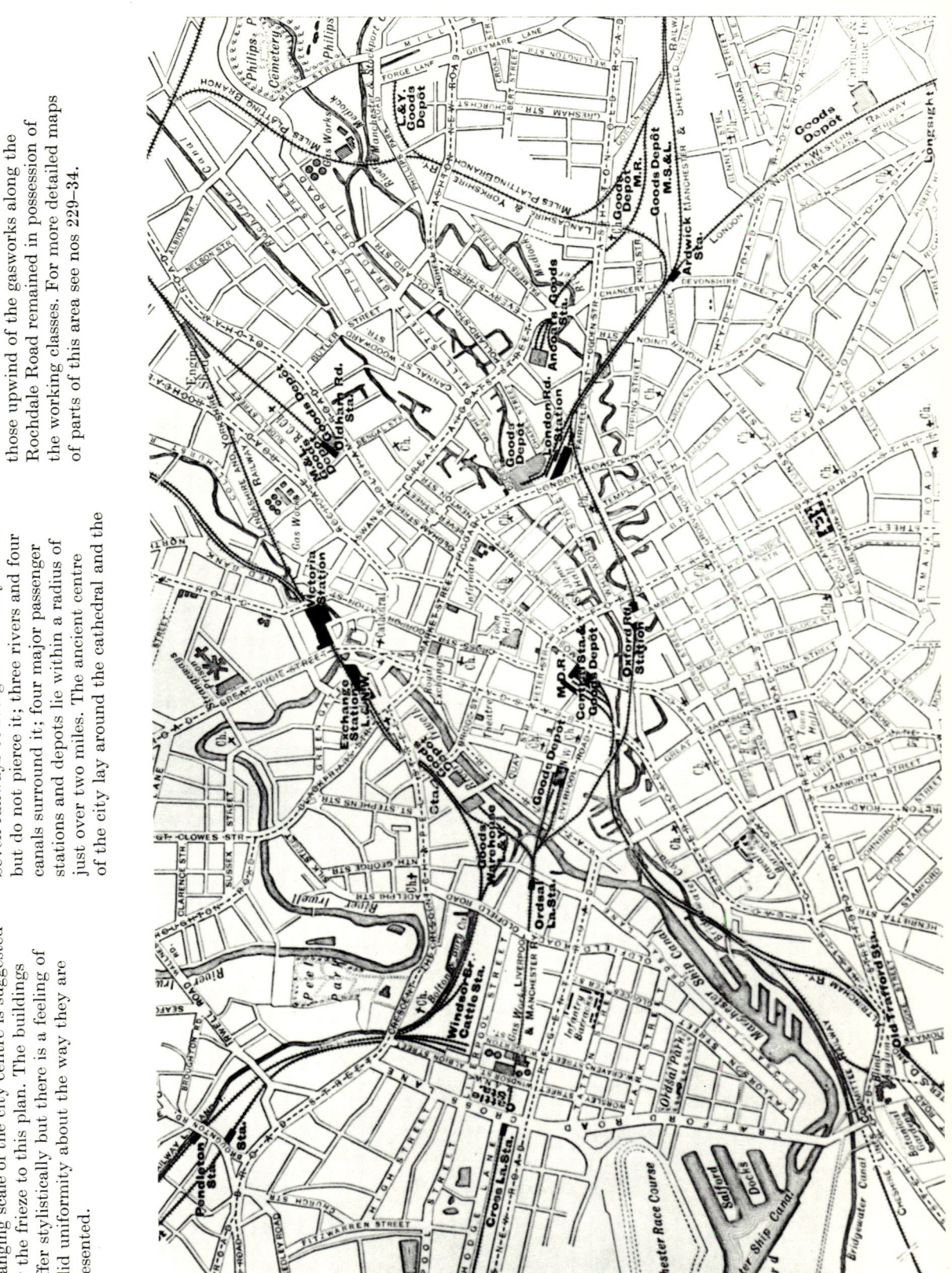

194 The Bridgewater Canal with the viaduct of the Manchester, South Junction & Altrincham Railway running alongside in the Knott Mill area. All the characteristic apparatus and signs of heavy wear and tear by industrial traffic. *Manchester Public Libraries*

195 River Irwell flowing under Regent Road. The last building in Salford on the left of the bridge is a brewery; the Hulme district of Manchester is on the right. This is part of the secret landscape of Manchester: its rivers functioned like sewers—out of sight and very largely out of mind.
Manchester Public Libraries

196 *below* London Road station, a place for loitering as well as dispatch. The rise to the station forecourt is steeper than it appears. A hotel surmounts the booking-hall.
Manchester Public Libraries

197 *above right* The southern approaches: Grosvenor Square on Oxford Road. The floor of the city was composed of acre upon acre of Penmaenmawr granite or Pennine gritstone.
Manchester Public Libraries

198 *below right* The London & North-Western Railway's Exchange Station. It overlooks the most ancient spot in Manchester. The Cathedral (so named in 1847) once stood on a mound sloping steeply to the Irwell, across which the station approach has since been thrown. The view of road and river on different levels below it is one of the grand sights of Victorian England, for the road into Salford is also concealed here. Cromwell did not arrive on the scene in the early seventies without a sectarian fight. The Greengate slum, which occupied much of this ground before the 1880s, did not go unresistingly.
Manchester Public Libraries

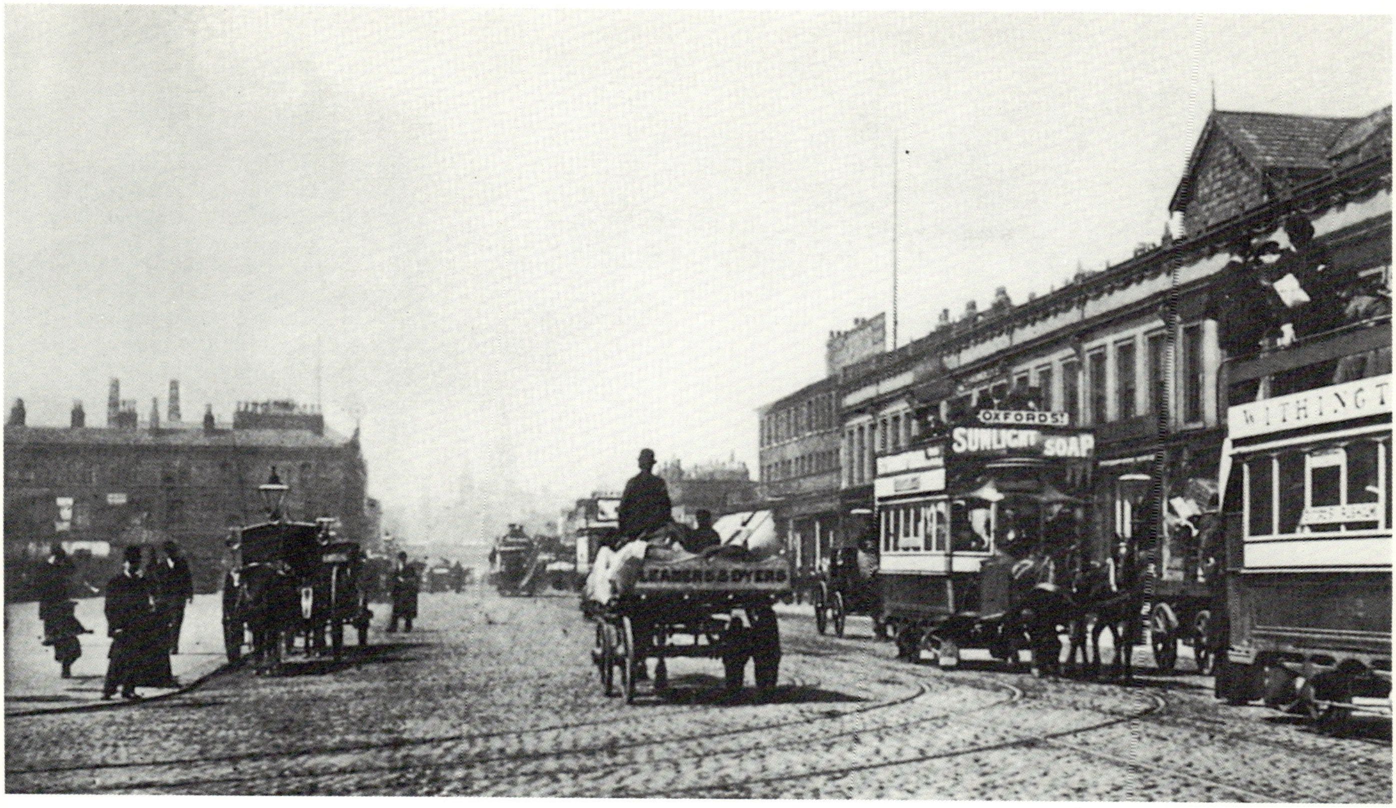
OXFORD ST
SUNLIGHT SOAP

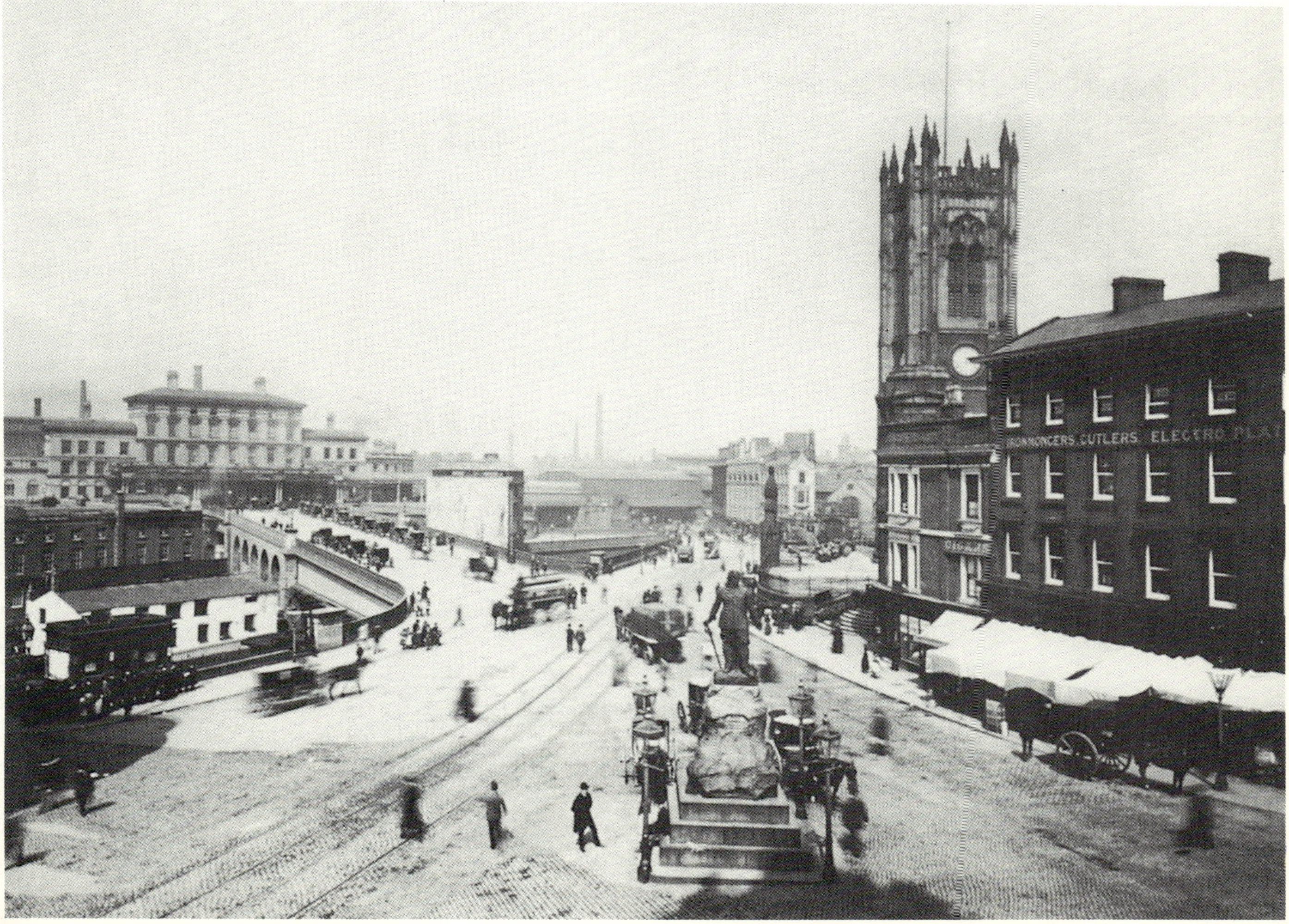
CUTLERS

199 Market Street, from Piccadilly looking west, the aorta of Manchester. It is early morning and the shops are still shut. The tallest buildings in the city line this street and comprise a veneer of commercialism uniformly undistinguished and graceless, except for the slender stems of the gaslamps (which have had to be protected from the traffic by wooden gaiters).
Manchester Public Libraries

200 *below* Market Street in full use, accumulating litter and loiterers.
Manchester Public Libraries

201 Market Street, looking towards Piccadilly: gritstone pavements, granite kerbs, setts laid at right-angles to the movement of traffic, ornate gaslamps, everything made to last.
Manchester Public Libraries

202 *below* The Town Hall, c.1885: Alfred Waterhouse's masterpiece erected between 1868 and 1877, in which economy and light were more telling in the public competition than in the elevation. Its outside walls blackened rapidly but the interior remained an archive of municipal images and architectural taste.
Photo: Valentine
Manchester Public Libraries

203 St Ann's Square, 1894. Here was the apotheosis of rich, comfortable, fashionable Manchester behind the Exchange. Richard Cobden is looking the other way.
Manchester Public Libraries

204 *below* Smithy Door market, *c.*1885. Places that have developed over many generations tend to have more human eccentricity than those laid out on a large scale. Manchester contained many irregular clusters of streets in the central areas, and many different aspects. This view is looking in the reverse direction to no. 205.
Photo: Valentine
Manchester Public Libraries

205 Smithy Door, 1875. The Royal Exchange is still being completed; the Vintners' Arms is being nibbled away.
*Photo: George Wardley
Manchester Public Libraries*

206 *below* The Shambles, c.1885. Twenty years on from no. 180.
*Photo: Valentine
Manchester Public Libraries*

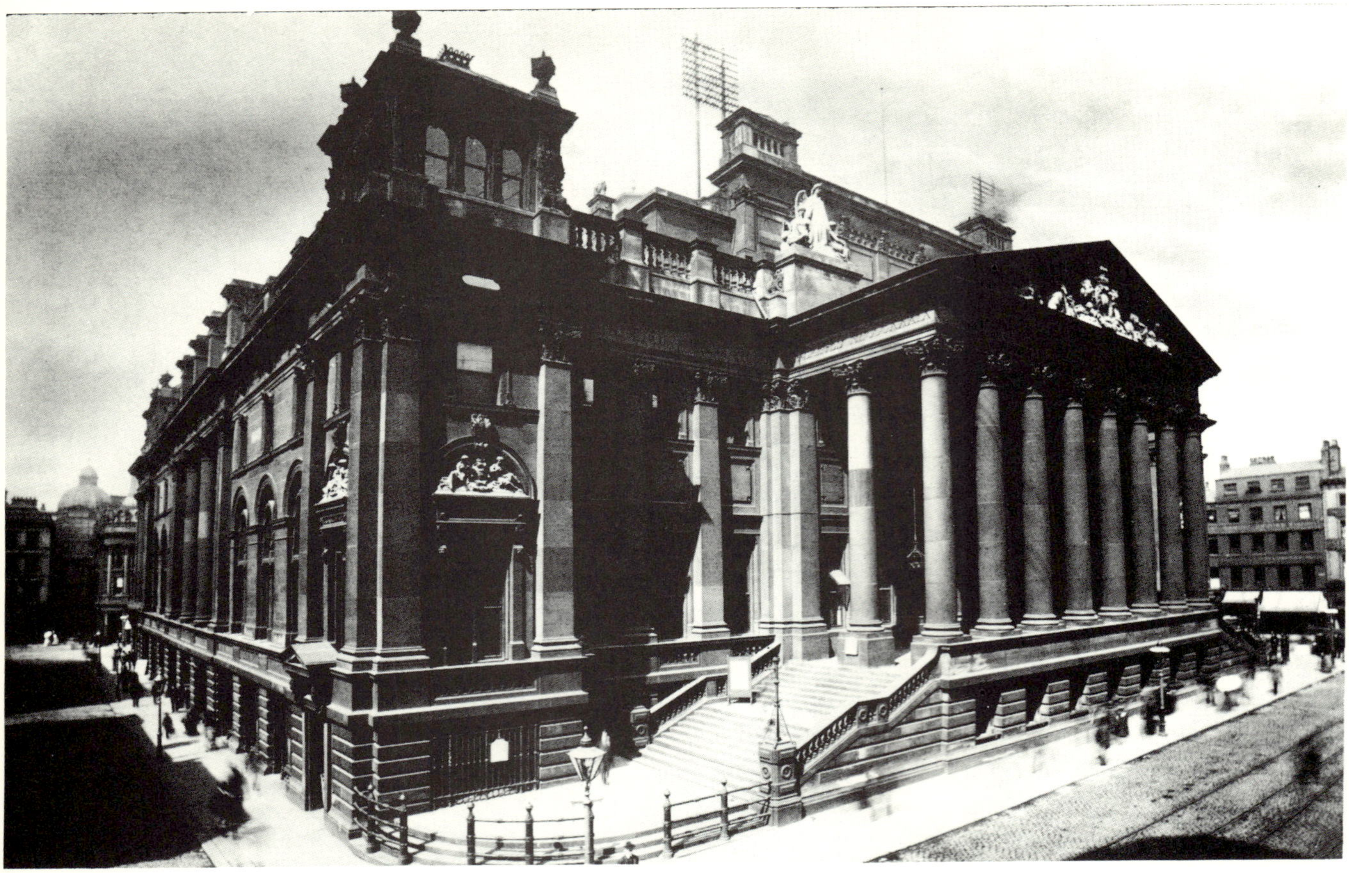

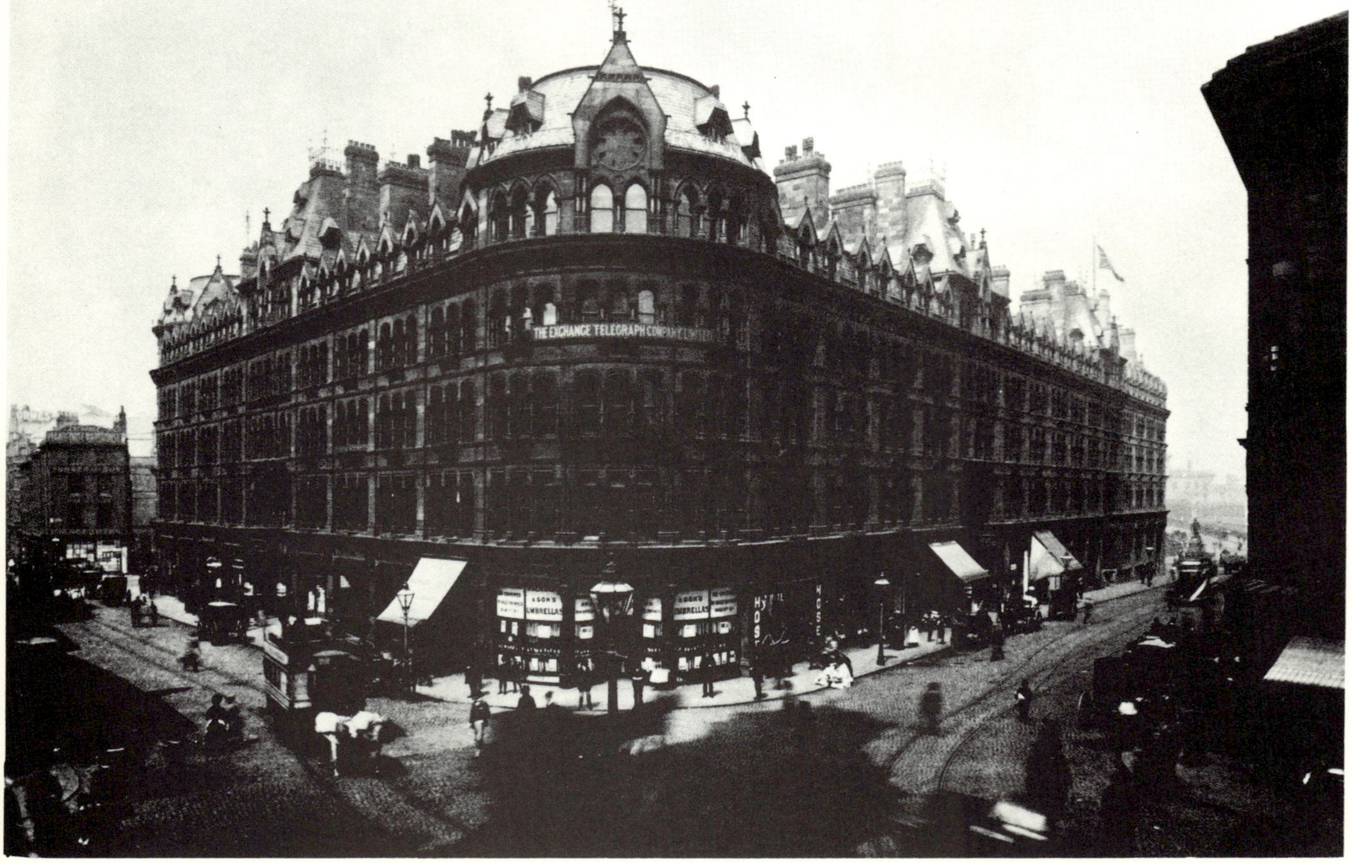
THE EXCHANGE TELEGRAPH
UMBRELLAS
UMBRELLAS

207 *left* The Royal Exchange, Market Street, c.1885. A self-important rather than distinguished building, this was the very heart of late-Victorian Manchester: every important firm in the city was represented here, and three-quarters of Britain's cotton yarn and woven cloth was marketed in it. The telephone, though catching on fast, had not yet made such human contacts a luxury.
Photo: Valentine
Manchester Public Libraries

208 *below left* Victoria Buildings, between Victoria Street, Deansgate, and St Mary's Gate, designed by William Dawes, a local architect, and erected in 1877. It contained 28 shops, 88 offices, and a hotel. The capitals to the columns at ground floor level are surmounted by carvings from Aesop's Fables.
Photo: Valentine
Manchester Public Libraries

209 *right* Daniel Lee & Company's warehouse, 1885. Here was an expression of commercial achievement and personal aspiration. Mr Lee was the most prominent Roman Catholic layman in the city in the fifties and sixties.
Photo: Valentine
Manchester Public Libraries

J.P. WESTHEAD
INDIAN BLEND 2d
LONGSIGHT
RUSHOLME · ALL SAINTS · OXFORD St

IMPORTERS
MEERSCHAUM
PIPES
BRIAR PIPES
CIGAR
TUBES
FRENCH
DUTCH
& IRISH
BODDINGTON AND LEICH'S CORN & FLOUR WAREHOUSE
W.H.PIERCE & Co CLAY PIPE WORKS IRK STREET
W.H.PIERCE & Co IMPORTERS OF TOBACCONISTS FANCY GOODS
W.H.PIERCE & Co CIGAR MERCHANTS
PIPES. CIGARS.
RALPH RENSHAW & SON
W.H.PIERCE & Co
RENSHAW'S
R. RENSHAW & SON. PROVISION

210 *left* Piccadilly, 1900: a succession of horse-trams. The trams were a major cause of traffic jams, especially on the narrower streets. Hackney carriages were much less conspicuous than in London at this time.
Photo: Valentine
Manchester Public Libraries

211 *below left* Shudehill Poultry Market. Such institutions tended to persist whatever changes occurred in the physical environment. Livestock are still sold on this spot.
Manchester Public Libraries

212 *right* Market Street. The social mix is striking, as is the filthy condition of the roadway. Shop windows were commonly clamorous.
Manchester Public Libraries

213 *below right* Between Cross Street and Princess Street.
Manchester Public Libraries

214–17 Rochdale Road. Life for the working classes was a matter as much of the street as of the home. This was a major thoroughfare but it suffered the penalty of containing the gasworks. The shops lining the road almost continuously from the city centre outwards were designed as such, and contained two storeys of accommodation over. The area round about was streaked by poverty and contained much housing built before by-laws became stringent regarding cellar dwellings.

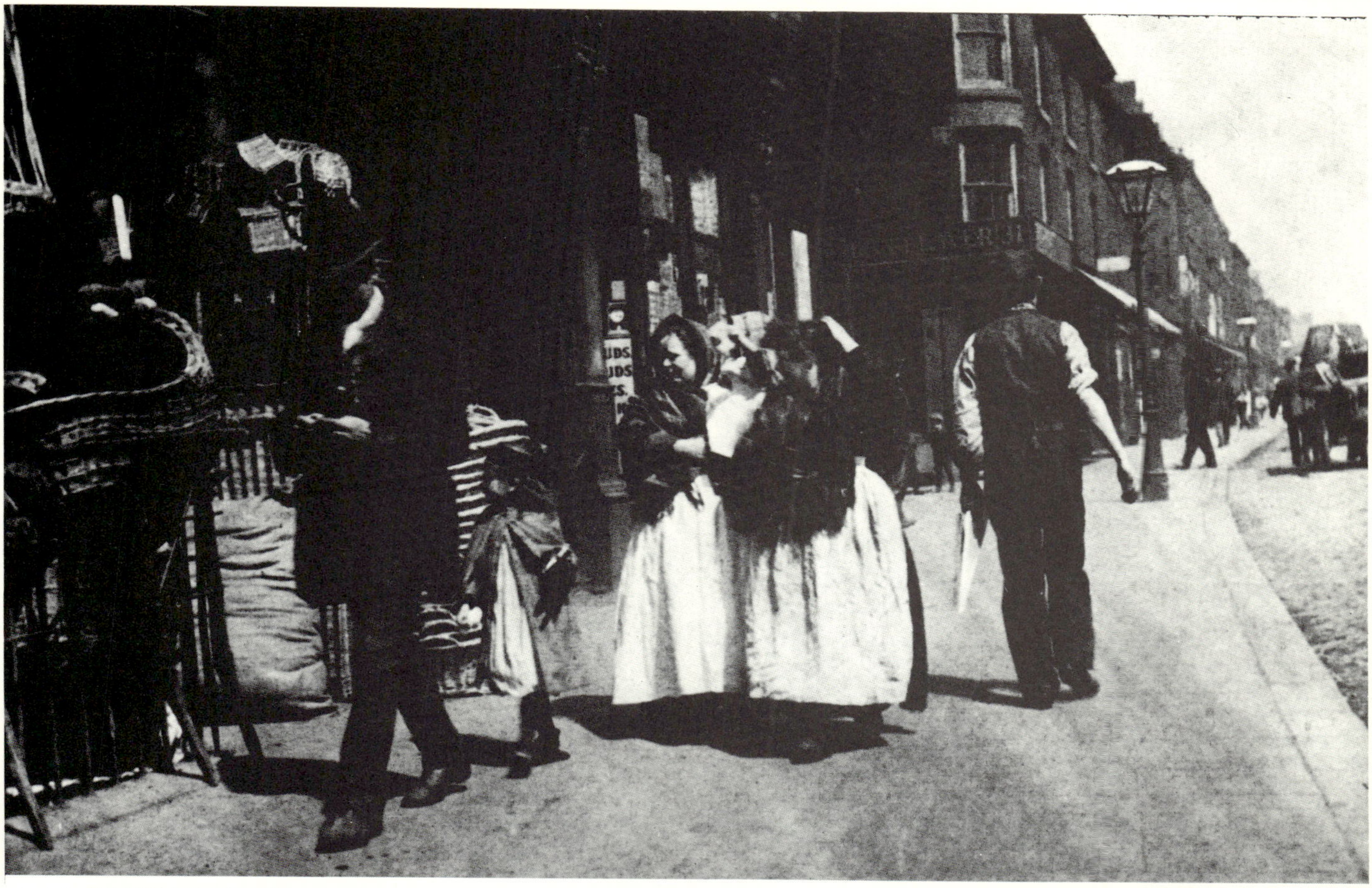

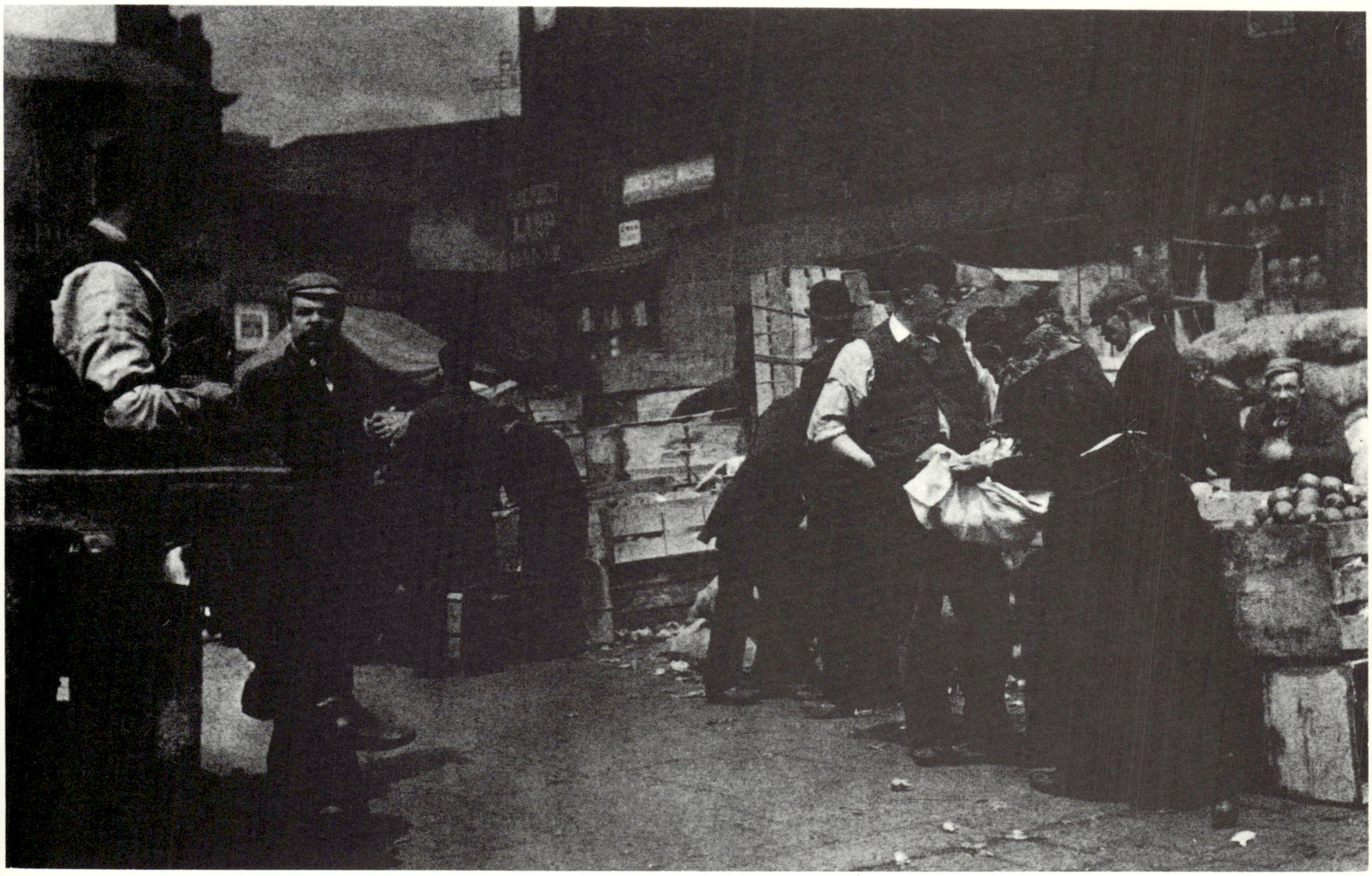

218 *left* Police and Sessions Court, Minshull Street, *c*.1877. The architect was a local man, Thomas Worthington, who liked to think that Manchester was the Florence of the nineteenth century. His Italian Gothic exercise of 1868 in red brick was too hemmed in by other brick-built warehouses and narrow streets to allow that to happen here, but the fine profusion of pointed arches and detailing round the entrances made it a proper seat of law.
Manchester Public Libraries

219 *below* Assize Courts, Great Ducie Street, *c*.1885. One of Waterhouse's earliest commissions, discharged under the influence of Ruskin in 1859, and reputedly the one building in the Venetian Gothic style that Ruskin approved of. Waterhouse went on to design Strangeways Gaol just beyond this site.
Photo: Valentine
Manchester Public Libraries

220 *right* Belle Vue Gaol, Hyde Road, 1890. An awesome pile that did not have a very long existence and was superseded by Strangeways.
Manchester Public Libraries

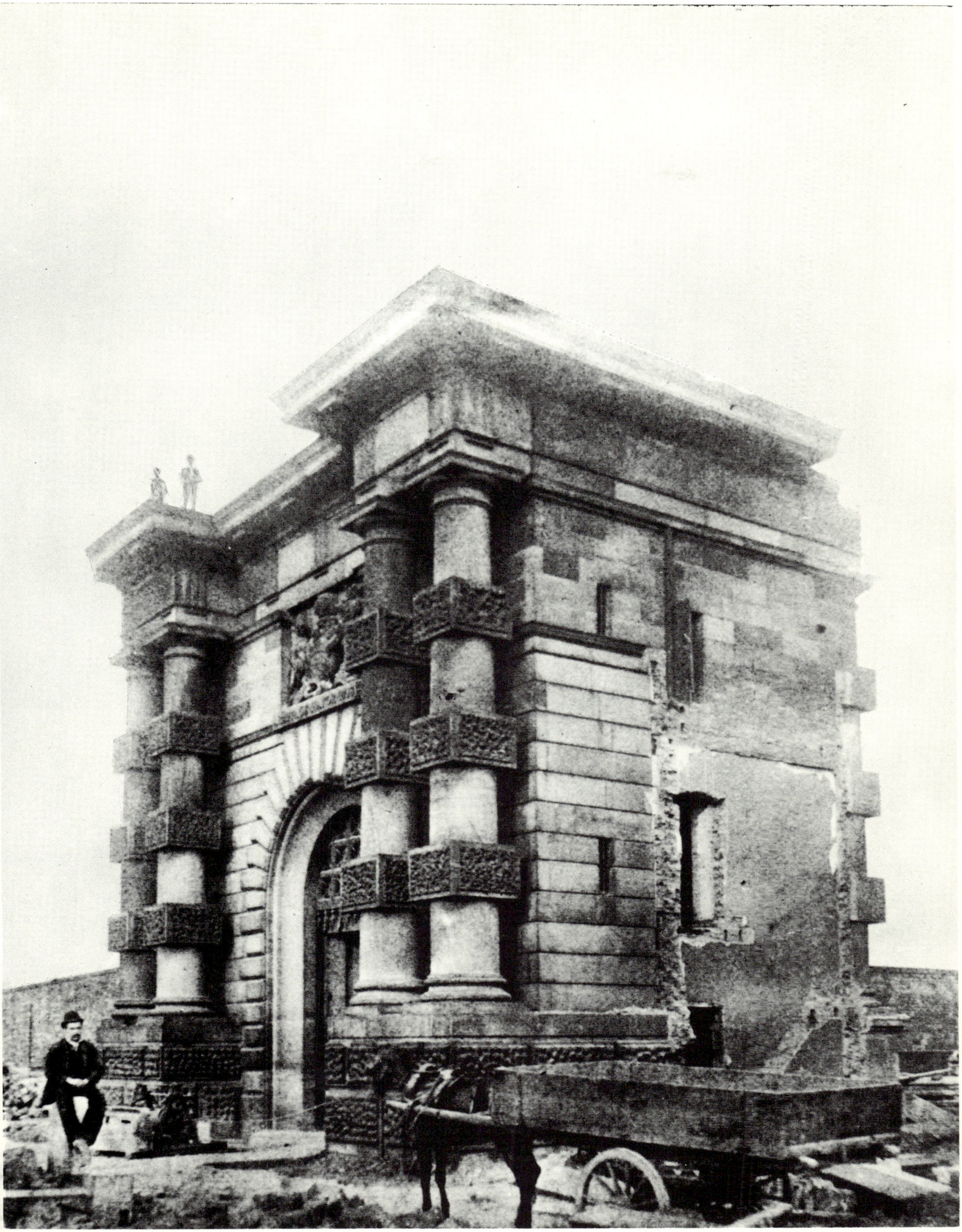

221 Palace of Varieties, Oxford Street. An aspect of urban enrichment largely unsuspected from the street. Electric lighting has been installed.
Manchester Public Libraries

222 Reference Library, Old Town Hall, King Street, c.1890. This building belongs to the Classical Revival, having been built in 1819 to the designs of Francis Goodwin. It was demolished about 1911 and its portico re-erected in Heaton Park. The adaptation of the main suite of rooms to the needs of a library was fittingly skilful though the original decorations had to give way to books. Contrary to the impression of studied indifference given here, the reference library was normally in heavy use. Before the present Central Reference Library was opened in 1934 readers had to use a series of temporary huts on Piccadilly.
Manchester Public Libraries

223 The Folly Theatre of Varieties, Peter Street, c.1890. A place for a sixpenny treat next door to the Free Trade Hall.
Photo: S. L. Coulthurst
Manchester Public Libraries

224 *above* Jersey Dwellings, Ancoats, 1897. The visitors to these model dwellings are ladies of the Manchester & Salford Methodist Mission, which was formed for such purposes in 1886.
Manchester Public Libraries

225 *above right* Strangeways Gaol entrance, 1900. Another activity of the Methodist Mission, performed daily. Prisons were places of private punishment and public admonishment. This was another of Waterhouse's designs.
Manchester Public Libraries

226 *right* The Whit Walk, Portland Street. Each denomination held its own on a different day of Whit week, the great festival that rivalled even Christmas in the nineteenth century. These were days for the throng, one's Sunday best, and protection against the rain.
Manchester Public Libraries

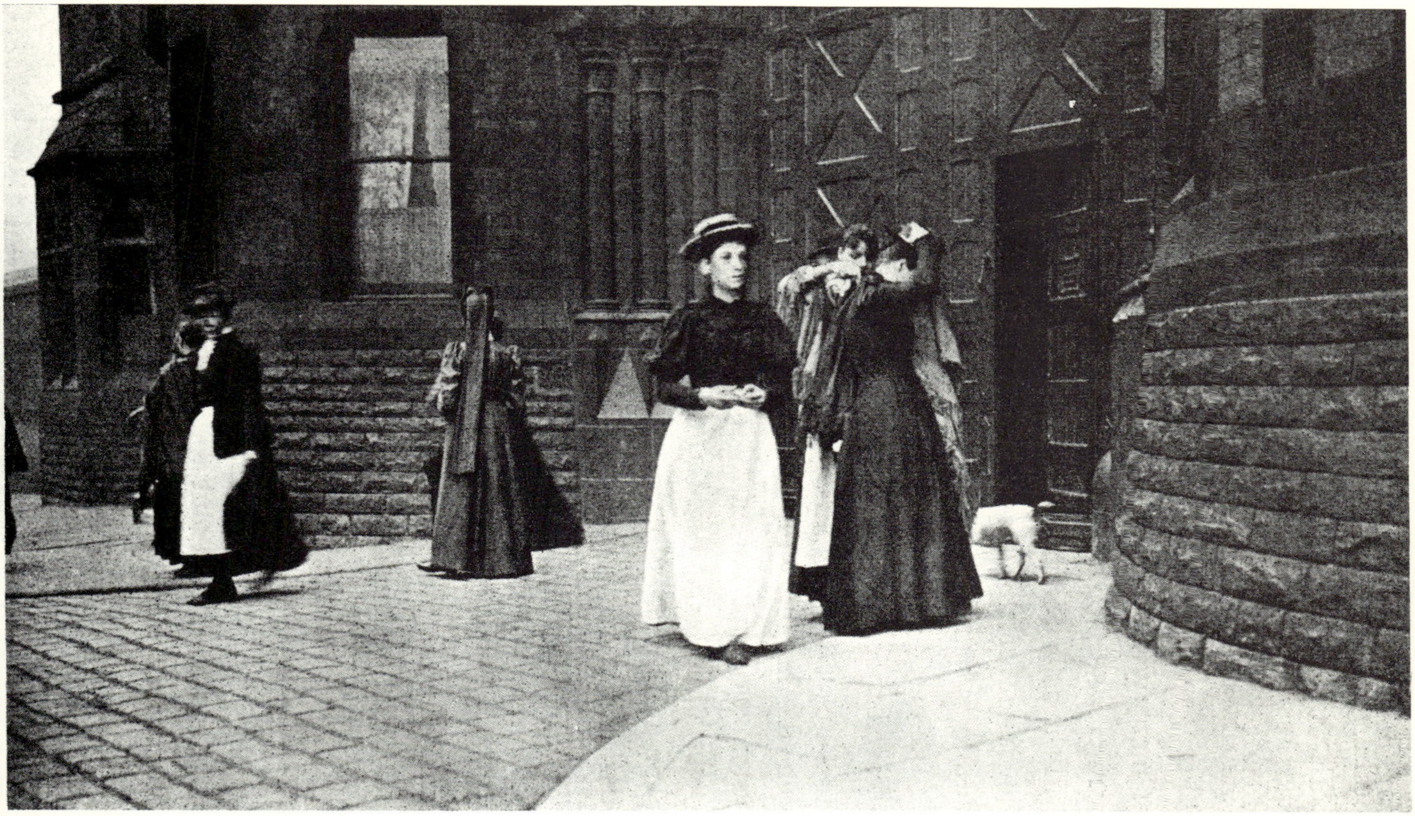

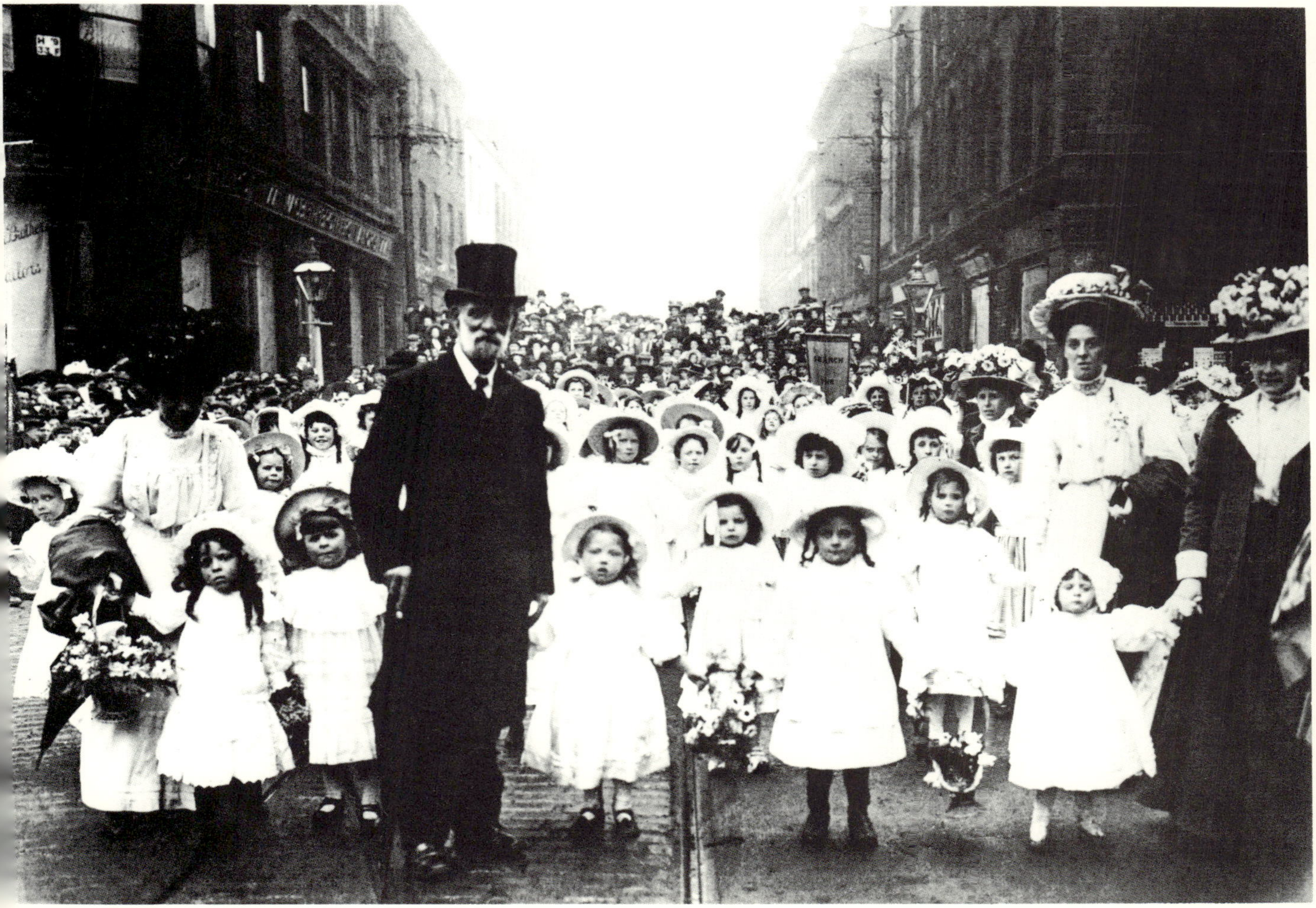

227 Botanical Gardens, Old Trafford. A place of beauty and recreation a tram-ride from the centre.
David Francis

228 The crematorium, three years after its
erection in 1892. Only 138 hearses arrived here
in the first ten years of its existence. The
Romanesque design was by Edward Salomons.
Photo: J. O. Pickard
Manchester Public Libraries

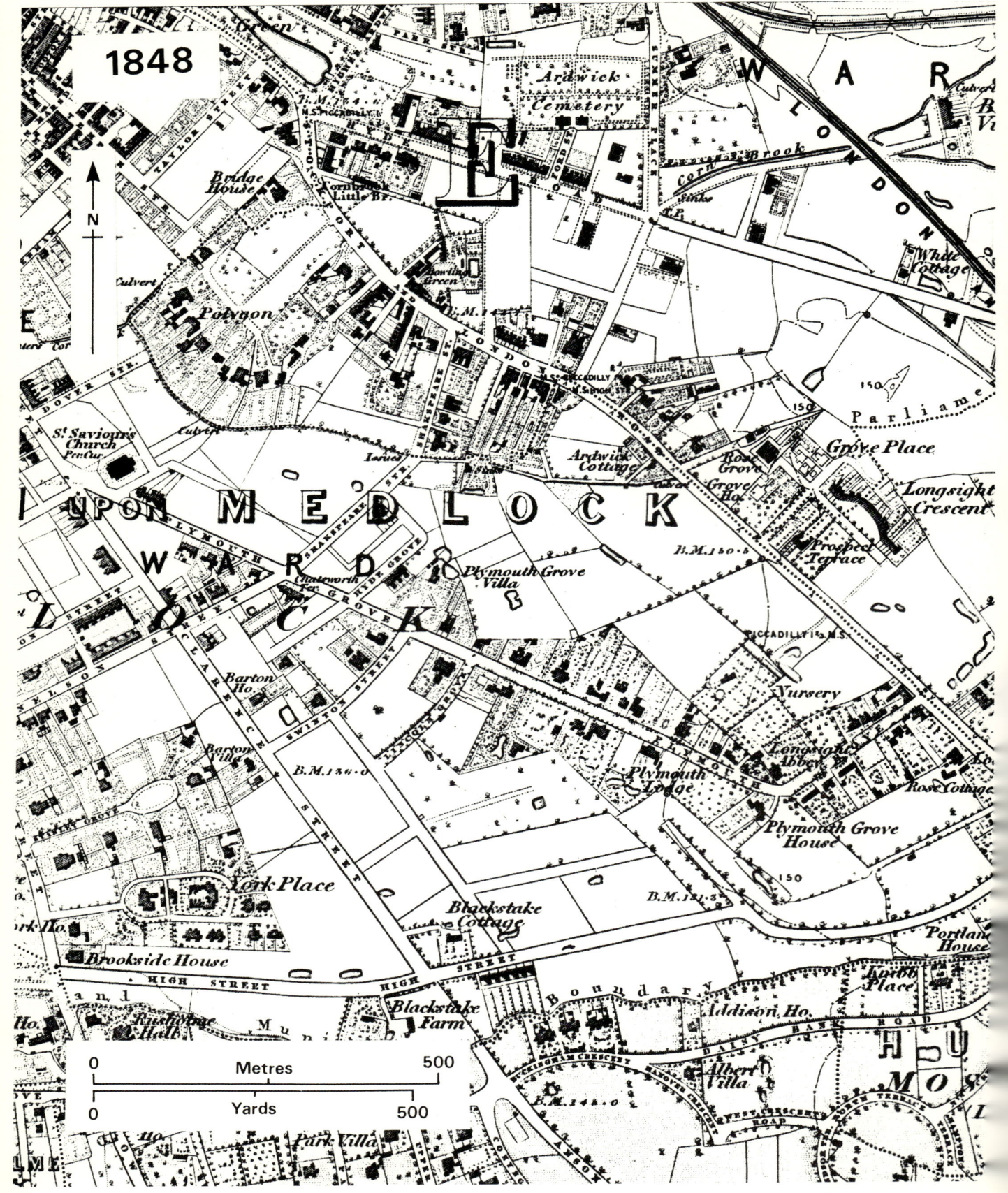

229–30 The south-eastern suburbs. Taken from the Ordnance Survey maps 1:10,560 of 1848 and 1895.

1895

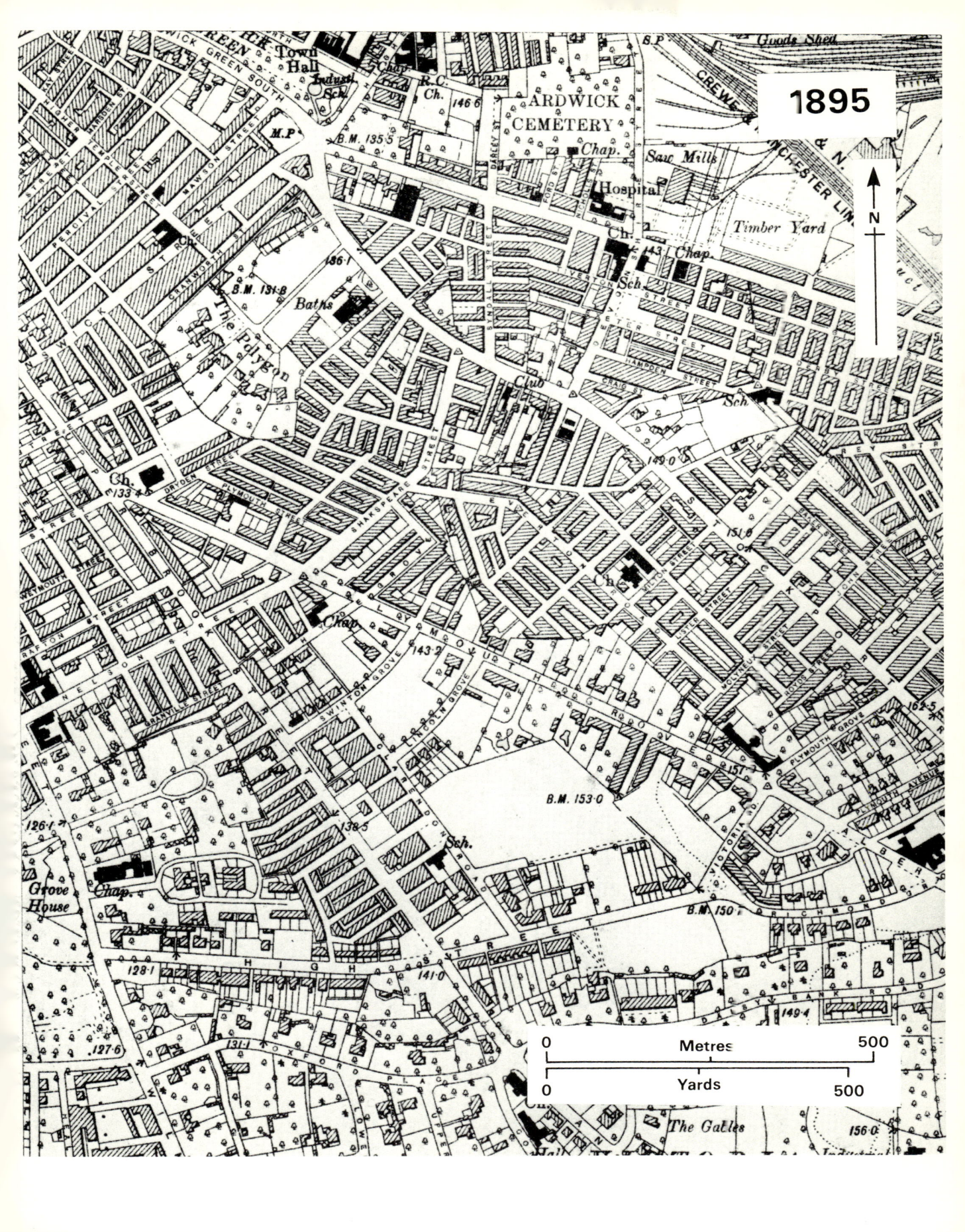

231–2 The central area. Taken from the Ordnance Survey maps 1:10,560 of 1848 and 1895.

1895

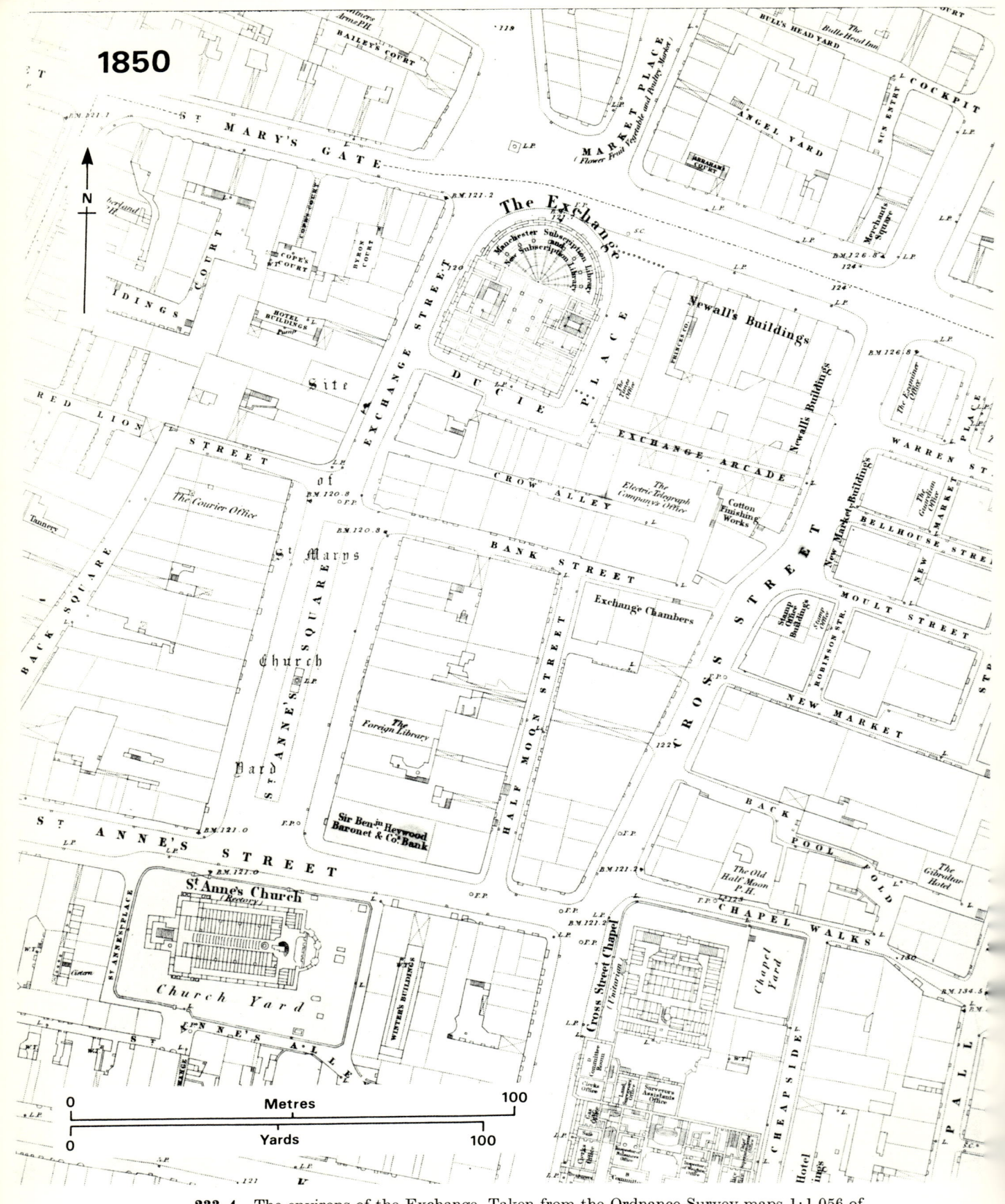

233–4 The environs of the Exchange. Taken from the Ordnance Survey maps 1:1,056 of 1850 and 1:500 of 1890.

1890

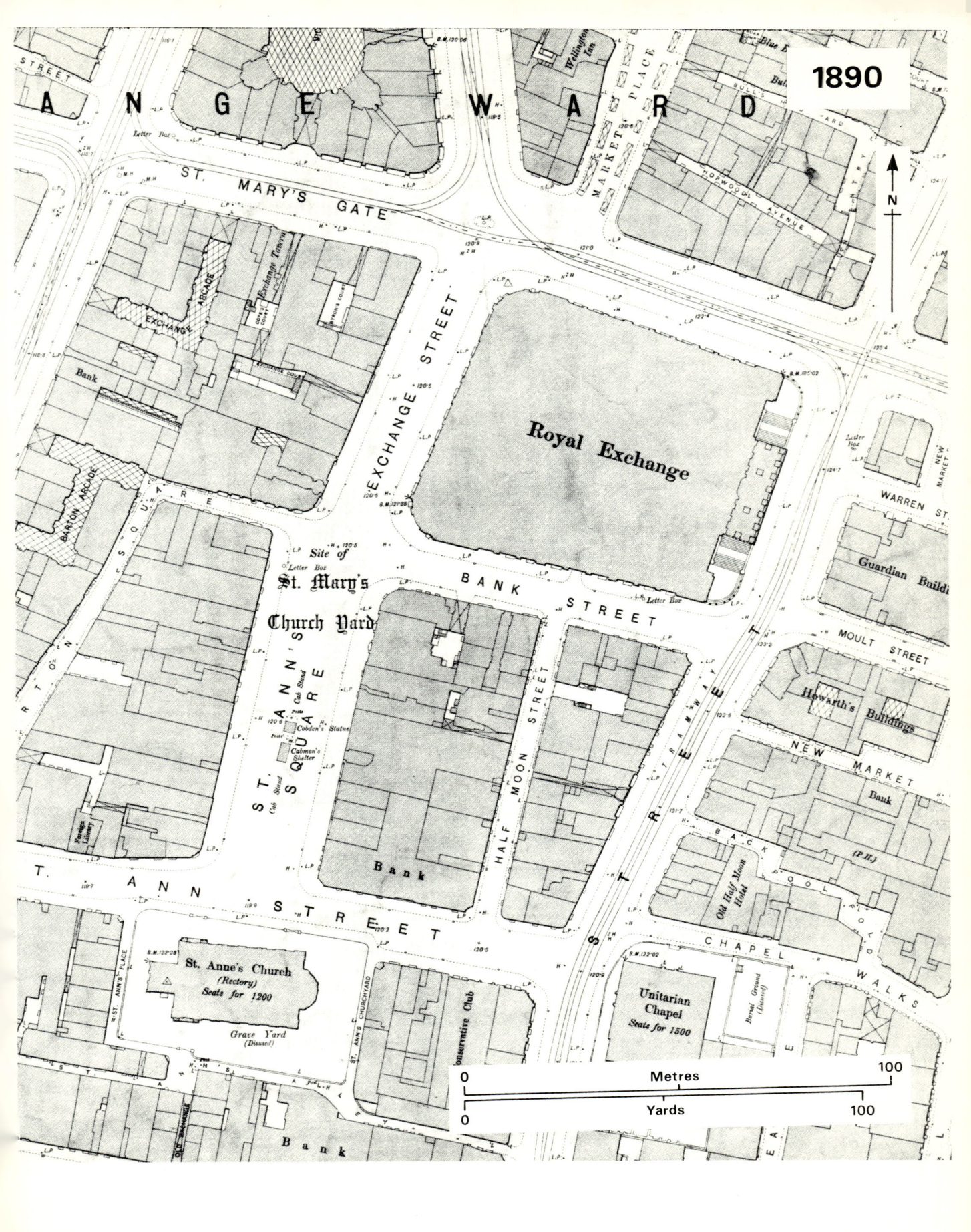

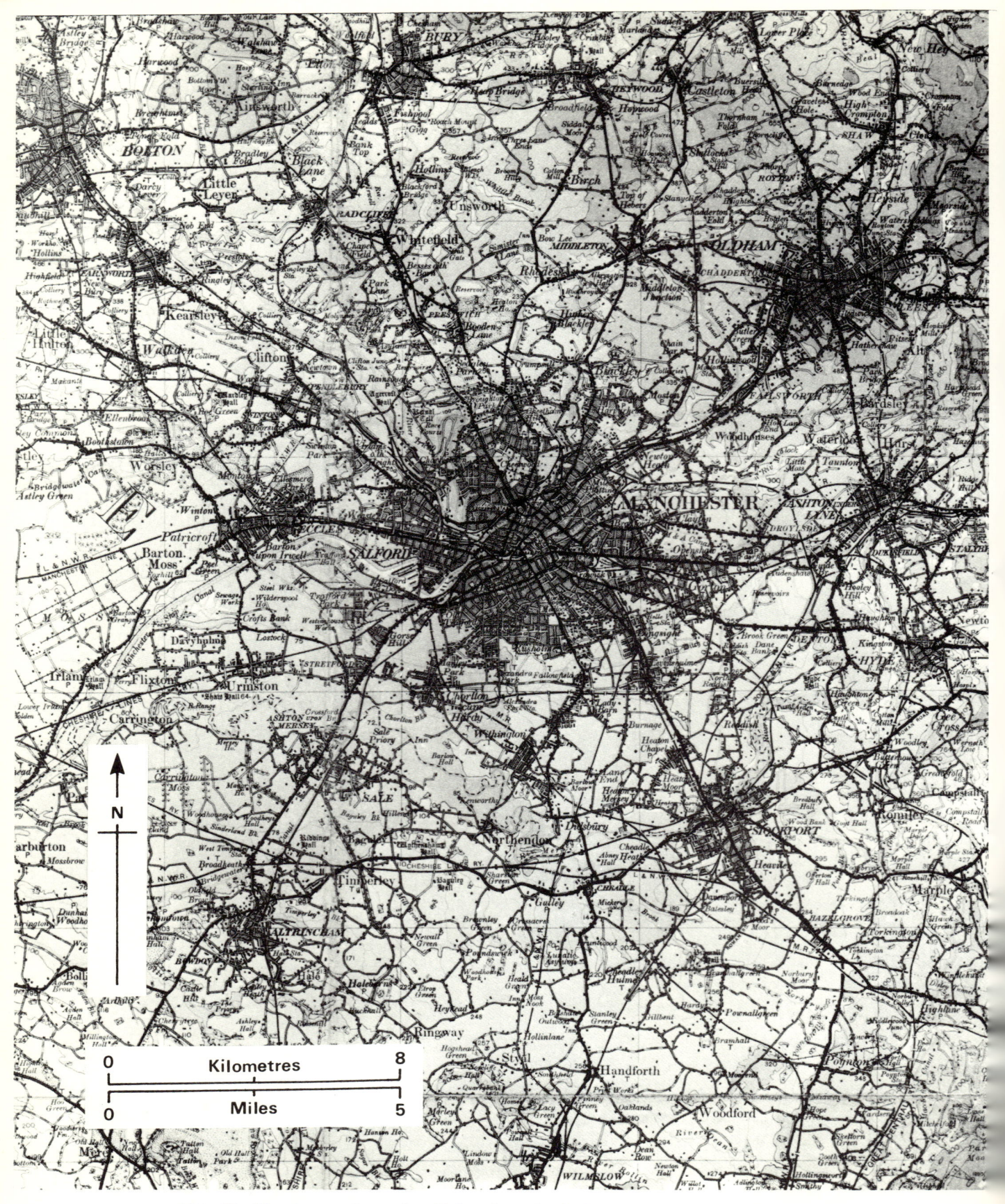

235 The Manchester city region. Taken from the Ordnance Survey map 1:126,720 of 1913, based on a survey of 1904.

11 Reading the Illegible

Steven Marcus

One of the chief components of the distress commonly felt by many people in modern cities is their sense that the city is unintelligible and illegible. The city is experienced as estrangement because it is not perceived as a coherent system of signs, as an environment communicating to us in a language that we know. After London, Manchester was the central site of that experience in Great Britain in the early nineteenth century, as Carlyle and Disraeli, Cooke Taylor and Kay-Shuttleworth, Faucher and de Tocqueville, among many others, have testified.[1] The discontinuities and obscurities, the apparent absence of large, visibly related structures, the disorganizations and disarticulations, seem to compose the structure of a chaos, a landscape whose human, social, and natural parts may be related simply by accidents, a random agglomeration of mere appearances.

It was into this prototypical anxiety-creating modern scene that Friedrich Engels, a young man sent by his father in Barmen to complete his business training in the family cotton-mills in Manchester, precipitated himself.[2] He arrived in Manchester at the very end of 1842 and remained twenty months. In that time he shunned polite circles and gave virtually all his spare time to 'intercourse with plain Working Men.' He wanted, he said in the dedication he wrote (in English) to 'the Working Classes of Great-Britain' in the original German edition of 1845 of his *The Condition of the Working Class in England in 1844*, 'to see you in your own homes, to observe you in your every-day life, to chat with you in your condition and grievances, to witness your struggles against the social and political power of your oppressors.' He wanted above all 'more than a mere *abstract* knowledge of my subject.'

Copyright 1973 Steven Marcus

The Victorian City: Shapes on the Ground

In the event, he learned to read Manchester with his eyes, ears, nose, and feet. Long before Ruskin declared that one has to read a building, he demonstrated that one had to read a city. He learned to read it with his senses, the chief inlets, if it is permissible to adapt Blake, of mind in the present age. What he experienced he also perceived as a total intellectual and imaginative structure. It was a vision sustained by applying and adapting the systematic, coherent, consequent, Hegelian style of analysis to a complex, apparently unsystematic, and possibly incoherent massive aggregate of experiences and materials of disparate orders; to an English social reality notorious for its capacity to withstand theoretical incursions. The result was an account of Manchester which became a central part of what I take to be the best single thing he ever wrote, the chapter on 'The Great Towns' in *The Condition of the Working Class in England in 1844*. His first words about the town itself, composed in a single paragraph that runs to almost three pages in length, remain one of the most enduring and important statements ever written about the modern city.[3]

> The town itself is peculiarly built, so that someone can live in it for years and travel into it and out of it daily without ever coming into contact with a working-class quarter or even with workers—so long, that is to say, as one confines himself to his business affairs or to strolling about for pleasure. This comes about mainly in the circumstances that through an unconscious, tacit agreement as much as through conscious, explicit intention the working-class districts are most sharply separated from the parts of the city reserved for the middle class. Or, if this does not succeed, they are concealed with the cloak of charity. [*Oder, wo dies nicht geht, mit dem Mantel der Liebe verhüllt werden.*]

If this is so, how then does the city work? Engels turns at once to a description which is simultaneously an explanation and analysis:

> In the center of Manchester there is a fairly extensive commercial district, which is about a half-mile long and a half-mile broad. This district consists almost entirely of offices and warehouses. Nearly the whole of this district is without permanent residents, and is forsaken and deserted at night, when only policemen on duty patrol its narrow, dark lanes with their bull's eye lanterns. This district is intersected by certain main thoroughfares in which an enormous volume of traffic is concentrated. The ground floors of the buildings along these streets are occupied by shops of dazzling splendor. Here and there the upper stories of such premises are occupied as residences, and these streets present a relatively lively appearance until late at night. With the exception of this commercial district, all Manchester proper, all Salford and Hulme, an important part of Pendleton and Chorlton, two-thirds of Ardwick, and certain stretches of Cheetham Hill and Broughton—all of these comprise a pure working-class district. This area [or district] extends around [or surrounds] the commercial quarter in a

> belt that is on the average one and a half miles in width. Outside, beyond this belt, live the upper and middle classes. The latter are to be found in regularly laid out streets near the working class quarter, in Chorlton and the lower-lying regions of Cheetham Hill. The upper middle class has situated itself in the remoter parts of Chorlton and Ardwick, or on the breezy heights of Cheetham Hill, Broughton, and Pendleton, where they live in villa-like houses surrounded by gardens.

And so on. There is the organization. What remains is to set it in motion. This is done in the first instance by the system of transport, the omnibuses which run every fifteen or thirty minutes and connect these outlying areas with the center of Manchester. And the beauty of it all, Engels continues,[4] is that the members of Manchester's monied aristocracy can now travel from their houses

> to their places of business in the center of town by the shortest routes, which run right through all the working class districts, without even noticing how close they are to the most squalid misery which lies immediately about them on both sides of the road. This is because the main streets which run from the Exchange in all directions out of the city are occupied almost uninterruptedly on both sides by shops, which are kept by members of the middle and lower-middle classes. In their own interests these shopkeepers should keep up their shops in an outward appearance of cleanliness and respectability; and in fact they do so. To be sure, these shops have none the less a concordant relation with those regions that lay stretched out behind them. Those shops which are situated in the commercial quarter or in the vicinity of the middle class residential districts are more elegant than those which serve to cover up [or as a facade for] the workers' grimy cottages. Nevertheless, even these latter adequately serve the purpose of hiding from the eyes of wealthy gentlemen and ladies with strong stomachs and weak nerves the misery and squalor that form the completing counterpart, the indivisible complement, of their riches and luxury.

He then proceeds to a concrete demonstration, and conducts the reader along a number of the main streets and simply and irrefutably shows how the changes in character of the buildings that front the street indicate what is to be found behind them. He is in fact charting one series of connected and stratified variables within the social ecology of the city.

By this means, Engels continues, 'it is possible for someone who knows Manchester to infer the social character of a district from the appearance of the main street that it adjoins. At the same time, however, it is almost impossible to get from these main streets a *real* view of the working class districts themselves.' He then moves towards his conclusion:

> I know perfectly well that this deceitful manner of building is more or less common to all big cities. I know as well that shopkeepers must in the nature

> of their business take premises on the main thoroughfares. I know that in
> such streets there are more good houses than bad ones, and that the value of
> land is higher in their immediate vicinity than in neighborhoods that lie
> at a distance from them. But at the same time I have never come across so
> systematic a seclusion of the working-class from the main streets as in
> Manchester. I have never elsewhere seen a concealment of such fine
> sensibility of everything that might offend the eyes and nerves of the
> middle classes. And yet it is precisely Manchester that has been built less
> according to a plan and less within the limitations of official regulations—and
> indeed more through accident—than any other town. Still when I consider
> in this connection the eager assurances of the middle classes that things
> are going splendidly for the working classes, I cannot help feeling that the
> liberal industrialists, the Manchester 'bigwigs,'[5] are not so altogether
> innocent of this bashful structural style.

In Manchester, as others had observed, the separation of classes had been driven to new extremes. What Engels added to this observation is that the separation had been built into the very structure of the city, and that this actual fabric both perpetuated such a condition and visibly expressed it. Even more, it had virtually effected the disappearance of one of the segregated classes, and the invisible poor of the mid-twentieth century was a reinvention of the invisible working classes of the mid-nineteenth. Yet how has this extraordinary phenomenon come about? In part, Engels says, it is through unconscious and unstated agreement, in part through deliberate intention—both of these, he intimates, applying to both groups, though not necessarily in equal proportions. The point to be taken is that this astonishing and outrageous arrangement cannot be fully understood as the result of a plot, or even a deliberate design, although those in whose interests it works also control it. It is indeed too huge and too complex a state of organized affairs ever to have been *thought up* in advance, ever to have preexisted as an idea.[6]

The city is a recognizably contemporary institution in other respects. It has already constructed for itself a central or inner core that is packed by day and deserted at night. This radical discrimination of function is equally borne out in the circumstances that those who during the day direct the city's workings from the center have moved away from there—where their homes originally were as well—and redistributed themselves as far from the center as the current means of transportation allow, on suburban heights. That is to say, as the city has rapidly and enormously expanded as an organization of production and a concentrated center of power and wealth, it has also deteriorated as a place in which to conduct other human and civilized activities. As the wealthier population have fled the center, they have leap-frogged over the vast working class and left them there massed, as it were, about that dense yet hollow core, in an immense unbroken belt in which they work and live. It is impossible at this point in history to disperse them, since they have to live within close walking-distance of their places of work. Despite all these sudden shiftings and evolutions—the fearful concentration of the working class, the flight from the center to the periphery—things have worked themselves out in what is apparently tidy and orderly detail. Indeed, the city reveals its social or class structure in each of its important spatial arrangements. These are most notable in those zones where the two extremes approach each other, or—to put the same idea in another way—at their lines of division, which is where the middle-middle and lower-middle classes come in. Their houses lie outside of the working-class belt, and between that belt and the more favorably situated suburban estates of the upper-middle class. And their shops and small businesses are located along the main thoroughfares—acting so to speak as insulators for the city's system of communication. Their location in space, therefore, is at both critical junctures an intermediate one, and their intermediary position is not merely structural but functional as well. They are acting as buffers between the antagonistic extremes.[7]

These streets, then, are Manchester's Potemkin villages. Yet behind the facades there lies not the nothingness of trans-Altaic wastes but a negative existence that is paradoxically a positive fullness, the indispensable creative source of all that positive wealth that lies beyond it. As Engels describes and analyzes them, those streets represent a collective effort of isolation, distancing, and denial, and work socially to the same ends as the same unconscious defensive processes work towards in individual persons. All he has neglected to add is that, when these symbolic aids were not available, it was possible to take to literal avoidance—the recourse adopted, for example, by the heroine of *North and South*:

> The side of the town [i.e., Manchester] on which Crampton lay was especially a thoroughfare for the factory people. In the back streets around them there were many mills, out of which poured streams of men and women two or three times a day. Until Margaret had learnt the times of their ingress and egress, she was very unfortunate in constantly falling in with them. (ch. 8)

These ways of dealing with experience were not confined to Manchester or the industrial towns or even to London. Some years earlier, in his essay 'Civilization,' John Stuart Mill had illuminated this subject from another point of view. 'One of the effects of civilization (not to say one of the ingredients in it),' he wrote, 'is, that the spectacle, and even the very idea of pain, is kept more and more out of the sight of those classes who enjoy in their fullness the benefits of civilization.' In the old cruel and heroic ages of Greece and Rome and of feudal Europe, this was not so. Today, however, all this has changed.[8]

> All those necessary portions of the business of society which oblige any person to be the immediate agent or ocular witness of the infliction of pain, are delegated by common consent to peculiar and narrow classes: to the judge, the soldier, the surgeon, the butcher, and the executioner. To most people in easy circumstances, any pain, except that inflicted upon the body by accident or disease, and the more delicate and refined griefs of the

imagination and the affections, is rather a thing known of than actually experienced. This is much more emphatically true in the more refined classes, and as refinement advances: for it is in keeping as far as possible out of sight, not only actual pain, but all that can be offensive or disagreeable to the most sensitive person, that refinement exists. We may remark too, that this is possible only by a perfection of mechanical arrangements impracticable in any but a high state of civilization.

It is a wonderfully intelligent passage and provides a useful context for Engels's remarks, as those remarks do in turn for Mill's. Both are aware that they are discussing a special modern development, and both regard this development with an equally critical, not to say jaundiced, eye. What is striking about Mill's statement is that, although it is written with the intention of the highest generality, the one thing that it fails to include is precisely the content of Engels's passage—the everyday world of labor and industry and the conditions of life in which it was sustained.[9] Furthermore, although Mill certainly sees the reciprocal and necessary connection between civilization or refinement and the concealment of pain, he does not press upon that connection with the same degree of force as does Engels. For what Engels is saying is that the riches and luxury are not only connected with the hidden suffering and squalor. They are connected in such a way as to be integral components of a unified-diversified phenomenon, whose very separation is the clue to their unity, both of these manifestly made visible in the structure of Manchester's streets.

Engels had demonstrated that a city could indeed be read. And yet, Engels says, doubling back upon himself, reading is *all* one can do from the main streets themselves. Those rows of shops, commercial buildings, pubs, warehouses, and factories, different and discriminable as they are, are still ultimately coverings. They function, depending upon the context in which we regard them, as appearances, symbols, or symptoms. They are the visible parts of a larger reality; they both reveal and conceal that reality; they are formations made up of displacements of and compromises between antagonistic forces and agencies. In order to get at the real things one has to go behind such appearances—no nineteenth-century novelist could have put his theoretical case more distinctly.[10]

But before he does this, Engels doubles back upon himself once more. He is aware of the fact that this kind of thing can be found in other cities, and he is acquainted with such innocuous 'explanations' of these arrangements as those that argue that the main streets are simply the places where shops and stores have to locate themselves, although there is some truth in these observations as well. Nevertheless, he has observed what he can only describe as the systematic exclusion of the working classes from the main streets of the city; at least, what he is describing —the informal ghettoization of the working classes—seems to act with systematic thoroughness and consistency, although the ghetto is itself the largest part of the town. But he knows, as well, that Manchester has been built up without plan, without explicit designs, and largely even through accident. Indeed, as he remarks later on: 'The big cities have sprung up spontaneously and people have moved into them wholly of their own free will' (p. 135). It is at the same time impossible to observe how this entire set of arrangements works to the advantage of the controlling class of industrialists, and to hear them speak of how the lives of the working class are improving—lives which have been in effect excluded from their sight—without suspecting that these industrialists are not so entirely innocent of it all as they appear; innocent in the sense of knowledge, of interest and implication, and of responsibility for behavior undertaken in the past, continued in the present, and projected into the future.

It is an invincible conclusion. It dramatizes the triumph of the intellect and experience of a single young man, but it represents as well the coming to fruition in him of a sovereign mode of thought, in the secularized application of its method to a new field of experience. What Engels has perceived and created is a general structure; its form is that of a coherent totality, a concrete, complex, and systematic whole, each of whose parts has a meaning, and more than one meaning, in relation to all the others. These meanings begin to come into view when we realize that this total whole is, naturally, more than the sum of its parts, that it is made of its parts and their histories, and that the entire elaborated and coordinated structure is in motion. For Engels is describing a process whose properties are unique. As I have remarked before, this process is neither mechanical nor organic; it partakes of both and yet is neither and something other as well. The only names we can assign to it are the human or the social, or, what is the same thing, the human-social, since the two cannot in fact or theory be held apart. It was this singular process that Hegel's style of thinking explicitly and preeminently dramatized. And yet there is something oddly moving and appropriate in the fact that this, one of the early successful adaptations of the method of this most difficult and obscure of all philosophers, should fall to the lot of a young man who considered himself to be a virtual auto-didact, who was the son of a factory owner, and that it should of all places have happened on the streets of Manchester.

Having thus constructively made out the macro-structure of the city, Engels girds himself to go behind those main street frontages, to enter that dark, dense belt formed by Manchester's working class and their dwellings, and to examine it. In this examination he has two purposes in mind: to describe it literally as it is, and to determine whether he can discover in it some kind of corresponding micro-structure. Before we follow him on this expedition, we may be prompted to ask ourselves a question. This question may appear superfluous and otiose in the very asking, and it may still seem so after we have tried to answer it. Why should Engels have followed this procedure? Why should he have begun with the buildings along the main streets of Manchester and then gone on to discuss the buildings in which the working class lived? The reader may be warranted in feeling that I have put the matter in this way because, like the character in the Talmudic story, I have an answer for which the question has to be invented; and he might very well reply, Why should he not have? He began with what was most strikingly evident to the senses; to begin anywhere else would have been tantamount to obscurantism. And moreover he was following the practice of other investigators, from Kay-Shuttleworth to Chadwick, who regularly opened their reports with an account of the conditions of the 'residences' of the laboring classes.[11] Still, the old saw—that children learn about the necessaries of life—runs 'food,' 'clothing,' and 'shelter,' in that order, at least in English. Is this only because that is the order of priorities for children, or is there some other meaning in the inversion practised by the investigators? In short, is there some unstated, perhaps unconscious, theoretical meaning at work? It seems hardly necessary to say that I think there is.[12]

In the houses men build for themselves they are expressing a behavior which binds them in common with other living creatures and species. The nests and burrows of animals represent efforts at altering the environment; they introduce something utterly new to it, something that did not exist before, but belongs to the creature who has made it. The purpose of this behavior is to create a surrounding suitable to the particular creature: shelter, protection, a stable micro-climate, a site for breeding, etc. These structures tend to be specifically characteristic, and scientists can distinguish between closely-related species, and even sub-species, of animals by examining these products of behavior that are in addition extensions of the creatures who made them. These structures are aboriginal artifacts; they are morphological embodiments of behavior, and represent in part the way in which particular creatures relate themselves to the world about them. In other species this behavior is genetically controlled or preprogrammed; in man it is not. It still holds, however, that in this fundamental artifact of a house or dwelling-place men express a good deal of how they relate themselves—or of how they are related—to the world in which they live. And in the collective human artifact—the settlement, town, or city—men as a group are expressing historically the character and quality of their existence, of the arrangements they have made, on the one hand, with the natural world, and, on the other, with one another. Engels is thus on sound theoretical grounds, when he remarks that the manner 'in which the need for shelter is satisfied furnishes a measure for the manner in which all other necessities are supplied' (p. 78).[13]

Engels takes the plunge. But where in this essentially doughnut-shaped belt is he to enter, and how is he to survey it with any coherence? His solution to this problem is as simple as it is sensible. At the northern limits of the commercial district— at about twelve on an imaginary clock— lying between that limit and the confluence of the rivers Irk and Irwell, and extending east along the banks of the Irk, is the Old Town, 'a remnant of the old, pre-industrial Manchester. The original inhabitants and their children have moved to districts that are better built, and have left these houses, which were not good enough for them, to a tribe of workers containing a strong admixture of Irish' (p. 57). Beginning with this section, which is conventionally, cartographically, and historically an admirable place for starting out, Engels proceeds to box the compass, to traverse the belt of working-class districts in a clockwise direction, describing what he has seen as he goes along. What he has seen and what he describes vary from section to section, but these variations fall within a uniform context of mass immiseration, degradation, brutalization, and inhumanization, the like of which had never before been seen upon the face of the earth.[14] We cannot follow him here in any detail but must select several of the more telling moments of his descriptive analysis.

As he stands in Long Millgate, one of the 'better' streets of the Old Town, Engels sees himself surrounded by houses that are old, dirty, and tumbledown; many are not even standing upright. Here, he says, one is 'in an almost undisguised working-class quarter,' since the keepers of the shops and pubs on such a street 'make no effort to give their premises even a semblance of cleanliness.' Yet these, too, are in a sense only facades, for they are as nothing compared to the alleys and courts that lie behind them, 'to which access can only be gained by covered passages so narrow that two people cannot pass' (p. 57). It is as if one were really penetrating into the heart of darkness. And once he is there, all Engels can do at first is to mutter 'the horror, the horror.' It is literally, he says, 'impossible to imagine for oneself'—or to represent—the chaos, confusion, density, cramming, and packing that exist in these spaces. Every available inch of ground has been built over, and the blame for this almost inconceivable overcrowding 'is not only to be ascribed to the old buildings surviving from Manchester's earlier periods.' It is in fact only quite recently, in modern times, that the practice has been followed of filling up every scrap of space that the old style of building had left. To prove his point, Engels then reproduces a 'small section of a plan of Manchester'—it is the subdistrict he has just been writing about—and adds 'it is by no means the worst spot and does not cover one-tenth of the Old Town.' It is a pretty enough section, but at least to my unpractised eye more or less unhelpful.[15]

But Engels is not to be stymied by the indescribable, and he takes us out of this set from an earlier version of *The Cabinet of Dr. Caligari* and into some other courts that branch off Long Millgate and are reached by covered passages that run down toward the banks of the Irk. On reaching these courts, he finds himself met with an assault of 'dirt and revolting filth, the like of which is not to be found...[and] without qualification the most horrible dwellings I have until now beheld.' He is going to say this more than once; he has run out of superlatives before he has barely begun; the language itself is giving out on him. In none of these inadequacies was he alone; indeed it may be suggested that one of the saving functions of language honestly used is that it should collapse before such realities, that it should refuse to domesticate these actualities with syntax and imagery, that it should compel writer and reader to at least a momentary extra-linguistic confrontation with such unspeakable man-made terrors, from which our man-made speech ordinarily protects us. In such a predicament, however, what is a writer to do? There is one course open, and Engels promptly takes it. He begins to specify. 'In one of these courts, right at the entrance where the covered passage ends is a privy without a door. This privy is so dirty that the inhabitants can only enter or leave the court by wading through puddles of stale urine and excrement (p. 58). This is the first of many such passages, and it will have to do service for almost all the others. It is at the same time difficult to know how and at what pitch of discourse literary criticism enters into such a scene. This difficulty has to do in the first place with what the intelligent student of literature can be expected to know. He may, I rashly speculate, be expected to know something about Swift and about Freud; he may have heard of Norman O. Brown's excremental vision, and will probably have read *Our Mutual Friend*; if he is a hard-working graduate student he will recall a scene in chapter six of *Mary Barton* that vividly, if gingerly, rehearses this material. That is, I believe, about it. The difficulty is further compounded by certain limitations that may be inherent in the historical imagination. How in fact does one reconstruct and apprehend the existential quality of such a situation? It may not be possible to do so. Or perhaps we ought to alter our terms, and recall Wallace Stevens's remark that 'in the presence of extraordinary actuality, consciousness takes the place of imagination.'[16]

What seems to have happened is that at about this moment in history, advanced middle-class consciousness—in which consciousness Engels may be regarded as representing the radical wing—began to undergo one of its characteristic changes. This consciousness was abruptly disturbed by the realization that, to put it as mildly as possible, millions of English men, women, and children were living in shit. The immediate question seems to have been whether they were not drowning in it. The catastrophe was worst in the great industrial towns where density and overcrowding went hand in hand with the interests of speculative builders, medieval administrative procedures and regulations—where they in fact existed—and produced a situation in which millions of grown people and their children were compelled to live in houses and neighborhoods that were without drains and often without sewers—and where sewers existed they took the run-off of street water and not waste from houses—without running water—sometimes not even in an entire neighborhood—and with one privy shared by who knows how many people: in some parts of Manchester over two hundred people shared a single privy. Such privies filled up rapidly and were cleaned out on an average of once every two years or so, which was as good as not at all. Streets were often unpaved, and where they were paved often no provision existed for their cleansing. Large numbers of people lived in cellars, below the level of the street and below the level of the waterline. Thus generations of human beings, out of whose lives the wealth of England was produced, were compelled to live in wealth's symbolic, negative counterpart. And that substance which suffused their existence was also a virtual objectification of their social condition, their place in society: that was what they were. We must recall that this was no Freudian obsessive neurosis or anxiety dream; but it is as if the contents of such a neurosis had been produced on a wholesale scale in social actuality. We can, then, understand rather better how those main-street palisades were functioning—they were defensive-adaptive measures of confinement and control. And we can understand what they were concealing: plenty.[17]

Engels tells the reader exactly where this court is, in case he is interested in confirming the truth of this account, and then takes us out of there and up onto Ducie Bridge, which spans the Irk at this point. Here he pauses to take a view, and composes a formal landscape in a style one wants to call authentic urban-picturesque:

> The view from this bridge—mercifully concealed from smaller mortals by a parapet as high as a man—is quite characteristic of the entire district. At the bottom the Irk flows, or rather stagnates. It is a narrow, coal-black stinking river full of filfth and garbage which it deposits on the lower-lying right bank. In dry weather, an extended series of the most revolting blackish green pools of slime remain standing on this bank, out of whose depths bubbles of miasmatic gases constantly rise and give forth a stench that is unbearable even on the bridge forty or fifty feet above the level of the water. ... Above Ducie Bridge there are tall tannery buildings, and further up are dye-works, bone mills and gasworks. The total entirety of the liquid wastes and solid offscourings of these works finds its way into the River Irk, which receives as well the contents of the adjacent sewers and privies. One can therefore imagine what kind of residues the stream deposits. Below Ducie Bridge, on the left, one looks into piles of rubbish, the refuse, filth, and decaying matter of the courts on the steep left bank of the river. Here one house is packed very closely upon another, and because of the steep pitch of the bank a part of every house is visible. All of them are blackened with smoke, crumbling, old, with broken window panes and window frames. The background is formed by old factory buildings, which resemble barracks. On the right, low-lying bank stands a long row of houses and factories. The second house is a roofless ruin, filled with rubble, and the third stands in such a low situation that the ground floor is uninhabitable and is as a result without windows and doors. The background here is formed by the paupers' cemetery and the stations of the railways to Liverpool and Leeds. Behind these is the workhouse, Manchester's 'Poor Law Bastille.' It is built on a hill, like a citadel, and from behind its high walls and battlements looks down threateningly upon the working-class quarter that lies below. (p. 60)

It is Tintern Abbey forty-five years hence, with the further exception that these ruins are densely inhabited.[18] Right up to the end, the representation is unusually telling. It sticks unwaveringly to the point, and, in the presence of catastrophe, consciousness, observation, *becomes* itself imagination. The juxtaposition of the paupers' cemetery and the railway stations is very good; but, it should be added, the work of synthesis is being selectively noted by Engels—it is already there. There is no evidence, in so far as I am aware, that Engels knew anything of Pugin, but this is *Contrasts* and more, and by a much more complex and civilized mind.[19] With the workhouse, Engels introduces some imagery and along with it a note of equivocation. The description of its situation and architectural style is accurate, and in so far as it is, communicates its own symbolic weight. But the representation is a trifle too 'medieval' in its suggestive associations and tones, and these do not really work well with the idea of the Poor Law Bastille. The current power of the phrase was contained in the threat implied, and that threat is the reverse of the one indicated by Engels. What the phrase brought to mind in these years of turmoil, protest, and crisis was less how the Bastille functioned in seventeenth- and eighteenth-century France than the fate it met in 1789. Still, on any reading, this is a very small falling-off on Engels's part.[20]

He moves back into Long Millgate and up along the Irk. If one turns left off the street here, he writes, 'he is lost. He wanders from one court to another. He turns countless corners, through innumerable narrow, befouled pockets and passageways, until after only a few minutes he has lost all direction and does not know which way to turn' (p. 61). This is certainly a sufficiently familiar report; to which we may add that it is also Todgers's without the fun and games.[21] Wherever one turns, Engels continues, there is filth—'heaps of rubbish, garbage, and offal.' Instead of gutters, there are stagnating pools, and the entire region is pervaded with 'a stench that would alone make it unbearable for any human being who was in some degree civilized to live in such a district.' The ambiguity pivots evenly and is going to take considerable working-out. Either those who live there are uncivilized, or those who live there do so under conditions which cannot be borne. If it is the latter, then a further series of contradictory alternatives come into play. Engels does not pursue these at this point. Instead, he continues to produce these large analytical observations.

> The recently constructed extension of the Leeds railway which crosses the Irk at this point has swept away some of these courts and alleys, laying others in turn, and for the first time, completely open to view. Thus, immediately under the railway bridge there exists a court that in point of filth and horror far surpasses all the others [there he goes again]—just because it was formerly so shut up, so hidden and secluded that it could not be reached without considerable difficulty. I thought I knew this entire district thoroughly, but even I would never have found it myself without the breach made here by the railway viaduct.

Readers will recall the epic passages in *Dombey and Son* (1846–8) to which this condensed and miniaturized passage has useful applications. Great projects like the railways led, among their many other inadvertent effects, to large-scale slum removal and clearance.[22] In addition, if one follows Engels in thinking of the city as a systematic, dynamic whole, the inherent course of its development brings that which has been pressed away and hidden into sight; it exposes itself by its own movement, the exposure being understood in a double sense. And the disarming, self-confident admission is in no way extraneous to the systematic effort of mind: self-moving intellect, self-realizing intention of will could never have made these discoveries by themselves. They are not victories of the study, wrested by rigor of method from the philosopher's teeming brain. Nor are they immediately victories of language, since social reality, on this side, expresses itself first in the concrete language of the nose, the eyes, and the feet. It is this developing group of signs that the young investigator must meet and follow and transform into the language of conceptualization. These are axioms of philosophy being proved upon the pulses, and elsewhere. It only remains to add what Engels could not then have known: that each of these new railway lines became itself the source of new divisions and demarcations; that for every slum destroyed in their construction a new one came into existence around them; that they became the new dialect lines of social distinction, having each of them a right and a wrong side, and serving in their finished state to restrict and confine as much as in their building they cleared out and exposed.[23]

Engels thereupon leads the reader into this micro-section of previously inaccessible slum.[24]

> Passing along a rough path on the river bank, in between posts and washing lines, one penetrates into this chaos of little one-storied, one-roomed huts. Most of them have earth floors; cooking, living and sleeping all take place in one room. In such a hole, barely six feet long and five feet wide, I saw two beds—and what beds and bedding—that filled the room, except for the doorstep and fireplace. In several others I found *absolutely nothing*, although the door was wide open and the inhabitants were leaning against it. Everywhere in front of the doors were rubbish and refuse. It was impossible to see whether any sort of pavement lay under this, but here and there I felt it out with my feet. This whole pile of cattle-sheds inhabited by human beings was surrounded on two sides by houses and a factory and on a third side by the river . . . a narrow gateway led out of it into an almost equally miserably-built and miserably-kept labyrinth of dwellings. (p. 63)

He has gotten almost to the center and the bottom, and what he finds there is that something has happened to the species. Men have gone back to living in holes (those thousands who dwelled in cellars were literally doing so). These nests or dens are virtual emptinesses, whose evacuated spaces are the counterpart of the densely packed humanity with which such absences are filled. The transformation may be taken one step further. Just as the pavement has disappeared beneath the accumulated wastes of the natural social life of the species, so too is man himself tending to vanish in certain of his distinguishing attributes. He has taken to living in cattle-sheds, or, what is the same thing, in shelters that cannot be told apart from them. It is quite impossible to know whether this is an image or a reality; that is its point.

The filth inside these hovels is in keeping with what is outside of them. Engels describes the condition of these interiors, which can only be compared to the pigsties that were frequently to be found in such quarters. Indeed, he remarks, it is impossible to keep such places clean or to keep oneself clean in them. The facilities for getting rid of wastes have already been mentioned; the only water available was from the

Irk itself, in which washing would only be a further form of pollution. It is at some such point in this account that the modern reader begins to realize that what Engels is embarked upon is a description of what has been called the Culture of Poverty, and that this is historically the first full-scale attempt at the representation.[25] Engels himself seems to have been struck by this awareness, and by the immensity and difficulty of the task he had undertaken. After conducting the reader through a number of other subsections of the Old Town, he pulls up, brings his account of this quarter to a close, and then turns to address the reader.

> This then is the Old Town of Manchester. On re-reading my description once more, I must admit that, instead of being exaggerated, it is by far not nearly strong enough. It is not strong enough to convey vividly the filth, ruination, and uninhabitableness, the defiance of every consideration of cleanliness, ventilation, and health that characterise the construction of this district, which contains at least twenty to thirty thousand inhabitants. And such a district exists in the very center of the second city of England, the most important factory town in the world. (p. 63)

Of all the characteristic utterances that issue from the humane consciousness of the middle classes—whether this consciousness be beleaguered, on the attack, or simply thunderstruck—during the decades around the middle of the century, this remains among the most poignant, the most authentic, and the most recognizably modern. One could compose an anthology of some decent size of such remarks and exclamations, whose authors represented every variety of political opinion and every substratum of middle-class life. Manchester itself, as we have seen, was often the source of such comments, but in this it was only hypertypical. The occasions were legion; among them were conditions that prevailed in all kinds of industrial work, in child labor, in agriculture, and everywhere in Ireland from 1846 onwards. And of course the twentieth century has been prodigal in the creation of such mind-stunning spectacles, mass events of such inhuman extremity that the only response to them is no response.[26] Out of all the responses from the earlier period, we can select only one more to represent the rest. After Lord Normanby accompanied Dr Southwood Smith on a tour through Bethnal Green, he wrote: 'So far from any exaggeration having crept into the descriptions which had been given, they had not conveyed to my mind an adequate idea of the truth.'[27]

In confessing that he was unable to represent the phenomenological quality of this reality, Engels was in effect coming very close to revealing its inner form. It was the dissolved and negated analogue of the unexampled transformation wrought by the industrial middle class upon the world.[28]

> The bourgeoisie ... has been the first to show what man's activity can bring about. It has accomplished wonders far surpassing Egyptian pyramids, Roman aqueducts, and Gothic cathedrals; it has conducted expeditions that put in the shade all former Exoduses of nations and crusades ... The

bourgeoisie, during its rule of scarce one hundred years, has created more
massive and more colossal productive forces than have all preceding
generations together.

Another part of that hegemony was revealed in the location of these working-class districts. They were at the very center of things, yet out of sight. To say that they were at once central and peripheral is to describe their contradictory existence in the structure of social consciousness of the time. It is also to define them in a rigorous and classical way as the hidden ground of things, a substructure. Nevertheless, to be at once central and peripheral is to occupy a place in society that is ordinarily relegated to crime and its attendant institutions—criminal sections, prisons, police-courts and police-stations. The wealth of contradictions implied in such relations did not escape Engels or other writers at the time.[29]

Engels proceeds to conduct the reader on a methodical survey of the other working-class sections. Beginning with the New Town, which is adjacent on the south and east to the Old Town, and then moving on around the compass to Ancoats and the area south of it, he analyzes the structural composition of each of the districts. These areas were built up later than the Old Town, and as he describes each section, Engels also describes the three different kinds of working-class cottages (all of them built in connected blocks or rows) that were to be found. Each style of building came into existence at a certain point in the Industrial Revolution and the expansion of Manchester, so that although sometimes the different methods of construction exist side by side, they are usually located in different parts of town and make it possible to 'distinguish the relative age' (p. 67) of each district. That is to say, even these working-class districts have their history, their industrial archaeology, their discernible meaning. Conditions in these districts are pretty much on a par with those already described. Some of the reasons for this have already been given, and include the want of public facilities for drainage and cleansing. Others have to do with methods of building and terms of land-holding and tenure, into all of which Engels goes in convincing detail. The outcome of this number of converging circumstances has been the building of working-class houses that last for about forty years: the houses themselves have their own life-cycle, which is on average only slightly longer than that of the industrial working people who inhabited them. This is another part of the structural composition of Manchester that signifies and not of Manchester or nineteenth-century cities alone.

As he systematically passes along this great circle, Engels drops pertinent observations about the ecological relations of the various parts of the district to the rivers, streams, and flats amid which they are situated; to the factories that surround them; and to the middle-class residential districts that begin at their outskirts. He takes us south of Great Ancoats Street, along the area crossed by the Birmingham railway, and through which the Medlock twists and turns. He then stops to represent in some detail the enclave of Little Ireland that lies in a bend of the Medlock.[30] This slum was described by Kay (later Kay-Shuttleworth) in 1831, and Engels remarks that in 1844 it is in a virtually unaltered state. It is surrounded by factories and embankments and is below the level of the river. Four thousand people live in it, most of them Irish; or, rather, they wallow in it, along with the pigs that thrive upon the garbage and offal in the streets. Large numbers live in porous cellars, and the density of habitation is ten persons per room. He moves on next to Hulme, to Deansgate and its rookeries, and finally to Salford, which faces Manchester in a bend in the Irwell. Conditions here in this district of some 80,000 people resemble those in the Old Town again, and Engels closes his discussion of this district and his entire survey of the belt of working-class quarters with the following observation:

> It was here that I found a man, who appeared to be about sixty years of age, living in a cow-shed. He had constructed a sort of chimney for his square-shaped pen, which had no flooring, no plaster on the walls, and no windows. He had brought in a bed, and here he lived, although the rain came through the miserable, ruined roof. The man was too old and too weak for regular work, and sustained himself by removing dung, etc. with his handcart. Puddles of excrement lay close about his stable. (p. 75)

We have at length reached the bottom, the bottom of the heap. The image has been actualized and become the literal reality—that is to say, in such an actualization the structure is openly disclosed. It is *Our Mutual Friend* without the Golden Dustman. And it is our urban pastoral again, but it also catches up another side of the micro-structure. The old man has been productively used up and discarded as refuse; accordingly in his old age he sustains himself from refuse. Yet he, too, is part of the life of the city and his life has a meaning, although the terms in which that meaning may be assessed will give comfort and peace of mind to almost no-one.[31]

Engels has brought to a close his first examination of what lies behind and between the network of main streets. It is Manchester itself in its negated and estranged existence. This chaos of alleys, courts, hovels, filth—and human beings—is not a chaos at all. Every fragment of disarray, every inconvenience, every scrap of human suffering has a meaning. Each of these is inversely and ineradicably related to the life led by the middle classes, to the work performed in the factories, and to the structure of the city as a whole. The twenty-four-year-old Engels has achieved a *tour de force*. I know of no representation of an industrial city before this that achieves such an intimate, creative hold upon its living subject. For anything that stands with it or surpasses it one has to go to the later Dickens, to *Bleak House, Hard Times, Little Dorrit*, and *Our Mutual Friend*.[32] But even to mention *The Condition of the Working Class* in the same sentence with such masterpieces suggests the quality of critical neglect which it has suffered and the misuses to which it has been put. And, one may add, in this last section it has *not* been Hegel's system that has been primarily instrumental in the creative achievement, but the courage and intelligence of a young foreign intellectual—and a businessman at that—who, during a twenty months' stay in Manchester, opened himself to its great and terrible realities and was not afraid of allowing himself to be overborne by them. He read the city well.

Notes

1. I have in mind, respectively: *Chartism* (1840) and *Past and Present* (1843); *Coningsby* (1844) and *Sybil* (1845); *Notes of a Tour in the Manufacturing Districts of Lancashire* (1842); *The Moral and Philosophical Condition of the Working Classes Employed in the Cotton Manufacture in Manchester* (1832); *Manchester in 1844* (1845); *Journeys to England and Ireland* (New York, 1968 edn).
2. I have considered the antecedents and meaning of this episode in his life and work in a book to be published under the title *Engels, Manchester, and the Working Classes*.
3. This last clause is given in Mrs Wischnewetzky's rendering, which retains the obscure referential character of the German. Since Engels read through and made revisions in her translation, it is to be presumed that he let this sentence pass as perspicuous. It is indeed highly suggestive and one could associate out from it in all directions, but it remains not exactly perspicuous. Henderson and Chaloner invent a sentence in its place: 'In those areas where the two social groups happen to come into contact with each other the middle classes sanctimoniously ignore the existence of their less fortunate neighbours' (Engels, *The Condition of the Working Class in England*, trans. and ed. W. O. Henderson and W. H. Chaloner (Oxford, 1958), p. 54). The references to this work in the present chapter follow the Henderson and Chaloner edition, though at certain points I offer an alternative translation.
4. 'Das ergänzende Moment' is what the old or former young Hegelian wrote. A beautiful illustration of Engels's thesis is to be found in the *English Note-Books* of Hawthorne, who came to Manchester in 1856 and then returned in 1857 for the Exhibition, rented a house first at Old Trafford and then on Chorlton Road, used the omnibuses exactly as Engels described, saw 'many handsome shops,' many fine paintings, and very little else—except for a sensational scene in Manchester cathedral. *Works* (Boston 1894), VIII, pp. 285ff, 517–45; also *Our Old Home*, *Works*, VII, pp. 169, 359ff.
5. In the original German edition Engels wrote 'big Whigs'; he later corrected this to read 'big Wigs,' a feeble enough pun in either form.
6. 'The first glance at history convinces us that the actions of men spring from their needs, their passions, their interests, their characters, and their talents ... Passions, private aims, and the satisfaction of selfish desires are ... tremendous springs of action. Their power lies in the fact that they respect none of the limitations which law and morality would impose on them ... [At the same time] human actions in history produce additional results, beyond their immediate purpose and attainment, beyond their immediate knowledge and desire. They gratify their own interests; but something more is thereby accomplished, which is latent in the action though not present in their consciousness and not included in their design.' G. W. F. Hegel, *Reason in History*, trans. Robert S. Hartman (Indianapolis, 1953), pp. 26, 35. (This is an augmented version of the Introduction to *The Philosophy of History*; an earlier translation of the whole work by J. Sibree remains widely available.)
7. Compare, with appropriate allowances, this passage from George Orwell: 'I was born into what you might describe as the lower-upper-middle-class ... a sort of mound of wreckage left behind when the tide of Victorian prosperity receded ... the layer of society lying between £2,000 and £300 a year: my own family was not far from the

bottom . . . In the kind of shabby-genteel family that I am talking about there is far more *consciousness* of poverty than in any working-class family above the level of the dole . . . Practically the whole family income goes in keeping up appearances . . . But the real importance of this class is that they are the shock-absorbers of the bourgeoisie. The real bourgeoisie, those in the £2,000 a year class and over, have their money as a thick layer of padding between themselves and the class they plunder; in so far as they are aware of the Lower Orders at all they are aware of them as employees, servants and tradesmen. But it is quite different for the poor devils lower down who are struggling to live genteel lives on what are virtually working-class incomes. These last are forced into close, and in a sense, intimate contact with the working class, and I suspect it is from them that the traditional upper-class attitude towards "common" people is derived.' *The Road to Wigan Pier* (New York, 1958 edn), pp. 153ff., 156.

8   *Essays on Politics and Culture*, ed. Gertrude Himmelfarb (New York, 1963), pp. 64f. See also the study of the development of the mechanism of death in the slaughtering and meat-packing industries in Siegfried Giedion, *Mechanization Takes Command* (New York, 1948), pp. 209–45.

9   I do not mean by this that the statements are opposed or contradictory of one another; they are not.

10 And no twentieth-century one, for that matter. See *L'Emploi du Temps* (Paris, 1957), trans. *Passing Time* (1960) by Michel Butor. Butor systematically uses the city in ways that are strongly reminiscent of Engels. He even provides a map that in its details resembles the maps supplied by Engels, and he traverses his northern English industrial town with the intention of finding the center of its 'complex net-work.' The main difference between the two is that Engels believes that through systematic study one can get to know the essence of the city, or of its reality, while Butor, one hundred and twenty years later, does not.

11 The first section of Chadwick's Report was titled 'General Conditions of the Residences of the Labouring Classes where Disease is Found to be most Prevalent.'

12 It may be noted in anticipation that for Rousseau's natural man, the three necessaries were 'nourishment, a female, and repose,' in *that* significant order: *The First and Second Discourses*, ed. Roger D. Masters (New York, 1964), p. 116.

13 In this paragraph I have summarized and drawn heavily on the writings of a number of behavioral scientists, such as N. Tinbergen, J. Calhoun, and J. Ralph Audy. I am in particular indebted to Audy's paper 'The Environment in Human Ecology: Artifacts—the Significance of Modified Environment,' in *Environmental Determinants of Community Well-Being*, Pan American Health Organization Scientific Publications No. 123 (Washington, 1965), pp. 5–16. The sense in which 'artifact' in this context is used is very close to a reversal of the usage of the eighteenth-century social and political theorists, who conceived of society as an artifact in the sense of a mechanical model. See J. H. Burrow, *Evolution and Society* (Cambridge, 1966), pp. 10ff., 26, and *passim*.

14 Immiseration is Schumpeter's apt translation of Marx's powerful coinage 'Verelendung.' The historical derivation of this word connects it inwardly with Marx's earlier, explicit writings about alienation, or 'Entfremdung,' and demonstrates, in terms of language, a genuine continuity in his thinking.

15 In *A Child of the Jago* (1896), Arthur Morrison tries to do the same thing for a small section of the East End, with relatively greater success—as far as the map is concerned.

16 *Opus Posthumous*, ed. Samuel Morse French (New York, 1957), p. 165.

17 The *locus classicus* for this subject is of course Edwin Chadwick's *The Sanitary Condition of the Labouring Population of Great Britain* (1842); two secondary works of interest and importance may be mentioned: S. E. Finer, *The Life and Times of Sir Edwin Chadwick* (1952); and R. A. Lewis, *Edwin Chadwick and the Public Health Movement, 1832–1854* (1952).

18 After Engels had visited Ireland in 1856, he wrote to Marx that ruins were characteristic of Ireland, 'the oldest dating from the fifth and sixth centuries, the latest from the nineteenth, with every intervening period included': *Werke*, XXIX, p. 56. Beginning in the eighteenth century, when artificial ruins were first constructed for taste and pleasure, civilization moved on in the nineteenth century to create instant, modern ruins—of which Ireland is a rural and Manchester an urban instance. The twentieth century has, in America at least, kept up the record.

19 Engels has in part introduced the cemetery and the railway stations in order to use them again, and to considerable effect, later on.

20 This does not of course mean that people did not feel threatened by the new Poor Law. They did in overwhelming numbers, even and particularly in places like Manchester where its most odious provisions were irrelevant and could not be enforced. For an account of how it was resisted in the North, see Cecil Driver, *Tory Radical: The Life of Richard Oastler* (New York, 1946), pp. 331–77.

21 See also *Oliver Twist*, ch. 50, for Dickens's similar description of Jacob's Island.

22 Further along, Engels notes that Liverpool railway has led to such a result in Salford (p. 74). In *Dombey and Son*, the pertinent subpassage in ch. 20 reads:

> Everything around is blackened. There are dark pools of water, muddy lanes, and miserable habitations far below. There are jagged walls and falling houses close at hand, and through the battered roofs and broken windows, wretched rooms are seen, where want and fever hide themselves in many distorted chimneys, and deformity of brick and mortar penning up deformity of mind and body, choke the murky distance. As Mr. Dombey looks out of his carriage window, it is never in his thoughts that the monster who has brought him there has let the light of day in on these things: not made or caused them. It was the journey's fitting end, and might have been the end of everything; it was so ruinous and dreary.

23 It was just such an experience that proved momentous in the career of Dickens's brother-in-law, Henry Austin. While working at surveying for the construction of the new Blackwall railroad, Austin first caught sight of what it was that the railway was clearing away and building through. The effect on him was to turn his interests permanently toward working for the alleviation of the living conditions of the laboring classes. See K. J. Fielding and A. W. Brice, 'Bleak House and the Graveyard,' in *Dickens the Craftsman*, ed. R. B. Partlow (Carbondale, Ill., 1970), pp. 115–39; 196–200.

24 The reader will recollect to what use Dickens put such properties of poor urban life as the posts and washing lines of the opening sentence. In *Little Dorrit* they become part of John Chivery's pastoral 'groves.'

The Victorian City: Shapes on the Ground

25 See Oscar Lewis, *La Vida* (New York, 1968), pp. xlii–lii for a summary discussion.
26 The literature on this subject is enormous. One may, however, point to the work of Robert S. Lifton, who has made a special study of responses to extremity. See in particular *Death in Life* (New York, 1967). The discussion of the relation of changing consciousness to such events is exceptionally tricky. It is certain that natural and social catastrophes on a mass scale existed before the nineteenth century; it is also certain that consciousness responded to these catastrophes differently before and after that time. Something happened to consciousness, but something was happening to society as well, to stimulate that consciousness and legitimize it. The tracing out of such intricately connected reciprocal developments is a study in itself.
27 Quoted in S. E. Finer, *The Life and Times of Sir Edwin Chadwick*, p. 161.
28 *The Communist Manifesto* (Penguin edn, 1967), pp. 82–5.
29 See J. J. Tobias, *Crime and Industrial Society in the Nineteenth Century* (1967), for an extensive discussion of this subject.
30 By this juncture, however, not quite willingly. 'If I were to describe all these separate parts in detail,' he conceded, 'I would never get to the end' (p. 71).
31 One makes wide exception for such sociological theorists as Edward Shils, in whose way of thinking such phenomena are part of the positive functioning of any society, and serve to accentuate its blessings. 'The humanitarian element in Marxism—its alleged concern for the poor—' he feelingly observes, 'can have no appeal where there are still many very poor people in Communist countries, and the poor in capitalist countries can now be seen not to be poor, not to be miserable, not to be noble—but to be as comfortable and as vulgar as, if not more vulgar, than the middle classes.' And again, 'Every society has its outcasts, its wretched, and its damned, who cannot fit into the routine requirements of social life at any level of authority and achievement... Those who are constricted, who find life as it is lived too hard, are prone to the acceptance of the ideological outlook on life.' Where would we be without such wisdom? 'Ideology and Civility: on the politics of the intellectual,' *Sewanee Review*, lxvi (1958), 450–80.
32 Mrs Gaskell's representations of Manchester in *Mary Barton* and *North and South* are very good—and, as has been mentioned earlier, are in accord with Engels's—but they do not have the generalizing and organizing power of Engels's account.

12 The Power of the Railway

Jack Simmons

'I perceive', wrote Carlyle in 1850,[1] 'railways have set all the Towns of Britain a-dancing. Reading is coming up to London, Basingstoke is going down to Gosport or Southampton, Dumfries to Liverpool and Glasgow; while at Crewe, and other points, I see new ganglions of human population establishing themselves, and the prophecy of metallurgic cities which were not heard of before. Reading, Basingstoke, and the rest, the unfortunate Towns, subscribed money to get railways; and it proves to be for cutting their own throats. Their business has gone elsewhither; and they—cannot stay behind their business! They are set a-dancing, as I said; confusedly waltzing, in a state of progressive dissolution, towards the four winds; and know not where the end of the death-dance will be for them, in which point of space they will be allowed to rebuild themselves. That is their sad case.'

As it often happened, what Carlyle saw was a quarter of the truth, or less; but his grasp of it was firm, and he expressed it with a memorable pungency. We are accustomed to think of the development of the railway as a potent force in the towns' growth and development. Carlyle, watching it at the very moment when it burst on the scene, saw only confusion: loss to the old-established towns, gain to the 'ganglions' he disliked, the Liverpools and Londons, the Crewes it had brought to birth. And in this he was wrong only in understatement: for the confusion, in the second half of the century, was greater than he foresaw.

With the proliferation of railways, Reading, Basingstoke, and Dumfries recovered themselves, to reach a new prosperity, founded largely on their position as

© 1973 Jack Simmons

junctions on the system.* It was not in the long run they, but towns of another kind, that lost. Nor did the ganglions gain, automatically or in equal measure. For those towns too, as they peered into the future, the railway could bring perplexity, the extremes of gloom and optimism. Here are two simple instances from Liverpool. Shortly before the opening of the Liverpool & Manchester Railway a correspondent of a local newspaper argued that if railway communication were extended to Birmingham and London, London would suck up Liverpool's trade. He assumed that when transport between two towns of unequal power was improved, the effect was to strengthen the stronger of the two; and he adduced the decline of Lancaster in relation to Liverpool, since the canal had been opened between them, as an example.[2] Five years later a much more sophisticated observer, de Tocqueville, prophesied that the development of railways would make Liverpool commercially more powerful than London.[3] Neither of these opposite forecasts turned out to be right. But there was some force in the reasoning behind both of them; and they illustrate very well the uncertainty that people felt as they tried to assess the effect of the railway on the towns they knew.

By the time Carlyle wrote those words the railway had arrived in every large town in the United Kingdom. The biggest that were not yet served by it were Worcester and Hereford.† In every town it touched it had begun its upheaval. It had all been the work of just twenty years. Before 1830 no major town in Britain had felt the impact of the railway to any significant extent. Primitive lines for the carriage of coal touched the fringes of a few of them—Newcastle, for example, and Edinburgh; the citizens of Swansea could make a jaunt along the shores of their Bay by coach on the Oystermouth Railway: but tiny developments like these were quite

---

* The elements of truth and error in Carlyle's assertions can be illustrated simply from the history of Reading. The railway arrived there in 1840, and in the succeeding decade the population of the town increased at a much slower rate than in any other of the nineteenth century. These are the percentage figures for the years 1821–81 (after which there was an alteration of boundaries):

1821–31 increase of	24·0
1831–41	19·6
1841–51	12·5
1851–61	20·6
1861–71	24·9
1871–81	30·1

At the same time it is worth noting that the critical expansion of the town's industries immediately followed the arrival of the railway. An 'extensive silk manufactory' appeared in 1841; the business of Messrs Huntley & Palmer dates from the same year. Sutton's emerged as seed-growers in the 1840s, conducting their trade largely with distant customers, by post and railway. W. M. Childs, *The Town of Reading during the Early Part of the Nineteenth Century* (Reading, 1910), pp. 23–4; idem., *The Story of the Town of Reading* (Reading, 1905), pp. 209–12, 228; *Victoria County History of Berkshire* (1906–27), I, p. 411; II, p. 243 (census figures). Reading was just over twice the size of any of the other Berkshire towns in 1801; by 1881 it was four times as big as Windsor, which stood second. It was the only town in the whole county that was placed on a main line of railway.

† Worcester got its first railway very shortly afterwards, on 5 October 1850; Hereford not until 1853.

marginal to the life of such places. Only two pairs of towns had yet been linked by rail: Gloucester and Cheltenham by a line authorized in 1809, which remained modest in scope, useful as a conveyor of coal from Gloucester Quay but otherwise barely perceptible as a factor in the economy of either town; Stockton and Darlington since 1825, but they were both towns of minor importance.

The Liverpool & Manchester Railway, however, opened up a new world. When its public traffic started on 15 September 1830 the railway began to exercise, for the first time, its full impact upon two of the great cities of Europe.* The railway was immediately successful, most conspicuously in the carriage of passengers. The number travelling between Liverpool and Manchester was at least twice as many as the coaches had conveyed before the railway was opened.[4] Nor was this a flash in the pan. The figure fell in 1832, when the novelty of railway travelling had worn off, but thereafter it rose steadily, to stand at 474,000 in 1835 and 609,000 in 1838.

Even before the railway was built, some landowners feared the effects it might have on their urban property, and with good reason. To the west of Deansgate in Manchester, for example, Henry Atherton had recently laid out a residential estate, in which trades and manufactures were prohibited. When the railway was first projected in 1825 his successors petitioned against it, on the ground that the locomotives would alarm the inhabitants, impair their comforts, and drive them away. This opposition was unavailing. When the Liverpool Road terminus was built, still more when it was extended in 1836, it exactly fulfilled the petitioners' fears, destroying the amenities of the Atherton estate, with the result that 'the establishment of the [station] . . . and the cushion of mean housing which surrounded it, tended to provide a westward barrier to the expansion of the business district.'[5]

The effect of the railway on the trade of Manchester and Liverpool can be inferred, though not precisely demonstrated. The railway was peculiarly well adapted to handling the commerce of Manchester, which was the metropolis of a large group of manufacturing towns, each playing its own part in the complicated production of textiles and some of the raw materials, like coal, that were consumed in it. The textiles were bulky to store, and subject in a high degree to the volatility of fashion. The railways at once speeded their concentration in Manchester and their distribution to clamorous customers, at the same time drastically reducing the space required for warehousing them.[6] For Liverpool the railways provided, what water communication could not, a system of rapid transit from the greatest English port facing out towards the Atlantic to the whole of the United Kingdom. 'There is now no town of any consequence in Great Britain', wrote Thomas Baines in 1852, 'with which the port of Liverpool has not constant communication by means of railway, and scarcely one which does not add something to the commercial prosperity of the port.'[7]

The physical impact made by the railway, both on Liverpool and on Manchester, was slight at first. The early passenger stations were placed at the very edge of the

* The Lyons–St Étienne line of 1829 made, by comparison, a trifling impact on Lyons.

built-up area. The inconvenience of this arrangement soon showed itself, and in Liverpool the passenger line was extended to a much better situated terminus in Lime Street, by the Haymarket. No such development occurred in Manchester, which came to be ringed round with a series of stations—Victoria and Exchange, London Road, Central—linked with one another by a series of not very convenient junction lines.

The system of unrestricted private ownership on which the British railways were constructed did much to determine the provision of major stations in the towns, which took on a pattern different from that usually found in Europe. The centre of London was encircled by a series of terminal stations unparalleled in number anywhere else in the world; the last of them—Marylebone, opened in 1899—was the fifteenth. The only other European capital with a similar multiplicity of stations was Paris, and that had no more than eight. By the end of the century the characteristic pattern in Continental cities was that of the single great station, the *Hauptbahnhof*. It was usually a terminus, and gathered to itself all the incoming lines, from whatever quarter they ran. It achieved its classic expression in Germany with the *Hauptbahnhof* of Frankfurt, built in 1879–88. By the end of the century almost all the great European cities had a station like this—Copenhagen, Amsterdam, Dresden, Marseilles, Milan, Rome—paralleled by the Union stations in North America.

In the comparable towns of Britain there are only three examples of this kind. Newcastle, Stoke-on-Trent, and Aberdeen alone concentrated the whole of their passenger traffic into one station. The single central station appeared also in Bristol, Hull, and Norwich; but in each of those towns there was also a small second terminus, belonging to a different company from that which owned the main one. In all the towns just mentioned, except Aberdeen, the central station was built by a railway company that enjoyed a preponderant share of the traffic in the district and could therefore mould the system, with comparative ease, so as to concentrate all its incoming lines at a single point. Elsewhere, each main company built its own station; and though sometimes, as in Leeds and Manchester, two or more companies might share a station between them, this still left a multiplicity everywhere: two in Birmingham, Sheffield, Edinburgh, Bradford, and Leeds;* three in Leicester,† Liverpool, and Belfast; three or four in Manchester;‡ four in Glasgow; five in Dublin. Aberdeen stood alone among all the large towns of Britain in the possession of a single central station, served by three companies.

The inconvenience of arrangements of this kind, in the passenger's eyes, does not need to be pointed out. The practice was even more tiresome for the trader and the manufacturer, for it was usual for each company to maintain at least one goods station of its own, sometimes more than one—those gaunt, austere brick buildings, many of which continue in use even under a unified railway system today.

* Three in Leeds if the New and Wellington stations, which were adjacent, are reckoned separately.
† Strictly speaking four, if the minor West Bridge terminus is included.
‡ According to whether Victoria and Exchange stations are reckoned as one, because they adjoined and were linked, or two.

Stations were usually placed well away from the centre. There were a few exceptions: the Central station in Glasgow, for example, was truly named. Only Birmingham, however, succeeded in getting the whole of its railway passenger traffic into the heart of the city. Its original stations, in Curzon Street and Lawley Street, were some distance out, and this was immediately complained of as a disadvantage.[8] In the 1850s the lines of all the three main railway companies were brought into two central stations (for a time it looked like being one,[9] but narrow gauge and broad gauge could not happily coalesce in this way); and the physical configuration of the town allowed both of them to be placed below the level of the ground, approached at each end through tunnels. In London there were eventually six terminal stations within the City proper;* but looking at the town as a whole, only Charing Cross could fairly be called central. The rest were all built on the perimeter. They have now been thoroughly engulfed by the spread of London; yet it should be remembered that the distance between Paddington and London Bridge is not far short of five miles.

There were those who advocated the construction of a *Hauptbahnhof* for London; they urged it again and again from the forties to the eighties. The most imaginative of them was Charles Pearson, the most persistent the *Builder*.[10] The more the termini multiplied, the more attractive the idea became: for the congestion of traffic in the streets rendered the transference from one station to another with each decade slower and more uncertain. The decisive arguments against any such proposal, however, were the still greater increase in street congestion that it would produce, concentrating the passengers on arrival and departure at one point rather than keeping them dispersed at a dozen, and the enormous capital cost of the land and construction. In the provinces, moreover, notorious difficulties of management had arisen in the big stations into which two or more companies ran their trains. London Road at Manchester provided perhaps the worst example.[11] But there were others: in Manchester, again, Victoria station,[12] Chester,[13] London Bridge, and King's Cross.[14] There were, it is true, other stations where sharing gave rise to few troubles: Carlisle (Citadel), for example, which eventually served as many as seven companies; Leeds (New);[15] those like Victoria in London, or Bristol (Temple Meads) in which the traffic of one of the sharing companies could be largely segregated from that of its partner. But one must admit that the chance of achieving the smooth joint management of a London station used by perhaps half-a-dozen companies was remote, and that the idea was, for this reason, rightly rejected. The *Hauptbahnhof* was more easily achieved on the Continent. There the cities were smaller, and the great stations were either under the management of a single company, as at Lille or Toulouse, or constructed under State control, as at Basle and Zürich and in Germany.

In London the alternative, adumbrated in 1846 and realized from 1863 onwards, was to allow multiple termini to develop, outside the centre, and to assist the movement between them by the building of urban railways. The first—and the most

* Broad Street, Liverpool Street, Fenchurch Street, Cannon Street, Holborn Viaduct, Blackfriars.

The Victorian City: Shapes on the Ground

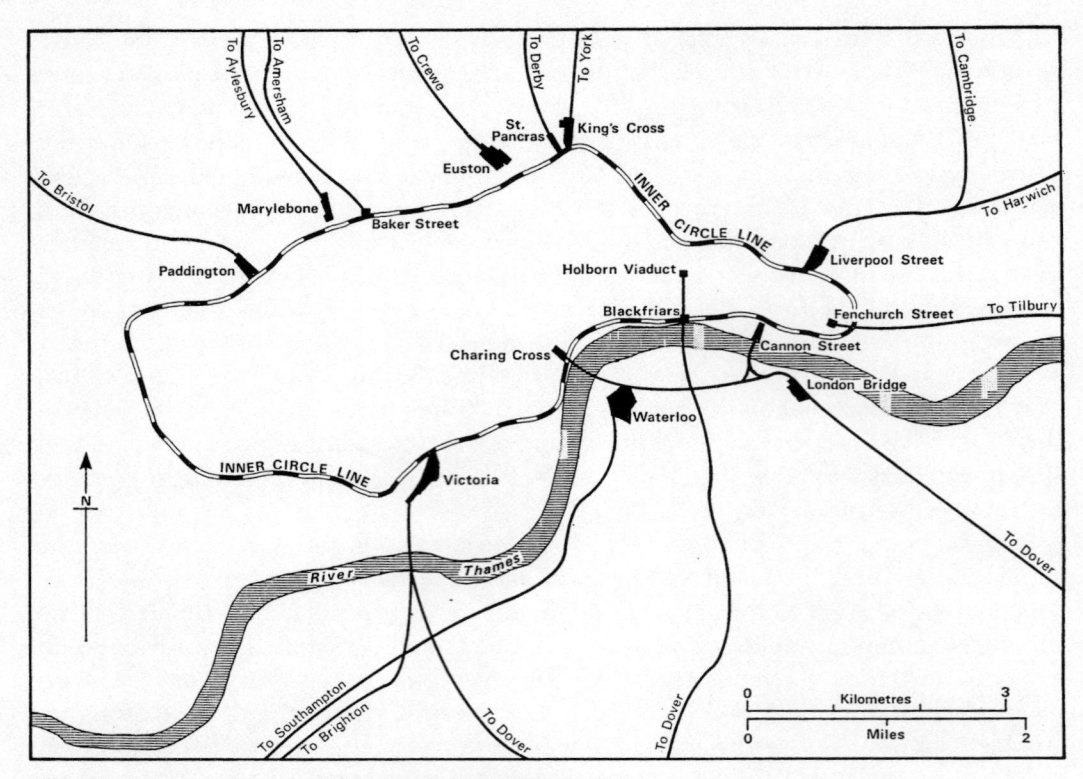

VII London terminal stations and Inner Circle line

important urban railway ever built anywhere, when one considers the consequences that sprang from it—was the Metropolitan Railway, opened from Paddington to Farringdon Street on 10 January 1863. In this too the far-sighted Charles Pearson had a hand. Its champions hoped that it would, at a stroke, solve the whole problem of urban congestion in the district through which it passed. They were too optimistic. Its immediate effect was to hit the buses and cabs very hard; but, once the novelty of travelling on it had worn off, the buses and cabs recovered their passengers. There seemed to be almost as much congestion as ever on the New Road, though the railway was *also* carrying over nine million passengers in its first year of operation.[16] The truth was that, here as elsewhere, the opening of the new railway had simply increased the number of travellers.[17]

The Metropolitan Railway was to be carried on eastwards to Moorgate, and under powers conceded in 1864 extensions were built to make the line into a complete circle. The circle (not completed for twenty years) developed in a way different from that originally intended. It remained primarily an urban railway, throwing out offshoots of its own—to St John's Wood and Harrow, to Ealing and Hounslow, to Whitechapel and Bow. It conveyed passengers between most of the main-line termini; only three—London Bridge, Waterloo, and Holborn Viaduct—lay off its route. But it did little to carry passengers through London without changing. Although trains from the Midland and the Great Northern lines ran on to it, they were mostly suburban trains, terminating at Moorgate. The London Chatham & Dover company, thrusting across the Thames by Blackfriars, linked up with the Metropolitan at Farringdon Street and thus gave central London in 1864 its solitary link from south to north. In the long run the greatest importance of this line was in the conveyance of freight; but it also carried a strange variety of passenger services, including some main-line trains from the Midlands and the North to the Kent coast. None of them was very successful. Nor were the outer suburban services: from St Pancras via Willesden to Earls Court; from the Great Western line—even, for a time, from Birmingham—to Victoria; from Croydon to Liverpool Street.

The truth was that the passenger in transit across London, from the north to the south coast, obstinately preferred to travel by the streets. Through services to the Channel ports, for the Continental steamers, were often discussed and occasionally provided.[18] But travellers generally chose, if they were passing through London, to stay at least a night there.[19]

The underground railways never made their shareholders' fortunes. The dividends of 5 percent that were usually paid by the Metropolitan Railway were derived in part from the Company's profitable surplus lands; the District had a struggle to keep up payment of any dividend at all, even on its Preference stocks.[20] These difficulties were partly due to the folly of the companies themselves in quarrelling rather than collaborating. But their chief cause lay in the conditions imposed upon them by landowners and local authorities, anxious to profit by the needs of the companies that arose from the special form the railways took. They were shallow lines, built directly below the surface. They interfered with sewerage; they required ventilating shafts, emitting fumes and smoke; even though they ran, as far as possible, underneath streets, their building involved the demolition of many houses; the Metropolitan Board of Works was determined to make them pay for a substantial part of the very expensive programme of street improvements that it rightly deemed necessary in London.

All these considerations are exemplified in the Metropolitan District Company's first Act.[21] There were to be no ventilating shafts on Lord Harrington's estate in South Kensington; special protection was accorded to the Duke of Norfolk and the Strand Estate when the Temple station was erected; no part of the Tallow Chandlers' Hall in Dowgate Hill was to be taken unless that Company gave its consent. Where the railway ran through the Inner and Middle Temple it was to be entirely covered in; and, except in emergency, 'the steam whistle of any locomotive engine . . . shall never be sounded within the prescribed limits', nor was steam ever to be blown off there. The District Company had to pay for enlarging Tothill Street to a width of 60 feet, and to construct, if so requested by the Metropolitan Board of Works, a low-level sewer, in the same trench with the railway, from Cannon Street to Trinity Square. And finally, on every weekday it had to provide a workmen's train in each direction at a maximum fare of a penny a journey.[22]

The last is an early example of an obligation that was placed on most London railways from 1860 onwards.[23] These workmen's services became numerous as time went on. In 1883 the number of trains required to be run by Act of Parliament was eleven; the number actually running was 107.[24] They were not always a cause of gratitude to the men who used them. In 1871, for example, a working man wrote to the *Builder* to complain of the overcrowding on the Metropolitan trains, which drove him to walk instead of using them, adding that he and his fellows would be great supporters of the new tramways as they were extended.[25] But Parliament kept the horse tramways out of the centre of London, and none of them ever provided an alternative to the Metropolitan Railway.[26]

The cost of building railways in the heart of London began to become prohibitive in the sixties. The Charing Cross and Cannon Street lines cost £4 million—that is, £2 million a mile, 'the most expensive piece of railway construction in the world'.[27] The rate for the Dover Company's Metropolitan Extension was £1½ million a mile.[28] When the Great Eastern first planned its line to Liverpool Street in 1862, the Company's Engineer, Robert Sinclair, estimated the cost at a little over £400,000;[29] the line was at last completed in 1875, and the price of it was in fact over £2 million.[30] A substantial proportion of this expenditure was attributable to large terminal stations. But the underground railways, which were not burdened in that way, were costing much the same. The bill for the first section of the District line, from South Kensington to Westminster, a little over two miles long, was £3 million.[31]

No wonder that the Circle line remained unique. Schemes for similar railways had burst out in profusion in 1863–4. All had failed, including the one for a line from Euston to Charing Cross, associated with a badly-needed new street, running southwards from the Tottenham Court Road. That project was authorized twice, in 1864 and 1871; but it was never realized. If more urban railways were to be built, it seemed that they must either be raised above ground level or tunnel much deeper in the earth. The first solution was adopted with success in New York and other American cities. It had supporters in this country,[32] and in the 1890s it was realized in Liverpool.[33] But that was a special case, the line running not through the centre of the city but along by the docks. Even then, the scheme was criticized for its unsightliness. It would have been much worse—indeed unthinkable, in terms of noise, dirt, and disfigurement—in the middle of London.[34]

The first experiments with tube railways, built at a deep level underground, and operated by 'atmospheric' (pneumatic) or cable power, had begun in the 1860s.[35] They failed; but when the Germans and Americans demonstrated the capabilities of electric traction, they were resumed, with much greater hope of success. The upshot was the City & South London Railway, the first electrically-operated underground railway in the world, opened to the public on 18 November 1890. A second line of the same sort—the Central London, running due east and west from the Bank to Shepherd's Bush—was sanctioned in 1891. In the following year half a dozen fresh projects for tube railways in London, to be worked by electricity or cable traction, were brought forward. Parliament referred them all to the consideration of a Committee of both Houses. Its report examined them from many different points of view—technical, commercial, economic. Some of the witnesses who appeared before it, looking far ahead into the twentieth century, considered these underground railways as agents of social change, an instrument of town-planning. The Committee accepted the necessity that these lines should be worked by electricity or cable power, that they should not be extensions of existing railways but independent, and that they should be constructed in tunnels of small bore, which excluded normal rolling stock. Parliament took the Committee's advice, and the London tube system, realized slowly and painfully in 1898–1912 and only slightly extended since that time, was the result.[36]

While that work was in progress the transport system of London came under the exhaustive scrutiny of the Royal Commission on London Traffic appointed in 1903. In its report it passed under review the whole course of legislation concerning railways in the capital, which it saw as a catalogue of almost uninterrupted error. The critical decision of the sixties, to keep railways out of the centre of London, had been wrong.[37] The tube railways then under construction were a grave extravagance; the provision of lifts had alone added 8 percent to the cost of building the Central London line. Shallow underground lines of the Metropolitan type (which required no such lifts) could by this time be built with comparatively little disturbance of the streets, as the Metro in Paris—then brand new—had demonstrated.[38] Lines of this sort should be preferred to deep-level tubes wherever possible. Almost the only point on which the Commissioners found themselves in agreement with current railway policy was in the tendency to favour electric traction.[39] (While their report was in preparation, the finishing touches were being put to the electrification of the Inner Circle, and the Brighton Company had begun the work of electrifying the South London line.) Yet though they were so much disturbed by the great expense of building tube railways, they ignored the financial problems that confronted some companies as soon as they began to contemplate electrification. The densest of the London suburban systems was that of the Great Eastern Railway, which was competently managed but indigent. Capital expenditure on the scale required for electrifying the lines out of Liverpool Street lay quite beyond its scope, without the benefit of State aid.

The Commission did a splendid work. The eight volumes of its report, evidence, appendices, and maps, constitute the most intelligent and thorough analysis of the problems of urban transport in the early twentieth century that any country can show, and they were studied with close attention on the Continent and in America. But the new Liberal government, which took office in December 1905, showed scant interest in the Commission's recommendations and implemented none of them. In the years that followed, the railways fought the electric trams and the motor buses all over London. It was a battle from which the State stood aside.

In considering the development of railways in London, one must be struck by the

The Victorian City: Shapes on the Ground

relatively small part that was played in it by the government of the city itself. This was in part because, at least until 1888, that government was dispersed among a large number of separate authorities—the Common Council of the City, the parish vestries, the Metropolitan Board of Works. In part, too, it was a consequence of the very size of London, of the inevitable difficulty of bringing public opinion to bear on matters of this kind. Something of the same sort can be seen in the biggest provincial cities too; but in those that were a little smaller the attitudes of the citizens could express themselves and might have a strong influence on the course of railway development. Let us consider two such towns: Sheffield and Nottingham.

Sheffield kept a particularly close eye on railway development. When the North Midland line, from Derby to Leeds, was projected in 1835–6, with George Stephenson for its engineer, it adopted a route from Chesterfield northwards that took it well to the east of Sheffield. Stephenson's stated reason for this choice was that he was thus

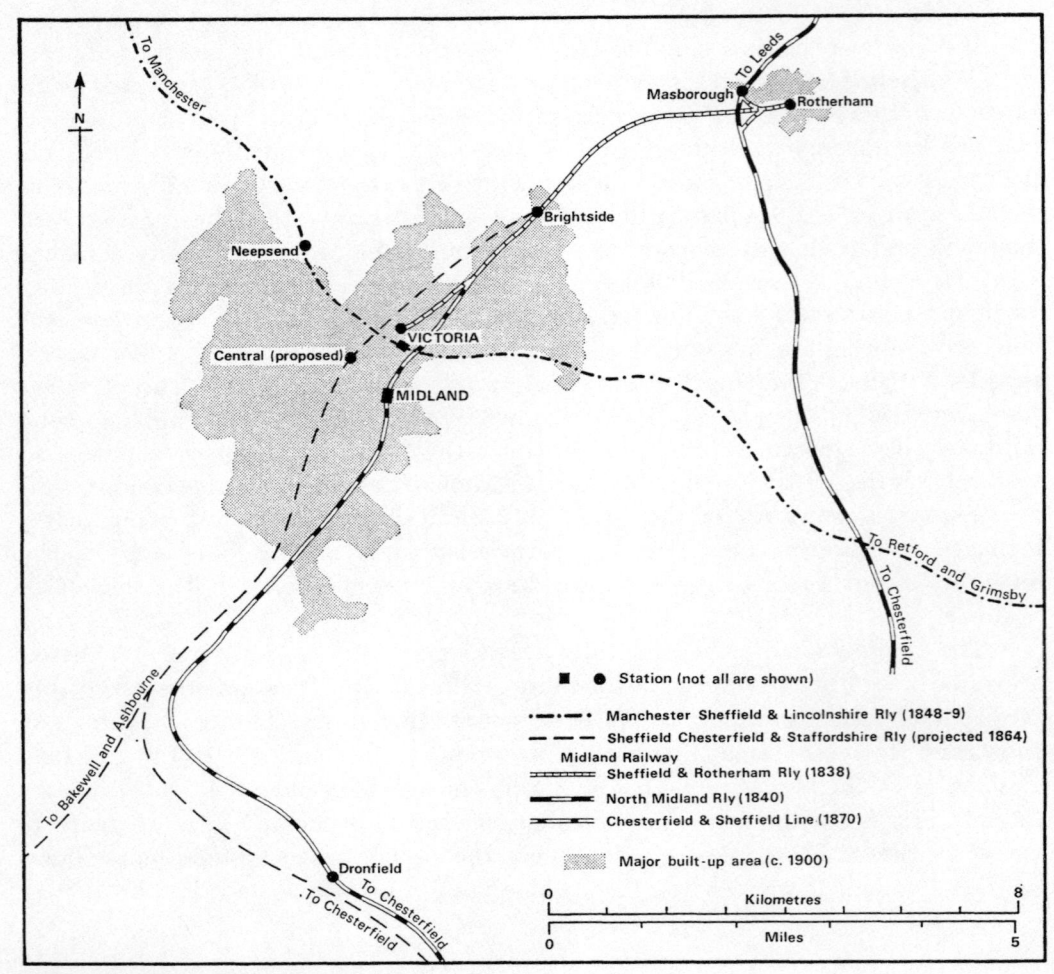

VIII Sheffield railway connections

enabled to keep to the high ground and avoid the heavy engineering works that would be necessary if the line took the direct route through the town, tunnelling at Dronfield and then running down the steep valley of the Sheaf. There was more to it than that, however. The Sheffield Coal Company and its landlord the Duke of Norfolk did not wish to stimulate the development of the competing Dronfield coal. When Stephenson came down to Sheffield to defend his plan his behaviour was arrogant and disingenuous. At one point he even went so far as to threaten the townspeople that if they succeeded in delaying the passage of the North Midland's Bill through Parliament 'Sheffield would be for ever deprived of any line whatever'. His performance gave great offence; but in the end, through apathy, the opposition to the Bill in the town collapsed.[40] In consequence, the line came no nearer to Sheffield than Masborough, five miles away to the north-east, whence a branch was built by a separate undertaking, the Sheffield & Rotherham Railway. Both projects were authorized by Parliament in 1836; they were in operation by 1840.

It quickly became clear that Sheffield had been wrong to acquiesce in Stephenson's plan. It was soon placed on what promised to be the chief east–west trunk line, from Manchester to Grimsby; but that accentuated the other failure, which prevented Sheffield from becoming one of the principal railway junctions of the North of England. During the 'railway mania' a succession of projects was brought forward with frantic insistence, aimed at giving Sheffield a direct route to the south: the Gainsborough Sheffield & Chesterfield, the Manchester Sheffield & Midland Junction, the Sheffield & Newark, the Boston Newark & Sheffield—all were launched in 1844–6, and all failed.[41] For fifteen years Sheffield then sank back into a tame acceptance of the existing situation, improved only, as far as passengers were concerned, by the opening of the Great Northern Railway and, after 1857, the running of a relatively brisk service via Retford to King's Cross.[42]

The Midland Railway built up for itself a large company of enemies in Sheffield. Its station, inherited from the Sheffield & Rotherham Company, was deplorable. Even Allport, the Company's General Manager, admitted that 'great inconvenience had arisen from [its] insufficiency';[43] at a meeting of the Sheffield Town Council a member observed that it was 'of such a character that even decent cattle might take offence at it'.[44] The coal and minerals of the Dronfield district remained inadequately exploited.[45] It was asserted by a Town Councillor—and he was not contradicted—that the price of coal in Sheffield was higher than it had been before the first railway arrived.[46] Iron and steel manufacturers were loud in their complaints against the Midland Company.[47]

The possibility of a new line to the south, joining the Midland main line at Chesterfield, reappeared in the autumn of 1861, and the Town Council appointed a committee to confer with the Midland directors. But the project moved forward at a snail's pace, and it began to be hinted that the Great Northern Company, one of the Midland's chief rivals, might interest itself in the matter. Next year the Midland revived its plans, and a new project also came forward, at first bearing the compendious title of the Sheffield Chesterfield Bakewell Ashbourne & Stafford & Uttoxeter

Junction Railway, later shortened to Sheffield Chesterfield & Staffordshire.[48] The chairmanship of the Company was accepted by the Mayor of Sheffield, the iron and steel manufacturer John Brown. In the Parliamentary session of 1864 it did battle with the Midland.

It seemed strong in its chairman and in the support of the Town Council, which petitioned in its favour. Yet it collapsed ignominiously. The chief promoter of this venture appears to have been William Field, one of the great railway contractors of the day, who had no local interest in it at all. He undertook to find £800,000 of the cost if a further £300,000 were forthcoming from Sheffield. Of that £300,000 the amount actually subscribed was rather less than £10,000.[49] So much for the urgent anxiety of Sheffield for the new railway. Worse still, the deposit required by the Standing Orders of the House of Commons was irregularly raised, not in the form of personal investment but through a loan from the General Insurance Company. When this came to light, the project was held to have evaded Standing Orders; and that settled its fate.[50] It was an interesting scheme. The line would have been very expensive to build and operate,* but from Sheffield's point of view it had important merits: it gave direct access to Chesterfield and the south, serving the intervening coalfield; it provided a valuable route to Staffordshire and South Wales; it offered a centrally-placed station in Sheffield;† and—a particularly interesting feature—it proposed to build six other stations within the borough boundary, linking them by a frequent service at fixed intervals, on the model of that run by the North Staffordshire Company in the Potteries.[51] Nevertheless, the Midland prevailed, securing powers in this session to build the line through Dronfield to Chesterfield.[52]

It was in no hurry, however, to exercise them. The Company was over-extended in many directions at the time, and hard hit by the Overend, Gurney crisis of 1866. The new Sheffield line took a low priority among its commitments. It was not opened until 1 February 1870.[53]

Sheffield had now at last achieved the objectives for which some of its citizens had been contending for thirty years: direct railway access to an under-exploited mineral field on its doorstep, a place on a trunk route from north to south, and a new railway station above the contempt of cattle. With this achievement it rested content —though the train service provided by the Midland and Great Northern Companies exhibited for a time the most laughable features of English railway competition.‡ But the Town Council still kept its eye closely on railway matters. In 1881 it gave

---

* The deposited plans (there is a copy of them in the Sheffield Central Library) show that, of the thirty-six miles of main line, thirty were graded at one in 130 or steeper, and over seven miles were in tunnel.
† This was to be at the junction of Townhead Street, Tenter Street, and Broad Lane (there is a roundabout there today), much closer to the middle of the town than any other railway station, then existing or subsequently built. It could not have been a satisfactory station, however, for it would have had to be crammed into a space of only 900 ft between a viaduct and a tunnel.
‡ Consider the down service provided by the two companies eighteen months after the opening of the Midland main line (*Bradshaw*, July 1871) as shown at foot of page 289.

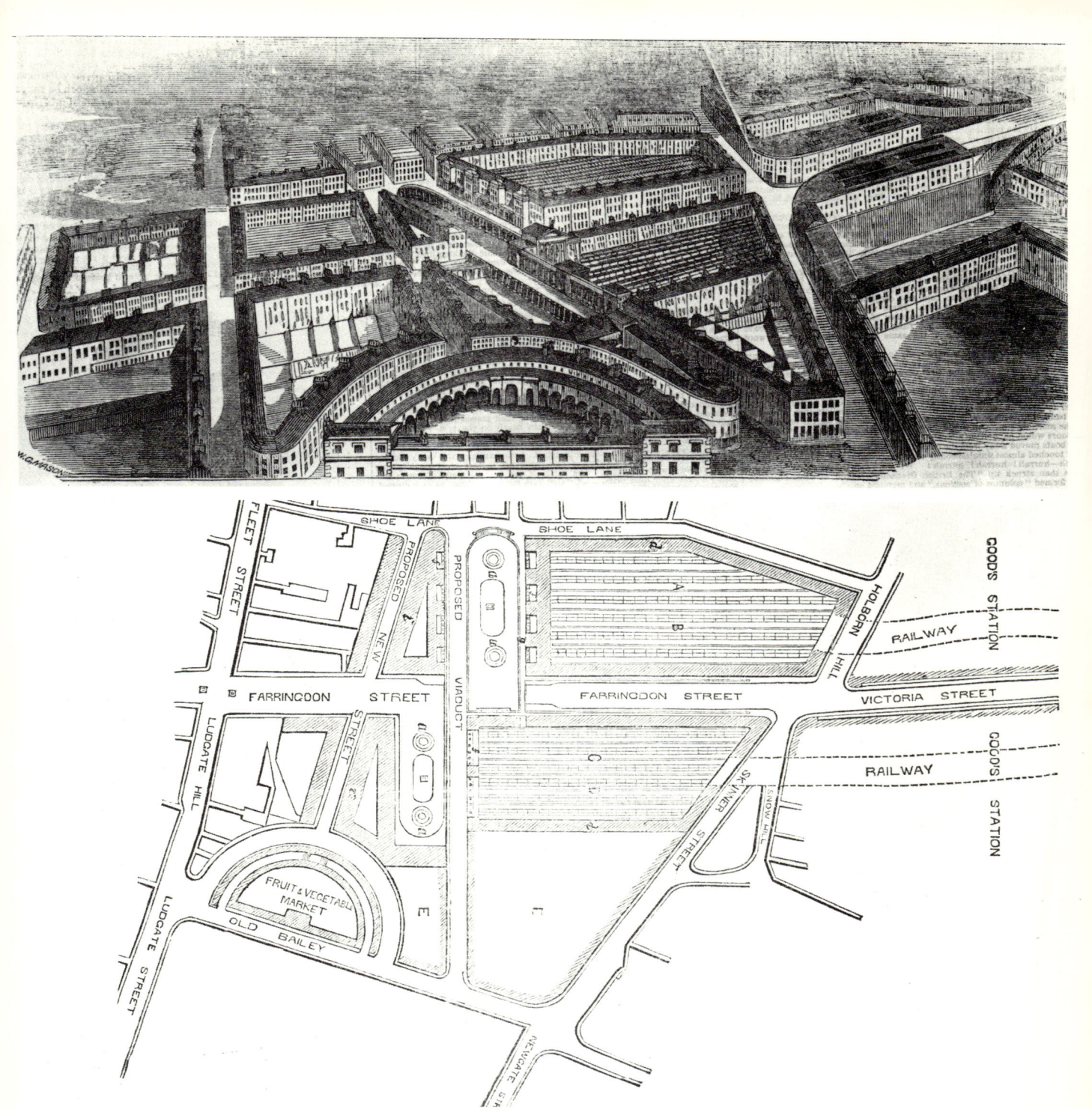

236–7 Charles Pearson's proposed Central Railway Terminus in London. The separate stations, for passengers and goods, to be used by a number of companies, appear in the upper part of the picture to the right with colonnaded entrances and forecourts. The plan shows the new streets he proposed, in conjunction with them, as a further London improvement. It shows stations for four companies (marked A, B, C, and D), and vacant sites (marked E) for one more. From *Illustrated London News*, viii (1846), 333–4.

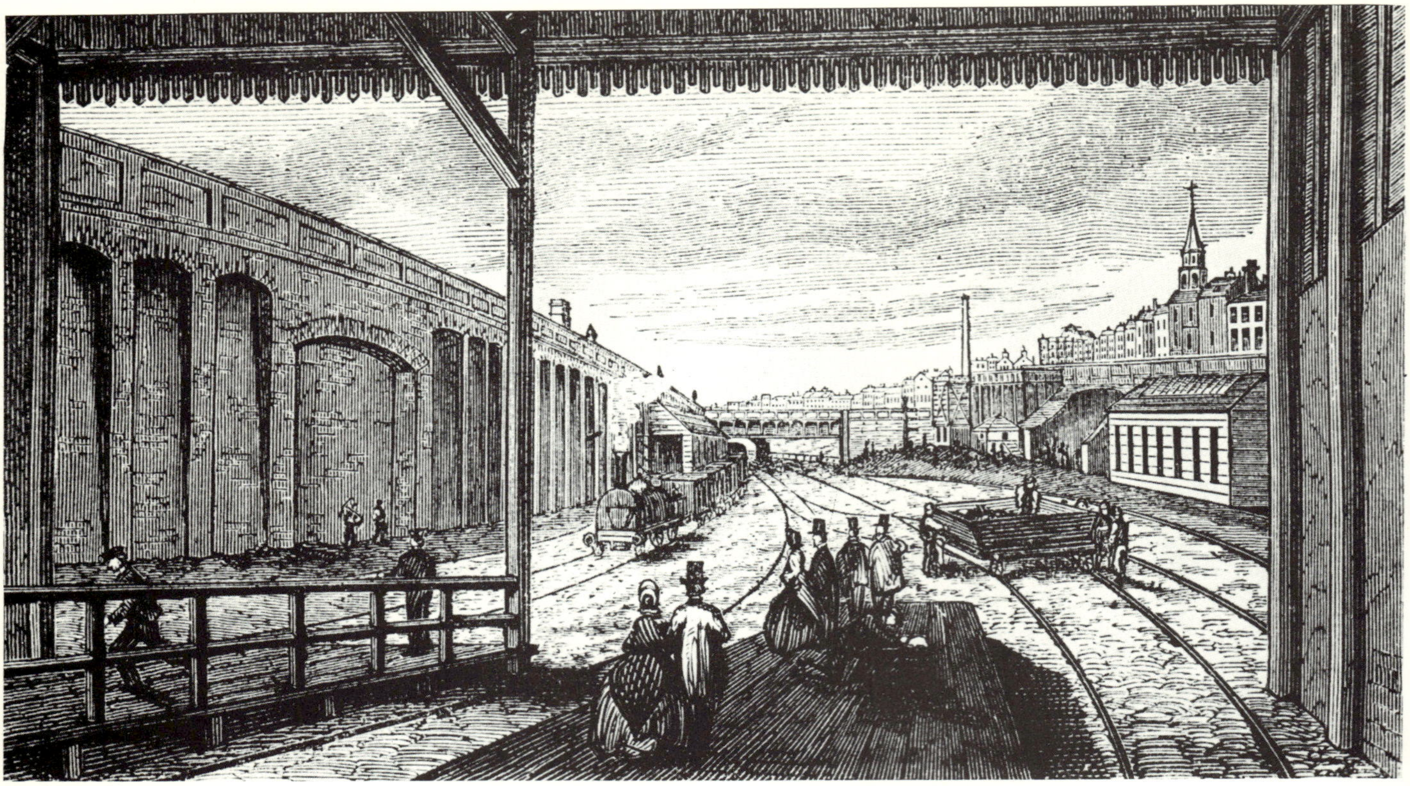

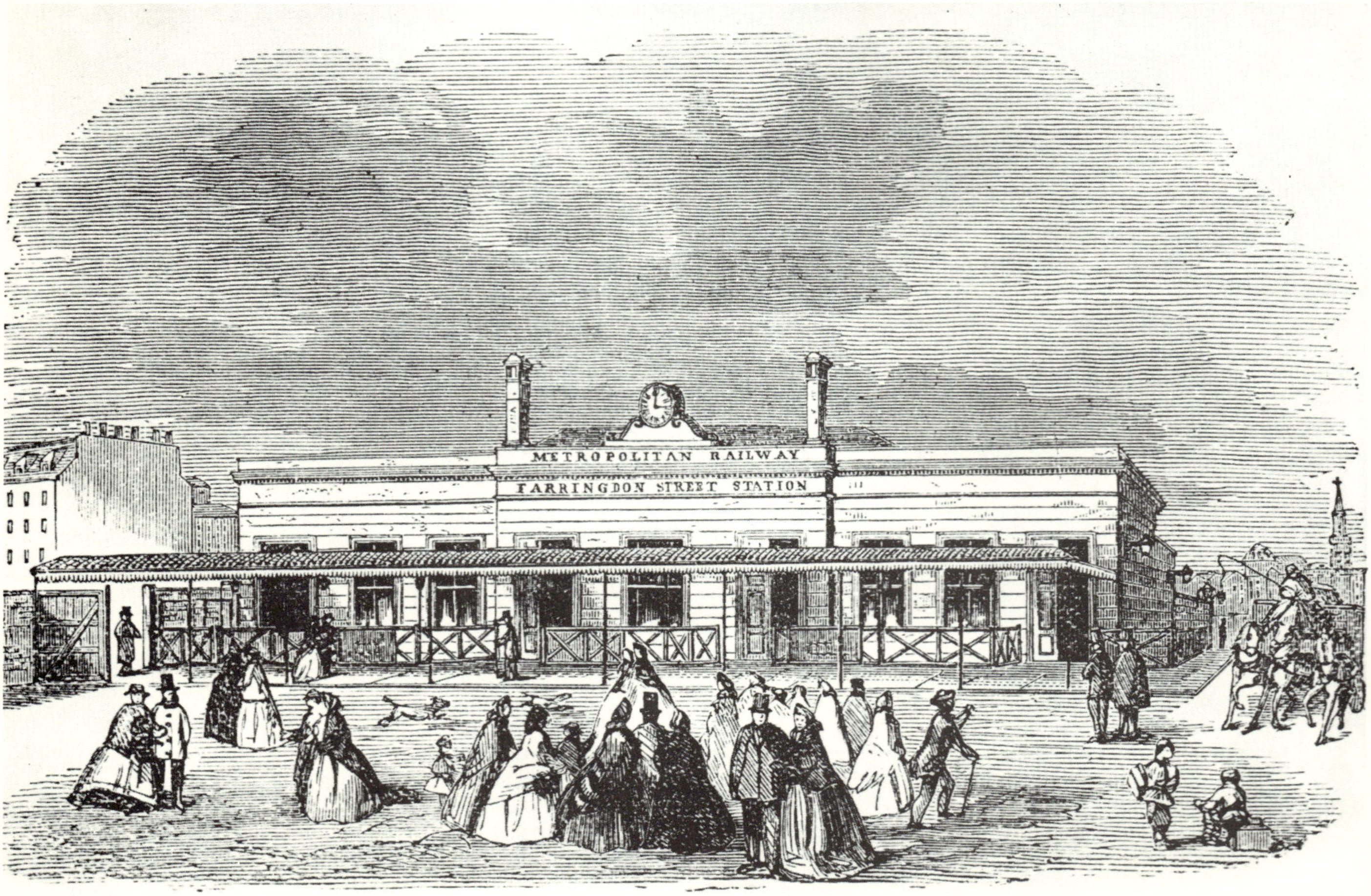
METROPOLITAN RAILWAY
FARRINGDON STREET STATION

238–9 *left* Metropolitan Railway: Farringdon Street station. The interior view looks east and shows the 'widened lines' to the Great Northern and Midland Railways on the left. From W. J. Pinks, *History of Clerkenwell* (2nd edn, 1880), pp. 357, 359.

240 *above* The Metropolitan Railway crosses the line coming in to join it from the Great Northern west of Farringdon Street station. The Metropolitan train is entering the Clerkenwell tunnel. The crossing was known as 'the gridiron'. This picture appeared immediately after the Great Northern passenger service on the line began, on 17 February 1868. From *Illustrated Times*, xii (1868), 137.

241 *above* The construction of a shallow underground railway: making the junction, close to Baker Street station, between the Metropolitan and Metropolitan & St John's Wood Railways. From *Illustrated Times*, xii (1868), 216.

242 *above right* The Thames Embankment under construction at Charing Cross. On the left the Metropolitan District Railway is seen in its tunnel; in the centre the sewers and a proposed 'pneumatic' railway intended to run from Charing Cross under the river to Waterloo, but never in fact built. From *Illustrated London News*, l (1867), 632.

243 *right* Blackfriars Bridge, London Chatham & Dover Railway: detail of the decoration. 'The entire invention of the designer [Joseph Cubitt] seems to have exhausted itself in exaggerating to an enormous size a weak form of iron nut, and in conveying the information upon it, in large letters, that it belongs to the London Chatham & Dover Railway': Ruskin, *Works* (ed. Cook and Wedderburn), XIX, p. 26. Engraving from W. Humber, *Record of the Progress of Modern Engineering, 1864* (1865), plate 27.

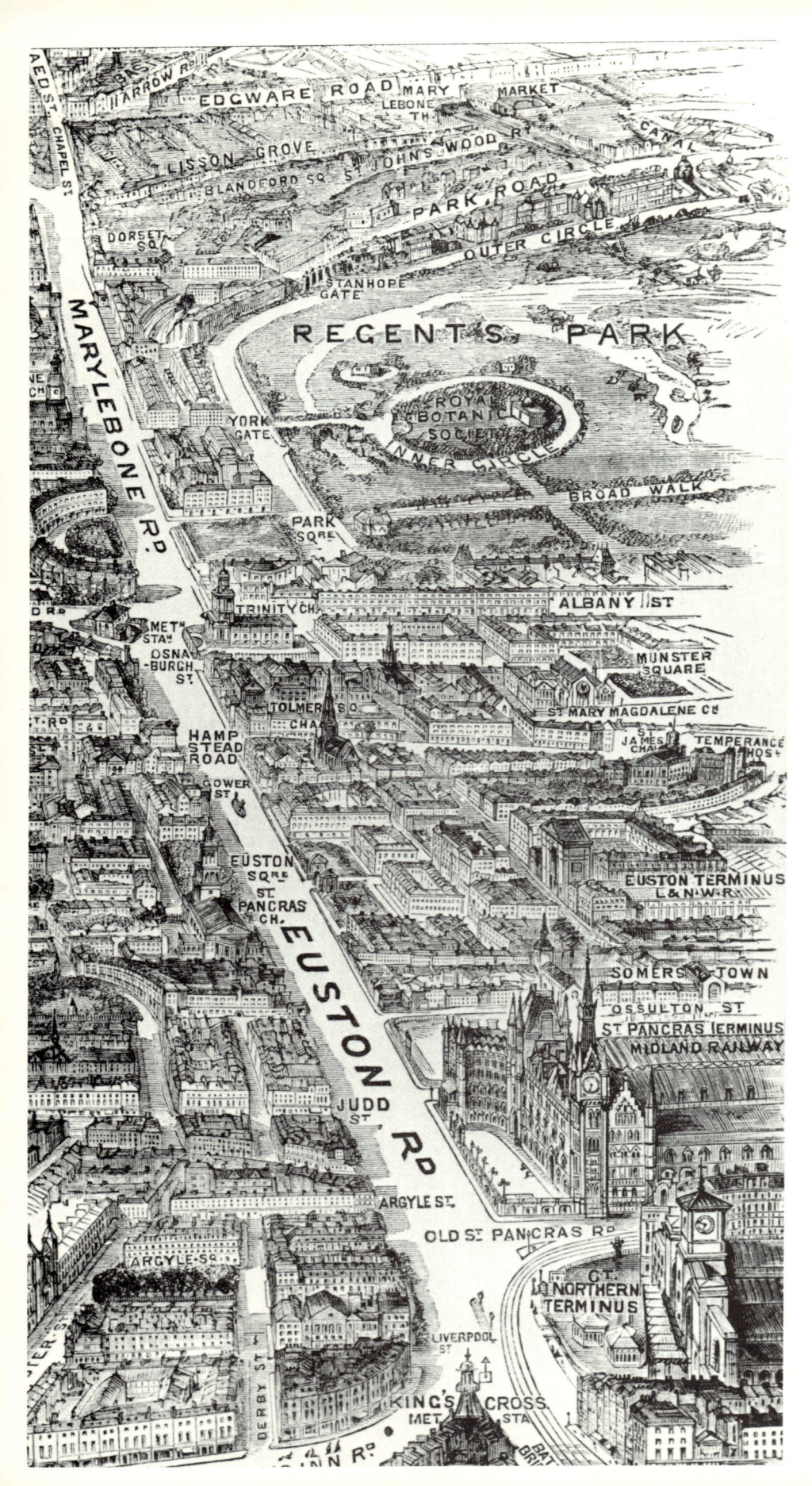

244 *left* Bird's-eye view of the Euston and Marylebone Roads, showing the four chief northern termini: King's Cross, at the foot, followed by St Pancras, Euston, and at the top of the picture Paddington. A fifth, Marylebone, was later to be built between Dorset Square and Lisson Grove. King's Cross station of the Metropolitan Railway is to be seen at the bottom; the line followed the course of the Euston and Marylebone Roads and Praed Street. Its Portland Road station, on an island site, has been exactly replaced by the modern Great Portland Street. From Herbert Fry, *London in 1885* (1885), plate xiv.

245 *above right* Bird's-eye view of the London & Greenwich Railway, rising above the houses on its continuous viaduct. The conspicuous church with an Ionic column for its spire is St John's, Horsleydown (now a ruin). The railway terminus is at Tooley Street—the nucleus of the modern London Bridge Station. From a lithograph by G. F. Bragg.
Photo: Science Museum, London

246 *right* St Pancras station under construction, 1868. The outer brick wall on the east side is nearly complete. Three of the spans of the roof have been erected, with the aid of the movable timber scaffolding that stands beneath them. In the foreground is an arch of the tunnel being built to contain the Fleet Ditch. From *Illustrated Times*, xii (1868), 149.

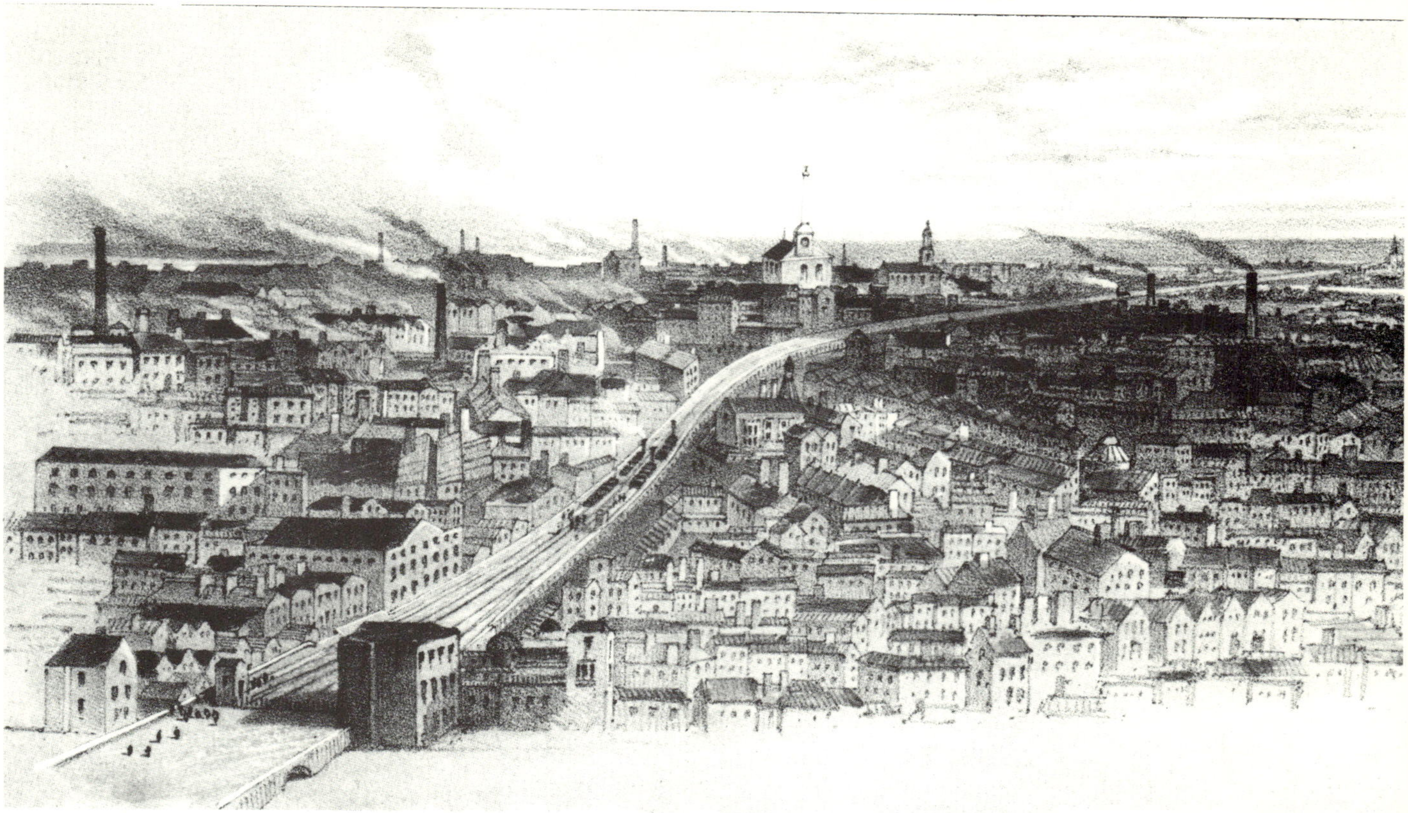

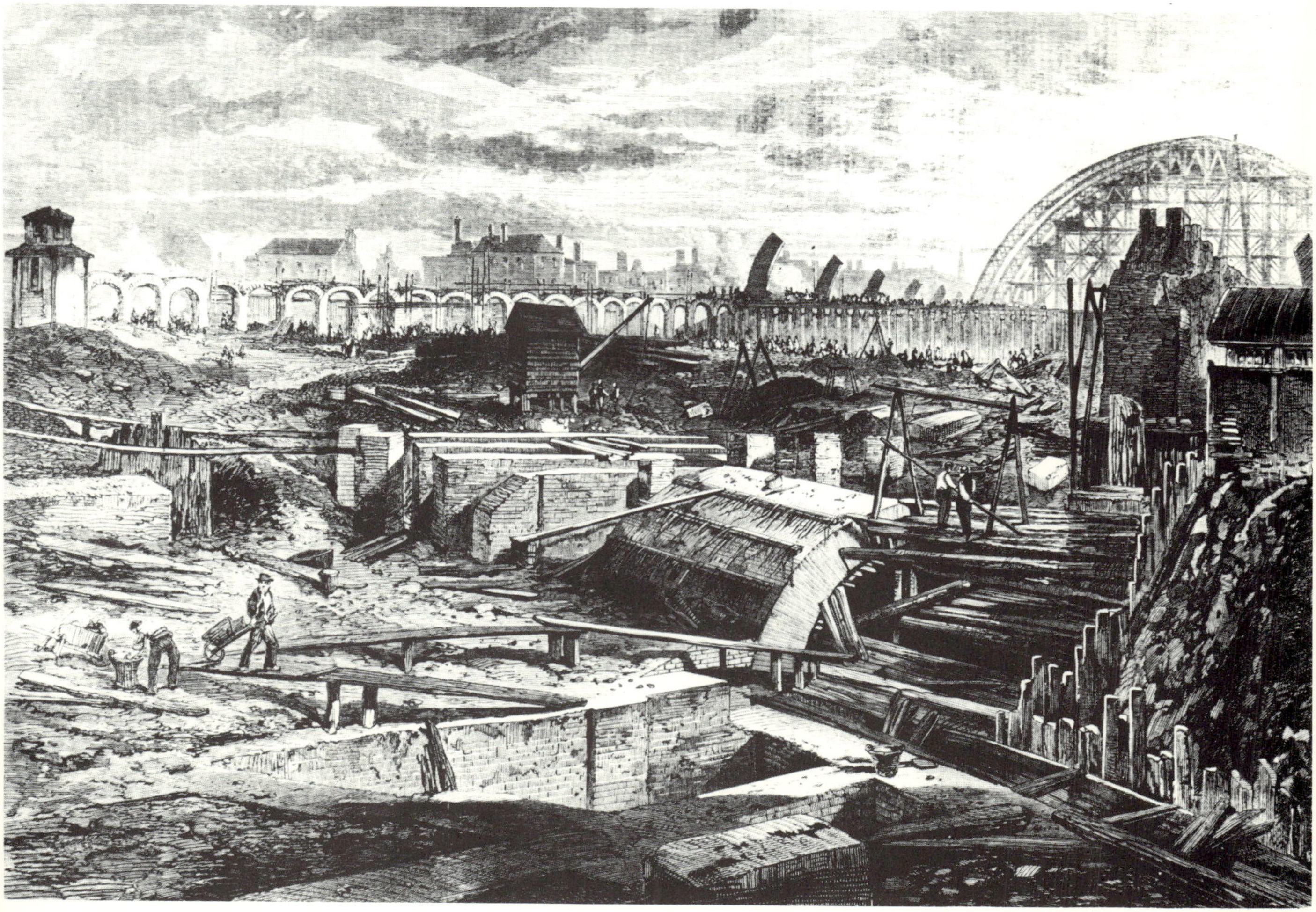

247 *above left* Hawkshaw's bridge and station at Cannon Street; London Bridge in the background.
below left The station roof under construction. From *Illustrated Times*, vi (1865), 99, 241.

248 *above right* The last fragment of the old Fleet Prison, on the east side of Farringdon Street close to Ludgate Circus; the London Chatham & Dover Railway above. From *Illustrated London News*, lii (1868), 261.
below right Ludgate Hill station. From *Illustrated Times*, vi (1865), 41.

REWARD
PARTINGTON & Co
CRYSTAL PALACE
NOW OPEN
DAILY TELEG
PARIS
MARGAT
RAMSGATE
SOLD
LUDGATE
HILL
VICTORIA
STATION
SALE

LONDON CHATHAM & DOVER RAILWAY
KINGS CROS

GREEN & McALLAN.
PRINTERS.
ART PRINTERS
HUTCHINGS & CROWSLEY
KING LUD.
OFFICES OF THE
GENERAL PRINTERS

249 *left*   The bridge over Ludgate Hill.
Photo: G. W. Wilson
Aberdeen University Library

250 *below*   The Great Northern Railway bridge over Friargate, Derby.
Derby Museum & Art Gallery

251 The North London Railway bridge over Church Street (now Mare Street). From a lithograph by C. J. Greenwood.
Photo: Science Museum, London

252 The Great Western Railway's entrance into Bath through Sydney Gardens, 1841. From a lithograph by J. C. Bourne.

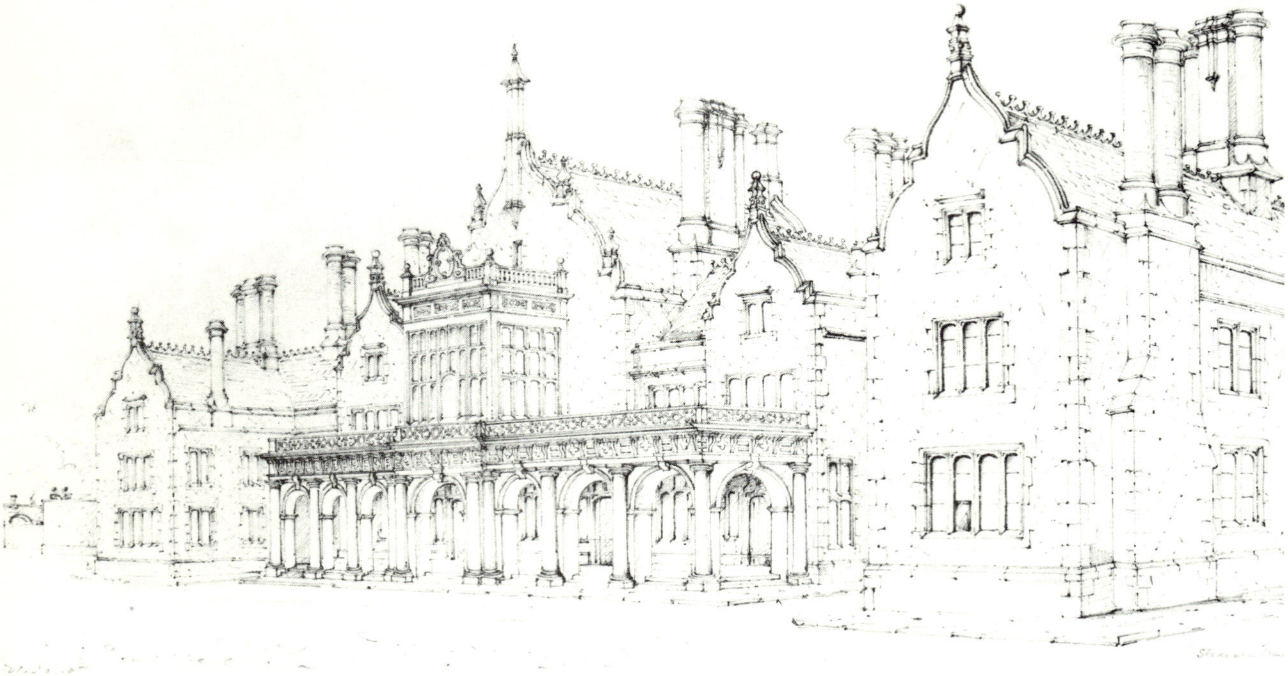

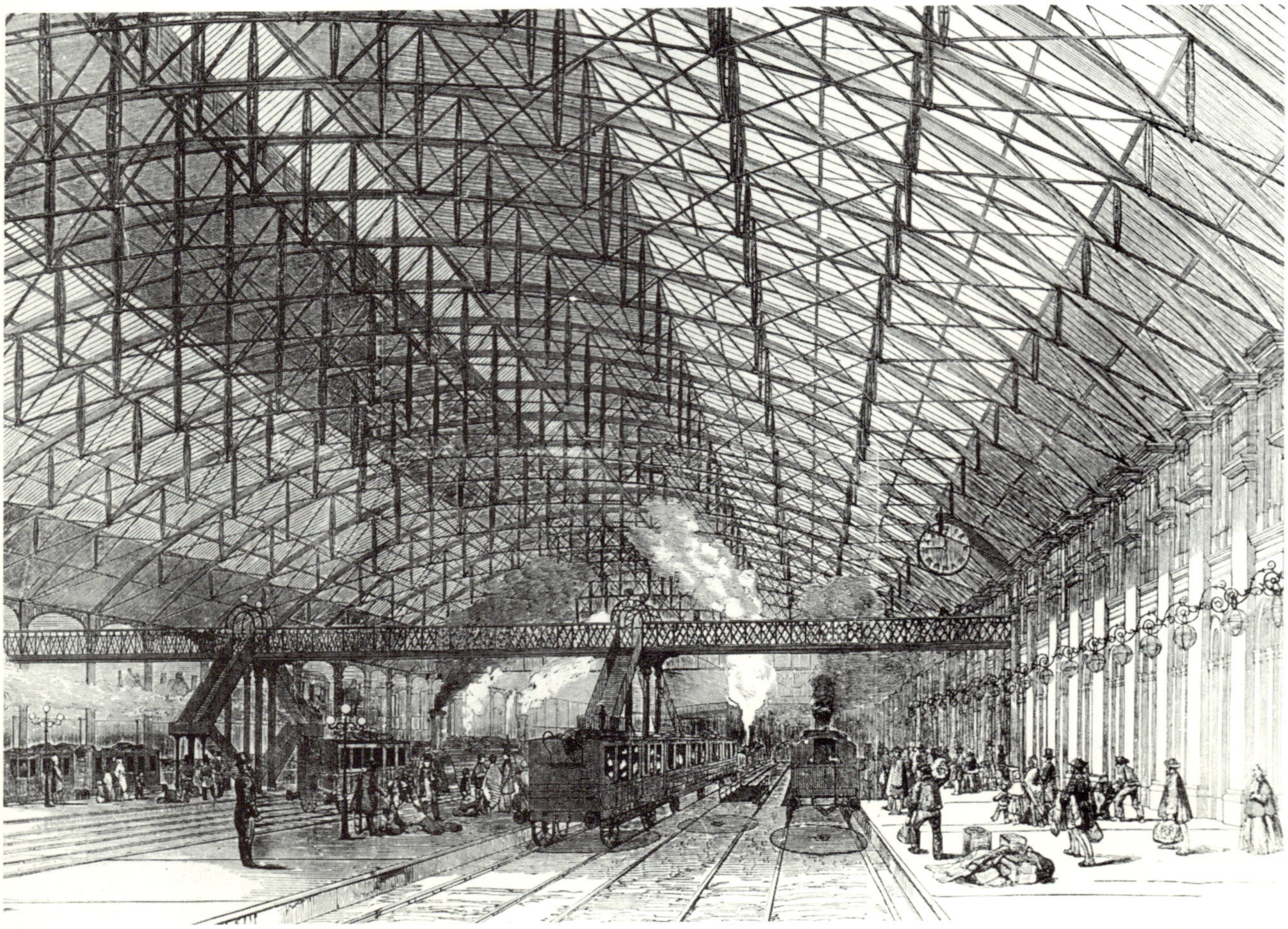

253 *left* Stoke-on-Trent station. Drawing by George Buckler: British Museum, Add. MS. 36387, fol. 232.

254 *below left* Birmingham, New Street station at the time of its opening. Its iron and glass roof, designed by E. A. Cooper, then claimed to be the largest single span in a station anywhere; its greatest width was 211 ft. From *Illustrated London News*, xxiv (1854), 505.

255 *below* Bristol, Temple Meads joint station (1871–8). The architect was Sir Matthew Digby Wyatt.
Photo: G. W. Wilson
Aberdeen University Library

256 *above* North end of the Royal Border Bridge, Berwick-on-Tweed. The railway cuts straight through the Castle. *National Monuments Record*

257 *below* The Great Central Railway striding across the Old Town of Leicester, on a viaduct ¾ mile long, built at the close of the railway age (1899). The road running across the lower part of the picture is aligned on the Roman Fosse Way; the Norman castle and its collegiate church stand towards the top left-hand corner of the picture. In the centre foreground is the yard of the Leicester & Swannington Railway, built in 1832 on the site of the Augustinian Friary. This photograph was taken in 1936. *Aerofilms Ltd*

evidence in the Parliamentary enquiry into railway rates; it petitioned the House of Commons in favour of the Railway & Canal Traffic Bill of 1888; two years later it urged the directors of the railway companies serving the town to provide a single central station.[54] Nor did it confine its attention to such large issues of general policy. It wrestled with the Manchester Sheffield & Lincolnshire Company for thirty years to secure a station to serve the populous north-western part of the town at Neepsend. Its pertinacity triumphed at last when the station was opened on 1 July 1888.[55]

Here was an attempt to improve transport to and from the suburbs, before the age of the mechanized road vehicle had arrived. So far as Sheffield was concerned it stood in isolation. But in Nottingham the Town Council took that matter seriously. It became interested in railways at an early date: chiefly because Nottingham, too, found itself placed on branch lines and feared the loss of its trade, in consequence, to its more happily-situated neighbours, Derby and Leicester. When the 'mania' came in 1845 the Council set up a special Railway Committee to examine proposals affecting the town. It looked into thirty-five in all.[56] The strongest support appeared for plans to build a line to link the Great Northern at Grantham, through Nottingham, with Manchester and Liverpool. This line eventually emerged in the hands of the Great Northern Company, which took the traffic east of Nottingham, and the Midland, which took it over to the west. Nottingham remained the focus of a series of branch lines until 1875–80, when the Midland Company at last opened a new trunk line to the north by Radford and to the south by Melton Mowbray and Kettering.

The interest taken by the Nottingham Council in railway development did not cease when the town had at last been placed on a main line. It had long had a particularly onerous and difficult task in promoting the development of inner suburbs, in order to relieve the dreadful overcrowding in the slums near the centre of the town.[57] That development could not begin until the common fields were enclosed, under an Act of 1845; and it was only in 1877, with a rational extension of boundaries, that the Council could grapple with Nottingham's urban and suburban problems as a whole. When, in 1885, local businessmen promoted a Nottingham Suburban Railway, the Corporation adopted an attitude of positive benevolence towards the scheme that was highly unusual among Victorian local authorities. The plan provided for a line 3¾ miles long, skirting the north-eastern side of the town, into which a large population was then moving. Its principal sponsor was Edward Gripper, who had established very large brickworks at Mapperley, in the district which the railway was

---

		a.m.					p.m.			
King's Cross (via Retford)	dep.	7.40		10.00			12.00	2.45		5.00
St Pancras	dep.		8.45		10.00	11.45		3.00		5.00
Sheffield	arr.	12.25	12.42	1.53	1.53	4.11	4.35	6.30	7.40	8.53	8.53

Thus of these ten trains, two pairs departed from adjacent stations in London and arrived in Sheffield at identically the same times, two other pairs left London within a quarter of an hour of each other. The responsibility for this farce rested with the Midland Company, for the Great Northern's timetable had been unchanged since before the Midland line opened.

The Victorian City: Shapes on the Ground

designed to serve.[58] The Council came forward with a grant towards the preliminary costs of promoting the scheme, and when Parliamentary sanction was granted in 1886 it was specially authorized to accept payment for land taken for the railway in the form of shares in the Company; this on the ground that the construction of the line would be 'of special benefit to the inhabitants of the town and borough of Nottingham'.[59] The line was opened in 1889.

The Council's efforts continued. When the Manchester Sheffield & Lincolnshire Company announced the plans for its London extension in 1890 it saw a fresh oppor-

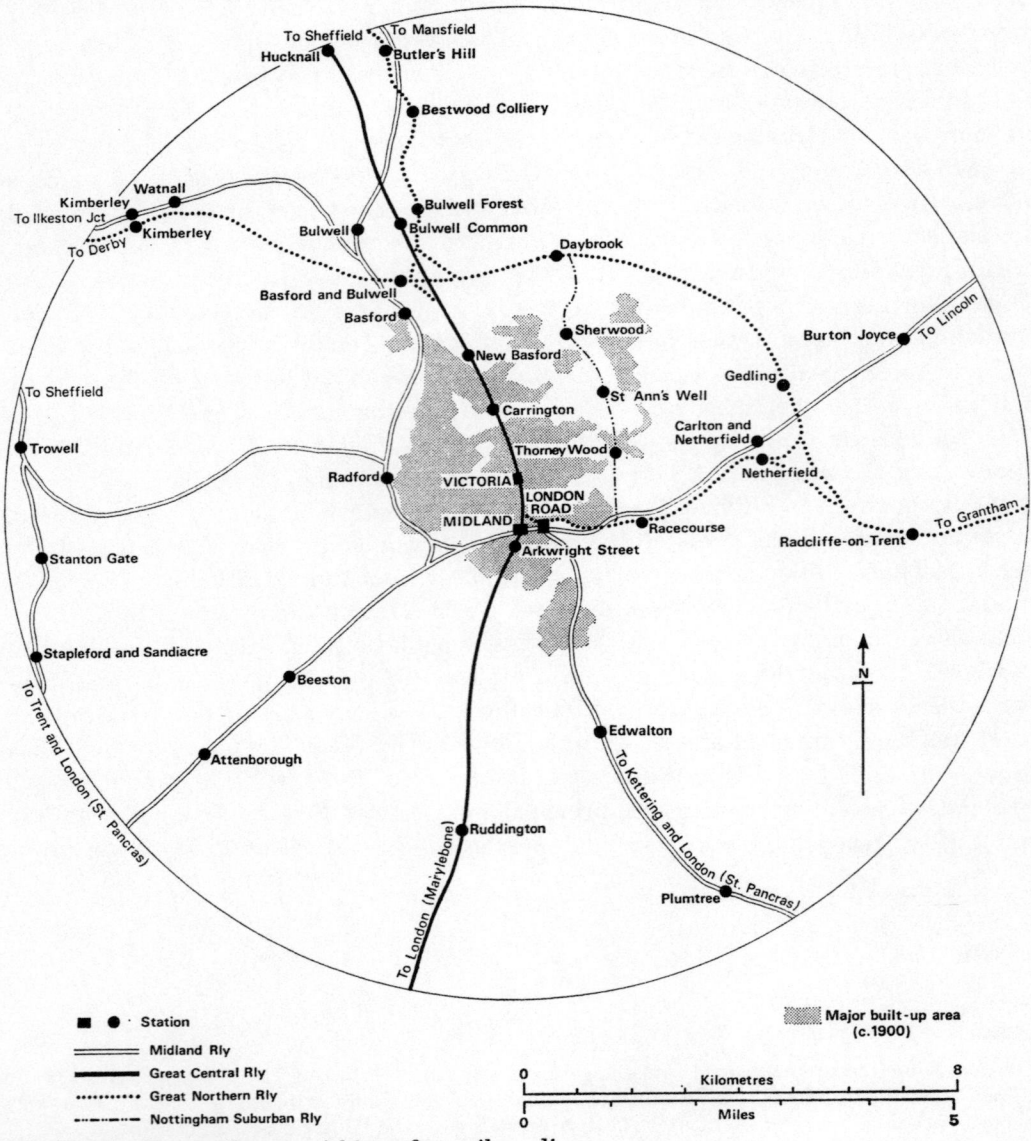

IX Nottingham railways within a five-mile radius

Simmons: The Power of the Railway

tunity. The Council was determined that this line too should help to solve the problems of suburban transport. It therefore insisted, successfully, that the railway company should be obliged to open four stations, in addition to the main one, within the limits of the borough. Their location was indicated in the Act of 1893, which required the provision of 'a reasonably effective service of local trains', stopping at each of these stations.[60]

By the time this railway was opened in 1899, the Nottingham Council had taken another decision, which diminished the value of such suburban railway services.

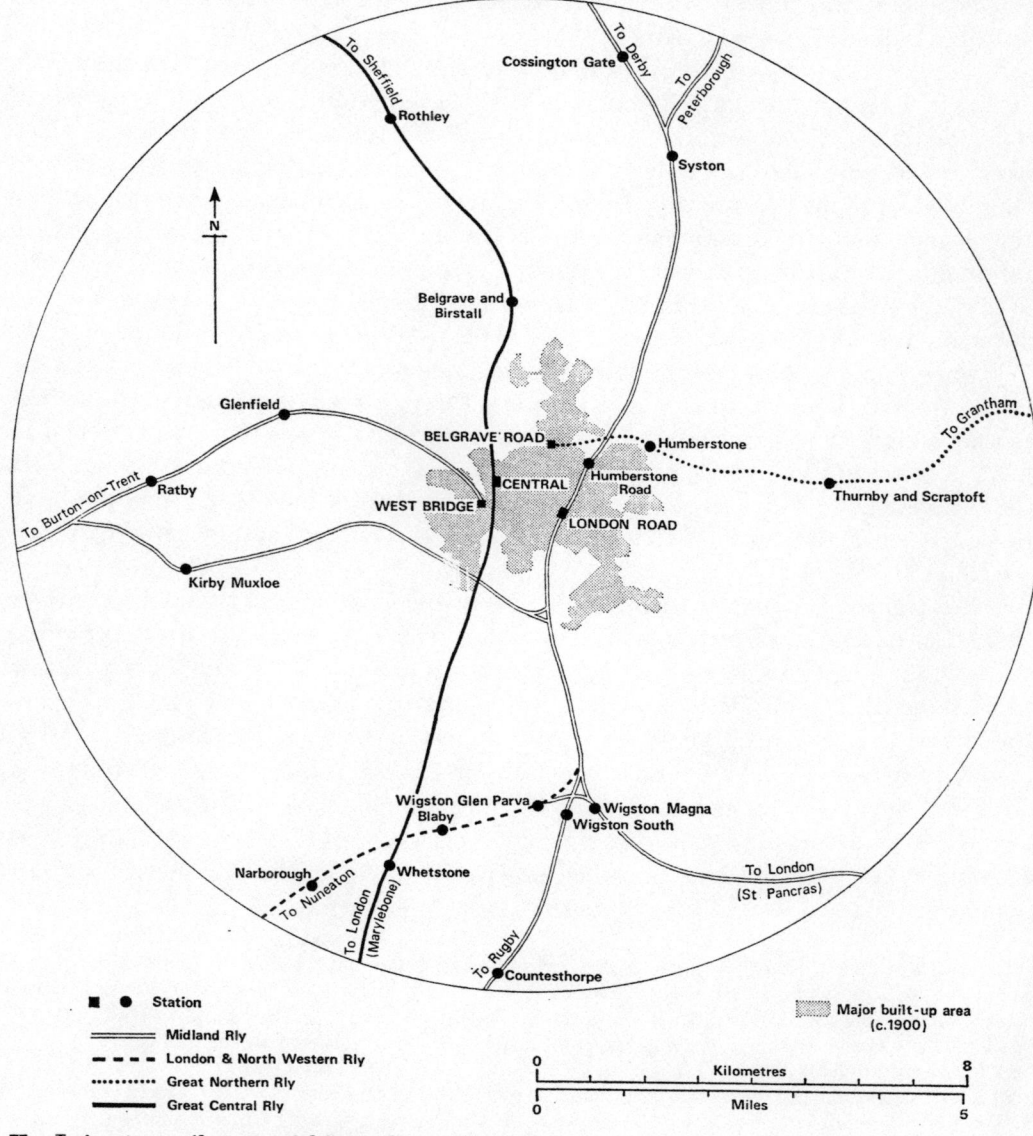

X Leicester railways within a five-mile radius

In 1897 it had bought out the privately-owned Nottingham Tramway Company, and in the following year it had decided to electrify the system.[61]

In Sheffield and Nottingham, then, the cities themselves were actively involved in railway development.* Though this book is concerned with the Victorian city, that is the large town, it is useful to glance briefly at what happened in small towns too: magnified as it were under a microscope. One can also see something of the power which the railway wielded in shaping its own course, regardless of the pleas or the hatred of the communities by which it passed. These towns had been there long before railways were thought of—prosperous markets, even centres of industry. How did they fare when the railway came?

Very variously. Some were fortunate or far-sighted enough to associate themselves with the railway: like Peterborough, which turned itself into a railway town, or Swindon, which found itself hitched up to a railway town adjoining it. Some, which were prospering before the railway developed, went on growing for a little and then stagnated, perhaps because they lay off a main line. Look at Kendal and Frome: two towns that were important enough to secure the right to separate parliamentary representation in 1832, and were deprived of it in 1885. Kendal was by-passed by the West Coast main line to Scotland—the two-mile branch from Oxenholme was a poor substitute. Frome's railway history was truly ironical. Placed on the meandering Wiltshire Somerset & Weymouth line, it found itself on a trunk route in 1906, when the Great Western inaugurated its new main line through Westbury—only to be relegated once more to a loop when a by-pass, running south of the town, was opened in 1921. Neither of these towns ceased to be industrial centres: K Shoes still come from the one, the high-class printing of Butler & Tanner from the other. But both have remained small, and for that their position on the railway system is in part responsible.

We can see the matter more clearly, however, if we look at towns smaller than these. Let us use Tewkesbury as an illustration. When Mr Pickwick stopped to dine there at the Hop Pole, it was a comfortable town of about 5,800 people, with a few small industries, profiting handsomely from its position on the great highway from the North to the West. Twenty-six coaches a day went from Tewkesbury to Worcester, twenty to Gloucester, besides others for Hereford and Malvern.[62] When the Birmingham & Gloucester Railway was first promoted, the Tewkesbury people complacently assumed it must pass through their town; but Cheltenham—a rapidly-growing place, already four times as large—insisted that the line should run by Ashchurch instead, two miles away to the east.† A public meeting of protest was held

---

* Other Corporations showed a no less anxious and long-continued concern in the matter. Railways took up a large part of the time of the standing Commerce of the Port Committee in Southampton; the Corporation subscribed to the Didcot Newbury & Southampton Railway in 1881 and gave it the most explicit and public support: Southampton Record Office, SC2/3/11, pp. 86, 96–7, 99, 112.

† Cheltenham's fear, no doubt, was that the line might follow the Severn valley closely through Tewkesbury (which would give a shorter route to Gloucester), serving Cheltenham by a branch. The line by Ashchurch was, however, superior from the engineers' point of view.

and two petitions submitted to Parliament. In vain. The railway company would only agree to build the line 'as close to Tewkesbury as practicable' and to throw off a branch running through the town to the Quay on the Severn.[63] After a little shilly-shallying on the Company's part, this was achieved. When the branch was opened in 1840 it was worked by horses; locomotives were not regularly used on it until 1844.[64] The trade of the town was now stagnant; drained away, one may fairly infer, by the railway to Cheltenham and Gloucester. The population rose, it is true, in 1851 to 5,878. But that was its peak. Thirty years later it was only a little over 5,000; it was 5,400 in 1901.[65] 'Whatever may be the ultimate effect of railroads', a local journalist remarked in 1841, 'it is evident that their introduction has hitherto been one of almost unmixed evil to the inhabitants of Tewkesbury.'[66]

This case stands by no means alone. There are many other examples of English towns like Tewkesbury: ancient urban centres in a rural society, which attained the limit of their growth in 1841–61 and then fell back for the rest of the century.[67] Such towns did not always stand on branch railways. Wellington had a station on the Bristol & Exeter line from its opening in 1843; but it reached its maximum size in 1851 and thereafter declined for forty years. Though Moreton-in-the-Marsh got its station on the Oxford–Worcester line in 1853, it was a smaller place in 1901 than it had been in 1851. It is, however, fair to say that the majority of the towns that went this way in the second half of the nineteenth century suffered from inferior railway communication or the total want of it.*

The railway had two outstanding duties to the towns it served: to promote their trade and industry and to make them more satisfactory places to live in. How did it perform them?

Nobody would dispute that railways literally created Crewe or New Swindon; or that they made substantial and old-established towns, like Doncaster and Darlington, into major centres of industry; or that they added a new element to places that were already industrial centres, by siting large works there, as at Derby and Wolverhampton. The equipment they needed was at first supplied almost wholly by private firms. The most successful of these firms grew steadily, to become important elements in the economic life of the towns in which they were situated. For some time they were on a relatively small scale. Robert Stephenson & Co. in Newcastle were employing only about 500 men in the late 1830s; they were building some thirty locomotives a year, and their turnover was about £60,000.[68] In the forties, however, the scale of operations was transformed. The rapid development of

---

* They are particularly numerous in the south-western counties, where the clothing trade was declining in these years in favour of Lancashire and Yorkshire. Wiltshire, for example, can show at least eight: Great Bedwyn, Highworth, Malmesbury, Market Lavington, Marlborough, Melksham, Ramsbury, and Warminster. Of these, Highworth and Malmesbury stood on branch lines; Great Bedwyn, Marlborough, Melksham, and Warminster on secondary cross-country routes; Market Lavington was not served by a railway until 1900; Ramsbury never at all.

the railway system set up a demand for plant of all kinds, from Britain and overseas. The railway manufacturing firms grew much bigger, partly as a consequence of amalgamations. At the same time the railway companies, competing with one another for the equipment they needed, often found it difficult to secure punctual delivery and began manufacturing on their own account. The companies' locomotive works at Crewe and Swindon were both established in 1843 and turned out their first engines two or three years later.[69] The Sheffield Ashton-under-Lyne & Manchester Railway* set up its works at Gorton in 1846–8. The choice fell on Gorton because it was far enough from the centre of Manchester 'to be clear of the heavy local taxes with which all such establishments in large towns are burdened'.[70] Those words were used by Richard Peacock, the Company's Locomotive Superintendent, and when he left the railway to set up in business on his own, in partnership with Charles Frederick Beyer, he selected a site for the new works just opposite those he had established in 1846. The relationship between the two companies was perfectly friendly; there was work enough for both of them. With Ashbury's carriage and wagon factory near by, the Lancashire & Yorkshire Company's works at Miles Platting, and other private locomotive builders such as Sharp Bros, Manchester became an important centre of railway manufacture. So did Glasgow. When the North British Locomotive Company had absorbed all its chief rivals, it was employing 8,000 men and building 700 locomotives a year.[71] The two biggest Scottish railway companies also had their works in Glasgow: the North British at Cowlairs, the Caledonian at St Rollox. Sheffield too, in spite of the unsatisfactory railway communication of which it had cause to complain, became a producer of railway goods—springs, rails, tyres, and axles—for almost the whole world.[72] The vast business of John Brown was founded largely on his invention and development of the conical steel spring for railway buffers in 1848.

The big railway companies' works were big indeed. In 1877 those at Crewe were employing 6,000 hands;[73] 14,000 were employed in 1905 at Swindon—it has been suggested that this was 'the largest undertaking in British industry at the end of the century, if not in Europe'.[74]

The influence of the railway is writ large in the history of commercial and industrial development in all Victorian cities. The part it played in assembling the raw materials of manufacture and in distributing the finished product was as evidently vital for West Riding woollens or Macclesfield silk or Birmingham guns as it was for the cottons of Manchester. Some of the evidence for this process is still plainly visible in the siting of factories by the side of the line: in the Lea Valley of north-east London, for example, by either route between Birmingham and Wolverhampton, or by the Midland line in Leicester and Nottingham.

The part played by the railway in transforming the economic life of cities is seen nowhere more clearly than in some of the great ports. Cardiff is a striking example. It grew into a great industrial city in the course of the mid-Victorian age. In the early nineteenth century it had been no bigger than its neighbour Newport, and much

* Later the Manchester Sheffield & Lincolnshire Railway, later still the Great Central.

smaller than Swansea. It decisively outpaced Newport in the fifties and Swansea in the seventies.[75] Its growth and prosperity were then so impressive that it seemed 'likely, at some future day, to outstrip Bristol'.[76]

It was the railway that made Cardiff into a coal-shipping port of world-wide fame. The work of bringing the coal down had been begun by the Glamorganshire Canal (completed in 1798) and the tramroads built in connection with it. In 1835 they were carrying 100,000 tons of coal annually into Cardiff; fifteen years later the Taff Vale Railway (opened in 1841) was carrying six times as much.[77] The opening of this railway gave Cardiff the advantage over Newport, which did not get a corresponding line for another ten years.[78]

The pre-eminent suitability of Welsh coal for use in steamships was now being demonstrated, in the thorough tests made at the instance of the Admiralty in 1847–51.[79] Henceforward, for the rest of the century, both supply and demand seemed boundless; the only limiting factor was transport. A virtual monopoly was exercised here for a long time by the Taff Vale Railway and its ally the Bute Docks Company. This brought great benefit to Cardiff, which became the one port of shipment for the unimaginable quantities of steam coal that poured out of the Rhondda in the seventies.[80] But the railway and the dock company overplayed their hand. Between 1874 and 1882, though the export of coal through Cardiff and Penarth rose by 170 percent, nothing was done to increase the facilities for handling it beyond the construction of one new dock. Congestion, both on the railway and at the docks, became intolerable. Having given repeated warnings, the coalowners took the sole measure open to them. They promoted new railways, providing alternative routes to the sea: first the Barry Railway, then others that connected the Rhondda to all Cardiff's rivals, Port Talbot, Swansea, and Newport. Cardiff therefore had to share with others in a trade that might have remained her own. But that trade was prodigiously valuable, and Cardiff's growth and prosperity continued uninterruptedly down to 1914. It was on the transport of coal that Cardiff emerged as the chief city of Wales.

The enduring symbol of that emergence is Cathays Park, laid out from 1899 onwards with imagination and a true civic pride. If the city fathers and their architects had had one touch more of imagination and piety, surely they would have mounted at its very centre, on a handsome plinth, a tank engine from the Taff Vale Railway.

Some of the benefits that accrued to the citizens of towns like Cardiff from these operations are plain enough. They brought in business to shopkeepers and bankers, to builders and professional men; they created much employment in the railways and docks. But how did they affect the whole lives of these people—not just their economic activities, the working day?

They unquestionably made things better for the comfortable classes. Indeed, they made possible a new kind of life—a life that could be lived in two quite different places at once. Almost as soon as the London & Brighton Railway was opened in 1841, it became practicable to live by the sea and work in London. The 8.45 express up from Brighton, and the 5 o'clock down from London Bridge at night, were already running in 1851, in an hour and a quarter each way.[81] They became a great institution, dignified with Edwardian pomposity by the name 'City Limited' in 1907, and still running at the same times (though to a faster schedule) even now. In the Victorian age they admitted first-class passengers only, the large majority of them holding season tickets, which cost about two shillings a day.[82] A similar express began to run to and from Southend in 1856.[83] In the North the merchants of Liverpool (more than their fellows of Manchester) were inclined to live a long way from their work;[84] and before the end of the century both Manchester and Liverpool men were building themselves 'palatial villas . . . on the shores of Windermere',[85] on the strength of business trains like those of Brighton and Southend.

When they took their families for a long journey, or visited a distant town, they might owe the railway more than rapid and easy conveyance. Many railway companies saw the need to provide hotel accommodation for their passengers and built hotels adjoining their stations, usually managed by lessees. The first in London was at Euston, followed by others at many termini, though not all. These hotels proved to be profitable ventures—so profitable that the railway companies themselves presently went into the business of management. The London & North Western Company, for instance, bought the Crewe Arms in 1876, took into its own hands the hotels at Liverpool and Birmingham in 1879 and 1882, and built a new Euston Hotel in 1881.[86] The Manchester Sheffield & Lincolnshire Company did likewise at Sheffield and Grimsby in 1885–90.[87]

The best of these hotels were very good indeed,[88] a real addition to the high-class amenities of London and the bigger provincial towns. Their contributions to urban commercial and social life were important, with the facilities they offered for public functions, for board and shareholders' meetings (it would be hard to calculate the number of those held in the Cannon Street Hotel alone in the course of a single year), for private discussions and family parties. Though they varied in merit, in most of the cities in which they were established—in Liverpool and Preston, in Hull, in Edinburgh and Glasgow—they set a new standard of comfort, solid, efficient, and dependable.

Such services benefited only the well-to-do. Except in so far as they created new employment, they were of no interest to the working classes or the poor. The railway, however, affected these people intimately, both for good and ill. At first the main thing it did for them was to help to lower prices, and especially the price of coal. The opening of the Leicester & Swannington Railway reduced the price of coal in Leicester at a stroke by 60 percent;[89] when the Great Northern Railway began to carry coal into London in 1851, in competition with that brought by sea, the price fell from thirty shillings to seventeen shillings a ton.[90]

Since in London (with the exception of the Blackwall line) the early railways stopped at the edge of the built-up area, or out in suburbs like Paddington, they did not ordinarily need to displace much of the population. The first notorious problems of that kind arose in provincial cities. The construction of the High Level Bridge in

Newcastle (opened in 1849) involved the removal of nearly 800 families;[91] more than 500 houses were pulled down to allow the erection of Tithebarn Street (later Exchange) station in Liverpool, which was completed in 1850.[92] When the plans for New Street station in Birmingham were first announced, it was remarked that with the completion of the undertaking 'the whole face of the present destitute-looking and filthy place will be changed'.[93] The building of the station (which was finished in 1854) was indeed historic, for it 'began the task of slum clearance in central Birmingham'.[94]

In London, while King's Cross station was being built in 1851, it involved the demolition of much of 'the awful rookery at the back of St Pancras Road'.[95] A lull then followed, for nearly all the rest of that decade, when no important railway construction was going on within the central mass of London; the lull was broken from 1859 onwards when the Metropolitan, North London, South Eastern, Dover, and Midland companies were engaged in fighting their way into the city. By this time a few consciences had been aroused by the suffering that evictions of this kind entailed. In 1853 Lord Shaftesbury persuaded the House of Lords to accept a Standing Order requiring the promoters of any scheme involving the demolition of thirty houses or more in the same parish to state the number and the occupants of them, and the steps they proposed to take to meet the consequent 'inconvenience'.[96] This did no more than oblige the promoters to supply information. The information was defective; but at least it provided some evidence, which could be well used by those who thought that the process should be regulated further. They were confronted by a phalanx of powerful opponents: not only the directors, managers, and shareholders of the railway companies, but *Punch* and *The Times*—with that insufferably sanctimonious devotion to *laissez-faire* which roused Matthew Arnold to a most noble wrath. It was only slowly that they battled through to reveal the full extent of the problem and to propound workable remedies for it.

The Demolition Statements tell us that between 1853 and 1901 sixty-nine schemes were put forward in London, requiring the displacement of more than 76,000 people. Nearly half these displacements were attributable to the great rush of metropolitan railway construction between 1859 and 1867.[97] The requirement to provide this information was only a step towards the re-housing of the displaced at the expense of the promoters. That notion made headway very slowly. It was not until 1874 that any provision of the kind was adopted, and then in a form that was totally ineffective. Even after numerous loopholes had been stopped up in 1885, this compulsion remained imperfectly enforceable. When the Great Central Railway built its London Extension at the end of the century, it did indeed re-house those who were displaced by its progress through Nottingham and Leicester and north-west London.[98] But 1,750 people, who lived on the site of Marylebone station, were not provided for, and they had to move out into the already overcrowded Lisson Grove.[99]

The Times and its supporters in the sixties, with a wilful ignorance, argued that those who were displaced would simply move further away from the centre of London, and that would be a gain all round. 'Thousands of cottages', the newspaper remarked, 'are springing up yearly in the suburbs.'[100] But of course this was airy and irresponsible nonsense. Of those who were losing their houses, the great majority walked to and from their work. Could they now suddenly afford substantial railway fares instead? And what of the casual labourers? Their chance of work depended first of all on their ability to be on the spot when work was going.

The first of these questions had already found its answer. (The second was never satisfactorily answered at all.) The requirement that was imposed on the London Chatham & Dover Company in 1860, to offer special fares to workmen, was rapidly extended to most of the railways that were building new lines into London.* In 1883 the Board of Trade was given a general power to compel companies to introduce workmen's fares wherever it considered them desirable.

This policy achieved its objective in part. It certainly facilitated the huge migration away from the centre of London that occurred in the last quarter of the nineteenth century. It contributed positively to the growth of working-class settlement within reach of the South London line[101] and, on a bigger scale, in north-east London and Essex, the preserve of the Great Eastern Railway, which was characterized in a report of a committee of the L.C.C. in 1892 as 'especially the workmen's London railway'.[102]

We have just looked at two extreme examples of suburban development: the growth of Brighton, Southend, and Windermere as settlements for the wealthy commuter, and of south and north-east London for the working man. But suburban development as a whole was not determined by people of either of these kinds. It was predominantly an affair of the middle classes—using that term in a broad sense. That is clear enough to any one today who walks round a suburb that was created in the Victorian age, using his eyes—in the way in which H. J. Dyos has taught us to use our eyes in south London.[103] The middle-class character of the suburb may well have changed since; it may have begun to fall into working-class occupation before the end of the nineteenth century. But most Victorian suburbs were originally intended to provide an airy, quiet, economical residence for middle-class family men. And for all of them alike it was an essential condition of living there that the suburb should be within easy travelling time of the man's place of work in the city.

That travelling might be performed in a number of different ways: on foot,† by bus or tram or bicycle or private carriage, as well as by railway. The railway's part

---

* But not to all. No such obligation was laid on the Midland Company in the Acts authorizing its London Extension in 1863–4.

† It was not only the very poor who walked to work. The middle-class family man who was economically minded did so too. My grandfather was appointed to a clerkship in the Admiralty in 1860. He lived in Stockwell, and for many years he walked to and from Whitehall daily, either at no expense at all or at the cost of a penny for tolls if he took the shortest route, over Hungerford Bridge, and back again. The practice of walking was noted as already on the decline when he began. 'The habit of walking is becoming less the rule; and the habit of riding less the exception,' said John Thwaites, Chairman of the Metropolitan Board of Works, in 1863. 'We find that parties instead of walking ride to the railway stations for the purpose of travelling by the line' (*P.P.*, 1863, VIII, S.C. on Met. Rly Comm., Q.1026). On the relation between different forms of suburban transport see G. A. Sekon, *Locomotion in Victorian London* (1938).

in suburban development varied greatly in importance. In some cities it was vital: in others it seems hardly to have mattered. The extremes are well represented by two neighbours, Nottingham and Leicester: towns closely comparable in size and not dissimilar in economic character. We have already seen an example in Nottingham of a railway built specifically to serve a new suburban area, and another of a burden laid upon a railway company to provide suburban stations with a satisfactory service. Nothing remotely like this occurred in Leicester. There the true suburban station was represented by two examples only, both on the Humberstone Road. Within a six-mile radius from the centre, there were thirty-two stations* in Nottingham, sixteen in Leicester. The Victorian expansion of Leicester was almost entirely independent of the railway. It followed the lines of the old main roads, and from the seventies onwards it was promoted by a service of horse trams. At the end of the century, when Leicester had become a city of 200,000 people, the Great Central Railway could sweep through it, opening only two suburban stations in a stretch of seven miles.

The railways themselves differed greatly in their willingness to provide suburban services. Such services were much less profitable than those running through on a main line, and some railways were frankly reluctant to provide them at all: the two earliest trunk lines opened into London, the London & Birmingham and the Great Western, for example.[104] The Great Northern began with a similar attitude and found itself thrust, step by step, protesting loudly, into the business of expanding its suburban services, ultimately in conjunction with tube railways running south-east and south-west from Finsbury Park.[105] The companies south of the Thames, on the other hand, with shorter main lines and much smaller freight traffic, were more interested in promoting suburban services, which became the life-blood of them all.

The pattern of development was not always like this, from shy beginnings into reluctant growth. Suburban services appeared in some cities that were in advance of their needs and had to be discontinued: in Hull, for example, one that made almost a complete circle was put on in 1853, and withdrawn for lack of patronage less than eighteen months later.[106] Here was an intelligent venture that failed. We are apt to overlook the tentative, experimental character of much of the railways' early work. They were feeling their way forward, and sometimes having to retract, with little experience to guide them and many local circumstances to be delicately appreciated and taken into account.

The expansion of the suburban railway system in the great cities is usually discussed in impersonal terms: the promotion of companies, the purchase of land, house-building, the train services provided. If we try to penetrate behind these large matters, to discover how the men and women who went to live in these suburbs felt, we are confronted with a great difficulty. The new suburban dwellers were the subject of no political or sociological investigations, nor did any important novelist except George Gissing take much interest in them. Yet by the seventies the railways had brought a whole new suburban world into being—or rather, two worlds: that of the closely built-up suburb, comprising miles of two-storeyed houses, mainly in terraces,

* Excluding the terminal or principal station on each line.

like Camberwell, Hornsey, or Kilburn, the world we associate with Mr Pooter; and the outer suburbs where the houses spread themselves out with sizeable gardens—which came to represent, for many of those they belonged to, the most passionate attachment of their lives. It was now possible for people of modest means to live in the country (an urban-dweller's country, at least) and to work in the heart of the city. 'Railways, and the untiring enterprise of suburban builders', said a writer in the *Cornhill* in 1874, 'have made it easy for them to do so if they like . . . As regards getting to business in the morning and home again in the afternoon, the train puts the dwellers in them nearly on a level with the inhabitants of Tyburnia or South Kensington.'[107] We can watch the process at work, almost photographically, in James Thorne's indispensable book *The Environs of London*, published in 1876.[108] But to Thorne, as a rule, it was just uglification.* Drab enough those houses may be to us now. To many of their first occupants they represented a Paradise. And yet, widespread though the experience was, where does it find expression? Very rarely, and more rarely still in terms of gratitude to the railway, which alone had made a life of this kind possible. Here is one of the rarities:[109]

Our Suburb

He leaned upon the narrow wall
That set the limit to his ground,
And marvelled, thinking of it all,
That he such happiness had found.

He had no word for it but bliss;
He smoked his pipe; he thanked his stars;
And, what more wonderful than this?
He blessed the groaning, stinking cars†

That made it doubly sweet to win
The respite of the hours apart
From all the broil and sin and din
Of London's damnèd money-mart.

The railway did not ordinarily present a handsome appearance as it made its way into the city. The passenger's usual view of it was well characterized by an American,

---

* Of New Barnet, for example, he says (p. 31): 'About the Barnet Stat. has sprung up, within the last few years, one of those new, half-finished rly. villages which we have come to look on as almost a necessary adjunct to every stat. within a moderate distance of London.' Of Buckhurst Hill (p. 64): 'The Stat. is at the foot of the hill, and about it a number of ugly houses have been awkwardly disposed.' In Redhill he sees (p. 481) 'a populous railway town of hideous brick shops and habitations, and around it a belt of ostentatious villas, comfortable looking mansions, and tasteful and ornate dwellings of many varieties, with a super-abundance of builders' detached and semi-detached malformations.' Cf. also his accounts of Croydon (p. 128) and Penge (p. 467).
† Underground; old style. (Author's note)

familiar with England, in the sixties: 'You enter the town as you would a farmer's house, if you first passed through the pig-stye into the kitchen. Every respectable house in the city turns its back upon you; and often a very brick and dirty back, too, though it may show an elegant front of Bath or Portland stone to the street it faces.'[110] And when the traveller emerged on to the street, if he looked back at the station he would seldom see anything very impressive.

The railway gave London a few major architectural monuments—the Arch and the Great Hall at Euston, the stations at King's Cross and St Pancras, the engineers' triumphs at Paddington and Charing Cross; some finely-treated works on the London & Birmingham line in masonry and iron, in the Camden Round-house and the portals of the tunnel under Primrose Hill; good brickwork on a vast scale, vermilion on the Midland line, Staffordshire blue on the Great Central. The provinces also had their imposing monuments. There were the great stations, at Bristol, Chester, Huddersfield, Newcastle; formidable displays of engineering skill at Birmingham, Liverpool, York, and Glasgow; the spacious and convenient stations built at the turn of the century, at Nottingham for example and Sheffield; some magnificent viaducts, striding over scores of houses, at Truro, Brighton, Folkestone, Mansfield, Stockport, Durham, Berwick.

In a few towns the railway showed pride in its achievement. When it approached a town of elegance, it often tried to make its works harmonize with the spirit of the place—at Bath, for instance (where the section through Sydney Gardens still shows something of its original charm), and at Tunbridge Wells. Yet not one major Victorian city used its railway station as a focal point in town-planning, as it was often used so handsomely on the Continent, at Geneva and Amsterdam, in Paris at the Gare de l'Est. The nearest approach to anything of the kind in Britain is to be found at Stoke-on-Trent. There the North Staffordshire Railway laid out its station in 1848 and opposite it, across a square, an hotel,[111] all in a Jacobean style, of red brick diapered with the local blue: a pleasing composition little altered today and still perhaps in the 1970s the most agreeable piece of coherent urban planning in the whole modern city.

Railway companies rarely collaborated with municipal authorities to make their stations into civic embellishments. The Town Council of Liverpool set an interesting precedent when it lent the services of its architect, John Foster the second, to the Liverpool & Manchester Company to design a stone screen to run in front of Lime Street station, facing across to St George's Hall, and contributed £2,000 towards the cost of erecting it.[112] But that example was not followed. The railways did indeed sometimes employ well-known architects of their time to design the façades of stations and their hotels: Tite, Dobson, E. M. Barry, Waterhouse, Gilbert Scott. But they rarely held any competitions for their buildings, even when that practice was fashionable.[113] They occasionally engaged other architects of merit, though no great fame: like Francis Thompson, who endowed Derby and Chester with stations of a true grandeur; David Mocatta, on the London & Brighton line;[114] Sancton Wood, who worked for the Eastern Counties Company in London and

Ipswich;[115] G. T. Andrews, the designer of the first stations in York and Hull;[116] J. T. Mulvany, and again Sancton Wood, who gave railway stations to the city of Dublin that worthily maintained its noble architectural tradition.[117]

In visual terms, however, it must be allowed that the railway generally damaged the city. There could be no doubt on the matter in London—certainly the Victorians themselves had none. Ruskin's denunciations of the railway bridges over the Thames[118] were not merely the raging of an affronted aesthete. On this matter he was at one with his middle-class contemporaries. 'The aggregate disfigurement of the metropolis by the London Chatham & Dover Railway, particularly by its viaducts and its bridges, is very great,' we read in an outspoken gazetteer of the late sixties. It goes on to characterize the newly-built line from London Bridge to Charing Cross in these terms:[119]

> A huge iron bridge goes over the roadway from the station [at London Bridge]: an enormous iron tube, long, high, and most ungainly, goes across Wellington Street, with severe injury to its formerly fine views; a struggling course follows, past the church of St. Mary Overy, across the Borough Market, and through dense back streets; another ungainly tube crosses the fine new street from Blackfriars into Southwark, utterly spoiling its handsome aspect; two more unsightly tubes cross Blackfriars Road, at awkward angles to each other and to the lines of houses; and another intersecting struggle through dense back streets goes onward to the site of the quondam beautiful Hungerford suspension bridge.

The *Builder* summed it all up in 1876 in a terse judgment on railways: 'No rural district we know of has suffered so much disfiguration from the structures connected with them as we have to complain of in London itself.'*

Nor were the visual disasters wrought by the railway in central London offset by the street improvements they brought with them; for here expectation far outran performance. The early underground railways made some contributions to this end: the Metropolitan to the widening of the New Road, the District to the Thames Embankment and to Queen Victoria Street. But the urgently-needed new street to run north and south, from Tottenham Court Road to Charing Cross, was delayed for years through the collapse of the Euston St Pancras & Charing Cross Railway

---

* *Builder*, xxxiv (1876), 847. Not all these bridges were universally condemned at the time. The Blackfriars railway bridge was pronounced 'successful' by William Humber: *Record of the Progress of Modern Engineering* (1864), p. 42. Hawkshaw, who was responsible for the bridges on the Charing Cross line, while admitting that they were 'ugly enough', went out of his way to explain to the Committee of 1863 that the conditions imposed by Parliament on their design precluded their being anything else (*P.P.*, 1863, VIII, S.C. on Met. Rly Comm., Q. 1105).

It is particularly unfortunate that all the railway bridges over the Thames in central London should have dated from the 1860s: a moment when structures of such exceptional strength could be produced only in cast iron in its heaviest and coarsest forms. The recent rebuilding of the bridge that carries the railway from Victoria to Battersea shows how simple and unobtrusive such a structure can now be, using the techniques of the mid-twentieth century.

project of 1871.[120] The Charing Cross Road was completed only in 1887, and then in complete dissociation from any railway.

The most fiercely hated of the London railway bridges, that across Ludgate Hill, exemplified all these matters to perfection. It was a means of street improvement: it afforded a pedestrian crossing over one of the busiest streets in London—for those who were willing to toil up it and down the other side; it made possible the construction of Ludgate Circus; and it was associated with the enlargement of Ludgate Hill to a width of 60 ft. But no one could regard it as anything other than an eyesore, a blow right across the face for St Paul's. Listen again to the gazetteer's judgment: 'That viaduct has utterly spoiled one of the finest street views in the metropolis; and is one of the most unsightly objects ever constructed, in any such situation, anywhere in the world.'[121]

If London suffered outstanding damage from railways in this sense, it did not suffer alone. A valiant battle had been waged in Edinburgh in 1838-42 to prevent the Edinburgh & Glasgow Company from extending its line from the Haymarket to the North Bridge through the Princes Street Gardens. At the end of the battle the honours were even. The railway got the powers it sought, but it agreed to disguise itself in a deep cutting, topped off by a high wall, made seemly to the residents of Princes Street by 'a profuse use of ivy, evergreens, and trees'.[122] The general convenience of the city demanded the extension of the railway from its western terminus. To put it wholly in tunnel would have been impracticable without a lavish provision of ventilators, which would have belched smoke like factory chimneys. In the course of time the traffic along the line grew, and it became the one offensive feature in the whole of that noble valley—until, with the disappearance of the steam locomotive in the 1960s, it has relapsed into an unobtrusive element in the townscape.

There were other railway bridges in provincial towns nearly as frightful as that over Ludgate Hill: the one erected by the Great Northern Company over Friargate in Derby in 1878, for example. Friargate is lined with exceptionally large brick houses: in its satisfying amplitude it is one of the grandest Georgian streets in England. The railway mauled it beyond repair, the bridge pinching in the street line to its own span and drawing attention to itself with gaudy paintwork and fussy decoration. It was a thoroughly offensive piece of commercial vulgarity, bawling away at the aristocratic company it kept. The harm the bridge has done is irreparable to the shape of the street and to the character it once bore as a residence.

The railway was, indeed, at many points in Britain, a vandal. It displaced medieval castles to establish stations and goods yards, at Berwick, Northampton, and Clare. It injured the castle and medieval walls of Conway—though here an enlightened architectural historian can recognize a 'skilful and considerate handling' of the problem which the railway builders faced.[123] It swept away Trinity Chapel in Edinburgh in favour of Waverley station.[124] It drove straight through the ruins of the Priory of St Pancras at Lewes.* At times its ambitions soared higher. A central

---

* This outrage provoked its response. It was directly responsible for the foundation of the Sussex Archaeological Society.

The Victorian City: Shapes on the Ground

station was projected for Dublin in 1876 near College Green. It was to be reached by continuing the Kingstown line from Westland Row station on a viaduct across Trinity College Park and the Fellows' Garden, incidentally demolishing the Provost's House. The advocates of this scheme believed that the compensation money payable to the College would silence its opposition, by enabling it to solve 'the vexed question of a fund for retiring pensions for Senior Fellows'.[125] But the Senior Fellows were denied their pensions. The railway was not built.

Sometimes railway companies were forced to accept a deviation, to avoid doing damage of this kind. The London & South Western Railway intended to cut through Maumbury Rings at Dorchester, but it was frustrated by a local antiquary, Charles Warne.[126] At Leicester a long battle was fought in the 1890s, chiefly by the Leicestershire Archaeological Society, to stop the Manchester Sheffield & Lincolnshire Railway, first from driving its London Extension through the Jewry Wall (one of the largest pieces of Roman masonry in England) and the medieval castle, and then from destroying a Roman pavement that had impertinently placed itself underneath the intended site of the station.[127] With the backing of the Town Council the battle was won. The pavement survives *in situ*, in a chamber of white glazed brick beneath the platform. And the Jewry Wall and the Castle laugh last, for the railway is now entirely closed.

Such battles had their effect. Protection was sometimes accorded to notable antiquities and historic buildings. Special arrangements for tunnelling within 200 feet of Westminster Abbey were among the financial burdens placed on the District Railway in 1864.[128] The Central London Railway was required in 1891 to preserve 'all objects of geological or antiquarian interest discovered by them in the execution of their works', and to deposit them in the Guildhall Museum.[129]

The railway obliterated the past as relentlessly as new roads, airports, and reservoirs obliterate it now. That is not to deny the improvement in life that the railway often brought. There is a proverb about omelettes and the breaking of eggs. But the steam locomotive, now passing into dim recollection, was a noisy machine[130] and could be a grave nuisance from the smoke it emitted. The railway offered new amenities to those who lived in the city. It also killed, or seriously impaired, a number of those they had enjoyed in the past.

And so we end, as we began, in a measure of confusion. For the country as a whole the building of railways in the Victorian age meant growth, progress at least in an economic sense, increasing power. Within cities it sometimes meant just that: the expansion and diversification of industry, the improvement of economic opportunity for most of the people living there. At other times it meant uncertainty or loss —the sudden removal of works elsewhere, to some more favoured site. In quality of life its effects were equally various. It offered access to recreation, to beauty and quiet, with undreamt-of ease. It also did much to destroy beauty and quiet within the city itself; and yet, on the other side again, it sometimes displayed a ruthless magnificence that was—and has remained—all its own. Anybody born about the time of the Battle of Waterloo, who grew up with the railway and watched its development closely, must have seen it as a complex and puzzling spectacle. Nowhere was it more complex, or harder to understand rightly, than in urban life. The railway did indeed set the towns of Britain a-dancing when it came. They were dancing to it still, though the tunes were different, when Queen Victoria died.

Notes

1 *Latter-day Pamphlets* (1858), p. 229.
2 *Liverpool Albion*, 26 April 1830.
3 *Journeys to England and Ireland*, ed. J. P. Mayer (1958), p. 112.
4 At the time of its opening it was reckoned that the greatest number of people who could be conveyed by the coaches between Liverpool and Manchester was 700 a day. If every coach had been filled to capacity, Sundays included (an impossible supposition), 255,000 passengers could thus have been carried in a year. In the course of 1831 the railway actually carried just over 445,000. J. Wheeler, *Manchester: Its political, social and commercial history* (Manchester, 1836), pp. 293–4.
5 J. R. Kellett, *The Impact of Railways on Victorian Cities* (1969), pp. 154–5, 164. This book, based on a comparative study of London, Birmingham, Manchester, Liverpool, and Glasgow, is by far the most important study of the subject.
It is especially valuable for its analysis of the dealings between the railway companies and the landowners in these five cities and the relations between the companies themselves.
6 Ibid., p. 173.
7 T. Baines, *History of the Commerce and Town of Liverpool* (1852), p. 828.
Cf. Braithwaite Poole's statement of the goods traffic in and out of Liverpool by the London & North Western Railway in 1851, analysed by towns and districts (ibid., p. 829).
8 J. A. Langford, ed., *A Century of Birmingham Life* (1868), II, p. 580.
9 *A Pictorial Guide to Birmingham* (1849), p. 162.
10 Pearson, who was Solicitor to the City of London, put forward two versions of his scheme. One was examined by the Commission on London Termini of 1846 (*Parliamentary Papers* (*P.P.*), 1846, XVII, QQ.2289–351, 2821–53), the other by the Select Committee (S.C.) on Metropolitan Communications in 1855 (*P.P.*, 1854–5, X, QQ. 1337–78). The *Builder* argued that the need for a great central terminus in London was 'obvious' as late as 1881: xli (1881), 195.
11 See Kellett, op. cit., pp. 161–3.
12 Ibid., p. 171.
13 E. T. MacDermot, *History of the Great Western Railway* (1927–31), I, pp. 356–7.
14 Kellett, op. cit., pp. 264, 266; J. Simmons, *St Pancras Station* (1968), pp. 16–17.
15 For the remarkable complexity of the railway system here see E. L. Ahrons, *Locomotive and Train Working in the Latter Part of the Nineteenth Century* (1951–4), I, pp. 59–61.
16 T. C. Barker and Michael Robbins, *A History of London Transport* (1963), I, p. 125.

17 *P.P.*, 1863, VIII, S.C. (H. of L.) on Metropolitan Railway Communications: Minutes of Evidence (500–II), 239.
18 A bizarre example is that which ran in 1899 from Liverpool (Central) to Folkestone, via the Mersey Tunnel, Birkenhead, and Reading—Liverpool dep. 8 a.m., Paris arr. 10.50 p.m.: G. W. Parkin, *The Mersey Railway* (n.d.), pp. 16–17.
19 It must moreover be remembered that until late in the Victorian Age all the express services for the Continent, except the night mails, left in the morning, too early to take passengers who had travelled any substantial distance from the North. See C. W. Eborall, Manager of the South Eastern Railway, in his evidence to the Committee of 1863: *P.P.*, 1863, VIII, S.C. on Met. Rly Comm., Q. 677.
20 Barker and Robbins, op.cit., I, pp. 74–5, 237, 271–2.
21 Local and Personal Act, 27 & 28 Vict., cap. cccxxii.
22 Sects 28, 45, 55, 36, 33, 76, 90.
23 Cf. C. E. Lee, *Passenger Class Distinctions* (1946), and H. J. Dyos, 'Workmen's Fares in South London, 1860–1914', *Journal of Transport History* (*J.T.H.*), i (1953–4), 3–19.
24 *P.P.*, 1883, LXI, Report to Board of Trade upon . . . Workmen's Trains, etc., 466.
25 *Builder*, xxix (1871), 16–17. Overcrowding on these trains was frequent: see, e.g., *P.P.*, 1883, LXI, Report on Workmen's Trains, 449, 451, 453; 1894, LXXV, Notes of Conference held at Board of Trade, etc., 836–7.
26 See the map in Barker and Robbins, op. cit., I, p. 185.
27 *Builder*, xxxiii (1875), 315.
28 Ibid.
29 British Transport Historical Records, GE1/1, 53–4.
30 C. J. Allen, *The Great Eastern Railway*, 1955, p. 58. No statement of the final cost was ever given to the shareholders; surprisingly, none of them seems to have demanded it.
31 Barker and Robbins, op. cit., I, p. 153.
32 Three of them spoke in favour of it to the Lords Committee of 1863 (long before the New York Elevated railway had appeared)—G. P. Bidder, John Parson, and William Baker—though the last two recognized that its unsightliness was a grave objection. (*P.P.*, 1863, VIII, S.C. on Met. Rly Comm., 88, 106, 154: QQ. 614, 802, 1139–41.)
33 The origins of this line go back to 1877, when the Mersey Docks & Harbour Board, having investigated the Elevated system in New York, determined to build such a line in Liverpool. They secured Acts for this purpose in 1878 and 1882, but did not carry out the work. The railway was eventually built by a separate company, incorporated in 1888; it was opened in 1893–1905. In one important respect it was an advance on all its American prototypes. It was the first such line to be electrically operated. C. E. Box, *The Liverpool Overhead Railway 1893–1956* (1962 edn), pp. 15, 18–19, 33.
34 The *Builder*, as early as 1881, was prepared to believe that electric traction might make the railway clean and relatively quiet, and in that case would perhaps have been prepared to accept the disfigurement (xli (1881), 196); but when the Liverpool

Overhead Railway got its second Act in the following year its comment was: 'These new works will, doubtless, afford many facilities, but will greatly damage the appearance of important parts of Liverpool. Everything is sacrificed nowadays to making money and "getting along"' (xlii (1882), 750).

35 For the pneumatic lines see H. Clayton, *The Atmospheric Railways* (1966), pp. 124–8; for the cable-operated Tower Subway see Barker and Robbins, op. cit., I, pp. 301–3.

36 The Committee's report and the minutes of evidence given before it are in *P.P.*, 1892, XII, Joint S.C. on Electric and Cable Railways (215–I), 1–172.
Cf. J. Simmons, 'The Pattern of Tube Railways in London', *J.T.H.*, vii (1965–6), 234–40.

37 *P.P.*, 1905, XXX, R.C. on the Means of Locomotion and Transport in London (Cd. 2597), 609.

38 Ibid., 621–2. The Metro had been the subject of a special investigation in 1902 by Lt-Col. H. A. Yorke, Chief Inspecting Officer of Railways at the Board of Trade: *P.P.*, 1902, XXIII, 479–93.

39 Ibid., 1905, XXX, R.C. on London Traffic, 623, 630.

40 The *Sheffield Mercury* (which was opposed to the North Midland's plan) gives the fullest account of these events: see especially the issues of 16 January, 19 and 26 March, and 9 April 1836. Its Whig rival the *Iris* supported the Sheffield & Rotherham Railway and paid little attention to the North Midland debates.

41 *Transactions of the Hunter Archaeological Society*, ix (1964), 16–19.

42 C. H. Grinling, *History of the Great Northern Railway* (1966 edn), pp. 162–3.

43 *Sheffield Daily Telegraph* (*S.D.T.*), 2 July 1864.

44 Ibid., 12 May 1864.

45 See, for example, the evidence of William Rangeley of Unstone before the Parliamentary Committee of 1864: *S.D.T.*, 2 July 1864.

46 *S.D.T.*, 10 October 1861. To be fair to the Midland Company, if this was true the blame rested also with the Manchester Sheffield & Lincolnshire and its ally the South Yorkshire.

47 Cf. Charles Cammell's evidence: *S.D.T.*, 1 July 1864.

48 *S.D.T.*, 10 October 1861, 13 November 1862, 10 September 1863.

49 Ibid., 6 July 1864.

50 The report of the Committee that enquired into the Company's promotion, and the evidence given before it, are in *P.P.*, 1864, X, S.C. on Standing Orders (510–I), 903–21.

51 *S.D.T.*, 16 October 1863.

52 For a clear account of the Midland Company's contest with its rival, from the Midland's side, see F. S. Williams, *The Midland Railway* (5th edn, 1886), pp. 135–43.

53 *S.D.T.*, 2 February 1870.

54 J. M. Furness, *Record of Municipal Affairs in Sheffield* (Sheffield, 1893), pp. 257, 409, 447.

55 Ibid., pp. 110, 119, 204, 253. Cf. *S.D.T.*, 13 December 1860, 13 June 1861, 12 March 1863.

56 Nottingham's place on the early railway system is discussed in R. A. Church, *Economic and Social Change in a Midland Town* (1966), pp. 170–4.

57 Church, op. cit., ch. 8.
58 For Gripper see Church, op. cit., pp. 229, 372, and Simmons, *St Pancras Station*, p. 53.
59 Local and Personal Act, 49 & 50 Vict., ch. xciv, sect. 34.
60 Local and Personal Act, 56 Vict., ch. i, sect. 21.
61 Church, op. cit., pp. 349–50.
62 *Tewkesbury Yearly Register and Magazine*, ii (1850), 45.
63 Ibid., i (1840), 289–91.
64 Ibid., ii (1850), 29, 157.
65 *Victoria County History (V.C.H.) of Gloucestershire* (1907), II, p. 187.
66 *Tewkesbury Register*, ii, 45.
67 Cf. *V.C.H. Essex* (1907), II, p. 137.
68 J. G. H. Warren, *A Century of Locomotive Building by Robert Stephenson and Co., 1823–1923* (1923), pp. 91, 93.
69 These were not the first locomotives to be built by a railway company in its own works. The Stockton & Darlington's Company's works at Shildon had turned out the *Royal George* in 1827: W. W. Tomlinson, *The North Eastern Railway* [Newcastle, 1915], p. 143.
70 G. Dow, *Great Central* (1959–65), I, pp. 106, 211.
71 *Murray's Handbook for Scotland* (8th edn, 1903), p. 148.
72 A. Gatty, *Sheffield Past and Present* (Sheffield, 1873), p. 311.
73 W. H. Chaloner, *The Social and Economic Development of Crewe 1780–1923* (Manchester, 1950), p. 74. The whole population of Crewe was then rather more than 20,000.
74 D. E. C. Eversley in *V.C.H. Wiltshire* (1959), IV, p. 215. Swindon's population was then about 47,000.
75 In 1840 the port of Swansea (including Neath) shipped 493,000 tons of coal, Newport almost exactly the same (489,000), Cardiff 166,000. By 1874 the corresponding figures were 1,043,000; 1,066,000; 3,780,000. (J. H. Morris and L. J. Williams, *The South Wales Coal Industry, 1841–75* (1958), p. 91.) In 1841 Cardiff and Newport both had populations (in round figures) of 10,000, as against Swansea's 25,000; by 1881 Cardiff had 83,000, Swansea 66,000, Newport 35,000.
76 *Murray's Handbook for South Wales* (1877 edn), p. 14.
77 D. S. Barrie, *The Taff Vale Railway* (2nd edn, 1950), pp. 6, 14.
78 Morris and Williams, op. cit., pp. 100–1.
79 Ibid., pp. 34–6. The North-country coalowners repeatedly tried to controvert this demonstration in subsequent years, but without success: pp. 36–41.
80 E. D. Lewis, *The Rhondda Valleys* (1959), p. 121. The increase was from 2,886,000 to 7,775,000 tons a year.
81 C. F. D. Marshall, *History of the Southern Railway* (1963 edn), p. 209.
82 Ahrons, op. cit., V, p. 28.
83 H. D. Welch, *The London Tilbury & Southend Railway* (1951), p. 25.
84 *Murray's Handbook for Lancashire* (1880 edn), pp. 115–16.
85 R. S. Ferguson, *History of Westmorland* (1894), p. 286.
86 W. L. Steel, *History of the London & North Western Railway* (1914), pp. 364–5, 375, 385, 388.

87 Dow, op. cit., II, p. 219.
88 Cf. Simmons, *St Pancras Station*, pp. 80–1.
89 A. T. Patterson, *Radical Leicester* (1954), p. 262.
90 Grinling, op. cit., p. 103.
91 J. Sykes and T. Fordyce, *Local Records* (Newcastle, 1867), III, p. 246.
92 *Builder*, viii (1850), 185.
93 *Birmingham Journal*, 9 January 1847, quoted in *Modern Birmingham*, ed. J. A. Langford (1873), I, pp. 5–6.
94 *V.C.H. Warwickshire* (1969), VIII, p. 10.
95 *London and its Vicinity Exhibited in 1851*, ed. J. Weale [1851], p. 811.
96 H. J. Dyos, 'Railways and Housing in Victorian London', *J.T.H.*, ii (1955–6), 13. This paper gives a very clear account of the problem and the remedies proposed and tried for it. See also Kellett, op. cit., pp. 325–36.
97 Dyos, ibid., p. 14. All these figures are certainly too low: see H. J. Dyos, 'Some Social Costs of Railway Building in London', *J.T.H.*, iii (1957–8), 23–5.
98 Dow, op. cit., II, pp. 244, 279, 287.
99 Dyos, 'Railways and Housing', 19.
100 Quoted in ibid., 14.
101 Dyos, 'Workmen's Fares', 15.
102 *Report of the Public Health and Housing Committee on Workmen's Trains* (1892).
103 H. J. Dyos, *Victorian Suburb: A study of the growth of Camberwell* (1961). See his remarks on the social character of suburbs, pp. 22–6.
104 M. Robbins, *Middlesex* (1953), pp. 78–80.
105 Grinling's *History of the Great Northern Railway* treats the suburban problem markedly better than any other history of a large railway company running into London.
106 K. Hoole, *Regional History of the Railways of Great Britain: North East England* (Newton Abbot, 1966), pp. 45–6.
107 *Cornhill Magazine*, xxix (1874), 603.
108 Reprinted by Adams & Dart in 1970.
109 Ernest Radford, *A Collection of Poems* (1906), p. 60.
110 Elihu Burritt, *A Walk from London to John o' Groats* (New York, 2nd edn, 1864), p. 2.
111 The architect was H. A. Hunt, of Parliament Street, London (not the 'otherwise unknown R. A. Stent'—H.-R. Hitchcock, *Early Victorian Architecture in Britain* (1954), p. 523), and he put the cost of the whole complex of buildings, including the passenger and goods stations, engine shed, workshops, and hotel, at £127,000: *Staffordshire Advertiser*, 4 August 1849. For accounts of the station when it was new see ibid., 14, 21 October 1848.
112 Hitchcock, op. cit., p. 497; J. A. Picton, *Memorials of Liverpool* (2nd edn, 1907), II, p. 187.
113 The most famous exception is the St Pancras Hotel, for which the competition was won by Scott in 1866: Simmons, *St Pancras Station*, pp. 48–9. Others include: the first station at Ipswich in 1846 where Sancton Wood was the successful competitor (*Builder*, iv (1846), 140; see also British Transport Historical Records, EUR 1/3, ff. 37–8, 41, 56); the Park Hotel at Preston, erected by the London & North Western and Lancashire & Yorkshire companies (competition won by

T. Mitchell of Oldham—*Building News*, xxxviii (1880), 248, 269); and Exchange Station, Liverpool (John West of Manchester—*Builder*, xli (1881), 222).

114 See *J.T.H.*, iii (1957–8), 149–57.
115 He designed for the Eastern Counties Company its second Shoreditch terminus in 1849 (Hitchcock, op. cit., p. 524); for Ipswich see note 113 above.
116 See *J.T.H.*, vii (1965–6), 44–53.
117 Mulvany was responsible for Broadstone station, for the Midland Great Western Railway; Wood for Kingsbridge (now Heuston), for the Great Southern & Western.
118 *Works*, ed. E. T. Cook and A. Wedderburn (1903–12), XVII, pp. 389–90; XIX, pp. 25–6.
119 J. M. Wilson, *The Imperial Gazetteer of England and Wales* [Edinburgh, 1866–9], II, pp. 167–8.
120 See the plan in the *Builder*, xxix (1871), 205.
121 Wilson, *Imperial Gazetteer*, II, p. 167.
122 D. Robertson, *The Princes Street Proprietors* (1953), pp. 37–46.
123 A. J. Taylor, *Conway Castle and Town Walls* (1956), p. 43.
124 The strange story of this demolition and its sequel is summarized in J. Grant, *Cassell's Old and New Edinburgh* [1884], I, p. 290.
125 *Builder*, xxxiv (1876), 1202. Dublin had long been interested in plans for linking its five terminal stations. Twelve years earlier the Dublin Trunk Railway Connecting Act had been passed, which authorized the construction of a tunnel under the Liffey. But it had a fatal drawback: it prohibited altogether the use of steam, atmospheric, and cable power on the line. (Local and Personal Act, 27 & 28 Vict., ch. cccxxi, sect. 19.)
126 *Murray's Handbook for Wilts. and Dorset* (1899 edn), p. 524.
127 For this struggle see *Transactions of the Leicestershire Archaeological Society*, vii (1893), 222, 273; viii (1899), 133, 134, 205, 375; ix (1904–5), 6, 15, 68, 114, 152. Public access to the pavement *in situ* was guaranteed under the Company's Act: Local and Personal Act, 56 Vict., ch. i, sect. 29 (17).
128 Local and Personal Act, 27 & 28 Vict., ch. cccxii, sect. 34.
129 Local and Personal Act, 54 & 55 Vict., ch. cxcvi, sect. 69.
130 Cf. for instance the agonized complaints of a correspondent of the *Builder* concerning the use of the steam whistle on the North London Railway near Dalston Junction: xxix (1871), 291.

13 London, the Artifact

John Summerson

It must be admitted that architecture in the nineteenth-century city is a somewhat abstruse agency of emotional effect. In any age, the 'language' of architecture is an almost exclusively professional affair; in the nineteenth century, when many architectural languages came to be spoken simultaneously, often in archaic dialects, with broken accents and much rhetorical improvisation, the situation is the equivalent of an exclusive babel, architects making speeches to each other or mumbling to themselves. Nevertheless, the babel makes itself felt. It has a telling effect within the mental image of the city. Anyone calling up an image of Victorian London will have a strong sense of this obscure and complicated dialogue.

Victorian London developed through sixty-four years, undergoing striking qualitative as well as quantitative changes in its fabric. With the quantitative changes we are here less concerned than with the qualitative. The enormous increment of suburbia will not be our concern at all. We shall deal only with those questions of architectural style and taste which so conspicuously transformed the central mass of the capital. These changes are not exclusively of an aesthetic kind. They are linked with economic considerations, such as increase of height on expensive land; socio-economic considerations like the arrival and increase of flats and office blocks; purely social considerations like the incidence of churches, schools, and infirmaries; and technological considerations, such as the exploitation of iron-frame construction, lifts, and plate-glass. All these helped to change the appearance and 'character' of

© 1973 Sir John Summerson

London between 1837 and 1901, but it is only in combination with factors of style and taste that the changes powerfully affect the image.*

In this process of change there are certain clearly marked phases of stylistic usage and they can be diagrammatically spaced at twenty-year intervals. Thus, from 1837 to 1857 there is post-Georgian eclecticism, characterized mainly by ornate Italian stucco streets, stone-built Italian club-houses, and Gothic churches. From 1857 to 1877 there is a phase of intense conflict between Italian and Gothic, accompanied by attempts to combine the two into a style appropriate to the age; stucco nearly disappears, brick, stone, and polished granite are the prevailing materials with some polychromy in tile and terracotta; street architecture becomes and remains chaotic. By 1877, the conflicts have relaxed and the next twenty years see a variety of stylistic enterprises, mostly based on English, French, Flemish, or other Renaissance schools; red brick is in high fashion, liberally combined with terracotta and a wide choice of stones. By 1897 a new phase is beginning, with the re-establishment of formal classicism on the one hand and new attempts at modernity of treatment on the other.[1]

These are some of the superficial indications of change. Before we pursue their meaning more deeply, however, there is one consideration requiring the utmost emphasis. *Victorian London was, to a very great physical extent, Georgian London.* That it was also, morphologically, Stuart and Tudor and medieval London is obvious. The point is that, whereas the buildings of those periods had mostly vanished or become curiosities, the buildings of the Georgian age survived in enormous bulk and in full use, to the extent that little of Georgian London was actually destroyed before the eighties. The only part of it which the Victorians did blot out was the City, where a steep rise in land-values made redevelopment pay. But the City was not in any case very extensively Georgian; much of it still dated from the rebuilding after the Great Fire. Further west, the great estates built up in the eighteenth and early nineteenth centuries preserved their integrity. The leasehold system put a brake on redevelopment and when ninety-nine-year leases did fall in they tended to be renewed. Only in streets where business interests strongly asserted themselves (Bond Street and Oxford Street, for instance) or where the raffish margin of an affluent estate stirred the interest of an ambitious landlord (as in Mount Street) did Georgian walls crumble,

---

* Advances in building technology made almost no visible difference to the London street-scene. The undivided plate-glass shop-window, increasingly common after 1870, was probably the most striking departure. Iron, although extensively used as an empirical substitute for brick or timber, and facilitating large uninterrupted floor-spaces, had little external effect. Building legislation consistently reflects a bricks-and-mortar mentality. The Metropolitan Building Act, 1855, insisted that openings and recesses in the fronts of buildings should not account for more than one-half the total area of the front above first floor level. This provision was carried over to the Act of 1894, thus perpetuating the idea of the solid masonry front. Iron buildings required special consent. On height, there was no restriction in the Acts of 1844 and 1855, the human capacity to mount stairs imposing a natural limitation. This being disposed of by the advent of lifts, the Act of 1894 imposed a height of eighty feet with two additional storeys in the roof and some liberty in respect of ornamental features.

and then only in the last three decades of the century. Victorian London *was* Georgian London: vastly extended, carved into by new highways, lanced by railways, elevated from its river by embankments, and sprigged with buildings new in scale and style. It was a city in which the ideas and products of one age were imposed upon and inserted into those of another.[2]

The Victorian attitude to the left-over Georgian environment needs to be defined. It was mostly and for most of the time one of utter disgust. The 'hole-in-the-wall' architecture of uniform eighteenth-century streets seemed to touch the very lowest architectural level conceivable and any opportunity of displacing some part of it was greeted with glee. Since it was rarely economic to do so, attempts were made to redeem these eyesores with adventitious ornaments. On the Bedford estate, Great Russell Street was richly stuccoed in the sixties[3] and the hated Gower Street was given heavy cement doorcases as leases fell in in the eighties.[4] When the leases of Adam's Adelphi Terrace fell in, around 1870, the opportunity was taken to add the coarsest kind of window-dressings and a gross pediment.[5] Some degree of kindness towards the earliest Georgian houses (usually identified as 'Queen Anne') does emerge with the new tide of taste in the seventies, and the cosmetic enrichment of Russell Square with terracotta dressings in 1900 is not entirely unsympathetic to the original structure. The objections to Georgian streets were their meanness, poverty of scale, and lack of 'character'. They were a standing incitement to be big, bold, and picturesque.

Notwithstanding the hatred of the Georgian, the principles of Georgian estate-development, planning, and design were for long inescapable. In fact, the whole architectural development of London from 1837 to 1857 shows so little radical change that it is more natural to think of it as a final chapter in Georgian urban history than as a period of innovation. The whole of Bayswater, from Marble Arch to Notting Hill and beyond to Shepherd's Bush; the whole of North Kensington, of South Kensington, Brompton, and Earls Court; the whole of Pimlico—all this is an overgrowth of late Georgian developed on the same leasehold system in big units and built by builders in the tradition of James Burton and of Thomas Cubitt (himself a major participant) or by ruthless captains in the building world like Sir Charles James Freake, William Jackson, and Charles Aldin, the creators of South Kensington.[6]

The Victorian houses are bigger than their Georgian equivalents but not bigger than those of Nash's Carlton House Terrace, the evident model for extreme affluence. Their layout in squares and streets and mews is no advance on eighteenth-century practice nor, in the relation of houses to their communal gardens, on what Nash had done in the grander Regent's Park terraces (Sussex or Cumberland, for instance). The ornaments, certainly, are different; they follow Sir Charles Barry's lead from severe neo-classicism to *cinquecento* revival. The fenestration of Barry's Reform Club echoes through Bayswater; the garden front of the Travellers' inspires terraces in the Ladbroke area. More lavish in the use of ornament than Barry's work ever was, these houses have nothing of his quality; with their serried ranks of windows and their ambitious height they betray too readily the builders' anxiety to exploit.[7]

Turning from estate development to 'street improvement', the impetus again is Georgian, and so indeed is the administration. Up to 1855 new streets were planned and their execution conducted by Her Majesty's Office of Works, just as Regent Street had been planned and administered by the Woods and Forests which the Office incorporated. James Pennethorne, Nash's protégé and professional legatee, was in charge. The façades in Cranbourne Street and New Oxford Street are a little different: there are fewer columns and more fancy window-dressings and sometimes a little Louis XV, even Elizabethan. But in principle they are the same.[8]

Even the railways started their careers in full sympathy with the Georgian framework. The Euston portico and lodges were nothing but an enlarged version of a type familiar as a park entrance. More than that, the portico was triumphantly placed on the great Bloomsbury axis established by an Earl of Southampton around 1660. Paddington, too, was carefully sited in 1838 in relation to the plans being prepared for the Bishop of London's estate in Bayswater. King's Cross, less happily sited, is Georgian at least to the extent that its frontispiece (sometimes mistaken for a piece of pure engineering) derives from a design in a French academic treatise.[9]

Lastly, churches. Here, certainly, there was an important displacement of Georgian practice by the new willingness to supersede state-aided church building by voluntary effort. But Bishop Blomfield, the pioneer of 'church extension', thought in statistical and political terms much as the old Commissioners had done; he even considered 'fifty new churches' as a goal, echoing a famous Act of Queen Anne.[10] Churches of the forties and fifties differ from churches of the twenties and thirties only in that the choice of style is always, instead of only sometimes, Gothic, and that the Gothic of the fifties has more of archaeology in it. The churches mostly stand on sites as well prepared for them on estate-development plans as any Commissioners' church by Nash, Smirke, or Soane.[11]

All this merely emphasizes the truth that early-Victorian London was a city which not only had a completely built-up Georgian body but also proceeded to expand from that body in a way which differed only in its superficial affluence from the practice of previous generations. Architectural innovation was in the air but scarcely yet on the ground. Architects talked of a new style but had little to show—nothing to compare in novelty with the Crystal Palace, which was not quite architecture. A few buildings, strange in style or structure, did make their appearance. In a cul-de-sac off Endell Street an obscure architect, James Wild, built a parish school in a severe version of Italian Gothic in 1849.[12] In the same year Butterfield started All Saints, Margaret Street, in red and black brick Gothic, strangely proportioned.[13] In 1851, the year of the Crystal Palace, I. K. Brunel and M. D. Wyatt tried consciously to evolve a new style for the sheds of Paddington Station.[14] There were other prophetic enterprises, but they were mere straws in the wind and London in the late fifties was still Georgian London—over-grown, over-ripe, over-rich in trite ornament. The Palace of Westminster, just approaching completion, was its great new symbol, but it was a symbol standing at the end, not the beginning, of an epoch. To despise it for its Palladian symmetry and multiplicity of corrupt ornament was a mark of sophistication. For a new epoch was already present in thought and word and in the years between 1857 and 1877 it exploded into architectural effect. The causes and nature of this explosion it is now our business to investigate.

The *causes* relate directly to the types of building on which people were prepared to spend and the opportunities available for siting such buildings. The *nature* of the explosion has more to do with architectural styles and the philosophies behind them. To draw a proper picture of Victorian London, it will not do to separate these issues. They must be taken together wherever the activities of investors, architects, and builders converge.

For concentrated redevelopment in the Victorian age there is nothing more striking than the virtual rebuilding of the central area of the City between 1857 and 1877. The transformation had indeed begun much earlier, the chief agents being the great banking-houses and, still more, the competing insurance companies, but it was around 1857 that the scramble for sites began. These businesses all required much the same sort of accommodation—and it was not very much. Vital, however, was that they should be well sited, and so the formula was quickly evolved of a large building on an important site (the sites, perhaps, of two or more old houses) in which the ground and first floor only would be occupied by the building-owner and the upper floors be profitably lettable for offices; there would be a housekeeper's flat in the roof, and boiler-room and strong-rooms or a lunch-room in the basement. The street front would be a lavish piece of Italian architecture, dedicated to the building-owner's sense of status. The interiors would comprise a handsome public hall and board-room and nothing else except space, to be subdivided at will. The formula combined convenience, prestige, and sound investment in a way which is still perfectly familiar. There still stand in the City buildings of the 1850s which differ little in type and style from their successors of the 1930s.

City architecture thus became a matter of competing, prestige-bearing façades. Many of the earlier buildings were architecture of the highest quality, like C. R. Cockerell's two works near the Bank of England.[15] Later, however, in the boom period, the City streets became a jungle of classical ornamentalism with all sorts of architects playing the Italian game with more gusto than sense. Barry's club style was the most commonly accepted basis, but devices from modern France, Sansovinesque essays in rustication, and extra flourishes of symbolic sculpture gave the City an aspect very different from that of serene Pall Mall.

Nor was the game always Italian. The new Gothic sophistication made a modest but not unnoticed entry when the Crown Life Office employed the Ruskinian partnership, Deane & Woodward, to design its headquarters in Bridge Street, Blackfriars, in 1859.[16] The lead was not immediately followed, but by 1868 the General Credit company was building in Lothbury, right opposite the Bank of England, a Venetian façade reflecting unambiguously Ruskin's plates in *The Stones of Venice*.[17]

Banks, insurance offices, and discount houses tended mostly to adhere to conventional Italian, accepted symbol of the aristocratic and secure. More latitude is noticeable in buildings put up by land-investment companies for the sole purpose of letting as offices. The 'office block' in this sense had emerged as early as 1823.[18] In 1844–5 came Royal Exchange Chambers whose architect, Edward I'Anson, was a leader in this field.[19] This was conventional enough, but he experimented with a nearly all-glass front in Fenchurch Street in 1857,[20] a fifteenth-century palazzo type in Seething Lane in 1859,[21] and a Venetian Gothic front in Cornhill in 1871.[22] Some daringly novel office blocks were built in Mark Lane in the mid-sixties, notably a smooth round-arched affair, with an all-iron interior, designed by George Aitchison, a future R.A.[23] Warehouses, too, sometimes ran to stylistic extravagance. The drapers and fancy warehousemen off Cheapside tended to prefer Italian of the most ornate kind.[24] A coarser type prevailed elsewhere, but in Thames Street, an old warehouse was re-fronted by the dedicated medievalist, William Burges, in 1866,[25] producing a type of warehouse imitated far and wide; while in Eastcheap, R. L. Roumieu built for a firm of vinegar manufacturers in 1867 a Gothic pile of extravagant absurdity and no little skill.[26]

It will be seen that from 1857 the stylistic character of City architecture began to change very fast. Before 1857 it could be said that there was scarcely a building in London which could not be accounted for by reference to some accepted school or tradition of architectural design. Within seven or eight years the whole tendency had changed. In 1864, George Gilbert Scott, surveying the scene with the lofty detachment of a Gothic master, delivered himself of the following:[27]

> There has . . . been no end to the oddities introduced. Ruskinism, such as would make Ruskin's very hair stand on end; Butterfieldism, gone mad with its endless stripings of red and black bricks; architecture so French that a Frenchman would not know it . . . Byzantine in all forms but those used by the Byzantians; mixtures of all or some of these; 'original' varieties founded upon knowledge of old styles, or ignorance of them, as the case may be; violent strainings after a something very strange, and great successes in producing something very weak; attempts at beauty resulting in ugliness, and attempts at ugliness attended with unhoped-for success.

Scott did not direct this diatribe especially towards the City, but for each of his gibes some City building, built before 1864, could be found to fit—not, perhaps, among the more costly buildings, in one of the better streets, or under the name of a known architect. An important aspect of Victorian architecture is the development of a class of architect with a temperamental reluctance to attach himself to any stylistic school and a certain aptitude for improvisation. There were a great many architects in this class, for which the politest name is the 'latitudinarian'.[28] They were not all bad architects, and their works constitute a very large proportion of the building product of the capital.

The boom period in the City's rebuilding tailed off in the late seventies and in its last years there was a slackening of stylistic competitiveness and a general acceptance of any style which any architect cared to bring along. This was partly due to the shock administered by one particular building—an office block in Leadenhall Street called New Zealand Chambers, designed by Richard Norman Shaw in 1873.[29] Here, for the first time, the turn of taste characteristic of the seventies intruded into the City. New Zealand Chambers went back to the rude vernacular of the early seventeenth century, with red-brick piers, peasant ornaments, and huge, luminous bay-windows. The architectural profession was shocked, but the well-lit floor-space let instantly at high rents, and Shaw's artistry soon reconciled the younger generation of architects to a brilliantly perverse choice of style.

New Zealand Chambers marks a turning-point in City architecture, and indeed in business architecture generally, because of its lack of stylistic seriousness; it was totally disloyal both to the Italian school and to the Gothic. Also, it came at a moment when classical banks and insurance offices were in sharp decline and City building consisted almost entirely of 'business premises' either for private firms or investment companies, mostly in the smaller streets. After Shaw, and in this context, all liberties could be taken. In 1878 gabled Flemish warehouses arrived in Wood Street[30] and in 1882 Ernest George & Peto were much praised for some brick buildings in Cheapside which would not have looked out of place in Bruges.[31] The *style François 1er* appeared in London Wall in 1880,[32] in Cornhill in 1881,[33] and was found extremely practical by the type of architect who believed that in City offices light was the biggest selling factor—which it obviously was.

In reviewing the rebuilding of the City we have been looking at the one really concentrated episode in the making of Victorian London—concentrated within a small area and within a period of no more than twenty years. There was no other concentration quite of this kind. To observe the progress of architecture outside the City the best course is to follow the formation of new highways. New highways necessarily brought new building-sites into the market and these sites were an invitation to business and residential projects of new kinds. The development of London architecture, both as to types and styles, is thus bound up with and richly illustrated by the buildings which these new thoroughfares attracted.

Victoria Street, Westminster, was the first new street created outside the control of the Office of Works.[34] It was formed by Commissioners under a series of Acts, the first of which was passed in 1845. Among the earliest buildings was a block of flats— the first middle-class flats in London—built by Henry Ashton for a Scottish developer and completed in 1853.[35] This pioneer block is, as one would expect from the date, much in the style of Bayswater or South Kensington and not bad of its kind. Unsuccessful at first in attracting tenants, it did succeed in planting the notion of flat-building firmly in the Victoria area of London; and there the idea grew. Belgrave Mansions, in a smart Parisian style, came in 1867,[36] with Albert Mansions rushed up alongside by 1870, in 'latitudinarian' Italian.[37] Hankey's shameless ten-storey enterprise, Queen Anne's Mansions, stood itself up in no style at all in 1874,[38] and from then on till the end of the century one block of flats followed another along the length of the street and in its lateral connections.

A characteristic of Victoria Street was the availability of sites of uncommonly long frontage. This facilitated economies of scale in various types of project. Thus, at the east end of the street a monster hotel and a monster block of offices were undertaken. The Westminster Palace Hotel[39]—the first of its size not projected by a railway company—was an immediate success; technically in advance of anything of its kind, its elevations were in a scraggy French style which the architect hoped 'would possess at least some indication of "high art"'. The long and monotonous office block—Westminster Chambers[40]—was financed by a tontine; though designed by one of Sir Charles Barry's sons, it displayed little more 'high art' than the hotel. Indeed, 'high art' was conspicuously absent from Victoria Street, a street of undertakings too anxiously calculated on the economic plane to admit serious thought about aesthetics.

The formation of new highways was vigorously pursued when, in 1855, the Metropolitan Board of Works took over from the Office of Works the responsibility for street improvements.[41] The responsibility was limited, for the Metropolitan Board of Works was never a planning body, nor did it ever assume more than a very superficial control of architectural design. So, as the new streets emerged and the lots of surplus land were taken up, strange assortments of façade architecture came upon the scene. In Garrick Street we have the Garrick Club,[42] in ponderous Italian, on one side of the street (1861) and, on the other side, a quaint Gothic stained-glass factory by A. W. Blomfield (1861).[43] In Southwark Street, cut through the poor district west of the Borough, we have the sensationally mannered Hop Exchange at the east end (designed by an architect who was one of the shareholders) followed by a long succession of warehouses, some very well designed, illustrating the whole gamut of brick polychromy.[44]

The M.B.W.'s most majestic achievement in the creation of new thoroughfares was, of course, the Thames Embankment.[45] This was not only a monumental achievement in itself but also the container for one of the cross sewers in Bazalgette's drainage system and the carrier of a major new highway. Furthermore, it led immediately to the construction of two other highways—one, linking the embankment to the Mansion House, called Queen Victoria Street; the other, linking the embankment to Charing Cross, called Northumberland Avenue. These three highways created together a new route from the City to the West End and, in the process, made available an amplitude of building sites, nearly all of which were occupied by the mid-eighties. The whole sequence, from Mansion House to Charing Cross, was, up to 1935, a continuously representative parade of Victorian architectural performance. The contents of the three highways are functionally very different. Queen Victoria Street, cutting through the City, necessarily attracted speculators in office accommodation. At the other end, Northumberland Avenue became almost entirely occupied by hotels and clubs. The Victoria Embankment developed miscellaneously. Here, the M.B.W. had no surplus land to dispose of, but by vastly enhancing the value of the land adjacent, it set the scene for a long riverside succession of expensive building enterprises.

If we glance down the lengths of these three connecting thoroughfares we get a pretty strong impression of what was going on in metropolitan architecture between 1870 (when the first buildings went up in Queen Victoria Street) and 1885 (the approximate date of completion of Northumberland Avenue). At the City end, in Queen Victoria Street, we find a few solidly competent blocks, characteristic of the sixties, alongside wildly experimental works, conditioned by a determination to combine optimum fenestration with vigorous stylistic exercises, both Gothic and Italian.[46]

The Victoria Embankment gives us a succession of monuments of wonderfully diverse sizes, shapes, and styles. No two of them seem to have anything in common. There are the academic French of the City of London School;[47] the English Perpendicular of Sion College;[48] the Elizabethan of the Astor Estate Office;[49] the Tudor of the houses on the Norfolk estate;[50] and, further west, the French Renaissance of Whitehall Court and the National Liberal Club;[51] the more or less pure François 1er of the St Stephen's Club;[52] and the French-Dutch-English mixture of New Scotland Yard.[53] All these are still standing. Missing are the School Board Offices (François 1er)[54] and the Hotel Cecil (modern Parisian).[55] These two, as it happens, were the earliest and latest in date (1875 and 1895 respectively). The whole assemblage is strangely unconvincing and raises all sorts of questions about the types of architects at large in London at this period and the nature of their stylistic persuasions and philosophies. When we come to Northumberland Avenue the same questions intrude themselves. The Society for the Promotion of Christian Knowledge building belongs to Italianism of the City type,[56] but next to it is a monster hotel in a very smart Parisian style.[57] Opposite, there stood until recently a large club in a felicitous blend of Jacobean and Flemish Renaissance.

If we go further in pursuit of new architecture in new streets we come to Shaftesbury Avenue, built up between 1879 and 1886. Here is a mixture of Renaissance improvisations in red brick, terracotta, and stone, exhibiting marked incompetence. The reasons for this are readily accessible. The M.B.W.'s policy of buying as little marginal land as possible had the effect of creating wholly inadequate sites; these were taken by wholly inadequate developers employing wholly inadequate architects. Before Shaftesbury Avenue was finished, the inadequacy of the M.B.W. itself had been exposed and it sank in a morass of scandal, to be superseded by the L.C.C.[58]

From Shaftesbury Avenue we may proceed to Charing Cross Road. Here we witness the effects of a crisis brought about by a statutory injunction on the M.B.W. to rehouse the victims of 'improvement' near the sites of their homes. The street was thus channelled for part of its course through meanly fenestrated cliffs of cheap working-class tenements. The remaining parts, shambling along into the nineties, are little improvement upon Shaftesbury Avenue.[59]

Much discredit attaches to the 'improvements' conducted by the M.B.W., an inadequate and, in its last phase, corrupt body. In contrast, the City Corporation conducted a series of operations within, and also without, its own boundaries which contributed honourably to the changing character of London. Within the City, King William Street and Moorgate were already appearing in well-dressed stucco at the beginning of the reign. Between 1846 and 1854 Cannon Street was extended westward. In 1847–9 came the Coal Exchange,[60] in 1850–3 Billingsgate Market[61] (rebuilt 1874), in 1867–8 Smithfield Meat Market[62] and, outside the City, in 1851–5, the Metropolitan Cattle Market.[63] All these were buildings of picturesque Italianate monumentality, the creation of the City's architects, J. B. Bunning and, after him, (Sir) Horace Jones. Other imposing City improvements were the construction, under the engineer to the Commissioners of Sewers, of Holborn Viaduct[64] and, concurrently, the rebuilding of Blackfriars Bridge by Joseph Cubitt.[65] Both were opened in 1869. Like so much Victorian engineering they were architecturally adorned by anonymous hands, the pillared piers of the bridge paying rich tribute to *The Stones of Venice*. Twelve years later the City was to sponsor Tower Bridge, magnificently if perversely engineered and clothed in absurd and outdated Gothic.[66]

The fact must be accepted that much of the architectural product of Victorian London was at a desperately low level. But whether the level of performance be low or high there is still the problem of the appearance and effect of these buildings—the styles they adopted, their sources, their varieties, and idioms. What are they and whence do they issue?

The answers must be sought in the new attitudes emerging in the architectural profession in the early seventies. These are attitudes both towards style and towards opportunity—attitudes adopted at first among the artistic élite of the profession and then spreading outwards towards its more or less illiterate fringes. Style and opportunity are of necessity closely linked. The Gothic Revival would not have succeeded to any extent had it not been linked with the High Church movement and with a widespread social and philanthropic drive towards the architectural rehabilitation of the established church. The momentum in this field was much reduced by 1870 and with it the chief incentives towards Gothic designing. Intelligent young architects of the 1870s saw ahead of them a widening range of opportunities in which churches played a diminishing part. The problem of Gothic for modern secular purposes was one which had never been happily solved; the competition for the new Law Courts in 1866–7 merely underlined the anomalies it involved, and the indirect outcome of that competition, Street's great Gothic composition in the Strand, was considered, even before it was finished, as the swan-song of the Revival.

The shift towards new stylistic attitudes is first observable in a particular sector of the domestic field—the house of the well-to-do artist, architect, or art-patron. We can witness the whole development in houses which are still for the most part standing, in Kensington and Hampstead. The house built for Val Prinsep by Philip Webb in Holland Park in 1864—a little brick house like a country vicarage—is said to have started the series.[67] Webb's house for George Howard in Palace Green (1868) still stands[68] but J. J. Stevenson's clever and important vulgarization of the same theme in Bayswater Road has gone.[69] Norman Shaw comes into the series with Lowther Lodge, Kensington Gore, in 1873, and from that date he is unquestionably the leading architect of the movement. His houses for artists in Melbury Road, Kensington, and Fitzjohn's Avenue, Hampstead, in Queen's Gate, Kensington, and on the

258 Threadneedle Street. On left, Imperial Insurance Office (John Gibson, 1848); in the centre, Bank of Australia (P. C. Hardwick, 1854). From *Illustrated London News*, xxvi (1855), 144.

259 National Discount Company, Cornhill (Francis Bros, 1858).
Photo: Timothy Summerson

260 City Bank, Threadneedle Street (Moseley, 1857).
Photo: Timothy Summerson

261 Nos 59–61 Mark Lane (George Aitchison, 1864).
National Monuments Record

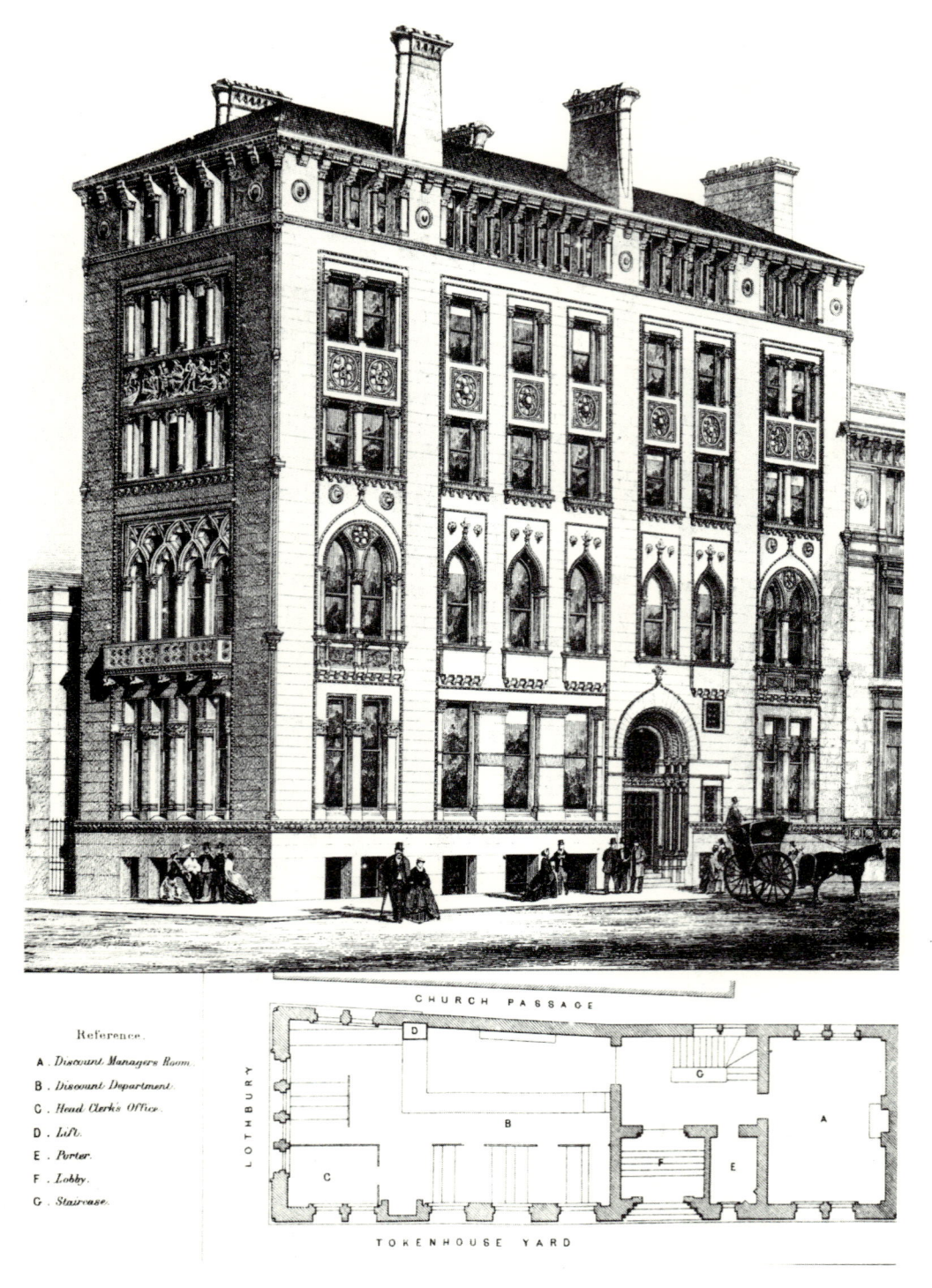

Reference.
A. *Discount Managers Room*
B. *Discount Department*
C. *Head Clerk's Office*
D. *Lift*
E. *Porter*
F. *Lobby*
G. *Staircase*

262 Design for General Credit Company, Lothbury (G. Somers Clarke, 1868), from *Building News*, xv (1868), 11.

HOUSES, VICTORIA-STREET, WESTMINSTER.

STAIRCASE WINDOWS.

UPPER STAIRCASE WINDOWS AND CORNICE.

PLAN.

ONE STORY DIVIDED INTO TWO SEPARATE DWELLINGS.

263 Flats, Victoria Street (Henry Ashton, 1853), from *Builder*, xi (1853), 722.

264 The London School Board Offices (G. F. Bodley, 1874), from *Building News*, xxix (1875), 14–15.

SAUNTERINGS IN SOUTHWARK-STREET.

AN impartial observer, a "looker-on in Vienna," if called on to choose a field for the display of collective architectural talent, would unhesitatingly select Southwark-street. The *tabula rasa* produced by the demolition of obsolete buildings, and the construction of the noble thoroughfare from Blackfriars Bridge to the Borough has offered a "fair occasion" for various architects to show their metal. That the metal should have, in all cases, the genuine ring is scarcely to be expected. There is alloy in all of the specimens, and pure pinchbeck, if the term be allowable, in too many. Asmodeus did not accompany us; nothing beyond the general appearance of the elevations can therefore be described in these notes. Crossing Blackfriars, and passing under the railway bridge, leaving the quaint old almshouses of good Charles Hopton on the left, the first building encountered is the immense store of Messrs. Tait and Co., army contractors. This building was illustrated in the BUILDING NEWS, No. 589, and will therefore be but cursorily noticed at present. The architect was Mr. R. P. Pope. It is distinguished by a liberal use of colour, and even gilding, but the effect is peculiar. Over the ground and first-floor windows are about eight courses of pseudo-reticulated glazed red and green bricks, and it is needless to say that the contrast is too violent for a building of otherwise sober colour. Again, the use of black—or, rather, artificially blackened—brick in the form of labels to windows and doors is most objectionable. In the present instance it makes the building appear as if in mourning, and the too free introduction of black tiles in the frieze of the cornice heightens the indications of woe. What connection there can be between an army contractor's establishment and a family mourning depôt we cannot see. It is true that a pursuit of reputation in the career of arms may occasionally be effectively checked, but from the time when the Venetians furnished supplies to the Crusaders to the days of Cloncurry—or, for that matter, to this year of grace—army contractors have had little occasion to make a moan. Why, then, this "suit of sables?" From the south-east a good view is obtained of the tower over the staircase, and from this point the building is seen to the greatest advantage, the management of the tower being really artistic. The material is, for the most part, yellow brick, with a little red Mansfield stone in the form of dressing. There is, however, an unreasonable mixture of noble with mean material. Common yellow bricks, Mansfield stone, glazed bricks, gilding, white stone—either Portland, white Mansfield, or some similar stone—encaustic tiles, polished granite shafts, elaborately carved caps, and a dismal bordering of black brick, a bordering such as the well-dowered widow orders to be put on her writing paper (see fig. 1), which shows one of the window heads.

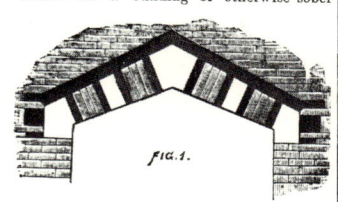

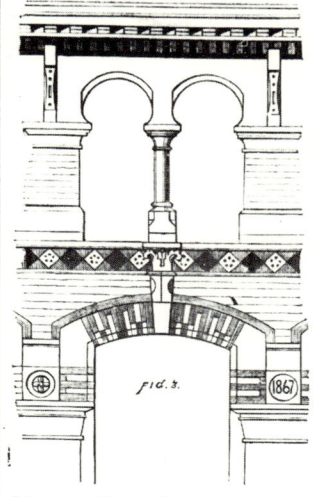

The next building on the same side of the street is the warehouse of Messrs. Causton and Sons, stationers. The utilitarian character of this building is so plainly indicated by its external appearance that criticism would be disarmed were it not that a display of festivity, in the guise of carving, has been attempted over the first-floor windows. Carving, if attempted, should be good of its kind. This is bad of its kind. Ill-designed and coarsely executed, it disfigures a building which should certainly not court attention. The jointing of the basement window heads is what we have ever protested against. A joint in the centre of a segmental or circular arch is neither conformable to recondite theory nor rule of thumb practice. The balcony over the entrance door, not yet fixed, may do something for the building, but the trusses already in place are of very uncouth form. The frontage is about 108 feet. Mr. Saunders, Finsbury Pavement, is the architect. On the same side of the way a warehouse is in course of erection for Messrs. Lawson and Co., seed merchants; Mr. J. Wimble, No. 2, Walbrook, being architect. So little progress has been made that beyond recording the fact that the first-floor joists are laid, nothing need be said at present. Immediately opposite, and in an unfinished state, is a warehouse for Messrs. Waite, Barnett, and Co., also seed merchants; Mr. Edis, architect. It is of yellow brick, and very plain in its present state, as the ornamentation is to be in stucco. The third-floor joists are not yet laid, so that it is not easy to say what will be its appearance when finished. It has a frontage of about 65ft., and a height of 50ft., enough for the display of a good deal of architectural skill. The contractors are Messrs. Sandon

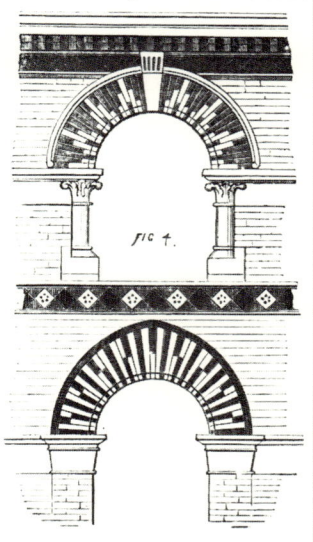

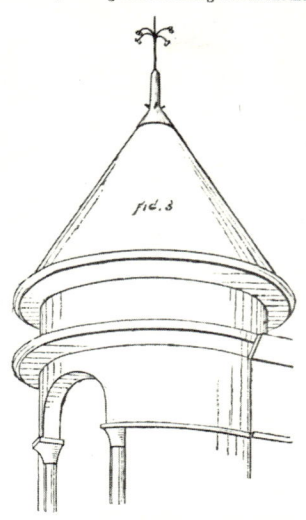

266 *left* New Zealand Chambers (R. Norman Shaw, 1873).
National Monuments Record

267 *above right* Northumberland Avenue. The S.P.C.K. building between the Hotel Metropole (left) and the Northumberland Avenue (later, Victoria) Hotel, 1883–4.
Photo: G. W. Wilson
Aberdeen University Library

268 *below right* Albert Buildings, Queen Victoria Street (F. J. Ward, 1871).
National Monuments Record

MULDIVO CALCULATING MACHINE Cº
CARTERS
TESTED SEEDS
CARTERS
SLATERS
RESTAURANT

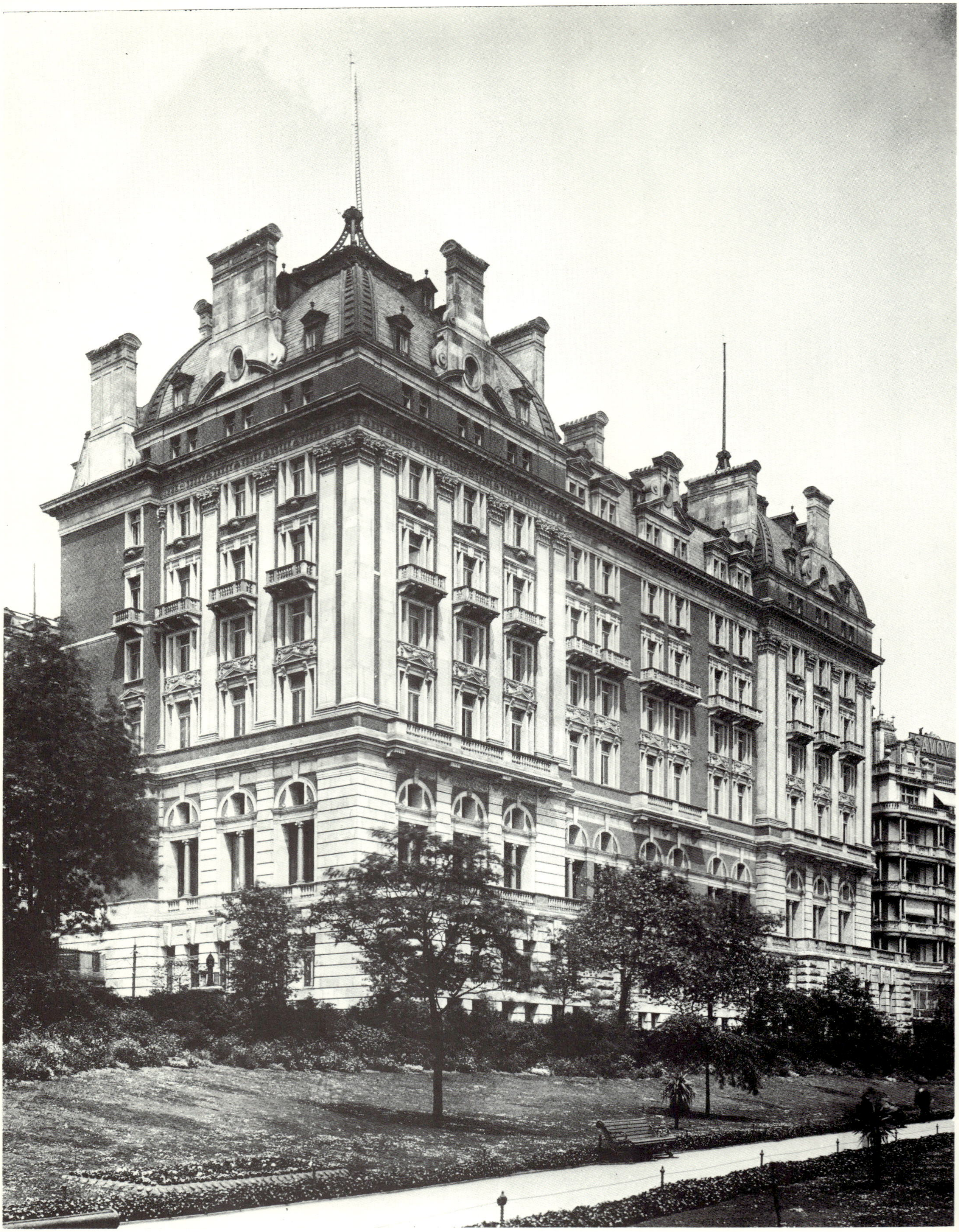
SAVOY

269 *left* The Hotel Cecil, designed by Perry & Reed and opened in 1886.
Photo: Bedford Lemere National Monuments Record

270 *right* The Red House, Bayswater Road (J. J. Stevenson, 1871).
National Monuments Record

A Plan of Lowther Lodge Kensington

CHELSEA FREE PUBLIC LIBRARY

271 *above left* Lowther Lodge, Kensington Gore (R. Norman Shaw, 1873), from *Building News*, xxviii (1875), following 716.

272 *left* Design for Chelsea Public Library (J. M. Brydon), from *Building News*, lvi (1889), 802–3.

273 *above* British Museum of Natural History (Alfred Waterhouse, 1873–80), from *Builder*, xxxi (1873), 10–11.

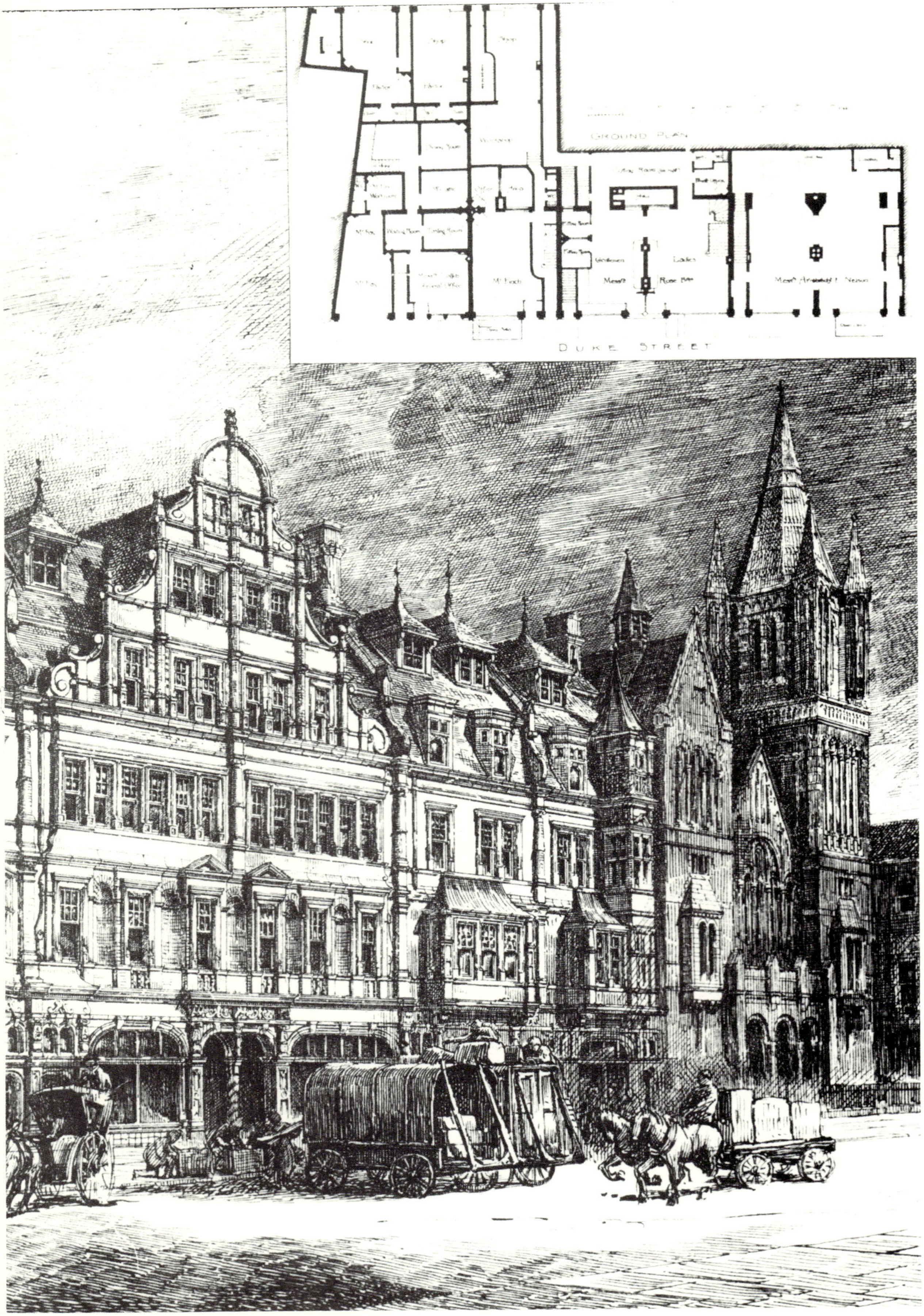

274 *left* Business premises, Duke Street (W. D. Caröe, 1891), from *Building News*, lx (1891), 238–9.

275 *right* No. 1 Old Bond Street (Alfred Waterhouse, 1881).
National Monuments Record

276 *below right* Grosvenor Hotel (James Knowles, 1860–1).
National Monuments Record

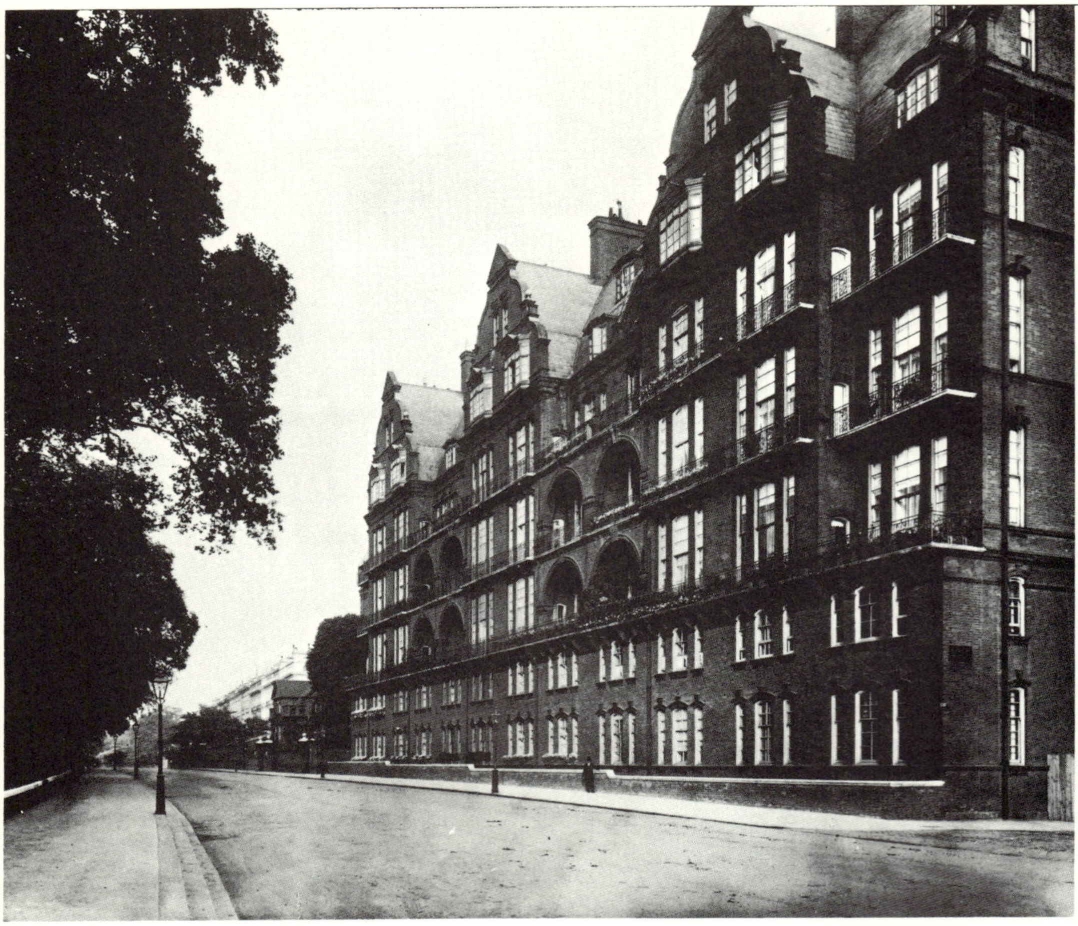

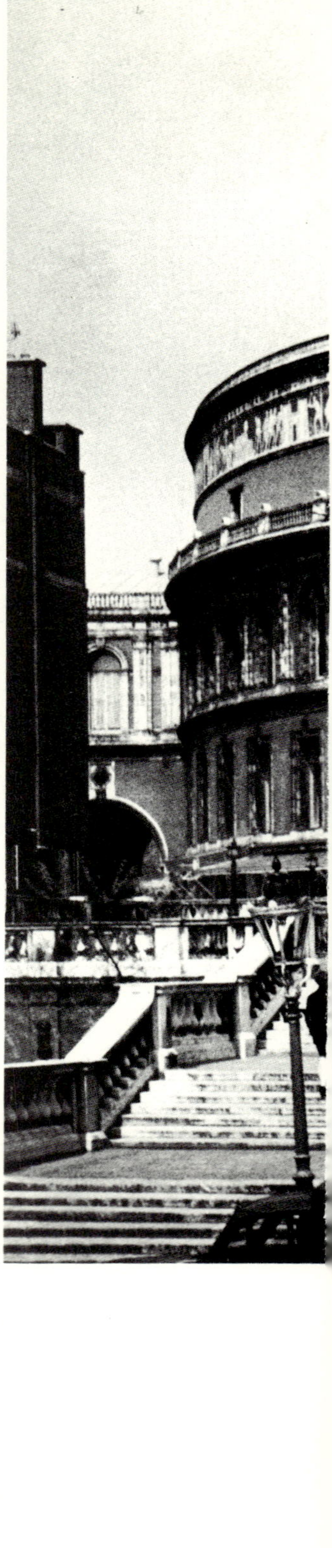

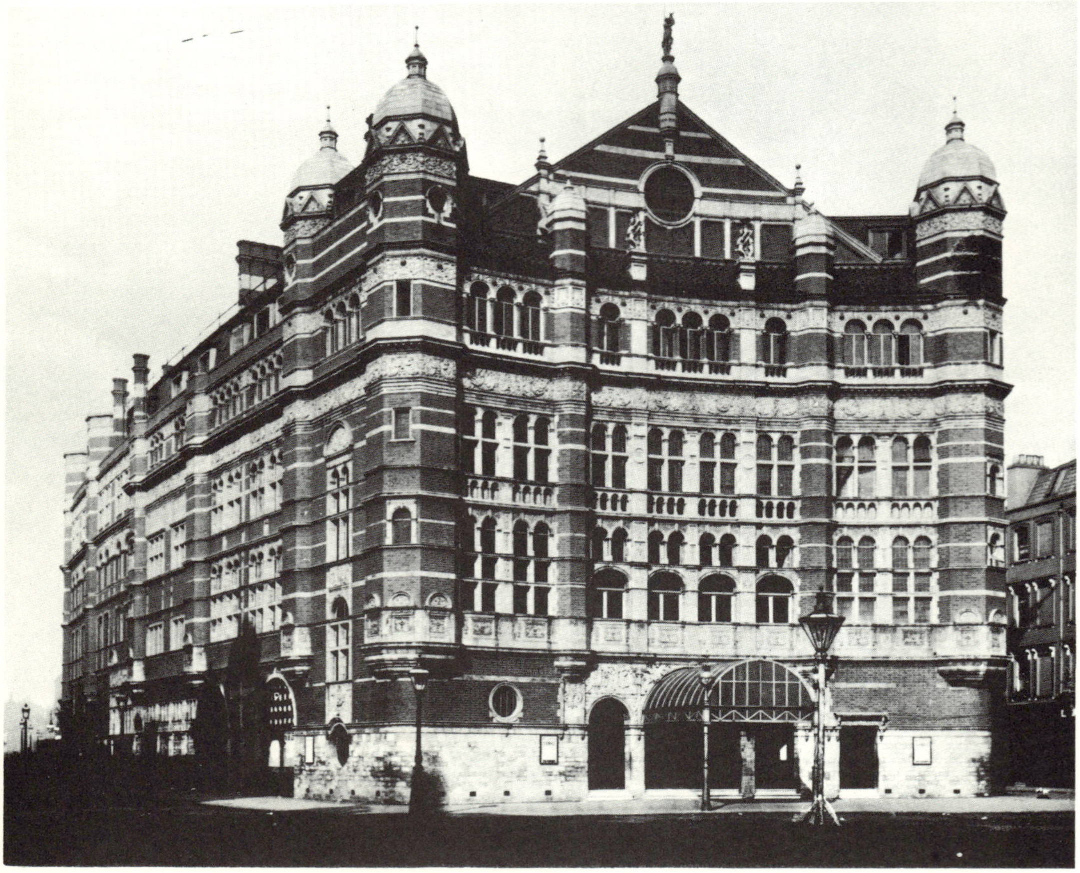

277 *above left* Albert Hall Mansions (R. Norman Shaw, 1881).
B. T. Batsford Ltd

278 *left* Palace Theatre, formerly British Opera House (T. E. Collcutt, 1891).
Photo: Bedford Lemere National Monuments Record

279 *above* Royal Albert Hall (Capt. Fowke and General Scott, 1867–71).
Photo: Timothy Summerson

280 Imperial Institute, South Kensington
(T. E. Collcutt, 1887–91).
National Monuments Record

Chelsea Embankment; his block of flats (Albert Hall Mansions); his famous office block in the City (New Zealand Chambers); his insurance office in Pall Mall—all these established him as by far the most influential architect of the last quarter of the century.[70] Just as Sir Charles Barry's Italian was the source of so much in the stylistic flavour of London of the forties, fifties and sixties, so the Renaissance of Norman Shaw pervades the London street-scenes of the later period.

What exactly was the nature of this revolution in taste emerging from the affluent circles of artists and art-patrons around 1870? It was a revolution away from the Gothic Revival and the 'battle of the styles'; away from church architecture as the central theme in architectural practice and towards domestic and social programmes; away from scholarly reference in design and towards complete freedom of handling. The new attachments to history were neither to textbook Gothic nor textbook Classical, but to the in-between vernaculars which had no textbooks. Indeed, a quite distinct new vernacular was created and given the curiously misleading title of 'Queen Anne'. From the early seventies 'Queen Anne' signified buildings usually of common brown and special red brick ornamented in the bourgeois style of the English seventeenth century and with gables and dormers moving in a Flemish direction. It was really a Renaissance revival—indeed, the term 're-renaissance' was suggested at the time as a more accurate designation than 'Queen Anne'—as indeed it would have been.[71] The metamorphoses of 'Queen Anne' are the clue to nearly all the architecture of London from the seventies to the nineties. The style incited no real antagonisms and the only architects who held away from it were those who felt a greater security (often for business reasons) in conventional Italianism and 'modern French'.

This revolution in taste coincides with changes in the responsibilities and opportunities open to architects brought about by the increased wealth of the professional classes and the tendency of the second generation of successful families to bring the arts into their lives and homes. It coincides also with a new proliferation of building types. Alfred Waterhouse, in a speech of 1883, observed that, whereas a few years earlier an architect's practice would embrace simply town and country houses, City buildings, churches, and the occasional public building, now, at the date at which he was speaking, it would embrace a whole multiplicity of types.[72] He gives a list, reflecting his own wide field of activity, but it falls far short of the reality as reflected, for instance, in the building journals of the time. Waterhouse very perceptively related the proliferation of types to the change in style which we have just been considering. He said that *style* (in the abstract sense associated with personal artistry) was now the thing, not *styles*. In other words, choice of this or that style from the past was less important than the way in which any style or mixture of styles was handled. Complexity of types demanded flexibility and richness in the architect's vocabulary, not rhetorical statement. Flexibility and richness are what the architecture of late Victorian London unambiguously reflects.

It was not only that building types were proliferating. The planning of each type was becoming more complex and so was the structure; so was the provision of services. Structural ironwork; lifts; fire-proofing the high buildings which lifts facilitated; the improvement and elaboration of sanitary equipment; heating and ventilation; and, latterly, electric lighting: the architect had to deal with all these, and in doing so saw himself as a different sort of professional from the man who built only churches, vicarages, and village schools, building them in the way in which they would have been built fifty or a hundred years before, and worrying mainly about pillar mullions and plate-tracery, symmetry, or the picturesque.

The proliferation of building types must not be taken to imply the introduction of positively new building-functions so much as the assumption by various classes of building of a new typological importance. Here, one of the most striking instances is that of the schools built by the London School Board after the Education Act of 1870. Schools of various kinds had been built in London for many years, mainly in association with churches. Their accommodation was primitive, their architectural style often picturesque. Until the School Board came into being it had been nobody's business to study the school as a type and to consider what principles were involved in, for instance, daylighting, heating, hygiene, and educational equipment. E. R. Robson, the Board's first architect, undertook to study these things and produced an admirable book on the subject in 1872.[73] He was also the designer, with his partner, J. J. Stevenson, of many of the early Board Schools. For these, the style of Stevenson's own Red House, in the Bayswater Road, was adopted. It was, of course, the 'Queen Anne' style, and the style characterized London Board Schools till near the end of the century. A 'Robson School' with its great windows, high Flemish gables, and dainty brickwork is still a familiarly repetitive London landmark. It was a new and exciting accent in the London sky-line of the seventies.

The Board Schools were immediate products of social legislation and in that respect are symptomatic of a tendency which is increasingly reflected in the London architecture of the eighties and nineties. The Metropolitan Poor Act of 1867 engendered grim, grey, towered infirmaries and led, in due course, to the great suburban hospital layouts for which the Sick Asylums Boards were responsible in the nineties. Architecturally and typologically more important are some of the building enterprises of the vestries. The vestries were not overloaded with building responsibilities, but there were buildings for which they had statutory powers to borrow money if prodded into doing so by their electors. From 1846 they had, like every authority in the country, powers to build public baths and wash-houses. The idea of public baths had been to raise the standard of hygiene among the grubbier sections of the population, but the notion that swimming could be a healthy recreation for all had reached England by the eighties. Lewisham built an establishment with first- and second-class swimming baths in 1885;[74] other vestries followed in rapid succession. They got bigger and bigger, and the Lambeth establishment of 1897, with four swimming-baths, ninety-six slipper-baths and provision for sixty-four washers was reckoned to be the biggest in Europe.[75] There was some fumbling with architectural style as the new type emerged. Lewisham's building was in the *démodé* Gothic of the seventies, Hampstead's was Queen Anne in front with a hammer-beam roof over the bath.[76]

But a handful of specialists soon evolved a satisfactory compromise between style and efficiency, just as Robson and his colleagues had done in the case of schools. And the style was always more or less 'Queen Anne'.

From 1850, the vestries could build free libraries, but built very few indeed until the nineties, when the new literacy of the first Board School generation required them and they came with a rush. Unlike public baths, they could not be made to pay and the reluctant vestries rarely proceeded without philanthropic subventions from such men as Henry Tate and John Passmore Edwards.[77] Architectural performance was variable but sometimes ran high, as in E. W. Mountford's library in Battersea,[78] J. M. Brydon's in Chelsea,[79] a neo-Georgian pioneer, and Henry Wilson's sensational free Gothic escapade in Paddington.[80]

The rising importance of architecture in the public sector is one of the significant features of the period after 1870. Its products are necessarily less conspicuous, however, than building in certain other sectors, especially those in which investment was the main incentive. What we may broadly call the investment sector includes a wide range of types: business premises of all kinds, blocks of flats, hotels, theatres and music-halls, restaurants, and public-houses. The religious and philanthropic sector is scarcely less important, with a vast production of churches and chapels on the one hand and the extensive development of philanthropic or semi-philanthropic industrial housing on the other, to which should be added a certain number of hospitals. Finally, the institutional sector includes clubs, the headquarters of professional bodies, and privately or corporately owned schools and colleges; the products of the Polytechnic movement of the nineties may appropriately be placed in this category.

Each of these many types has its own architectural history and its own peculiar engagement with the question of style. Rarely indeed is there any consistent association of type and style such as we found in the case of the Board Schools and such as we should find again if we were to look at London churches after 1870. On the other hand the general stylistic character of London advances with the development of the more important types, and to some of these we may give cursory attention before proceeding to general conclusions.

Clearly, it is in the investment sector that the ubiquitous and most manifest changes occur, blocks of flats, business premises, and hotels being of outstanding effect. Of flats and their adolescence in and around Victoria Street something has already been said. The first architecturally original block of flats was Norman Shaw's brilliant, but perhaps extravagant, Albert Hall Mansions of 1881, with its great Dutch gables and recessed balconies, and dwellings ingeniously planned on two levels.[81] Thereafter, flats were rarely placed in the hands of architects of major talent and reputation. The first large block after Shaw's was Whitehall Court (1884) where the firm of Archer & Green looked to early French Renaissance (the lost Château de Madrid in particular) for a way of controlling a huge cellular mass and giving it animation and a silhouette.[82] But the later monster blocks are, for the most part, either mechanical solutions made more or less acceptable by a conventional distribution of cornices, or vulgar concoctions by architects of a very low type working with builders who were often themselves the developers. Stylistic improvisations based on 'Queen Anne' with strong infusions from Norman Shaw's private vocabulary are the typical product. One trouble about flats was that they depended for success on a good address. Suburban flats were unknown, except in Battersea, till after 1900.[83] So investment in flats meant the grabbing of central sites and the piling on to them of as much as the Building Acts would permit. In the nineties, Cavendish Square, Hanover Square, Berkeley Square, and Kensington Square were all broken into by blocks of flats. The destruction of Georgian London had begun.

'Business premises', that vague denomination which usually signifies a combination of lettable offices and shops, but may also include flats or multi-storey show-rooms, began to change the character of the West End in the seventies. Successful shopkeepers, content before with show-windows in an old house or a clutch of old houses, now tended to rebuild their entire premises with ample provision for office- or flat-letting as an investment and, very naturally, in as fashionable a style as possible. Marshall and Snelgrove rebuilt a large block in Oxford Street for their drapery business as early as 1878, but drearily, in bad French;[84] 'Queen Anne' and the gay red of Farnham bricks had not yet arrived. Oxford Street only changed colour in the eighties. In Bond Street, the quaint windows of Norman Shaw's New Zealand Chambers found a ready imitator in 1876,[85] and in the same year Robert Edis rebuilt a shop at the corner of Brook Street with a tile-hung gable, the first of its kind in Central London.[86] Here were country manners coming shockingly into the smart West End. In Piccadilly, which up to the seventies had had a soberly institutional character, Waterhouse's shop at the corner of Bond Street[87] must have dazzled in 1881 with its cheerfully decorative terracotta. Wigmore Street was transformed by Collcutt, Ernest George, and others;[88] while on the Grosvenor estate 'Renaissance' flourished as nowhere else, with Ernest George's François 1er, Robert Edis's Dutch, and 'Queen Anne' mixtures by Thomas Verity, Wimperis, and others making a regular gallery of re-births in Mount Street,[89] while in Duke Street the blocks flanking Waterhouse's King's Weighhouse Chapel show W. D. Caröe at work with what is perhaps Danish Renaissance (1891).[90] The quality architecture of the Victorian West End cannot be better appraised than by probing this concentrated collection.

Among the most substantial and visually prominent enterprises in Victorian London were the hotels. The large hotel was the creation of the railways, and Paddington was the first (1852–3).[91] Incorporating some French ideas in the programme, the architect gave it a stucco front on the lines of a *hôtel de ville*. The next monster was the French-styled Westminster Palace Hotel, already mentioned as an important component of the new Victoria Street.[92] This was followed in 1860–1 by the London Chatham & Dover Railway's Grosvenor Hotel at the other end of the street. By this date the railway hotel was seen to have become a portent; it was 'one of those striking conceptions which distinctively mark the civilization of the age' and the Grosvenor with its bigness of scale and liberality of ornament (with busts of the Queen's prime ministers) played this role with conscious success.[93] In the early sixties came two more railway hotels, at Charing Cross[94] and Cannon Street,[95] and also the Langham, on the site of an old house and grounds in Langham Place.[96] The emerging type was a massive and ornate brick-and-stone block with high French roofs (for staff accommodation) and corner pavilions. At St Pancras (1865–72)[97] the Midland departed into something different at the Gothic hands of Sir Gilbert Scott, but the Holborn Viaduct Hotel strongly re-established the 'modern French' idea in 1874.[98] The same architects (Isaacs & Florence, the latter Paris-trained) designed the best of the three great hotels in Northumberland Avenue.[99] Again, in the Carlton[100] and the Cecil (1895)[101] the French idea prevailed, but the last years of Victoria's reign saw free Renaissance at the Great Central[102] and, in Russell Square, two exuberant Renaissance outbursts—the Russell and the Imperial, the latter a Baltic fantasy with a strong infusion of *art nouveau*.[103]

Among types in what may broadly be called the investment sector none had anything like the size or architectural effrontery of the hotels. The creation of 'theatre-land', by the exploitation of sites offering themselves during the formation of Shaftesbury Avenue and Charing Cross Road, is a phenomenon more striking socially than visually. Theatre designing was in the hands of a few specialist architects (notably C. J. Phipps and Walter Emden) who contributed little to the street-scene beyond weakly conventional 'modern French' or commercial 'Queen Anne'. The one remarkable late Victorian theatre is Collcutt's British Opera House (now the Palace Theatre) in Cambridge Circus (1891), a Plateresque study in brick and terracotta;[104] but Collcutt did not plan the theatre—merely overlaid the structure with his idiosyncratic art. Restaurants make a number of sharply accented contributions without developing a type. There is Pimm's Restaurant (now no longer such), a ludicrously mannered front in Poultry[105]—the first rationalized City chop-house (1870); and in Piccadilly still stands the Criterion (1874), a highly elaborate enterprise of Spiers and Pond where Thomas Verity enclosed behind his *cinquecento* façade a whole world of dining-halls, grills, bars and buffets, dives and saloons, with an art gallery at the top and a theatre in the basement.[106] The great restaurants of the eighties and nineties—the Holborn, Romano's, Frascati's, and the Trocadero—never quite competed with this, though each had its stylistic singularity. Descending from restaurants to public-houses, we can observe departures in various directions from the pre-1870 convention that a public-house was simply an over-decorated version of a common London house, preferably on a street corner. The pubs which today are regarded as being enjoyable by reason of their uninhibited vulgarity mostly belong to the eighties and nineties; their exteriors are, in fact, largely made up of assorted thefts from Norman Shaw and 'modern French'.

This almost entirely random inspection of types within the investment sector alone is perhaps sufficient to indicate the kind of complexity with which an analyst of late Victorian London's visual image has to deal. The question now arises whether the apparent chaos of the stylistic scene can be reduced to some coherent order. If it can, the main clue to it is to be found in the study of the architectural profession. From being a fairly small body in the 1830s, numbering probably fewer than a hundred, the profession in London was said to number 1,114 in the census of 1851 and had reached nearly 2,000 by 1871. At this date it was a grossly inflated, ill-defined, and badly under-educated professional category, dependent for leadership on a few brilliant men who formed that élite to whose reforming agency reference has already been made. The élite was a constellation rather than a group, and from the major figures there was constant devolution to lesser men and lesser constellations. Furthermore, there were architects who stood away from the main sources of influence, holding to old-fashioned ideas and bringing them out in new guises. And there was always the rabble of hacks, ghosts, and incompetents of all sorts, cribbing and corrupting the performances of their betters.

Philip Webb,[107] J. J. Stevenson, Norman Shaw, and Waterhouse were the rising constellation of the seventies. All born in 1830 or 1831, they powerfully changed the face of London. They changed its style through their new attitudes to the past. They changed its colour by changing its materials: from brown brick to brown and red, then to red; from Portland stone to granite or Mansfield or to red or buff terracotta. A second constellation followed when men born around 1840 came into full practice in the eighties. In the hands of Ernest George, R. W. Edis, and T. E. Collcutt, the 'Queen Anne' liberalism was diffused into a variety of northern Renaissances—French, Flemish, and Dutch, and even Spanish Plateresque, while J. M. Brydon turned 'Queen Anne' into a truly historical George I, thus promoting the return to Wrennian and Gibbsian classicism which was one of the directions taken at the end of the century.

An analysis of the whole architectural content of London as it stood in 1901 would show an overwhelming indebtedness to these names. Furthermore, if the work and influence of the 'constellations' could be detached from its context and considered in isolation, we should have before us an entirely consistent pattern of stylistic growth and performance. London, however, gave room and opportunity for much else. Conspicuously, there is the 'modern French' school allied with the Italian conservatism of Barry's time (mainly represented in the work of his sons, Charles and Edward M.), operating at a somewhat conventional, though usually affluent level. And in the church-building world there are still the protagonists of Gothic, like G. F. Bodley and A. W. Blomfield and, later, J. P. Sedding and J. F. Bentley, running parallel with our two constellations and, in their secular works, showing no aversion to 'Queen Anne' and her progeny. Indeed, Bodley's headquarters for the London School Board, an early and well-informed interpretation of François 1er, was instanced by J. J. Stevenson, in a lecture of 1874, as a good example of the 'Queen Anne' school!

At this stage, it may well have occurred to the reader that in the course of a fairly lengthy discussion of London types and styles not a single building of national status, built by the central government or built to incorporate national sentiment, has received more than passing mention. The omission is natural and reflects an important truth about Victorian London. Whatever may have been the case in provincial towns, where vast town-halls and exchanges arose amid general applause, in the capital the history of major public buildings is, to an astonishing extent, a history of governmental parsimony, ill-conducted competitions, and results which were either roundly condemned or soon forgotten. The loss of esteem of the Palace of Westminster was noted earlier. In 1856 came the fiasco of the Government Offices competition, and if, in the outcome, a reasonably good work of the Italian school emerged at the hands of Gilbert Scott, it is not one that counts for much in the history of English architecture. The Law Courts competition was a notorious administrative disaster and Street's eventual building, cramped in site and cost, makes a sad cadence to the Gothic adventure.[108] The Admiralty competition of 1884 was badly assessed, the award going to architects of slight calibre and a design outside the main creative currents of the time.[109] Meanwhile, the Government did achieve an important architectural result when it directly commissioned Waterhouse to design the British Museum of Natural History in South Kensington.

South Kensington, rather than Whitehall, is the district where one must look for the expression of Victorian architectural ideas at a national level. The particular area is that between Kensington Gore and the Cromwell Road, bought with the surplus arising from the Great Exhibition of 1851.[110] The first buildings to be put on the site were arcades and a conservatory for the Horticultural Society, which leased part of the ground: slight buildings, but already, in 1861, announcing that the South Kensington site would be no place for the normal Italian or Gothic convention.[111] The Albert Hall, promoted by a private company to honour the memory of Prince Albert, was built on the northern part of the site in 1867-71[112] and, on another part, the nucleus of the museum later to become the Victoria and Albert was begun in 1865.[113] Under the influence of Henry Cole these buildings were placed in the hands of military engineers.[114] The Albert Hall, based on a Roman theme and daringly contrived, was decorated by an architect in a north Italian Renaissance style, appropriate to brick and terracotta. The north court of the Museum, in the same materials, wears the same dress. The results are a special South Kensington product, rather outside the general stylistic flow.

The British Museum of Natural History is another matter. The Government commissioned Waterhouse directly, without competition, and the building was begun in 1873.[115] At a moment when his contemporaries were moving forward from Gothic to early Renaissance vernaculars, Waterhouse moved dramatically backward to a hard, rhythmical, Germanic Romanesque. The colossal emphasis of Waterhouse's building posed a problem for the next comer to the South Kensington site. T. E. Collcutt won the competition for the Imperial Institute in 1887.[116] It was to be on a site to the north of the museum (the old Horticultural Garden) and on the same axis. It must either agree with or protest against Waterhouse. It did, in fact, agree in its general disposition but disagreed in the stylistic handling. If Waterhouse's building is rooted in hard mid-Victorianism, Collcutt's was rooted in the gentler 'Queen Anne' movement of the seventies and owed much to the compositional devices of Norman Shaw. Its extreme elegance of detail and subtle combinations of brick and stone were the exact contrary of Waterhouse's tough, positive handling in terracotta.

The Victorian City: Shapes on the Ground

It is sad that we have lost the Imperial Institute (except for its slim Sevillian tower) because that building and the Natural History Museum together represented, as no other London buildings do or did, the power of expression of mid-Victorian and late-Victorian architecture at the level of national monumentality.

South Kensington is a good point at which to leave the subject of Victorian London as artifact. From Scott's Albert Memorial some of the most significant phenomena are within easy range—the stucco terraces of the fifties; the rich houses of Webb, Stevenson, and Shaw; and the monumental museums, with the steeples of Kensington and Lancaster Gate peering over the trees. But, having said that, it becomes obvious that to consider London as a visual entity is to entertain a plurality of images. Many suggest themselves, but perhaps five have crucial relevance: the rebuilt City; the trail of new buildings in new streets; the products of domestic affluence; the products of religion, philanthropy, and social progress; and South Kensington. From sixty-four years of prodigious expansion and energetic innovation no unique totality, no single architectural image emerged to typify the London of Victoria. To the twentieth century, the London legacy was incomprehensible chaos. Only today, with architectural horizons stretching to irredeemable monotony, does the Victorian city, by the very virtue of its complexity, once again strike the imagination.

Notes

1 The history of style in early Victorian architecture is admirably expounded in H.-R. Hitchcock, *Early Victorian Architecture in Britain* (New Haven, 1954). There is no comparable study for the later decades, but on a small scale the English sections of the same author's *Architecture: Nineteenth and Twentieth Centuries* (1958) are valuable.
2 Dickens's London is essentially Georgian London, even when he sets his scene in the railway age. The new London of his time was 'Stucconia' where the Veneerings resided (*Our Mutual Friend*, ch. 10) and that was perhaps Belgravia or Tyburnia.
3 The houses on the corners of Great Russell Street and Museum Street were rebuilt or remodelled by an architect called Trehearne in 1862 (*Builder* (*B*), xx (1862), 416).
4 This was part of an endeavour on the part of the Bedford estate to preserve Bloomsbury from 'lodging-house dry-rot'. See D. J. Olsen, *Town Planning in London: The eighteenth and nineteenth centuries* (New Haven, 1964), pp. 174–7.
5 London County Council, *Survey of London*, XVIII: The Strand (1937), p. 103, plate 71b.
6 There are, as yet, few studies of Victorian estate development in London. The following are important: Dorothy Stroud, *The Thurloe Estate, South Kensington* (published for Thurloe Estates Ltd by Country Life Ltd, 1965); I. Scouloudi and K. P. Hands, 'The Ownership and Development of Fifteen Acres at Kensington Gravel Pits', *London Topographical Record*, xxii (1965), 77–125; D. A. Reeder,

'A Theatre of Suburbs: Some Patterns of Development in West London, 1801–1911', in *The Study of Urban History*, ed. H. J. Dyos (1968). For the Radcliffe Estate, Brompton, see *B*, xxvi (1868), 201; for South Kensington developments, *B*, xxvii (1869), 629.

7   The planning and equipment of the London house after the Georgian period has never been studied, partly because of the obvious difficulty of obtaining access to private houses. The building periodicals give few town-house plans.

8   For the various metamorphoses of the Office of Works, see *B*, xxxv (1877), 897. For early Victorian street improvements see 'A Quarter of a Century of London Street Improvements', *B*, xxiv (1866), 877, and J. W. Penfold, 'On Metropolitan Improvements', *B*, xxiii (1865), 427–8. The best sources for descriptions and illustrations of the new streets are the *Companion to the Almanac* and the *Illustrated London News*.

9   See T. C. Barker and Michael Robbins, *A History of London Transport* (1963), I; John Summerson, *Victorian Architecture: Four studies in evaluation* (New York, 1970); E. A. Course, *London Railways* (1962).

10 Alfred Blomfield, *A Memoir of C. J. Blomfield, Bishop of London* (1863), pp. 233–5.

11 T. F. Bumpus, *London Churches, Ancient and Modern* (2nd series: classical and modern) [1908]; B. F. L. Clarke, *The Building of the Eighteenth-Century Church* (1963); M. H. Port, *Six Hundred New Churches: A study of the Church Building Commission, 1818–56, and its building activities* (1961).

12 J. Summerson, 'An Early Modernist: James Wild and his Work', *Architects' Journal*, lxix (1929), 57–62.

13 Paul Thompson, 'All Saints, Margaret Street, Reconsidered', *Architectural History*, viii (1965), 73–94.

14 H.-R. Hitchcock, 'Brunel and Paddington', *Architectural Review*, cix (1951), 240–6.

15 The London and Westminster Bank, Lothbury (with William Tite), 1837–9, was demolished in 1928; the Sun Fire Office, Threadneedle Street, was demolished in 1970 but had previously been much altered. See A. E. Richardson, *Monumental Classic Architecture in Great Britain and Ireland during the Eighteenth and Nineteenth Centuries* (1914), p. 78 and plate xxxvii. Cockerell's employment in these instances was probably due to his position as architect to the Bank of England, where he succeeded Soane in 1833.

16 *Building News* (*BN*), iv (1858), 723; details, xii (1865), 447.

17 *BN*, xv (1868), 11, 29, 147. For critical comments on this and many other city buildings of the sixties see three leading articles in *B*, xxiv (1866), 641, 677, 792.

18 E. I'Anson, 'Some Notices of Office Buildings in the City of London', *Trans. Royal Institute of British Architects* (*R.I.B.A.*), xv (1864–5), 25–36.

19 *Illustrated London News*, vii (1845), 215–16. H.-R. Hitchcock, *Early Victorian Architecture*, II, plate xii, p. 2.

20 *BN*, iii (1857), 1125.

21 Exhibited R.A., 1860. See *B*, xviii (1860), 290.

22 *B*, xxix (1871), 187.

23 *BN*, xi (1864), 134–5. H.-R. Hitchcock, 'Victorian Monuments of Commerce', *Architectural Review*, cv (1949), 61–74.

24 A specially elegant example was Munt & Brown's warehouse, Wood Street. See *BN*, iii (1857), 68, 261, 285.
25 *B*, xxiv (1866), 851; *BN*, xiii (1866), 780.
26 *B*, xxvi (1868), 749.
27 G. G. Scott, *Personal and Professional Recollections*, ed. G. G. Scott, Jr (1879), p. 210.
28 The term is Robert Kerr's. Summerson, *Victorian Architecture*, pp. 8–9.
29 Destroyed by bombs: *B*, xxxi (1873), 358, 607, 632. Commenting on this building, shown at the R.A. in 1873, the *Builder* asked why Shaw should 'affect so unnecessarily the manner of a bygone age ... entirely contradicting the tone and feeling of his own day'. The aged Professor T. L. Donaldson could not 'conceive what motive could have induced [the architect] to rake up a type of the very lowest state of corrupt erection in the City of London, of a period that marks the senility of decaying taste.' R. Blomfield, *Richard Norman Shaw, R.A., Architect, 1831–1912* (1940), p. 50, recalled that 'even ten years later ... people were not quite sure whether they should regard the building as a freak or as a work of genius.'
30 The warehouse of Rylands & Sons by John Belcher: *BN*, xxxvi (1879), 17.
31 The premises of Cow, Hill & Co.: *BN*, xlii (1882), 25, 121.
32 The Submarine Telegraph Co.'s offices by John Norton: *B*, xxxix (1880), 617.
33 The London & Lancashire Assurance Office, by T. Chatfield Clarke, now called Yorkshire House: *B*, xli (1881), 174–5.
34 The early history of the street is summarized in J. Penfold, 'On Metropolitan Improvements', *B*, xxiii (1865), 427. See also *B*, xix (1861), 459.
35 Plan and details: *B*, xi (1853), 721–2. The flats, now let as offices, are entered from Nos 137, 147, 157, 167 Victoria Street.
36 *B*, xxv (1867), 121–3.
37 Priscilla Metcalf, 'The Rise of James Knowles, Victorian Architect and Editor' (unpublished Ph.D. thesis, University of London, 1971), pp. 359–61.
38 *B*, xxxiii (1875), 924. See also [Sylvain Mangeot], 'Queen Anne's Mansions', *Arch. & Building News*, 13 January 1939, 77–9.
39 Fully described by its architect, Andrew Moseley, in *B*, xx (1862), 165–7. The building still stands though with its mouldings shaved off and the walls rendered white. The staircase hall survives.
40 *B*, xix (1861), 570; xxi (1863), 490.
41 Percy J. Edwards, *History of London Street Improvements, 1855–97* (1898).
42 *BN*, xi (1864), 464–5.
43 *B*, xxii (1864), 901.
44 Southwark Street was largely destroyed by bombs in World War II. For the Hop Exchange, damaged but still (1972) standing, see *B*, xxv (1867), 731. For notes on the street, see *B*, xv (1857), 121; xx (1862), 870; xxv (1867), 348; 'Saunterings in Southwark Street', *BN*, xiv (1867), 707.
45 See Edwards, op. cit.
46 The first building in the street was I'Anson's British and Foreign Bible Society, still standing: see *B*, xxiv (1866), 447. Also still standing are the two remarkable office blocks designed by F. J. Ward for the developer, Major Wieland, standing

opposite each other at the eastern part of the street: Albert Buildings (see *B*, xxx (1872), 187) and Imperial Buildings (see *B*, xxxii (1874), 461).

47 By Davis & Emmanuel, 1880–2: *B*, xxxviii (1880), 600–5; xlii (1882), 495.
48 By Sir Arthur Blomfield, 1880–6: *B*, xxxviii (1880), 538, 541–2.
49 Now Incorporated Accountants Hall. By J. L. Pearson, 1895.
50 By John Dunn, 1889–94. Now (1972) almost entirely demolished.
51 Whitehall Court is by Archer & Green: *BN*, xlvii (1884), 90–1. The National Liberal Club, forming the eastern part of a nearly symmetrical river-front, is by Alfred Waterhouse: *BN*, xlviii (1885), 165, 174–5.
52 By J. Whichcord: *B*, xxxii (1874), 309.
53 *Architect*, 5 June 1891; *BN*, lviii (1890), 654.
54 By G. F. Bodley: *BN*, xxix (1875), 22.
55 *Book of the Hotel Cecil*. Publicity brochure with illustrations by J. Pennell, S. Reid and T. R. Davison [1895].
56 By John Gibson: *B*, xxxvii (1879), 1153.
57 The Northumberland Avenue Hotel, by Isaacs & Florence: *B*, lii (1886), 640, 642; liii (1886), 664 (sculpture). For Northumberland Avenue generally, see Edwards, op. cit.
58 *Survey of London*, XXXI, pp. 71–4.
59 Ibid., XXXIII, pp. 297–300.
60 H.-R. Hitchcock, 'London Coal Exchange', *Architectural Review*, i (1947, i), 185–6.
61 *B*, x (1852), 9.
62 *B*, xxiv (1866), 956–7; xxv (1867), 261.
63 *Companion to the Almanac*, 1855, p. 229.
64 *B*, xxvii (1869), 320–1; sections 321, 326.
65 *B*, xx (1862), 732–3.
66 Theo Crosby, *The Necessary Monument* (1970).
67 *BN*, xxxix (1880), 504. The house was greatly enlarged by Webb and subsequently much altered.
68 W. R. Lethaby, *Philip Webb and his Work* (1939), pp. 87–91, and plate opp. p. 20.
69 *BN*, xxvii (1874), 342, 351.
70 See R. Blomfield, *Richard Norman Shaw, R.A.* (1940).
71 Important for the beginnings of 'Queen Anne' are papers read by J. J. Stevenson at the Architectural Conference in 1874 (*B*, xxxii (1874), 537; *BN*, xxvi (1874), 689–92), and at the Architectural Association in 1875 (*B*, xxxiii (1875), 179–80).
72 *BN*, xliv (1883), 245.
73 E. R. Robson, *School Architecture* (2nd edn, 1877).
74 H. Percy Adams, 'English Hospital Planning', *Journal R.I.B.A.*, xxxvi (1929).
75 By Wilson, Son & Aldwinckle: *BN*, xlix (1885), 98.
76 By A. Hessell Tiltman: *BN*, lxxii (1897), 434.
77 By Spalding & Ould: *BN*, lii (1887), 78.
78 T. Greenwood, *Free Public Libraries* (1887).
79 *BN*, lvi (1889), 806–7.
80 H. S. Goodhart-Rendel, *English Architecture since the Regency* (1953), pp. 197–8.
81 Blomfield, pp. 39–41. See also *BN*, xli (1881), 526 ff.

The Victorian City: Shapes on the Ground

82 See above, p. 319.
83 S. Perks, *Residential Flats of all Classes, including Artisans' Dwellings* (1905).
84 *B*, xxxvi (1878), 753.
85 Clifford Chambers, by T. H. Watson & F. H. Collins: *BN*, xxxiii (1877), 240.
86 *BN*, xxxii (1877), 90.
87 *B*, xl (1881), 126.
88 Nos 42, 44, 46 by Ernest George: *BN*, xlv (1883), 50; Nos 7, 9 by A. Payne & E. M. Elgood: lvii (1889), 152; No. 40 by Collcutt: lxi (1891), 10, 718.
89 Nos 126–9 by W. H. Powell: *BN*, liii (1886), 420. See also lvii (1889), 586; lxiii (1892), 805, 824; lxxii (1897), 91, 92.
90 *BN*, lx (1891), 229–30; lxx (1896), 563, 565.
91 *Companion to the Almanac*, 1853, p. 253; *Illustrated London News*, xxv (1854), 217.
92 See above, p. 318.
93 *B*, xviii (1860), 755; xix (1861), 375.
94 *Illustrated London News*, xliv (1864), 563–4.
95 *B*, xxiv (1866), 758.
96 *B*, xxi (1863), 533.
97 See Jack Simmons, *St Pancras Station* (1968).
98 By Isaacs & Florence: *B*, xxxii (1874), 212, 591.
99 Northumberland Avenue Hotel, by Isaacs & Florence: *BN*, xliv (1883), 254. The Grand and Metropole Hotels (*B*, xxxvii (1879), 344–5; *BN*, xliv (1883), 830) are both by F. & H. Francis.
100 *BN*, lxxvii (1899), 8, 9.
101 See note 54.
102 S. Hamp, 'Hotel Planning', *Journal R.I.B.A.*, xiv (1907), 405–7.
103 N. Taylor, 'Doll's Palace', *Architectural Review*, cxl (1966), 451–4.
104 *BN*, lx (1891), 80, 196.
105 *B*, xxviii (1870), 407.
106 *BN*, xxiv (1873), 270–1, 330.
107 All the architects named in this and the following paragraph are the subjects of biographical notices in the *Dictionary of National Biography*.
108 Summerson, *Victorian Architecture*, pp. 113–17.
109 See *BN*, xlvii (1884), 153; winning design, 168–9.
110 For the early development of the site, see *B*, x (1852), 677; xi (1853), 681; xiv (1856), 263; xv (1857), 45, 357; xvi (1858), 137–8; xvii (1859), 456–7; xviii (1860), 312, 836–7; xix (1861), 173, 213.
111 *B*, xix (1861), 497.
112 *B*, xxv (1867), 365, 368; xxviii (1870), 977, 986, 1046; xxix (1871), 249; *Architect* (1869), opp. p. 294.
113 *B*, xxviii (1870), 467.
114 *Fifty Years of Public Work of Sir Henry Cole* (1884), II, pp. 296, 305.
115 *B*, xxxi (1873), 10–11.
116 *BN*, lv (1887), 10, 12; unsuccessful designs, 92–110.

14 House upon House*
Estate development in London and Sheffield

Donald J. Olsen

'They were not so much towns as . . . the barracks of an industry,' wrote the Hammonds of the new industrial cities of the Midlands and the North. 'These towns reflected the violent enterprise of an hour, the single passion that had thrown street on street in a frantic monotony of disorder . . . these shapeless improvisations . . . represented nothing but the avarice of the jerry-builder catering for the avarice of the capitalist.'[1] 'Truly the state of London houses and London house-building, at this time,' exclaimed Carlyle in 1867; 'who shall express how detestable it is, how frightful!'[2]

However earnest our attempts to avoid imposing our own moral and aesthetic standards on the past, and to look at Victorian cities through Victorian eyes, it is hard not to think that they could have made a better job of it than they did. The pretentious gentility of the Victorian middle-class suburb is a denial of the urban values implicit in Bedford Square and Regent's Park; while the cramped, dark, unventilated, undrained hovels in which millions of the poor, deserving and undeserving alike, spent their lives were a standing reproach to the society that permitted their existence. Although we have reached the stage of being able genuinely to admire particular pieces of Victorian architecture, it is less easy to admire the Victorian townscape as an aesthetic whole—even the Victorians could rarely bring themselves to do that—still less to admire the quality of life imposed by the urban environment on the Victorians themselves.

* I wish to thank the John Simon Guggenheim Memorial Foundation for awarding me a fellowship for 1967–8 which enabled me to do the greater part of the research for this chapter.

Copyright 1973 Donald J. Olsen

The Victorian City: Shapes on the Ground

Was the Victorian city perhaps a monstrous blunder, as remote from the nineteenth-century urban ideal as it is from ours? Whether or not the Victorians got the cities they deserved, did they get the cities they desired? The answer is to be sought in the mediocre, representative acres of middle-class villas and working-class back-to-back cottages as much as in their self-conscious aspirations towards urban art or magnificence: in the anonymous suburbs of North London and the grimy terraces of the West Riding as much as in Prince Albert's South Kensington or Cadbury's Bournville.

It is arguable that the Victorian building industry, whatever the aesthetic and sanitary defects of its products, on the whole responded well to the challenge of a population explosion, rapid urbanization, a revolution in transport, technological upheaval, and changing notions as to what the home and the city ought to be. Before attempting to account for some of the flaws in the Victorian city and its suburbs, it might be appropriate to suggest how it was that they were built as well as they were.

The nineteenth-century suburb clearly responded to some fundamental cultural need, for, however much aesthetes and planners may deplore it, it provided the model for the environment in which most English-speaking people choose to live even today. It is harder to praise Victorian housing for the working classes. Even the model dwellings seem almost as repellent as the speculative housing whose evils they were designed to alleviate. With respect to sanitation, the private producers of housing did little beyond cooperate reluctantly with the efforts of the public authorities. Nor can working-class housing be much admired for structural soundness, architectural excellence, or imaginative layout. Where the Victorian building industry did succeed was on the quantitative plane. With notable exceptions in places like central London, it managed for the most part to keep up with the massive growth in population and with the expectations of that population for rising standards of accommodation. It provided more middle-class housing than the middle classes could absorb, and a supply of working-class housing that, if of minimal standards of amenity, was in most cases adequate for the needs of the people, at a price that the great majority of the working classes could afford. The very poor and those forced to live near their employment in areas of abnormally high land values formed a special, if deplorable, case.[3]

Building-land on the outskirts of Victorian cities was plentiful, cheap, and not subject to significant fluctuations in price. In no town does there appear to have been a shortage of building-land, freehold or leasehold, once one left the central area: landowners consciously kept both freehold prices and ground rents low in order to attract purchasers and builders to their estates. The practice of charging peppercorn rents for the first years of a lease, or accepting payment in instalments meant that the speculator himself often paid little or nothing for the land.

Nor was there any shortage of speculative builders. Entry into the business required no specialized skills and little or no capital; the smallest of firms was able to coexist perfectly satisfactorily with the largest.[4]

To an abundance of landowners eager to participate in the unearned increment that urban growth offered them, and an abundance of builders ready to risk their all in covering their land with houses, there was added a complementary abundance of investors, virtually forcing their money on the builders and developers. The buoyant supply of capital for house-building matched the supply of land and labor. G. Calvert Holland attributed what he thought the over-production of houses in Sheffield to 'the petty capitalist . . . desirous of realizing a handsome per centage . . .' as much as to the landowner, 'naturally anxious to appropriate his land to building purposes,' and the 'pennyless speculative builder.'[5] John Nash had earlier charged 'attornies with monied clients [with] facilitating, and indeed putting in motion, the whole system . . .' by which London was artificially extended, 'by disposing of their clients money in premature mortgages, the sale of improved ground rents, and by numerous other devices.'[6] With opportunities for investment far more limited than today, the thrifty Victorian who wished a greater return than he could earn from Consols without undue risk was almost forced to invest in the housing industry.[7] 'The supply of capital for house-building certainly ebbed and flowed,' H. J. Dyos has concluded, 'but there is no clear evidence that it was ever checked in such a way as to impede development at all seriously; there is, on the contrary, rather more evidence of over-building in periods of easy money than of under-building when money was tight.'[8]

An examination of how two urban estates, the Eton College estate at Chalcots, in Hampstead, and the Norfolk estate in Sheffield, were developed for building-purposes suggests not only the variety of ways in which the Victorian city took shape, and that its history was not entirely one of lost opportunities and the complacent creation of the unsanitary and the ugly, but also what prevented the urban land-owner from doing very much to alter, for better or worse, the form that the new city took.

These estates exhibited vast differences but had two qualities in common: mediocrity and success. As to the first, there can be little argument. Neither Eton College nor the Duke of Norfolk ever did anything to startle the expectations or outrage the sensibilities of the builders or tenants on their estates. It is the unremarkable quality of the management and the indifferent nature of its results that give the estates their particular interest, just as second-rate books often have more to tell the intellectual historian than do masterpieces.

The success of the estates was clearly not an artistic one. No one would challenge Sir John Summerson's characterization of the architecture on Chalcots as 'catch-penny,' the result of a 'commonplace . . . chain of events . . . a process which comes as near as possible to complete anonymity in its results, for it can truthfully be said that not one solitary soul was ever really interested in what the physical visible results would be.'[9] Hugh C. Prince has described the houses of Samuel Cuming, the principal builder on the estate in the forties and fifties, as 'nondescript . . . inoffensive . . . only remotely classical yet no more than vaguely romantic. They were built

The Victorian City: Shapes on the Ground

to please respectable but undiscerning clients.'[10] (Plates 288 and 290 are elevations of Cuming's houses; Plates 289 and 291 are of residential and commercial property put up by other builders in the fifties.) Kate Simon has recently called the productions of William Willett, whose late Victorian and Edwardian houses were among the last to be built on the estate, 'dour, red-brick, lightless, airless, insane excrescences.'[11]

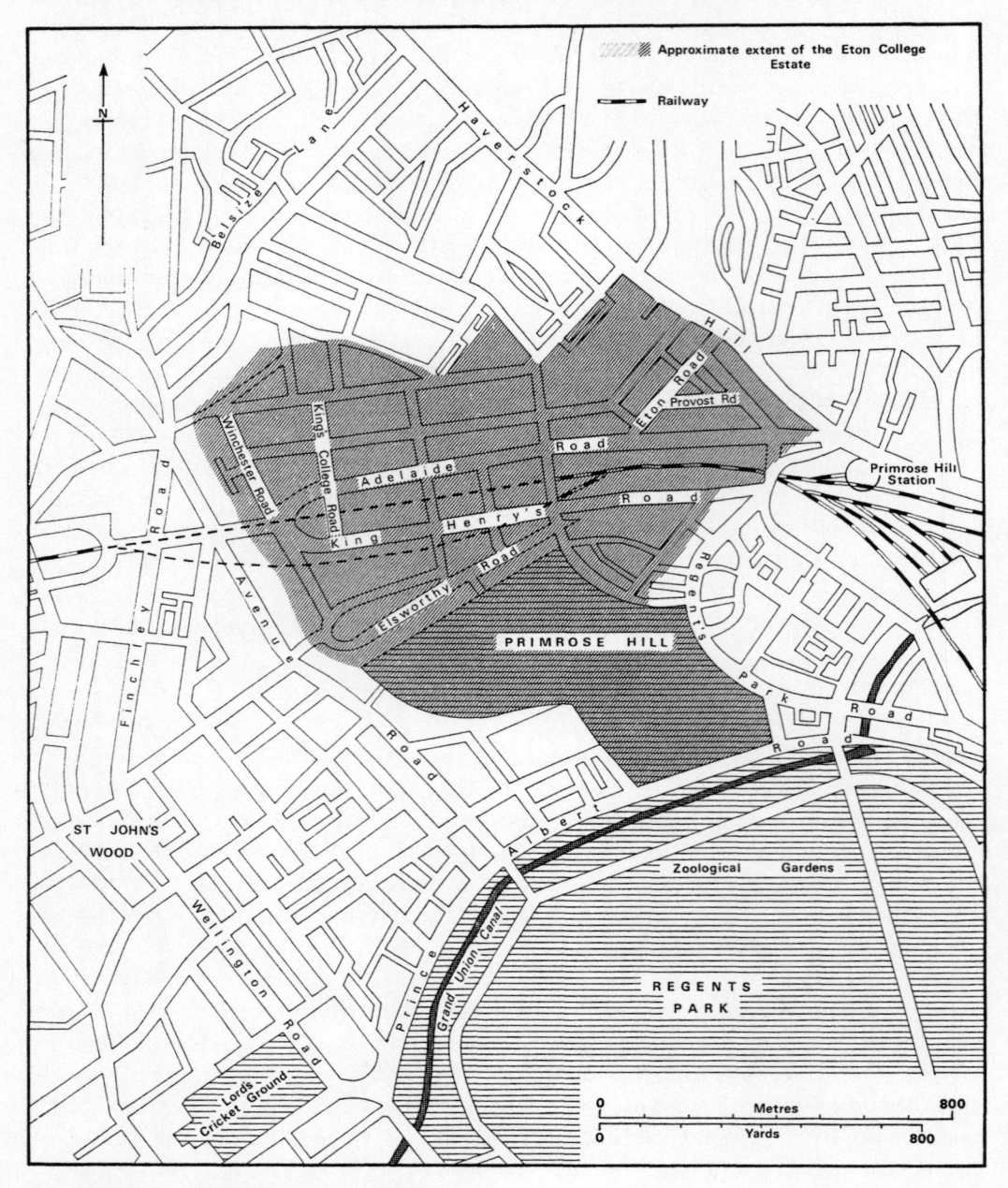

XI Position of the Eton College Estate

Nor can Victorian Sheffield, either those parts owned by the Duke of Norfolk or elsewhere, be easily described as other than a blot on an otherwise magnificent landscape. 'We have surveyed Birmingham, Stafford, Wolverhampton, Newcastle-upon-Tyne, Hull, Shrewsbury, and other towns; but Sheffield, in all matters relating to sanitary appliances, is behind them all,' reported the *Builder* in 1861.[12] In 1848 James Haywood and William Lee described the leasehold houses on the Norfolk estate in Sheffield Park as 'very low, consisting of one or two stories, and ... crowded together in the most irregular manner, forming angular yards of the most inconvenient description.'[13] The *Sheffield and Rotherham Independent* over twenty years later found 'the Park estate of the Duke of Norfolk almost unimproved ... The buildings consist mostly of narrow rows of houses, along narrow streets north and south, at different elevations, forming a series of terraces. In many of these there have been no changes during the last fifty years, except that the houses have become dilapidated, and arrangements for light, air, and cleanliness seem never to have been thought of.'[14]

And yet Chalcots and Sheffield reflect as much of what was admirable about the Victorian city as of what was deplorable. In a period characterized by a chronic oversupply of middle-class housing, Chalcots attracted and continued to attract the kind of resident for whom it was designed. While the Bedford and Foundling Hospital estates were fighting their losing battles to keep their genteel tenants from deserting their elegant Georgian squares, the 'respectable but undiscerning clients' who constituted the bulk of the middle-class house-market moved into the anonymous Italianate villas of Chalcots and, as a class, remained there. Sir John Summerson has remarked on the unusual degree of social continuity in the neighborhood. After describing the inhabitants of 1851, according to the census records of that year, as 'a mixed lot—small manufacturers, solicitors, a Congregational minister, an architect, a painter or two, widows bringing up families on small incomes,' he pointed out that 'its character ... remained much the same and so far as I can ascertain has always done so. Allowing for the changes which have taken place in the structure of society as a whole one can say that these houses are now inhabited by exactly the same type of people who lived in them a hundred years ago.'[15] There was a perceptible social decline by the early twentieth century. One resident, writing in 1918, observed that 'the character of the population has changed. This is no longer, to anything like the former extent, a neighbourhood wherein rising, or successful, professional and business men set up households and bring up families.'[16] Yet an examination of the *Post Office London Directory* shows the decline to have been far from catastrophic, certainly far less than occurred in Bloomsbury. What generations of architects, surveyors, solicitors, and stewards strove vainly to achieve in Bloomsbury, Eton College seems to have managed without really trying.

For the records of its estate show little evidence of serious planning, even of the landlord holding reluctant builders to higher standards of design, construction, or layout than they would themselves have wished. The history of the estate is one of mediocre management and mediocre builders proceeding according to an accepted pattern, a standard way of going about such things. If Bloomsbury is a magnificent failure, Chalcots is an undeserved success.

The Duke of Norfolk did even less to control or direct the building development on his estate. Michael Ellison, the agent from 1819 to 1860, intelligent, conscientious, energetic, and wholly devoted to what he saw as the best interests both of the Duke of Norfolk and the town of Sheffield, left most of the superintendence of the leasehold development of the estate to the steward, Marcus Smith, a man of ordinary intelligence and limited attainments. London building-leases varied considerably in stringency, but none with which I am familiar left so much to the discretion of the tenant as did the standard Norfolk leases for nineteenth-century Sheffield. Yet in spite of this, both the quantity and quality of working-class housing in Sheffield compared favorably with that of most other English towns.

Given the level of wages and incomes for both the very poor and the industrious artisan in Victorian England, their chief requirement as to housing was that it be cheap and that it be densely enough built so that everyone could live within walking distance of his work. The landowners, investors, and speculative builders of London were unable to satisfy such requirements. The size and wealth of the metropolis made land values in the center—where most jobs were to be had—such that the construction and maintenance of working-class dwellings became increasingly uneconomical. The operations of the London Building Acts and the further requirements imposed by the landlords of leasehold estates raised the cost of building to a level higher than that of the provinces, while a variety of considerations encouraged both landowners and builders to concentrate disproportionately on middle-class housing. Speculative working-class housing on the cheaper land available in the suburbs did not, until perhaps the very end of the century, provide an adequate substitute, because of the insufficient service of workmen's trains, the casual nature of much employment, and a level of incomes too low to permit the very poor to pay the economic costs of decent housing anywhere.[17]

Sheffield, although it reached a population of 411,000 in 1901, remained physically compact. 'The population of Sheffield is, for so large a town, unique in its character,' Dr Frederick W. Barry reported in 1889, 'in fact it more closely resembles that of a village than of a town, for over wide areas each person appears to be acquainted with every other, and to be interested with that other's concerns.'[18] By the time it expanded beyond walking limits, cheap, municipally-owned electric trams brought the outer suburbs within the reach of the working classes.[19]

Despite the crowding of houses on the ground and the primitive state of sanitation for much of the century, most contemporaries were agreed that the Sheffield worker was comparatively well housed. John Parker, M.P., a member of the Select Committee on Public Walks, told his colleagues in 1833: 'Generally, in Sheffield, the average of the comfort of the lower classes is above that of most other places; we have not yet got into the abominable way of cellars, or of many families living in the same house.'[20] There was comparatively little overcrowding throughout the century: the average dwelling was the single-family three- to four-room cottage, and the average number of persons per house less than five.[21] Haywood and Lee, in their generally critical account of sanitary conditions in Sheffield, found housing as such perfectly adequate:[22]

> Notwithstanding all the evils we have seen during our inspection of the town, we do not hesitate to say, generally, that the *construction* of the houses occupied by the working classes in Sheffield is better, and the rental more moderate, than in almost any other town in the kingdom. The great amount of sickness, and the high rate of mortality here arises, so far as the dwellings are concerned, more from an inadequate supply of water and inefficient drainage, than from defective structural arrangements.

There were many attacks on the speculative builders of Victorian Sheffield, but no one suggested that they produced too few houses for the needs of the population. Rather, the complaint was that they built too many houses, of too flimsy a character.[23] 'Sheffield,' a local solicitor told the Select Committee on Town Holdings, 'is undoubtedly an over-built town.'[24]

'In building as in other things,' Marian Bowley reminds us, 'quality as well as quantity costs more. Anyone providing houses for the great mass of working-class families in the nineteenth century had to provide them at a price that they were willing or able to pay . . . The lower the income group to be served, the cheaper the buildings in terms of quality and quantity of materials, workmanship, fittings and amenities . . . Building which is frequently referred to as shoddy and nasty is not necessarily dishonest. It may be a thoroughly honest job for the price.'[25]

The ability of Chalcots to meet so precisely the wishes of its middle-class tenants, and of Sheffield to house, at whatever low standards of comfort, its vast artisan population may have come about not so much in spite of the passive policies of the ground landlords, as because of them. The Georgian tradition of town planning, the only one available to the Victorian landowner, was irrelevant to most of the needs of the nineteenth-century city.

What values underlay a 'good' eighteenth- or early nineteenth-century town plan—in Bath or Dublin or the New Town of Edinburgh? Coherence and uniformity: uniformity of facade, of design, of the social status of the occupants. This required segregation, with the garden squares and principal streets reserved for the better sort of resident, the back streets for the middling sort, and the courts and mews for the lower orders, decently screened from view. Georgian town planners were not able to achieve the single-class neighborhood that so distresses sociologists but so delights estate agents and building societies, but it was not from any belief in the virtues of social integration. There was less mixture of classes in eighteenth-century Bloomsbury than in seventeenth-century Covent Garden, more social uniformity in Belgravia in the 1830s than in Mayfair in the 1770s. Chalcots, though it discarded the outward appearance of Georgian urbanity, did achieve one of its principal aims in becoming an exclusively middle-class dormitory.

Along with social segregation went segregation of occupation. The ideal street to the Georgian planner was one of gentlemen's private residences, whose quiet and uniformity were undisturbed by trade, manufacture, or through-traffic. Shops, offices, and, above all, manufacture, were to be either hidden from sight or excluded altogether. Noise, smoke, and noxious odors were rigorously proscribed in all respectable leases. Such a policy swept the dust under the carpet, and helped members of the middle and upper classes to ignore the existence of poverty, filth, and even the necessary operations of the economy. Here again, Chalcots restricted shops to specified streets, and managed to hide even the London & Birmingham Railway in a tunnel. But the very nature of Sheffield's economy was such as to frustrate the most well-meaning ground landlord in this respect.

Chalcots succeeded without taking too much thought for success—in strict eighteenth-century terms. Its history suggests that, given sufficient size and a favorable location, a London estate planned and managed itself. The history of the Norfolk properties in Sheffield suggests that the techniques of town planning available to an eighteenth- or nineteenth-century landowner were of little use in the West Riding of Yorkshire.

'Irrespective of the sort of house adapted for the estate, and the area allowed to each,' advised the *Building News* in 1860, 'much depends on the "laying out": a greater income and more general success will occur from one form of planning roads, etc... than another.'[26] Even in Georgian London the extent of control that landlords exercised over builders had varied greatly. '... Some of the Proprietors,' wrote the architects to the Crown estate in 1811, 'have confined the Builders ... to a certain Plan as to the general distribution of the intended Street, Squares, etc. without reference to the particular Class of Houses; others have gone further, stipulating the Class of Houses, and the numbers of them; others again have suffered the Builders ... to distribute the Streets, and cover the Ground as they thought proper.'[27]

The instrument for putting such schemes into effect was the building agreement. The enforcement of restrictive covenants depended not only on the skill with which they were drawn up at the outset, but also on the perseverance of the ground landlord. He had every temptation not to employ his legal powers. J. Wornham Penfold remarked in 1858 that it was 'the builder's object and interest to build as cheaply as he possibly can,' while the landlord was reluctant to enforce the covenants, being 'oftentimes more anxious about what seems to be his present interest than about what is best for the future; and if too stringent conditions are placed on either design or construction, the estate will get a bad name among the speculators.'[28] The Select Committee on Town Holdings agreed: 'If the ground landlord makes it his first object to get the land covered with houses, and then secure to himself the commercial ground-rent ... it is improbable that he will make the stipulations of the building agreement press too heavily upon the speculative builder.'[29]

In the latter years of the eighteenth century the 9th and 11th Dukes of Norfolk, and their agent Vincent Eyre, had attempted to astonish the North by adorning Sheffield with a succession of terraces whose architecture would not have been out of place in Bath or Dublin. They commissioned the architects James Paine and Thomas Atkinson to design the façades for new streets to be built on the former Alsop Fields, and the resulting elevations would have graced the capital of an enlightened despot of a German principality. The gridiron network of Norfolk, Arundel, Surrey, and Eyre Streets still stands out in contrast to the narrow, winding, and irregular pattern of both the older and newer parts of the town, and a fragment of one of Atkinson's designs can still be seen in Norfolk Street. But it is unlikely that any substantial portion of the planned buildings was ever erected: if it had been, some Man of Taste on a Picturesque Tour would have noticed it. For whatever reason, the 11th Duke sold the whole neighborhood in the early nineteenth century.

The disposal of the Alsop Fields development marked the abandonment by the estate of the Georgian approach to town planning. Regular street patterns, unified facades enforced by architectural controls written into the leases, and coherent neighborhoods in which street names, layout, and physical appearance expressed the wealth, taste, and public spirit of the landowner, however appropriate for a capital city or a fashionable watering place, evoked little enthusiasm from the independent cutlers of Sheffield.

Sheffield was a wealthy town, but not one whose wealth was of the sort to encourage good architecture or good urban design. The quality of life of a Sheffield artisan in his ugly back-to-back cottage may on balance have been preferable to that of the occupant of an exquisitely designed Georgian slum in Dublin. Certainly the principles and techniques of urban planning available to the Duke of Norfolk, however suitable to conditions in London, had nothing to contribute towards solving the problems of Sheffield. This is partly a question of social class, but more one of the realities of economic geography.

Sheffield never was, and probably never could have become, a place where people of wealth, leisure, and taste would willingly congregate. It was not a York, a Harrogate, or a Scarborough. Nor did its industries—at least until the great expansion of the steel industry in mid-century—produce the great concentrations of wealth that were to be found in Leeds or Manchester. Finally it was, to quote Horace Walpole, 'one of the foulest towns in England in the most charming situation.'[30]

Noisy, smoky, and loathsome, Sheffield was surrounded on all sides by some of the most enchanting countryside to be found on this planet. The flight to the suburbs and beyond was an early and understandable phenomenon for all who could afford the move. 'Within the past few years,' wrote G. Calvert Holland in 1841,[31]

> the town has extended widely in all directions . . . The same change presents itself in the picturesque sites of the immediate vicinity. There is no richly clothed hill or attractive valley but what is embellished by the tasteful decorations of art . . . All classes, save the artisan and the needy shopkeeper, are attracted by country comfort and retirement. The attorney,—the manufacturer,—the grocer,—the draper,—the shoemaker

and the tailor, fix their commanding residences on some beautiful site, and adorn them with the cultivated taste of the artist.

Segregation of function was a basic principle of Georgian planning. But the cutlery industry, on which Sheffield's prosperity was based, remained until the present century a matter of small workshops around which the cutlers and their employees lived. Later, the smoke and noise of the steel industry became at once Sheffield's greatest nuisance and the foundation of its greatness. 'A thick pulverous haze is spread over the city, which the sun even in the dog days is unable to penetrate, save by a lurid glare . . . and a buzz, softened down from the first clanging and clashing utterance of machinery, into a hum as of a swarm of bees, rises into the air and is distinctly audible,' reported the *Builder* in 1861, as much in admiration as in condemnation.[32] The Norfolk estate did from time to time send orders to tenants to abate the smoke nuisance, but, given the state of Victorian technology, to have abolished industrial nuisances would have been to abolish Sheffield.

The principal Norfolk holdings were to the east and north of the town. They were gradually covered with steel-mills along the valley of the Don, and with densely-built terraces of working-class cottages up the surrounding hills. Massive postwar slum clearances have obliterated most of what went up on the estate in the nineteenth century, but little that remains is calculated to excite admiration for the Duke's agents as town planners. Yet these terraces were no worse than what was going up elsewhere in Sheffield, and superior to much working-class property in other towns.

The building agreement, which in London was a lengthy and detailed document, precisely prescribing the nature and quality of the houses to be erected, setting forth the provisions of the leases to be granted when the houses were near completion, was in Sheffield no more than a verbal arrangement by which the future lessee agreed to put up buildings, and the landlord agreed to grant a lease at so much per square yard.[33] The leases required that the premises be kept in repair, but lacked the machinery for enforcement contained in the ordinary London lease. Instead of the provision, customary in London, that the premises not be used, without specific license from the freeholder, other than as a gentleman's private residence, there was merely a prohibition of any trade which the agent of the Duke of Norfolk might deem offensive. Such general covenants were notoriously hard to enforce.[34]

While the general tendency in London was for landlords to add more and more restrictive covenants to their leases as time went on, Sheffield leases granted in the nineteenth century were in some respects less stringent than those dating from the 1780s and 1790s. The printed form of lease in the time of Vincent Eyre and the 11th Duke included a specific list of offensive trades—soap-boiler, distiller, melting-tallow-chandler, and so on—unlike the later general covenant against nuisances. More important, it required that the lessee submit a plan and elevation to be approved by the Norfolk agent for any buildings 'at any Time hereafter to be erected and which shall front upon or adjoin to any public Street or Road in or near the said

Town of Sheffield.'[35] By 1800 such provisions had disappeared from most leases. Nothing, moreover, ever prevented the lessees from filling in the courts behind the houses with the meanest of erections, which was precisely what they proceeded to do.

Any attempts to introduce stricter controls brought immediate opposition. James Sorby wrote to Michael Ellison in 1835 to protest against a covenant that would have been taken for granted in any London lease:[36]

> A Draft of a Covenant which you propose to introduce into the Lease from the Duke of Norfolk to Mr. Alfred Sorby, has been handed to me for perusal... My Brother ... certainly will not agree to so unexpected a clause, being inserted in the Lease, neither do I think that you can, in fairness expect him to agree to it, as it is a very novel idea, to restrict a Tenant from laying out Money upon the property of his Lessor, & occupying it to best advantage ... it is ... much out of the common way. My Brother has not any intention ... of building either Workshops or Dwellinghouses upon the Land in question, but he naturally wishes to have it in the most marketable state, in case he should, (after expending so much Money upon the place,) be driven away by the nuisances, which surround him.
>
> I hope you will reconsider the matter, & grant a Lease, on the usual Terms.

The lease in question was for Park Grange, the mansion that Sorby was to build for himself south of the Farm, later the ducal residence. The houses that were erected on this part of the property in the thirties and forties were by far the most pretentious ever built on the estate. Yet even here, a 'gentleman's private residence' covenant such as would have been found in the leases in any genteel London street seemed an outrageous innovation.

William Fowler, of Birmingham, surveyed the operations of the Sheffield office in 1861. His penciled marginal comments on a 'Draft Proposed Form of Lease for 99 Years' indicates that an outside observer did not regard the degree of control permitted by the covenants as at all insufficient: if anything, he thought the leases erred on the side of severity. He wrote 'very stringent covenants,' beside the provisions against nuisances and assignments without license, and one stating that the lessee

> shall [not] use or exercise ... upon the said premises any trade or manufacture which shall be declared by such Agent [of the ground-landlord] to be offensive or a nuisance to the neighbourhood without ... consent in writing ... Or [a provision for forefeiture] if the buildings erected ... shall be suffered to be dilapidated or out of repair and the same shall not be repaired within one year after notice.

He commented on the provision permitting the agent to make a schedule of fixtures on the premises any time during the last five years of the lease, 'A power probably never exercised.'[37]

Building-leases remained in general loosely drawn on all estates in Sheffield throughout the century.[38] Such covenants as there were were rarely enforced. It would be fair to say that so long as the ground rent was regularly paid and no flagrant nuisance committed—and in Victorian Sheffield a nuisance would have to be flagrant indeed to be noticed—the landlord would not interfere with the leaseholder's interest. 'I have never heard of the Duke [of Norfolk] interfering at all with the property until it got to the end of the term,' the mayor of Sheffield told the Select Committee on Town Holdings.[39] His impression finds confirmation in penciled memoranda in the margin of the questionnaire sent by the Committee to the Norfolk estate office. Alongside question five, 'Is it your experience that ground landlords habitually abstain from enforcing the covenants to paint and otherwise keep in repair universally inserted in building leases?' is written, 'No [sic]—Ground Landlords only require Buildings to be maintained at a certain value & to be insured & do not notice painting or ordinary repairs.' In answer to question eight, 'Is it your experience that the system of building leases is conducive to bad building, to deterioration of property towards the close of the lease, and to a want of interest on the part of the occupier in the house he inhabits?' is the cautious opinion, 'Probably no worse than on freeholds.'[40]

The result was what one would have expected: building of the cheapest possible character densely crowded upon the ground; streets often left unpaved and undrained—in Sheffield (London varied somewhat), builders were not obliged to form roads and sewers at the same time as erecting the houses; property in a ruinous state at the expiration of the original lease. The preservation of the reversionary value of the leasehold property, on a London estate the continual preoccupation of the ground-landlord, rarely seemed a consideration in Sheffield.

It would be a mistake to blame Victorian Sheffield on apathy, or mismanagement in the Norfolk office. The surviving correspondence relating to Sheffield in the Arundel Castle MSS. is incomplete, but those years for which there is relatively full documentation—chiefly from the twenties to the fifties—bear witness to the consistent and well-informed interest taken in Sheffield affairs by both the Duke and his London agent. The Sheffield agents—in particular Michael Ellison and his son M. J. Ellison—displayed an obvious sense of responsibility both for the interests of the Norfolk estate and for the prosperity of Sheffield as a whole. Finally, the Duke of Norfolk as a London landowner—he possessed a considerable estate between the Strand and the Thames—seems to have behaved like any other enlightened metropolitan ground landlord at the same time that he was being so permissive in Sheffield.[41]

One explanation is found in a letter of 1840, which the Sheffield agent wrote to Edward Blount, then responsible for the management of all the Duke's properties, about the problem of determining policy with respect to the renewal of leases in Sheffield, particularly as to fines. 'The rule observed in such cases in London,' he argued, 'will not do for this place: the object there being for the Lessor, to exact the utmost amount, without reference to any other consideration. Here, we must take care not to destroy confidence, but on the contrary deal liberally with parties, and hold out encouragement to others, to embark capital in buildings on the Duke of Norfolk's Land.'[42] Whether or not one accept's Ellison's view of the principles governing estate management in London, it is significant to see the stress he placed on doing everything possible to encourage speculative building.[43] Actually anyone involved in the management of a London estate would have been equally convinced that his chief aim should be to make conditions attractive for the speculative builder. The difference lay in the builders and speculators in the two places: the builder in Sheffield was simply not in an economic position to bear the weight of the quality of building demanded in London.[44]

Land values, and hence ground rents, were not consistently lower on the outskirts of Sheffield than they were on the outskirts of London. What made Sheffield attractive to the builder of minimal or non-existent capital was that he could build as cheaply as the market permitted. Sheffield, like most provincial towns, had nothing comparable to the London Building Acts until 1867. Building agreements on well-managed metropolitan estates, by adding to the minimum standards of the Building Acts, necessarily increased costs. In Sheffield, the agreements did not even specify a minimum expenditure by the builder. The absence of any obligation to construct the streets brought costs even lower.

As a result, house-building proved irresistibly attractive to men of small capital —with whom Sheffield abounded—who built cheaply and in large quantities. Any attempt by the landowners to impose metropolitan notions of estate management would have reduced the quantity of working-class housing, not only by raising building costs, but also because a fundamental principle of Georgian town planning was to concentrate on building for the upper and middle classes. But nothing the Dukes of Norfolk could have done would have made central Sheffield attractive to the already suburbanized middle classes of the town. With the relative failure of the Alsop Fields development as a cautionary example, they did nothing because there was nothing to be done.

There was much more that could be done in London, at least in certain parts. The Select Committee on Town Holdings pointed out that it was 'easier to enforce salutary regulations . . . when there are special advantages attaching to a particular neighbourhood, and marking it off for the residence of a certain class.'[45] Chalcots had such advantages.

Its situation enabled Chalcots to absorb some of the prestige of Regent's Park to the south, St John's Wood to the west, and Hampstead to the north. The estate failed to develop a distinctive character of its own, and lacked even a name, for 'Chalcots' had meaning only for the antiquary.[46] The neighborhood was variously identified in whole or in part as Haverstock Hill, Primrose Hill, St John's Wood, and

South Hampstead, and its anonymity reflected the derivative nature of both its layout and its architecture.

In order to derive full advantage from its promising site, Eton College did no more than any other large, responsible London landlord did, and far less than some. It hired a succession of surveyors, who proposed, modified, and abandoned a succession of spacious but unoriginal street plans; it had its solicitors draw up versions of standard London building agreements and leases. It negotiated with a succession of developers and builders, who proceeded to erect detached and semi-detached villas in whatever style was currently in fashion. (See Plates 288, 290-1 for the fashions of the forties and fifties.) For amenities it encouraged the building of churches and an adequate number of respectably managed public houses. The houses were decently designed and substantially built, in accordance with the building agreements: not so much in consequence of their provisions, as because the builders and developers were accustomed to put up houses of that sort. One would imagine that the more respectable and better-established builders would be attracted to a sizeable, conservatively managed estate such as Chalcots. The dilatory behavior of the Provost and Fellows, and particularly the bursar and registrar, with respect to the management of their London property—leading especially to delays in the execution of leases—did nothing to encourage builders and, it might be argued, made it impossible for any but a builder of some standing and considerable capital resources to indulge in the luxury of an Eton College speculation.

Unlike many ground landlords, the college offered no financial assistance to builders in the form of loans, purchases of improved ground rents, or contributions towards the expense of roadways and sewers. The builders did not seem to find the controls and covenants in the agreements irksome, although they inevitably increased their costs. There were disagreements as to the rate of rent per acre, the amount of time during which the full rent would be replaced by a peppercorn and a 'grass' rent, and the time allowed for the houses to be built; but there is no evidence that the builders were either trying to scamp on the dimensions or quality of materials, or that they were trying to crowd more and smaller houses onto their plots than the agreements allowed—both recurring problems on the Bedford and Foundling estates.[47] From the point of view of the developers and builders, the situation called for precisely the large—but not too large—substantial houses on a spacious plan that the agreements called for. None of this is surprising, since there is every indication that most of the initiative for the layout and quality of construction came from the building developers themselves. The surviving estate plans that did originate with the landlords and their agents either proved impractical or were modified out of recognition in the negotiations with the developers.

The leasehold covenants served not so much as obstacles to bad building as encouragements to good, and an implied guarantee that the rest of the estate would be built to similar standards. What mattered was less the conscious policies of the ground landlord than the continuing existence of a market for the kind of houses being erected on Chalcots, the willingness of substantial builders to speculate on the estate, and the seemingly inexhaustible pool of capital available to support middle-class housing developments in London.

The earliest plan for Chalcots would have made it a northward extension of Regent's Park; the actual development began as an eastward extension of St John's Wood, and ended as a southern continuation of Hampstead. Perhaps its success derived from its ability to reflect the aesthetic and social character of all three.

The Crown architects had in 1811 suggested that what was to become the northern portion of Regent's Park, land adjacent to the Eton College estate at Primrose Hill, could be 'most advantageously disposed of for Villas, having each an allotment of from two to five, ten, or a greater number of Acres.'[48] John Nash similarly proposed that the whole of the Park 'be let in parcels of from four to twenty acres, for the purpose of building villas, and so planted that no villa should see any other, but each should appear to possess the whole of the park.'[49] The handful of villas actually built in the park is only a fragment of the original conception.[50]

The first plan for Chalcots, dated by Noel Blakiston 1822, proposed a similar layout. It argued that 'the Estate from its proximity to London and from its elevated situation is so well adapted for the erection of Villas with a small Quantity of Land attached thereto that it would be eagerly sought after by the wealthy Citizens of the Metropolis—and builders would readily speculate in taking the plots.' It proposed that the whole estate of about 230 acres, except for fourteen acres on Primrose Hill itself, be divided into seventy-seven building lots, all larger than half-an-acre.[51] Two years later John Jenkins produced a building plan (Plate 281). It divided the whole of Chalcots, including Primrose Hill, into seventy-five large plots, to be served by four new roads.[52] Primrose Hill was never developed, and was in 1842 granted to the Crown in exchange for property in Eton.[53]

By 1826, when the college obtained a private Act authorizing it to grant ninety-nine-year leases on Chalcots, the great building boom of the twenties had collapsed, and it was becoming evident that Regent's Park was not going to become a garden suburb for the very rich. More modest and realistic plans for Chalcots were thus called for, and the estate brought in the architect John Shaw (1776–1832), surveyor to Christ's Hospital and to the Eyre estate in St John's Wood, to supervise operations. In March 1827 he prepared a scheme for the small part of the estate immediately adjacent to Haverstock Hill (see Plate 282). It provided for thirty-four detached villas set in plots of at least half-an-acre. None of the villas was to face directly on Haverstock Hill, along which trees were to screen its traffic; instead three new roads were to be formed. Each house was to contain at least twelve 'squares' of building, i.e., 1,200 square feet, well above the minimum nine squares for first-rate houses specified by the London Building Act. Although a notable retreat from the grandiose intentions of the earlier plan, Shaw's scheme contemplated far larger houses than were ultimately built on the site.[54]

A later map, dated 1829, shows a less ambitious street plan and a smaller number of villas, fourteen in all, on either side of the new road paralleling Haverstock Hill (Plate 283). Later additions in pencil show two alternative routes for a projected road bisecting the estate and linking Chalk Farm with the Finchley Road, together with the line of the London & Birmingham Railway and the portal of the Primrose Hill tunnel.[55]

In the same year printed 'Proposals for Building,' with a colored plan on the back, were prepared (see Plate 284). They called the attention of builders to 'this very desirable Property, which is too well known to render necessary any description of its eligibility, in all respects, for Villas and respectable Residences, combining the advantages of Town and Country.' The college proposed to offer 'in the first instance . . . that part of the Estate adjoining the Hampstead Road at Haverstock Hill . . . containing about Fifteen Acres, in lots of not less than half an Acre, for the erection of single or double detached Villas.' The college would form the roads, which were intended, '(should the Buildings go on) to be continued and connected with other Roads, particularly with the new Turnpike Road from Mary-le-bone to Finchley, now in progress.' Even without such connections, 'the Roads at present proposed will afford very desirable Frontages for Buildings, having the advantage of adjoining the main Hampstead Road, and being at the same time secluded from its publicity.' There was no mention of the dimensions of the houses or their rate of building, except that they were to be 'substantial and respectable private Houses,' whose plans, elevations, and structural specifications would require the approval of Mr Shaw.[56]

The result must have been disappointing to all concerned. As Sir John Summerson puts it, 'some rather wretched pairs of villas sprang up on one side of the estate. Nothing more happened until 1842.'[57]

The younger John Shaw (1803–70) succeeded his father as surveyor in 1832, but it was 1840 before he presented the college with his own comprehensive plan of development. His proposals are interesting both as an expression of the ideas of suburban planning held by an experienced early-Victorian architect, and as an indication of how hard it was for any landowner rationally to direct the growth of London. For no portion of Shaw's plan was put into effect, since it proved to be as unacceptable to speculative builders as the plan of 1822 had been. For it was the builder, and behind the builder the investing public and their estimate of the market for houses, who ultimately determined the shape and character of the expanding English town. The landowner in London had greater power than landowners in most provincial towns, but he could exercise it only within the limits imposed by geography, fashion, the structure of the building industry, and the state of the money market.

Shaw was able to report that building on the portion of the estate fronting the Hampstead Road was virtually complete. 'Before making any further advance,' he wrote the bursar, 'it is highly important to consider . . . what will probably be the most beneficial mode of appropriating the Estate generally to the purposes of building, so that whatever may be done shall be upon a definite plan.' To that end he enclosed a plan containing a comprehensive pattern of streets and building lots (Plate 285). He surprisingly chose to locate a major roadway directly above the line of the railway tunnel. Negotiations were already in progress for the sale of Primrose Hill, but the plan showed that part of the estate marked out in building plots like the rest. So long as arrangements were made 'for definite and sufficient access into the Estate' from Regent's Park, Shaw thought the preservation of Primrose Hill as a public open space would 'materially benefit the Estate, while by securing the Roadways I have alluded to, it will not lead to any alteration in the design of the *residue* of the Estate.'

The plan represents an intermediate stage between the extremely spacious scheme of 1822 and the relatively compact layout that the building developers ultimately gave to the estate. In laying down the lines of roads, Shaw 'endeavoured to suggest the best Communications to the most important points, and to secure double frontages to each line as far as practicable.' He also provided for deep building lots, considering 200 feet the minimum that ought to be permitted. In fact, most plots turned out to be less than 150 feet deep. There were to be no more than three major east–west roads, compared with the five actually built, and two north–south roads, where there were eventually to be four. Apart from the comparatively narrow building plots for the houses already completed along Haverstock Hill, the plan provided for a total of no more than about 215 lots, nearly three times the number contemplated in 1822, but far fewer than there were eventually to be.[58] The accompanying map shows how Chalcots was actually to be developed.

The two indispensable amenities for any respectable housing estate were a church and a public house. William Wynn, the first substantial developer on the estate, gave it its first public house, the Adelaide Tavern—at the eastern end of Adelaide Road in 1842. In requesting the assistance of the college in his application for a license, Wynn pointed out that there was up to then 'not 1 public house [on the estate] . . . and taking the improvements altogether with the new park and Adelaide Road being a principal thoroughfare to the park the house will be an improvement to the estate.'[59] Three years later Shaw authorized Samuel Cuming to build a second 'Hotel or Tavern' in Adelaide Road, 500 yards distant from the Adelaide Tavern:

> its being restricted from being inferior to the *1st* Rate Class of Building will I think [Shaw wrote to the bursar] insure its respectability and Mr. Cuming who would be most interested in maintaining this respectability (considering his large Building Engagement) thinks such an house essentially necessary for the convenience of the large neighborhood which will be established: He is also desirous of ascertaining if the College would be willing to give a site for a Church on the new Line of Road . . . he tells me there is very great want of a place of worship and . . . he feels confident he could obtain the necessary funds for its erection.[60]

Shaw had given considerable thought to the need for a church in his 1840 report:[61]

> From frequent applications made by the Residents on the Property, and in the Neighborhood, I find that there is a general demand for a Chapel on the Estate, there being no place of worship within a very considerable

> distance; Mr. Wynn the builder of many of the houses and other persons have led me to believe that it would be profitable even as a speculation to establish one, and that if the object could be entertained or promoted by the Provost and College such a subscription would be made by the neighborhood as would with their aid accomplish it.
>
> Undoubtedly the existence of such a building on the Estate would most materially lead to the formation of a neighborhood around it . . . I have . . . in the Plan shown where a Church or Chapel might be advantageously placed, either at C or D.

The church that was finally erected, St Saviour's, was located at neither of the sites that Shaw suggested, but instead in the triangular enclosure, north of Adelaide Road, formed by Provost and Eton Roads and Eton Villas. Samuel Cuming was the largest of the original subscribers, pledging £200, and the college donated the site.[62] The church, in the Early English style, was completed in 1856, except for the tower and spire, which were added later.[63] In 1870 the college conveyed a freehold site at the junction of King Henry's and Elsworthy Roads to the Ecclesiastical Commissioners for the erection of another church, St Mary's, completed in 1873.[64]

Leases generally prohibited the use of houses as other than gentlemen's private residences, except in streets designed from the outset for shops, notably King's College and Winchester Roads, and England's Lane. Plate 289 is an elevation of four shops and a public house built along the east side of King's College Road by Robert Yeo; their lease was granted on 27 May 1858.[65] Yeo's lease for the terrace of six shops opposite the one illustrated prohibited the lessee from permitting 'any open or public shew of business . . . on that side of the house which abuts on the Adelaide Road,' or allowing 'the private door or entrance to be used for the purposes of any trade either for the reception or delivery of any goods and merchandize . . . nor [to] permit . . . any . . . of the said messuages or tenements to be converted into . . . a public house tavern or Beer shop nor in any way alter or add to the said six . . . buildings . . . nor destroy the uniformity of the said premises.'[66]

The initiative for setting out the actual pattern of streets lay with the building developers, notably Samuel Cuming. Cuming, who dominated the operations on the estate from the mid-forties through the fifties, continued to build until his death in 1870. Plate 286 shows how far the building had progressed by November 1849.[67] Plate 287 shows the houses, mostly along Adelaide Road, built by Cuming under his first five agreements.[68] Most of the houses shown in the earlier plan that are not in Plate 287 were built by Wynn or his subordinates, of whom Cuming was one, before he set out on his own.

The difference in the scale and density of construction between the plans of the ground landlord and the proposals of the building contractor reflected the customary conflict of interest between landlord and speculator, whereby the latter, with his better knowledge of the house market, would scale down the ambitious designs of the former.[69] Even so, the actual development carried out the surveyor's basic intention: that it consist of streets of villas, of the same order as those that were making the Eyre estate such a speculative and social success.

St John's Wood, 'resorted to by dissipated men of affluence for the indulgence of one of their worst vices,' already had a reputation for impropriety. Yet most of its inhabitants were 'opulent and industrious professional men and tradesmen,' and it set a pattern of suburban development that Chalcots wished to emulate. Whatever the 'dark, demoralizing scenes' that were supposed to take place in the 'handsome residences' behind the five-foot brick walls which were the 'peculiar characteristic' of the district, the estate found no difficulty in attracting residents of the utmost respectability.[70] Even Lady Amelia de Courcy, in *The Small House at Allington* (1864), was perfectly content to move to St John's Wood on her marriage, and her sister, Lady Alexandrina, would have been happy to do likewise had her husband not refused to live north of the New Road.

Shaw described the Eyre estate in 1845 as 'one of the most important and valuable Properties connected with the Metropolis, especially that part which adjoins Chalcots, the Roads being wide and the houses of a very superior Class.'[71] He therefore considered communications with the estate of the utmost importance, and laid great stress on pushing a road across Chalcots to join the Finchley Road. For this reason the central portion of the estate was developed several decades before either the northern or southern section.

The printed building-proposals of 1829 had promised through communications with St John's Wood. Ten years later negotiations were in progress with William Kingdom, 'a Gentleman ... of great respectability, and considerable Property,' who had recently engaged in a speculation on Lord Holland's estate in Kensington, to construct the entire road and to build at its two ends, in 'a style of Building somewhat similar to the Terraces of the Regent's Park, or of the Oxford and Cambridge Terraces [Sussex Gardens, Paddington]; there can be little doubt of the new Road,' which would be called Eton Terrace, 'becoming a splendid addition to the projected improvements of the neighbourhood.' Shaw was willing to reduce Kingdom's ground rent from £35 an acre to as low as £20 if he would at his own expense build the road, 'intended to be 400 feet from the Center of the Railway and Tunnel and extending the whole length of the North side of it,' and from which 'branch roads may be formed ... in many directions,' since it would 'so materially enhance the value of the Estate generally.'[72]

The negotiations with Kingdom fell through, but the following January Shaw was proposing to 'communicate with Colonel Eyre with reference to the access proposed to the St. Johns Wood Estate ... which ... will afford a mutual benefit to the *two* Estates.'[73] On 14 July 1845 Shaw reported that Adelaide Road had by then been formed 'to the extent of about 1500 feet,' one third of the way to the Finchley Road, and that arrangements had been made with Colonel Eyre for the junction of the two roads. He urged that the road and sewer 'be formed *at once*, & not progressively ... and in this Condition Mr. Cuming acquiesces; indeed it is to his own interest with reference to his selling or letting the houses now building upon the Land he has taken, as there naturally exists a feeling in the Public, either that the Road may not be continued, or that many years may first elapse.'[74]

He wrote confidently the following January, 'when this new line is . . . opened I feel assured that the remaining frontages of the College Land will be advantageously disposed of, and the whole Estate greatly increased in value.'[75] He contemplated two more roads running west from the Hampstead Road north of Adelaide Road, but did not intend to extend the Private Road (later College Road) farther northward. There was also to be a road 'on the Chalcotts Estate running parallel with the Marylebone & Finchley Road at about the same distance from the boundary of the two Estates,' which became Winchester Road.[76]

The building agreements with Cuming, like those with later developers on the estate, while immensely more restrictive than a Sheffield builder would have tolerated, left him a fair amount of freedom to determine the details of his operations. For example, in his first agreement, he contracted to build 'not less than 8 single or double detached houses of not less than the 2d. rate on that part of the plan marked A,' with similarly loose specifications for the other four plots. 'Plans of each house and a general Specification of the same [were] to be submitted to, and to be subject to the approval of the Surveyor to the estate before it is commenced and no house [was] to be built inferior as respects materials and workmanship to those now building by Mr. Cuming [under an agreement made previously with William Wynn] on the estate.'[77] A standard covenant required that 'a space of at least 15 feet . . . be reserved between the main walls of each single or double detached House.'[78]

The surveyor found nothing to object to in Cuming's operations. In July 1845 he wrote to the bursar: '. . . I passed some hours yesterday in going over it [the Chalcots estate], and was very much pleased with the New line of Road which is now complete up to Mr. Cuming's last take, and with the houses he has built, & is building, which are of a superior description as he proceeds.'[79] Late in 1850 he reported that although 'there has been generally a cessation of building during the last two years . . . what has been done at Chalcot I am happy to say has been well done, and the Houses are of a respectable class & character.'[80]

Both the Chalcots estate and the Norfolk estate were developed on ninety-nine-year leases, but in every other way the policies and practices of the landlords and the developing builders on the two properties were totally different. Neither the Provost and Fellows of Eton College nor the Duke of Norfolk were consciously pursuing idiosyncratic methods of estate management, but they were instead responding to the wholly different sets of social and economic circumstances operating in North London and in a West Riding industrial center. Eton College developed Chalcots as an exclusively middle-class residential suburb; the Norfolk estate had an overwhelmingly working-class and industrial character. Eton College imposed strict regulations on its builders and tenants; the Norfolk estate exercised the mildest of

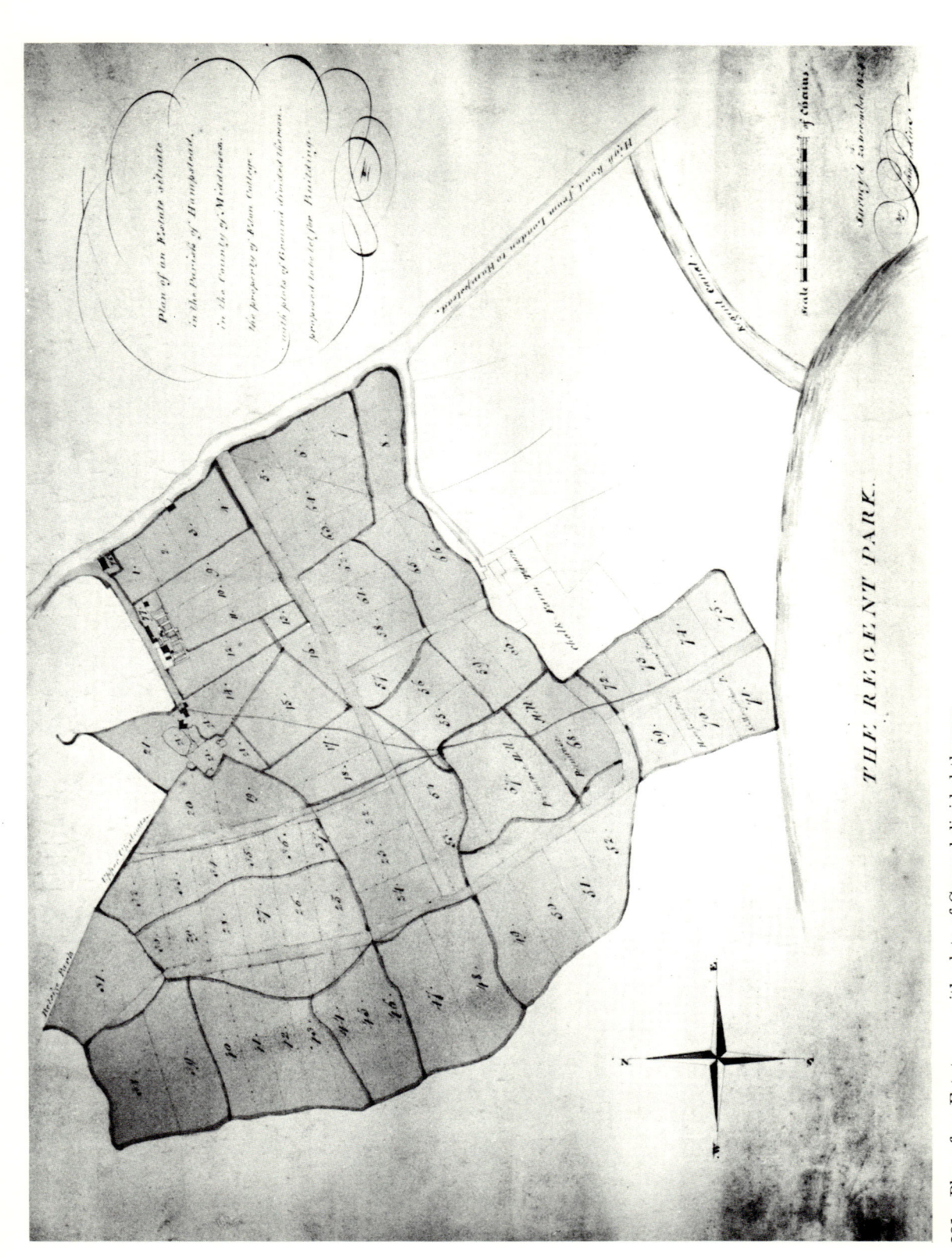

281 Plan of an Estate ... with plots of Ground divided thereon proposed to be let for Building. Surveyed 20 December 1824, by John Jenkins. Eton College Records, vol. 51/13.
Photos: S. G. Parker-Ross
Eton College

282 Plan of part of the Estate of Eton College . . . showing a Design for building thereon, March 1827 [by John Shaw]. E.C.R. vol. 51/17. *Eton College*

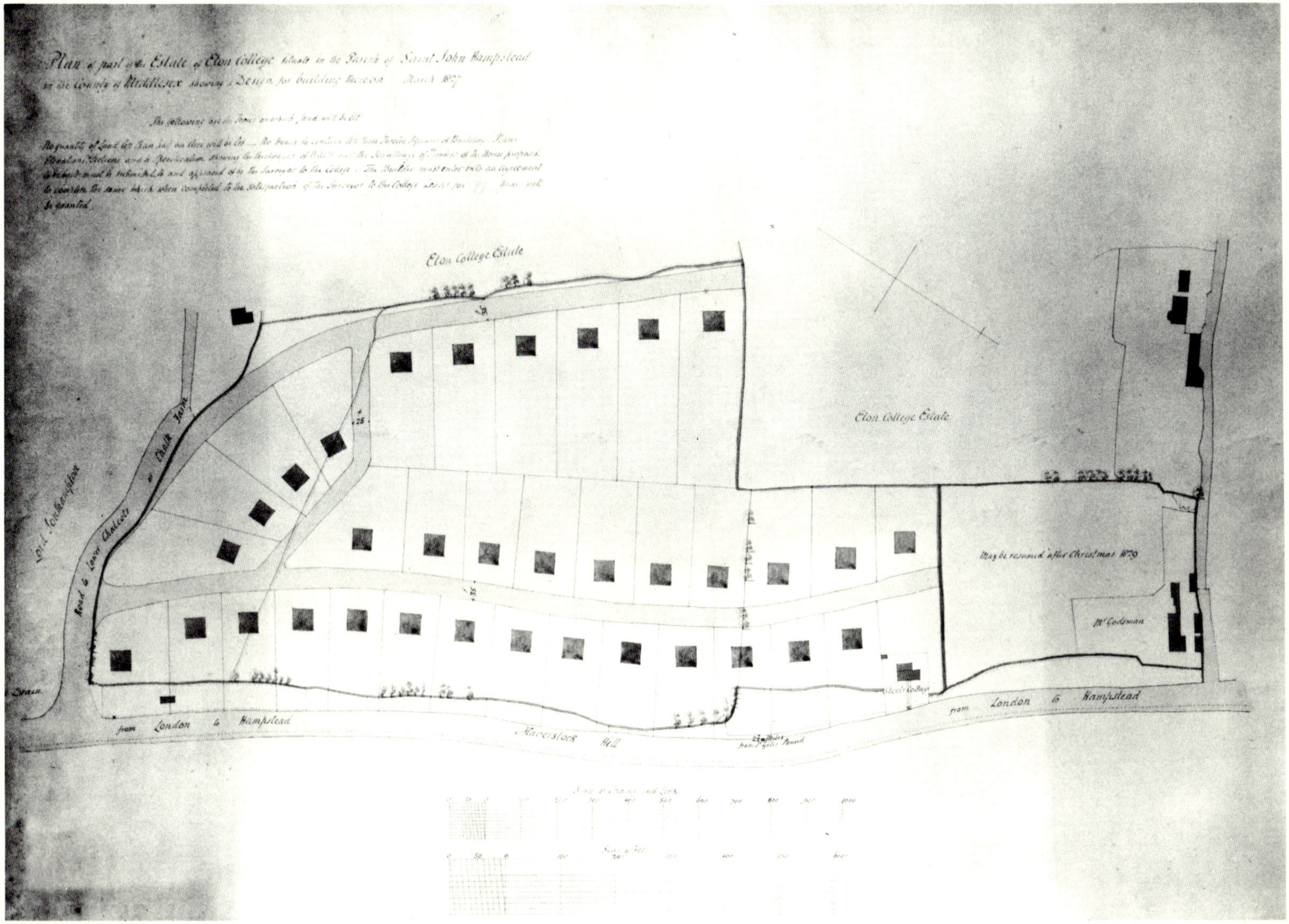

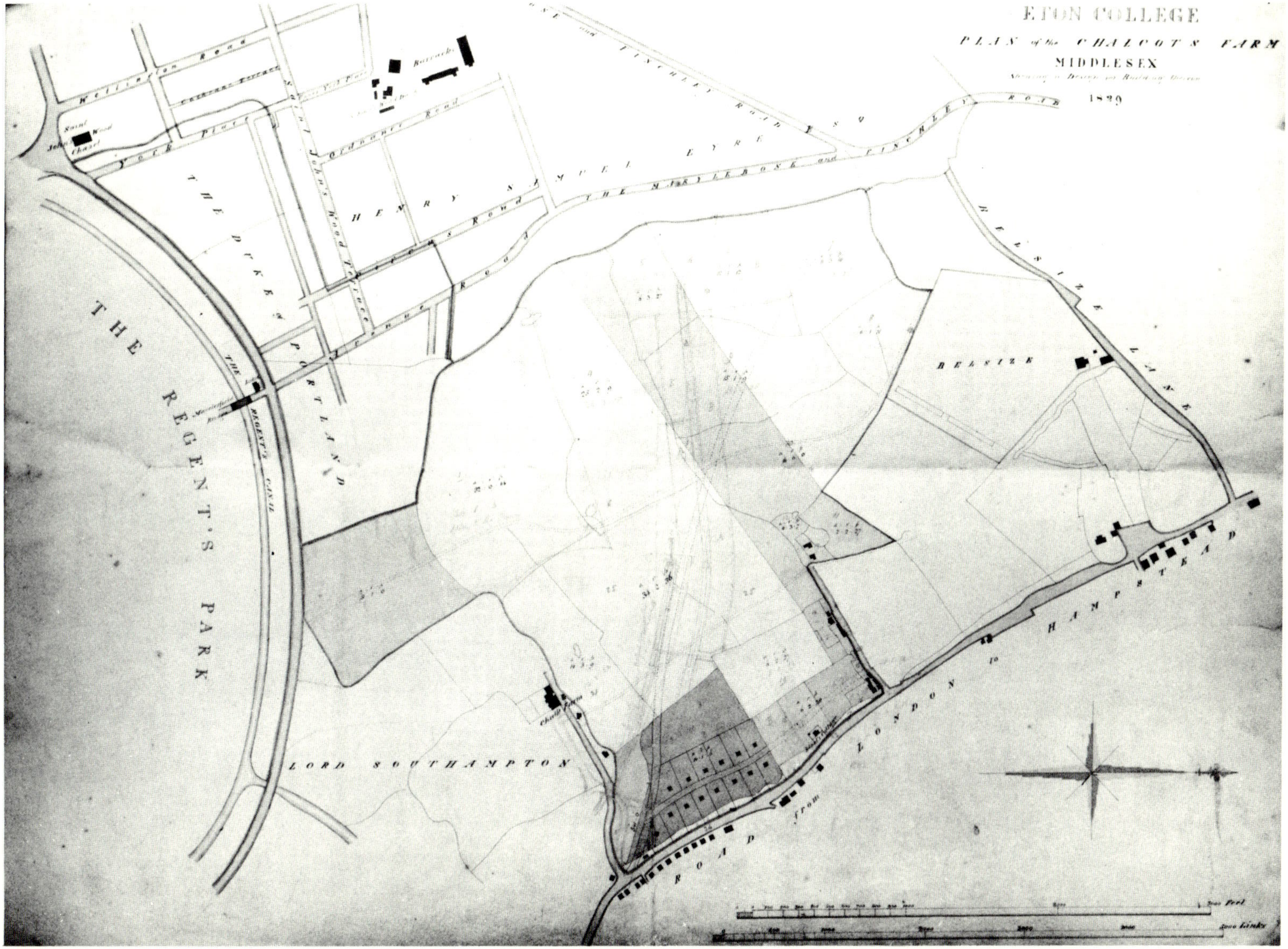

283 Plan of the Chalcots Farm ... showing a Design for Building thereon, 1829. E.C.R. vol. 51/19. *Eton College*

284 Eton College Estate, Chalcots, Hampstead. Proposals for Building, 1 May 1829. E.C.R. vol. 49/27. *Eton College*

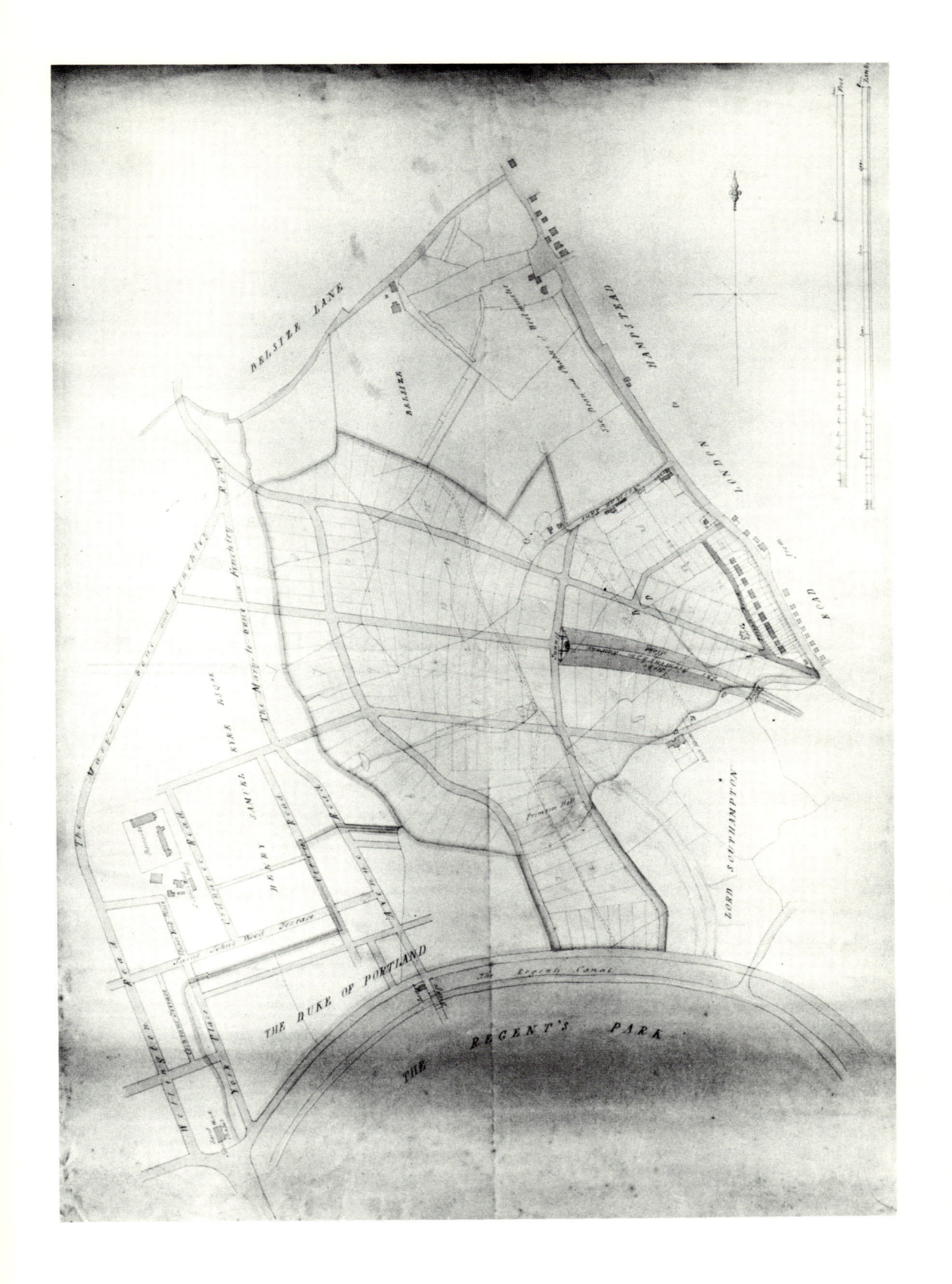

285 Chalcots estate [January 1840]. E.C.R. vol. 51/25.
Eton College

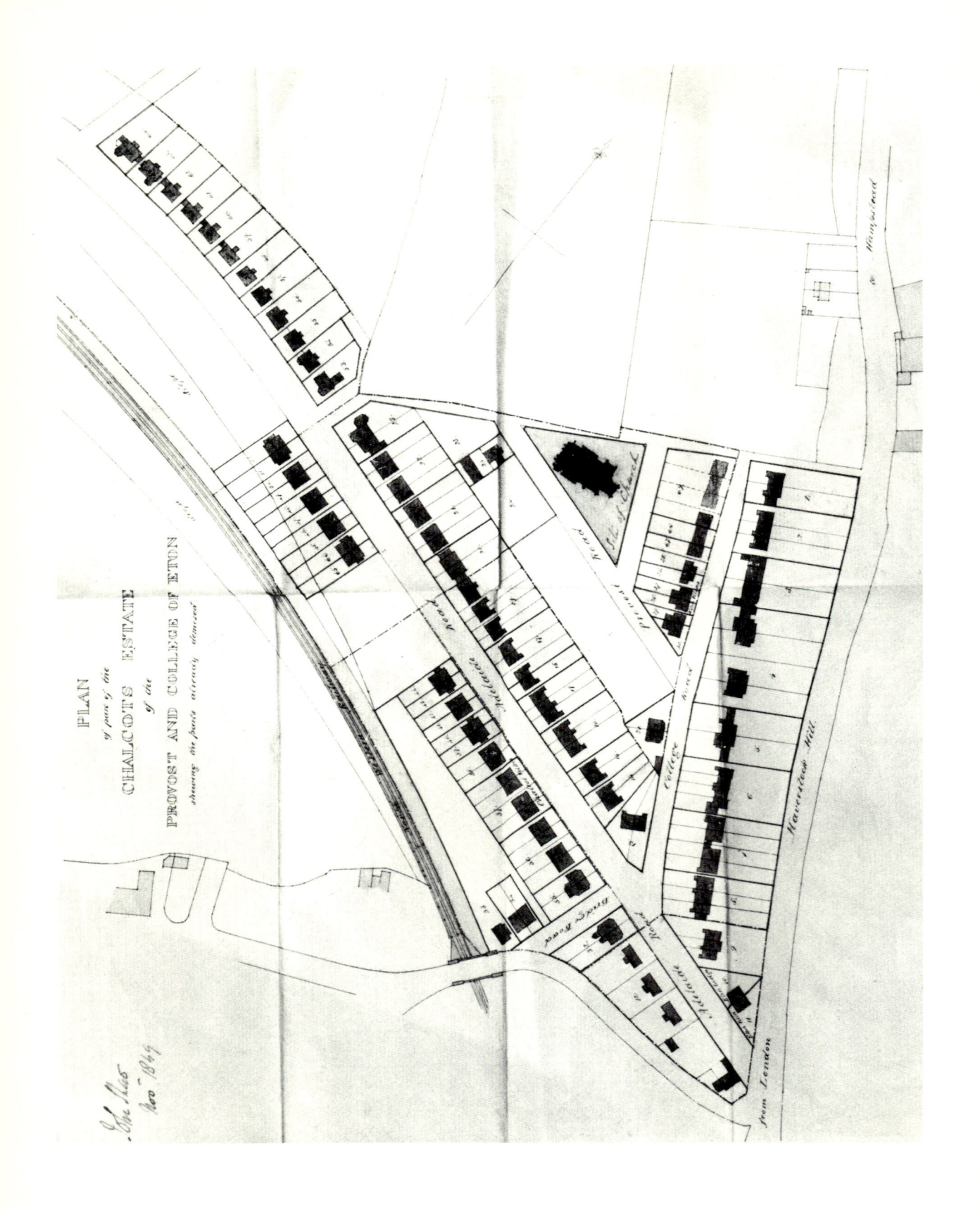

PLAN
of part of the
CHALCOTS ESTATE of the
PROVOST AND COLLEGE OF ETON
shewing the parts already demised

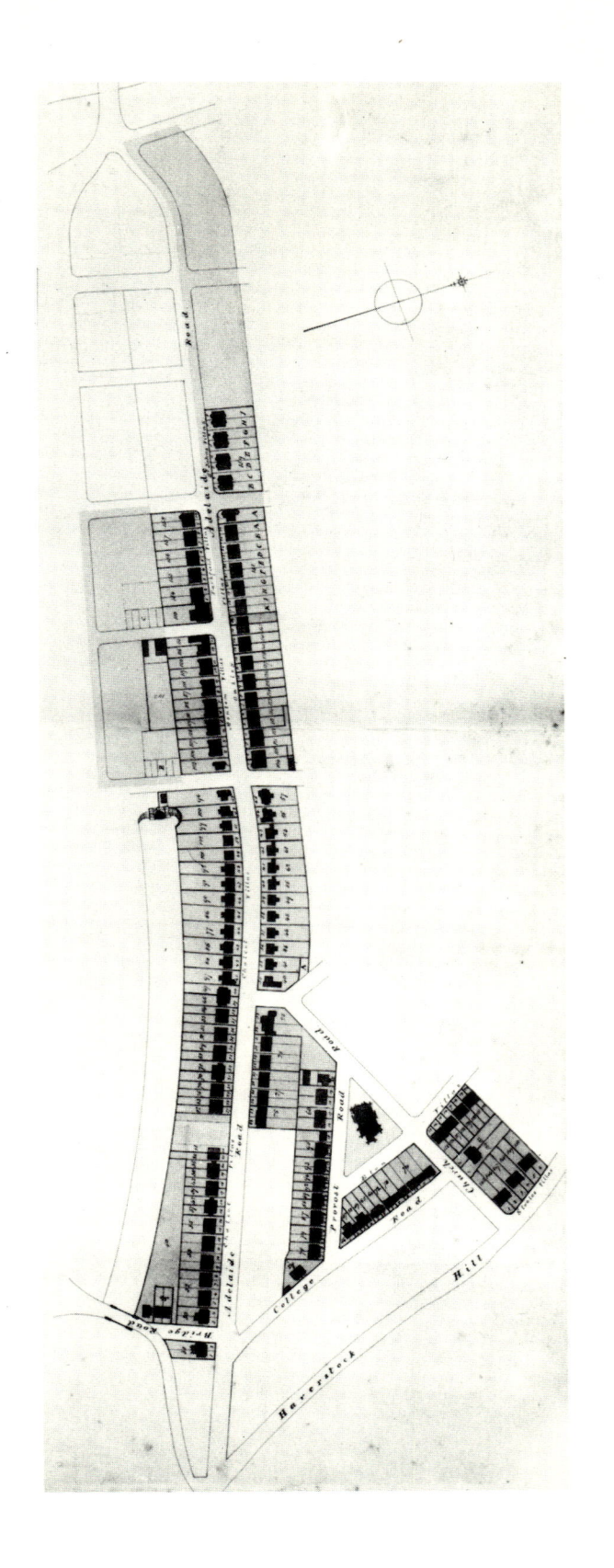

286 *above* Plan of Chalcots Estate showing the parts already demised. November 1849, John Shaw. E.C.R. vol. 49/131.
Eton College

287 *below* Plan of building on the estate to illustrate five building agreements ... with Samuel Cuming, 4 January 1858. E.C.R. vol. 51/40.
Eton College

288 *left* Elevation for 1 Eton Villas, 28 March 1849, Chalcots Leases no. 1, fol. 56. *Eton College*

289 *right* Elevation, four shops and public house, Nos 1, 3, 5, 7, 9 King's College Road, 27 May 1858, Robert Yeo, Chalcots Estate. Leases, no. 1 (9 March 1836–26 September 1860), fol. 160. *Eton College*

290 *below* Elevations for 17–26 Oakley Villas, Adelaide Road, 2 May 1856, Samuel Cuming. Leases no. 1, fol. 151. *Eton College*

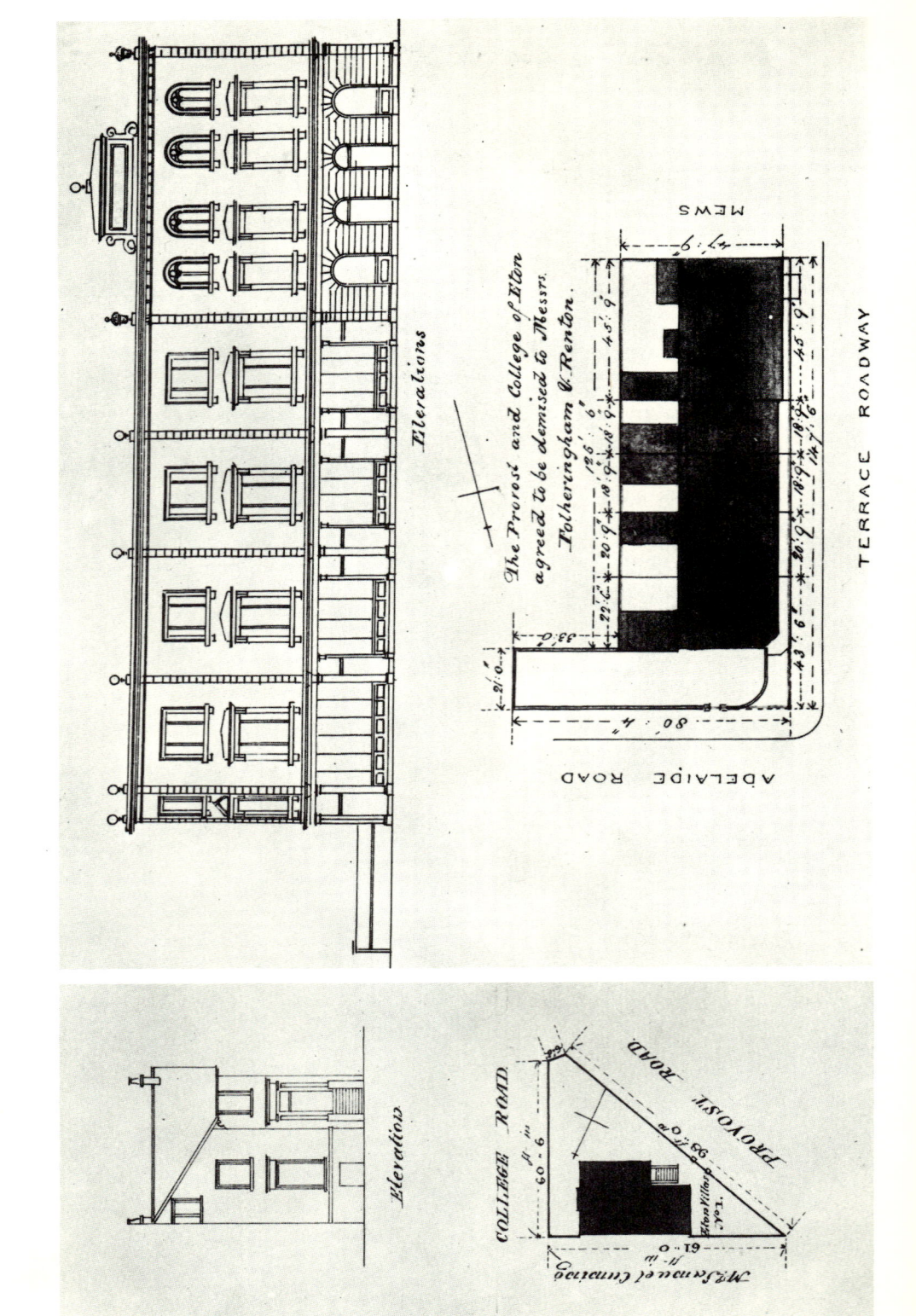

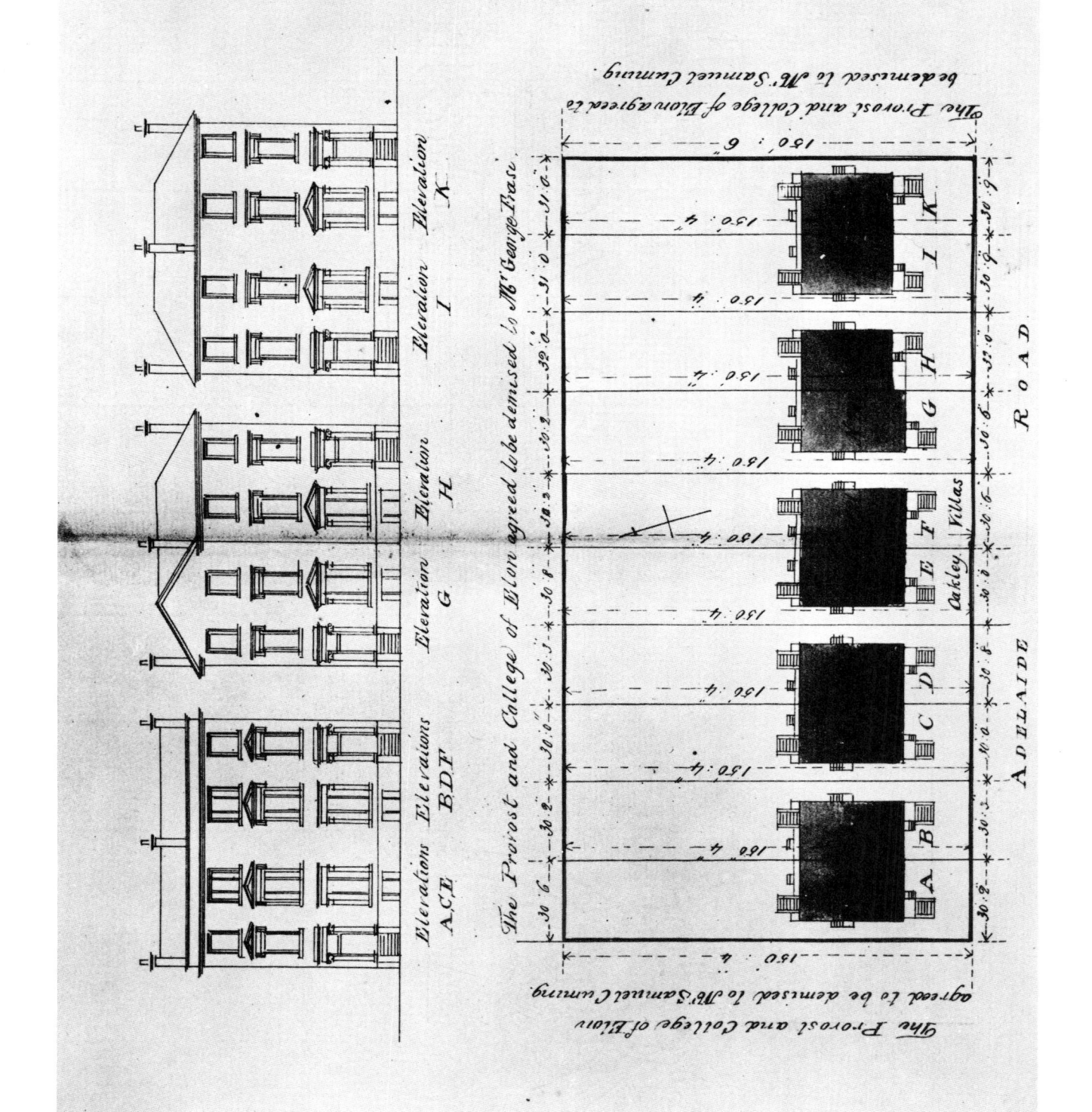

Elevations Elevation Elevation Elevation
A,C,E   G   H   I   K
B,D,F

The Provost and College of Eton agreed to be demised to Mr. George Frase

The Provost and College of Eton agreed to be demised to Mr. Samuel Cuming

ADELAIDE ROAD

Oakley Villas

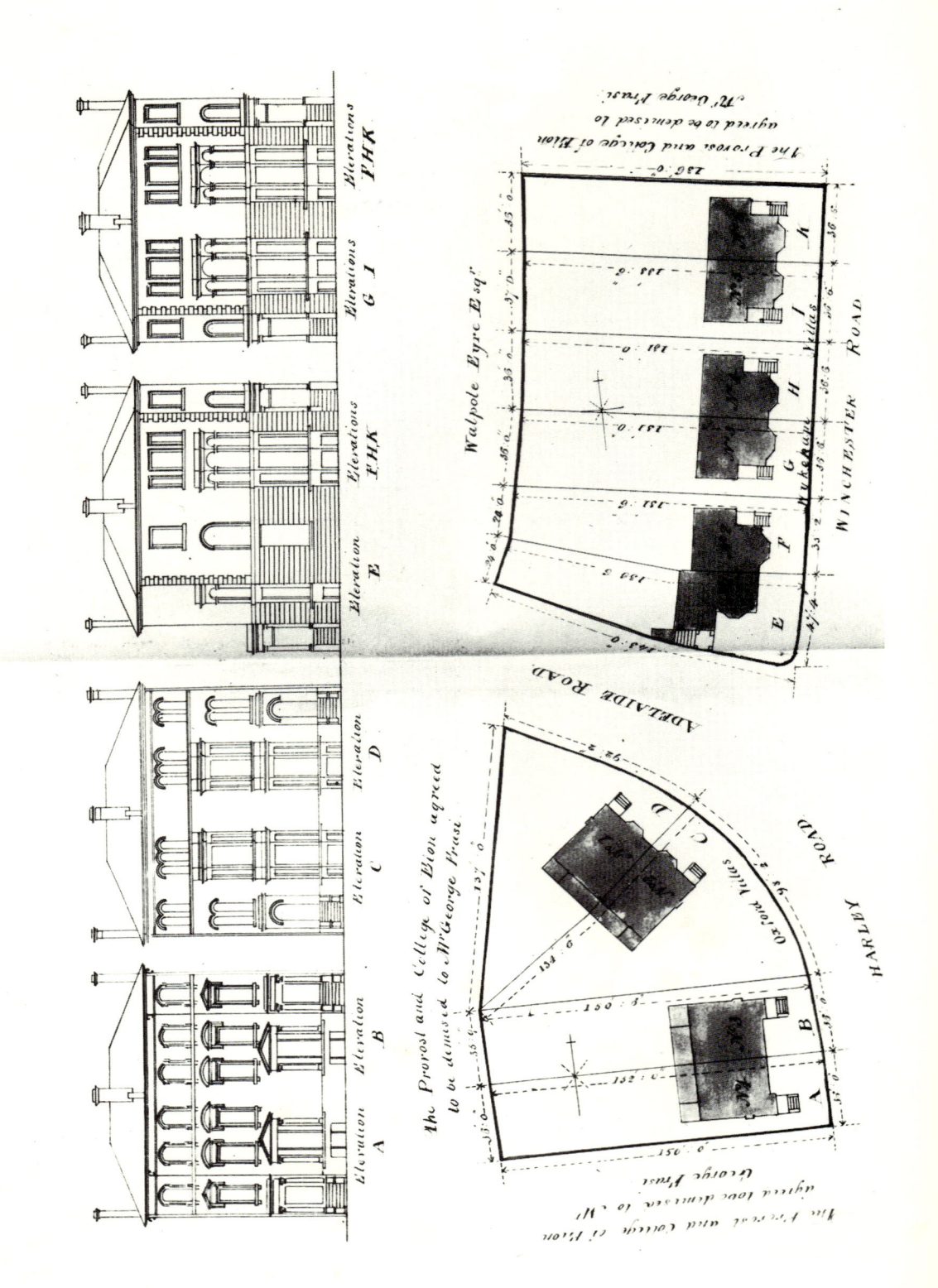

291 Elevations for 1–4 Oxford Villas, Harley Road; and 1–6 Wykeham Villas, Winchester Road, 2 May 1856, Frank Clemow and Mary Anne Angell, lessees Leases no. 1, fol. 152. Eton College

controls. Chalcots was a low-density neighborhood of detached and semi-detached villas; the Norfolk estate was a mixture of cutlers' workshops, steel mills, and low rows of cottages, back-to-back in the early Victorian years, fronting on narrow streets or courts. Chalcots was developed by comparatively large-scale builders and contractors; Sheffield by small speculators, with a minimum of capital. Speculators at Chalcots were themselves usually professional builders; speculators in Sheffield— at least until the 1850s—were more likely to be investors outside the building industry. In Chalcots, the responsibility for providing roads lay with the builder; in Sheffield, with the ground landlord. Before building could begin in Chalcots, carefully drafted articles of agreement had to be executed; in Sheffield, a verbal understanding was all that was required. For all its superior social aspirations, Chalcots is as much a denial of the basic principles of Georgian town planning as is Sheffield. As an exercise in urban design it pales in comparison with the near-contemporaneous development of the Bedford estate in northern Bloomsbury. Thomas Batcheldor and Samuel Cuming make a poor showing if set beside Christopher Haedy and Thomas Cubitt. Whatever the leafy charms of Eton Villas today or Adelaide Road until recently, they cannot compare in architectural excellence with Tavistock or even Gordon Square. Figs Mead, which was being developed in the forties and fifties at exactly the same time as Adelaide Road, and for a markedly lower class of intended resident, shows immensely greater imagination in layout.[81] Yet the same London & Birmingham Railway that frustrated the intentions of the Bedford Office to make Ampthill Square a center of gentility did no apparent damage to the houses in Adelaide and King Henry's Roads through whose back gardens it ran.[82]

Like the Norfolk estate in Sheffield, Chalcots is less a consciously imposed town plan than a rational response to a particular housing market. Sheffield demanded the maximum number of the cheapest possible cottages, crowded as closely together as possible around places of employment. The narrow streets and confined courts that hugged the slopes of Sheffield's hills did just that, reconciling the single-family house with a compact, high-density community. Chalcots demanded villas, separate and semi-detached, solid and respectable but not unduly costly, a church, trees, private gardens, seclusion; above all that the lower orders be kept at a distance. By co-operating with, rather than opposing, social and economic realities, the two ground landlords adopted unheroic roles for themselves. The real key to what happened lies with the builders, developers, and investors who embodied these realities.

Notes

1 J. L. and Barbara Hammond, *The Town Labourer 1760–1832* (1917), pp. 39–40.
2 Thomas Carlyle, 'Shooting Niagara: and After?' *Macmillan's Magazine*, xvi (1867), 332.
3 See A. S. Wohl, 'The Housing of the Artisans and Laborers in Nineteenth Century London, 1815–1914' (unpublished Ph.D. dissertation, Brown University, 1966);

'The Bitter Cry of Outcast London,' *International Review of Social History*, xiii (1968), Part 2, 189–245; and 'The Housing of the Working Classes in London, 1750–1914,' in Stanley Chapman, ed., *The History of Working Class Housing* (Newton Abbot, 1971).

4   H. J. Dyos, 'The Speculative Builders and Developers of Victorian London,' *Victorian Studies*, xi (1968), 641–90; Marian Bowley, *The British Building Industry* (Cambridge, 1966), pp. 337–40.

5   G. Calvert Holland, *Vital Statistics of Sheffield* (1843), p. 56.

6   John Nash in J. White, *Some Account of the Proposed Improvements of the Western Part of London* (2nd edn, 1815), Appendix, pp. xxvii–xxviii.

7   A. K. Cairncross, *Home and Foreign Investment 1870–1913* (Cambridge, 1953), p. 84.

8   Dyos, 'Speculative Builders,' p. 663. See also John R. Kellett, *The Impact of Railways on Victorian Cities* (1969), pp. 412–13.

9   Sir John Summerson, 'Urban Forms,' in Oscar Handlin and John Burchard, eds, *The Historian and the City* (Cambridge, Mass., 1963), p. 174.

10 Hugh C. Prince in J. T. Coppock and Hugh C. Prince, eds, *Greater London* (1964), p. 105.

11 Kate Simon, *London : Places and pleasures* (New York, 1968), p. 293.

12 *Builder (B)*, xix (1861), 641.

13 James Haywood and William Lee, *Report on the Sanatory Condition of the Borough of Sheffield* (2nd edn, Sheffield, 1848), p. 10.

14 *Sheffield and Rotherham Independent*, 13 January 1872.

15 Sir John Summerson in Handlin and Burchard, op. cit., p. 175; and in 'The Beginnings of an Early Victorian London Suburb,' II, a lecture delivered at the London School of Economics, 27 February 1958, the MS. of which Sir John has kindly allowed me to read.

16 Reginald J. Fletcher, *St. Saviour's Church South Hampstead, A Retrospect* (1918), pp. 6–7.

17 See the works of A. S. Wohl cited in note 3.

18 *Parliamentary Papers (P.P.)*, 1889, LXV, Report on an Epidemic of Small-Pox at Sheffield during 1887–88; by Dr Barry (C. 5645), p. 286.

19 Sidney Pollard, *History of Labour in Sheffield* (Liverpool, 1959), pp. 185–6.

20 *P.P.*, 1833, XV, Select Committee (S.C.) on Public Walks: Minutes of Evidence (448), Q. 899. See also ibid., 1833, VI, Select Committee on Manufactures, Commerce, and Shipping: Minutes of Evidence (690), Q. 2917.

21 Sidney Pollard, op. cit., pp. 100–1. See also G. Calvert Holland, op. cit., pp. 29, 46, 69; 'Plans generally adopted in the Town of Sheffield for Cottage Houses,' by William Flockton, in *P.P.*, 1845, XVIII, Royal Commission (R.C.) on the State of Large Towns and Populous Districts: 2nd Report, Part II, Appendix (610), p. 347.

22 James Haywood and William Lee, op. cit., p. 122. See also *P.P.* 1888, XXII, S.C. on Town Holdings: Minutes of Evidence (313), QQ. 859–62.

23 Ibid., Select Committee on Manufactures, Commerce, and Shipping (690), QQ. 2884–9; G. Calvert Holland, op. cit., pp. 56–8; *P.P.*, 1884–5, XXX, R.C. on the Housing of the Working Classes: Minutes of Evidence (Cd. 4402–I), Q. 10,795.

24 Ibid., 1888, XXII, S.C. on Town Holdings: Minutes of Evidence (313), Q. 720.
25 Marian Bowley, op. cit., p. 360.
26 *Building News*, vi (1860), 871. See also 'Report of Messrs Leverton & Chawner...' in J. White, *Some Account of the Proposed Improvements*, Appendix, p. vi; *Architectural Magazine*, i (1834), 115; James Noble, *The Professional Practice of Architects* (1836), pp. 91–3; Donald J. Olsen, *Town Planning in London: The eighteenth and nineteenth centuries* (New Haven, 1964), pp. 14–20.
29 'Report of Messrs Leverton & Chawner,' pp. xi–xii.
28 *B*, xvi (1858), 177. See also Olsen, op. cit., pp. 34–5, 224–7.
29 *P.P.*, 1889, XV, S.C. on Town Holdings: Report (Cd. 251), p. 85.
30 Horace Walpole to George Montagu, 1 September 1760, in W. S. Lewis and Ralph S. Brown, Jr, eds, *Horace Walpole's Correspondence* (New Haven, 1941) IX, p. 295.
31 Holland, op. cit., pp. 50–1.
32 *B*, xix (1861), 641.
33 Sheffield City Libraries, 'Applications for Building Ground and Premises, 1824–1850 A,' passim, Arundel Castle MSS S384. All references to the Arundel Castle MSS in the Sheffield City Libraries are by the kind permission of His Grace, the Duke of Norfolk, E.M., K.G., and of the City Librarian, John Bebbington, F.L.A.
34 'Ratification by Henry, 15th Duke of Norfolk, of building leases in a schedule,' 24 December 1868, Arundel Castle MSS–S.D. 546. Alfred Emden, *The Law Relating to Building, Building Leases, and Building Contracts* (2nd edn, 1885), pp. 79, 87, 293–4.
35 Printed form of lease (dated 6 May 1788), Arundel Castle MSS–S.D. 868/184.
36 James Sorby to Michael Ellison, 12 September 1835, Arundel Castle MSS–S478 (xiii).
37 Duke of Norfolk's Trust (Sheffield Estate), 'Draft Proposed Form of Lease for 99 Years—As perused by Mr. Fowler, 1861,' pp. 3–6; Arundel Castle MSS–SP38.
38 *P.P.* 1888, XXII, S.C. on Town Holdings: Minutes of Evidence (313), QQ. 729–30.
39 Ibid., Q. 3517.
40 Marginal memoranda, *Queries Prepared with a View of Obtaining Evidence on the Matters Referred to the Committee*, S.C. on Town Holdings, 1886, Arundel Castle MSS–S439.
41 See Few & Co. to H. W. Beavan, 25 August 1836 (copy), Arundel Castle MSS S478 (xiv); for the policies on the London estate in the eighties, see *P.P.* 1887, XIII, S.C. on Town Holdings; Minutes of Evidence (260), QQ. 11,958–12,360.
42 Michael Ellison to Edward Blount, 20 April 1840, Arundel Castle MSS–S478 (xviii).
43 In this connection, see also Henry Howard, Auditor, in 'Sheffield Order Book,' 11 and 12 July 1815, Arundel Castle MSS–S391; Edward Blount to John Housman, 10 August 1819, Arundel Castle MSS–S478 (ii); Michael Ellison, Diary, 26 April 1853, Arundel Castle MSS–S523.
44 For the small resources and limited scale of operation of the Sheffield builders, see 'Applications for Building Ground and Premises, 1824–1850 A,' Arundel Castle MSS–S384; 'Applications for Building Ground, and premises, From 29th September

1865 to 31st December 1867,' Arundel Castle MSS–SP38; Holland, op. cit., pp. 55–9, 66; *P.P.*, 1833, VI, S.C. Manufactures, Commerce, and Shipping (690), QQ. 2884–9; Ibid., 1884–5, XXX, R.C. on the Housing of the Working Classes: Minutes of Evidence (C. 4402–I), Q. 10,795; Pollard, op. cit., p. 101.

45 *P.P.*, 1889, XV, S.C. on Town Holdings: Report (251), p. 85.
46 *B*, xxxiii (1875), 274.
47 Olsen, op. cit., pp. 20, 33–4, 81–3, 224–7.
48 'Report of Messrs Leverton & Chawner,' p. xv.
49 John Nash, op. cit., pp. xxxiv–xxxv.
50 John Summerson, *Georgian London* (1945), p. 164.
51 'Proposed plan to divide the Land belonging to Eton College situate at Primrose Hill . . . into plots of ground to be let for Building,' Eton College Records (hereafter cited as E.C.R.) vol. 51/12. All MS. references to material on Chalcots in the Eton College Library are by the kind permission of the Provost and Fellows of Eton College.
52 'Plan of an Estate . . . with plots of Ground divided thereon proposed to be let for Building,' 20 December 1824, E.C.R. vol. 51/13. See also George Bethell, 'Observations heard at the meeting relative to the waste,' 1825, E.C.R. vol. 49/22.
53 *An Act for effecting an Exchange between Her Majesty and the Provost and College of Eton*, 5 & 6 Vict. cap. lxxviii.
54 'Plan of a part of the Estate of Eton College . . . showing a design for building thereon,' March 1827, E.C.R. vol. 51/17.
55 'Plan of the Chalcots Farm . . . Shewing a Design for Building thereon,' 1829, E.C.R. vol. 51/19.
56 Chalcots, Hampstead. Proposals for Building, 1 May 1829, E.C.R. vol. 49/27.
57 Summerson, 'Urban Forms,' p. 174.
58 'Chalcots estate plan' [January 1840], E.C.R. vol. 51/25; John Shaw to George Bethell, 14 January 1840, E.C.R. vol. 49/84.
59 William Wynn to Thomas Batcheldor, 12 December 1842, Chalcots, Box 1.
60 John Shaw to George Bethell, 8 January 1846, E.C.R. vol. 49/119.
61 Ibid., 14 January 1840, E.C.R. vol. 49/84.
62 Henry Bird to George Bethell, 7 September 1846, E.C.R. vol. 49/128; *Proposed District Church, Chalcott's Estate, Haverstock Hill, Hampstead Road*, 5 September 1846, E.C.R. vol. 49/128; Two plans 'of part of Fourteen Acres Field inserted in a conveyance from the Provost and College of Eton as a site for a Church . . .' 17 December 1846, E.C.R. vol. 51/34; Tooke Son and Hallowes to the Provost of Eton, 10 July 1847, E.C.R. vol. 49/130; John Shaw to George Bethell, 21 November 1850, E.C.R. vol. 49/137.
63 *B*, xiv (1856), 390; Stephen Buckland, 'The Origin and Building of S. Saviour's Church,' *S. Saviour, Eton Road, Hampstead, Parish Paper*, October 1956.
64 'The Provost and College of Eton and The Revd. Chas. James Fuller and others, Agreement for Conveyance of Land . . . for a Site for a Church,' 20 July 1870, Chalcots, Box 6.
65 'Chalcots Estate Leases No. 1 (9 March 1836–26 September 1860),' Lease no. 160.
66 Lease, Robert Yeo, 13 June 1859, 'Chalcots, Leases, January 1859–July 1865,' p. 59.

67 E.C.R. vol. 49/131.
68 'Plan of building on the estate to illustrate five building agreements ... with Samuel Cuming,' 4 January 1858, E.C.R. vol. 51/40.
69 See, for instance, Olsen, op. cit., pp. 20, 35, 224–7.
70 Alfred Cox, *The Landlord's and Tenant's Guide* [1853], pp. 231–2.
71 John Shaw to George Bethell, 22 July 1845, E.C.R. vol. 49/116.
72 William Kingdom to John Shaw, 22 March 1839, E.C.R. vol. 49/81; John Shaw to George Bethell, 28 March 1839, E.C.R. vol. 49/82.
73 Ibid., 14 January 1840, E.C.R. vol. 49/84.
74 Ibid., 14 July 1845, E.C.R. vol. 49/114.
75 Ibid., 8 January 1846, E.C.R. vol. 49/119.
76 Ibid., 5 March 1846, E.C.R. vol. 49/123.
77 'Particulars of a proposed letting of part of the Chalcott's Estate to Mr. Saml. Cuming ... Builder,' 5 August 1844, E.C.R. vol. 49/111.
78 'Particulars of a proposed letting ... to Mr. Saml. Cuming ... Builder,' 1 January 1845, E.C.R. vol. 49/112.
79 John Shaw to George Bethell, 22 July 1845, E.C.R. vol. 49/116.
80 Ibid., 16 November 1850, E.C.R. vol. 49/136. See also *Land and Building News*, ii (1856), 251.
81 Olsen, op. cit., pp. 63–73.
82 Ibid., pp. 150–1.

15 Slums and Suburbs

H. J. Dyos and D. A. Reeder

Victorian London was a land of fragments. Yet the immensity of its mass and the bewildering tissue of its human associations argue for something quite different. To all appearances, its streets abutted each other with a stumbling logic and the names of the neighbourhoods they enclosed unrolled with a continuous rhythm across the map. By day, the great improvised structure could be seen to heave into operation like a piece of fantastic machinery set going by an invisible hand. By night, its operatives could be seen submitting with equal discipline to a different spatial logic as they redistributed themselves for sleep. The impression created by this daily act is of a kind of social contract being discharged by a society held together by a common purpose and conscious to some degree of being a community. Such a symmetry was not a complete illusion even in Victorian London. People commonly congregate in cities because they look for mutual support and they accept un-thinkingly the commercial ethos ruling there as the first organizing principle of their working lives. The rule of the market was the original and most natural means of settling rival claims to all things in short supply, not least to space, and the population reshuffled itself by day and by night in tacit acceptance of the prevailing values laid upon the land. Whatever space they occupied was everywhere contracted for in some terms, whether by duly attested indentures or merely a nod upon the stairs, and this great interlocking bargain came daily into play. It was a contract of a kind.

But the irrigations of commercial capital that sustained it even in the smallest channels were not capable of keeping a larger sense of community alive when it was in danger of being crushed by sheer force of numbers. Whatever sense of that kind

© 1973 H. J. Dyos and D. A. Reeder

could be said to have lingered within living memory at the start of the Victorian period had been fractured by just such growth, and the animus that was replacing it already looked more like an instinct for survival. London was too vast, and the consciousness of the crowd too immanent, to admit the intimacy of a single community for the whole. There was instead the beginnings of a fragmentation that has never been reversed. This was only partly geographical. It was also social and psychological. The characteristic tensions being produced were not so much, perhaps, between class and class as between the individual and the mass, and between the individual's inner life and his outward behaviour. The characteristic shapes which these produced on the ground—the realities composed from their images, then and now—were of suburbs and of slums.

In the crudest model of their development with which we might begin, it is here that urban society most visibly diverged. Centrifugal forces drew the rich into the airy suburbs; centripetal ones held the poor in the airless slums. But the compelling pressures of expansion caused ripples of obsolescence, which overtook places once dancing with buttercups and left them stale as cabbage stalks. Suburbs begat or became slums, rarely if ever the reverse, and the two never coalesced. Whole districts lost ordinary contact with their neighbours, and London became, in the most indulgent terms, an island of villages; in the most heartless, a geographical expression of increasing vagueness. Yet this disintegration was not a disconnected process. The fact of the suburb influenced the environment of the slum; the threat of the slum entered the consciousness of the suburb.

We want in this chapter to look at some of the ways in which their individual characteristics were mutually determined. It is confined to London because the relationship between these two ways of life was not exhibited so clearly anywhere else in Victorian Britain. We concern ourselves chiefly with the narrow focus of the economic forces which were at work simply because we believe that, apart from the multiplication of the population, these were the most fundamental. And we have not the space to go beyond them.

We can see this London quite literally, then, as a great commercial undertaking. Here, even in 1815, was a million-peopled city articulated by commerce into a single gigantic enterprise, the supreme money market of the trading world and the most commanding concentration by far of people, industry, and trade to be found anywhere. This commercial metropolis was presently made more impregnable still by the concentrative power of the railways at home and the gains of imperialism abroad, and its capacity to stimulate new demands within the national economy was enhanced by its indomitable leadership in wealth and fashion. It was an enterprise supported by a labour force that was docile, abundant, and cheap, and by a capital supply that seemed more and more mobile and less and less exhaustible. Here was a formula for inexorable growth—of numbers, output, distances, elevations, mass.

It was a headlong business. What mattered most to a commercial metropolis was commercial success, whether on some remote frontier where money mattered less than existence or on its own doorstep, where it sometimes mattered a good deal more. Building and extending its own plant were inseparable parts of the whole field of operations, and investment in the necessary urban equipment—in houses and domestic amenities no less than in the means of locomotion or places of work and the storage of goods—obeyed as it could the first rule of the market place: buy cheap, sell dear. The housing of the poor, as of the rich, was an item of real property, capable of providing titles to wealth for a whole series of property holders; it was expected to yield profits commensurate with whatever were regarded as the risks of investing in it; and it was held to be quite as much subject to the ordinary pressures of the market as any other commodity put up for sale. As the manipulators of capital, the middle classes helped to make possible the expansion of house-building in the suburbs, the parts of the city in which they were shaping their own environment, but in diverting resources for these purposes they also helped to determine the environment of those left behind in the city centre.

It must be remembered here that the distribution of personal incomes and social disabilities in any society helps to determine not only the way in which power is used at the top but also the way in which weaknesses are shared at the bottom. The condition of the housing of the poor was a step through the looking-glass of the rich—a reflex of the allocation of political power and economic resources in society at large. It was no accident that the worst slums were generally found in places where large houses were vacated by the middle classes in their trek to the further suburbs. Such property could only be occupied economically by lower classes by being turned into tenements, but the rent for a whole floor or even a whole room was often too much for those eventually in possession, and the sub-divisions of space that followed usually meant the maximum deterioration in living conditions. It must also be recognized that the resources which might have made such a transition less dramatic were being ploughed heavily back into the commercial machine instead of being distributed in higher wages. It was tacitly accepted that, if better houses had been built to house industrial workers during the nineteenth century, the higher wages paid out to make this possible would have raised the costs of exports and reduced the capital being sent abroad, which would in turn have held back the growth of exports. More sophisticated economic reasoning might suggest that, so long as there were few substitutes for British goods abroad, paying higher wages and charging higher prices would actually have increased these receipts, but it did not look that way at the time. More matter-of-fact logic demanded that labour costs be cut to the bone. The slums were part of this argument for the economy of low wages, and one of their practical functions was therefore to underpin Victorian prosperity. The truth of the matter now seems to be that they embodied some of the most burdensome and irreducible real costs of industrial growth that might have been imagined.[1] When Henry Jephson wrote his classic study, *The Sanitary Evolution of London* (1907), it was, he said, 'the all-powerful, the all-impelling motive and unceasing desire' for 'commercial prosperity and success' that provided London's motive power. 'That indisputable fact must constantly be borne in mind as one reviews the sanitary and social condition of the people of London . . .' (pp. 7–8).

London's appetite for people had always been immense, but its commercial aggrandizement made it even more voracious. From just under a million inhabitants at the beginning of the century, its requirements had grown, by its close, to about four and a half millions within its own administrative boundaries alone, and a further two millions were distributed, less congestedly but within easy daily reach, in an outer belt of the conurbation—known for statistical purposes after 1875 as Greater London. London grew by sucking in provincial migrants because jobs were either better paid there or thought to be so; it also offered a more liberal array of charities, richer rewards for crime, a more persuasive legend of opportunity than could be found anywhere in the country.[2] So the net migration into London during the 1840s resulted in the addition of about 250,000 inhabitants, or almost one-fifth of its mean population for the decade—a rate of intake which declined appreciably over the next two decades while its absolute level continued to climb, but which surged up again in the 1870s, when almost 500,000, or over fifteen percent, were added to the natural increase. By 1901 Greater London contained one-fifth of the entire population of England and Wales, and in the preceding decade it absorbed, one way or another, one-quarter of the net increase in population of the whole country.[3]

How this provincial tribute was gathered and where it was harboured we cannot tell for certain. The paths of migration into Victorian London have not been traced sufficiently systematically, and whole decades of coming and going, of movements by the million, have dropped completely from our view. The traditional place for the stranger, the poor, the unwanted—for any threatening presence—was at the city gates, and the motley colonies that had accreted there since the Middle Ages remained the receptacles for the sweepings of the city until the space available within the walls was exhausted by the growth of urban society proper in the sixteenth and seventeenth centuries. The suburbs were the slums. The process of accretion that was already in motion around London at the beginning of the nineteenth century was quite different from this. The colonization of London's nearby villages by merchants and men of affairs in the eighteenth century had pre-empted a new middle-class ring of suburbs which, as it spread, congealed in the nineteenth century into a great continent of petty villadom. The new suburb had leap-frogged the older slum.[4] Before 1861 or so the growth of these outer districts, though often dramatic topographically, was not the most rapid or most important. The central districts, which had always belonged to both rich and poor, remained the chief focus of growth. It was here that the main influx of newcomers congregated, and it was here too that the slum was first given its modern name.

It was originally a piece of slang which meant among other things a room of low repute—a term which Pierce Egan extended in his *Life in London* (1821) to 'back slums', defining it in one place as 'low, unfrequented parts of the town'.[5] Dickens used it so when he wrote a letter dated November 1841: 'I mean to take a great, London, back-slums kind of walk tonight.'[6] Presently it took the form of 'slums' and began to pass into everyday use, though it took another forty years for the inverted commas to disappear completely. It was perhaps always an outsider's word, the lack of which did not necessarily imply that the thing it named was too esoteric for ordinary speech, but one too readily accepted to require a general pejorative term. That did not come about before the 1880s—when the assumed place names of the suburbs were first swept aside by the collective gibe of suburbia—and the all-too-real housing 'problem' began to take shape as a public issue. Precisely what a slum ever meant on the ground has never been clear, partly because it has not developed that kind of technical meaning which definition in an Act of Parliament would have given it: indeed, it still lacks this kind of precision and tends to be defined even nowadays in terms of the number of obsolescent houses which a local authority can clear rather than the kind of houses involved. This vagueness has in practice been aggravated by the failure of parliament or the courts to define at all clearly the chief characteristic of a slum, namely its overcrowding, or the actual basis of the medical judgment that a house was 'unfit for human habitation'.[7]

The implication of this is that, like poverty itself, slums have always been relative things, both in terms of neighbouring affluence and in terms of what was tolerable by those living in or near them. Such a term has no fixity. It invokes comparison. What it felt like to live in a slum depended in some degree, for example, on what it might feel like to live in a suburb. Yet there was no simple polarity. There were degrees of slumminess just as there were degrees of suburban exclusiveness, and there were many irregularities in the declension between them. To Patrick Geddes the whole lamentable process of city-building in the nineteenth century was of a piece: 'slum, semi-slum and super slum,' he wrote, 'to this has come the Evolution of Cities.'[8] The old order of slum had certainly insinuated itself as readily into the central purlieus of wealth in St James's as the new order was doing in the suburban ones of Bayswater or Camberwell.[9] Slums as well as suburbs had their moving frontiers and to try to define either too concisely makes very little sense. In terms of human values, the slums of Victorian London were three-dimensional obscenities as replete as any ever put out of sight by civilized man; in terms of the urban economy, they were part of the infrastructure of a market for menial and casual labour; in terms of urban society, they were a lodging for criminal and vagrant communities; in terms of real property, they were the residue left on the market, the last bits and pieces to command a price; in terms of the dynamic of urban change, they were the final phase in a whole cycle of human occupation which could start up again only by the razing of the site. The making of these slums was a process that began far beyond the reach of the slummers who packed into them. It is important to recognize that the unfortunates who occupied the central slums, and those only slightly more fortunate, perhaps, who occupied the slums of the inner suburbs or the embryo slums of the outer districts, were more than anything else merely the residuary legatees of a kind of house-processing operation which was started by another social class with little or no idea or concern as to how it would end.

We cannot here explain in detail what one reviewer of Mayhew called 'this localicity of pauperization'.[10] The most general explanation for slum tendencies in particular places is that, without the kind of general control on the spatial development of the city that might have been given, say, by a rectilinear grid, there were bound to be innumerable dead-ends and backwaters in the street plan. A glance at Booth's maps shows how often these introspective places were seized by the 'criminal classes', whose professional requirements were isolation, an entrance that could be watched, and a back exit kept exclusively for the getaway. They were not difficult to fulfil in scores of places in every part of Victorian London. A more careful reading of Booth's maps would show how some additions to the street plan—a dock, say, or a canal, a railway line, or a new street—frequently reinforced these tendencies. What often made them more emphatic still was the incense of some foul factory, a gas-works, the debris of a street-market, or an open sewer. They all acted like tourniquets applied for too long, and below them a gangrene almost invariably set in. The actual age of houses seldom had much to do with it, and it was sometimes possible to run through the whole gamut from meadow to slum in a single generation, or even less. Animal husbandry survived, of course, in caricature in some of these places, where pigs, sheep, cows, and other livestock were still being slowly cleared in the 1870s from the slums they had helped to create. One man who had spent twenty years visiting the poor of a riverside parish could write as late as 1892 of the donkeys, goats, and chickens which had as free a run of some of the houses by night as they had of the streets by day.[11]

It is possible, too, to trace the origins of slumminess in buildings scamped, whether legally or not, on inadequately settled 'made ground' or by virtue of some builder's sanitary blunder. The history of building regulations is a tale of the regulators never quite catching up with the builders, and of the piecemeal enlarging of the statutory code so as to reduce risks from fire and to health. The machinery for approving street plans and drainage levels took time to evolve and an incorruptible corps of local government officers to administer it.[12] No one can say with real confidence that the many hands required in building and rebuilding slum-prone neighbourhoods were under full and proper control in the public interest before the last decade of the century. Whatever was physically substandard was inclined to become socially inferior, too: the slummers themselves, though often adding the penalty of their own personal habits to the descending scale, seldom *created* such slums as often as they confirmed the builders' and others' mistakes. When this social descent began it was seldom, if ever, reversed, but went inexorably on—respectability taking itself a little further off, the sheer durability of such houses visiting the sins of their builders on the third and fourth generations of those that occupied them.

The more fundamental explanation of the slums, as of the distribution of all urban space before the era of zonal controls, rests, however, on the more basic commercial concept of supply and demand—of housing itself and of the capital and land which were its vital elements. Take first the dominant influence, the supply of housing at the centre. Here, as compared with the outer districts, were much stricter economic limits to an increase in house-room at low rents, because the alternative uses of the land were capable of bearing a heavier charge for it. Economizing on land by building high was moreover inherently more costly per unit of space provided. Private capital was therefore shy of building flats on central sites for the working classes, and the working classes themselves had a horror of anything so undomesticated as a street standing on end. There was no real possibility of enlarging the housing capacity of the central districts. Indeed, it was impossible even to maintain it. The conversion of houses into offices and warehouses within the City alone began to force its total population down after 1861.[13] But the most draconian changes felt all over central London arose, as they had always done, in making way for a greater traffic in merchandise and fare-paying passengers, whether by sea, road, or rail; and the docks, street improvements, and railways being built in this period set off a whole series of detonations which could be felt not only on the spot but also in long chains of reactions which reached even to the suburbs.[14]

The connections between all this activity and the supply of living room became a commonplace among common people even though the scale of operations was never—nor perhaps ever can be—measured with real accuracy. In the history of the slums of Victorian London these 'improvements'—to use the generic term—had a special irony. They had always been hailed as the means of clearing the slums, though they had hardly ever failed to aggravate them, for their effect always was to reduce the supply of working-class housing, either absolutely or in terms of the kind of houses which those turned out of doors by their operations could afford or wish to occupy. 'Here was a continual pushing back, back, and down, down, of the poor', one slum clergyman told his lecture audience in 1862, 'till they were forced into the very places which were already reeking with corruption.'[15] This had been so, long before the Victorian period. It is possible to see, allied to the commercial zeal for wider streets and larger openings in the City since the eighteenth century, a restless opportunism for the demolition of the ugly, unhealthy, overcrowded—above all, commercially unjustifiable—bits of the ancient city.

Street improvement could even be justified on such grounds alone, for the 'perforation of every such nest', wrote one enthusiast for improvement in 1800, 'by carrying through the midst of it a free and open street with buildings suitable for the industrious and reputable orders of the people, would let in that *Eye* and observation which would effectually break up their combinations.'[16] Such a point of view was expressed literally over forty years later, when James Pennethorne pushed Victoria Street through the backstreets of Westminster along a line he chose solely for its effectiveness in puncturing Pye Street and its neighbouring slums, known locally as Devil's Acre. Not every pair of dividers that stepped across the map of London in this long age of street improvement between the 1780s and 1840s was guided solely by the topography of the slums: the Temple Bar improvers seem to have been practically oblivious of them; John Nash deliberately skirted them with his Regent Street; and to the arch-improver, James Elmes, the rightness of the line was settled primarily by reference to architectural principles. Yet by 1840 no stronger supporting claim for a scheme could be made than that it improved an unhealthy district; of marginally slighter importance only was the aim of the 'melioration of the moral conditions of the labouring classes closely congregated in such districts'.[17] Scarcely a scheme of street improvement in London failed to respond in some degree to this call of duty, or expediency, over the next sixty years. 'Courts, close, crooked, and ill-looking are in plenty; swarms of children . . . costermongers . . . the low night lodging-houses and ugly dens of Golden Lane,' was how one man saw such an opportunity as late as 1877. 'Let a good, broad, straight thoroughfare be cut through it from North to South: the whole district is at once opened up and plenty of elbow room given to the City.'[18] It was a deepening and pitiless irony that the 'moral condition' of those affected tended to deteriorate in processes like these.

London was being dug and re-dug with restless thoroughness by railway navvies, too, and would have been even more trenched by them, especially during the 1860s, if the struggle between the railway companies had not cancelled out some of the schemes. It was above all the railways' demand for land to carry their lines into or under the central districts that had the most direct effect on the dwindling supply of living room there, especially as they had, like the street improvers, every financial reason for choosing the poorest districts for their lines wherever they could. Until the 1870s or even the 1880s, their demolitions tended to evoke a mixture of barely modified censure and almost unstinted praise, except on the part of those directly affected, or their few champions. Scores of writers described what happened as the railways burrowed their way into the centre and swept aside whole neighbourhoods of densely packed houses: some were merely stupefied by engineering marvels or the new vandalism; some reflected on their unerring aim at working-class districts; some thankfully remarked on the supposed benefits of destroying slums so efficiently; some tried to tot up the real costs of operations which defied all normal accounting methods. Watching the trains come by became an almost routine assignment for journalists, ticking off the days 'to get out', picking their way over the rubble, counting the houses that had to come down, following in the footsteps of those turned out of doors to see where they had gone, estimating the further overcrowding of the already overcrowded. From the fifties to the seventies the press was full of this kind of thing.[19]

It is almost impossible to believe that the repercussions of all this random slum-clearance were not recognized at the time. 'Carts of refuse turn down one street and dirty families another,' explained one observer in the early 1860s in a book suggestively entitled *The Hovel and the Home*, 'the one to some chasm where rubbish may be shot, the others to some courts or fallen streets, making them worse than they were before.'[20] The notion that slummers turned out of doors could take flight for the suburbs, or that the shock of flattening acres of crowded houses could be absorbed by the surrounding areas without difficulty, was contradicted by brute facts so often, in every kind of newspaper, periodical, pamphlet, and public utterance, that one is driven to conclude that there was some deliberation in this kind of permissiveness. William Acton wrote: 'This packing of the lower classes is clearly not yet under control.'[21]

That is clear from the more or less bland acceptance, by select committees on railway bills and street improvement schemes before the late 1870s, of easy assurances by the promoters that no difficulty would be encountered by those displaced and of arguments which hinged on the facility of taking slum houses rather than factories or warehouses. At the back of this lay an over-tender but sharply legal regard for property compulsorily acquired, and a readiness of the courts to settle handsomely for property owners. Thus their lordships were told by the Metropolitan Board of Works' Superintending Architect, when examining the Bill for Southwark Street in 1857, that the course given to it enabled them to skirt several expensive properties (including Barclay Perkins's Brewery—'That was a property which it was desirable to avoid') and save £200,000, while displacing fourteen hundred people living in the slums of Christchurch and St Saviour's. 'No inconvenience is anticipated and the Bill does not contain any provisions,' ran the written statement. 'The surrounding district is much peopled with workmen engaged in the different factories and the Artizans are migratory and there is great accommodation for Artizans in the Borough—but the houses are dense and mostly too crowded for health, comfort and convenience.'[22]

Demolitions for docks, railways, and new streets added immeasurably to the slums that were spared, and exacerbated a problem they were powerless to solve without the most elaborate rehousing. The complete failure to do this was the prime reason why the West End Improvements—designed to carry Charing Cross Road and Shaftesbury Avenue, among other streets, through some of the worst slums at the centre—should nearly have been frustrated by a parliament which was coming to realize the need to rehouse those displaced. This was the first major scheme of its kind, outside the City, in which a proper attempt was made to prevent the multiplication of slums by the provision of alternative housing for those displaced, or at least for others who could afford the higher rents.[23] The failure even here to give permanent homes to the lowest grade of slummers displaced was also the major defect in the activities of the numerous charitable bodies which laboured to increase the supply of working-class housing in or near former slums. Their activities, puny as they were in relation to the problem, scarcely touched the principal classes involved.[24]

The other side to these market forces was the working man's inability to transform his need for living room at the centre into an effective demand for it. He could not outbid his employers in the rent they were prepared to pay for their premises where he went to work, nor could he follow them into the suburbs where they went to live. The reasons were almost entirely economic. Though we know far too little about family budgets, it is possible to relate slum or slum-prone housing to low and irregular incomes in a very general way. It is clear, for example, from a survey taken by the Registrar-General of about thirty thousand working men in different parts of London in March 1887,[25] that the scantiest living accommodation was occupied by the families of the lowest-paid and most irregularly-employed men and that the amount of living room increased roughly in step with money wages and security of employment: half the dock labourers, at one extreme, occupied a single room or part of one, while only 1 percent of policemen occupied less than two rooms. As the proportion of men who were married did not vary significantly between occupational groups—the mean was 82 percent—and the size of families appears to have been pretty constant, it seems reasonable to connect type of employment directly with room density. It is also fairly clear that, in general, the less a man earned the lower the rent he paid, the two sums bearing a rather surprisingly fixed relationship to each other, with rent coming out at around one-fifth of the income of the head of the family.[26] The picture is made still sharper by taking account of liability to unemployment in different groups. Very roughly, one may say that at one extreme (St George's-in-the-East) was a situation in which almost half the working men and their families in a whole parish occupied single rooms or less, with over a third out of work and over a quarter earning less than nineteen shillings a week; at the other extreme (Battersea) less than a fifth were unemployed, and around two-thirds occupied three rooms or more and earned over twenty-five shillings a week. This suggests that crowded living conditions were related to the general structure of the labour market.

So long as employment remained on a casual basis, with the number of available jobs fluctuating violently from day to day or hour to hour, not only for unskilled but also for some of the most skilled trades, working men were obliged to live within reasonably close walking distance of their work. The distance considered practicable varied between trades and might stretch to three or four miles, but there are many signs that working men often felt chained more closely to their workplaces than that.[27]

> I am a working man [explained[28] one factory worker, who had a regular job].
> I go to my factory every morning at six, and I leave it every night at the same
> hour. I require, on the average, eight hours' sleep, which leaves four hours for
> recreation and improvement. I have lived at many places in the outskirts,
> according as my work has shifted, but generally I find myself at Mile End.
> I always live near the factory where I work, and so do all my mates, no matter
> how small, dirty, and dear the houses may be . . . One or two of my
> uncles have tried the plan of living a few miles out, and walking to business
> in the morning, like the clerks do in the city. It don't do—I suppose because
> they have not been used to it from boys; perhaps, because walking
> exercises at five in the morning don't suit men who are hard at work with their
> bodies all day. As to railways and omnibuses, they cost money, and we
> don't understand them, except on holidays, when we have got our best
> clothes on.

The circle of knowledge of what work was going was in some trades a narrow one, and the prospect of making a journey for, rather than to, work shortened the commuting radius. Sometimes it was vital to be literally on call. In the docks or some of the East End trades, the connection between the worker's home and workplace sometimes had to be more intimate still, and it was in these circumstances of sweated labour and of insecure and poorly-paid employment that creeping congestion either made a district ready for the complete descent into slum or more indelibly confirmed a condition that had already been sketched in. 'A slum', in a word, 'represents the presence of a market for local, casual labour.'[29]

The circle that closed over so much of the labouring mass was a spring for the middle classes. The wealth that was created in the commercial metropolis benefited them first, and they used it quite literally to put a distance between themselves and the workers. Whatever proportion of their earnings remained in their hands, once the capital requirements of commercial enterprise itself were satisfied, was invested in the manufacture of the city itself. Augmented by a certain amount of capital drawn in from outside, these resources were committed to a kind of self-generating expansion, a re-investment—to use economists' language—in the social overhead capital that was needed not only for the conduct of the enterprise City of London Ltd, nor even for so many safe-as-houses additions to personal portfolios, but also for making available to themselves those suburban parts of the city in which it was now thought desirable to live. Here was one of the most beautiful parts of the metropolitan mechanism. There were the business openings themselves: turning fields into streets brought large speculative gains to nimble dealers in land; keeping the suburbs supplied with building materials, fuel, and provisions was full of promise for the railways and the new multiple stores; transporting the commuters lifted the ceiling for street transport as well; buying and selling the property was endlessly rewarding for solicitors, auctioneers, banks, and building societies. It was for many a bonanza. These things apart, the middle-class suburb was an ecological marvel. It gave access to the cheapest land in the city to those having most security of employment and leisure to afford the time and money spent travelling up and down; it offered an arena for the manipulation of social distinctions to those most conscious of their possibilities and most adept at turning them into shapes on the ground; it kept the threat of rapid social change beyond the horizon of those least able to accept its negative as well as its positive advantages.

Within the analysis we are offering here, these suburbs were above all the strategic component in the housing of the whole urban community. It was the pace of their development and the amount of capital resources they consumed which determined not only the general scale of provision that could be made for the housing of the working classes but also the actual dimensions of their houses. In this sense the slums were built in the suburbs, and some of them actually were. These financial and logistical influences required their own kind of infrastructure—a new social order capable of transmitting by imitation the habits and tastes of the middle classes through their intervening layers to the upper strata of the working classes so as to form a continuum—in a word, suburbia. This had become a geographical reality by the last quarter of the century and the economic and social processes that sustained it have never lost their momentum. We must recognize that beckoning the middle classes and their imitators in their flight to the suburbs were images of many kinds. We are not disposed to succumb to their siren calls here, but it is important to see how beguiling they were.[30] The 'suburban quality' they sought, according to Henry

James, was 'the mingling of density and rurality, the ivy-covered brick walls, the riverside holiday-making, the old royal seats at an easy drive, the little open-windowed inns, where the charm of rural seclusion seems to merge itself in that of proximity to the city market.'[31] For what filled their sails may well have been trade winds of irresistible reality, but what took hold of the helm were dreams of aspiration and even romance. It was the Englishman's practicality that found in the suburbs the solution to the essentially middle-class problem of escaping the snares of the city without losing control of it. It was his romantic idyll of pastoral bliss that wove in and out of all his plans for taking to the suburbs.

It was not merely that the central areas were becoming so clogged by commerce and infested by slums. These things assailed his sight, his touch, his smell. They undermined his health and his property. But what also became fixed in his mind was the realization that density itself spelt death and depravity on his own doorstep. The convulsions of the city became symbolic of evil tendencies in the best and in the worst writing of the day, while the serenity of the suburbs became a token of natural harmony. The undrained clay beneath the slums oozed with cesspits and sweated with fever; the gravelly heights of the suburbs were dotted with springs and bloomed with health. It did not greatly matter to the individual that neither the slum nor the suburb conformed in every case to its stereotype. What mattered to him was that here was his own way out of the urban mess, a protection for his family, a refreshment for his senses, a balmy oasis in which to build his castle on the ground.[32]

To return, however, to our own ground, we can see now that the middle-class suburb was both an invention for accentuating and even refining class distinctions, and a means of putting off for a generation or two the full realization of what was entailed in living in a slum. C. F. G. Masterman knew and shared these middle-class feelings, and his powerful imagery of the nether world across which the middle classes were carried daily on their railway viaducts towards the heart of the commercial metropolis conveys this well. The image of the working classes storming up the garden path was no more of a caricature to them than it had been, half a century before, to Ruskin, horrified at the cockneys tearing down the apple blossom and bawling at the cows.[33] To a society in which landed wealth, drawn though it increasingly was from urban revenues, was still underpinning great embankments of privilege and political leverage, as well as the smaller domains of parochial aristocracies, the slum people were like a sleeping giant. Reports of its size or its hideous appearance—even cautiously conducted inspections on foot—were as entertaining as a menagerie.[34] But when the thing stirred or broke loose it was as if those watching had seen a ghost. 'We are striving to readjust our stable ideas', wrote Masterman after one of these brief gestures of power by the multitude. 'But within there is a cloud on men's minds, a half stifled recognition of the presence of a new force hitherto unreckoned; the creeping into conscious existence of the quaint and innumerable populations bred in the Abyss.' That was written in 1902.[35] It was a premonition that the great days of the middle-class suburb were numbered. There were men in their fifties in certain suburbs who could not remember the time when workmen's tickets were not in use, and for twenty years this traffic had been planting new colonies of working-class commuters—at the behest of the Board of Trade if need be —around scores of suburban stations.[36] What had begun as a fugitive solution to the problem of rehousing slummers, turned out of doors where costs of new accommodation were prohibitive, was fast becoming by common consent the urban ethos of the twentieth century. It was now the turn of the suburbs to lose their immutability and for distance to lend less enchantment. To flight had been added pursuit.[37]

The means of making this mass exodus was a belated and perhaps dubious gift of the commercial system itself. Taine's native wits told him that the suburban trend of the 1860s implied 'large profits from quick turnover, an opulent free-spending middle class very different from our own'.[38] An earlier estimate had put commuters' incomes in 1837 at between £150 and £600 a year—'provided their business did not require their presence till 9 or 10 in the morning'.[39] What brought such giddy expectations nearer for the masses was not cut-price fares alone but advancing real incomes arising more than anything from the great fall in world commodity prices in the last quarter of the century, and by shorter working hours and securer jobs.[40] The employment of clerks, for example, rose geometrically in London after mid-century, and they went to live in their favoured inner suburbs—22,000 of them, one in eight of all London's clerks, were to be found in Camberwell, for instance, by 1901, when around 5 percent of London jobs were in offices of some kind.[41]

It should not be overlooked, of course, that neither these plebeian places nor their more patrician counterparts were inhabited solely by commuting workers. Once established, these suburbs put on natural growth—perhaps accounting for up to a third of their overall expansion in the second half of the century—and they created employment locally mainly in the service trades. The longer the lines of suburban communications were stretched the more transport and distributive workers were needed to keep them open. This may help to explain how it was that by the 1860s outer suburbs beyond the reach of buses were growing markedly more quickly than their railway commuting services.[42] We might also add that when this tendency began to be augmented at all substantially by a growing volume of visitors to London on business or pleasure—something plainly happening from the 1850s—it helped to create a disproportionately large number of relatively badly-paid jobs in London as a whole, and this reinforced the inherently depressing forces making the inner ring of places into potential slum.

It occasionally happened that this convoluted influence of the suburb on the structure of the city, and in particular on the slum, uncoiled itself as if to demonstrate beyond a doubt what was going on. Look for a moment at North Kensington, a district lying perhaps more off the line of fashionable progress beyond the West End than on it, but containing some of the more sublime as well as some of the more ridiculously ambitious suburban neighbourhoods to be found among the developments of the 1840s, 1850s and 1860s.[43] Among these were some where their builders had badly overreached themselves and raised reasonably prepossessing houses for single families that never came, and which had had instead to be sub-divided immediately into tenements. Worst of all, as one of Dickens's reporters explained to his readers in *Household Words* in 1850, 'in a neighbourhood studded with elegant villas ... is a plague spot scarcely equalled for its insalubrity by any other in London: it is called the Potteries.'[44] Here was a custom-built slum of seven or eight acres, nurtured by expediency, occupied by people made homeless by improvements in the West End or in the process of shifting in or out of London altogether, and sustained by a kind of bilateral trade with its affluent neighbours.[45] Its topography included a septic lake covering an acre, open sewers and stagnant ditches galore, a puzzle of cul-de-sacs and impenetrable settlement (especially on its most fashionable quarter), and a heap of hovels numbering under two hundred by mid-century. It was exceptionally low-lying, but made more of a gully for surface water from the surrounding district by having been extensively dug for brick-clay for the mansions round about, and left unpaved and unmetalled. In 1851 the population was returned as 1,177; pigs outnumbered people three to one. It was a nauseous place, but from it came a substantial supply of good quality bacon raised on middle-class swill, a profitable flow of rents for middle-class landlords, and a large pool of adult and juvenile labour for middle-class households—domestic servants, cleaners, wet-nurses, prostitutes, laundry-women, needlewomen, gardeners, night-soil men, chimney-sweeps, odd-job men, and builders. The return flow had to be reckoned not only in wages and charitable coppers but in things pilfered or pockets pinched. It was an impressive trade. Nor was it an isolated one. Agar Town in St Pancras and Sultan Street in Camberwell are similarly documented, and it is hard to believe that this uneven enterprise between suburb and slum was not being carried on in some degree in places barely known to historians even by name.

We have been speaking so far as if these migratory movements into the suburbs and the slums were taking place within closed frontiers, but it is clear that they were not. London, we noted earlier, was drawing off people from the rest of the country as well as redistributing them itself. The question naturally arises, where did these provincial migrants to London go? Were the inhabitants of the slums provincial in origin, drawn in perhaps at different times from overmanned farms or from underdeveloped or technologically obsolescent industrial towns? The evidence for the 1870s—to take a decade of high growth—does not suggest this. It is not possible to discover, by looking merely at the numbers enumerated at a particular census, which of those born elsewhere had arrived in the last ten years, nor is it possible to discover how many came and went. However, what the birthplaces of those enumerated on census night in April 1881 suggest is that, at a time when not much over one-third of London's population had been born elsewhere, the movement into London was producing in the most rapidly expanding suburbs a larger proportion of provincials than were to be found in districts nearer the centre. Mayfair was an exception in drawing practically 60 percent of its population from outside London. Bethnal Green, by contrast, which had been three parts slum when Hector Gavin had rambled over it in the late 1840s,[46] and was by now one of the most extensive congeries of slum in London, contained little more than 12 percent who had been born outside London, and the whole area of Whitechapel and St George's-in-the-East surrounding it did not raise the figure above 20 percent; Seven Dials itself, plumb centre, had less than half as many inhabitants born outside London as had the most affluent parts of the same West End.[47]

This general pattern corresponds very closely with that prevailing in the Notting Dale Potteries as compared with the surrounding parts of Kensington. If we look simply at the birthplaces of heads of households in 1861, we see that just over half of them were born in London, whereas less than one-quarter of those in the entire district were—a disparity that appears to have increased as the area was being settled in the course of the preceding decade. Similarly, if we look simply at servant-keeping households for the same date in two long-developed districts of Bayswater—a very affluent quarter—we find the corresponding figures are 28 percent and 35 percent; whereas if we do the same for a district of quite new and less affluent settlement—to be quite precise, having exactly half the servant-keeping establishments (1·9 domestic servants per household as compared with 3·6 in Bayswater)—what we see is that the proportion of London-born was just over half.[48] By this date the proportion of London-born in London as a whole was dwindling below the half-and-half division of 1851. Charles Booth's calculations for the 1890s suggest that there was here an inverse ratio between the proportion of provincial immigrants and the poverty of a district, and if we look finally at the proportions of cockneys to provincials in what he regarded as one of the vilest slums he knew in London—Sultan Street in Camberwell—we can see just how true this is.

The birthplaces of the heads of the 104 households and their wives enumerated there in 1871, when the whole area had just started its social descent, divided in the proportion of six Londoners to four provincials and Irish, a ratio which widened considerably over the next three censuses: in 1881, 36 percent of heads and wives were born outside London; by 1901 the figure had fallen as low as 26 percent. Interestingly enough, the ratio of London-born in Sultan Street increased very markedly during this period as compared with that in Camberwell as a whole, even though by 1901 this larger community was showing signs of deterioration in social class and, along with this, an increase in the proportion of Londoners to the rest. That only 8 percent of the children enumerated in Sultan Street in 1871 should have been born outside London (and less than 2 percent in 1901) supports the evidence of the more general statistics that the slums of Victorian London were mostly occupied by second or later generation Londoners, and that the suburbs were the ultimate destinations of the incoming provincials. The slums of Victorian London are therefore more properly thought of as settlement tanks for submerged Londoners than as settlement areas for provincial immigrants to the city.[49]

We get here another glimpse of that centripetal and downward movement in the development of Victorian London that needs to be set in relation to the more familiar one of the centrifugal deployment of another instalment of population into the suburbs. Here were two social gradients that sometimes intersected and sometimes even changed direction for a time—it all depends, as it commonly does, on where one

The Victorian City: Shapes on the Ground

is standing and how far one can see, and especially on when one is looking. In as general terms as we dare use, what we see is one such slope leading upwards and outwards, and the other leading downwards, if not inwards. The cultural slopes of Victorian London, which must be mapped one day more carefully than anyone has yet attempted, made no doubt an undulating plain of dull mediocrity—which is perhaps the natural condition of human society—relieved only occasionally by peaks or precipices of dramatic dimensions.[50] Moving up and down these slopes was always an exertion of some kind, whether to cut a better figure or to tap fresh credit, to escape the rent-collector, or even the police. How it all happened on the way up we know well enough: Wilkie Collins, Chesterton, Galsworthy, the Grossmiths, Thackeray, Trollope, Wells, describe some of the traverses and ascents; and Keble Howard, Pett Ridge, Shan Bullock, and unsung authors like Mrs Braddon and William Black, whose tedious novels were set in the suburbs because they were designed to be read there, echo some of the chatter that took place on the way.[51] Fewer people—Arthur Morrison, Israel Zangwill, Walter Besant, Gissing, and Kipling certainly among them—have told how it was on the way down or at the bottom.[52] Before we take a last look in a moment at the suburbs and the commercial ascendancy to be found there, it is worth measuring as dispassionately as we can the human ingredients of a slum as its slumminess intensified.

We must return to Sultan Street. Here was a street of around seventy six-roomed stock brick houses arranged on three floors, which had been virtually completed between 1868 and 1871. It was built, with an adjoining street, on a small plot of cow-pasture which for forty years had been completely enclosed by a heap of clap-board cottages on two sides, a row of modest villas on another, and by the long back-gardens of a decent Georgian terrace on the fourth. What gave to this ineligible building land still greater insularity was a railway viaduct, pierced by two bridges. The Herne Hill & City Branch of the London Chatham & Dover Railway was taken clear across the back-gardens of the terrace between 1860 and 1864 as part of the larger strategy that that company was using to get the better of its rival, the South-Eastern Railway, in its drive for the Continental traffic to Dover. The building of Sultan Street was therefore specially speculative, as nothing makes plainer than the speed with which the improved ground-rents were passed from hand to hand. The local influences at work on this place correspond closely with what was said earlier. Cowsheds and piggeries squeezed up with the surrounding houses, and a glue factory, a linoleum factory, a brewery, haddock-smokers, tallow-melters, costermongers keeping their good stuff indoors with them while leaving rotting cabbage-stalks, bad oranges, and the like on the street, created between them an atmosphere which, mingled with household odours, kept all but the locals at bay.[53]

Sultan Street was *not* badly built, but some of its houses became slums almost at once and the rest followed inexorably. By 1871 these seventy or so houses were packed out by 661 persons and almost a quarter of them were being made to hold between thirteen and eighteen occupants each, or more than two to a room. However, just over half of all the houses contained, in 1871, between seven and twelve occupants apiece, a situation which could not be described as one of overcrowding. In ten years the whole spectrum had shifted. The total number living in the street had grown by more than half as much again (57·4 percent) to 1,038, over half of whom were now to be found thirteen to eighteen per house, while a further 14 percent were living nineteen or more per house; only one-third of the inhabitants of the street were living, on the average, less than two to a room. The net effect of what had happened was that another forty-three families (197 persons) had come in and lifted the mean figure of persons per house from about nine in 1871 to fifteen in 1881. How many families had come and gone in the interval there is no way of telling precisely, but it seems clear enough from a statistical analysis of age-patterns at successive censuses that there was a big turnover in inhabitants, a process that produced by the 1890s a community that was distinctly older than it would have been if the younger people who had gone there at the start had stayed: the tendency was for older people to move in as the houses themselves aged.

As this was happening, the structure of households was changing appreciably. One noticeable feature was that relatives and lodgers were disappearing from the street. That lodgers should be declining is probably more revealing than that relatives should be growing scarcer, because families too poor to put up a nephew or a parent might be supposed to have taken in a lodger, provided room could be found: few indexes of poverty and overcrowding could conceivably be more significant than the inability to sub-let even sleeping-room. In 1871 the proportion of households with lodgers, almost 15 percent, had been practically the same as the average for Camberwell as a whole, but by 1881 it had fallen to just over 4 percent. By this date the whole pattern of Sultan Street society was beginning to diverge with statistical significance from that of the surrounding population in Camberwell as a whole. Three features make this plain: the marital status of the heads of households; the size of families; the occupational and industrial distribution of wage-earners. In all three respects Sultan Street society had in 1871 corresponded quite closely with that of Camberwell as a whole. By 1881, and still more markedly by 1891, there were signs that it was tending to become more matriarchal: relatively more widows, more married women whose husbands were 'away', and more single women as heads of households. So far as the size of both the household and the family are concerned, there was again virtually no difference between Sultan Street and Camberwell in 1871. By 1881, the average size of household had begun to diverge appreciably and before long the same was true of the size of families: both were larger in Sultan Street. Rather interestingly, although there seem to have been no significant differences in the number of children to be found in families of the different social groups in Sultan Street or outside it, in 1871, both skilled and unskilled workers in Sultan Street in 1901 had appreciably more children than the average for their class in Camberwell as a whole. The street seemed to be somewhat deficient in fathers and abundant in children, a situation which was liable to be open to only one interpretation. So far as jobs were concerned, the Sultan Street community already had many more labourers and domestic servants in 1871 than were to be found in Camberwell as a whole

(80 percent against 44 percent), though the distribution between skilled and unskilled scarcely varied. It was this distribution which now proceeded to vary: the proportion of skilled to all labourers in Camberwell fell only slightly (62 percent to 54 percent between 1871–1901), but in Sultan Street it was halved in these thirty years (62 percent, 46 percent, 29 percent, 32 percent); actually, the disparity was at its greatest between 1881 and 1891, for Sultan Street was slightly redeemed in the nineties, whereas Camberwell at large deteriorated a little more rapidly than it had become accustomed to do. There were by now enough other slums in it to offset the effect of its middle-class residents in keeping its average status high. Across the northern reaches of Camberwell, at least, the menace of more general social decline was beginning to show. The suburb that had at the beginning of the nineteenth century contained 'few poor inhabitants and not many overgrown fortunes'[54] contained too few fortunes of any size and too many poor by its end. For it, as for other suburbs on the original frontier of London's expansion, the tide of middle-class settlement had rolled on. The erstwhile suburb had to take increasing care not to become a slum.

The dynamics of urban growth by suburban accretion had just such an irresistible momentum. The spread of the built-up area gave every appearance on its leading edge of an unstoppable lava flow; the encrustations that eventually cooled behind it seemed to take on the dull unworkability of pumice. The inhabitants of large cities begin to accept as part of the ground of their being the obduracy of these forces and see them almost as urban nature's way. And when we look at the suburban flux in the full heat of its making, there do seem to be relentless pressures demanding a release. It was there that money not only talked but also lived and moved almost under its own compulsion. The supply of middle-class housing in the suburbs was not simply a reflex of the demand for somewhere to live. It was the active quest for its own outlets that kept the supply of capital on the move. This was not the operation of some anonymous, inscrutable financial wizard who knew some special incantations. Anyone with a bit to spare could open sesame. Lending on mortgage was the passive 5-percent way of taking part; going into the land market, getting some suburban land ripe for development, making it go, was the active, speculative, all-or-nothing way of doing so.[55] 'The formation of ground rents', one auctioneer's notice blandly proclaimed in 1856, 'has been the study and occupation of many of the most intelligent men of the day, and is accomplished by the purchase of freehold land, to be let on building leases.'[56] To this should be added innumerable supporting enterprises which were also creating their own demand, adding their weight to the engine of suburban economic growth: drainage schemes, gas and water undertakings, shops, schools, pubs, music-halls, parks, bus and jobmaster concerns, tramway and railway companies—the whole intricate web of agencies without which a suburb would have been a castle in the air.

The opportunities for this speculative enterprise were too numerous and too diverse, and their interpenetrations too tangled, to be described in any detail here. Nor can we draw up a balance sheet which would show their debit and credit sides. What mattered most was the *expectation* of gain. The tide ran too fast for the experience of loss to impede the flow. In fact, many were disappointed. Main-line railway companies in particular, compelled by their shareholders to get a share of the suburban market, had many regrets over their passenger accounts when they understood the economics of the rush-hour better.[57] Most of the money to be made out of suburban enterprise went to the men who dealt in land, who were first on the scene, who leased or bought before the rise, who developed or sold it on the very top of the tide. What evidence we have about land values in the Victorian city suggests that land prices in the outer suburban districts appreciated by ten- to twenty-fold in the thirty years after 1840.[58] In this the railway played the role of inflating land values *en route* and offering windfalls to those quick or knowing enough to pick them up while still ripe. The Engineer of the Acton & Hammersmith Railway remarked in 1874: 'The moment a line is deposited and there appears a chance of carrying on, the speculative builders of London all rush to the ground to cover it with houses. Many cases have arisen where the Act has not gone through one committee before the builders were on the spot and commencing to sell.'[59] Those who courted ruin either moved in too soon or arrived too late, and occupied some salient easily outflanked by more desirably placed estates. These were the men who committed themselves to unsound schemes in the belief that they would share what others were known to have reaped, who kept doggedly on when their returns failed to rise, or tried to bolster the market by taking shares in railway companies when all their capital was locked up in bricks and mortar.[60]

Such men did not lose their nerve; nor did they normally run out of capital. By failing to sell their houses they fell short on the payment of interest on their loans and their creditors foreclosed on their security—the houses that had dragged their mortgagees down. There were too many undertakings, and too much was expected of them individually in the short run. In terms somewhat alien to the times, such investments looked too much for growth and too little for income. In the long run they made collectively very handsome holdings indeed, but the whole atmosphere in which they were created was charged with speculation and the operations themselves were hobbled by their scale. The supply of capital was never seriously depleted by other demands being made upon it—partly because the money market was more specialized than is often supposed—and even the counter-attractions of overseas investment merely had the effect of making funds more readily available when they were not going abroad, not of restricting them when they did.[61] The most palpable consequence of this in the suburbs was the sight from time to time of new untenanted streets of houses, a temporary fall in house prices, and an increase in bankruptcies among builders. The long-term trends in population growth and the supply of houses in London as a whole, on the contrary, scarcely wavered or diverged. All that these marginal alternations in the supply of capital did in the suburbs was to fix the reckoning day and to award the prizes. The estates that suited the tastes and the pockets of their middle-class adjudicators went on; those that did not were put on one side for more vulgar approval later on; and the housing industry re-formed its battalions for another campaign.[62]

The bowler-hatted field-marshals commanding these operations were the speculative builders—if by this term is meant not only those directly involved in erecting houses but also anyone capable of remaining solvent long enough to raise the carcass of a house and to find a tenant or a buyer for it. These were functions as readily discharged by men totally inexperienced in building as by men that were. The Forsytes' fortunes were based reputably enough on the success of a mason from Dorset, but there were speculative builders who had been tailors, shopkeepers, domestic servants, publicans: one District Surveyor complained in 1877 of a clergyman-cum-speculative builder whom he remembered only too well because he had built several hundred houses and left the district without paying him his fees.[63] The big men were veritable entrepreneurs as much at home among the scaffolding as in the chambers of money-lending solicitors, men who were not only scrupulous businessmen but discreet vulgarizers of the fine arts, whose work was the stuff of dreams that would not fade, and whose plumbing was impregnable. But there were also others, not necessarily dishonest or incompetent men, whose business methods precluded such attention to detail or a guarantee of the best materials being used—men more skilled very often in raising money than raising houses. In any rapidly growing suburb of size there would be scores of such builders rubbing along on precarious credit from timber-merchants or brickmakers (something harder to obtain, incidentally, from the 1850s), mortgaging their houses floor by floor or pair by pair or terrace by terrace to a building society, a solicitor, or even to a bank. If they were lucky they would survive with virtually no stock-in-trade nor capital of their own, and skip from one take to another by finding buyers for their houses at the eleventh hour and transferring to their clients the mortgage they had raised on the house when the roof went on, in time to finance the first stages in the next operation. There were in this category some speculative builders who were little more than hirelings of the ground-landlord or the developer who had first laid out the estate, and who came almost inevitably and often very quickly into financial servitude to more powerful men. There were also builders' merchants and building societies which had to step in to complete a job on which they had been forced to foreclose. It was the kind of enterprise which tempted many men, not least the long-established family concerns of jobbing builders—father and son, brothers and brothers-in-law—who, when new houses did not seem to hang on the market above a week or two, could be seen measuring the leap into full-scale speculative building, and taking it.

These suburban outlets for capital were like a delta of a great river system, made fertile by deposits collected in distant places and carried along irreversibly by superior force. Landowners of London estates made advances in cash and kind—the prevailing leasehold system itself was but the means of transferring capital assets on credit—and were doubtless as ready to tap the resources of their country estates as to return them the proceeds of their urban ones. Cash left on trust in the country made its way into solicitors' hands in London, and professional men in particular knew such channels well. The intermediaries themselves also became the principals, and the legal system—despite its obscurity in the annals of economic history—became the axis for many of these movements of capital, especially after 1872, when advances to speculative builders by building societies began to flag. Thus the savings of the farmers and retired gentlefolk of Chippenham were channelled during the 1880s into the making of a seedy suburb in Paddington, and in 1874 some of Mr Speaker's money was combined in Lincoln's Inn with that of a reverend gentleman in Southampton to finance the building of a terrace in Walworth.[64] The most conservative of City institutions, the Royal Exchange Assurance, and several other insurance companies, along with two leading London banks, engaged in this disposal of loanable funds too, doing so in such a way as to accentuate the uncertainties of speculative building and, arguably, to make the employment of this capital more wasteful.[65] The standard instrument for all these transfers was the mortgage deed, the main outlet for surplus funds in the nineteenth century, but it was used to special effect by the building societies. Whether as channels for funds for occupying owners or for speculative builders, these highly parochial institutions performed the prime function of releasing the capital resources of one suburb for the development of a neighbouring one, just as the Church building societies were the means of transforming that wealth into places for prayer apart.[66] The crucial role of all these institutions was to convert the savings of the suburban communities into funds for re-investment in their own structure, into a hoopsnake of incomes and satisfactions that recycled what was earned in the tangible means of enjoying it.

But what of the slums? The mechanism that activated and populated the suburbs did not function for them in the same way. In the suburbs, a correspondent to the *Builder* was commenting as early as 1848, it was[67]

> as though one-half of the world were on the lookout for investments, and the other half continually in search of eligible family residences ... There is a leaven of aristocracy in the parlour with folding-doors ... The villa mania is everywhere most obtrusive ... But the poor want dwelling-places. Whilst we are exhausting our ingenuity to supply our villas with 'every possible convenience,' we are leaving our working-classes to the enjoyment of every possible inconvenience, in wretched stalls to which men of substance would not consign their beasts of burden.

It is indeed an interesting question whether the flow of private capital into suburban house-building has not always tended to be at the expense of investment in lower-grade housing, unless moderated by public subsidies for working-class housing. The whole ethos of the 5-percent philanthropy idea which was developed in the 1850s and 1860s to get private capital for this type of investment really evolved from a situation in which the returns on suburban house-building were not only setting the pace but also making the *idea* of securing comparable returns on housing the lower working classes, as distinct from the 'aristocracy', largely nugatory.[68]

The Victorian City: Shapes on the Ground

Worse than this, it cannot be taken as an open question whether the people of the slums were not actually contributing to the environment of the suburbs. The lubrications of the money economy had made even these rusty parts revolve. 'There are courts and alleys innumerable called by the significant name of RENTS,' explained one sociologist, in all but name, of the 1840s. '... they are not human habitations ... They are merely so many man-traps to catch the paying animal in; they are machines for manufacturing rent.'[69] It was profoundly true. The urban landlord already appeared to have become a kind of dinosaur across the water who took his tribute in flesh and blood. Whereas in the old pre-urban days, when the rent-receiver was often visible as employer or social better within a tolerable frame, and when the rent-night gathering might sometimes almost be said to have been celebrated 'as if the thing, money, had not brought it there',[70] the relations between landlord and tenant in the city were characteristically impersonal, conducted through agents called rent-collectors, and the rent itself seldom bore any demonstrable connection with the human container itself. The money that passed was the sum necessary to deny the space on the ground to some other use, and it soon bore, in the central districts, little relationship to the amenities available in the shape enclosing it. One enthusiast for garden suburbs for the masses, long ahead of his time, calculated in 1846 that a four-roomed cottage could be provided for half the rent of six shillings a week paid by working men for two rooms in town.[71] Little wonder, perhaps, that these extractions were so painful. 'Absentee-landlordism, subleases, rack-rents,' wrote Henry Lazarus in a diatribe hurled at the landlords of London in 1892, 'here is the trinity of England's land curse.'[72] Between tenant and landlord often stretched a whole chain of shadowy intermediaries, held in their contracted order by a series of subleases which divided responsibilities for the upkeep of the property and inflated the rents paid for it.

The direction in which these financial obligations led could scarcely fail to be outwards towards the suburbs. And when George Bernard Shaw first tried his hand as a playwright it was to these calculated connections between slums and suburbs that he turned. 'In *Widowers' Houses*', he explained after it was put on in 1892, 'I have shewn middle class respectability and younger son gentility fattening on the poverty of the slum as flies fatten on filth.'[73] The rent-collector describes his client's business:

LICKCHEESE. ... I dont say he's the worst landlord in London: he couldnt be worse than some; but he's no better than the worst I ever had to do with. And, though I say it, Im better than the best collector he ever done business with. Ive screwed more and spent less on his properties than anyone would believe that knows what such properties are. I know my merits, Dr. Trench, and will speak for myself if no one else will.

COKANE. What description of properties? Houses?

LICKCHEESE. Tenement houses, let from week to week by the room or half room— aye, or quarter room. It pays when you know how to work it, sir.

	Nothing like it. It's been calculated on the cubic foot of space, sir, that you can get higher rents letting by the room than you can for a mansion in Park Lane.
TRENCH.	I hope Mr. Sartorious hasnt much of that sort of property, however it may pay.
LICKCHEESE.	He has nothing else, sir; and he shews his sense in it, too. Every few hundred pounds he could scrape together he bought old houses with—houses that you wouldnt hardly look at without holding your nose. He has em in St. Giles's: he has em in Marylebone: he has em in Bethnal Green. Just look how he lives himself, and youll see the good of it to him. He likes a low death-rate and a gravel soil for himself, he does. You come down with me to Robbins's Row; and I'll shew you a soil and a death-rate, so I will! And, mind you, it's me that makes it pay him so well. Catch him going down to collect his own rents! Not likely!
TRENCH.	Do you mean to say that all his property—all his means— come from this sort of thing?
LICKCHEESE.	Every penny of it, sir.

That scene was enacted in 'the library of a handsomely appointed villa at Surbiton'. It conveys, as the most meticulous property ledger perhaps never will, the underlying logic of what another earlier Victorian commentator described as the 'terrible physiology' of the map of London.[74] The movements of capital in the making of the metropolis determined the social space available to its different users, and hence the relative wealth or poverty of its different districts. This is not to say that the living conditions which were produced by these commercial discriminations were necessarily inferior to, or their limits more extreme than, what had gone before. What was different was the geographical scale and the fact that with virtually no curbs placed on the way in which capital was allowed to spend itself—beyond the rather narrow limits set by legislation for the public health—the poor were quite literally starved of such finance. The curse of these poor was no greater than the poverty of any other, but it produced more visible ironies. The wealthiest parishes of London had the lowest municipal rates.[75] It was difficult at first to reverse the capital flows we have been describing within the boundaries of any one of them. Kensington, the wealthiest of them all, for example, reacted to efforts to do so on behalf of the Potteries as a healthy body does to some foreign matter embedded in it—by trying to evict it.[76] The effort to redress such unbalances, to equalize the rates, to steer capital where it would not normally go, was what the municipal socialism that found its voice with the establishment of a single government for London in 1889 was all about.[77] It was about the need, among other things, to make transfer-payments between suburbs and slums, to give the community a conscience, to put together again the fragments into which it had been shattered by the impact of its own growth. It was the start of a road that would never end.

The Victorian City: Shapes on the Ground

Notes

1 This argument was originally advanced in a paper from which some of the material for this chapter has been drawn: see H. J. Dyos, 'The Slums of Victorian London', *Victorian Studies*, xi (1967–8), 5–40. It has benefited since from the comments of Professor Lionel Needleman, to whom we are grateful. His own work, *The Economics of Housing* (1965), deals more theoretically with the subject, especially in regard to the supply and demand factors influencing housing.
2 For a discussion of the relations between London and the rest of the country, see H. J. Dyos, 'Greater and Greater London: Notes on Metropolis and Provinces in the 19th and 20th Centuries', in *Britain and the Netherlands. IV: Metropolis, Dominion and Province*, eds J. S. Bromley and E. H. Kossmann (The Hague, 1972), pp. 89–112.
3 H. A. Shannon, 'Migration and the Growth of London, 1841–91', *Economic History Review*, v (1935), 79–86. These and other estimates of population growth given in this chapter have been calculated from the census returns.
4 For an account of this, see H. J. Dyos, 'The Growth of a Pre-Victorian Suburb: South London, 1580–1836', *Town Planning Review*, xxv (1954), 67–78.
5 Pierce Egan, *Life in London* (1821), pp. 274, 288, 343, 345–6.
6 To Daniel Maclise, 20 November 1840. For other contemporary definitions of the slum, see J. H. Vaux, *Flash Dictionary* (1812); Jon Bee, *A Dictionary of the Turf* (1823), p. 161; Henry Mayhew, *The Great World of London* (1856), p. 46; J. S. Farmer and W. E. Henley, *Slang and Its Analogues* (1890–1904), see 'slum'; *The Times*, 17 January 1845.
7 'It may be one house', wrote Robert Williams in *London Rookeries and Colliers' Slums* (1893), p. 13, 'but it generally is a cluster of houses, or of blocks of dwellings, not necessarily dilapidated, or badly drained, or old, but usually all this and small-roomed, and, further, so hemmed in by other houses, so wanting in light and air, and therefore cleanliness, as to be wholly unfit for human habitation.'
8 Quoted by Lewis Mumford, *The City in History* (1961), p. 433.
9 See, for example, Rev. Thomas Beames, *The Rookeries of London: Past, present, and prospective* (1851); Cardinal Wiseman, *An Appeal to the Reason and Good Feeling of the English People on the Subject of the Catholic Hierarchy* (1850), p. 30. Bayswater and Camberwell are dealt with below.
10 'Ragged London', *Meliora*, iv (1862), 300. There is a large bibliography of contemporary writings on the slums in Dyos, 'Slums of Victorian London'.
11 [Thomas Wright], *The Pinch of Poverty* (1892), p. 187.
12 The only available history of building regulations in London during this period is C. C. Knowles, 'A History of the London Building Acts, the District Surveyors, and their Association', an unpublished MS. dated 1947 in the Members' Library, Greater London Council, County Hall, Westminster Bridge.
13 Corporation of London, *The City of London: A record of destruction and survival* (1951), p. 165 ff.
14 The large displacements in the vicinity of the new Waterloo station in 1858–9 for the Charing Cross Railway were augmented by the actions of landowners seizing the opportunity of re-shuffling their tenantry in the vicinity. The only step-by-step

reconstruction of such a process is H. C. Binford, 'Residential Displacement by Railway Construction in North Lambeth, 1858–61' (unpublished M.A. thesis, University of Sussex, 1967).

15 Rev. G. W. M'Cree, *Day and Night in St. Giles* (Bishop Auckland, 1862), p. 6.
16 [C. G. Stonestreet], *Domestic Union, or London as it Should Be!!* (1800).
17 For a more extended account, see H. J. Dyos, 'Urban Transformation: the objects of street improvement in Regency and early Victorian London', *International Review of Social History (I.R.S.H.)*, ii (1957), 259–65.
18 Henry Chevassus, *Overcrowding in the City of London* (1877), p. 10.
19 See the references in H. J. Dyos, 'Railways and Housing in Victorian London', *Journal of Transport History*, ii (1955), 11–21, 90–100.
20 Ebenezer Clarke, Jr, *The Hovel and the Home; or, Improved dwellings for the labouring classes, and how to obtain them* (1863), p. 31.
21 William Acton, *Prostitution . . . in London and Other Large Cities* (1857), p. 180.
22 Quoted from the Demolition Statement submitted with its Bill by the M.B.W. under the provisions of a House of Lords Standing Order adopted in 1853 (House of Lords Record Office).
23 See Dyos, 'Urban Transformation', *I.R.S.H.*, 261.
24 See the references in J. N. Tarn, 'Housing in Urban Areas, 1890–1914' (unpublished Ph.D. thesis, University of Cambridge, 1961); C. J. Stewart, ed., *The Housing Question in London, 1855–1900* (1900); E. R. Dewsnup, *The Housing Problem in England: Its statistics, legislation and policy* (Manchester, 1907); J. S. Nettlefold, *Practical Housing* (Letchworth, 1908).
25 *Parliamentary Papers*, 1887, XV, Tabulation of the Statements made by Men living in Certain Selected Districts of London in March 1887 (C. 5228). The areas covered were the registration sub-districts of St George's-in-the-East, Battersea, Hackney, and Deptford. Despite the warning given that the details of the returns were 'of very small statistical value', there is a very general pattern discernible in them.
26 For a discussion of house rents in London in relation to the problems of housing the working classes, see A. S. Wohl, 'The Housing of the Working Classes in London, 1815–1914', in *The History of Working-Class Housing* (1971), ed. Stanley D. Chapman, pp. 15–54.
27 See E. J. Hobsbawm, 'The Nineteenth-Century London Labour Market', in *London: Aspects of change* (1964), Report No. 3, edited by the Centre for Urban Studies [University College, London], pp. 3–28.
28 John Hollingshead, *Today: Essays and miscellanies* (1865), II, p. 306.
29 B. F. C. Costelloe, 'The Housing Problem', *Transactions of the Manchester Statistical Society*, 1898–9, 48. The economic forces at work here have been brilliantly analysed by Gareth Stedman Jones, *Outcast London: A study of the relationships between classes in Victorian society* (Oxford, 1971), especially in part I.
30 These images can be discerned best from the numerous guides to the politer parts of the metropolis and the advertisements accompanying the 'well-advertised' building estates. See, for example, the Homeland Reference Books, *Where to Live Round London (Northern Side)*, ed. Freeman Bunting (1897, 1908).
31 Henry James, 'The Suburbs of London', *Galaxy*, xxiv (1877), 778.
32 For an explanation of these generalized statements, see H. J. Dyos, *Victorian Suburb: A study of the growth of Camberwell* (Leicester, 1961), especially pp. 20–33.

33 C. F. G. Masterman, *The Condition of England* (1909), p. 72; John Ruskin, *Praeterita* (Orpington, 1886), I, p. 70.
34 See Dyos, 'Slums of Victorian London', 11–24.
35 [C. F. G. Masterman], *From the Abyss* [1902], p. 4.
36 The background will be found in H. J. Dyos, 'Workmen's Fares in South London, 1860–1914', *Journal of Transport History*, i (1953), 3–19.
37 For some discussion of the dynamics of suburban change in Victorian London, see D. A. Reeder, 'A Theatre of Suburbs: Some patterns of development in West London, 1801–1911', in *The Study of Urban History* (1968), ed. H. J. Dyos, pp. 253–71; and for an outline account of the rise and decline of a suburban district, see his description of Fulham in P. D. Whitting, ed., *A History of Fulham* (1970), pp. 150–64, 275–90.
38 H. Taine, *Notes on England* [1868–70], translated by Edward Hyams (1957), p. 14.
39 *Penny Magazine*, 31 March 1837, quoted in T. C. Barker and Michael Robbins, *A History of London Transport* (1963), I, p. 36.
40 William Ashworth, *The Genesis of Modern British Town Planning* (1954), ch. 6.
41 Dyos, *Victorian Suburb*, p. 62.
42 For a discussion of the relationships between suburban rail-travel and suburban development in Victorian London, see John R. Kellett, *The Impact of Railways on Victorian Cities* (1969), especially pp. 365–87, 405–19. Some account has been taken of the growth of low-paid jobs in the service industries by Mary Waugh, 'Suburban Growth in North-west Kent: 1861–1961' (unpublished Ph.D. thesis, University of London, 1968).
43 Reeder, 'Theatre of Suburbs', pp. 255–6, 264.
44 *Household Words*, i (1850), 463.
45 We are grateful to a former student for permission to use her work on this suburban slum. See Patricia E. Malcolmson, 'The Potteries of Kensington: a study of slum development in Victorian London' (unpublished M.Phil. thesis, University of Leicester, 1970).
46 Hector Gavin, *Sanitary Ramblings: being sketches and illustrations, of Bethnal Green. A type of the condition of the metropolis* (1848). See also R. J. Roberts, '*Sanitary Ramblings* ... by H. Gavin', *East London Papers*, viii (1965), 110–118.
47 Stedman Jones, op. cit., ch. 6, deals more extensively with distribution of provincial immigrants in Victorian London.
48 These few calculations are derived from a larger study being undertaken by D. A. Reeder due to be published under the title *Genesis of Suburbia* by Edward Arnold in 1974.
49 These figures are also taken from a larger piece of work on the social organization of Victorian Camberwell, by H. J. Dyos. For a discussion of the methods being used to calculate these data, see H. J. Dyos and A. B. M. Baker, 'The Possibilities of Computerising Census Data', in *The Study of Urban History*, ed. Dyos, pp. 87–112.
50 For an attempt to work out some of the broad contours of this map for the mid-nineteenth century, see F. Bédarida, 'Croissance urbaine et image de la ville en Angleterre au xixe siècle', *Bulletin de la Société d'Histoire Moderne*, third series, No. 1 (1965), 10–14; also by the same author, 'Londres au milieu du xixe siècle: une analyse de structure sociale', *Annales*, March–April 1968, 268–95. Interesting work has also been done on the internal movements of particular and easily identifiable ethnic

groups. For example, Lynn Lees, 'Patterns of Lower-Class Life: Irish slum communities in nineteenth-century London', in *Nineteenth-Century Cities*, ed. Stephan Thernstrom and Richard Sennett (1969), pp. 359–85. See, too, Vivian D. Lipman, 'The Rise of Jewish Suburbia', *Transactions of the Jewish Historical Society of England*, xxi (1968), 78–103.

51 For example, W. M. Thackeray, *Vanity Fair* (1848); Anthony Trollope, *The Three Clerks* (1858); Wilkie Collins, *Hide and Seek* (1861); George and Weedon Grossmith, *The Diary of a Nobody* (Bristol, 1892); G. K. Chesterton, *The Napoleon of Notting Hill* (1904); W. Pett Ridge, *Mrs Galer's Business* (1905); Keble Howard, *The Smiths of Surbiton* (1906); Shan Bullock, *The Story of a London Clerk* (1907); H. G. Wells, *Ann Veronica* (1909).

52 See P. J. Keating, *The Working Classes in Victorian Fiction* (1971).

53 Dyos, *Victorian Suburb*, pp. 109–13.

54 J. C. Lettsom, *Village Society* (1800), p. 5.

55 For more extended accounts of these business operations and of the financial interests involved, see H. J. Dyos, 'The Speculative Builders and Developers of Victorian London', *Victorian Studies*, xi (1968), 641–90; D. A. Reeder, 'Capital Investment in the Western Suburbs of Victorian London' (unpublished Ph.D. thesis, University of Leicester, 1965).

56 From the notice of sale of the Gunter Estate including building land in Chelsea and Fulham (G.L.C.R.O.), 1856.

57 The economics of railway operations in Victorian London and other cities are discussed in: Barker and Robbins, op. cit., pp. 208–40; Kellett, op. cit., pp. 60–99, 388–405; H. J. Dyos and D. H. Aldcroft, *British Transport: An economic survey from the seventeenth century to the twentieth* (Leicester, 1971), pp. 215–19, 147–75.

58 Reeder, 'Capital Investment', pp. 104–9.

59 Minutes of Evidence, House of Lords Committee, Acton & Hammersmith Railway Bill, 1874, p. 208. For a discussion of the timing of estate development in outer west London after mid-century and for some examples of how development prospects were frequently over-estimated, see M. A. Jahn, 'Railways and Suburban Development: outer West London, 1850–1900' (unpublished M.Phil. thesis, University of London, 1970).

60 There was, for instance, the case of Charles Henry Blake, a retired Indian civil servant, who put savings of £116,000 into building thirty-six houses on the Kensington Park estate in the 1850s, lent heavily to unreliable builders, and plunged into railway speculation, before being hauled back from ruin by his solicitors in time to catch a boom that brought financial success. His large collection of papers are in G.L.C.R.O.: Ladbroke Estate.

61 The most comprehensive statements on investment in house-building in the nineteenth century are: H. J. Habakkuk, 'Fluctuations in House-Building in Britain and the United States in the Nineteenth Century', *Journal of Economic History*, xxii (1962), 198–230, and J. Parry Lewis, *Building Cycles and Britain's Growth* (1965). Among other writers whose work is indicated in Professor Parry Lewis's book, E. W. Cooney was the first to discuss building fluctuations in London and S. B. Saul offers an explanation nearest to the interpretation offered by us. See, respectively, 'Long Waves in Building in the British Economy of the Nineteenth

Century', *Economic History Review*, xiii (1960), 257–69; 'House Building in England, 1890–1914', *Economic History Review*, xv (1962), 119–37.

62 From the 1840s to the 1870s, 80 percent of house-builders undertook six houses or fewer per annum, and very few built more than fifty. Following the boom that peaked in 1880–1, relatively depressed conditions forced the industry to rationalize somewhat, and at the next peak in 1899 a mere seventeen firms (3 percent of the total) were building over 40 percent of new houses in London; even so, 60 percent of all builders were still undertaking six houses or fewer in a year, and they accounted between them for about one-fifth of all new houses (G.L.C.R.O., Monthly Returns of the District Surveyors, 1872–99).

63 *Builder*, xxxv (1877), 42.

64 Details of the Paddington estate loans are contained in the St Peter's Park estate records of the ground-landlords, the Dean and Chapter of Westminster, now the Church Commissioners; details of the Walworth transactions are contained in the property ledgers of Edward Yates, a speculative builder in South London, the only such business records known to us to have survived as historical records.

65 See P. G. M. Dickson, *The Sun Insurance Office, 1710–1960* (1960); T. E. Gregory, *The Westminster Bank through a Century* (1936); Barry Supple, *The Royal Exchange Assurance. A history of British insurance, 1720–1970* (1970).

66 The operations we are describing here have not been sufficiently researched as yet but we are confident from the two most helpful collections of building society records that we have examined—those relating to the Temperance Permanent and the West London Permanent Building Societies—that there is much more to be done by way of detailed investigations in this field. The best available general histories of the building society movement in the nineteenth century are: Sir Harold Bellman, *Bricks and Mortals* (1949); J. Seymour Price, *Building Societies, their Origins and History* (1958); E. J. Cleary, *The Building Society Movement* (1965).

67 'The Building Mania', *Builder*, vi (1848), 500–1, quoting the *Morning Herald*.

68 See David Owen, *English Philanthropy, 1660–1960* (1964), pp. 372–93.

69 William Howitt, 'Holidays for the People. Michaelmas', *People's Journal*, ii (1846), 171.

70 Ibid., 170.

71 Andrew Winter, 'Country Houses for the Working Classes', *People's Journal*, ii (1846), 135.

72 Henry Lazarus, *Landlordism: An illustration of the rise and spread of slumland* (1892), p. 46. For the rise of urban land-reform movements at this time, see D. A. Reeder, 'The Politics of Urban Leaseholds in Late Victorian England', *I.R.S.H.*, vi (1961), 1–18.

73 *Plays Unpleasant* (1926 edn), p. xxv. The following excerpt comes from Act II (pp. 33–4).

74 Winter, op. cit., p. 134.

75 For 'rates made' see the annual volumes of *London Statistics* from 1890 onwards.

76 Malcolmson, op. cit., ch. 4. For the law on this, see G. St Leger Daniels, *A Handbook of the Law of Ejectment* (1900), which was designed to be 'specially useful to those landlords and their agents who often find the "getting rid of tenants" an unpleasant necessity'.

77 J. F. B. Firth, 'London Government, and How to Reform It', in *Local Government and Taxation in the United Kingdom*, ed. J. W. Probyn (1882), pp. 147–269; S. and B. Webb, *The London Programme* (Fabian Society Tract, 1891).

IV A Change of Accent

16 Another Part of the Island
Some Scottish perspectives

G. F. A. Best

One of the things that must strike anyone reading this book thus far is how much of what we know about the Victorian city is subject to illusion. We see it for the most part on a darkened stage in which pools of light are cunningly thrown by the illusionist. But like any small boy at a production of this kind we suspect that there is much more to be seen. That is a revelation not easily granted. If we were to be allowed behind the scenes we might find a wholly different paraphernalia of stage furniture and backcloths stored away—the scenery for another part of the magic island. But out front the view we have of the stage itself is governed to an unsuspected degree by our own optical and psychological readiness to see what is not there.

It is possible, for example, to be so stunned by a sense of the speed of physical change in the twentieth-century city that one may slip into thinking of this or that Victorian city as a relatively fixed, solid thing, that emerged, dingily, from some Georgian chrysalis and only changed to become the less fixed twentieth-century thing we know. This illusion is perhaps promoted by the masters of the British topographical illustration school, who left such attractive, clean, clear-cut vistas (no doubt somewhat idealized) of the later-Georgian and very early-Victorian city and who had no immediate successors, until photography for popular commercial purposes brought the sight of the Victorian city back to us most vividly in its very late-Victorian form. This shortage of visual information helps to produce the erroneous impression that the Victorian city was more fixed than the twentieth-century one. The fact is that prosperous cities of that time experienced physical flux no less than those of our own. The railways and the new roads wrought great changes on the face

© 1973 G. F. A. Best

of the Victorian city. 'Improvements', as they were almost always called, were going on from the very early nineteenth century, and the ones we have heard most about cannot have been more than a fraction of the whole. The clearance of old city-centre inconvenience and squalor to make way for handsome business buildings, which achieved such flattering publicity when Chamberlain did it in Birmingham, was going on as best it could in every city where business boomed and local pride surged. Everyone knows about the city centres' characteristic loss of resident populations as their concerns became dominantly civic and mercantile, but it is less easy to bear in mind the physical change that accompanied this demographic and functional one. Birmingham men returning after fifteen years' absence in 1890 and 1970 respectively might be equally perplexed to find their way from New Street station. In the most prestigious streets of the City, façades may well have changed several times in the course of the century. Mayfair's rebuilding as the ground-leases fell in during the eighties and nineties totally changed its appearance from the Mayfair which Palmerston knew.

Another illusion of the Victorian city tends to be its size. Here was the most urbanized country in the world, with the biggest city in the world and far and away the biggest cluster of large-sized cities to support it; with, moreover, continual complaint all through the century from administrators and magistrates about the unmanageability of cities of these sizes and from spokesmen of the labouring classes about the difficulty of ever getting out of them. The paradox is that these allegedly unmanageable and prison-like cities were not all that large by the genuinely unmanageable and prison-like standards of our own day. London was indeed a monster by 1870 and no description of the green suburban belt will persuade anyone that mid-Victorian London did not present a problem to get out of for those who did not have several hours to spend, together with the energy to sustain them. But even so, it cannot have taken a Westminster pedestrian more than two hours to get well into the fields beyond Fulham or Hammersmith, or for one from Shoreditch to get into the countryside beyond Clapton and Leytonstone. These are not prohibitive times and distances. Proportionately still less so are the distances between the hearts of provincial towns and their green hinterlands. At the beginning of Victoria's reign Manchester and Leeds were physically as easy to walk out of as Winchester or Stirling are now. So far, then, as the Victorian city was difficult to get out of, or likely to inspire feelings of enclosure and imprisonment, the cause was not sheer physical immensity. Perhaps the 'countryside' was made unwelcoming by gamekeepers, guard-dogs, policemen, railings, fences, and intimidatory notices to keep the gentle traveller in mind of the abundance of laws for the defence of property. Whatever the reason for that constantly alleged divorce of townee from countryside, it cannot have lain in the factor of mere size.

Then there are the dangers of 'London fixation': of interpreting provincial matters in metropolitan terms. Henry James's remark that 'all England is in a suburban relation' to London[1] was less true in the nineteenth century than it has become in the twentieth. The bright lights of London are dangerous. The world city, the city where the wealth and display of Britain were gathered, the city where contrasts were most startling and thought-provoking, the city which so many different kinds of writer liked to explore and write about is such an attractive and absorbing subject that the historian is in real danger of judging provincial urban matters by London standards. London's differences went beyond the sufficiently obvious ones of being the world city and a socialite one. Wages did not go nearly so far in London as they did in Glasgow or Middlesbrough. There was in London a spectacular mass of destitution and pauperism which called into existence the most elaborate philanthropic apparatus that we can trace in the world in the nineteenth century, and yet defied that philanthropic apparatus to abolish or much diminish it. London had an 'entertainment' quarter so notorious that one is forced to conclude that London was the seedy vice capital of the western world. In these and many other respects London was a different kettle of fish entirely from the other cities of the land, and yet historians, I think, often slip into the trick of describing this or that attribute of nineteenth-century urban society in terms of the London society that is easiest to find out about, and that we know the most about, and perhaps (because of the literary quality of the evidence) most enjoy finding out about.

And, last of these preliminary points, there is the probability that the Victorian city is less of a type than would be the Victorian city of this or that region. I know of no attempt as yet to determine what these provincial patterns were and how far they spread. The regions to study would be of course the standard ones: Tees–Tyne, West Riding, Lancashire–Cheshire, Black Country, South Wales; and—call it regional or national—Scotland. So to the main business of this essay: to investigate some of the aspects of the Victorian city as it developed in Scotland.[2]

The big cities of Scotland at the 1901 Census were a small group, in this order of size:[3]

Glasgow (excluding Govan, Partick, and Pollockshaws)	762,000
Edinburgh (excluding Leith)	317,000
Dundee	161,000
Aberdeen	154,000
Paisley	79,000
Leith	77,000
Govan	77,000
Greenock	68,000

Glasgow's pre-eminence had been undisputed since quite early in the century. Edinburgh (I am now perforce including Leith, as, in respect of Glasgow, I shall include Govan, etc.) had actually been larger by 6,000 in 1801, by 2,000 in 1811. Thereafter Glasgow leapt ahead. By 1851 only Dublin was larger in the U.K.; by 1871 Glasgow was next after London; Liverpool, Manchester, and Birmingham were above Dublin, in that order.

The first trap awaiting the incautious historian who lumps them in with the English cities for common study is that the statutory framework which controlled the Scottish cities (so far as any statutory framework did control Victorian cities, which was not greatly, before the 1870s) was not coincident with that controlling the English ones. The quantity of statutes clearly labelled 'Scotland' indeed strongly suggests this, but they do not tell quite the whole story. For one thing, many public and general statutes right through the century were supposed, at the time of their passing, to work on both sides of the border. Some indeed were able to do so, but others were not. The Board of Supervision (Scotland's central health and local government authority, such as it was) complained to the Lord Advocate in November 1866 that it had tried and failed to work the Nuisance Removal (Scotland) Act of 1856, Lindsay's Burgh Police Act of 1862, the 1865 Sewage Utilization Act, and the 1866 Sanitary Act; besides being largely irreconcilable with one another, the two last-named Acts simply did not fit Scottish law and circumstances.[4] The same hard truth was, about the same time, being discovered independently in Whitehall. Certain leading inhabitants of Uddingston, Lanarkshire, took up the 1866 Act's suggestion that the Home Secretary should be memorialized to put pressure on their laggardly local authority. The official endorsements of the Uddingston memorial tell their own story: '[Tom] Taylor to inquire S.H.W. [Spencer Walpole]'; 'Does the Act apply to Scotland?'; 'I do not find any provision to the contrary but ask Mr. Thring.' Henry Thring, the Home Office Counsel, soon came to the lame conclusion that the Act (Part I of it, anyway) was apparently meant to be applicable to Scotland, but that it was no use applying it there because the ultimate means of enforcement was the exclusively English court of Queen's Bench.[5]

The series of statutes designed to assist good-quality working-class housing, known after their promoters as the Torrens and Cross Acts, are another case in point. The Torrens Acts (1868, 1879, and 1882) were said to be 'practically a dead letter' north of the border;[6] Cross's Acts, however, *were* applied, presumably because Cross had the sense to get a special version tailor-made to suit Scottish conditions.[7] A generation earlier, a similar tailoring had done the same for the Lands Clauses Act of 1845, which cheapened and simplified the compulsory acquisition of property by public authorities.[8] But public and general statutes did not absolutely have to adopt Scottish costume in order to survive north of the border. The 1890 Housing of the Working Classes Act and its lineal successors functioned everywhere, as apparently did the Small Dwellings Acquisition Act of 1899 and the Housing and Town Planning Act of 1909.

The preliminary difficulty, however, is not yet fully exposed. It goes beyond the facts that some general Acts did, and others did not, work in Scotland, and there is no cut-and-dried means of telling which is which. There is the further difficulty that in some of its branches the course of Scottish social legislation ran much nearer to the English than it did in others. This fact will not upset a purely Scottish historian, to whom the course of purely English legislation is of no interest. But it can upset the historian who seeks to bring shared English and Scottish concerns within a single

'British' view. He will (quite rightly) be on his guard, for instance, about such matters as education, for there the differences between the two nations are celebrated and obvious. In the cases of legislation about primary local authorities and the Poor Law the historian will also know from the start that he is dealing with two perfectly independent sets of institutions, and he will not be surprised, therefore, if their legislative histories are out of phase with one another.

In respect of public health, however, the case is surely rather different. Here was a nineteenth-century innovation, a brand-new problem demanding legislative intervention to cope with problems much the same wherever they cropped up. One might expect the English and Scottish legislative and administrative histories of public health to be coincident and concurrent. But they actually appear more different, both in content and chronology, than any of those three parallel cases already mentioned. And they are the more confused and confusing, because it seems to have taken some years before the inapplicability to Scotland of English-framed general public health legislation was clearly realized. Only in the 1860s did the general government of Scotland attain, in respect of public health, the state which England had been in since 1848, and only in the 1890s did the Scottish laws of public health really catch up with the English.

The Scottish equivalents of the 1848 Public Health Act and its lineal successors were Lindsay's Burgh Police Act of 1862 and the Public Health (Scotland) Act of 1867. Lindsay's Act enabled communities which chose to adopt it to make the sort of building and sanitary by-laws which were the very foundations of early-Victorian public health; the 1867 Act established a Scottish equivalent to the central sanitary office which had been functioning in England (under changing titles) since 1848, and (ominously foreshadowing the English development of 1871) lumped it in with the central poor-law authority, the Board of Supervision. It was no more than a pale copy of the English office already established. As one of its more intelligent critics, the Sheriff-Substitute of Lanarkshire, W. C. Spens, pointed out ten years later: 'that the duties under the Act were not supposed to be very onerous may be gathered from the fact that the remuneration which was proposed to be allowed to the legal members of the Board [i.e. the three Sheriffs] ... was £50 a year.' None of that Board's honorary members regularly attended. The Chairman, W. S. Walker of Bowland (he had been Secretary from 1852 to 1868), was, Spens conceded, active and capable; but he said, 'although I always am prepared to stand up for my profession, I certainly am not prepared to hold that the supervision of public health matters in Scotland should be entrusted to a committee of advocates.' The Board had no medical member, and only one medically-qualified officer—and he a part-timer, Dr H. D. Littlejohn, the Edinburgh Medical Officer of Health, who was prevailed upon to accept 'the huge salary of £200 per annum ... upon the understanding that for that salary he shall undertake all investigations within ... a day's journey of Edinburgh'.[9]

Nor was this feeble Board's powers *vis-à-vis* the local sanitary authorities at all adequate to the situation.[10] It had only three full-time Inspectors for Poor Law and Public Health functions combined, and their salaries—'fixed by the Treasury'—

were a good deal lower than those of their English and Irish counterparts. Its English counterpart—in the sixties the Local Government Act Office—drew some of its large and ever-growing measure of control and influence over local boards of health from its powers to approve sets of by-laws and to issue time- and money-saving Provisional Orders. The Board of Supervision had no such powers.[11] Most burghs that thought fit to make by-laws under Lindsay's Act did not even need to get the approval of that Pooh-Bah of Scottish local administration, the Sheriff. It was nobody's official business, until the nineties, to make sure either that the by-laws which were needed were made, or that, having been made, they were enforced. Regulations made under the Scottish Public Health Act came under the Board's purview, but, again, the Board had no power to see that they were made. These provisions for urban public health were not impressive even at the dates of their introduction, and they remained substantially unchanged for over twenty years.[12]

They were not unchanged because nobody saw their failings. Repeated attempts at improving them foundered, it seems, on rocks painfully familiar to the well-intentioned Scottish legislator: shortage of parliamentary time; lack of interest or lack of intelligence in Lords Advocates and, after 1885, Scottish Secretaries; selfish and stupid opposition from the very local authorities in Scotland whose performance most needed improvement. And when, at last, in the nineties, substantial improvements were made, there was not, in the Burgh Police (Scotland) Act of 1892 which replaced Lindsay's Act of thirty years before, that clear and helpful distinction of health from police and building functions which the reformers desiderated. The Public Health (Scotland) Act of 1897 extensively reorganized its administration. From that time on, Scottish and English public health law marched more in step, but Scotland was, by then, a long way behind. It was not in every respect so far behind in the big cities, which had been steadily looking after themselves under their own local Acts for anything up to a hundred years. 'The truth is', said Chairman Walker to the Dilke Commission,[13] 'that we have no great anxiety about the large towns. We are aware that their staff is ample . . . and they are intelligent and active in the executing of the [Public Health] Act'; but it was lamentably behind in the smaller urban areas and in the country towns and villages, where the cause of public health, lacking the steady pressure of a vigilant and sufficient central authority, seemed still at the turn of the twentieth century to be in its infancy, and where only the closer proximity of the countryside sometimes made insanitary horrors less horrific.

We turn back from the unlovely spectacle of late-Victorian Coatbridge, Motherwell, Blantyre, and Port Glasgow to the major cities, where urbanization was being experienced at its most intense. These big six—Glasgow, Edinburgh, Leith, Aberdeen, Dundee, and Greenock—preferred to keep themselves to themselves and to work under their own local 'Police Acts', as did some of the greater English cities. On both sides of the border, motives of local pride and independence often urged municipalities to shape and pay for their own statutes. Manchester men firmly believed (reasonably or not) that its 1844 Improvement Act was Lord Morpeth's main model for his 1848 Public Health Act, and Glasgow men believed that R. A. Cross got the idea for his first, 1875, Artizans' and Labourers' Dwellings Improvement Act, from its own pioneer Improvement Act of 1866. Provincial citizens argued that they had to be left free to attend to particular local needs, which were bound to vary from place to place. This may well have been true; yet the mainspring of their action was surely the determination not to admit central authorities to any share in the management of their local affairs. Such a measure of local pride, independence, and, perhaps, obstinacy, was one of the many shared characteristics of English and Scottish Victorian cities, itself neither surprising nor discreditable, but worth attention because it appears that its Scottish form was rooted in a civic tradition and an urban outlook so different in one important particular that it merits closer examination.

Scottish cities in general were accustomed to a civic government much firmer and more positive, in some respects, than were English cities. Evidence of many varieties suggests that Scottish city government was in the last century in some ways tougher, and in others more enterprising, than English. Consider, first, Edinburgh, in its golden age. Was it absolutely by chance, or was it partly because it was set in a Scottish tradition of civic authoritarianism, that Edinburgh was in fact the only city in eighteenth- or early-nineteenth-century Britain officially to develop itself to such a remarkable extent, and that, as A. J. Youngson points out,[14] 'all Scottish towns of any consequence undertook building programmes in these decades'? John Nash's metropolitan magnificence was a special case: the beautification of the imperial capital under the passionate stimulus of the monarch and with a good deal of Crown property in hand to start with. Dublin's great array of public buildings was an even more special case: Dublin was the Georgian New Delhi. The splendour of Georgian Bath was entirely the creation of a family of gifted property developers; the municipal contributions to the amenities of Bath were consequent upon the initiatives of the Woods. Early Victorian Newcastle, for a tragically short while Britain's most handsome industrial city, was principally the achievement of a gifted property developer, though Robert Grainger was certainly much helped by his friendship with the Town Clerk. Beyond Newcastle there is nothing to match the City of Edinburgh's conception and development of its New Town. All the rest—Bloomsbury, Belgravia, Brighton, Bournemouth, Birkenhead, *et hoc genus omne*—were so much the creations of private landowners and/or speculators, that the role of the public authority was in comparison quite negligible.

The episode of Edinburgh's New Town may the more plausibly be set into a context of civic authoritarianism because several nineteenth-century Scotsmen thought this existed, and there is some evidence to suggest that they were right. For instance, the Edinburgh lawyer John Hill Burton—whose summary of *The State of the Law as regards the Abatement of Nuisances and the Protection of the Public Health in Scotland* for the benefit of Edwin Chadwick's famous Sanitary Inquiry is one of the clearest pictures we have of this side of 'the condition of Scotland question'—said in 1840 that 'formerly any public regulations or restrictions which the courts of law approved of were very arbitrarily enforced, without much respect to individual rights'. Perhaps too much could be made of this reference to a strict regime which had, on Burton's own plentiful evidence, shrunk from its earlier grandeur; but only fifteen years later allegations of civic authoritarianism were made by another citizen of Edinburgh, one Robert Mason of Meadowbank. Complaining to the Home Secretary about the proposed amalgamation of the southern suburbs with the City, he concluded by denouncing the Council for already possessing a bigger empire than they could efficiently handle. 'Our present Lord Provost [Sir John Melville, Writer to the Signet] is a great admirer of the abilities and management of the Emperor of the French; and although he may possess much unheard-of wisdom, he is not appointed to be hereditary Provost of Edinburgh.'[15] It is interesting that, whereas English Mayors held office for a year at a time, Scottish Provosts served for three years at a stretch.

Though, perhaps, no ratepayer of an English city in the fifties would have felt moved to express himself thus strongly about his mayor, it is very doubtful whether any English city was yet doing anything spectacular or expensive enough to provoke such passionate reactions. Similar things were probably being said in Birmingham about Mayor Chamberlain in the later seventies, and in the County of London about Chairman, the Earl of Rosebery, in the nineties. Each presided over a regime which showed three unmistakable marks of the collectivist beast: willingness to spend ratepayers' money for the common good; positive anxiety to spend it for the benefit of those who were too poor to pay rates; and a Germanic idea of civic magnificence and subsidized culture. Birmingham and the London County Council regularly appear in the accounts as shining lights of later-Victorian civic enterprise, and Liverpool's record seems to have been extraordinarily impressive, though its full measure has not yet been taken. But a fourth British city ought to be put beside them: a city which may well not have needed to acquire notions of civic enterprise, because they were already in its national tradition—Glasgow.

Glasgow certainly took over from Edinburgh as the show-city of Scotland in the second half of the nineteenth century, and, long before it became controlled by the Labour party, its collectivist enterprise was provoking unsympathetic observers to the same sort of alarm and disapprobation that it now provokes as a model of socialist civic planning. A commentator in normally sober legal circles angrily called Glasgow 'the oppressor of the West'.[16] Anger and fear informed his whole piece, which was printed in 1905, when Glasgow had been in the van of municipal enterprise for at least fifty years. By 1888 Glasgow's Medical Officer of Health, the great Dr J. B. Russell, was already complaining that 'The public of Glasgow trust too much to authorities and officials for the solution of their social difficulties—more, I think, than any other community'. Proud of Glasgow's sanitary and housing work, he regretted the city's relative backwardness in the ' "Open Spaces and Playgrounds' Associations"... "Artizans' Dwellings Companies" and the like, which unite the business capacity and Christian sympathy of the citizens of so many other cities in successful labour for the common good.' Why, he concluded, 'have we not an Octavia Hill in Scot-

292 An intersection in the St Leonard's district of Edinburgh (St Leonard's Hill—Dumbiedykes Road—Carnegie Street—Heriot Mount) at the time of demolition, *c.*1960. This crossing was complete, in this shape, by 1867. At exactly what dates through the preceding fifty years each separate block was erected, I cannot judge.
Scotsman Publications Ltd

293 *above* Amphion Place, Calton Road. Built about 1825. The photograph (1958) shows the railway into Waverley from the east tunnelling under the Calton Hill. From a photograph in the Local History Collection.
Edinburgh Central Public Library

294 *right* Lower Viewcraig Row, in the Dumbiedykes district, 1957; the terraces are on a hillside descending to the valley between Holyrood Palace and Salisbury Crags.
Photo: Adam H. Malcolm

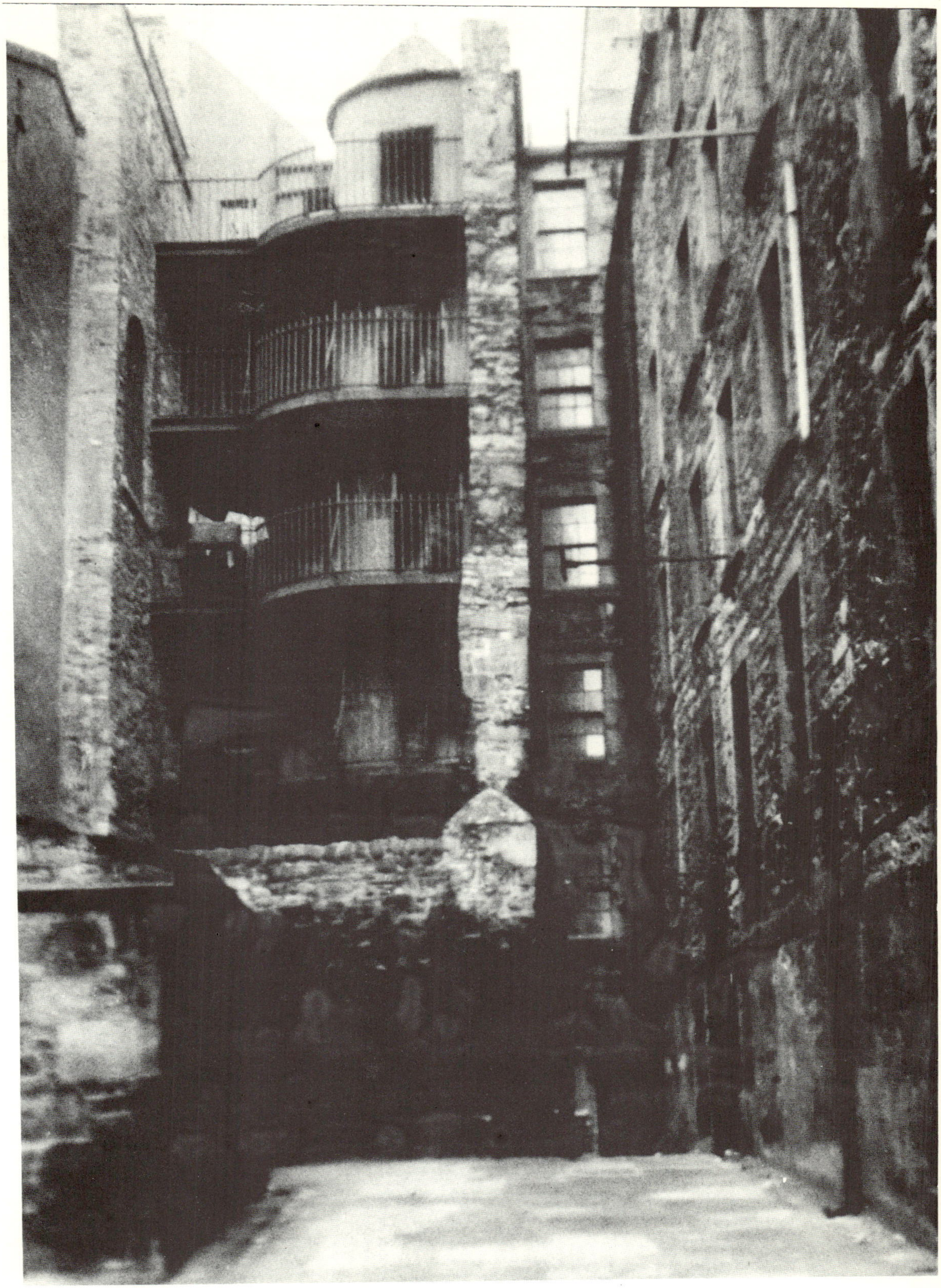

295 Tweeddale Court, off the High Street. Undated, but c.1950, showing a characteristic scene in the Old Town as it was inhabited through the nineteenth and into the early twentieth century.
Edinburgh Central Public Library

land?'[17] Dr Russell's complaint about a certain lack of *voluntary* citizens' enterprise towards better urban life was echoed by others; by, for instance, the Edwardian Royal Commissioners on the Poor Law, who, finding it remarkable that there were in the big Scottish cities no medical Provident Dispensaries of the kind by then quite common in English cities, said they thought it 'regrettable that so many persons in the large centres of population in Scotland appear to be willing to accept charitable medical relief instead of adopting some form of providence or thrift in order to provide themselves with medical attendance', and urged the Scots to 'vindicate the national character' and establish some.[18]

Glasgow was not a 'model' city in all respects. The extent of its municipal enterprise was not unlimited. But it was, by Victorian standards, both extensive and early; and Glasgow's record seems more impressive than—at any rate—Birmingham's, and deserves to be much better known. It shows that Glasgow was at least as active as the other early enterprising English municipality, Liverpool—at least according to one American through whose eyes, as yet undazzled by Brummagem glitter, we might look at the Glasgow of 1889.[19]

The Municipality of Glasgow 'transformed the Clyde from a mere rivulet with mud banks into an artificial channel for navigation by the largest vessels'. The Loch Katrine water scheme, completed in 1859, was, with the possible exception of Liverpool's, the first of the great municipal long-distance water projects, and the only cities in the United Kingdom that got their water more cheaply were Dublin and Edinburgh. The city bought out the two existing private gas companies in 1869, and in 1893 was supplying a higher candle-power gas than any other city except Liverpool, at a price cheaper than most. With Dundee, it pioneered the public lighting of private courts and common stairs. On its own it led the way in the municipal leasing and hire-purchase of gas-stoves and fires. When tramways suddenly became the rage, Glasgow took a course in the early seventies which London and other less collectivist-minded cities soon came to regret not having taken: it built the tramlines itself and leased them to private companies to work; thus it retained control of the amenities of the highways and of the development of the system, and was the more easily able to municipalize them later on. It was not the Council's fault that the obstinate long refusal of the circumambient little burghs (all by nature parts of 'greater Glasgow') to come within the municipal boundary, reduced the efficiency of the tramway system's suburban services, just as they impeded the efficiency and growth of other municipal services. Glasgow owned and managed all its markets and slaughter-houses—nothing exceptional about that—and owned what the American observer called 'a magnificent system of public baths and wash-houses'.

But the most impressive of Glasgow's early municipal enterprises was in something much bigger and more basic than abattoirs and laundries: it was in housing and town-planning. The three big Scottish cities led the way in slum-clearance. Their own Improvement Acts gave to Dundee in 1871, Edinburgh in 1867, and Glasgow in 1866 the powers to purchase, clear, and redevelop central slum areas which the Cross Acts, from 1875 onwards, made generally available. These pioneer improvements are obviously of great interest, and one student has summarized the Glasgow property development in terms of 'municipal socialism' thus: 'By 1902 the Corporation owned, among other things, 2,488 houses, 78 lodging-houses, 372 shops, 86 warehouses and workshops, 12 halls, 2 churches [and] a bakehouse.'[20]

This venture into 'council-housing' was quite remarkable but too much should not be made of it: first, because the city only went into it *faute de mieux*, after failing to find private speculators to put up the much-needed houses; second, because there is no adequate material for comparing it with the municipal housing achievements of other cities; and third, because after all the significant thing was not that some quantity of housing was municipally erected and leased but that it was the biggest Scottish city that undertook 'the first massive municipal intervention [in Britain] to sweep away the most insanitary and dilapidated and archaic central urban areas and to replan them on a modern basis'; it was Glasgow that first recognized 'that a free market, and private philanthropy and public health regulations could not provide an adequate solution to these problems: that the City Fathers must at least supervise and plan redevelopment';[21] and it was Dundee and Edinburgh, not Birmingham or anywhere else, that quickly followed suit.

Now why were the Scottish cities the pioneers in this business of slum clearance? It is not enough to say simply that they happened to have worse slums than anywhere else. Probably their slums were the very worst—at least, equal with the very worst—in Britain; but surely history and contemporary experience alike suggest, however unhappily, that social remedial measures are not necessarily taken most decisively by the cities that most need them. Perhaps there was something in the Scottish local government tradition which encouraged an in some ways tougher handling of the problems of poverty.

Dr H. T. Hunter—one of Dr John Simon's brilliant band of investigators in the sixties—whose *Report on the Housing of the Poorer Parts of the Population in Towns*[22] is among the most important social enquiries of that inquisitive decade, included the memorable fact that 'At Glasgow the question of over-crowded houses has attained such magnitude of importance as to compel the abandonment of all prejudice, and the peremptory defence of society against individual aggression, by the enforcement of a fixed ratio of capacity and inhabitation.' This was the origin, in the sixties, of the famous Glasgow system of 'Ticketed Houses':[23] registration and labelling, by tinplate 'tickets' firmly fixed on the outer wall, of private houses of three or fewer rooms in which the aggregate volume of air-space (not including lobbies and recesses) did not exceed two thousand cubic feet. The 'ticket' stated the cubic content and the number of occupants allowed at the low rate of three hundred cubic feet for each adult. But this was not all. The police as well as the sanitary inspectors had the power to enter and inspect 'ticketed houses' at any time of day or night—and it was almost invariably at night that they did enter, since it was only then that the overcrowders could be caught at it.

This was indeed a 'peremptory defence of society against individual aggression'! There was nothing like it in England at that date (the registration and inspection of common lodging-houses was quite a different thing), nor anything else yet in Scotland. But in the course of the next fifty years it became known in all the big Scottish cities, and it was made generally available by the 1903 Burgh Police Act. The only English cities in which anything like it could be discovered by the 1917 Royal Commission on Housing in Scotland were Bradford, Birmingham (where a stricter sense of propriety confined inspections to Sunday mornings), and Liverpool, where it was worked with full Glaswegian vigour.[24]

The fact of its flourishing in Liverpool so much more than anywhere else south of the border rather suggests that by 'individual aggression' Dr Hunter may actually have meant 'Irish pauperism'. The extreme nastiness of the slums inhabited by Irish immigrants was a constant theme with Victorian social and sanitary reformers, and close investigation would very probably discover public authorities and police to have become tougher, and more emphatic in their use (or abuse) of discretionary powers, when they were dealing with the 'paddies'. Glasgow and Liverpool of course were worse hit by Irish immigrants than anywhere else: T. H. S. Escott observed in 1879, without a flicker of surprise, that 'the repressive measures enforced by the Liverpool magistrates are exceptionally severe, and that the police often apprehend upon charges which would be deemed trivial elsewhere'.[25] Glasgow's statistician, James Cleland, thought there were about 35,500 Irish natives in Glasgow in 1831, nearly 18 percent of the total population, and all of it crowded into the worst slums. J. H. Clapham wryly comments that 'the West Scottish towns [were not] hospitable to the newcomers', and that by some unspecified ' "moral compulsion" . . . 1,517 Irish were shipped from Paisley in 1827', and even more from Glasgow.[26] Again, it may have been because the denizens of the fearful Sandgate slums were largely Irish that the two police inspectors in charge of Newcastle's common lodging-houses (under Shaftesbury's Act of 1851) went beyond their legal brief and in practice inspected, between ten and two at night, 'all houses let in single-room' dwellings, as well as common lodging-houses proper. Dr Hunter was rather astonished at their audacity, but reported that the sanitary and moral results had been excellent, and, 'such [was] the tact of the officers, that no anger or resistance seems to have been produced'.[27]

The foreign character of the slum-dwellers of the Scottish cities may have contributed to bringing out strong authoritarian streaks in public officials. But it does seem likely that some national characteristic in the authorities governing those cities was also operating. The Poor Law Commission of 1909 discovered that some Scottish poorhouse authorities had gone a long way beyond what was legally permissible in their liberty-restricting rules and regulations. They also reported that 'It has long been the practice in Scotland to remove children from vicious parents'. In Glasgow, they said, the Parish Council retained a special officer for prosecuting the parents of deserted or 'separated' children.[28] And John Naismith startled the National Association for the Promotion of Social Science in 1874 with his account of the ways in which Glasgow magistrates, Justices of the Peace, procurators-fiscal (i.e. public prosecutors), and police between them so operated the local Police Acts as to perpetrate petty injustices without possibility of appeal.[29] One of his instances was of a prosecution, under the Police Act, of a workman for loitering in and obstructing the street.

> Not only was the offence not proved, but it was directly negatived by the police officers called to prove it. The sitting bailie, however, convicted the accused, but dismissed him with an admonition. The worthy magistrate afterwards explained to me that he did so because there had been grievous complaints from the shopkeepers in the neighbourhood of the annoyance caused to themselves and their customers by the workmen loitering about at meal times, and that it was necessary that the practice should be put a stop to.

The practical operation of these Scottish City Police Acts invites scrutiny. Mayhew printed on the wrappers of his weekly issues of *London Labour and the London Poor* in early 1851 a number of letters from Edinburgh correspondents reporting a regular crusade against or, from the opposite angle, persecution of, street-sellers, asserting that one magistrate had announced his refusal to go along with his colleagues in thus enforcing the City Police Bill. One gets a strong impression that Scottish towns were 'better' policed, earlier, than English towns. Did Scottish municipal authorities enjoy under their Police Acts larger powers of interference in citizens' lives than normally did English ones? That they did so is suggested by Sheriff James Watson's address to the Social Science Association in 1877. He said that the Police Acts and magistrates' 'provisional orders' (by-laws?) had raised 'a vast number of frivolous acts ... into criminal offences punishable by a caution to keep the peace, a small fine, or a few days' imprisonment'.[30] Did magistrates friendly with shopkeepers, on no matter which side of the border, connive with police to keep the lower classes a-moving along respectable pavements? Perhaps; and yet I should like to close this section of my argument with a quotation from an embattled Glasgow pawnbroker, Mr William McKay of 99 Cowcaddens Street, who complained of police oppression (under the local Police Act) in a letter to the *Glasgow Herald* of 15 November 1860. In England, he alleged, referring to a Commons debate on 4 July 1860, such police powers were not tolerated. But in Glasgow, the police enjoyed and freely used power to enter and inspect his premises at any hour of day or night—'up out of your bed, away from drill, or from any other evening engagement, to accommodate a tyrannical set who, if they got us under their heel, would oppress us like the worst Bomba that ever swayed a sceptre'.

There is then some circumstantial evidence to suggest that the Scottish municipal approach to the manifold problems of industrial urbanization may have been essentially different from the English one. As to the problems themselves, which of them took on, north of the border, a distinctive shape and colour?

It is difficult to avoid the impression that Scottish city slums had, as indeed they still seem to have, some peculiar nastiness about them. Whether or not they were absolutely 'the worst' in Britain, they did appal observers.[31] And that their constant major characteristic was the density of their overcrowding, there can be no doubt.

Professor W. P. Alison, whom Professor Flinn allows to have occupied 'a similar position in Scotland . . . to that held by Chadwick in England',[32] dwelt significantly on a description of life in old Glasgow's wynds given to the 1839 Handloom Weavers Inquiry.[33] It was from Alison's writings, apparently, that Thomas Carlyle drew both his warning parable of 'the poor Irish widow'—conceivably the germ of that theme of providential retribution soon afterwards exploited by Kingsley in *Cheap Clothes and Nasty* and by Dickens regularly from *Dombey and Son* (ch. 47) onwards—and his summary of the Scottish wynds as 'scenes of woe and destitution and desolation, such as, one may hope, the sun never saw before'.[34] Edwin Chadwick appeared to be equally shocked. 'The most wretched account of the stationary population of which I have been able to obtain any account, or that I have ever seen, was that which I saw in company with Dr. Arnott, and others, in the wynds of Edinburgh and Glasgow.' And again, his summary: 'It might admit of dispute, but, on the whole, it appeared to us that both the structural arrangements and the condition of the population in Glasgow was the worst of any we had seen in any part of Great Britain.'[35]

This bad eminence of the Scottish slums continues to be proclaimed throughout the nineteenth century and into our own. Dr Hunter indeed neglected to mention any Scottish city when he deliberately described 'life in parts of London and Newcastle [as] infernal', but his colleague Dr George Buchanan, whom Simon had set to investigating the epidemic typhus of Greenock, found its population almost as densely crowded as Liverpool's (the densest in England) 'and considerably denser than in the poorest parts of London'. 'In Greenock', he said, 'no less than 2,747 persons . . . are living under conditions which would not be permitted in the worst parts of London.' About five hundred souls were living or partly living in a manner 'to which', he thought, 'there can be but few parallels and which represents people living day and night in a space about the size of a street cab apiece.'[36]

Ten years later Chadwick, in jolly mood, told the Social Science Association that if he had to construct a city sacred to Siva the Destroyer, with a guaranteed death-rate of over forty per thousand, he would 'copy literally and closely the old parts of Whitehaven, those of Newcastle-upon-Tyne, and the wynds of Glasgow and Edinburgh'.[37] Chadwick was going, as he had always gone, by mortality figures, which had always been worst in the cities. The sanitary reformers' campaigns had therefore been directed mainly at the cities, with the gratifying result that, by the turn of the century, the killing powers of the cities were much diminished. This was true of the Scottish cities as well as of the English ones.[38] Their overall death-rate figures of course concealed sharp variations between the better and the worst localities; thus in Glasgow about 1900, the overall death-rate of 20·7 per thousand concealed the gap between Kelvinhaugh's 15 and Brownfield's 36;[39] but the improvement in the general urban death-rates during the past thirty years was undeniable, and a comparable and quicker improvement in the urban infant death-rates was about to begin. Sanitarians' attention began to turn from the cities proper, where the battle seemed to have been won, to the rough frontier-towns of industrialism—above all, the mining communities of Glamorgan, the Midlands, County Durham, Lanarkshire, and Fife. In such forlorn industrial districts the infant death-rate could be nearly as bad—over a larger population, and without a fraction of the administrative excuse—as it still was in the worst quarters of Glasgow,[40] and it was to those hitherto neglected places that the national battle for public health now moved. But to one feature of big city life the attention of sanitary and social reformers alike remained riveted—a lingering and loathsome feature, for which the Scottish cities held and hold an unenviable palm: overcrowding.

It was for the peculiar character of their overcrowding that the Scottish cities were notorious. They had always been so to some degree, but they became increasingly so as Victoria's reign wore on, because there were elements in the nature of Scottish overcrowding which prevented its amelioration during years which witnessed its diminution elsewhere. So, always absolutely worse, it seemed relatively much worse, too, as time went by. Density of population had always meant public ill-health and the reformers had been quite clear about this since Alison's and Chadwick's heyday: but the mid-Victorian public health experts had learnt to distinguish between one sort of density and another, and had come to appreciate that some sorts of density were more dangerous than others. Dr J. B. Russell of Glasgow, faced with density problems equalled only by those of Liverpool, conclusively demonstrated that the most lethal sort was that measured (and it required more sensitive instruments to measure it), not in persons per acre, but in persons per room.[41] More people lived in one- and two-room houses in Scotland than anywhere else in Britain: in 1911, 47·9 percent, as against England's 7·5 percent.[42] These one- and two-room houses, moreover, if they were in cities (and most, though not all of them were), were almost always in tall, deep tenements which presented reformers with other special problems.

The origins of the Scottish urban tenement and the one- or two-room home (which together gave the Scottish Victorian city much of its distinction) were often discussed, as reformers tried to accustom their inhabitants to higher standards of comfort and space. Earlier in Victoria's reign, reformers supposed that, once three- or four-room 'houses' or independent two- or three-room 'cottages' were available, the poor one- and two-room families would voluntarily and gladly move into them. Such 'houses' and 'cottages' did become available from the sixties onwards. The poor did not rush into them. Sympathetic reformers attributed this to the fear which the uncertainly-employed poor had of getting into a home the higher rent (or mortgage repayment) of which they could not afford in times of slack trade or family misfortune; unsympathetic reformers attributed it to the slum-dwellers' debased preference for dirt and drink. But by the early years of our present century it was clear that the complete explanation of the persistence of one- and two-roomed housing had to include some recognition of the ghastly fact that experience of it had made it seem a natural thing to the third and fourth generations. It was, for many of them, a way of life.[43]

When and how came about the urban Scot's induration to the overcrowded life of the tenement? Russell's predecessor, the first Glasgow Medical Officer of Health, Professor William Tennent Gairdner, fancifully supposed (as from time to time have others) that the tenement plan came to Scotland from France, a disagreeable by-product of the Auld Alliance.[44] Clapham comes much nearer the mark when he strikingly says, 'The Scottish way of living had been got by ancient country habit out of ancient town necessity, and it was a one- or two-room way, very tough and indestructible, as were the eighteen-inch or two-foot stone walls of the houses which encased it.'[45] The plentiful supply of stone must have had something to do with it, and the example of Edinburgh, for so long the biggest Scottish city besides being the capital, must have had some influence, too. The tenements of its Old Town remained into the nineteenth century the national archetypes of one of the two main species of low-class housing: the nearly indestructible vast old tenement, never undercrowded even in the better days which it had usually known, now utterly gone to seed and divided and sub-divided again and again—'made down' was the expressive technical term—until it had become a warren of one- and two-room dwellings. London had its equivalents in the 'rookeries' of old Westminster, Holborn, Jacob's Island, and other places well known to Dickens. One hears less often of similar places in Liverpool, Nottingham, and Newcastle. But whereas even in London they were regarded (and feared) as exceptional, in Edinburgh and Glasgow they were the hearts of the cities: the wynds which trickled narrowly down each side of the Royal Mile in Old Edinburgh, which sneaked at right-angles off the main grid thoroughfares of Old Glasgow. 'End-on to the building line of these streets', explained Dr Russell,

> we see long narrow strips, extending in what was the natural direction of the growth of a house placed endwise to the street—*backwards*, looking on the maps like sections of geological stratification, with cracks or flaws in between . . . parallel intervals left between tenements simply for convenience of access, only wide enough to permit two persons to pass, or perhaps a barrow or a cart. Each proprietor was bent on covering every inch of his grounds with his building, and the only function exercised by the Dean of Guild Court was that expressed in the phrase, which is still in use, 'to grant a lining'—that is, to see that if he built *up to*, he should not *build over* the line of his building.

The end-result of decades of 'making-down' in such places was house-addresses like 'Bridegate, No. 29, back land, stair first left, 3 up, right lobby, door facing'.[46]

Such were the typical pre-nineteenth-century tenements with which the ancient Scottish cities were burdened. More modern tenements, however, were not necessarily better. They began to appear in Glasgow late in the eighteenth century, as that city's pent-up prospering population, like Edinburgh's at much the same time, burst its seams and sought better-style housing in the then suburbs: Hutchesontown, Tradeston, Cowcaddens, Laurieston, Gorbals. Some of it was in villas or terraces of the common English style, but most of it was in three- or four-storey common-staired blocks, rectangularly lining the gridded streets beloved of the Georgian estate developer. From these initially not unsavoury blocks of flats there soon—sometimes very soon[47]—came unmistakable evidence of social decline. The speculation had gone sour; the more moneyed citizens were not moving in as briskly as they should; some neighbouring development—some factory, canal, or wagon-way, perhaps—had drastically damaged the amenities. In financial desperation or as a calculated switch from a respectable to a disreputable means of making a profit, the ground-owner filled the hollows behind the first façades with other tenement buildings, of meaner cast. The streets and stairs became noisy with a lower class of more numerous residents than had been originally foreseen; the hammer and chisel of the 'maker-down' sounded in these newish tenements as they had long done in the old, and the second main species of low-class Scottish Victorian housing had arrived. Its style was, from whatever causes, hypnotic. Later tenement blocks, of similar external appearance and basic common-stair design, needed no 'making-down', for they were of one- and two-room dwellings from the start; though it should be observed that the one- and two-room dwellings newly built as such would usually be of quite large rooms, and equipped with sleeping-closets and storage-space in plenty. Of the 26,794 dwellings built in Glasgow between 1866 and 1874, half were of two rooms and a quarter of only one;[48] in 1911, 20 percent of Glasgow's families were still living in one room, 46·3 percent in two rooms; of those one-room families, 57 percent were living three or more to the room—not counting lodgers.[49] Glasgow and Dundee were respectively about one-half and one-quarter as bad again, in percentages of population living more than two to a room, as the *worst* London boroughs, Bethnal Green, Stepney, Shoreditch, and Finsbury—the populations of which were, of course, nothing like so large. On the other hand it may reasonably be suspected that a Scottish family in a *new* 'single-end' was likely to be immeasurably better off than any English family in a made-down single room.

It remains only to answer two obvious questions about this tenement-type housing which was characteristic of the poorer parts of the Scottish Victorian city. Why did it, with all its evident nastiness, become characteristic? And—keeping a constant eye on the English comparison—was it in practice even more difficult a staple form of slum for Scottish municipalities to deal with than any which English ones faced?

To the presumed causes already suggested for this slum style's virtual universality in nineteenth-century Scotland must be added the workings of the feuing system. The independence of Scottish from English common law no doubt made innumerable practical differences to the business of local government and property ownership—differences which might well repay detailed investigation. The biggest and basic difference, however, was the Scottish landowner's normal method of developing his estates not (as in England) either by outright sale of the 'freehold' or by a lease of the ground for a stated term of years, but by a sort of compromise between these two normal English modes—a sale of the piece of land in perpetuity on condition of a perpetual payment of a 'feu-duty' in respect of it. Like the English landowner who issued a ground-lease, the Scottish could, if he were so minded (and if his bargaining power were strong enough), stipulate certain conditions for the ground's development; unlike the English, he (or rather his successors in the title) would never be able to resume absolute control of it. All owners of property in Scotland own it absolutely, in the sense that their right to it is inextinguishable except by their own voluntary cession of it; but no Scottish property owners, except the inheritors of property so ancient and secure as never to have been itself feued, are free from some obligation to pay a feu-duty on it; and that includes the owners of flats above the ground. Some feu-duty was bound to have become attached to each flat or, in Scottish parlance, 'house' that anybody owned, with the extremely complicated result that modern processes of sub-infeudation could give to each owner of each 'house' in a many-staired four-storey tenement block an absolute right to a proportionate interest in the piece of ground above which his 'house' stood. John Hill Burton did his best to explain this 'peculiar' tenure, as he called it, in his 1840 report for the benefit of Chadwick: 'should the tenement be destroyed, [the "house" owners] retain their right, though it can have no physical representative, till the proprietors who held beneath them have, by rebuilding, made as it were a pedestal for the real property to be erected on.'

The obstacles that such properties could put in the way of authorities seeking substantially to improve or to demolish them can be easily imagined, and constituted a weighty item in the list of Scottish city singularities. The feuing system made such transactions even more complicated than they were bound to be anyway. But—to revert to the question of Scottish slum origins—how far was the feuing system itself to blame for the staple Scottish slums being there at all? The extent of its responsibility can be exaggerated. For instance, the 1917 Royal Commission on Housing in Scotland, after carefully describing it, concluded that it had 'in many cases hampered the free development of land for building purposes' and encouraged landowners to feu land for beetling tenements, rather than for pleasanter, lower-level buildings, because they could thus get a higher aggregate of feu-duties.[50] No doubt this happened.[51] But the Scottish feuing system cannot have been much more to blame than the general British systems of private landownership and, through most of the nineteenth century at any rate, feeble or non-existent public controls of land-use and housing standards. Property owners south of the border also knew the greater financial returns to be gained, where demand was insistent enough, by similarly intensive exploitation of the housing needs of the poorer classes, who, no less in Glasgow than in Liverpool and London, absolutely had to live close to their places of (potential) employment, and hence on land the value of which was likely to be enhanced by its alternative suitability for industrial or commercial uses. If it is argued that, nevertheless, the Scottish estate developer did in his tenements generally crowd more persons on to an acre than did his English counterpart in his back-to-backs, it is also arguable that regional building and mortgaging customs were partly responsible, and English speculators, if they had known about tenements and understood how to build them, would certainly have done so. But in one respect, certainly, the feuing system did tend to encourage this tenemented exploitation of the land. The Scottish proprietor, or, to use the legal term, 'superior', parting with his land, knew that he would never recover it for redevelopment as would the Englishman developing it by building-leases. For the Englishman, especially if he was (as in the older cities he often was) of a hereditary land-owning family, accustomed to 'think big' and over a long term, immediate gains were not everything. An estate that was relatively lightly built on for the present was often thought to have a better chance of retaining its amenities within the growing city and of guaranteeing richer rewards when the ninety-nine-years leases expired, than one which was from the start exploited to its uttermost. The English suburban landowner was likely to have more of a choice; if he cared for his descendants' prospects, he would keep his development classy, and unconcernedly accept a financially lower return than was offered by immediate intensive development. The Scottish suburban landowner had no such choice. His national system compelled him, willy-nilly, to get the most he could, straight away. So up went the blocks; into them moved the house-owners or tenants; into the superior's coffers flowed the feu-duties; and, unless the district was one which exceptionally retained a high initial social rating, into the municipal offices sooner or later came intimations of social and sanitary nastiness.

In two respects this nastiness was different in kind from any that could bother an English municipality. First, it might have to do with that Scottish peculiarity, the common stair and the 'lobby' leading to it, which, being everybody's business to keep clean, became in effect nobody's. Its nastiness—'an upright street, constantly dark and dirty'—was specially mentioned in the famous *Proposals* of 1752, the blue-print for the making of classical Edinburgh. Both of Glasgow's Victorian Medical Officers of Health specified the common stairs of the tenements as one of their worst features. W. T. Gairdner remarked that in England there was always some ultimate owner of a tenement who was responsible for the stairway, and could be compelled to clean it; in Scotland it was simply an untended 'receptacle for foul air'.[52] J. B. Russell thus memorably summarized it:[53]

> The common stair of a Scotch flatted tenement is the analogue of the English court, not only as the means of access to the houses, but especially in the old buildings, in respect that it contains the common jawbox—the representative of the English gully-hole—and the common water-tap . . .; and in modern buildings in respect that it contains the common water-closet, the representative of the privy or trough water-closet, which stands at the head of the English court. Yet with all this similarity of function, the English court is at the worst a box open above to the free air, while the Scotch common stair is *at best* a longer, narrow box, fully open only at the lower end, with or without certain mockeries of ventilators at the upper end, and with windows at intervals, which may admit light, but are never opened, and serve no useful purpose for ventilation until by a providential accident, or a merciful exhibition of malice, the

panes of glass are smashed. In their *worst* form it is hard to say what the Scotch common stair is but a dark noisome tunnel buried in the centre of the tenement, and impervious both to light and air, excepting the fetid air which is continuous and undiluted from the house along the lobbies and down to the close, from which you start on your perilous and tedious ascent.

The second respect in which the social and sanitary problems of the Scottish slum were characteristically different from those of the English one had to do with the practice of 'making-down'. Victorian legislators, local and national alike, were (understandably enough) slower to authorize public interference with the insides than with the outsides of private dwellings, and the early public-health reformers were only able to get at insanitary interiors during cholera emergencies, which by general agreement justified exceptional powers. During the forties and the fifties building regulations began to be enforced by the better English local authorities—there were prescriptions of minimum volume, lighting, solidity, sanitation, and so on—but such regulations applied only to *new* buildings, and left the interior conditions of already existing ones subject merely to such limited interferences as could be justified under the Nuisance Removal Acts (1855 and 1858) and the 1848 Public Health Act.[54] In Scotland the situation was even less satisfactory, partly because the jealously guarded ancient jurisdiction of the Dean of Guild Courts—which, intelligently extended, might have been the Scottish slums' salvation—was, until very late in the century, concerned only with structural strength and *external* amenity (thus, for instance, Dr Littlejohn was once obliged to complain, that the city officials had no power to prevent the installation of defective drains, or the ventilation of internal water-closets into bedrooms; they could do nothing until, a building being furnished and occupied, its occupants formally complained):[55] and partly because that wholly internal 'making-down', which was so much more a Scottish problem than an English, produced slums in exactly the way which was most difficult for the Victorian legislator to envisage tackling. Edinburgh—ahead, for once, of Glasgow in a housing context—seems to have got these powers for itself in its Police Act of 1879;[56] Glasgow and the rest of Scotland had to wait until after 1890.[57]

The chief impression we are bound to take back from this limited though outlandish excursion to the most northerly Victorian cities is one of contrast and variety. The Scottish cities themselves plainly differed from one another in the ways in which they met or did not meet the challenges that peculiar combinations of urban circumstances threw up in one place or another. These were differences of response, as palpable in their way as the differences of accent, to which long force of social circumstance had given rise in every place. There could on the face of it be no justification for a Scottish stereotype of the Victorian city. Yet the more one looks the more one feels justified in identifying a very general type of Scottish Victorian city, characteristically unlike the English one. To discover the whole extent of its singularity would naturally require a far more searching expedition than the one I have been conducting. It would have to be a full-scale undertaking, concerned not only with superficial differences of urban form or the means of civic identification but also with a whole range of matters—beginning perhaps with climate and geology, crossing the whole terrain of the arts and sciences, and ending with Calvinism and the national character. None of these considerations need—or should—obliterate the particular characteristics of the individual case, for cities are human aggregates and cannot conceal their human inclinations. Whenever Englishmen think of Scotland they think, as often as not, of 'Caledonia stern and wild'. They never lose sight of the contrasts between their own landscape and that of the Scots. The point of this essay is to suggest that the contrasts of townscape, if not of the urban lives they shaped, were as great.

Notes

1. 'London', in *English Hours* (1905), p. 34 [first published in *Century Magazine* in 1888].
2. What follows is an amended version of my paper, 'The Scottish Victorian City', *Victorian Studies*, xi (1968), 329–58.
3. *Parliamentary Papers* (*P.P.*), 1902, CXXIX, Eleventh Decennial Census of the Population of Scotland: Appendix VII, 'Municipal and Police Burghs' (Cd. 1257), pp. 937ff.
4. Public Record Office (P.R.O.), Home Office Papers (H.O.), 45/7940. The Board's complaint led to the appointment of George Munro, the Sheriff of Linlithgow, to draft the bill which became the Public Health (Scotland) Act of 1867.
5. P.R.O., H.O., 45/7933/22–4.
6. The only attempt to apply this legislation was in Leith. See *P.P.*, 1884–5, XXXI, Royal Commission (R.C.) on the Housing of the Working Classes: Scotland [Dilke Commission] Minutes of Evidence (C. 4409), QQ. 18,470–1, 19,923–8.
7. Cross's English Acts were 38 & 39 Vict. cap. xxxvi [1875], 42 & 43 Vict. cap. lxiii [1870]; the Scottish one was 38 & 39 Vict. cap. xlix [1875].
8. The Lands Clauses Consolidation (Scotland) Act, 1845: 8 & 9 Vict. cap. xix.
9. W. C. Spens, 'On the Necessity of a General Measure of Legislation for Scotland with regard to Public Health', *Proceedings of the Philosophical Society of Glasgow*, xi (1877–9), 129–43.
10. See Dilke Commission, Q. 18,371.
11. See Royston Lambert, 'Central and Local Relations in Mid-Victorian England: The Local Government Act Office, 1858–71', *Victorian Studies*, vi (1962), 121–50.
12. Various piddling amendments to the Public Health (Scotland) Act were passed in 1871, 1879, 1882, and 1891. They are conveniently summarized by a complacent lawyer, T. G. Nasmyth, 'The Public Health (Scotland) Act, 1897', *Juridical Review*, x (1898), 12–24.
13. Q. 18,363.
14. *The Making of Classical Edinburgh 1750–1840* (Edinburgh, 1966), pp. 50–1.
15. Letter of 12 May 1856, P.R.O., H.O., 45/6365.
16. *Scottish Law Review*, xii (1905), 223–30.

17 'On the "Ticketed Houses" of Glasgow, with an Interrogation of the Facts for Guidance towards the Amelioration of the Lives of the Occupants', *Proc. Phil. Soc. Glasgow*, xx (1888–9), 1–24; the article also appears in a memorial volume of his writings edited by his successor, A. K. Chalmers, *Public Health Administration in Glasgow* (Glasgow, 1905).

18 *P.P.*, 1909, XXXVIII, R.C. on the Poor Law: Report on Scotland (Cd. 4922), p. 263.

19 Albert Shaw, 'Municipal Socialism in Scotland', *Juridical Review*, i (1889), 33–53. I have added some information given by William Smart, 'The Municipal Industries of Glasgow', *Proc. Phil. Soc. Glasgow*, xxvi (1894–5), 36–53.

20 C. M. Allan, 'The Genesis of British Urban Redevelopment with special reference to Glasgow', *Economic History Review*, new ser., xviii (1965), 608.

21 Ibid., 613.

22 *P.P.*, 1866, XXXIII, Eighth Report of the Medical Officer of the Privy Council: Appendix No. 2 (3645), pp. 421ff.

23 25 & 26 Vict. cap. cciv [1862]. 'Ticketing' is mistakenly dated from 1866 in *P.P.*, 1917–18, XIV, R.C. on the Housing of the Industrial Population of Scotland, Rural and Urban: Report (Cd. 8731), p. 345, para. 794.

24 Ibid., paras 790–3. It was introduced into Liverpool in 1866. By 1888, 18,967 'houses' were under inspection. (Glasgow had 23,288.) See E. W. Hope, *Health at the Gateway* (Cambridge, 1931), pp. 150–2. Somewhat similar powers were given to metropolitan local authorities by section 35 of the 1866 Sanitary Act, but only two of them, Chelsea and Hackney, seriously sought to use them. See C. J. Stewart, ed., *The Housing Question in London 1855–1900* (1900), p. 71.

25 *England: Its People, Policy and Pursuits* (1879), I, p. 153.

26 *An Economic History of Modern Britain* (Cambridge, 1930), I, pp. 61–2.

27 Hunter, *Report on the Housing of the Poorer Parts of the Population in Towns*, pp. 145–6. On pp. 150–7, he lists the names of one-room dwelling proprietors who were found to be keeping lodgers. Most of them certainly sound Irish.

28 R.C. on the Poor Law, 1909, pp. 199, 201, 230, 231.

29 'Suggestions with regard to the Summary Jurisdiction of Magistrates in Scotland', *Transactions of the National Association for the Promotion of Social Science* (*T.N.A.P.S.S.*), 1874 [Glasgow meeting], pp. 230–3.

30 'Crime in Scotland', *T.N.A.P.S.S.*, 1877 [Aberdeen meeting], pp. 312ff.

31 See Edwin Muir, *Scottish Journey* (1935), p. 116.

32 See his introduction to his edition of Edwin Chadwick's *Report on the Sanitary Condition of the Labouring Population of Great Britain* (Edinburgh, 1965), p. xxiii. The judicious William Smart described Alison as 'neither a sentimentalist nor a doctrinaire'. See Smart's impressive 'Memorandum on the History of the Scots Poor Laws prior to 1845', appended to the Report on Scotland (p. 313) of the R.C. on the Poor Law, 1909.

33 *P.P.*, 1839, XLII, Reports from Assistant Hand-Loom Weavers' Commissioners (159), pp. 565–6. Alison concludes: 'It is my firm belief that penury, dirt, misery, drunkenness, disease and crime culminate in Glasgow to a pitch unparalleled in Great Britain.'

34 *Past and Present* (1843), Bk III, ch. 2; Bk I, ch. 1.

35 Flinn, op. cit., pp. 97–9.

36 Appendices to Eighth Report of the Medical Officer of the Privy Council, 1866, pp. 62, 213.
37 'Address on Health', *T.N.A.P.S.S.*, 1877 [Glasgow meeting], p. 100. He added 'some edifices in Paris and Berlin, and some tenement-houses and crowded slums reported of in New York and Boston'.
38 Clapham, op. cit., III, pp. 451–5.
39 A. K. Chalmers, *The Health of Glasgow* (Glasgow, 1930), opposite p. 76.
40 Cf. the figures given by Chalmers, op. cit., p. 193, with those in Clapham, op. cit., III, p. 455.
41 See especially the following papers: 'On the Ticketed Houses of Glasgow' (no. 17); and 'The House in Relation to Public Health', *Trans. Glasgow Insurance and Actuarial Society* (1877), series 2, no. 5 (repr. Chalmers, ed., *Public Health Administration in Glasgow*, pp. 170–89). He also gave, in *Life in One Room* (Glasgow, 1888), an unforgettable impression of its accumulated nastiness (originally a lecture delivered to the Park Parish Literary Institute, Glasgow, 27 February 1888; repr. Chalmers, pp. 189–206).
42 Clapham, op. cit., III, p. 462.
43 This conclusion is based partly on my reading of the evidence given to the R.C. on Housing (Scotland), 1917–18, pp. 345ff. Clapham (III, p. 463) seems to say the same.
44 See his paper, 'Defects of House Construction in Glasgow', *Proc. Phil. Soc. Glasgow*, vii (1870–1), 245ff. He may have been wrong about this, but he was ahead of his time in the clarity of his recognition of the force of habit and custom.
45 Clapham, op. cit., II, p. 495.
46 'On the Immediate Results of the Operations of the Glasgow Improvement Trust at May 1874, as Regards the Inhabitants Displaced, with Remarks on the Question of Preventing the Recurrence of the Evils which the Trust seeks to Remedy', *Proc. Phil. Soc. Glasgow*, ix (1873–5), 214.
47 See John R. Kellett, 'Property Speculators and the Building of Glasgow, 1780–1830', *Scottish Journal of Political Economy*, viii (1961), 211–32.
48 Russell, 'On the ... Operations of the Glasgow Improvement Trust', *Proc. Phil. Soc. Glasgow*, ix (1873–5), 225.
49 Tables 69 and 73 in *Glasgow* ('Third Statistical Account of Scotland'); Clapham, op. cit. III, p. 460; R.C. on Housing (Scotland), 1917–18, pp. 345ff., paras 724–51.
50 *Report*, para. 1511.
51 Kellett, op. cit., p. 212.
52 See note 44.
53 'On the ... Operations of the Glasgow Improvement Trust', *Proc. Phil. Soc. Glasgow*, ix (1873–5), 217.
54 Dr Hunter gives a detailed description of how this was done in his report, already referred to in note 22. On p. 58 he sums up thus: 'The administration of [these Acts] by the local authorities varies in efficiency in various places with the strength of the bye-laws, the skill, industry and independence of the officers, and lastly with the presence or want of a M.O. in places where such an appointment is desirable.'
55 In his *Report on the Sanitary Condition of the City of Edinburgh* (Edinburgh, 1865), pp. 118–19.

56 See Russell, ed. Chalmers, *Public Health Administration in Glasgow*, pp. 224–5; evidence of James Gowans (Edinburgh's Dean of Guild Courts) to Dilke Commission, QQ. 18,848–50; and, in general, Dr Littlejohn's evidence to the same, Q. 18,939ff.
57 See G. W. Barras, 'The Glasgow Buildings Regulations Act (1892)', *Proc. Phil. Soc. Glasgow*, xxv (1893–4), 158–9. The relevant legislation was the Glasgow Police (Amendment) Act of 1890, and the Glasgow Buildings Regulations Act of 1892; respectively, 53 & 54 Vict. cap. cci (Local) and 55 & 56 Vict. cap. ccxxxix (Local).

17 Metropolitan Types
London and Paris compared

Lynn Lees

How can we reconstruct the Victorian city? Pictures, newspapers, and reminiscences tangibly recreate the 'light and shade'[1] of the Victorian environment and thus present the historian with indelible images of a vanished urban world. Visual and personal links to the nineteenth century are indeed indispensable, but a note of caution must be sounded. The human memory is selective, and artists and journalists are involuntarily drawn to the picturesque, to the different. We should not neglect those sources which describe the typical as well as the unique. We should turn more often to the work of those bureaucrats and amateur statisticians who tried through their reports to keep pace with the urban world exploding around them. Although these men were sometimes too close to their subjects to do more than record and classify, the historian can return to their voluminous tabulations and turn them to his own purposes. That is, he can go beyond the numbers to discover the half-hidden structure of the nineteenth-century city.

The search for the Victorian city will be both simplified and enriched if the historian turns not only to the English past but also to the foreign analogues of his chosen subject. International comparisons of urban structures can help to establish the distinctively English qualities of Victorian cities. Moreover, they force the historian to refine his vocabulary and his analytical approach. To use the term, 'Victorian city,' is to suggest that national differences, rather than functional ones, provide the most useful way of classifying cities in the modern period. But the evidence for this assumption has not been adequately examined. If our use of the Victorian label is to have anything more than sentimental significance, we must be prepared to

The Victorian City: A Change of Accent

justify its application to English towns and to specify how and why these cities in the nineteenth century differed from hundreds of others located abroad. The search for variations on an English type should not deflect historians from the task of defining the place of English cities generally in the international urban spectrum.

Before proceeding any further with this design for research and analysis, I should add that this chapter will not attempt either of these projects with reference to 'the Victorian city,' but instead will compare two specific cities—London and Paris—during the nineteenth century. The aim of the comparison will be to illustrate certain similarities and differences of structure and of function between these capitals. Finally, I hope to justify the use of the category 'metropolis' as an urban type and to specify ways in which the English variant was unique. Whatever national characteristics link Victorian cities, it is a mistake to neglect international uniformities that can help to explain the structure of cities and the consequences of urbanization at specific points in time.

Cities such as London and Paris had a special identity in the minds of contemporaries. Cobbett's description of London as 'the Great Wen' was only one of many that marked the metropolis off from other English cities. Calling the capital 'Babylon' or simply marking it off from 'the provinces'—as was also done in France—indicates an awareness that these cities were extraordinary. They were, in fact, viewed in a highly ambivalent manner, as if they possessed both the best and the worst that the nineteenth-century city had to offer. Part of their symbolic value as prototypes of urban good and evil came from their attraction for outsiders and from their unique sizes. Indeed, their extraordinary growth, which set them apart from many other cities, was one of the primary determinants of those changes in urban geography that marked their evolution during the Victorian period. Between 1801 and 1851, Paris doubled while London almost tripled in size as a result of growth rates that ranged from 10 percent to over 25 percent per decade. Consequently, the Victorian metropolis passed the million mark by 1811, and Paris did the same thirty-five years later. By 1891, over 4,000,000 people lived in London and 2,500,000 in Paris.[2] Growing far faster than its national population, each reinforced its long-established dominance as the country's major city over the very weak claims of rivals who lacked the political and cultural importance of a capital.

In societies not accustomed to the galloping expansion of 'megalopolis,' the immense size of these cities seemed extraordinary, particularly in France where far more of the population lived in rural areas. Even in the much more urban English society, the size of London was clearly atypical. As was the case with many large English cities at the time, migration into it was a flood rather than a slowly moving stream; during the 1840s, over 300,000 people settled in the metropolis, and during the 1870s, an equivalent number moved into Greater London. Even if London's rate of expansion was slow when compared with that of northern industrial cities, its growth remained largest in absolute terms. Across the Channel, the preponderance of Paris was equally marked, and the element of imbalance even more striking, although both the rate of increase and the absolute level of migration into the French capital remained far lower than in London. (For example, about one-third as many people moved into Paris during the 1840s as settled in the English capital.) Parisian growth, nevertheless, has been called 'monstrous' and 'aberrant'; during the years between 1831 and 1851, when the French population grew only by 9 percent, Paris grew by 55 percent.[3] Not only did the capital expand enormously at a time when the national rate of population increase was slowly declining, but also it regularly attracted more newcomers, both relatively and absolutely, than the manufacturing and commercial cities which were directly affected by industrialization.

In addition to this internal expansion, both capitals induced intensive migration into nearby areas—into Greater London and the remainder of the Seine Department —producing a fringe of urban districts. During the second half of the nineteenth century, these areas grew at faster rates than did the center cities and quickly became part of the gargantuan urban regions which today dominate England and France. Already in 1851, 13 percent of the population of England and Wales lived in metropolitan London, but at that time the northern industrial region provided a second center that attracted men, money, and attention. Since 1900, however, when the economic expansion of the northern textile and metal industries slowed, London has had little competition which could redirect migrants elsewhere. Movement into the south-east intensified, producing by 1961 a metropolitan region of 722 square miles and over 8,000,000 people, itself dwarfed by the London planning region that extends roughly forty miles in all directions from Charing Cross. In France, the attraction of Paris has been even more consistently unchallenged, with the result that the expanding ring of industrial suburbs around the city proper spread throughout the Seine during the nineteenth century and beyond the department by 1918. Again, planners and administrators have been forced to redirect their attention from city to region, to an area of several thousand square miles where 18·5 percent of the French population lives.[4] The distinction between city and country no longer applies because the metropolis has constantly expanded beyond the formal limits set by its governors.

Even before suburban London and Paris engulfed the nearby countryside their links with neighboring counties and departments were close. Arthur Redford has described the 'drift' of new settlers from farm to village to metropolis that characterized migration in England during the early nineteenth century. Subsequent studies of the birthplaces of Londoners show that about half of the migrants to the capital during the Victorian period came from bordering counties and other areas of the east, south-east, and south Midlands. The Scots and the Irish were the only exceptions to the rule that movement into London decreased rapidly as the distance to travel increased. Over 50,000 Celts (one migrant in six) arrived and stayed in the city during the 1840s, although the number dropped to 25,000 during the next decade. Paris, too, recruited most of its migrants from bordering departments and the north-east part of France. Except for a stream of men coming from the Massif

The Victorian City: A Change of Accent

Central, a legacy from a time of heavier seasonal migration, Paris in 1850 held far less attraction for those who had a long distance to travel.[5]

Although the growth of both capitals depended to a large extent upon regional migration, it should be noted that the reliance of Paris upon outsiders was much more extensive. A comparison of the birthplaces of Londoners and Parisians makes this quite clear. (See Table 17.1.) Approximately 40 percent of all Londoners during the mid-nineteenth century were born elsewhere. In Paris, migrants made up an even larger share—50 percent to 60 percent—of the capital's population. Indeed, Adna Weber has calculated that only 15 percent of the city's expansion between 1821 and 1890 resulted from natural increase. In contrast, London's growth during the second half of the century depended very heavily upon an excess of births, despite

Table 17.1 *Birthplaces of inhabitants, London, 1841–91 and Paris, 1833–96*

London	1841	1861	1891	Paris	1833	1861	1896
	(percent)				(percent)		
London born	64·9	62·1	65·0	Paris born	50·0	36·1	36·8
				banlieue	—	—	3·0
Other counties—							
England & Wales	28·9	30·4	29·4	Other departments	43·0	58·7	50·1
Scotland	1·3	1·3	1·1	Alsace-Lorraine	—	—	3·7
Ireland	7·6	3·8	1·1	Abroad	4·0	5·2	6·4
Abroad	1·8	2·4	3·4	Unknown	3·0		

Sources: H. A. Shannon, 'Migration and the Growth of London, 1841–1891,' *Economic History Review*, v (1935). L. Chevalier, *La formation de la population parisienne au xixe siècle* (Paris, 1950), p. 46.

Table 17.2 *Mortality and fertility in London and Paris, 1821–81*

	London		Paris	
	Births per 1,000	Deaths per 1,000	Births per 1,000	Deaths per 1,000
1821–30	—	—	37·4	32·6
1841–50	30·0*	26·1*	31·2	29·3
1861–70	35·4	24·4	30·5	26·4
1881–90	33·2	20·4	26·3	23·7

* 1838–41 only.
Sources: Jacques Bertillon, *Des recensements de la population de la nuptialité, de la natalité, et de la mortalité* (Paris, 1907), pp. 17–18. *Annual Reports of the Registrar-General* (1841–81).

the high level of migration. Weber attributed this difference between the two capitals simply to the difference between their birth- and death-rates. (See Table 17.2.) Mortality in Paris was substantially higher, and the birth-rate lower, than in the English metropolis. To some extent, these cities were merely reflecting national demographic patterns which produced a much slower rate of population growth in France than in England at this time, but the specific effects of extreme overcrowding, poverty, and inattention to sanitation, problems which seem to have been relatively more acute in Paris, should not be overlooked.[6]

Differences in the capacity of these two capitals to house their thousands of new citizens adequately can help to account for the more dismal Parisian demographic situation. London had one substantial advantage, that of space. Although Paris remained surrounded by the *octroi* wall of Louis XVI until 1861 and by the fortifications of Louis Philippe until the First World War, London had spilled over its Roman wall centuries earlier, gradually absorbing the small settlements in the area. When reached by speculative builders and the transportation network, outlying parishes became part of London's suburban sprawl and joined the metropolitan district for all practical purposes. Thus shaped by the availability of transportation to the center and by the entrepreneurial expertise and interest of large landowners rather than by administrative fiat, London continued its haphazard development throughout the Victorian period.[7] Because its physical area was almost ten times the size of Paris, the city avoided, for the most part, those extremes of concentration that transformed blocks of gracious *hôtels* into rabbit warrens. While most of London's outside ring of districts contained fewer than twenty-five persons per acre in 1851, the lowest density in a Parisian *arrondissement* at this time was three times that level. (See Maps XII and XIII.) Although a few of the more notorious workers' quarters in central London sheltered over 300 persons per acre, these areas were very small in comparison with the large sections of the center city that were much less densely populated. Moreover, the average density per dwelling in the most tightly-packed parishes did not exceed fifteen persons per house, and rarely did it rise over a rate of ten. On the other hand, in central Paris virtually all of the central *arrondissements* north of the river sheltered more than 200 persons per acre, and fewer parks and open spaces existed to relieve overcrowded areas. As early as 1826, the III[e] *arrondissement* had an average density of thirty-five persons per house. Paris was a concentrated city whose growth both upward and outward was confined by government regulations limiting the heights of buildings and preserving the eighteenth-century boundary of the commune. Only in 1861 were adjoining suburbs annexed, and even then the area of the city was little more than doubled. This style of growth contrasts directly with the London pattern.[8]

The urban landscapes of London and Paris were, nevertheless, evolving in the same direction. By mid-century, rapid migration into the historic cores of both cities had slowed decisively. Although newcomers still settled in the center cities, their arrival was balanced by the regular exodus of others. The City of London's population, which had remained stable for the first half of the nineteenth century, declined rapidly thereafter in size. Adjoining areas in both the east and the west

The Victorian City: A Change of Accent

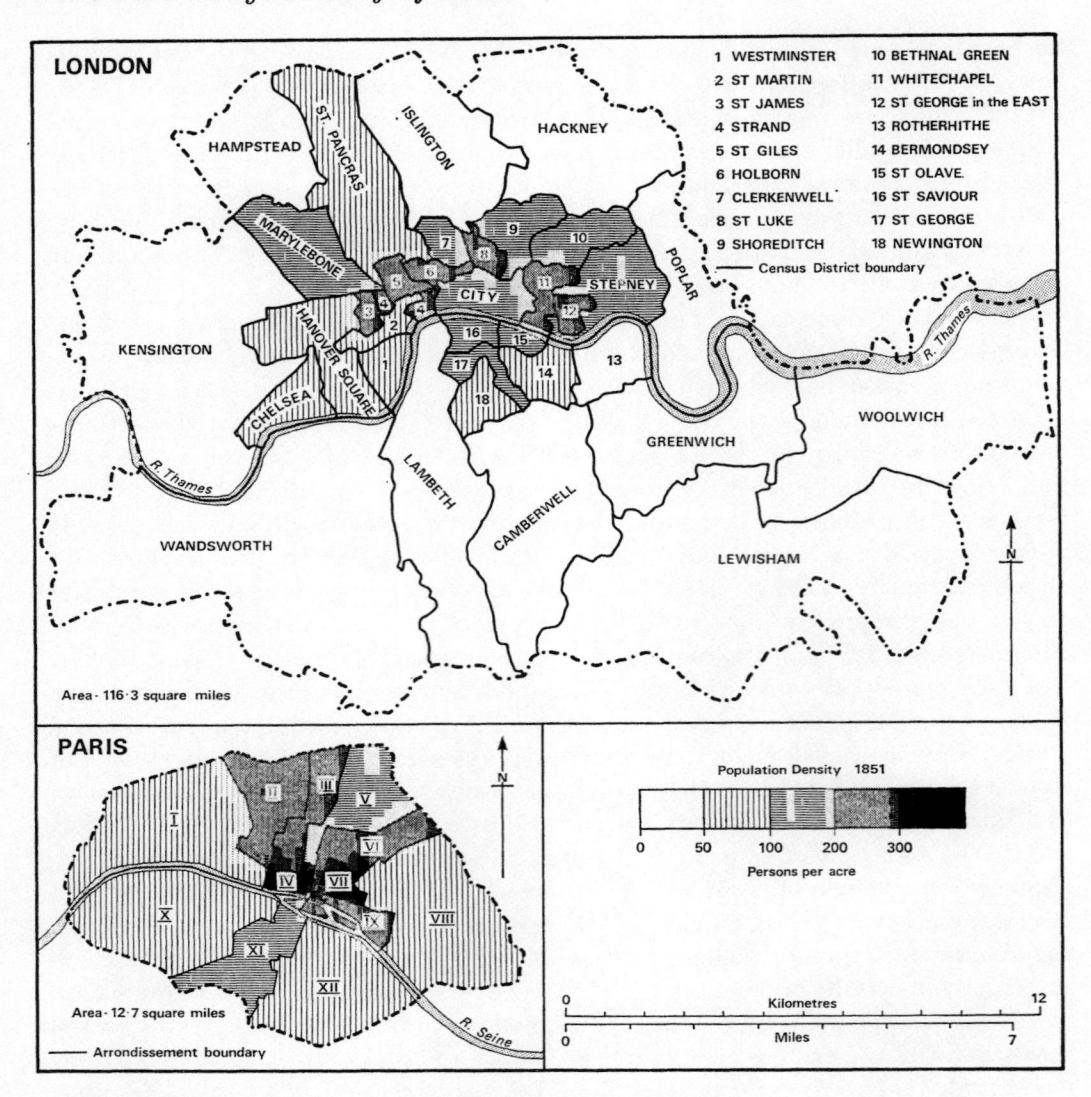

XII and **XIII** Population densities: London and Paris, 1851

underwent a similar change around 1851: net migration into St Giles, the Strand, parts of Holborn, and Whitechapel virtually stopped, while movement into the outer ring of districts intensified. Similarly, the growth of the central *arrondissements* in Paris declined and sometimes stopped during the 1830s and 1840s in periods when the city was expanding less rapidly. After 1861 the central ring lost population regularly to the area beyond the boulevards under the impact of Haussmann's razing and an accentuated tendency to move out of the overcrowded core to areas of lower rents and lower density.[9] In both capitals, this centrifugal movement intensified the separation of social classes in residential areas which, in London, was already well advanced.

418

The cost of transport into the city effectively insulated the outer ring of suburbs from most working-class residents until the 1870s, and thereafter the availability of cheap trains directed them into certain areas. Meanwhile, many in the London middle class had deserted the center of town for the villas and wide streets of Kensington or Camberwell, and the Parisian bourgeoisie marked out for its own the expanding quarters that flanked the Seine in the west. The much-lauded Parisian system of mixed housing, where class divisions operated in vertical rather than horizontal space, broke down during the July Monarchy and the Second Empire. The whirlwind of renewal that changed the city geographically also changed it socially, destroying the older pattern of settlement along with the houses that had maintained it. As in London, areas in the east became much more exclusively inhabited by workers, while the middle and upper classes settled in the west. The communes of the soon-to-be-annexed Parisian *banlieue*, classic examples of the industrial suburb, became and have remained working-class areas.[10]

These currents of internal migration helped not only to differentiate areas from each other socially, but also to accelerate changes in the capitals' economic geography. As pre-industrial patterns of production broke down and fewer and fewer households lived, made goods, and sold them in the same building, these activities separated into different streets and parts of the city. Particularly in the case of capitals like London and Paris, which performed complex economic functions for the entire country, the specialization of their economic geography by region was inevitable. Each was its country's largest center of manufacturing and distribution; each provided a multitude of services and products obtainable elsewhere only with difficulty. In addition, London served as the chief port and international entrepôt for England's commercial empire. As these activities expanded, much of the manufacturing that had taken place in the centers of the capitals became unprofitable and was displaced. The watchmaker of Clerkenwell and the masons of the Place de Grève had to move elsewhere in order to make room for new banks, department stores, and offices.[11]

By the middle of the nineteenth century this transition was under way, but by no means complete. Although the secondary and tertiary sectors provided a growing share of the jobs available, industry still employed the largest share by far of the labor forces. (See Table 17.3.) London and Paris were advancing at different rates, however, in this transformation of their economies. At first glance, their industrial structures in 1850 look surprisingly similar; a variety of artisanal trades and light industries were located in both capitals, and several of these trades employed almost the same proportion of the male labor force.

Yet the more advanced state of the English economy can be seen in the greater development of several important sectors. The existence in London of substantial service and transport industries, as well as the higher proportion employed in commerce and finance, testifies to the more complex and diversified economic development of the English metropolis.[12] In Paris, each of these activities occupied a smaller percentage of the work force, in part because the city did not provide as favorable a physical or financial climate for international trade and business transactions. By

The Victorian City: A Change of Accent

Table 17.3 *Occupations of inhabitants, London and Paris, 1851–4*

	London, 1851		Paris, 1854	
	Males	*Females*	*Males*	*Females*
	(percent)		(percent)	
Agriculture	2·1	0·1	0·2	0·1
Industry	35·3	12·2	50·1	30·0
Transport	8·1	1·0	3·0	0·1
Service	3·6	17·0	6·1	13·7
General labor and unclassified occupations	6·0	2·4	5·1	3·9
Commerce and finance	4·2	0·5	2·7	0·8
Professions, administration, education	6·4	1·1	10·1	5·3
Outside labor market	34·3	66·4	23·4	47·5

Sources: *Census of Great Britain, 1851*, 'Occupations of the People, London.'
Seine, *Recherches statistiques sur la ville de Paris et le département de la Seine*, VI (Paris, 1860).

1850, French merchant bankers had not yet diverted large amounts of energy into the joint stock investment companies that would soon export French capital and entrepreneurial skills to Central and Eastern Europe, and their methods of financing trade were more stringent than those of the English. The heyday of the French Bourse and bond market that drew foreign governments in need of cash into Paris, like pins to a magnet, lay in the second half of the century.[13] But even apart from differences in banking systems, Paris simply could not handle the volume of foreign trade that regularly passed through the Victorian metropolis. The French shipping industry was no match for the English; the *quais* along the meandering Seine bore little resemblance to the giant wet docks of the Thames, built to hold ocean-going ships. As a result, Paris was less involved with international commerce.

Yet when all parts of these urban economies are taken into account, it is clear that we are dealing with basically similar industrial landscapes. Both London and Paris were centers of artisanal production before the Industrial Revolution and remained so long afterward. Despite the general shift to a machine-dominated technology, the transition to factory production was a slow process in both cities. But each was an important center of craft production. The size of the cities meant that a local mass-market for simple necessities existed, and their status as national capitals with comparatively large, resident upper classes also provided important markets for handmade luxury goods. 'Cheap and nasty' versions of the same items could be easily produced for those who would emulate the style of the rich without their incomes. In both places, the tradition of skilled handlabor remained strong. Therefore, artisans and journeymen—tailors, shoemakers, printers, butchers, and bakers—served both the rich and the poor; they, rather than proletarians, formed the major part of the urban working class in London and Paris before 1850. In both cities, the small workshop dominated the process of production. In 1851, only 21 percent of all industrial establishments in London employed over four workers, and a mere 14 percent had ten or more employees. In Paris at mid-century, 18 percent of industrial firms employed ten workers or more. While there were many small factories in both cities, which combined work by skilled laborers and simple machinery driven by steam power, large establishments were concentrated in a few trades. The capitals could boast large building and engineering firms, and also a few big gas-works, metallurgical plants, and refineries. A small but growing number of wage laborers worked in these trades and in the transport sector, particularly on the docks or on the railroads.[14]

The English example of this artisanal-city-in-transition was, paradoxically enough, less directly affected by industrialization than its French counterpart. Initially, rather than stimulating the growth of either the textile or metal trades in London, the Industrial Revolution dampened their progress. Except for a few residues of the silk industry, textile production virtually ceased in the metropolis by 1850, and the amount of heavy industry was quite limited. London specialized in other things, in the manufacturing and finishing of many items of clothing, machinery, consumer goods, and in trades where large-scale factory production emerged only gradually. Its vast and constantly expanding home market meant that local demand for food, clothes, and housing was exceptionally strong and that other industries which had been solidly established in the eighteenth century remained in the capital. Despite pressures to lower costs and to expand the supply of goods, the London labor force maintained its primacy in many areas. Large contingents of highly skilled residents produced the finest jewelry and surgical instruments in the country, and an unceasing supply of cheap low-skilled workers were recruited into the sweated clothing trade and into jobs such as the making of boxes or matches. Many types of manufacturing, therefore, continued in the metropolitan area in spite of competition from other cities.[15]

Paris, on the other hand, had much more heavy industry. Several large metallurgical and engineering firms were solidly based in Paris, and the suburban areas that were annexed in 1860 had become by that time minor industrial areas sheltering railroad yards, chemical plants, and a variety of small factories. Although many of these plants had been driven out of the city by high rents and high wages, the pressures of decentralization drove them no further than the city wall. Yet the economic growth of Paris, which encouraged these establishments, undermined them at the same time. The strength of Parisian industry lay in the high quality of its techniques and in the abundance of skilled labor available locally. But costs of production were comparatively high. As soon as firms began to produce for a mass-market and to mechanize, they found it more profitable to move elsewhere. By the 1830s, the Parisian textile industry had become increasingly decentralized and had turned from cotton spinning to the more specialized, luxury branches of the trade—to silk, cashmere, and fine wools. Firms which needed a great deal of space moved into the suburbs or the provinces; even luxury workshops decentralized their production, allowing the routine stages of work to be done outside the city and leaving only the difficult tasks to Paris artisans. Operations which required highly skilled labor could still flourish in the capital, but industrial progress elsewhere in France contributed to the relative economic decline of the city. Over the long run, Paris, like London, would turn more and more to the provision of services.[16]

If we turn away from the economic structures of these capital cities to the structure of their labor forces, we can explore the operation of two urban societies, translating their economic functions into human terms. The first point to be made is that, during the middle of the nineteenth century, most adults considered that they had an occupation, and indicated this to the census-taker when he enquired. Whether or not they were currently employed, the vast majority of men in both capitals and a sizeable percentage of women identified themselves with one specific trade. The proportion of both men and women in the labor force, however, differed significantly between the two capitals. (See Table 17.3.) Comparatively more Londoners were supported by their families and did not work themselves. The difference in the position of women is especially marked: only one-third of London women had an occupation, while in Paris the proportion was one-half. In this case, however, the statistics are deceptive, for they do not take into account the substantially different demographic profile of the two cities. Part of this difference between the two proportions of employed women results from differences in the age structures of the two populations. In London in 1851, 32 percent of the residents were under the age of fifteen, while the corresponding figure for Paris was only 20 percent. In as much as there were substantially fewer children in the French capital, average Parisian families in 1851 were small relative to those in London. (Compare a range of 2·01 to 2·87 persons per family in the French capital with mean family sizes between 3·5 and 4·5 persons found in London.) Death and estate records for the year 1847 in Paris indicate that only one-third of the workers' or servants' families had more than one surviving child. While bourgeois families in Paris tended to be larger, only in a few occupations did the proportion with more than one child rise over 50 percent. Local studies of London working-class families show, however, that most had between two and three children.[17]

Part of the difference, therefore, between the proportions of people in the London and the Parisian labor markets resulted from the simple fact that there were fewer children in the French capital. But two other factors, the availability of education and the frequency of child labor, also affected the rate. In England, children had a greater opportunity to attend school. One child in three under the age of fifteen living in the metropolis was listed in the 1851 census as 'scholar.' Enrollments in Paris in 1846, however, were limited to one child in five under the age of fourteen. There is also evidence to indicate that the level of child employment was higher in the French capital around 1850. If information contained in the English census is used as evidence of participation in the labor market, 10 percent of residents under the age of fifteen claimed to have had an occupation; at the same time, 11·4 percent of Parisian children of fifteen and under were employed in the primarily skilled jobs surveyed by the Chamber of Commerce. Although various kinds of factory jobs were included, many of the unskilled, casual jobs that children would have held were not covered by the enquiry. The most plausible conclusion, therefore, is that comparatively more children had jobs in the French capital.[18]

Even though demographic factors help to account for differences in these metropolitan labor forces, other pressures helped to produce this outcome. A comparison of social structures will clarify some of these forces. Statistics on the social structures of London and Paris can be obtained indirectly from the censuses of 1851 and 1854. I have reclassified information on employed residents by placing each occupational category into one of the five status groups outlined by the English Registrar-General in 1950. Comparisons between the two capitals are possible because of the degree of specificity in the occupational listings of both censuses. The report for Paris in 1854 is more general than the English tally but many of the categories used correspond exactly. In addition, distinctions made in the French census between *patrons*, *employés* or *ouvriers*, and *domestiques*, separate the Parisian labor force automatically into social groups. This tripartite scheme does not permit the division of French workers into detailed categories according to the nature of their jobs, but the lines of demarcation between the middle class and workers are clearly drawn. Although detailed problems of classification abound and certain trades are suspiciously hard to find among the supposedly exhaustive published lists, the French categories can be compressed into the English ones.[19]

Both London and Paris around 1850 can be described as hierarchical societies which consisted of comparatively small upper and middle classes completely outnumbered by a working class of artisans, laborers, and domestic servants. The proportion in each capital that should be counted as working class falls between 75 percent and 80 percent. The remainder of the people can be divided between two categories that include upper- and middle-class occupations (*grande* and *moyenne bourgeoisie, petite bourgeoisie*). (See Table 17.4.) In both countries, the elite composed of professionals, bankers, large employers, and administrators was quite small. Yet the top category of Parisian society appears to have been significantly larger than its London counterpart. This disproportion results primarily from the position of two groups, the bureaucrats and the *rentiers*. For example, over 46,000 persons were classified as *rentiers* or *propriétaires* in Paris; they constituted over half of category I. The figure for London's men and women of independent means and house or land proprietors was only 14,920. Even if part of this disproportion arose from different official definitions, the nature of social values and of social reality was responsible for much of it. Many English property owners also had occupations and listed them in the census, thus placing themselves in other categories; in France, a generalized bourgeois prejudice against commerce and industry probably produced the opposite effect. It also led families to retire and live on unearned income in greater proportion than in England at this time, thereby swelling the ranks of group I in Paris. This difference in the position of the *rentier* illustrates two dissimilar standards of social

Table 17.4 *Social structure of London and Paris, 1850*

	Social groups	Percentage of employed population	
		London	Paris
I.	Capitalists, professionals, administrators, *rentiers*	4·3	10·4
II.	Lower ranking professionals and administrators, *patrons* and small employers, shopkeepers	16·6	13·0
III–V.	Working Class—all levels	79·1	76·1
	Skilled labor	39·7	—
	Domestic service	17·9	11·0
	Other semi-skilled labor	6·6	—
	Clerks	2·0	—
	Unskilled labor	10·6	—
	Unclassified	2·3	—

Sources: *Census of Great Britain, 1851*, 'Occupations of the People, London.'
Seine, *Recherches statistiques sur la ville de Paris et le département de la Seine*, VI (Paris, 1860).

behavior and patterns of behavior. Another obvious difference in French and English social structure can be seen in the larger size of the French bureaucracy. Approximately 19,000 Parisians were employed in *administrations publiques*; an equal number worked in local and national government posts in London. (These figures include the civil service as well as the police, customs, judiciary, and local administration of the capitals.) Since Paris was half the size of London in 1850, the group in government employ in France was proportionately twice the size of the one in London. But the representation of other occupations, such as doctors and lawyers, was relatively stronger in the English metropolis. When both categories I and II are added together, therefore, the upper and middle classes of the capitals seem to have been equivalent in size, even though not identical in structure.[20]

At the opposite end of the social spectrum were the workers, who can be divided into several groups according to the level of skill and security of their jobs. Craftsmen and other skilled workers made up the largest share. Some of these men can be called the 'aristocracy of labor,' a group which Eric Hobsbawm estimates to be no more than 15 percent of the working class although it varied from trade to trade, but others were far less favored. As might have been expected as a result of London's industrial structure, the skilled—whether well or badly paid—far outnumbered those employed in semi-skilled or unskilled jobs (the proportion of men with unspecified casual jobs having been 5 percent of the working class). This division into three categories is also useful for analyzing the Parisian working class. Unfortunately, neither the Paris Chamber of Commerce nor the census divided those employed within certain industries by the nature of the job that each man did. Until other sources are discovered, a statistical comparison of the structure of the Parisian working class with that of London will not be possible. The only group that can be isolated with some assurance is the domestic servant. The number employed in this trade in the English capital was proportionately much larger than in Paris. (See Table 17.4.)[21]

Some persons were better fitted than others to rise within this hierarchical structure. The lack of access to training and to education clearly blocked many from joining the favored few in the 'aristocracy of labor' or in the ranks of the middle class. The London Irish illustrate an extreme case of a group that clustered at the bottom of the society.

About half of the employed Irish-born males living in five London parishes of heavy Irish settlement in 1851 were either employed as general laborers or had similar jobs in the transport industry, principally on the docks. Many of the rest earned their livings as construction workers, street hawkers, tailors, or shoemakers. Fewer than 2 percent could be called 'middle class,' and under 10 percent had skilled jobs of any sort; moreover, half of this latter group worked in the worst of the sweated trades and can scarcely be counted among the 'aristocrats' of labor. Charles Booth's data on the position of migrants in a wide range of London trades clearly show that most of the jobs demanding skilled labor were dominated by the London-born. Except for some branches of the building, clothing, and engineering trades which were common in the provinces and, therefore, recruited heavily from them, the list of jobs held primarily by working-class migrants is weighted towards semi-skilled and low-skilled occupations. Where the rewards were low, so were the barriers to entry. Respectable branches of artisanal trades as well as the better paid jobs on the docks, where men were protected somewhat from the vagaries of fortune and economic changes by trade unions or friendly societies, could easily take on the character of closed shops, where entry was limited to a specific number of trained apprentices, often the relatives of those already in the trade. The power to exclude outsiders could offer protection in precarious times, and newcomers tended to be those excluded.[22]

Similarly in Paris, migrants were concentrated in the lowest ranks of the working class. A survey based upon the city's death-records in 1833 shows that of the migrants from twenty-six departments (73 percent of whose occupations could be traced), 16 percent were professionals, *propriétaires*, or *rentiers*, while 77 percent worked in industry or had a salaried position. Of this latter group, over half were porters, domestic servants, unskilled laborers, tailors, shoemakers, or laundresses. Almost certainly migrants were overrepresented in these trades. In addition, Adeline Daumard has shown that the Parisian bourgeoisie, with the exception of government employees, were more likely to have been born in Paris or to have resided there for over twenty years than were men employed as domestic servants or laborers, jobs which recruited much more heavily from the provinces. Despite the fact that migrants formed the majority of the Parisian population, the bourgeoisie was dominated by native Parisians or long-time residents. Booth's data on London middle-class occupations indicate that in the English capital migrants dominated education, the civil service, the ministry, and literature. Yet without information on length of residence in London of the men surveyed, it cannot be determined whether this result is incompatible with Daumard's findings.[23] Very little is known about social mobility among migrants; until more specific studies are made, it is premature to make a definite pronouncement upon their place in urban societies of the nineteenth century. The most plausible conclusion appears to be, however, that more of them entered and probably remained on the bottom levels of the society than their numbers warranted. This tendency was particularly strong in the case of identifiable ethnic groups like the Irish.

Up to this point in the analysis the structural similarities of London and Paris have outweighed the differences between them. Both capitals expanded enormously during the nineteenth century under the impact of heavy migration, each becoming the center of a vast metropolitan region. At the same time the geography of both cities underwent similar transformations, which were closely linked to wider changes of economic structure triggered by the Industrial Revolution and by their position as national capitals. Each acquired a variety of national and international economic functions which entailed the development of complex links with the provinces and with other parts of the world. On another level, the social structures of these urban societies resembled each other closely. Both were composed of similar groups with almost identical proportions in the middle class and the working class.

But a detailed comparison of the two capitals has also revealed important differences between them. The composition of the middle class, for example, differed substantially. In Paris, the elite was dominated by those claiming to live on unearned income and by men in the employ of the government, while the representation of several independent professions and of commercial occupations was much stronger in London. In keeping with England's economic supremacy, her upper and middle class was much more deeply involved in business and in other kinds of productive employment. The evidence presented in this essay indicates the presence of greater wealth and probably a more even income distribution in the English capital around 1850. Not only did proportionately more Londoners have domestic servants but also there were fewer people in the labor force and more children in school. To be sure, relative poverty within the working class is hard to measure. But the fact that Londoners had more children and lived longer than their Parisian counterparts offers circumstantial evidence that Londoners had a higher standard of living.

The most obvious difference between London and Paris, of course, lay in the realm of political structures. Paris was closely controlled by the central government, through the Prefect of the Seine and the Prefect of Police. The city was thought to be too dangerous to rule itself, and every new outbreak of revolutionary violence has reconfirmed this feeling. Even its Municipal Council had very limited powers: during the Second Empire, it was appointed from above. London parishes, on the other hand, possessed the right of self-government almost up to the point of urban anarchy:

no effective government beyond that of the City of London was installed until 1889 when the London County Council came into being. In neither case were these arrangements typical. The capitals' political structures were sharply differentiated from those of other cities. The English and the French alternatives, however, pointed in different directions. London was administratively ill-equipped to handle the day-to-day crises of metropolitan growth and change. Paris had a strong authority with easily expandable rights and powers that could, in theory at least, cope with urban problems. Yet with the exception of Baron Haussmann's concerted attack on Paris during the Second Empire, both cities moved only slowly and often ineffectively to improve the conditions of urban life. To borrow the words of one historian, 'Street improvements appeared to be a panacea for every urban problem.'[24] In fact, the status of London and Paris as capitals worked against any opportunities for constructive change and adaptation. So immense, expensive, and difficult to solve were their problems that even the halting steps toward reform taken in other cities were often denied them. As urban giants they could not be given the tools of mortal men.

The position of London and Paris as capitals decisively shaped the institutional framework within which their economic and social functions could operate. Without further research and statistical comparisons it is impossible to say whether each was structurally more like its international rival or its provincial siblings, but the special conditions under which each developed placed Paris and London in a category by themselves. The metropolis was distinct not only in size and splendor from the ordinary city but also in its institutions and its opportunities for development. Although the Victorian metropolis was unique in several ways, it should be seen as a national example of an international urban type.

Notes

1. Gustave Doré and Blanchard Jerrold, *London. A Pilgrimage* (1872), p. 2.
2. Adna F. Weber, *The Growth of Cities in the Nineteenth Century* (New York, 1899), pp. 46, 73.
3. H. A. Shannon, 'Migration and the Growth of London, 1841–1891,' *Economic History Review*, v (1935), 81; Charles Booth, *Life and Labour of the People in London*, first series: *Poverty* (1902), III, p. 124; Charles H. Pouthas, *La population française pendant la première moitié du xixe siècle* (Paris, 1956), p. 174.
4. Peter Hall, *The World Cities* (1966), pp. 30–1, 59, 68–9; Weber, op. cit., p. 47; Donald Read, *The English Provinces, 1769–1960. A study in influence* (1964), pp. 271–3.
5. Arthur Redford, *Labour Migration in England, 1800–1850* (Manchester, 2nd edn, 1964), pp. 62–6; Shannon, op. cit., 80–1; Louis Chevalier, *La formation de la population parisienne au xixe siècle* (Paris, 1950), pp. 162–7.
6. Shannon, op. cit., 80–2; Chevalier, op. cit., p. 45; Weber, op. cit., p. 240; Seine (département), *Recherches statistiques sur la ville de Paris et le département de la Seine*, V (Paris, 1840), Tables 118, 119.
7. Steen Eiler Rasmussen, *London: The unique city* (1937), pp. 23–4.
8. H. Price-Williams, 'The Population of London, 1801–1881,' *Journal of the Royal*

    Statistical Society, xlviii (1885), 401–13; Louis Chevalier, *Classes laborieuses et classes dangereuses* (Paris, 1958), p. 218; Pouthas, op. cit., pp. 158–64.
9 Price-Williams, op. cit., 389–99.
10 Chevalier, op. cit., pp. 240–1; Harold Pollins, 'Transport Lines and Social Divisions,' in Centre for Urban Studies, *London: Aspects of change* (1964), pp. 41–6.
11 Booth, op. cit., second series, V, pp. 59–60.
12 François Bédarida has calculated that in 1851 the tertiary sector employed over 50 percent of the active population. See 'Londres au milieu du xixe siècle: une analyse de structure sociale,' *Annales: Économies, Sociétés, Civilisations*, xxiii (1968), 278.
13 Rondo Cameron, *France and the Economic Development of Europe, 1800–1914* (Chicago, 2nd edn, 1961), passim.
14 George Dodd, *Days at the Factories; or, The manufacturing industry of Great Britain described. Series I-London* (1843); *Parliamentary Papers* (*P.P.*) 1852-3, LXXXVIII, *Census of Great Britain, 1851:* Pt I, pp. 28–9; Chambre de Commerce de Paris, *Statistique de l'industrie à Paris* (Paris, 1851), pp. 38–48; Chevalier, op. cit., pp. 107–11.
15 Booth, op. cit., second series, V, pp. 86–95.
16 Maurice Levy-Leboyer, *Les banques européennes et l'industrialization internationale* (Paris, 1964), pp. 116–18, 345–7; Chevalier, op. cit., pp. 108–17.
17 *Census of Great Britain, 1851*, Pt I, pp. 10–15; Chevalier, op. cit., p. 264; Adeline Daumard, *La bourgeoisie parisienne de 1815 à 1848* (Paris, 1963), pp. 9, 337; Lynn H. Lees, 'Social Change and Social Stability among the London Irish, 1839–1870' (unpublished Ph.D. thesis, Harvard University, 1969), pp. 94–5, 188–9; *P.P.* 1863, LIII, *Census of Great Britain, 1861*, Pt I, p. 11; R. W. Rawson, 'Results of some Inquiries into the Condition and Education of the Poorer Classes in the Parish of Marylebone in 1838,' *Journal of the Statistical Society of London*, vi (1843), 44.
18 *Census of Great Britain, 1851*, Pt I, pp. 10–15; Chambre de Commerce de Paris, p. 48; M. Gréard, *L'instruction primaire à Paris et dans les communes du département de la Seine en 1875* (Paris, 1876), p. 33.
19 More specific discussions of problems arising from the classification of occupations listed in the English census can be found in Bédarida, op. cit.; H. J. Dyos and A. B. M. Baker, 'The Possibilities of Computerising Census Data,' in H. J. Dyos, ed., *The Study of Urban History* (1968), pp. 87–112, and W. A. Armstrong, 'Social Structure from the Early Census Returns,' in E. A. Wrigley, ed., *An Introduction to English Historical Demography* (1966), pp. 209–37. Many of the same cautionary notes apply to the procedure of fitting French occupations into English categories.
20 Seine, *Recherches statistiques*, VI, pp. 626–51; *Census of Great Britain, 1851*, Pt I, pp. 10–15; David S. Landes, 'French Entrepreneurship and Industrial Growth in the Nineteenth Century,' *Journal of Economic History*, ix (1949).
21 E. J. Hobsbawm, *Labouring Men* (1967), pp. 328–36; Lees, op. cit., p. 160.
22 Booth, op. cit., second series, V, pp. 29, 131; Lees, op. cit., p. 152; J. C. Lovell, 'Trade Unionism in the Port of London, 1870–1914' (unpublished Ph.D. thesis, Cambridge University), pp. 97, 119–20, 127–31.
23 Daumard, op. cit., p. 9; Seine, *Recherches statistiques*, V, Tables 118–19; Booth, op. cit., second series, V, p. 29.
24 Anthony Sutcliffe, *The Autumn of Central Paris* (1970), p. 27.

Index

Index

Figures in italic type indicate illustrations; those in bold type indicate principal entries. Superior figures refer to notes.

A

Aberdeen, 139
  infant mortality, 649
  overcrowding, 23
  Police Acts, 394
  railway station, 280
  size, 391
  University library, 233
Abergavenny, infant mortality, 650
Abraham, A. P., 244[21]
Abraham, G. D., 244[21]
Accum, F., 680, 689[63]
Ackerknecht, E. H., 689[63]
Ackermann, R., *306*; 463
Acorn, G., 152, 160[239]
Act for Preventing the Adulteration of Articles of Food and Drink (1860), 660
Act of Separation of Free Church of Scotland (1843), 231
Acton, W., 188[20], 366, 383[21], 658, 700, 705[15]
Acts of Parliament, *see* separate titles
Adam Bede, 523
Adam, R., 432, 438
Adams, H., 551
Adams, W. E., 151, 160[238]
Adamson, R., 231
Adcock, A. St J., 599
Adderley, J., 592, 599, 600
Addison, J., 222
Adelaide Road (Hampstead), *290*; 349, 351, 353
Adelphi Arches, sleeping rough in, 128–9, 458, 573
Adelphi Terrace, 313
Adelphi Theatre, *53*, *54*, *64*; 220, 224
Adgey, R. J., 811[58]
Adickes, F., 56[66]
Adshead, S. D., 432
Africa, South, rate of urban concentration, 41 (table)
Africa, urban population growth, 34
  urbanization, 33
After London, 535
Ahrons, E. L., 305[15], 308[82]
Ainsworth, H., 739
Aintree, *158*
Aitchison, G., 316, 443
Akroyd, E., 882, 883
Akroydon, 883
Albert, Prince Consort, *82*; 233, 327, 634
Albert Hall Mansions, *277*; 321, 323
Aldcroft, D. H., 385[57]
Alden, P., 866
Alderson, F., 245[28]
Aldgate, Butcher's Row, *69*
Aldin, C., 313
Aldwych, *175*
Alexander, D. A., 440
Alford, H., 865
Alhambra Theatre of Varieties, *56*
Alice in Wonderland, 535
Alison, Dr, 624[77]
Alison, W. P., 401, 409[32,33]
All Men are Liars, 599
All Sorts and Conditions of Men, 589, 590, 592
All the Year Round, 576[5]
Allan, C. M., 409[20]
Allen, C. J., 306[30]
Allen, D. E., 190[44]
Alma-Tadema, L., 453
Alpert, H., 120[23,33]
Altick, R. D., 575[1]
Alton Locke, 535, 586
Amato, P. W., 55[59]
Amberley, Viscount, 781
America, Latin, agglomeration of population, 9

431

Index

America, Latin—*continued*
 child-woman ratio, 37
 demographic characteristics of immigrants, 37
 fertility, 37
 migration of foreigners, 15
 mortality rates, 34
 population concentration, 8, 41
 urban population growth, 34
 urbanization, 32, 33 (table)
America, North, concentration of population, 8
 density of population, 26
 effect of births on natural resource consumption, 43
 Union stations, 280
 urbanization, 32, 33 (table), 34
 see also United States
Amsterdam, railway station, 280, 301
Anderson, A. M., 157[118]
Anderson, M., 122[71]
Anderson, R., 444
Anderson, W. J., 850[30]
Andover, 444
Andreas-Salomé, L., 684[12]
Andrews, G. T., 302
Anerley, 126
Angrist, S. W., 686[25]
Anson, P., 851[55]
Answer to Davenant's Preface, 519
Anti-Corn Law League, 161, 176
Apology for the Revival of Christian Architecture in England, 443
Archer & Green, 323
Archer, J. W., 473[55]
Architecture, 311–28, 431–46, 903–4
 changes in investment sector, 323–4
 complexity of Victorian, 328
 concept of the Beautiful, 431–2
 concept of the Picturesque, 432–4, 903
 concept of the Sublime, 434–46
 connection with religious feeling, 320, 443–4
 dockland, 440–1
 ecclesiastical, 443–4
 governmental control, 326–7
 grotesque style, *151*
 growing importance of style, 321
 hotel, 324
 importance in public sector, 323
 industrial, 440–2
 individuality of, 433
 influence of Georgian, 312
 'latitudinarian', 316
 office block, 442
 profession of, 325–6
 reasons for changes in style, 311–12
 relation of High Church Movement to Gothic Revival, 320
 relation to new streets, 317
 Renaissance inspiration in Victorian, 432
 revolution in taste, 321
 rhetorical, 443, 444
 variety of styles in City, 315–17
 Victorian attitude to Georgian, 313
 see also individual architects
Architecture of Humanism, The, 434
Argentina, annual rate of urban concentration, 5 (table)
 child-woman ratio, 37
 concentration of population, 10, 34
 rate of urban concentration, 41 (table)
 vital statistics, 35 (table)
Argyll Rooms, 695, 699, 700
Aris's Birmingham Gazette, 828
Arkwright, Sir R., 438
Armstrong, D. L., 813[129]
Armstrong, W. A., 97, 104[64], 428[19]
Armytage, W. H., 888[15], 889[20,23]
Arnold, Matthew, 297, 434, 485–6, 496, 547, 556[13], 572, 581[58]
Arnott, N., 630, 661[20]
Arsenic Act (1851), 660
Art, 449–68
 acceptable subjects of, 451
 'Aesthetic Movement', 445, 490
 aspects of industry as subject of, 450, 470[23]
 campaign against artificial standards, 467
 ceremonial occasions, 464
 degradation of urban life shown in, 458
 development of flamboyant taste, 249
 French portrayal of London life, 460, 466
 increasing documentary realism, 459–60
 Irish migrants as subjects of, 459
 moral significance in, 456, 457
 Pre-Raphaelites, 452–3
 railways as subject of, 470[15]
 rejection of Academy authoritarianism, 468
 rejection of satire, 452
 relation to photography, 230–1, 235
 reluctance to depict social and urban realities, 450–1, 452, 457, 465
 theme of fallen women, 457
 topographical, 462–3
 use of genre, 454–6
 working-class activity depicted, 456, 457
 see also Architecture; Photography
Art and Socialism, 510
Art Journal, 449, 469[2]
Art of the People, The, 510
Art Union, 229, 232
Artisans, Labourers and General Dwellings Company, 183
Artists' response to city, 449–74
Artizans' and Labourers' Dwellings Acts (1868, 1879) ['Torrens Acts'], 392, 611, 613, 614
Artizans' and Labourers' Dwellings Improvement Acts (1875, 1879, 1882) ['Cross Acts'], 392, 395, 397, 568, 613, 614, 617
Arundel and Surrey, Countess of, 840
Arundel and Surrey, Earl of, 848
Ashton, H., *319*; 317
Ashton, J., 208[6,9,19,20], 209
Ashton, T. S., 101[12], 102[15]
Ashton-under-Lyne, 127, 128
Ashworth, H., 103[34]
Ashworth, W., 54[51], 119[10], 691[72], 889[24]
Asia, agglomerations of populations, 9
 demographic characteristics of migrants, 37
 urban population growth, 34
 urbanization, 32, 33 (table)
Associations of Medical Officers of Health, 608, 610
 see also Health, Medical Officers of
Astley's Theatre, 212, 549
Astor Estate Office, 319
Atget, E., 245[32]
Athenaeum, 576[5], 728
Athens, as primate city, 39
Atherton, H., 279
Atkinson, T., 341
Auden, W. H., 732

Index

Audy, J. R., 274[13]
Aurora Leigh, 90, 483–4
Australasia, annual rate of urban concentration, 5 (table)
  concentration of population, 8, 10
  population growth, 34
  urbanization, 32, 33 (table)
Australia, urban transformation, 42
  population, 8
  rate of urban concentration, 41 (table)
  vital statistics, 35 (table)
Austria, annual rate of urban concentration, 5 (table)
  marriage rates of urban population, 18
Autobiographical Sketches, 476
Aves, E., 135
Axon, W. E. A., 758[19], 759[40], 761[52]
Azad, Q., 56[67]

B

Bache, S., 820
Bagehot, W., 539, 541, 550, 555[4,7], 560, 576[4]
Bagshawe, J. R., 147, 159[204], 160[213]
Baines, E., 765, 776
Baines, M. T., 771, 776, 783
Baines, T., 279, 305[7]
Bakan, D., 684[5], 686[33]
Baker, A. B. M., 104[65], 384[49], 428[19]
Baker, E. A., 534, 536[16]
Baker, R., 768, 772, 782
Baker, Sybil, 789–814
Baker, W., 306[32]
Balfour, J., 172
Ballard, E., 640
Balston, T., 472[43]
Baltimore, population, 9
  slum conditions, 25
Baly, Dr, 644
Balzac, H. de, 223, 479, 481, 489
Bamford, S., 739, 741, 756[2], 757[5,6,11], 758[16,18], 760[45]
Bamforth, J., 239, 240
Banbury, 127
Bank, architecture, 315
Bank Buildings, 437
Bank, City, *260*
Bank, London and Westminster, 323
Bank, Manchester and Salford, *191*
Bank of Australia, *258*
Bank of England Branch—Manchester, *187*

Bank station, 284
Banks, J. A., **105–22**, 121[57,58,59], 122[64,65], 685[20]
Banks, Mrs Linnaeus (Isabella Varley), 745, 756
Banks, Olive, 121[58], 122[64,65]
Banton, M., 190[45]
Barber, M. and S., 735[68]
Barbican (London), 858
Barchester Towers, 518, 524, 525, 527, 528
Barge, T., 843
Barker, T. C., 103[34], 305[16], 306[20,26,31], 329[9], 385[57], 689[61], 691[73]
Barley, M. W., 690[66]
Barlow, H. C., 631, 661[25]
Barlow, T., 666[94]
Barlow, W. H., 440
Barnaby Rudge, 547
Barnes, G. N., 156[72]
Barnes, H. E., 120[32]
Barnes, W., 876
Barnett, S., 586, 592, 595
Barracks, Wellington and Knightsbridge, pubs near, 166, 172
Barraclough, G., 48[1]
Barran, J., 773
Barras, G. W., 411[57]
Barrie, D. S., 308[77]
Barrow-in-Furness, 125
Barry, Sir Charles, *324*; 255, 313, 315, 321, 326, 432
Barry, E. M., 301, 326
Barry, F. W., 338
Barry, J., 438
Bartlett, W. H., 473[49]
Basingstoke, railway, 277
Basle, 281
Bass, Ratcliff and Gretton, records of, 145
Batcheldor, T., 353
Bateson, J., 766, 777
Bath, building, 301
  railway architecture, 301
  uniformity of town plan, 339
Battersea, 467
  flats, 324
  gypsies, 130
  library, 323
  off-licensed premises, 168
  Shaftesbury Park, 183, 568
  unemployment, 368
Battiscombe, G., 662[46]
Baudelaire, C., 467, 479, 489, 493[12], 577[11]
Baumeister, R., 54[51], 56[66]
Bavaria, annual rate of urban concentration, 5 (table)
  infant mortality, 25
Baxter, R., 817

Bayswater, 313, 314, 317, 363
  migrants, 373
Bayswater Road, Red House, *270*; 322
Bazalgette, Sir J., 318, 468
Beale, C. H., 834[7]
Bealey, R. R., 757[5], 761[53]
Beames, Rev. T., 382[9], 735[68]
Bean, W. W., 787[65]
Beardsley, R. R., 690[64]
Beaumont, J. B., 590
Beckett, Sir J., 781
Beckett, J. C., 814[143]
Beckwith, J., 769, 770
Bédarida, F., 384[50], 428[12]
Bedford, Bishop of, 95
Bedford, Duke of, temperance policy, 165, 166, 167, 181
  estate of, 313
  see also Figs Mead
Bedford, F., 243
Bedford Park, 434
Bedford Square, 333
Bedfordshire, marriage-rates, 17
Bedingfield, Lady, 848
Bee, J., 382[6]
Beer Act (1830), 169, 170
Belcher, T., 173
Belfast, *427–34*; **789–809**
  Ancient Order of Hibernians, 799, 808
  attitude to Home Rule Bill, 799, 800, 803
  attitude to Repeal of the Union, 795
  Catholic Association formed, 799
  Catholic stronghold in the Pound, 795, 796, 797, 807
  Catholic strength, 796
  cholera, 789
  concept of territoriality, 809
  conflict over 1832 election, 790–3
  Conservatism, 790
  discrimination against Catholics, 801–2, 803
  distribution of Catholic population, 794 (map)
  drunkenness, 807
  education, 804–5, 808
  emigration from, 789, 795
  growth of built-up area, 791 (map)
  house-wrecking, 797
  housing, 804
  importance of ethnicity, 801
  influence of lodges, 808
  introduction of army to control conflict, 799
  introduction of guns in conflicts, 797, 807

433

Index

Belfast—*continued*
　lack of recreational activity, 806–7
　'March to Hannahstown', 798
　migration to, 13, 789, 796, 802
　Nonconformity, 790, 803, 805
　occupations in different areas, 794
　Orange Order, 789, 798
　overcrowding, 23
　Party Processions Act (1832), 790, 796, 798
　police, 798, 799, 805, 806
　population, 789
　processions, 799, 800
　prosperity, 793, 802
　Protestant Operatives Society, 795
　Protestant strength, 796
　railway stations, 280
　Ribbonmen, 790
　Sandy Row, 791, 795, 796, 797, 798, 807
　typhoid, 804
　Unionist party, 808
Belgium, annual rate of urban concentration, 5 (table)
　mortality, 27, 35 (table)
　vital statistics, 35 (table)
Belgravia, 395, 486
　scarcity of pubs, 166
　social uniformity, 339
Bell, Colin and Rose, 889[25]
Bell, G. D., 57[71]
Bell, Lady, 125, 154[18]
Bell, R., 619[12], 620[15], 623[69]
Bellman, Sir H., 386[66]
Bell's Life in London, 565, 566
Beloff, H., 685[14]
Bence Jones, H., 653
Bennett, A. R., 187[13]
Bennett, J., 760[40]
Bennett, J. H., 656
Bennett, Mary, 469[12]
Bentham, J., 87, 676, 875
Bentley, J. F., 326, 445, 447[15]
Beresford, A., *334*
Beresford, M. W., 785[3]
Beresford, W., 834[15]
Berger, B. M., 56[66]
Berkeley Square, 324
Berkshire, migrants from, 133
Berlin, 440
　age structure of migrants to, 17
　as agglomeration of suburbs, 29
　birth-rate, 19
　crude mortality, 21 (table)
　infant mortality, 24
　mortality, 20
　population, 9
　as primate city, 39

Bermondsey, 139, 538, 848
　Jacob's Island, 403, 540, 708, 717
Bernard, W. B., 216
Berry, B. J. L., 55[62]
Berthoff, R. T., 153[5]
Bertillon, J., 416n.
Berwick, *256*; *301*
　barracks, 436
　castle destroyed by railway, 303
Besant, Annie, 594
Besant, W., 160[236], 374, 585, 586, 589, 590, 591, 595, 596, 597, 600[1]
Bessbrook, pubs banned, 166
Best, G. F. A., **389–411**, 689[59], 833[2]
Bethnal Green, 270, 486, 573, 586
　density of population, 26
　employment, 134, 137
　labour-yard of Employment Association, *10*
　licensed premises, 164 (map)
　migrants, 152, 372
　mortality, *415*
　overcrowding, 23, 404
　poverty, 596
　sanitary inspector for, 607
Betjeman, Sir John, 245[32]
Bett, W. R., 665[84], 66[95]
Bettelheim, B., 685[12], 686[31], 688[55]
Bevan, G. P., 52[32]
Beverley, 6
　political conflict over Poor Law, 771
Beyer, C. F., 294
Bidder, G. P., 306[32]
Billingsgate Market, 320, 505
Binford, H. C., 383[14]
Binny, J., 734[53]
Birch, A. H., 785[2]
Bird, H., 356[62]
Birkenhead, 395
Birmingham, appearance, 440
　ascendancy of Evangelicals, 828, 829
　Bishop's House, 442
　change of standards in ministry, 828–9
　changing attitudes of churches to society, 833
　children's hospital, 653
　church attendance, 829–30, 855
　church building, 823–7
　church encouragement of social welfare, 832
　clearance of slum houses, 390
　Cobden Hotel, *81*

　compared to Sheffield, 337
　County Council, 396
　crude mortality, 21 (table)
　difficulty of creating new parishes, 827
　diocese, 818
　diseases, 638
　Dissenters, 819–21, 830
　effect of commerce, 874
　Great Western Arcade, *309*
　influence of bankers and merchants, 819
　influence of Quakers, 820
　influence of Unitarians, 820, 829
　migration through, 117
　mortality, 20
　municipal honours, 775
　music halls, 174
　New Street station, *254*
　Night Refuges, 142
　police inspection of housing, 399
　population, 9, 391
　Primitive Methodism, 858
　prosperity, 818, 819
　pub debates, 179–80, 185
　Radicalism, 819
　railway hotels, 296
　railway stations, 280
　religious denominations, 824 (map), 825 (map)
　Roman Catholics, 821
　social divisions, 819
　Society, 86
　Ten Churches Scheme, 824–6
　see also Church of England; Edgbaston
Birth control, 114–15
　infrequent use, 701
　as an urban development, 114
　in working-class families, 115
Birth-rate, *see* Fertility
Bishopsgate, slums, 461
　transport-inns, 162
Bitter Cry of Outcast London, The, 589, 592
Black, W., 374
Blackburn, fair, 131
　infant mortality, 650
Blackfriars, 315
　architecture, 443
　Bridge, *243*, *426*; 320
　station, 283
Blackpool, *165*
Blackwood's Magazine, 576[5]
Blaise Hamlet, 433
Blake, N. M., 52[32]
Blake, W., 6, 522
Blakiston, N., 347
Blantyre, 394
Blatchford, R., 879, 886

Index

Blaug, M., 101[9], 734[40]
Bleak House, 103[31], 235, 272, 458, 507, 509, 530, 531, 542, 550, 563
Bleicher, H., 51[28]
Blenheim Park, 437
   Woodstock Manor, 433
Bloch, M., 92
Blomfield, Alfred, 329[10]
Blomfield, A. W., 318, 326
Blomfield, Bishop C. J., 314, 857
Blomfield, Sir Reginald, 330[29], 331[70,81], 432
Bloomsbury, 337, 338, 339, 395, 433, 568
   Bedford estate, 353
   licensed premises, 167 (map)
   scarcity of pubs, 166
Blount, E., 344
Blum, H. F., 686[24]
Blunt, R., 155[63]
Blyth, A. W., 623[70]
Bodley, G. F., *264*; 326, 444
Bodmin Gaol, 444
Boeckh, R., 13, 50[14]
Bogan, B., 850[25]
Bolton, 740
   church attendance, 855
   link between pub and stage, 175
   pub density, 169
Bolton, A. T., 446[4]
Bolton, T., 578[21]
Bombay, 9, 36
Bond Street, 321, 324
Bond Street, Old, No. 1, *275*
Boni, A., 245[32]
Bonomi, I., *304*; 440
Bonomi, J., 443
Booth, C., 14, 50[16,19], 51[23], 92, 93, 94, 96–7, 103[48,50], 112, 120[38], 121[52], 126, 130, 136, 140, 154[12ff], 155[50ff], 156[69ff], 157[111ff], 158[152], 159[206,209,210], 160[216ff], 164, 165, 168, 169, 184, 186[2], 187[9], 364, 373, 426, 427[3], 428[12,15], 586, 592, 595, 600, 862, 865, 866, 867, 868[6], 871[50,55,58,61]
   importance of his work, 596–7
Booth, J. B., 160[231]
Booth, M. R., 211–24
Booth, General W., 155[52], 592, 595
Bose, A., 56[66]
Boston, as agglomeration of suburbs, 29
   birth- and death-rates, 14
   foreign-born population, 50[15]
   infant mortality, 25
   migration to, 14
   population, 9, 14
Boucicault, D., *49, 62, 65*; 219, 223
Bough, S., 465
Boulding, K. E., 57[74]
Boulton, J. T., 437, 446[3]
Bourne, J. C., *355*; 439, 463
Bournemouth, 395
   overcrowding, 22
Bournville, 334
   development, 884
Bow Common, 151
Bow station, 282
Bowden, J. E., 850[11]
Bower Theatre, 212
Bowley, Marian, 339, 354[4], 355[25]
Box, C. E., 306[33]
Box railway tunnel, 439
Boyd, A., 814[148]
Boyd, R., 664[66]
Boyd-Orr, Lord, 57[69]
Boys, T. Shotter, *354*; 463
Brace, C. L., 659
Bradbury, F. C. S., 642, 664[53]
Braddon, Mrs M. E., 374
Bradford, 99, 883
   dram-shops, 178
   German families, 441
   Liberalism, 781
   mortality, 20
   police inspection of housing, 399
   Primitive Methodism, 858
   railway stations, 280
   warehouses, 441
Bradford Observer, 171
Bradlaugh-Besant trial, 115
Bradley, J., 732
Bradshaw's Journal, 746
Brady, M., 232
Braithwaite, C., 690[63]
Bramley, F., 468
Bramwell, Baron, 183
Brand, Jeanne, 619[5]
Brayshaw, J., 599
Brazil, housing deficits, 37
   rate of urban concentration, 41 (table)
   rate of urban growth, 37
   urban natural increase, 37
Breese, G., 190[45]
Brenner, C., 686[30]
Breslau, overcrowded housing, 24
Breslaw, J., 57[73]
Brett, J., 474[59]
Brewster, Sir D., 233
Brice, A. W., 275[23]
Bridgewater Canal, *194*; 250, 457
Brierley, B., 742, 756, 757[5], 760[44,45], 761[53]
Journal, 753
   as spokesman for working class, 753, 755
   sympathy with unemployed, 754
   writing in dialect, 753–4
Briggs, A., 6, 83–104, 103[44], 104[59,71], 119[11], 122[67], 686[36], 687[40], 689[60,62], 690[65], 691[74], 756[1], 785[2], 834[4,5], 888[2]
Bright, J., 176, 198, 717, 868
Brightfield, M., 624[92]
Brighton, 125, 139, 395
   Chain Pier, *85*
   children's hospital, 653
   church architecture, 444
   deaths from phthisis, 647
   launching the *Skylark*, *161*
   lodging-houses, 127
   railway viaduct, 301
   return of migrants, 144
Brindley, J. M., 835[34]
Bristol, carriage factory, *331*
Britannia Theatre, *61*; 212, 224
British and Foreign Medical Review, 632
British Association, statistical section, 90
British Medical Association, 608
British Medical Journal, 607
British Museum of Natural History, 273; 327
British Towns: A statistical study of their social and economic differences, 98
Britton, J., 439, 463
Brix, J., 53[38]
Brixton, Christ Church, *334*
   No. 17 Tulse Hill, *321*
   St Matthew's Church, *315*
Broadsides, *see* Street literature
Brockington, C. F., 619[5,8], 624[87], 662[42], 688[49]
Brodbeck, M., 104[62]
Brodrick, C., *311, 318*; 437, 439
Bromley, 433
Bromley, J. S., 382[2]
Brompton, 313
   hospital, 647
Bromyard, 138
Brontë, Charlotte, 452, 532, 535
Brontë, Emily, 525, 527
Brontë, P., 647
Brook, J., 770
Brook, W., 772
Brougham, Lord, 90
Brown, G. H., 57[70]
Brown, J., Mayor of Sheffield, 288, 294
Brown, J., 809[5,6,11]
Brown, N. O., 266, 688[54]
Brown, P. E., 689[63]

435

Index

Browne, H. K., 458
Browning, B., 444
Browning, Elizabeth Barrett, 90, 483–4
Browning, Robert, 483, 492, 553
Brownlee, J., 642, 662[50]
Bruckner, N., 51[23, 26]
Brunel, I. K., 314, 439, 454
Brunelleschi, F., 431
Brussels, as agglomeration of suburbs, 29
  crude mortality, 21 (table)
  population, 9
  as primate city, 39
Bryant & May, strike by match-girls, 594
Brydon, J. M., 272; 323, 326
Bryson, R. A., 57[72]
Buchanan, G., 401, 639
Bücher, K., 122[69]
Buckingham, migrants from, 133
Buckingham, J. S., 881
Buckland, S., 356[63]
Buckle, H. T., 101[4], 524, 536[9]
Budapest, population, 9
  as primate city, 39
Budd, W., 635, 636, 637–8, 662[37]
Buenos Aires, effect of population concentration, 38
  population, 9
  primacy, 38
  rate of population growth, 37
Builder, 281, 284, 302, 305[10], 306[34], 330[29], 337, 342, 379, 460
Building, agreements, 342–3, 344
  capital available for, 335
  cost, 338
  effect of leasehold system, 312
  land for, 334
  London Building Acts, 338, 345, 347
  and population growth, 324
  regulations, 364, 407
  see also Housing; Leasehold system
Building News, 421–2; 340
Bulgaria, rate of urban concentration, 41 (table)
Bull and Mouth (Queen's Hotel), 27; 162
Bull, G. S., 830
Bullock, S., 374, 385[51]
Bulstrode, H. T., 663[52]
Bulwer, E., 534
Bumpus, T. F., 329[11]
Bunning, J. B., 320, 437, 439, 443
Bunting, F., 383[30]
Bunyan, J., 518, 519, 520, 522, 641
Burchard, J., 56[66], 57[73], 354[9, 15]

Burges, W., 316, 447[15]
Burgess, E. W., 102[22]
Burgh Police Act (1903), 399
Burgh Police (Scotland) Act (1862), 392, 393, 394; (1892), 394; (1903), 399
Burial places, *169, 228, 337, 413*
Burke, E., 431, 442
  on the Sublime and Beautiful, 434–6
Burleigh, J. H. S., 835[19]
Burlington, Lord, 432
Burn, W. L., 689[63], 692[75, 76]
Burn, William, *303*; 440
Burne-Jones, E. C., 453, 469[13]
Burnett, J., 667[105]
Burritt, E., 309[110]
Burrow, J. H., 274[13]
Burton, D., 443
Burton, J., 313
Burton, J. H., 395, 405
Burton-on-Trent, migrants to, 143, 144–5
  Suffolk maltsters, *20*
Bussey, Peter, 179
Bute Docks Co. (Cardiff), 295
Butler, S., 532, 534
Butler, E. M., 704[4]
Butor, M., 274[10]
Butterfield, W., *316*; 314, 444, 445, 447[15]
Buttery, J. A., 765, 766
Byron, H. J., 212
Bythell, D., 757[8]

C

Cabinet of Dr Caligari, The, 265
Cadbury, G., 884
Cadogan Place (London), 549, 550
Cailler, P., 471[24]
Cairncross, A. K., 50[12], 120[18], 354[7]
Cairo, population, 9
Calcutta, infant mortality, 36
  population, 9
Calvert Holland, G., 335, 341–2, 354[5, 21, 23]
Camberwell, 300, 363
  changing structure of households, 375
  migrants, 373
  overcrowding, 374–5
  relation between slums and suburbs, 372–6
  as a suburb, 371, 419
  Sultan Street, 372–6
Cambridge, University of, 483, 605
Cambridge Circus (London), 325
  pubs, 166

Camden Pratt, T., 160[228]
Camden Town, 463, 538
Camera obscura, 228, 230
  influence on art, 459
  see also Photography
Cameron, Sir C. A., 810[23]
Cameron, R., 428[13]
Cameron, W., 192, 208[4]
Campbell, A. V., 216
Campbell, C., 432
Campbell, T. J., 812[87, 92]
Campden Hill, 444
Canada, annual rate of urban concentration, 5 (table)
  child-woman ratio, 36
  migrants to, 15, 789
  rate of urban concentration, 41 (table)
  vital statistics, 35 (table)
Canaletto (Antonio Canale), 228, 462
Canals, *194*; 250, 254, 295, 457
  see also individual canals
Canning Town, 143
Cannon Street, 320
  station, *247*; 283, 284
Cantril, H., 691[73]
Caracas, effect of population concentration on, 38
Carceri d'Invenzione (Piranesi), 437
Cardiff, as coal-shipping port, 295
  Cathays Park, 295
  influence of railways, 294–5
  Tiger Bay, 123
Caricature, 567
Carlisle, 281
Carlisle, Lord, 102[31]
Carlton House Terrace, 313
Carlyle, Mrs Jane, 848
Carlyle, Thomas, 89, 180, 257, 353[2], 401, 456, 457, 486, **495–502**, 511–12, 513, 514, 521, 886
  as an allegorist, 498
  ambivalent attitudes towards social change, 877–8
  on housing, 333
  on railways, 277
  on Reason and Understanding, 480
  rejection of city, 496, 878
  rejection of his age as 'mechanical', 498
  social background, 498
  view of working class, 727
  vision of the city, 498–502
Carmarthenshire, winter migrants in, 143–4
Carnegie, A., 31
Caröe, W. D. *274*; 324
Carpenter, E., 879

Index

Carpenter, Mary, 735[78]
Carter, R. M., 772
Cassedy, J. H., 690[64]
Castle Howard, 436
Casual wards, *9*, *392*; 128, 135, 142
    payment for, 142
Catnach, J., *41*, *46*; 199–200, 201, 202, 577[14]
Cavendish Square (London), 324
Cemeteries, *see* Burial places
Census, when instituted, 94
    taking the, *396–400*
Census of England and Wales
    (1831), 626
    (1851), 4, 241, 326, 337, 424, 710
    (1851, Religious), 821, 829, 830, 855, 857
    (1861), *396–400*; 573
    (1871), 373
    (1881), 7, 372, 373
    (1891), 17
    (1901), 241, 373, 391
    (1911), 114
    (1931), 6
Ceylon, infant mortality, 36
Chadwick, Edwin, *411*; 87, 88, 89, 96, 101[8], 264, 274[11], 275[17], 395, 401, 409[32], 604, 627, 628, 635, 643, *669–83*, 689[61], 690[64, 65], 735[78]
    achievements, 683
    Chartist attacks, 102[27]
    compared with Mayhew, 721
    influence of Bentham, 676
    influence of mother, 675
    opposition to, 682
    proposal to appoint medical officers, 677
    as Royal Commissioner, 677, 718, 719
    on sewage utilization, 679, 682
    vision of inter-connecting sewage schemes, 678, 681
Chadwick, Esther A., 759[34]
Chadwick, G. F., *247–56*
Chadwick, O., 836[46], 859, 861
Chalcots (Eton College Estate), *289–91*; *335–40, 345–53*
    architecture, 335
    attitude of landlords, 346, 352
    attractions of positions, 346
    importance of church and pub, 349–50
    initiative of builders, 346, 350
    position, 336 (map)
    restrictions on shops, 340, 350
    social continuity of residents, 337, 339
Chalmers, A. K., 409[17], 410[39, 40, 41], 411[56]
Chalmers, T., 828

Chaloner, W. H., 121[51], 273[3], 308[73], 691[73], 757[6]
Chamberlain, J., *81*
Chamberlain, W. H., 390
Chambers' Journal, 576[5]
Chance, W., 831
Chancellor, E. B., 557[20]
Chancellor, Valerie, 869[18], 870[41]
Channing, W. E., 87, 88, 102[21]
Chant, Mrs Ormiston, 703, 704
Chapman, S. D., 354[3], 383[26], 622[50]
Characteristic Sketches of the Lower Orders, 565
Characteristics, 498
Charing Cross, 318
    hotel, 325
    railway project for, 302–3
    railway station, 281, 284, 301
    Road, 303, 319, 325, 367
    tramps, 129
Charles, C., 849[3]
Charlton, 129
Charrington, F. N., 169, 175
Chartist movement, 161, 179, 185, 198
    in Leeds, 772
    public meeting rooms denied to, 175
Chase, Ellen, 139, 143, 158[141], 159[179]
Chatham, 436
Chaussier, F., 228
Cheadle, W. B., 655, 656
Cheap Clothes and Nasty, 401, 645
Cheap Trains Act (1883), 284
Cheapside (London), *12*; 316, 317
Checkland, S. G., 119[12, 114]
Chelsea 467, 843
    Hospital, 437
    library, *272*; 324
    off-licensed premises, 168
    registration of houses, 611
    see also Embankment
Cheltenham, church architecture, 443
    railway, 279
Chemnitz, infant mortality, 24
    overcrowded housing, 24
    urban illegitimacy, 19
Cheshire, migration through, 117
Chesney, K., 188[22]
Chester, railway station, *308*; 281, 301, 439
    the Rows, *115*
Chesterfield, 286, 287, 288
Chesterton, G. K., 374, 385[51], 489–90, 529, 536[13, 15]
Chevalier, L., 92, 97, 104[63], 187[18], 416n., 427[5, 6], 428[8π], 735[78]

Chevalier, S.-G., *see* Gavarni
Chevassus, H., 383[18]
Chicago, as agglomeration of suburbs, 29
    migration to, 14
    mortality, 27
    population, 9
    slum conditions, 25
Chichester, Lord Arthur, 791, 792
Child, I. L., 684[11]
Child of the Jago, A, 598
Children, activities, 61, 62, 68, 69, 73, 74, 78
    attitudes to, 63, 64, 66, 67
    diseases, 628, 651–2, 653–4, 655
    in factories, 86, 676, 677, 806, 856
    games, *97*, *98*; 60, 61, 72
    hospitals, 653
    mortality, 16, 23–7, 401–2, 650–1
    school attendance, 421
    schools for pauper, 653
    in Sunday School, 867
Children of the Ghetto, 599
Chile, annual rate of urban concentration, 5 (table)
    child-woman ratio, 36, 37
    rate of urban concentration, 41 (table)
    rate of urban growth, 37
    urban natural increase, 37
    vital statistics, 35 (table)
Chimes, The, 539, 555[5]
Chippenham, 379
Chirk, 438, 450
Chislehurst, 445
Cholera, *406*, *418*; 88
    in Belfast, 789
    deaths caused by, 636
    epidemics, 604, 636, 659, 680, 682, 708, 717
    theories of causation, 637, 680
Chorlton-upon-Medlock, 249, 254, 259
    disease, 638
Christaller, W., 56[65]
Christmas Carol, A, 542, 555[5]
Church Building and New Parishes Acts (1818–84), 827
Church Building, Parliamentary Commissioners, 823
Church, R. A., 307[56], 308[58], 688[51], 691[73], 787[62]
Church, Nonconformist, denominational structure of, 833
    organization of, 816
    Primitive Methodism, 857, 858, 862, 864

437

Index

Church, Nonconformist—*contd.*
  reaction to growth of cities by, 815–16, 906
  see also Belfast; Birmingham; Leeds
Church of England, attendance, 857, 860
  in Birmingham, 822–7, 829, 830, 832, 833
  education encouraged by, 832, 856
  excessive work for incumbents, 832
  growth of indifference towards, 816
  limited success of reform movement, 856
  need for new churches, 823, 856
  outdated organization, 822
  parochial system, 823, 832–3
  reaction to growth of cities, 815, 869[10]
  renewal in life of, 816–17
  renting of pews, 822, 830, 869[10]
  rural bias, 817
  Temperance Society, 161
  see also Religion
Church Pastoral Aid Society, 829
Church, Roman Catholic, 837–49
  achievements of missionary activity, 849
  class segregation, 841
  emotional mass conversions, 838–40
  formation of confraternities, sodalities and guilds, 846
  need for popular urban ministry, 837
  opposition to missions, 844
  pew rents, 842
  reaction to growth of cities, 815
  tasteless adornment, 846
  temperance associations attached to, 848
Church, Roman Catholic, Irish in, 837–49
  confraternities and societies, 847
  education, 843
  enjoyment of spectacle, 846
  need of mission, 837
  need to transform churches, 840
  segregation, 841, 842
  temperance pledge taken by, 848
Churches, Victorian, architecture, 314, 432

in City, 433
importance on housing estates, 349–50
Churchill, Lord Randolph, 174
Cities, affinities with villages, 898–9
  ambivalence of Victorian response, 877
  complexity, 247
  concept, 815
  definition, xxvii–xxviii
  dependence on migration, 112
  development, 31–2
  dominance of large over small, 39
  fear of, 478–9, 483, 501, 903
  growth, 4–15, 389
  influence of modern, 893
  meaning underlying geography, 900–1
  need for 'qualitative' and 'quantitative' evidence for study of, 83
  new ideas originating from, 111
  nodality, 40
  numbers, 99
  quantitative growth of wealth and population, 31
  repression of knowledge of, 479, 903
  sense of community, 899
  size, 390
  social divisions, 896
  social progress revealed, 895
  structure, 262–3, 264
  study of society based on, 93
  Victorian attitude to growth, 84, 873–4, 884
  see also England; Urbanization; individual cities and countries
Cities, Victorian, anonymity, 897
  effect of introducing industry into, 898
  as hybrid growth, 898
  influence on modern cities, 894
  levelling processes of, 897
  need for further study of, 907
  as phenomenon of modernity, 896
  qualitative difference between eighteenth-century and, 479
  see also England and Wales, growth of cities
Citron, P., 481, 493[12]
City Development. A Study of Parks, Gardens, and Culture Institutes, 31
City Sewers Act (1848), 604
'Civilization', 261–2

Clapham, J. H., 113, 120[39], 121[46], 399, 403, 410[38π]
Clapham, M. H., 113, 120[39], 121[46]
Clapham Road (London), *118*
Clapton, 390
Clare, castle destroyed by railway, 303
Clare-Market (London), 586
Claremont, 436, 437
Clark, J. M., 441
Clarke, A. M., 850[13]
Clarke, B. F. L., 329[11]
Clarke, G. S., 443
Clarke, J. E., 383[20]
Class, social, acceptance of distinctions in, 745
  fear of conflict between, 593
  in relation to fertility, 12, 28–9
  in relation to religion, 860
Classes, middle, attitude towards children, 63, 64, 66, 67
  attitude towards destitute, 65, 68
  attitude towards slums, 370
  attitude towards working classes, 63, 65, 66, 67, 830
  children's activities, 63, 64, 65, 66, 67
  employment of servants, 63, 65, 66, 67
  fear of city, 574
  leisure activities, 63–4, 67
  in London and Paris, 426–7
  moral attitudes, 64, 67
  occupations, 64, 65, 66, 67
  Romantic and Utilitarian attitudes, 875
  see also Suburbs
Classes, upper, attitudes to working classes, 830
  East End pleasures enjoyed by, 587–8
  meeting with lower classes in pubs, 172–4
  portrayed in late Victorian drama, 224
  poverty among, 70
  pubs banned near residences, 182
  servants, 70–1
Classes, working, apprenticeship of, 77
  attitude of upper classes to, 67, 68, 74–5, 830
  attitude to children, 61, 62, 69–71, 73, 74, 77
  attitude towards neighbours, 72, 75, 78
  attitude to police, 75, 78, 80
  attitude to religion, *see* Religion

attitude to wealth, 62
begging, 80
children's activities, 61, 62, 68, 69, 73, 74, 78
and criminal class, 75–6, 77
effect of railways on, 296, 297
family occupations, 61, 67, 71, 72, 73, 76
fighting as pastime, 62, 77
gulf between city and country, 69
housing, 334, 367–8
illiteracy, 61, 73, 77
interviews, **60–80**
literature written by, 753–4
living conditions, *375, 389–90, 393–4*; 72, 206, 264–72, 660
in Manchester, *214–17*; 258–72
manners, 62, 73, 74, 76, 77
meeting with upper classes in pubs, 173–4
mortality, 249
necessity of living close to work, 368–9, 405
in Paris and London, 424–6
portrayal in the press, 565–70
profligacy, 731, 896
and prohibitionism, 181
rising wages, 660
and temperance movement, 183
thieving, 78, 79, 80
treatment in literature, 748
see also Housing; Occupations; Poverty; Slums
Clay, Rev. J., 649–50
Clayton, E. G., 667[105]
Clayton, H., 307[35]
Cleary, E. J., 386[66]
Cleckheaton, 443
Cleland, J., 399
Clerkenwell, Italians in, 135
St John's Gate, *67*
workhouse infirmary, 647
Cleveland, migration to, 14
population, 9
size of families, 19
Cleveland, H. W. S., 54[52]
Clinard, M. B., 684[2]
Clough, A. H., 484–5, 494[20]
Clutton, H., 447[15]
Coaching inns, 162
modernization, 163
Coal Exchange (City of London), 320, 437
Coal Hole, 172
Coalbrookdale, 438, 450
Coale, A. J., 51[23]
Coatbridge, 394
Cobbett, W., 134, 157[102], 521
Cobden, R., 172, 868
Cockerell, C. R., *187, 323*; 315

Coe, W. E., 810[41]
Coffee-shops, 127, 128
Coger's Hall, 179
Cole, G. D. H., 121[44]
Cole, G. V., 473[58]
Cole, H., 327
Cole, W., 865
Coleman, T., 155[54]
Coleridge, H. J., 849[1]
Coleridge, S. T., 522, 876
Collcutt, T. E., *278, 280*; 324, 325, 326, 327
Collett, E. B., 52[36]
Collier, Frances, 688[48]
Collins, J., 189[29]
Collins, P., 245[31], **537–57**
Collins, W., 374, 385[51], 531
Colombia, foreign migration to, 37
rate of urban concentration, 41 (table)
urban growth, 37
urbanization and per capita income, 37
Colombo, child-woman ratio, 36
Coming Race, The, 534
Cominos, P. T., 705[21]
Common Lodging Houses Act (1851), 615
Communications, *see* Transport
Communist Manifesto, 728, 729
Computers, use in quantitative analysis, 98
Comte, A., 110
Conder, C., 492
Condition of England, The, 96
Condition of the Working Class in England in 1844, The, 257, 272
Connan, D. M., 622
Conrad, J., 523, 531, 535–6
Constable, J., 464, 465, 473[52]
Constantinople, population, 9
Contagious Diseases Acts (1864, 1866, 1886), 116, 658, 697
Continued Fevers of Great Britain, 637
Contrasts, 267
Conway, castle injured by railway, 303
railway bridge, 439
Cook, E. T., 187[16], 310[118], 470[17], 494[22], 556[11], 888[1]
Cook, M., 785[8]
Cook, T., 113
Cooke, Rev. Dr, 790, 803
Cooke Taylor, W., 5, 10, 257, 757[4]
Cooley, C. H., 29
Cooney, E. W., 385[61]
Cooper, D., 467, 473[59], 474[60, 64]
Cooper, E. A., 254

Co-operative societies, 113
Copenhagen, 18, 280
Copenhagen Fields, pubs, 166
Concanen, A., 472[35]
Conservative party, *see* Tory party
Corbett, Raby & Sawyer, *312*
Corbridge, Sylvia L., 760[40]
Corcoran, B., 200
Cornhill, 316, 317
National Discount Company, 259
Cornhill Magazine, 300, 576[5]
Cornwall, economic decline, 109
Methodism, 857
Corot, J. B. C., 465
Costelloe, B. F. C., 383[29]
Cotman, J. S., 450
Couling, S., 188[24]
Coulsdon, overcrowding in, 99
Coulson, Colonel, 792
Coulthart, Dr J. R., 128, 155[38, 40]
Coupe, W. A., 207[1]
Courbet, G., 473[59]
Courthion, P., 471[24]
Covent Garden, *140*; 339, 538
architecture, 432
gin-shops, 586
market, 100, 439
prostitutes removed from theatre, 695
pubs, 166
sleeping rough, 128
theatre, 212
Coventry, Butcher Row, *70*
diseases, 638
infant mortality-rates, 650
migrants, from, 133
Cowan, R., 626, 660[2], 661[18]
Cowlairs, 294
Cowper, W., 476, 477, 480, 486
Cox, D., 450
Crabb, J., 156[88]
Craig, W. M., 565
Craven, J., 770
Crawford, J. H., 15
Creese, W., 882, 889[26, 28, 30, 31]
Creighton, C., 639–40, 652, 660[3], 661[19], 662[29, 31, 32], 665[82, 85]
Cremorne, 695
Cresy, E., 155[37]
Crewe, employment on railways, 294
locomotive works, 294
railway, 277, 293
sanitation, 683
Crime, growth, 107
housing of criminals, 364
incidence, 713
in relation to poverty, 716

439

Index

Crime—*continued*
  as subject of street literature, 198–9
Criminal Law Amendment Bill (1882–5), 696, 697
Cripplegate, Night Refuge, 142
Cromford Mill, *301*; 438
Cronkhill, 437
Cronne, H. A., 810[27]
Crosby, T., 331[66]
Cross, I. B., 810[48], 814[161]
Cross, R. A., 394, 616
Cross Acts, *see* Artizans' and Labourers' Dwellings Improvement Acts
Crossley, Sir F., 882, 883
Crossley, J., 750
Crossley West Hill Park Estate, 883
Crowe, E., *346*; 457
Croydon, migrants from, 133, 138
  railway, 238
  tramps, 129
  typhoid, 640
Cruikshank, G., *39, 338*; 451, 565, 566, 571, 580[39]
Cruikshank, P., 188[24]
Crum, F. S., 50[15], 51[25, 28]
Crystal Palace, *21, 82*; 174, 233, 240, 314
Cubitt, J., *243*; 320
Cubitt, L., 439
Cubitt, T., 313, 353
  employment of migrants, 146
Cumberland, R., 222
Cuming, S., 335, 336, 349, 350, 352, 353
Cunnison, J., 814[146]
Currie, Sir E. H., 592
Currie, R., 868[7]
Curtis, C., 822
Curtis, J. S., 787[63]
Curtis, L. P., Jr, 581[50]
Cyder Cellars, *32*; 172

D

Dagenham, 99
Daguerre, L. J. M., 229, 230, 240
Daily Telegraph, 577[5]
Dale, A. W. W., 834[5]
Dale, R. W., 820
Dalton, J., 457
Dalziel, E. and T., 564, 578[26]
Dance, G., the Younger, *296, 298*; 437
Dangerous Classes of New York, The, 659
Daniel Deronda, 529, 530, 535
Darby, A., 438
Darlington, railway at, 279, 293

Darrah, W. C., 244[15], 245[32]
Darwin, Charles, 725
Daumard, Adeline, 425, 428[17, 23]
Daumier, H., 474[59]
Davenant, Sir W., 519
David Copperfield, 458
Davidson, M., 664[67]
Davies, E. T., 833[3]
Davies, G., 834[11]
Davies, J. B., 649, 665[75]
Davis, K., 33n.
Davis, W. A., 686[33]
Davy, Sir H., 244[3]
Dawes, W., *208*
Dawson, G., 832, 833
De Courcy, Lady Amelia, 351
De Loutherbourg, P. J., 438, 450, 469[4]
De Quincey, T., 476, 547, 587
De Tocqueville, A., 6, 49[5], 257, 278, 730, 757[4], 899
Deane, Phyllis, 49[10], 50[13], 52[37], 55[61], 687[40]
Deane & Woodward, 315
Dearle, N. B., 159[194]
Death-rate, *see* Mortality
Debenham, Sir E., 445
Defoe, Daniel, 547
Delamotte, P. H., *82*
Delaroche, H., 230
Dell, R. S., 836[36]
Demography, *see* Population
Denison, E., 586, 590, 592, 593, 595, 596
  life, 591
Denison, G. A., 865
Denison, L. E., 871[52]
Denmark, marriage-rates of urban population, 18
  rate of urban concentration, 41 (table)
  urban birth-rate in, 19
Dent, R. K., 834[13, 23]
Denvir, J., 138, 154[14, 17], 157[126, 129]
Deptford, 143, 538, 848
  Mill Lane, 126
Derby, church architecture, 444
  effect of railways, 293
  lodging-houses, 127
  railway bridge, 303
  railways, 286, 289
  station, 301, 439
Derbyshire, migrants, 144
Derry, overcrowded housing, 23
Description of Modern Birmingham, 819
Descriptive and Statistical Account of the British Empire, 84
Destitute Sailors' Asylum, 142, 146

Detroit, 9, 14
Developers, *see* Building; Leasehold system
Deverell, W. H., 459
Devlin, J., M.P., 808
Devon, Methodism in, 857
Dewar, M. W., 809[5, 6, 11]
Dewhirst, R. K., 889[27]
Dewhurst, K., 256[1]
Dewsnup, E. R., 22, 23, 52[33], 383[24]
Dexter, W., 554, 557[20]
Dialect, formation of society to preserve, 750
  poetry, 751, 903
Dickens, Charles, 84, 85, 86, 91, 102[31], 243, 372, 401, 403, 434, 451, 457, 458, 459, 479, 502, 507, 509, 518, 523, **537–54**, 563, 569, 578[19], 581[46], 584, 613, 625–6, 660[1], 698, 708, 717, 735[67], 903, 905
  attraction of city at night, 540
  'attraction of repulsion', 537, 540
  choice of locations, 550–1
  favourite locales, 538
  importance of London to, 544–545, 553
  knowledge of London, 545, 550
  possible debt to Mayhew, 727
  wit and imagination, 546
Dickinson, G. C., 186[6]
Dickinson, H. W., 689[63]
Dickson, P. G. M., 386[65]
Dictionary of National Biography, 609
Dighton, E. W., 571
Dingle, A. E., 189[37], 853[97]
Diocesan Report on the Condition of the Bristol Poor, 858
Diphtheria, 628, 651, 655
  deaths from, 653, 666 (table)
  treatment of, 653
Disdéri, A., 234
Disease, 625–60
  diarrhoeal, 638–9
  dysentery, 628
  immunization against, 655
  medicine used to treat, 650
  venereal, 628, 658–9, 697
  see also Health; Sanitation; and individual diseases
Disraeli, Benjamin, 87, 121[51], 198, 257, 535, 539, 549, 556[16], 717, 882
Division of Labour in Society, 110
Djakarta, population of, 9
Dobson, A., 489
Dobson, J., 301, 439

Docks, London, architecture, 440–1
  construction, 587
  Mayhew's description, 724–5
  strike, 595
Doctor Marigold, 548
Dodd, G., 428[14]
Dodds, E. R., 812[101]
Dodgson, C. L., 232
Dodington, 'Bubb', 436
Dollard, J., 685[13,16]
Dombey and Son, 268, 401, 463, 538, 905
Domestic service, employment of country girls in, 115
  in London and Paris, 425
Dominican Republic, child-woman ratio, 37
Doncaster, effect of railways, 293
  lodging-houses, 293
Donegall, Marquis of, 791
Donno, E. S., 690[68]
Dorchester, railway diverted from, 304
Doré, G., *3, 42, 57, 347;* 427[1], 441, 461, 472[41], 554, 556[20]
Dostoevski, F. M., 479
Dou Chang, Sen, 49[9]
Douglas, Mary, 673, 686[27]
Dover Wilson, J., 581[58]
Dow, G., 308[70], 309[87,98]
Dowgate Hill, Tallow Chandlers' Hall, 283
Downing, A. J., 29, 54[48]
Doyle, L., 813[126]
Doyle, R., 452
Drama, commercial, 222, 223
  high moral tone, 218
  life in London as theme, *48–51, 62–4;* 212, 215–17, 222, 223
  temperance, 217–18
  themes, 213–18, 222–4
  symbol and verisimilitude in, 216
  upper-class life as theme, 224
  see also Theatres
Drescher, S., 735[80]
Dresden, 280
  urban illegitimacy, 19
Drew, Rev. Dr, 803
Drink, argument for free trade, 175
  off-purchase, 167
  prohibition, 181–3
  restriction of hours, 181
  and sporting activities, 173–3
  as theme of Victorian melodrama, 217
  and women, 78, 168
  see also Drama; Gin-palaces; Pubs; Temperance movement

Driver, A. H., 834[9]
Driver, C., 275[20]
Dronfield, 288
Dronsfield, J., 761[55]
Drummond, A. L., 446[14]
Drury Lane, 568, 569, 843
  gin-shops, 586
  prostitutes removed from theatre, 695
  soup-kitchens, 123
  theatre, *63;* 212
Drysdale, Dr G., 701
Du Maurier, G., 452, 455, 569, 570
Dublin, Catholic predominance, 793
  growth-rate, 793
  incidence of fever, 804
  overcrowded housing, 23
  population, 391
  public buildings, 395
  pubs, 807
  railway stations, 280, 302, 304, 310[125]
  Sackville Street, *74*
  scarlet fever epidemic, 652
  uniformity in town plan, 335
Dubos, R. J., 43, 48, 57[71]
Duby, G., 493[2], 494[18]
Dudley, 138, 440
  disease, 638
Duke Street (London), *274;* 324
Dulwich, 433
  Art Gallery, 437
Dumfries, fair, 131
  railway, 277
Duncan, R., 120[19]
Duncan, T., 444
Duncan, Dr W. H., 604, 610, 611
Dundee, 394
  Cox's stack, *302*
  Improvement Act (1871), 397
  municipal intervention, 398
  overcrowding, 23, 404
  public lighting, 397
  size, 391
  whalers, *24*
Dunfermline, 31
Dunlop, J., 163, 186[6], 187[13]
Dunnett, H. McG., 187[15]
Dupee, F. W., 556[18]
Durand-Ruel, P., 465
Durham, overcrowding, 23
  railway viaduct, 301
Durkheim, É., 109–10, 111–12, 117, 120[20,29,34], 122[68]
Dyos, H. J., **359–86**, 54[49], 98, 104[64,65,67], 186[1], 187[9], 190[47], 298, 306[23], 309[96ff], 329[6], 335, 354[4,8,10], 382[1,4], 383[17,23,32], 384[34,37,41,49], 385[53,55,57], 428[19], 684[1], 687[42], 691[73], 732[19], 733[19], 786[33], 893–907

E

Ealing, station, 282
Earls Court, 313
  architecture, 432
  Exhibition, *310*
  station, 283
East, T., 830
East End Idylls, 599
East London, 590, 595
Eastcheap (London), 316, 443
Eberth, C. J., 640
Ecology, human:
  abuse of human ecosystem, 42–8
  application of cost-benefit analyses to ecosystem, 47
  future plans for environment, 48
  welfare-accounting technique applied to environmental policies, 47
  see also Pollution
Economist, 418; 576[5], 728
Ecuador, vital statistics, 35 (table)
Eddison, E., 776
Eder, J. M., 244[26], 245[32]
Edgbaston, 433, 821, 822, 826
Edgeworth, F. Y., 94
Edinburgh, architecture, *73, 292;* 432, 443
  children's hospital, 653
  effect of migration, 13
  gasworks, *303;* 440
  Improvement Act (1867), 397
  Lower Viewcraig Row, *294*
  municipal intervention, 398
  New Town, 395, 503
  Night Refuge, 142
  Old Town, *75, 295*
  overcrowding, 23
  photography, 232
  Police Acts, 394, 407
  population, 391
  railway, 278, 280
  railway hotel, 296
  railway stations, 280
  Ruskin's attitude to, 503
  uniformity of town plan, 339
  water systems, 397
Edinburgh Medical Journal, 693
Edinburgh Review, 89, 101[13], 576[5], 728
Edis, R., 324, 326
Edmonds, E. L. and O. P., 101[5]
Edmonds, T. R., 627, 628, 660[6], 661[11]

Index

Education, provision of, 86
  see also School
Education Act (1870), 322, 570, 773, 784
Edward, Prince of Wales, 634, 860
Edwards, G., 870[39]
Edwards, H. W. J., 187[17]
Edwards, J. P., 323
Edwards, P. J., 330[41, 45]
Edwards, R. D., 120[17]
Effingham, 213
Egan, P., 202, 362, 382[5], 451, 547, 579[31], 589
Egg, A., 457, 458
Eginton, F., 579[27]
Egley, W. M., *343*; 454, 470[18]
Egypt, rate of urban concentration, 41 (table)
  urbanization and per capita income, 37
Ehrenzweig, A., 685[18]
Ehrlich, P., 48, 57[75]
Elements of Social Science, 701
Elephant and Castle, Metropolitan tabernacle, *313*
  pubs, 166
  theatre, 224
Eliade, M., 475, 871[57]
Eliot, George, 518, 521, 522, 524, 527, 533–4, 535, 629, 717
Eliot, T. S., 492, 493
Ellegård, A., 577[5]
Elliott, C. M., 785[9]
Ellison, M. J., 344
Ellison, Michael, 338, 343, 344, 355[42]
Elmes, J., 365
Elson, G., 156[85]
Elton, Sir A., 469[2], 579[27]
Embankment, Chelsea, 321
Embankment, Thames, *132*, *138*, *242*, *373*
  homeless and destitute on, 458
  scarcity of pubs, 166
  as system of drainage, 318, 537, 558
  tramps on, 129
  winter employment, 146
Embankment, Victoria, *76*; 165, 321
Emden, A., 355[34]
Emden, W., 325
Emerson, P. H., 241
Emerson, R. W., 480, 546, 556[12]
Emigration, *131*; 880
  see also Migrants
Émile, 477
Employment, industrial, competition with agriculture, 133
  Christmas increase in, 149

growth of regular, 153, 371
lack of in late winter, 150
for migrants, 148
slackness of, 137
women's, 639, 650
see also Occupations
Emmons, R., 474[71]
Encyclopaedia Britannica, xxvii, 92
End and the Means, The, 511
Engels, F., 3, 121[51], 275[18, 19, 22], 717, 731[1], 856, 857, 868[1], 869[13], 900
  in Manchester, 257–72
  opinion of the East End, 593, 595
  on working-class religious attitudes, 855
England, annual rate of urban concentration, 5 (table)
  home ownership, 25
  increase in town and country population, 14, 15
  industrialization of labour force and product, 11 (table)
  internal migration, 15
  marriage-rates for urban population, 18
  migration to towns, 13
  mortality, 20, 24
  overcrowding, 23
  population concentration, 11 (table), 12
  population, 6, 8
  sex and age structure of population, 16–17
  size of households, 25
  size of rural population, 12, 15
  typhus epidemic, 633
England and Wales, growth of cities, 39
  mortality-rates from cholera, 636
  mortality rates from diarrhoeal diseases, 638–9
  mortality rates from typhoid fever, 635
  nineteenth-century growth of population, 105, 362, 574[1]
  number of serials current in, 575 (table)
  see also Census; Great Britain
English Common Reader, The, 575[1]
English Dialect Dictionary, 750
English Dialect Society, 750
English Traits, 546
Entertainment: balloon ascents, 746
  travelling circuses, 131
  see also Fairs; Music halls; Pubs; Theatres

Environs of London, The, 300
Epsom Downs, Derby Day at, *5*, *159*; 124
Erewhon, 534
Erikson, E. H., 685[15, 21]
Ernst & Co., *192*
Escott, T. H. S., 399
Essay on Population, 723
Essay on the Picturesque as compared with the Sublime and the Beautiful, 437
Essex, migrants to, 135
Esther Waters, 600
Eton College Estate, see Chalcots
Ettlinger, L. D., 446[9]
Europe, concentration of population, 8
  increase in urban and rural populations, 14
  scarlet fever, 652
  tuberculosis, 642
  urban birth- and death-rates, 14
  urbanization, 32, 33 (table)
Euston, Arch, 439
  hotel, 296
  pubs, 163
  railway portico, 314
  station, *244*, *305*; 284, 301, 302
Evans, G. E., *20*; 132, 144, 156[86], 159[182, 187, 191]
Evans, J., 742, 758[14, 15], 759[37]
Evans's Cave of Harmony, 172
Evers, G., 769
Eversley, D. E. C., 308[74]
Everton, gypsies at, 129
Ewing Ritchie, J., 157[126], 180, 188[19]
Exeter, 681
Eyes of the Thames, 137
Eyre, V., 340, 342

F

Faber, F. W., 839, 845
Factory Act (1867), 646
Factory and Workshop Act (1901), 646
Fairbairn, Sir A., 773, 774, 777, 782
Fairbairn, P., 766, 777, 782
Fairfield, C., 189[42]
Fairs, autumn, 140
  spring migration of showmen to, 131
  summer migration to, 136
Falkner, G., 746, 759[27]
Family, changes in relationships, 60, 114
  growing economic independence of children, 114

442

migration, as escape from, 113–14, 121[51]
size, 61
standards of behaviour, 62
Fanger, D., 557[20]
Farington, J., 450
Farmer, J. S., 382[6]
Farr, W., 95, 103[55], 607, 617, 626, 627, 628, 660[5,12], 665[73]
Farringdon Street station, *238–239*; 282, 283
Fast, L. S., 246[32]
Fatal Reservation, A, 599
Father Hilarion, 599
Faucher, J., 52[38], 257
Faucher, L., 757[4], 856–7, 869[11]
Fawcett, H., 91
Fayers, T., 157[108]
Fenchurch Street (London), 316
Fenichel, O., 688[54]
Fenton, R., 232
Ferenczi, S., 685[16], 686[32]
Ferguson, R. S., 308[85]
Ferguson, T., 664[71], 665[72,74]
Fertility, child-woman ratio as measure of, 36
class differences, 53[44]
highest in mining and agricultural families, 114
in London, 49[11]
negative correlation with socio-economic status, 27, 40
parity between mortality and, 34
relation between size of agglomeration and, 19
rural compared with urban, 51[29]
in urban populations, 18, 114
Festive Wreath, The, 746
Fever, *see* Diseases
Fiction—Fair and Foul, 508
Field, W., 288
Fielden, K., 691[75]
Fielding, H., 547
Fielding, K. J., 102[31], 275[23], 688[57]
Figs Mead, 353
Fildes, Sir L., *350*; 452, 459, 472[31], 568, 573, 580[44]
Finchley, 538
Common, 150
Road, 348, 351, 352
Finer, S. E., 275[17], 276[27], 619[2], 686[36], 687[39], 687[44,45], 689[58,61], 690[64]
Finland, child-woman ratio, 36
rate of urban concentration, 41 (table)
Finnie, E., 770, 773

Finsbury, gypsies in, 130
overcrowding, 22, 23, 404
Firth, J. F. B., 386[77], 621[32]
Fisher, R. A., 94
Flanagan, J., 126, 154[21]
Fleet Prison, 248
Fleet Street, *121*; 696
Fleetwood, shipping trade in, 146
Fleming, Rev. Canon W., 850[29]
Fletcher, I., 494[31]
Fletcher, J., 101[1]
Fletcher, R. J., 354[16]
Flinn, M. W., 101[8], 401, 409[35], 619[4], 660[4], 661[20], 689[61], 734[37]
Foard, J. T., 760[44]
Fogel, R. W., 103[56]
Folkestone, 301
Folkingham House of Correction, 444
Food, adulteration of, *408*; 680
Forbes, S. A., 468
Ford, B., 102[24]
Ford, G., 101[4]
Fordyce, T., 309[91]
Fors Clavigera, 502, 503
Forster, J., 537–8, 539, 546, 554, 555[11]
Forster, W. E., 781
Forsyth, R. A., 876, 877, 888[11]
Fortnightly Review, 96, 465, 490, 576[5]
Foster, D. B., 880
Foster, J., 186[1], 190[47], 786[33]
Foster, J., II, 301
Foster, M. B., 473[47]
Foundling hospitals, 80, 546
Fowke, Captain, *279*
Fowler, C., *307*; 439
Fowler, W., 343
Fox, Celina, **559–81**
France, age and sex structure of population, 51n.
annual rate of urban concentration, 5 (table)
concentration of population, 4
death-rate in cities, 13
decreasing internal migration, 15
effect of 1830 Revolution on intelligentsia, 481
effects of migration on cities, 13
mortality, 20, 35 (table)
rate of urban concentration, 41 (table)
research on urbanization, 97
vital statistics, 35 (table)
writers' attitude to city, 480, 481–3
see also Paris

Francis, D., **227–46**
Frankenberg, R., 120[24], 685[20]
Frankfurt-am-Main, age structure of migrants to, 17
railway station, 280
Fraser, D., **763–88**, 787[67]
Fraser's Magazine, 88, 728
Frazer, W. H., 620[15], 621[36], 622[42], 624[88]
Frazer, W. M., 665[72,73,84], 666[91], 691[73]
Freake, Sir C. J., 313, 432
Fredur, T., 157[103]
Free Libraries (Ireland) Act (1855), 804
Freeman, A., 156[7]
Freeman, T. W., 54[53], 119[7]
French, S. M., 275
Freud, Anna, 689[63]
Freud, S., 266, 536[11], 684[9], 685[17,18], 686[30], 688[45], 702, 705[18]
theories, 67-2, 673, 674
Friedmann, J., 55[59]
Frith, W. P., *50, 342*; 454, 471[27], 474[59,63]
Frome, 292
Fromm, E., 684[5]
Froude, J. A., 888[13]
Frye, R. M., 690[68]
Fulham, 390, 538
off-licences, 168
Fun, 376; 560, 569, 576[5]
Furness, J. M., 307[54]
Furniss, Fr J., 839

G

Gaiety Theatre (London), *59*
Gairdner, W. T., 403, 406
Gale, F., 173–4, 188[22]
Galsworthy, J., 374
Galton, F., 27, 121[54]
Gamble, W., 577[12]
Gamson, W. A., 691[71]
Garbett, J., 834[16]
Garden cities, advantages, 31
development, 887
proposals for, 886
see also Howard. E.
Garrick Club, 318
Garrick Street (London), 318
Garrick Theatre (London), *58*
Gaskell, Mrs, 86, 88, 276[32], 530, 717, 739, 753, 756
achievements, 748–9
Gaskell, Rev. W., 757[11]
Gateshead, 776
overcrowding, 22, 90
Poor Law administration, 771
Gatty, A., 308[72]
Gautier, T., 490

Index

Gavarni (S.-G. Chevalier), *348*; 460, 461, 472[38], 578[22], 579[30]
Gavin, H., 372, 384[46], 603, 621[40], 690[64], 735[68]
Gay, J., 476
Geddes, P., 31, 48, 54[52], 363
Gee, S., 655, 656, 666[93]
General Medical Council, 605
Geneva, 301
  Ruskin's opinion of, 503
Genoa, Dickens in, 539, 540, 554
Gentleman's Magazine, 67; 520
Genzmer, F., 53[38]
George, M. Dorothy, 579[31], 580[40], 581[53], 851[41]
George, Sir E., & Peto, Morton, 317, 324, 326, 434
Géricault, T., 465
Germ, 452, 486
Germany, effect of migration on growth of cities, 13
  marriage-rates of urban population, 18
  railway stations, 280, 281
  sex and age structure of urban population, 16–17
  vital statistics, 35 (table)
Gernsheim, Helmut and Alison, 244[1]ff, 245[30, 32]
Ghetto Tragedies, 599
Gibbon, Sir G., 619[12], 620[15], 623[69]
Gibbon, S., 158[157]
Gibson, J., 432
Giedion, S., 274[8], 676
Giffen, Sir R., 93
Gilbert, J., 563, 564
Gill, C., 691[74], 785[2], 834[4]
Gill, J. C., 835[34]
Gilley, S., **837–53**
Gillfillan, J. B. S., 814[146]
Gillray, J., 451, 452, 567
Gin-palaces and gin-shops, *29, 33*; 170, 586
Gingell, W. B., *322*
Giorgione, 504, 506
Girtin, T., 463
Gissing, G., 299, 374, 489, 518, 530, 739
Gladstone, W. E., 163, 180, 185, 190[48], 198, 717, 727
  grocers' licences, 167, 168
  refreshment house licences, 165, 168
Glamorganshire Canal, 295
Glasgow, architecture, *123*; 432, 442, 443
  collectivist enterprise, 396
  children's hospital, 653
  concentrations of population, 38
  drunkenness, 184
  effect of commerce, 874
  Egyptian Halls, *328*; 442
  extent of disease, 629–30
  Free Presbyterian churches, 443
  housing, 397–9, 402–7
  Improvement Act (1866), 395
  infant mortality, 649
  Irish, 399
  lack of voluntary enterprise, 397
  Loch Katrine water scheme, 397
  mortality, 20, 401
  overcrowding, 22, 23, 404
  in poetry, 485
  population, 9
  Police Acts, use of, 394, 399–400
  public lighting, 397
  pubs, 807
  railway, 277
  railway hotel, 296
  railway manufacture, 294
  railway stations, 280, 281, 301
  sanitation, 397
  size, 391
  suburbs, 403
  'Ticketed Houses', 398
  tramps, 129
Glass, D. V., 52[31, 37], 121[53]
Glass, Ruth, 732[19], 888
Gloucester, 292
  children's hospital, 653
  lodging-houses, 127
  railway, 279
Gloucestershire, economic decline, 109
Godalming, 445
Godfrey's Cordial, 650
Godstone, migrants to, 133
Godwin, E. W., *331*; 434
Godwin, G., 460, 472[35], 689[63], 735[68]
Golden, H. H., 33n.
Goldwater, Senator B., 803
Gombrich, E. H., 567, 577[10, 17], 580[38, 41], 581[48]
Gomme, A., 442, 446[11]
Good, J. W., 814[153]
Goodall, A. L., 661[24, 26]
Goodhart-Rendel, H. S., 331[80]
Goodman, G., 771, 775, 777–8, 782
Goodman, J. B., 870[41, 47]
Goodwin, F., *222*
Gordon Square (London), 353
Gorham, M., 187[15]
Gorton, 294
Gosport, railway, 277
Gott, B., 764
Gough, A. D., *317*; 442
Govan, 391
Gower, H. D., 246[32]
Gower Street (London), 313
Gowing, L., 464
Graham, J. Q., 103[56]
Grainger, R., 395
Grant, T., 839
Grantham, 291
Graphic, 385, 388; 459, 466, 472[31], 560, 563, 566, 572, 573
Graves, R. J., 661[16]
Gravesend, migration from, 138
  migration to, 140
Gray, B. Kirkman, 102[17]
Gray, D. J., 581[48]
Gréard, M., 428[18]
Great Britain, birth-rate, 10, 11, 35 (table)
  concentration of population, 41
  death-rate, 10, 35 (table)
  effects of urban transformation, 10
  Gross National Product, 11
  growth of cities, 8
  income per capita, 11
  internal migration, 15
  level of urbanization, 35 (table)
  mortality, 27, 35 (table)
  population, 6, 10
  prices, 10
  see also England and Wales
Great Exhibition (1851), 198, 204, 233, 234, 327
Great Expectations, 521, 524, 539, 554
Great Packington, 443
Great Russell Street (London), 313
Great Transformation, The, 874
Great Unwashed, The, 860
Great Yarmouth, sailors' home, 78
Greaves, W., *360*; 468
Grecian Theatre (London), 60
Greece, rate of urban concentration, 41 (table)
Green, A. L., 691[71]
Green, J. R., 592
Green, T. H., 178
Greenhow, Dr E. H., 605, 638–641, 644, 650, 664[58], 666[88]
Greenock, 391, 394, 649
  overcrowding, 401
Greenwich, migration to, 140
  pier, *160*
  poverty, 597
  Royal Observatory, 432
Greenwood, J., 124, 153[6], 154[24], 159[176], 160[235], 188[19] 587 589

Index

Greenwood, T., 331[78]
Greg, W. R., 86, 101[13]
Gregan, J. E., *329*; 441
Gregory, Canon, 861
Gregory, R., 209[24]
Gregory, T. E., 386[65]
Gregson, J. S., 740, 747–8
Grenville Murray, E. C., 243
Grey, Sir G., 173
Grimsby, 287
   railway hotel, 296
Grimshaw, J. A., *359*; 469, 468
Grindon, L. H., 760[40]
Grinling, C. H., 307[42], 309[90, 105]
Gripper, E., 289
Groome, F., 154[9]
Grossmith, G. and W., 385[51]
Grosvenor Hotel (London), *276*
Guardian Society, 698, 704
Guillou, J., 50[18]
Gutteridge, J., 858, 863, 867
Guy, W. A., 603, 643
Gwynn, D., 810[40]
Gypsies, and the law, 130
   hop-picking, 138
   in and near London, *4, 5, 13, 16*
   on outskirts of towns, 142–3
   outward movement from London, 132
   sites of encampments, 129–30
   see also Migrants; Migration; Tramps

H

Habakkuk, H. J., 385[61]
Habits and Customs of the Working Classes, 860
Hackney, church architecture, 443
   off-licensed premises, 168
   registration of houses, 611
   skating on marshes, 151
   temperance campaigns, 848
   upper-middle class, 596
Hackney Wick, 865
   gypsies, 130
Hadley, A. T., 28, 29, 54n.
Haedy, C., 353
Haeser, H., 665[85]
Hagen, E. E., 35n.
Haggeston, 865
Haines, J. T., 215
Haiti, child-woman ratio, 37
Haldane, Elizabeth, 759[34]
Halévy, E., 483, 875, 888[7]
Halifax, 855, 882
   Town Hall, *324*
Hall, Sir B., 608
Hall, E. T., 685[20]
Hall, H., 765
Hall, P., 427[4]

Halliday, A., *50*; 220
Hambridge, C., *321*
Hamburg, birth-rate, 19
   population, 9
Hamerton, P. G., 466
Hammersmith, 390
   House of the Good Shepherd, 844
   off-licensed premises, 168
Hammond, J. L. and Barbara, 119[12], 353[1]
Hamp, S., 332[102]
Hampden Club, 175
Hampshire, migration via, 117
Hampstead, 345
   architectural styles, 320, 322
   Garden Suburb, 445
   Heath, *154*; 150
   medical officer, 609
   off-licensed premises, 168
Handlin, O., 56[66], 57[73], 354[9, 15]
Hands, K. P., 328[6]
Hanham, H. J., 785[6]
Hanover Square (London), 324
Hankey, H. A., 317
Hanly, M., 175
Hanna, Rev. H. 803
Hanna, W., 835[25]
Hansen, G., 27
Hansom, J. A., *314*; 443
Harcourt, W. V., 178
Hard Times, 84, 85, 86, 272, 544
Hardie, M., 473[56]
Hardwick, P., *305*; 441
Hardy, F. D., 455
Hardy, T., 175, 187[16], 521, 522, 903
Hare, A., 900
Harington, Sir J., 681–2, 690[68]
Harland, J., 749, 750, 757[5, 9], 759[28]
Harpenden, 445
Harper's Weekly, 597
Harrington, Lord, 283
Harris, A. E., 621[37]
Harris, B., 38, 55[60]
Harris, C. D., 56[67]
Harris, E. S. F., 850[19]
Harris, E. V., *335*
Harris, J. R., 689[61], 691[73]
Harrison, A., 121[40]
Harrison, B. H., **161–90**, 189[37], 853[97]
Harrison, J. F. C., 189[30], 889[22, 23]
Harrison, T., *188*
Harrold, C. F., 516[2]
Harrow, station, 282
Hart, E., 607, 620[21]
Hart, Gwen, 691[73]
Hart, J. P., 218
Hart, P. D., 664[53]

Hartley, J., *326*; 441
Hartrick, A. S., 563, 577[15]
Hartwell, R. M., 868[7]
Hartwell, Robert, 172
Harvey, J., 444
Harvey, W., 563
Hassall, A. H., 667[105], 689[63]
Hastings, *162–3*
Hastings, Marquis of, 173
Hatt, P. K., 187[15], 190[45]
Hauser, P. M., 33n., 52[29], 55[56, 58, 59], 56[64], 104[61]
Haussmann, Baron, 418, 427
Haverstock Hill, 345, 347, 349
Havighurst, R. J., 686[33]
Haw, G., 599
Hawkshaw, J., *247*
Hawley, A. H., 54[53]
Hawthorne, N., 450, 551
Hayford, H., 493[7]
Haymarket (London), 173
   prostitutes, 696
   theatre, *57*; 212
Hayter, Alethea, 662[46]
Haywards Heath, 445
Haywood, J., 337, 339, 354[13, 22]
Hazlitt, W., 577
Headlam, S., 595
Health, advances in bacteriology, 95
   demand for national policy, 96
   effects of improvements in medicine, 10, 30
   medical officers, *see below*
   sea-bathing for, *166*
   see also Disease; Sanitation; Water systems; and individual diseases
Health, medical officers of, **603–618**
   achievements, 616–18
   appointment, 608–9
   assistance given by sanitary inspectors, 606–7
   condemnation of overcrowding, 611–15
   duties, 605–6
   house inspection, 610–11
   petition for better housing, 614–15
   prevention of disease as main function, 606
   proposed by Chadwick, 677
   qualifications, 621[29]
   reports, 94, 610
   salaries, 609–10
   working-class opinion of, 755
Health of Towns Association, 95, 664
Heaps, C., 769
Heart of Darkness, 523
Heasman, Kathleen, 836[37]

Index

Heath, W., 570
Heathcote, 445
Heberle, R., 120[32]
Hegel, G. W. F., 272, 273[6], 730
Heimann, P., 684[8]
Henderson, P., 516[13]
Henderson, W. O., 121[51], 273[3]
Hendrick, I., 685[15]
Henley-on-Thames, 238
Henley, W. E., 382[6], 489
Hennock, E. P., 104[67], 186[1], 689[60]
Henriques, F., 705[21]
Henry, Rev. Dr, 799
Henry, T., 810[47]
Hepler, L. G., 686[25]
Herbert, G., 817
Herdman, W. G., 464
Hereford, 278, 292
Herefordshire, level of nuptiality, 17
   migrants, 137, 138
Hertford, employment of migrants, 144
Hertfordshire, 22
Herzfeld, H., 54[49]
Hewitt, M., 665[77]
Hey, W., 771, 782
Hibbert, C., 557[20]
Hick, B., 438
Hicks, G. E., *343*; 455
Higgins, T. T., 666[89]
Higgs, Mary, 42
Highgate, 538
   architecture, 443
   Archway, *306*; 438
Hilda: A study in passion, 599
Hill, A. H., 160[231]
Hill, D. O., 231
Hill, Octavia, 26, 139, 396
Himes, N. E., 56[66], 121[58]
Himmelfarb, Gertrude, 274[8], **707–36**
Hindley, C., 154[10], 156[75,78], 158[151], 160[224], 193, 208[4,6,7,8,9], 577[14], 579[19], 579[28]
Hirsch, A., 665[85]
Hirsch, W., 122[65]
History and Description of the Great Western Railway, The, 439
History of England, 100
Hitchin, tramps, 126
Hitchcock, H.-R., 309[111,112], 328[1], 329[14,23], 331[60], 446[10]
Hobbes, T., 519–20
Hobsbawm, E. J., 108, 119[13], 120[15], 121[62], 383[27], 424, 428[21], 576[3], 688[52], 868[1]
Hobson, J. A., 28, 53[46], 172, 187[19], 188[20]

Hobson, Joshua, 772
Hocking, J., 599
Hodgkinson, Ruth G., 653–4, 666[90], 667[98]
Hodgson, J., 838, 844
Hofman, W., 470[23]
Hogarth, W., 451, 454, 458, 547
Hogben, L., 121[53]
Hogg, J., 187[14]
Hoggart, R., 581[51]
Holbeach, migrants to, 133
Holborn, *122*; 865
   gin-shops, 586
   migration to, 418
   'rookeries', 403
   station, 282
   Viaduct, *425*; 320, 325
Holden Pike, G., 159[202]
Hole in the Wall, The, 598
Hole, J., 889
Holl, F., *351*; 452, 459, 580[44]
Holland, N., 888[12]
Holland Park, 320
Hollar, W., 462
Hollingshead, J., *59*; 141, 158[164], 383[28], 555[11], 735[68]
Hollis, Patricia, 576[2]
Holmes, G. K., 53[41]
Holmes, O. W., *90*; 233
Holmfirth, 239
Holyoake, G. J., 121[44], 206
Home Rule Bill, *see* Belfast
Homestead Act (U.S.A., 1862), 108
Hone, W., 188[24], 197, 579[31]
Hongkong, population, 9
Hood, T., 488–9, 547, 709
Hook, W. F., 766, 817
Hoole, K., 309[106]
Hooligan Nights, The, 600
Hooper, W., 92
Hop-picking, 22
   see also Migrants
Hope, E. W., 409[24]
Hopkins, A. A., 557[20]
Hopkins, A. B., 759[34]
Hopkinson, J., 863, 864
Hopper, T., 438
Hornsey, 300
Horowitz, M. J., 685[20]
Horsfall, T. C., 52[38]
Horsley, J. W., 126, 154[28]
Horsley, J. Callcott, 455
Hotels, growth of, 163, 324–5
   in London, 324–5
   in Manchester, 255
   railway, 296
   see also individual names
Hotel Cecil, *71*, *269*; 319, 325
Houghton, A. B., *344*; 455
Houghton, W. E., 576[3], 705[21]
Houndsditch, slums, 461

Hounslow, station, 282
Hours with the Muses, 744
House, H., 102[31]
Household Words, 173, 372
Housing, **333–53**
   availability of land for, 334
   in central districts, 364–5
   contribution of medical officers to improved, 616–17, 618
   division into tenements, 361
   effects of improved transport, 365
   effects of railways, 250, 296, 297, 366
   extent of new, 334
   importance of cheap, 338
   inadequacy, 106, 250
   as an increasing urban problem, 22
   investment in, 334–5, 369, 376–9
   as an item of real property, 361
   legislation on, 616
   middle-class, 259, 338, 346, 376
   overcrowding, *see below*
   registration of houses, 611
   in relation to wages, 367
   in United States, 25–7
   working-class, 338, 615
   see also Building; Lodging-houses; Scotland; Slums; Suburbs; and individual cities
Housing, overcrowding in, 22, 611–16
   aggravation by house destruction, 614
   causes, 25–7
   consequences, 612
   effect on health, 651
   effects on morality, 612–13, 657
   statistics, 623[59]
Housing and Planning Act (1909), 392
Housing of the Working Classes Act (1890), 392
Houseless Poor Asylum, 143
Hove, 144
Hovel and the Home, The, 366
How, Bishop Walsham, 592
How the Poor Live, 589
Howard, Diana, 214n.
Howard, E., 30–1, 885–7, 888[4], 889[36]
Howard, G., 320
Howard, K., 374, 385[51]
Howard, R. B., 626, 660[3]
Howarth, E. G., 158[159]
Howell, G., 180

Howells, W. D., 243
Howitt, W., 386[69]
Hoxton, 192–3, 213
  Theatre, *61*
Hoyle, W., 166–7, 169
Hoyt, H., 56[66]
Hsu, F. L. K., 685[22]
Huddersfield, lodging-houses, 127
  railway station, 301
Hueffer, F. M., 470[21]
Hughes, A., 453, 457
Hugill, S., 155[57]
Hull, compared to Sheffield, 337
  railway hotel, 296
  railway station, 280, 302
  suburban railway, 299
  waterfront industries, 147
Hume, J., 781
Humphry Clinker, 520
Hungary, annual rate of urban concentration, 5 (table)
  migration from, 15
Hungerford Market, *307*; 439, 504
Hunt, H. A., 309[111, 112]
Hunt, Leigh, 547
Hunt, W. Holman, *353*; 453, 473[53]
Hunter, A., 771, 782
Hunter, Henry J., 656
Hunter, H. T., 398, 399, 401, 409[27], 410[54]
Hunter, J. W., 761[51]
Huntly, W., 656, 666[95]
Hussey, C., 434
Hustings, *168*
Hutchins, Barbara L., 121[40]
Hutchinson, T., 536[10]
Hutchison, T. W., 103[47]
Huxham, J., 629, 661[15]
Huysmans, J.-K., 552, 556[19]
Huyton-with-Ruby, population, 99
Hyams, E., 188[22]
Hyde Park (London), *153*
Hyndman, H. M., 92
Hypatia, 534

I

I'Anson, E., 316, 329[18], 330[46], 441
Ikin, J. A., 776
Illustrated London News, 362–3, 368–9, 371, 373, 375, 381; 466, 570, 577[5]
  advertisement, *87*
  changes in, 572
  city life as seen by, 572, 573
  conservatism, 561, 564, 573, 574
  lack of authenticity in illustrations, 563
  metropolitan readership, 560
  treatment of the poor, 565
  use of half-tone blocks, 237
  use of photographs, 234–5
  ventures into working-class districts, 567
Illustrated Times, 366, 372, 386–9, 391–4, 396–400, 416, 420; 560, 563, 568
  social realism, 573
  street scenes, 565
Imperial Institute, *280*; 327, 328
In Darkest England and the Way Out, 592
In the Image of God, 599
India, effect of births on natural resource consumption, 43
  housing deficits, 37
  migration to cities, 41
  mortality-rates, 34
  urbanization and per capita G.D.P. in, 38
  vital statistics, 35 (table)
Indonesia, migration to cities, 41
Industrial Revolution, effects of, 4
  on the Church, 816
  on housing, 271
  on population, 11
  on textile and metal trades, 421
  on the working classes, 108
Industrial-urban transformation, *see* Urban transformation
Industrialism, features of, 10, 11, 248, 646, 904
Inglis, K. S., 122[74], 835[27], 836[46], 855, 868[4, 5], 870[48], 871[59]
Ingram, H., 563
Ingram, W., 578[20]
Innes, J. W., 121[55]
Inns of Court, Dickens in, 538, 539
International Encyclopedia of the Social Sciences, 94
International Exhibition (1862), 233, 449
Interviews, *see* Oral history
Intuitionism, 902
Ipswich, 302
  Customs House, 441
  English-born population, 118
Iraq, rate of urban concentration, 41 (table)
  urbanization and per capita income, 37
Ireland, 270
  annual rate of urban concentration, 5 (table)
  famine, 796
  migrants from, 108, 795
  overcrowded housing, 23
  scarlet fever epidemic, 652
  see also Belfast; Dublin; Irish
Irish, concentrations of, 165, 399
  as hay-makers, 134
  as hop-pickers, 125
  in market gardens, 135
  as migrants, 124, 125, 133, 134, 138, 149, 373, 399, 415, 459, 801, 837
  occupations, 425
  seasonal movements, 140
  as subject of Victorian art, 459
  women migrants, 137
  see also Church, Roman Catholic
Ironside, R., 471[29]
Irvine, 443–4
Isaacs & Florence, 325
Islington, *335*; 538, 568
  Milner Square, *317*; 442
  Roman Catholics, 842, 845
  tramps, 126
  'World's Fair', 141
Italians, 166
  as migrants, 135, 149
Italy, decreasing internal migration, 15
  effects of migration on growth of cities, 13
  mortality 27, 35 (table)
  vital statistics, 35 (table)
Itinerant Traders of London, The, 565
Ivins, W., 564, 578[24, 25]

J

Jack the Ripper, 586, 594–5
Jackson, J. A., 122[75]
Jackson, Mason, 564, 578[23], 580[36]
Jackson, W., 313, 473[54]
Jacob, H., 56[63]
Jacobs, A., 192
Jaffe, A. J., 51[29]
Jahn, M. A., 385[59]
James, H., 93, 100, 104[72], 369–70, 383[31], 390, 469[13], 489, 551
James, J. A., 820, 834[6]
Jane Eyre, 523
Japan, migration to cities, 14
  mortality in cities, 36
  population restriction, 36
  rate of urban concentration, 41 (table)
  urban demographic transition, 35

Index

Japan—*continued*
  urbanization, 34, 35 (table)
  vital statistics, 35 (table)
Jarrow, 6
Jeaffreson, J. C., 556[19]
Jefferies, R., 535
Jeffery, J. R., 119[5]
Jenkins, E., 619[4], 661[26]
Jenner, W., 632, 654, 662[28]
Jephson, H., 361, 619[2], 620[21, 24], 622[45, 47, 57], 660[9], 662[30]
Jerningham, the Hon. S., 848
Jerrold, D., 215
Jerrold, W. B., **3**; 208[2], 427[1], 461, 556[20], 565, 579[29], 581[47]
Jesmond, 433
Jessopp, A., 862, 870[38]
Jevons, W. A., 157[104]
Jews, exclusion by street-gangs, 76
Jewsbury, Maria J., 757[11]
John, A. H., 159[184]
Johnson, E. D. H., **449–74**
Johnson, Edgar and Eleanor, 556[15]
Johnson, L., 492–3
Johnson, Dr S., 476, 520, 536[5]
Johnson, S. C., 120[16, 17]
Jones, E., 684[7]
Jones, Ernest, 179, 180, 198
Jones, F. M., 684[1]
Jones, G. R. J., 785[3]
Jones, G. W., 785[2]
Jones, Sir H., 320
Jones, I. G., 190[46]
Jones, Inigo, 432
Jones, J., 757[11]
Jones, O., 439–40
Jonson, B., 520, 536[4]
Journal of the Statistical Society of London, 628
Journalism, *see* Press
Joy, G. W., 470[18]
Joyce, James, 493
Jude the Obscure, 521, 524, 535
Judy, 383; 560, 570, 572
Juillard, E., 56[65]
Jullien, L. A., 202

K

Kampf, L., 757[10]
Kane, Rev. Dr, 803
Kant, E., 480
Kaplan, A., 104[62]
Kate Hamilton's, 695
Kay, J., 457
Kay-Shuttleworth, Sir J. P., 90, 257, 264, 271, 688[48]
Kayne, G., 663[52, 55, 70]

Keating, P. J., **585–602**, 385[52], 581[47]
Keeling, Bassett, 444
Keene, C., 452, 567, 580[37]
Keith-Lucas, B., 689[60]
Kellett, J. R., 98, 104[66], 155[54], 186[1, 6], 305[5, 11, 14], 354[8], 385[57], 410[47, 51]
Kelly, J. J., 848
Kelly, W., 842
Kelso, 443
Kempley, 445
Kendal, 292
Kendall, M. G., 103[54]
Kenilworth, migrants to, 133
Kennedy, D., 814[143]
Kenney, R., 155[57]
Kennington, turnpike gate, 120
Kensington, architecture, 320
  attitude to slums in, 381
  Charity Estate, 432
  medical officer, 609
  off-licensed premises, *168*
  Park Gardens, 112
  scarcity of pubs, 166
  Square, 324
  as a suburb, 419
  workhouse infirmary in, 647
Kensington, North, 313
  suburbs becoming slums, 371–372
Kensington, South, 313, 317, 334
  station, 283, 284
  Victorian architectural ideas, 327, 328
Kensington, West, gypsies, 130
Kensington Gore, Lowther Lodge, *271*; 320
Kent, J., **855–72**
Kent, migrants to, 132, 135, 138
Kent, W., 431, 442
Kentish, J., 820
Kettering, 289
Kettle, A., 88, 102[24]
Kettle, J., 54[52]
Kilburn, 300
Killen, W. D., 812[83]
Kilmarnock, overcrowding, 23
Kilsby, tunnel, 439
Kilvert, Rev. Francis, 865
King, J., 769
King, K. D., 599
King William Street (London), 319
Kingdom, W., 351, 357[72]
King's Cross, design of station, 314
  Fair, 133
  pubs, 163
  railway station, *244*; 281, 287, 301
  railway station shed, 439
King's Lynn, fair, 132

King's Weston, 436
Kingsbury & Neasden, station, *133*
Kingsley, C., 29, 84, 90, 101[2], 401, 530, 586, 645, 664[61], 728, 729, 735[75], 885, 889[32]
Kingston, 433
Kipling, R., 374, 597
Kira, A., 689[61], 692[75]
Kirwan, D. J., 128, 155[45], 472[35]
Kitson Clark, G., 106, 119[9], 684[3], 689[60, 61]
Kitton, F. G., 556[12]
Kittrell, E. R., 688[52]
Klebs, E., 653
Klingender, F. D., 450, 461, 469[2], 473[48], 579[27]
Knight, C., 186[5], 561, 562, 568, 577[8], 578[20], 580[42]
Knoepflmacher, U. C., **517–36**, 888[10]
Knowles, C. C., 382[62]
Knowles, J., **276**
Koch, R., 637, 641
Kodak Company, counting house, *147*
  head office, *150*
Köllman, W., 50[19]
Königshütte, overcrowded housing, 24
Korea, rate of urban concentration, 41 (table)
  urbanization and per capita income, 37
Kossman, E. H., 382[2]
Kris, E., 567, 580[38, 41], 581[48]
Kubie, L. S., 685[12], 686[29]
Kuczynski, R., 51[22]
Kuper, H., 190[45]
Kuznets, S., 56[68]
Kyne, J., 843, 844, 845, 851[54]

L

La Barre, W., 685[19], 688[53]
La Nouvelle Héloïse, 477
Labour Church Movement, 864, 866
Labour force, growth of, 49n.
Lackington, T., 55[59]
Laforgue, J., 482
Lamb, C., 489, 547, 902
Lamb, H. A. J., 690[68]
Lambert, B., 592
Lambert, F. C., 244[19]
Lambert, Royston, 103[53], 104[58], 613, 619[2, 7], 624[90], 662[34], 666[91], 689[63], 691[73]
Lambert, W. R., 187[18]
Lambeth, 538
  baths, 322

Index

half-gypsy colony, 141
music hall, *36*; 174
police station, 300
Lami, E., 460
Lampard, E. E., **3–57**, 49[3], 52[29], 55[60, 65, 66], 56[64], 104[61]
Lampen, H., 769
Lancashire, administration of Poor Law, 771
Authors' Association, 745
Chetham Society, 750
cotton famine, 633
drunkenness, 184
effect of industry on appearance, 247–8
high marriage-rates for employed women, 18
infant mortality-rates, 24
migrants from, 137
nuptiality for urban and rural adults, 17
pub distribution, 167
typhoid fever, 640
see also Liverpool; Manchester
Lancaster, E., 213
Lancaster Gate (London), 432
Lancet, 604, 607, 608, 628, 638, 694, 699, 804
Lanchester & Rickards, *336*
Lander, H., 599
Landes, D. S., 428[20]
Landlords, attitude to tenants, 78, 353, 380–1
conflict of interest with speculative builders, 350–1
little interference with leaseholders, 344
in Scotland, 404–5
see also Bedford, Duke of; Norfolk, Duke of
Lands Clauses Act (1845), 392
Langford, J. A., 305[8], 309[93], 834[4, 10]
Langford, M., 692[75]
Langley, A. S., 834[8]
Lankester, Edwin, 607, 612
Lankester, E. R., 703
Lasker, B., 134, 157[105]
Laski, M., 662[46]
Latter-Day Pamphlets, 495, 501
Latter-Day Saints (Mormons), 858
Lausanne, Dickens in, 539–40, 555[6]
Lauter, P., 757[10]
Lautréamont, Comte de, 482
Lavoie, E., 56[63]
Law Courts (London), 146
Law, C. M., 575[1]
Lawrence, C. E., 188[28]
Lawrence, D. H., 880
Lawton, R., 119[4, 6]

Layard, G. S., 577[18]
Lazarus, H., 380, 386[72]
Le Breton, J., 599
Le Play, F., 110
Lea Valley, 294
Leadenhall Street (London), 317
Leader, 700
Leasehold system, building agreements, 340, 342, 345, 346, 352
conflicts of interest, 350
effects on quality of building, 344
estate management, 334–57
finance of builders by landlords, 346, 378
ground rents, 374, 376, 380
landlordism as product, 380–1
ninety-nine year leases, 352
peppercorn rents, 334
provision of roads, 353
restrictive covenants, 342–4, 346, 350, 352–3
role of landlords, 352–3, 380
role of solicitors, 378–9
speculative investment, 374, 376–9
see also Building; Chalcots; Landlords
Lecky, W. E. H., 702, 705[19]
Ledoux, C.-N., 444
Lee, C. E., 306[23]
Lee, J. M., 785[2]
Lee, W., 337, 339, 354[13, 22]
Leech, J., 452, 569
Leeds, abolition of church rates, 766
ballot-rigging, 769
character of merchant class, 764
church attendance, 855
churchwarden elections, 765–6
as commercial centre, 764
conflict over workhouse board, 768
Conservative seats in municipal elections, 799 (table)
Conservative share of poll at elections, 781 (table)
Corn Exchange, *311*; 437
Council membership, 782 (table)
disease, 638
Dissenters, 766, 771, 773, 774
economic structure, 764
economic status of Leeds wards, 780 (table)
effect of commerce, 874
effect of migration, 13
highway surveyors, 772–3
Improvement Acts, (1824) 771, (1842) 772, (1866) 778

Liberal influence, 768, 769, 771, 773, 774, 776, 779, 781
Marshall's Mills, 304, 440
mortality, 20
municipal government, 775–780, 784
music halls, 765
out-townships, 765
parliamentary elections, 780–781, 784
political agitation, 783, 784
political composition of Bench, 776
political conflict over parochial administration, 764
politicization of Poor Law, 768, 769, 770, 771
population, 9, 764
Primitive Methodism, 858
public health, 778
railway stations, 280, 281
rates, 778
religion, 784
St Thomas's church, *316*
school board elections, 773, 774
size, 390
suffrage, 783
town council conflicts, 775
township and parochial administration, 764–74, 784
township and wardships, 766, 767 (map), 770
Tory influence, 766, 768–9, 771, 776, 779, 780, 781
Tramways Order, 163
urban politics, **763–84**
Voluntaryism, 784
water supply, 771, 778
Leeds Express, 773
Leeds Intelligencer, 765, 769
Leeds Mercury, 784
Lees, Lynn **413–28**, 428[17], 732[19], 851[65]
Legoyt, M. A., 51[21]
Leicester, Baptist Chapel, *314*; 443
County Gaol, *297*
crude mortality, 21 (table)
Domestic Mission Society, 113
English-born population, 118
expansion independent of railways, 299
infant mortality-rate, 649
market place, *80*
medical officers, 604
mortality, 20–1
overcrowding, 22
railways, *257*; 289, 304
railways in area, 291 (map)
railway stations, 280
rates, 778

449

Index

Leicester Square, 696
  employment of boardmen, 150
  soup kitchen, 123
Leighton, F., 453
Leiper, W., 443
Leipzig, birth-rate, 19
  illegitimacy, 19
Leisure Hour, 577[5]
Leith, 391, 394
Lemaitre, H., 474[65]
Lemon, R., 207[1]
Lemoisne, P.-A., 472[36]
Leningrad, migration, 14
  population, 9
Leno, J. B., 176
Leopold of Saxe-Coburg, Prince, 462
Leslie, H., 219
Letchworth, 887
Lethaby, W. R., 331[68], 445
Letheby, H., 624[89]
Lettsom, J. C., 85, 101[7], 385[54]
Levasseur, Émile, 13, 50[14], 51[27]
Lever, W. H., 884
Levi, Leone, 159
Levine, G., **495–516**
Levy-Leboyer, M., 428[16]
Lewes, Mrs C. L., 689[59]
Lewes, G. H., 223
Lewes, priory destroyed by railway, 303
Lewin, B. D., 689[63]
Lewis, E. D., 308[80]
Lewis, O., 276[25]
Lewis, R. A., 102[19], 275[17], 619[2], 686[34], 687[39], 688[56], 689[63]
Lewis-Faning, E., 56[66]
Lewisham, 538
  off-licensed premises, 168
  swimming baths, 322
Ley, J. W. T., 554
Leyton, 77
Leytonstone, 390
Liberation Society, 161
Libraries, provision of, 113, 146, 323
Licensed premises (maps)
  Bethnal Green and Spitalfields, 164
  Bloomsbury, 167
  dockland, 166
  Strand, 165
  see also Pubs
Lichfield, 818
Liddle, J., 603, 607, 620[19], 622[54]
Life and Adventures of a Cheap Jack, The, 149, 192
Life and Labour of the People in London, 595
Life in London, 202, 362, 451, 587
Life in West London, 600

Lifton, R. S., 276[26]
Lille, 281
Lillie, W., 691[73]
Lillo, G., 222
L'Illustration, 460
Limehouse, 538
  Charlie Brown's Railway Tavern, 175
  Strangers' Home, 2
Limerick, overcrowding, 23
Lincoln's Inn Fields, 843
  Great Wild Street, 845
Lincolnshire, migrants to, 133, 134, 143
Lindley, K., 577[13]
Linsley, G. A., 778
Lipman, Vivian D., 385[50]
Lister, R., 469
Lithgow, R. A. D., 758[20,23,25]
Little Dorrit, 272, 537, 548, 551, 554
Littlejohn, H. D., 393, 407, 411[56], 609, 616, 620[20]
Liverpool, *96, 114, 117*; 106
  Academy of the Arts, 453
  Brown Library, *173–4*
  building trade, 146
  children's hospital, 653
  church attendance, 855
  civic enterprise, 396
  commuter travel, 296
  dockland architecture, *326*; 441
  emigration from, *131*
  English-born population, 118
  Exchange Buildings, *149*
  extension of passenger lines, 280
  Irish, 118, 399
  lodging-houses, 127
  mortality, 20, 21 (table)
  overcrowding, 403
  overhead railway, 284
  police inspection of housing, 399
  population, 9, 13, 391
  Primitive Methodism, 858
  pubs, 807
  railway, 277, 279
  railway architecture, 301
  railway hotels, 296
  railway stations, 280, 301
  Sanitary Act (1846), 604
  shipping trade, 146
  Toxteth Park cemetery, *169*
  tramp shelters, 129
  waterfront industries, 148
Liverpool Street station, 283
Livesay, A., 444
Livius, T., 539, 849[5]
Liza of Lambeth, 600
Llewellyn Smith, Sir H., 50[19], 94, 103[50], 112, 120[38]

Lloyd, A. L., 156[72]
Lloyd, J., 156[76]
Lloyd's Weekly, 577[5]
Local Government Act (1888), 605, 606
Local Government (Ireland) Act (1898), 802
Lockwood & Mawson, 441
Lodging-houses, conditions, 127–128, 725
  desertion at hopping time, 138
  pleas for inspection, 726
  tramps' use, 126, 127
Loeffler, F., 653
Logsdail, W., 473[58]
London, as agglomeration of suburbs, 29
  birthplace of inhabitants, 416 (table)
  birth-rates, 49n.
  as centre of drama and theatre, *48–65*; 212, 224
  cholera deaths, 636
  churches, *313, 315, 334*
  City, *see below*
  city-size distribution, 39
  coffee shops, 127
  compared with Paris, **413–28**, 483
  complexity of economic development, 419
  concentration of population, 38–9
  conditions in underworld, 460
  Corresponding Society, 175
  cost of living, 391
  County Council, inception, 94, 427; civic enterprise, 396
  County Hall, *336*
  crude mortality, 21 (table), 49n.
  density of population, 26, 106, 417, 418 (map)
  East End, *see below*
  effects of migration, 13, 362
  'entertainment' quarter, 391
  Fever Hospital, 633, 637
  fog, *382*; 466, 541, 555[8]
  Guildhall, 437, 438
  hospitals, 441, 605, 637, 653
  inns, *25, 26, 27, 72*
  lodging-houses, 127
  Mansion House, 238, 318
  Metropolitan Board of Works, 318–19
  middle-class interest in business, 426
  migration from, 133, 139
  migration to, 49n., 139, 141, 414, 415, 416

450

mortality and fertility, 416 (table), 417
music halls, *361*; 174, 175
Night Refuges, 142
occupation of inhabitants, 420 (map), 421
paintings, *341–4, 349–55, 357–358, 360–1*; 228
population, 4, 9, 362, 414, 422
Port of, 419, 420
prisons, *248, 296, 298*; 444
prostitution, 694, 701
rates, 381
sleeping rough, 128
social structure, 423–6, 424 (map)
Society for the Protection of Young Females, 696
standard of living, 476
Statistical Society, 86–7
structure of working class, 424–5
tradition of skilled labour, 420–1
typhus deaths, 633
van-dwellers at Agricultural Hall, *16*
Working Man's Association, 176
see also Architecture; Docks; Drama; Housing; Prostitution; Pubs; Railways; Slums; Street life; Suburbs; Theatre; and individual institutions and placenames.
London, City of, xxvii
churches, 433
Dickens and, 538
medical officer of, 609
population, 4, 7–8
portrayed in drama, 222, 223
pub density, 163
sanitary inspection, 607
school, 319
terminal stations, *247*; 281
Theatre, 212, 221
transport inns, *26–7, 72*; 162
Victorian buildings, *258–62, 266, 268, 323*; 312, 315
London, East End of, 585–600
attitude of wealthy to, 180
cholera, 600[2]
churches, 447[15]
contemporary view of monotony of, 590, 591, 598, 600
deficiencies of contemporary writing on, 599–600
migrants from, 135
novelists and, 589, 597–9
Ruskin's attitude to, 503
Sailors' Home in, *1*
statistics of poverty, 595–7

as symbol of urban poverty, 585, 589
visits of aristocratic youth, 587–8
worst areas, 587–8
London: A pilgrimage, 3; 441, 461; *see also* Doré, G.
London and its Environs in the Nineteenth Century, 463
London Bridge, *135, 247*; 238
pubs, 163
railway station, 281, 282, 296
sleeping rough near, 123, 128
London Films, 243
London Labour and the London Poor, 86; 400, 460, 708, 710, 715; *see also* Mayhew, Henry
London Shadows: A glance at the homes of the thousands, 460
London Wall, 317
London's Shadows, 727
Long, S. E., 809[5, 6, 11]
Long Millgate (Manchester), poets meeting at *182*; 745
Longmate, N., 689[62]
Longstaff, G. B., 50[16, 18]
Los Angeles, migration to, 14
population, 9
Lothbury, *262, 323*; 315, 443
Loudon, J. C., 685[23]
Louse infestation, 631
Love, D., 192, 208[4]
Lovett, W., 183, 189[29, 42]
Low, S. J., 54[49]
Lowder, C. F., 592
Lowe, R., 180
Lowery, R., 176
Lowestoft, 140
migrants to, 143
Lowry, H. F., 536[1]
Loyal and Patriotic Fund, 176
Lubbock, B., 24
Lubove, R., 53[41]
Luccock, J. D., 775
Lucky Bargee, 599
Luddites, 178
Ludgate Circus, 303, 461
Ludgate Hill, 121
railway bridge across, *249*; 303
station, *248*
Ludlow, J. M., 728, 729
Lumley, W. G., 122[73]
Lupton, D., 776, 782
Lutyens, Sir E., *337*; 445–6
Lyceum Theatre (London), 65
Lyle Cummins, S., 664[54]
Lynch, K., 104[70]

M

Maas, J., 469
Macarthy, D., 151

Macaulay, T. B., 100, 101
McClelland, D. C., 692[75]
McClelland, V. A., 852[89]
M'Connell, J., 798
M'Connell, W., 472[35]
McCord, N., 786[26]
M'Cree, Rev. G. W., 383[15]
McCulloch, J. R., 84, 85
MacDermot, E. T., 305[13]
MacDonagh, O., 108, 120[17]
M'Donnell, T. M., 821
McDougall, J. B., 664[54]
McDowell, R. B., 810[24]
Macduff, W., 458
McGee, G. W., 180
McGregor, O. R., 624[91], 705[21]
McGrew, R. E., 689[62]
McKay, W., 400
McKendrick, N., 687[40]
Mackenzie, J. S., 6, 49[5]
Mackenzie, R., 732
McKenzie, R. D., 102[22]
McKeown, T., 642, 662[51]
Mackonochie, A. H., 866
Maclagan, W. D., 859, 869[21]
Maclaren, J., *302*
MacLeod, R. M., 690[63]
Maclise, D., 382[6]
Macmillan's Magazine, 597
Macnamara, C., 689[63]
MacNeice, L., 804, 812[101]
Macready, W., 695
M'Tear, T., 814
Madness of David Baring, The, 599
Madox Brown, F., *345*; 453, 456, 465, 471[30], 486
Madras, population, 9
Madrid, population, 9
Magazine of Art, 449
Magazines, *see* Press
Mairet, P., 54[52], 57[76]
Maitland, F. W., 243
Malcolm, A. J., 812[102]
Malcolmson, Patricia E., 384[45], 386[76], 620[24]
Malignant Cholera: Its mode of propagation and its prevention, 637
Malthus, Rev. T. R., 86, 101[11], 723, 724, 725, 730, 734[57]
Malton, Thomas, father and son, 463
Malvern, 292
Manby Smith, C., 140, 158[150]
Manchester, 247–72
agglomeration of suburbs, 29
architecture, 252, 253
Art Treasures Exhibition (1857), 233, 449, 450
avarice of inhabitants, 740
Barton's Arcade, *312*

Index

Manchester—*continued*
 building trade, 146
 chapels and churches, 250, 255
 church attendance, 855
 commercial district, 258
 control over weavers, 741
 Courts, *218–19*
 crematorium, *228*
 crude mortality, 21 (table)
 disease, 638
 effect of commerce, 874
 fair, 131
 Fish Market, *179*
 Free Trade Hall, *84, 183*; 254;
 gaols, *220, 225*
 Grosvenor Square, *197*
 growing acceptance of city by writers, 756
 growth of industry, 250, 255
 growth of railways, 249, 268, 279
 growth of suburban areas, 249, 254
 growth of tramways, *177*; 252, 253, 255, 259
 hotels, *189*; 255
 housing, 22, 250, 265, 266, 271, 272
 infant mortality-rates, 24
 Infirmary and Lunatic Asylum, 251
 Irish, 264, 272
 King Street, *186*
 Literary Club, 752–3
 Little Ireland, 271–2
 Market Street, *199–201, 212*
 meaning underlying street geography, 900
 medieval street and building patterns, 251
 Methodist mission, *224–5*
 middle-class districts, 259, 260, 261
 mortality, 20
 Mosley Street, *364*
 music halls, 175
 Newall's Buildings, *184*
 Night Refuges, 142
 Old Shambles, *180, 206*
 Old Town, 264–70
 Old Town Hall, *222*
 photographic survey, 236
 Piccadilly, *210*
 pictorial history, 564
 politics of churchwardens, 766
 Polygon, 249, 253
 population, 9, 13, 391
 Primitive Methodism, 858
 pub density, 169
 pubs, *181*
 Queen's Hotel, *189*
 railway manufacture, 294

 railway stations, *196, 198*; 250, 280, 281, 287
 Rochdale Road, *214–17*
 Royal Exchange, *188, 205*; 207
 St Ann's Square, *203*
 sanitation, 683
 segregation of classes, 258–61
 Ship Canal, *167*
 Shudehill Poultry Market, *211*
 size, 250, 252, 390
 Smithy Door, *204–5*
 social background of writers, 742
 squalor, 267–8
 start of commuter travel, 296
 Statistical Society, 86, 88, 739
 street literature, 740, 741, 757[10]
 street plan, *193*
 subjects of poetry, 746
 terraced houses, 253, 254
 theatres, *221, 223*; 254
 Town Hall, *202*; 255, 443
 tramps, 128, 141
 use of canals, 250, 254
 use of rivers, *195*
 Victoria Buildings, *208*
 warehouses, *187, 209, 327, 329*; 440–1
 weaving and factory verse, 740, 741, 742, 748–9, 757[9]
 Whit Walk, *226*
 workhouse, 267
 working-class districts, 258, 259–60, 262, 264–72
 writers and poets, **739–56**
 writers' attitude to city, 744, 747, 752
 see also Engels; Gaskell
Manchester City News, 750
Manchester Examiner and Times, 91
Manchester Guardian, 91, 146, 746, 749
Mander, R., 188[25, 26]
Mandron, R., 493[2], 494[13]
Manet, E., 457, 465, 471[24], 474[59]
Mann, H., 855, 857, 858, 868[3], 869[9]
Mann, P. H., 121[56]
Manning, Cardinal, 175, 181, 182, 595, 848
Mansfield, 301
Mapperley, 289
Marble Arch, 313
Marcus, S., 257–76, 667[99], 690[66], 736[82]
Marey, J. É., 237
Margate, 125
Marius the Epicurean, 534
Mark Lane (London), *261*; 316, 443

Marlow, W., 462
Marmor, J., 691[71]
Marriage, advocation of postponement, 28
 chances of, in towns, 113
 demographic consequences of early, 114
 fluctuations in marriage-rates and population growth, 91
 rates, 17–19
Marris, P., 190[45]
Marseilles, 280
Marsh, C. M., 835[23]
Marshall, A., 27, 28, 30, 54[50], 93, 103[47], 887
Marshall, C. F. D., 308[81]
Marshall, H. C., 777, 782
Marshall, J. G., M.P., 782, 783
Marshall, L., 887, 889[38]
Marshall, T. H., 660[4]
Marshall, W., 158[140]
Marshall & Snelgrove, 324
Martin, G. H., 227–46
Martin, J., 438, 461–2
Martin, P., 244[22], 245[32]
Martin Chuzzlewit, 542
Martineau, Harriet, 551, 556[18]
Martineau, R. B., 471[27]
Martley, 138
Marwood (public executioner), 174, 188[23]
Marx, K., 3, 110, 274[14], 471[23], 510, 712, 717, 728
Marx, L., 888[10]
Mary Barton, 65; 88, 101[13], 266, 276[32], 748, 749, 902
Marylebone, railway station, 280, 297
 workhouse, *9*
Masborough, 287
Mason, R., 396
Massachusetts, birth-rates, 19
 foreign-born population, 50[15]
 high marriage-rate, 18
 infant mortality, 25, 26
 mortality, 20
Master Humphrey's Clock, 569
Masterman, C. F. G., 96, 104[60], 370, 384[33]
Masters, D. R. D., 274[12]
Mathew, Father T., 845, 848
Matthiessen, F. O., 103[49]
Matz, B. W., 162, 186[3], 555[3]
Maugham, S., 600
Maurice, F. D., 456, 486
Maurice Quain, 599
May, H., 432
May, P., 452
Mayer, H. M., 246[32]
Mayer, J. P., 305[3], 757[4]
Mayfair, 339, 372, 390
Mayhall, J., 787[65]

Index

Mayhew, A., 150
Mayhew, Henry, 88, 89, 101, 102[23,28], 124, 125, 132, 143, 154[11,25], 156[74]ff, 157[109,124], 158[144]ff, 159[173,180], 160[224, 225, 234], 169, 173, 188[21], 192, 199, 200, 201, 208[2,3], 209[17], 234, 363–4, 382[6], 400, 459, 530, 581[49], 586, 657, 658, 679, 684[1], 688[54], **708–32**, 734[53], 735[67], 858, 869[17], 900, 904
  articles in *Morning Chronicle*, 709
  attitude to street-folk, 711–12, 721
  compassion, 720
  criticism of, 716
  definition of the poor, 709, 900
  dramatic treatment of material, 714
  errors, 713
  failure of contemporaries to mention, 717
  limited influence, 717
  limited section of population described by, 710
  obsession with statistics, 713
  opinion of philanthropists, 722
  opposition to contemporary reforms, 723
  opposition to Malthusianism, 724
  understanding of environmental factors, 721
Mazee, H. E., 660[3]
Meacham, S., 833[3]
Mead, Margaret, 684[2]
Meadows, K., 563, 565, 566
Mechanics' Institutes, 113
Mechie, S., 833[3]
Medical Directory, 609
Medical Times, 618
Meek, G., 139, 158[142]
Mehta, S. K., 55[61]
Melbourne, population, 9
Melton Mowbray, 289
Memoirs of a Social Atom, 151
Menai Straits, railway bridge, 439
Mennie, H., 255
Menninger, W. C., 684[7]
Mercer, E., 761[55]
Meredith, G., 518, 535
Meredith, R., 244[13]
Merrie England, 879
Merseyside, effect of industry on appearance, 248
  see also Liverpool
Merthyr Tydfil, 'China' in, 126
  disease, 638

infant mortality, 650
Ironworks, 438
proletarian control, 772
Merton, infant mortality, 99
Metcalf, Priscilla, 330[37]
Metropolis Local Management Act (1855), 604, 605, 606, 607
Metropolitan Board of Works, *see* London
Metropolitan Cattle Market, 320
Metropolitan Improvements, 463
Metropolitan Poor Act (1867), 322
Meuriot, P., 51[22]
Mexico, rate of urban concentration, 41 (table)
  rate of urban growth, 37
  urban natural increase, 37
  vital statistics, 35 (table)
Mexico City, population, 9
Miall, E., 835[27]
Microcosm of London, 463
Middle classes, *see* Classes, middle
Middle East, urban population in relation to gross domestic product, 37–8
Middlemarch, 533–4, 629
Middlesbrough, age of population, 99, 100
  as boom town, 125
  'moral density', 117
  sex of population, 116
Middlesex hospital, 605
  migrants to, 137
  wages, 133
Middleton, T., 520
Middleton, W., 770
Midwinter, E. C., 691[73], 786[26], 814[144]
Migrants, long-term: adaptation to town life, 118
  age, 112
  contact with villages of origin, 117–18
  cultural shock felt by, 184–5
  economic opportunities, 108–9
  French, 138
  habitations, 372–3
  Irish, 108
  living conditions, 217
  in lowest ranks of working class, 425
  marriage opportunities, 113
  in middle-class occupations, 426
  Scots, 415
  sex-distribution of, 112
  Welsh, 138
Migrants, seasonal: as agricultural itinerants, 132, 133, 134, 136, 137

  attracted by pubs, 172
  in breweries, 44
  Christmas increase in employment, 149
  cultural shock felt by, 184–5
  decline, 153
  different classes, 124, 134
  hay-making, 134
  hop-picking, 125, 133, 134, 135, 136, 137, 138, 139
  industrial employment, 148
  influence of weather, 131, 150–1, 152
  lack of employment, 7; 150
  legislation against, 153
  in market gardens, 135
  occupations, 124, 134, 135, 136, 145
  relation to urban social economy, 15
  respectability of, 152
  sailors, 146–8, 151
  sleeping places, 127–8
  street-trading, 149
  wages, 133
  work for women, 133
  see also Irish; Italians; Occupations; Tramps
Migration, age of population at time of, 16–17
  as component of urban increase, 12–17
  consequence, 625
  effect on urban theatre, 213
  effects on American cities, 14
  effects on European cities, 13
  effects of railways, 298
  international, 14–15
  results, 643
  rural–urban, 105, 108–9, 112–118, 139, 143
  seasonal nature, 139
  undertaken in stages, 117, 142
  urban–rural, 136–7
Milan, 280
Mile End Old Town, 596
Miles Platting, 294
Milford Haven, shipping trade, 147
Mill, J. S., 48, 87, 88, 91, 261–2, 717, 724, 876, 887
Mill on the Floss, The, 521, 524
Millais, J. E., 453, 457, 458
Millais, J. G., 471[25]
Millbank Penitentiary, 444
Miller, D. P., 156[78]
Miller, D. R., 686[33]
Miller, J. C., 829, 830–1, 833, 834[18], 835[28,35], 836[45]
Miller, J. Hillis, 542–3, 555[9]
Miller, N. E., 685[13,16]
Miller, T., *416*; 128, 155[46]

453

Index

Milligan, H., 244[20]
Mills, H. V., 879
Mills, J. S., 472[32]
Millthorpe, 879
Milner, G., 760[44, 47]
Milner, Marion, 686[31], 688[53]
Milner and Sowerby's 'Cottage Library', 191
Milton, John, 518, 519, 520, 521, 522, 742
Milwaukee, mortality, 27
Mining towns, infant mortality, 24, 402
  typhoid fever, 639
Minneapolis, migration to, 14
Mint, the, 589
  pubs near, 166
  tramps near, 166
Minto, C. S., 245[32]
Mishan, E. J., 57[73]
Mission and Extension of the Church at Home, The, 829
Mitchell, B. R., 49[10], 50[13], 52[37], 55[61], 118[3], 119[4]
Mitchenson, J., 188[25, 26]
Mocatta, D., 301
Modern London, 565
Modern Painters, 502, 503–4, 506
Mole, D. E. H., 815–36, 835[35]
Molesworth, Sir W., 536[3]
Moll Flanders, 520
Mols, R., 50[11]
Moncrieff, W. T., *48, 53*; 215, 219
Monet, C. J., 465, 466, 474[59]
Monk, S. H., 446[3]
Monmouthshire, winter migrants 143
Monod, S., 101[4]
Montague, C. J., 209[22]
Monthly Magazine, 463
Moody, T. W., 814[143]
Moody and Sankey, 866
Moonshine, 560
Moore, F. F., 811[56], 814[145]
Moore, G., 600
Moore, Fr John, 846
Moore, T., 812[91]
Moorfields (London), 843
Moorgate, 319
  office blocks, 442
  station, 282, 283
Mord Em'ly, 600
Morden, 99
Moreton-in-the-Marsh, 293
Morgan, D., 160[240]
Morgan, J., 200
Morgan, J. Minter, 881
Morgan, M., 573
Morland, G., 454
Morning Advertiser, 561

Morning Chronicle, 88, 645, 708, 709, 710, 712, 728
Morocco, rate of urban concentration, 41 (table)
Morpeth, Lord, 394
Morris, J. H., 308[75, 78]
Morris, May, 516[12]
Morris, O. J., 244[27], 245[32]
Morris, R. N., 102[22]
Morris, William, 434, 487, 488, 495–8, 509–15, 886
  attitude to middle classes, 512
  attitude to pleasure, 514
  attitude to women, 515
  concern with working men, 510
  debt to Carlyle, 511
  rejection of city, 496, 510–11
  rejection of profit motive, 497
  socialism, 879
  vision of London, 513–14
Morrison, A., 275[15], 374, 586, 591, 597–8
Morse, R., 39, 56[64]
Mortality, and environment, 19–23
  and fertility, 34–6
  as indicator of community health, 650
  declining rate, 27, 95–6
  effects of decline in urban, 27–8
  infant, 16, 23–7, 53[43], 95, 249, 401–2, 649, 650–1
  in London, 49[11]
  in relation to sickness, 627
  in Scotland, 401
  of men, 16
  see also Population
Mortality, causes of:
  diarrhoea, 636
  diphtheria, 666 (table)
  arising from female factory employment, 639, 650
  phthisis, 663 (table)
  scarlet fever, 665 (table)
  typhoid, 635
  typhus, 633
Morton, C., 174
Morwood, V., 155[67]
Moscow, migration to, 14
  population, 9
Moseley, A., 330[39]
Moseley, T., 827, 828
Moser, C. A., 98–9, 100, 104[69]
Mosley, Sir Oswald, 188[23]
Mother-Sister, 600
Motherwell, 394
Mottram, R. H., 689[62]
Mouat, F. J., 103[36]
Mount Street (London), 312, 324
Mountford, E. W., *176*; 323

Mozley, J. K., 850[9]
Mudd, J., *84*
Mudie-Smith, R., 858, 866, 869[19], 871[59]
Muir, E., 409[31]
Muir, P. H., 209[15]
Mulock, Dinah, 705[13]
Mulready, A. E., 455, 458
Mulready, W., 454
Mulvany, J. T., 302
Mumford, L., 31, 54[51], 382[8], 476, 690[66]
Munich, overcrowded housing, 25
Munn, P. S., 450
Murchison, C., 637
Murdock, K. B., 103[49]
Murphey, R., 55[61]
Murphy, Dr S., 612, 617
Murray, J. F., 54[49], 155[44]
Music halls, *36*; 695
  evolution, 174
Muthesius, S., 446[13]
Muybridge, E., 234, 237
My Secret Life, 173
Mystery of Edwin Drood, The, 472, 538, 543, 544, 586

N

Nadelhaft, Janice, 555[18]
Nagle's, employment by, 150
Naismith, J., 399
Napier, Sir C., *37*; 176
Naples, population, 9
Nash, J., *306*; 313, 314, 347, 357[49], 365, 395, 433, 438
Nasmyth, T. G., 408[12]
National Association for the Promotion of Social Science, 90, 91, 399, 401, 604, 607, 618, 619[4], 831
National Film Archive, 238
National Liberal Club, 319
National Political Union, 176
National Review, 196, 206
National Union of the Working Classes, 175, 176
Navy, methods of recruitment, 79
Neff, Wanda F., 664[57]
Negretti & Zambra, 233
Neison, F. G. P., 627, 660[7, 8]
Nelson, H. S., 735[67]
Netherlands, Dutch, infant mortality, 25, 35 (table)
  mortality, 20, 35 (table)
  population, 4
  vital statistics, 35 (table)
Nettlefold, J. S., 383[24]
Neuburg, V. E., **191–209**, 209[13, 21]

Index

Neumann, E., 688[53]
Nevinson, C. R. W., 472[39]
Nevinson, H., 592
New, H., 834[7]
New Delhi, Viceroy's House, 445
New England, reasons for growth, 14
New Lanark, model factory community, 881
New Orleans, mortality, 27
New Scotland Yard, 319
New South Wales, sex ratio of population, 16
New York, Booth's attitude to, 94
   city-size distribution, 39
   crude mortality, 21 (table)
   infant mortality, 25, 26
   'metropolitan district', 52[30]
   mortality, 20, 24, 26, 27
   population, 9
   railways, 284
   Ruskin's attitude to, 503
   size of families, 19
   slum conditions, 25–6
New Zealand, different pattern of urban transformation, 42
   rate of urban concentration, 41 (table)
   urbanization and per capita gross domestic product, 38
   vital statistics, 35 (table)
New Zealand Chambers, *266*; 317, 321, 324
Newark, employment of migrants, 144
Newbigging, T., 761[55]
Newcastle-upon-Tyne, *330, 356*; 401
   appearance, 395
   architecture, 432
   artistic activity, 464–5
   cholera, 636
   church attendance, 855
   compared to Sheffield, 337
   crude mortality, 21 (table)
   employment provided by railways, 293
   High Level Bridge, 456
   'hopping', 131
   local elections, 765
   pictorial history, 564
   pubs, 176
   railway, 278, 280
   railway station, 301
   railway-station shed, 439
   slums, 22, 403
   tramps, 129
Newgate, Oxford Arms, *72*
Newgate Gaol, *296, 298*; 437
Newhall, B., 244[3,7], 245[32]

Newman, F., 885, 889[33,34]
Newman, Sir G., 612, 619[11], 622[56], 624[77,88]
Newman, J. H., 821, 845
Newmarch, W., 91
Newport, 294, 295
News from Nowhere, 510, 511, 513
Newsholme, A., 22, 52[33], 102[18], 641, 648, 664[70], 665[85], 666[87]
Newspapers, *see* Press
Newton, E., 432
Newton, Mary P., 119[5]
Newton, R., 691[73], 785[2], 787[62]
Nicholas, H. G., 812[95]
Nicholas Nickleby, 540, 547, 549, 554
Nicholson, R., 172
Nicol, E., 459
Nicolle, C., 633
Niepce, Isidore, 229
Niepce, J. N., 228–9, 230
Night Side of London, 180, 727
Nightingale, Florence, 635, 636
Nine Elms, 865
Nisbet, R. A., 120[25]
Nixon, President R., 57[70]
No. 5 John Street, 600
No Thoroughfare, 546
Noakes, D. W., 240, 241
Noble, J., 355[26]
Nodal, J. H., 750
Noel Park, pubs excluded from, 183
Noel, R., 478–9
Nonconformist Church, *see* Church, Nonconformist
Norfolk, Duke of, 283
   as enlightened landlord, 344
   estate, 319
   lack of control over building development, 338
   as owner of Sheffield Coal Company, 287
   as owner of Sheffield estate, 335, 344
   terraces commissioned by, 341
Norfolk, migrants from, 143
Normanby, Lord, 270
North and South, 261
Northampton, 303
   migration through, 117
Northampton, Marquis of, temperance policy, 181
Northern Star, 88
Northumberland Avenue (London), *267*; 318, 319
   hotels, 325
Norway, annual rate of urban concentration, 5 (table)

Norwich, children's hospital, 653
   English-born population, 118
   railway station, 280
Norwood, 538
Nossiter, T. J., 785[2]
Notting Dale, gypsies, *4*; 130
   migrants, 373
   migrants' return, 140
   tramps, 126
Notting Hill, 313
Nottingham, children's hospital, 653
   disease, 638
   English-born population, 118
   infant mortality, 650
   municipal elections, 765, 779
   overcrowding, 22, 289, 403
   railway development, 286, 289–91
   railway station, 301
   railways in area, 290 (map)
   tramps, 125
Nottinghamshire, typhoid fever, 640
Novelists, attitude to city, **517–536**
   attitude to city, in eighteenth century, 520
   challenge of city not met by, 529
   city men in country viewed by, 524–7
   city viewed as mysterious cosmos, 531
   compromise with urban realities, 533, 535
   debt to Protestant allegorists, 518
   emphasis on orphaned children in city, 522
   inconstancy of city life seen as threat, 521
   Romantic view of city, 522–3
   urban information acquired from, 902
Novello, V., 841
Nuisance Removal Acts (1855–8), 22, 407, 610, 611, 613, 659
Nuisance Removal (Scotland) Act (1856), 392
Nunns, T., 830, 836[40]
Nyren, R., 173

O

Oakley, F., 842
Oakum-picking, *170*
Obermeier, O., 632
O'Connell, D., 793, 821
O'Connell, J., *409*

455

Index

O'Connor, J., 468
O'Connor, T. P., 118
O'Connor, W., 844
Occupations, in Belfast, 794
   change from indoor to outdoor, 132
   depicted in theatre, 221
   of inhabitants of London and Paris, 420 (map)
   injury to health in indoor, 645
   largest occupational groups, 710
   of migrants, 124
   number of new, 109
   restrictions in leasehold agreements, 342
   of street-sellers, 193
   of slum-dwellers, 372
   winter-time, 139–40, 143–4, 145–51
   working conditions in different, 644
Oddy, D. J., 103[34]
Odessa, migration to, 14
Ogle, W., 51[24], 91, 121[49]
Ohlin, G., 57[69]
Okun, B., 51[29]
Old Curiosity Shop, The, 547
Old Trafford, Botanical Gardens, 227
Oldham, 248, 740, 772, 855
Oliphant, Mrs, 697–8
Oliver Twist, 540, 547, 548, 698, 708
Olsen, D. J., 189[39], 328[4], **333–57**, 355[26, 28], 356[47], 357[69, 81]
Omnibus, introduction of, 6
O'Neill, G. B., 455
On Architecture and Buildings, 437
On the Mode of Communication of Cholera, 636
Ong, W., 877, 888[12]
Oppé, A. P., 469[6]
Oral history, possibilities of, 60–80
Orchard, B. G., 190[46]
Orchardson, W. Q., 455
Ord, W. M., 645
Ordish, R. M., 440
Ordnance, Survey (1850), 248 (1891), 252
Orel, H., 536[6]
Organ-grinders, 136
Orthodox London, 865
Orwell, G., 273[7], 581[52]
Osaka, infant mortality, 36
   population, 9
Osborne, F. J., 54[51]
Osler, W., 635
Oslo, as primate city, 39

Our Mutual Friend, 266, 272, 328[2], 541, 542, 543, 554, 583, 587
Our New Masters, 860
Overcrowding, *see* Housing, overcrowding
Overend, Gurney crisis (1866), 288
Ovsjannikov, Y., 208[1]
Owen, David, 386[68]
Owen, D. J., 810[40]
Owen, R., 176, 881
Owenites, 87
Oxenholme, 292
Oxford, pubs, 176–8
Oxford, University of, 483
Oxford House, 592
Oxford Street (London), 312, 324
   Pantheon, 443
Oxfordshire, wages in, 133

P

Paddington, hotel, 324
   library, 323
   off-licensed premises, 168
   Frith's painting of station, *342*; 454
   sanitary inspectors, 607
   siting of railway, 314
   station, *50*, *244*; 282, 301, 439
   suburb, 379
Paddock Wood, 139
Page, J. K., 57[72]
Pahl, R. E., 888[2]
Pakistan, effects of births on natural resource consumption, 43
   migration to cities, 41
   vital statistics, 35 (table)
Paine, J., 341
Paisley, 391
Palace Journal, 597
Palace Theatre (London), *278*
Palgrave, F. T., 454, 456
Pall Mall, 321
   scarcity of pubs, 166
Pall Mall Gazette, 607, 697
Palm, T. A., 656, 66[96]
Palmerston, Lord, 432
Paradise Lost, 518–19, 520
Paraguay, founding of Cosme Colony, 879
Paris, 94, 181, 414–27
   age and sex structure, 51[29]
   attitude of Victorian poets, 478
   birthplace of inhabitants, 416 (table)
   central government control, 426–7

   centrifugal tendency, 418, 421
   child labour, 423
   crude mortality, 21 (table)
   density of population, 26, 417, 418 (map)
   Gare de l'Est, 301
   heavy industry, 421
   industrial employment, 419
   lack of middle-class interest in business, 426
   migration to, 415, 416
   migration to nearby areas, 415
   mortality and fertility, 20, 416 (table), 417
   occupation of inhabitants, 420 (table), 421
   population, 9, 171–2, 414, 422
   as primate city, 39
   regulations governing growth, 417
   Rue de Rivoli, 439
   social segregation, 418–19
   social structure, 423–6, 424 (map)
   structure of working class, 424–5
   tradition of skilled hand-labour, 420–1
   underground railway, 285
Park, R. E., 102[22]
Parker, J., M.P., 338
Parkes, J., 775
Parkes, L., 624[75]
Parkin, G. W., 306[18]
Parkinson, Rev. R., 757[11]
Parkman, F., 551
Parliamentary Candidates' Society, 176
Parry Lewis, J., 385[61]
Parson, J., 306[32]
Parsons, T., 120[23]
Parsons, W., 297
Partlow, R. B., 275[23]
Party Processions Act (1832), 790; (1850), 796; (1872), 798
Pask, A. T., 137, 157[123]
Passavant, J. D., 449
Past and Present, 89, 456, 878
Pater, W., 534
Patterson, A. T., 121[42], 309[89], 691[73], 785[1], 787[59]
Paul, B. D., 691[71]
Paul-Dubois, L., 814[144]
Pavilion Theatre (London), *55*; 212, 224
Paxton, Sir J., 440
Payne, E. A., 833[1]
Payne, J. H., 213
Peabody Trust, 26
Peacock, A. J., 189[35]
Peacock, R., 294

Pearl, C., 705[21]
Pearson, C., *236–7*; 281, 282
Pearson, J. L., 443
Pearson, K., 94
Peebles, J. H., 661[21]
Peel, A., 833[1]
Peel, Mrs C. S., 121[47], 690[68], 705[8]
Peel, F., 189[34]
Peking, population, 9
Pelling, H., 785[2]
Pencil of Nature, The, 230, 231
Penfold, J. W., 329[8], 330[34], 340
Penge, tramps, 126
Pennell, E. R., 474[65]
Pennell, J., 331[55], 474[65]
Pennethorne, J., 314, 365
Penny Illustrated Paper, 560, 563, 568
Penny Magazine, 561, 578[20]
Penrhyn, 438
Pentonville, 538
People's Palace (London), 589, 592, 593, 594
People's Paper, 179
Percier & Fontaine, 439
Percy Adams, H., 331[74]
Perkin, H. J., 101[6], 735[79]
Perks, S., 332[83]
Perloff, H. S., 56[65]
Perring, R., 765, 769, 777
Perry, R., 631, 661[24]
Persius & Hübsch, 442
Perth, 31, 444
  infant mortality, 649
Peru, housing deficits, 37
  rate of urban concentration, 41 (table)
  urbanization and per capita income, 37
  vital statistics, 35 (table)
Peterborough, 292
Pett Ridge, W., 374, 385[51], 599, 600
Petre, Mrs E., 848
Pettenkoffer, M. von, 640
Pevsner, Sir N., 433, 441, 445, 447[16], 451
Philadelphia, as agglomeration of suburbs, 29
  population, 9
  slum conditions, 25
Philanthropic Society, 698
Philip, R. W., 648
Phillips, G., 467
Phillips, W., *51*; 215–16
Philosophical Enquiry into the Origin of our Ideas of the Sublime and Beautiful, A, 434
Phipps, C. J., 325

Photography, amateurs' enthusiasm for, 241
  as reflection of contemporary taste, 241
  contemporary appreciation, 243
  developments, *see below*
  early achievements, 229–30
  for educational purposes, 234, 239, 241
  effect of roll-film camera, 237
  formality of urban, *94*
  inclusion of figures, 242
  newspaper use, 237–8, 562–4
  photographic firms, 236, 239
  portraiture, 231
  as record of architecture, 241–242
  as record of city life, 237, 238, 240
  in relation to graphic art and painting, 230–1, 235
  slum life as subject, *91–2, 106, 134*
  technical standards, 238
  temperance lecturers' use of, 239, 240
  of town and country, 236
  use in historical research, 227–8, 242–3, 906–7
  wars as subject of, 232
  see also Press
Photography, developments:
  calotype, 231
  camera, 228
  carte de visite, *88*; 234
  daguerreotype, 229–32, 578[21]
  dry-plate process, 235
  stereoscopic view card, *83*; 126, 233, 239
  techniques, 229–34, 243
  wet-plate process, 232–3
Phthisis, deaths from, 647, 663 (tables)
  incidence, 645
Piccadilly (London), architecture, 324
  prostitutes, 696
  scarcity of pubs, 166
  theatre, 325
  transport inns, 162
Pickering, W. S. F., 835[27], 868[4]
Pickwick Papers, 162, 540, 547
Picton, J. A., 309[112]
Pictorial Relics of Ancient Liverpool, 464
Pictorial Times, 395, 401–5; 560, 564
  social realism, 573–4
Picture of Dorian Gray, The, 529–30
Pierson, S., 873–89

Pilgrim's Progress, The, 518–19
Pimlico, 313
Pimlott, J. A. R., 121[41], 186[4], 594
Pinchbeck, Ivy, 121[50]
Piranesi, G. B., 437, 439
Pissarro, C., 465
Pitts, J., 200
Pittsburgh, population, 9
Pixham, 445
Plint, T., 783
Plomer, W., 871[53]
Plumstead, A. W., 493[7]
Plumstead, gypsies, 130
Plymouth, King William Victualling Yard, 441
  migration through, 117
Poets and writers, 475–93, 876, 902
  anti-urbanism among, 477–8
  attitude to city among American, 480–1
  attitude to city among French, 481–3
  attitude to city of eighteenth century, 476
  causes of anti-urbanism among, 478–9
  changing attitude in Aesthetic Movement among, 489–92
  city as stimulus to, 482
  conservatism, 483
  exceptions to general attitude, 488–9
  unsuccessful treatment of city as subject, 484–5, 486
  see also Carlyle; Dickens; Morris; Novelists; Ruskin
Polanyi, K., 874
Police, attitude to prostitutes, 703
  attitude to tramps, 129
  living conditions, 368
Political Economy Club, 87
Pollard, A., 102[13]
Pollard, J., *26*
Pollard, S., 354[19, 21], 576[3], 687[40], 691[71, 73], 868[8], 881, 889[21]
Poll-books, use of, in assessing political motivation, 98
Pollins, H., 428[10]
Pollution,
  of air, in Sheffield, 342
  of air, in United States, 43
  laws against, 47–8
  reduction, in air, 30
  see also Thames
Poole, J., 547
Poor Law, Commission (1834), 603
  composition of workhouse board, 768
  condemnation of, 723, 725

457

Poor Law—*continued*
  distinction between paupers and independent labourers, 728
  effects, 726
  Report (1834), 719, 730
  workhouse infirmaries, 647, 653, 656
  workhouses, *10, 401–3*; 136, 138, 267
  see also Poverty
Pope, A., 431, 476
Poplar, 139
Popplewell, F., 156[83]
Population, age structure of urban, 16–17, 99
  agglomeration, 7, 8
  concentration, 7, 8, 13
  density, 109
  effects of increase in urban, 28, 30, 107
  fecundity, 13, 17, 36
  growth, 7, 105
  growth of world and world urban, 33 (table)
  increase in 'surplus' women, 116
  and pub density, 171
  in relation to health, 402
  in relation to housing, 334
  sex structure of urban, 16–17
Porden, C. F., *315*
Pornography, 896–7
Port, M. H., 329[11], 834[14]
Port Glasgow, 394
Port Sunlight, 884
Port Talbot, 295
Porter, G. R., 84, 93
Porter, J., 812[96]
Portfolio, 449, 466
Portland Place (London), 696
Portlock, J. E., 101[10]
Portman, D., 690[67]
Portman Square (London), Calmel Buildings near, 845
Portsmouth, 125
  migrants from, 139
Portugal, annual rate of urban concentration, 5 (table)
Potter, Beatrice, *see* Webb
Poultry (London), Pimm's Restaurant, 325
Pound, Ezra, 493
Pouthas, C. H., 427[3]
Poverty, among migrants, 143
  attitude of wealthy to, 730
  causes, 59
  connection with ill-health, 626, 643
  culture of, 270, **707–32**
  distinction from pauperdom, 719, 720, 723

  European compared with British, 730
  homeless, *6, 9, 14, 350, 351, 384–5, 387, 392, 400, 404*
  national policy needed, 96
  in relation to pub density, 168
  statistics, 596
  study of, 595–6
  unofficial enquiries into, 94
  in upper classes, 70
  of working-class families, 61–2
  see also Slums; Class, working; Crime; Mayhew; Prostitution; Poor Law
Poverty, 96
Powell, T., 861
Power, W. H., 640
Poynter, E. J., 453
Prague, density of population, 26
Pratt, E. A., 186[6]
Pre-Raphaelite Brotherhood, 453, 486
Press, **559–74**
  circulation figures current in England and Wales (1781–1900), 575–6, 576–7[5]
  greater freedom of humour magazines, 565–6, 567
  introduction of news illustration, 562
  lack of authenticity in illustrations, 563–4
  power printing, 578[20]
  role of journalism in urban culture, 559, 900
  social and urban problems not faced by, 561
  as source of contemporary opinion, 559
  street life depicted, 570
Prest, A. R., 104[66]
Prest, J., 691[73]
Preston, 296
  health conditions, 649
Price, H., 777, 782
Price, J. Seymour, 386[66]
Price, Sir U., 436
Price-Williams, H., 427[8], 428[9]
Priestley, J. B., 6, 7, 21, 49[6]
Priestley, Joshua, 551, 556[18]
Primrose Hill, 301, 345, 347, 348, 349
Prince, H. C., 335, 354[10]
Prince, J. C., 743–5, 752, 756, 757[11], 759[27]
  attitude to literature, 743–5
  rejection of urban environment, 744
Prince Miller, D., 156[78]
Princess's Theatre (London), 224
Principles and Practice of Medicine, 635

Pringle, K. C., 836[37]
Prinsep, V., 320
Prior, E. S., 445
Pritchard, T. F., 450
Process of Calotype. Photogenic Drawing, The, 231
Procter, R. W., 743, 758[14,15], 759[26]
Progress of the Nation, 84
Prohibitionism, *see* Drink
Proposals (1752), 406
Prosser, T., 439
Prostitutes, **693–704**
  attractions of London to, 701
  domestic servants becoming, 657
  in literature, 698
  numbers, 693
  in pubs, 173
  as subjects of paintings, 457, 458
  temporary migration to suburbs, 700
  in theatres, 694
  travelling, 142
Prostitution, *374*; 116, **693–704**
  attitude of Parliament, 697
  decline, 704
  encouraged by contemporary morality, 701–2
  and poverty, 700, 731
  press interest in reform, 698–700
  as an urban phenomenon, 657
Prowse, R. O., 599
Prussia, annual rate of urban concentration, 5 (table)
  internal migration, 15
  mortality, 20
  population, 4, 10
  rural fertility, 18
Psychological interpretation of public health movement, **669–92**
Public Baths Act (1847), 804
Public Health, 610
Public Health Act (1848), 393, 394, 407, 604
Public Health Act (1875), 605, 606
Public Health (London) Act (1891), 611
Public Health (Scotland) Act (1867, 1879), 393, 494
Public House Closing Act (1864), 173
Pubs, **161–90**
  appearance, *28, 30, 34*; 170, 171, 325
  in Belfast, 807
  as centre for public meetings, 175

Index

children in, *156*
and commuter travel, 163, 165
debates held in, 179–80
density, 162, 163–7, 172
dependence of migrants on, 171, 172
discussion groups, *38*; 179–80
entertainment, *31, 35, 37*; 174
exclusion from building estates, 183
increasing size, 171
Irish clubs, 847
lack of class distinction, 172
licences, 167
in London underworld, 192
number in relation to churches, 184
political tendencies, 185
and railway travel, 162, 163
as recreational centres, 168, 169, 171
revolutionary activities accredited to, 178
in rural society, 161
similarity of role to temperance movement, 184–5
siting, 164n., 165–9
and sporting activities, 173–4
as transport centres, *26*; 162
in urban society, 904
variety of, 176
working-class, 166, 167
   see also Drink; Licensed premises; Temperance movement; and individual cities
Pückler-Muskau, Prince, 694
Pudney, J., 121[45], 690[66]
Pugh, E., 600
Pugin, A. C., 267, 442, 443, 444, 463
Pulpit Photographs, 832
Punch (London Charivari), *6, 374, 378–80, 382, 390, 406–10, 417*; 22, 85, 222, 231, 297, 555[8], 556[16], 576[5], 709
blandness, 452
city life as seen by, 573, 574
greater freedom, 567
loss of radicalism, 572
metropolitan readership, 560
satirical treatment of statistics, 89
street life depicted, 571–2
treatment of the poor, 568–70
Purcell, E. S., 852[73]
Purleigh, Tolstoyan colony, 880
Putney, 538
Pye, C., 819
Pyne, W. H., 469[4]

Q

Quarterly Review, 539
Queen Anne's Mansions (London), 317
Queen Victoria Street (London), *268*; 318, 319
  warehouse, 441
Queen's Park, pubs excluded from, 183
Quennell, P., 732
Quinlan, M. J., 705[21]

R

Radcliffe, N., 640
Radford, 289
Radford, E., 309[109]
Ragged London, 727
Railway and Canal Traffic Bill (1888), 289
Railway, Metropolitan, *238–41*; 146, 282, 283
  Central Line, 284
  Circle Line, 284
  District Line, 284, 302, 304
  effect on streets, 302
  provision of employment, 153[4]
Railway companies, ownership of stations, 280, 281
  Acton & Hammersmith, 377
  Barry, 295
  Bolton & Manchester, 250
  Brighton, 285
  Caledonian, 294
  Cheshire Lines, 254
  Dover, 284
  Eastern Counties, 301
  Great Central, 297, 299, 301
  Great Eastern, 285, 298
  Great Northern, *250*; 287, 288, 289, 291, 296, 299, 303
  Great Western, *252*; 283, 299, 439
  Lancashire & Yorkshire, 294
  Leicester & Swannington, *257*; 296
  Liverpool & Manchester, 250, 278, 279, 301
  London & Brighton, 295–6, 301
  London Chatham & Dover, *248*; 283, 298, 302, 324, 374, 461
  London & Greenwich, *245*
  London & Manchester, 746
  London & North Western, 249, 255, 296
  London & South Western, 304
  Manchester South Junction & Altrincham, 194
  Manchester Sheffield & Lincolnshire, 249, 251, 289, 290, 291, 296, 304
  Midland, 283, 287, 288, 289, 301
  North British, 294
  North London, *251*
  North Midland, 286, 287
  North Staffordshire, 301
  Oxford–Worcester line, 293
  Oystermouth, 278
  Sheffield Ashton-under-Lyne & Manchester, 294
  Sheffield Chesterfield & Staffordshire, 288
  South Eastern 374
  South London, 284, 285
  Taff Vale, 295
  Wiltshire Somerset & Weymouth, 292
Railways, **277–310**
  arches used as shelters, 129
  architectural monuments, 301
  companies, *see above*
  damage done to cities, 302
  development of manufacturing firms, 293–4
  effect on other transport, 162, 282, 284
  effect on town development, 292, 293, 366, 538
  effect on trade, 279, 291, 292, 294
  houses pulled down for, 297
  as imagery in street ballads, 195–6
  lack of government control over, 286
  land values inflated by, 377
  in Leicester, 291 (map)
  in London, *see below*
  Metropolitan, *see above*
  mobility increased by, 107, 117
  in Nottingham, 290 (map)
  prices reduced by, 296
  private ownership, 280
  proposed terminus, *236–7*
  and pubs, 162
  results of expansion, 304–5
  siting of factories next to, 294
  slum clearance resulting from building, 268
  stations, and migrants, 125
    arguments for 'Hauptbahnhof', 281
    in Europe, 125
    see also individual stations
  statistics of urban, 98
  and suburban commuters, 163, 296, 299
  as subject of paintings, *342, 355*; 470[15]

459

Index

Railways—*continued*
  underground, *240–1*; 283–5
  viaducts, *245*, *248*, *257*; 438–9
Railways in London, cost, 283, 284
  development of multiple termini, 281–2
  electrification of underground, 284, 285
  map of terminal stations and Inner Circle, 282
  stations, *133*, *238–9*, *244*, *248*, *342*; 280, 281
  underground, *240*; 283–5
  workmen's services, 283, 284, 419
Raleigh, J. H., 518, 536[2], 888[10]
Rambler, 838
Rammell, T. W., 154[33]
Ranger, W. C. E., 154[35]
Ranyard, Ellen, 158[156]
Raper, J. H., 182
Rasmussen, S. E., 427[7]
Rasselas, 520
Ratcliff, Sir J., 821
Ratcliffe Highway (London), 123, 142, 173, 588–9
  drunkenness, 181
  gin-palaces, 170
  murders, 587
  squalor, 587
  visits of aristocratic youth, 587–8
Rauchberg, H., 50[19], 51[26]
Ravenstein, E. G., 50[17], 100, 116, 117, 122[66,70,72]
Rawlinson, J., 691[74]
Rawson, Sir Rawson W., 101[1], 103[41], 428[17]
Rayner, S., 579[28]
Read, D., 427[4], 759[39]
Read, N. F., 557[20]
Reading, 445
  fair, 132
  migrants from, 139
  photographic printing works, 229
  railway, 277, 278n.
Réclus, E., 888[4]
Record, R. G., 642, 662[51]
Record of Badalia Herodsfoot, The, 597
Redford, A., 122[70], 415, 427[5], 684[1], 691[73]
Redgrave, R., 458
Redistribution Act (1885), 802
Reeder, D. A., 54[49], 328[6], **359–86**, 384[37,43,48], 385[55,58]
Reform Act (1867), 801
Reform Club, 313
Refuges, Night, 128, 150

closing down, 131, 132
for the destitute, 143
financing, 142
opening, 142
Regent Street (London), 100, 313, 365, 696
  Nash's Quadrant destroyed, 703
Regent's Park, 333, 345, 347, 433, 443, 566, 696
Registrar-General, the, 19–20, 24, 367, 633, 641, 654
Reid, F., 578[26]
Reid, J. C., 579[31]
Reid, J. S., 812[83]
Reid, W., 189[44]
Reilly, Sir C., 432
Rein, S. R., 49[9]
Reiss, A. J., 187[15], 190[45], 888[8]
Rejlander, O., 235
Religion, middle-class attitude to, 63, 64, 66, 67, 68, 858–9
Religion, working-class attitude to, 61, 69, 71, 72, 73, 77, 80, 821–2, 829, 830, 831, **855–68**
  attendance at church, 860
  belief in value of ritual, 864–5
  contemporary explanations, 857–8
  effect of social divisions, 860, 861
  effect of urban growth, 859, 862
  failure of Anglican church to change, 856–7
  indifference to institutionalized religion, 866, 867
  see also Church
Rendle, T. McD., 188[19]
Rennie, Sir J., 441
Report on the Housing of the Poorer Parts of the Population in Towns, 398
Report on the Sanitary Conditions of the Labouring Population, 395, 604, 627, 718–19
Repton, H., 433
Restaurants, architecture, 325
Reynolds, G. M. W., 87
Reynolds, Graham, 455, 469[2]
Reynolds, R., 690[66]
Reynolds' Weekly, 577[5]
Rhys, E., 556[14]
Ricardo, D., 85
Ricardo, H., 445
Rice, J., 160[236]
Richard, H., 172
Richardson, A. E., 329[15]
Richardson, Hamilton, 776
Richardson, H. H., 442
Richardson, J., 776, 782

Richardson, Fr R., 848, 849, 853[94]
Richardson, S., 520
Richardson, T. M., 464
Richmond, Surrey, 538
Rickets, 655–6
Ridings, E., 75[11,12]
Riesman, D., 684[5]
Rilke, R. M., 6
Rimbaud, A., 482
Rimmer, W. G., 785[5]
Rio de Janeiro, population, 9
Ripley, W. Z., 28, 53[47]
Ritchie, J. E., 735[68]
Robbins, M., 305[16], 306[20,26,31], 309[104], 329[9], 385[57]
Roberts, D., 119[11], 464, 692[72]
Roberts, M., 599
Roberts, R. J., 384[46]
Roberts, W., 579[28]
Robertson, D., 310[122]
Robertson, G., 450
Robertson, J., 438
Robertson, W. B., 446[14]
Robinson, H. A., 814[158]
Robinson, H. Peach, 235
Robson, E. R., 322, 331[73]
Roby, J., 749
Rochdale, infant mortality rates, 99
Rochdale Canal, 250, 254, 255, 256
Rochester, Dickens in, 538, 544, 554
Rodgers, B., 103[35]
Rodney, H., 599
Roebuck, Janet, 621[25]
Rogers, F. W., 786[27]
Rogerson, J. Bolton, 743, 745, 746–7, 756, 757[11], 759[30]
Roll, Sir E., 687[40]
Rolt, L. T. C., 186[5]
Roman Catholic Church, *see* Church, Roman Catholic
Rome, 280
  crude mortality, 21 (table)
Roker, 445
Romola, 534
Rook, C., 600
Rooker, M. A., 450
Rookeries of London, The, 727
Rose, M. E., 786[26]
Rosebery, Earl of, 396
Rosen, G., **625–67**, 661[10,14], 664[58,71], 665[84]
Rosenberg, C. E., 689[62]
Rosenberg, J. D., 732
Rossetti, D. G., 486–7, 564, 578[26]
Rossetti, W. M., 474[65]
Rossiter, C., 474[59]
Rotherhithe, 123, 129

460

Roumieu, R. L., *317*; 316, 442, 443
Rousseau, J.-J., 477, 486
Rouvray, F. G., 664[67]
Rowe, R., 153[3]
Rowlandson, T., 451, 452, 463, 565
Rowney, D. K., 103[56], 104
Rowntree, B. Seebohm, 93, 96, 97, 103[44], 134, 154[16], 157[105], 168, 187[11], 189[32], 691[73]
  drink map of Oxford, 177
Rowntree, J., 171, 187[13,17], 189[43]
Royal Academy, 231
Royal Albert Hall, *279*; 240, 327, 328
Royal College of Physicians, 614
Royal Commission (R.C.) on Children's Employment (1861), 646
R.C. on Elementary Education (1847), 126
R.C. on Elementary Education Acts (1886–8), 861
R.C. on Employment of Children in Factories (1833), 676
R.C. on Housing in Scotland (1917), 399, 405
R.C. on the Housing of the Working Classes (1884–5), 611, 613
R.C. on Labour (1892–4), 132, 134
R.C. on Liquor Licensing (1897), 182
R.C. on London Traffic (1905–6), 285
R.C. on Magisterial and Police Jurisdiction (1865), 798
R.C. on Municipal Corporations (1835–7), 765
R.C. on Poor Laws (1834), 603, 719–20
R.C. on Poor Law (1909), 397, 399
R.C. reports, 718
Royal Exchange Buildings (London), 442
Royal Sanitary Institute, 618
Royal Statistical Society, 91, 92
  see also London, Statistical Society
Royal Victoria Dispensary for Consumption, 648
Royston Pike, E., 199[9]
Rubin, M., 51[26]
Rudd, R., 55[55]
Rudé, G., 189[34]
Ruhrgebiet, population of, 9
Rural life, as theme of nineteenth-century drama, 216–17
  considered superior to urban life, 108–11
  difference between urban and rural population, 69–70, 100
  nineteenth-century desire for, 874
  restorative powers, 886
Ruskin, John, *219*; 29, 89, 103[33], 171, 258, 302, 315, 370, 384[33], 434, 442, 455, 457, 489, 495–8, 502–9, 510, 511, 512, 513, 514, 555[11], 590, 593, 873, 877, 886
  on architecture, 487–8, 502–3
  attitude to Turner, 504–7
  concept of city life, 508–9
  Guild of St George formed by, 878–9
  opinion of Carlyle, 502
  rejection of city by, 496, 502, 503, 507, 878
Russell, G. W. E., 871[56]
Russell, Dr J. B., 396–7, 402, 403, 406, 410[41,48]
Russell Square (London), 313
  hotels, 325
Russia, mortality, 27
  sex ratio of population, 16
  see also U.S.S.R.
Russian Empire (Europe), annual rate of urban concentration, 5 (table)
  growth of cities, 8
Rutherford, M., 534
Rutland, level of nuptiality, 17
Ryle, Annie, *46*

S

Sabine, E. L., 690[68]
Sachs, H., 686[32]
Safdie, M., 54[52]
Saffron Hill (London), 539, 845
St Andrews, 232
St Bartholomew's Hospital, 441
St George's Circus (London), 166
St George's-in-the-East, 586
  migrants, 373
  unemployment, 368
St Giles's, 418, 586
  Dickens's view, 538
  migrants' return, 140
  tramps, 126
  workhouse, 138
  workhouse infirmary, 647
St James's Park, frost fair, *11*
  tramps, *14*
St John-Stevas, N., 555[4]
St John's Wood, 345, 347, 352, 433
  reputation, 351
  station, 282

St Louis, population, 9
St Martin's-in-the-Fields, 442, 607
St Marylebone Infirmary
St Olave's, 607
St Pancras, Agar Town, 372
  hotel, 309[113], 325, 440, 443
  pubs, 163
  sanitary inspector, 607
  slums and suburbs, 372
  station, *244, 246, 333*; 283, 301, 302
  train shed, 440
St Paul's Cathedral, *136*; 538, 696
St Petersburg, crude mortality, 21 (table)
  population, 9
St Rollox, 294
St Stephen's Club, 319
St Thomas's Hospital, 637
  Lectureship in Public Health, 605
Sala, G. A., *88*; 472[35], 545, 555[11]
Salford, 272, 740
  deaths from phthisis, 647
  disease, 638
  Flat-Iron Market, *141*
Salisbury Cathedral, 444
Salmon, Rev. W., 851[48]
Salomons, E., *228*
Salt, Sir T., 882, 883
Saltaire, development, 882–3
  Lockwood & Mawson's buildings, *320*; 441
  prohibition of public houses, 882
  pubs banned, 66
Salvation Army, 80, 142, 704, 866, 867
  founding, 592
Samuel, R., **123–60**, 160[240]
Samuelson, P. A., 57[74]
San Francisco Bay, migration to, 14
  population, 9
Sandby, T. and P., 463
Sandford, J., 829
Sanger, 'Lord' G., 132, 141, 151, 156[80,89], 158[165], 160[237], 192, 208[4], 655
Sanitary Act (1866), 392, 611, 613
Sanitary Condition of the Labouring Population of Great Britain, The, 718, 719
Sanitary Evolution of London, The, 361
Sanitary Ramblings, 727
Sanitation, *406–7, 409–26*
  advances, 318, 683, 905–6

461

Index

Sanitation—*continued*
  decline of disease in relation to improvement, 633
  doctors' insistence on improved, 638, 639
  effect on health of poor, 339
  effects on growth of London, 14
  lack of satisfactory techniques for sewage treatment, 678, 681, 683
  lack of provision, 106–7, 271
  need to make water closets compulsory, 683
  in new housing, 334
  open sewers, 372
  pride in improvements, 85
  provision of public baths, 322
  public ignorance of conditions, 718
  sewage, *365, 419–20, 424*
  statistical knowledge, 95
  task of sanitary reformers, 669, 673, 674
  technological advances, 679, 680
  in United States, 25
Santiago, effect of population concentration, 38
São Paulo, population of, 9
Sartor Resartus, 499
Saturday Review, 539, 546, 576[5], 696, 699, 700, 751
Saul, S. B., 385[61]
Saville, J., 118[2]
Saxony, annual rate of urban concentration, 5 (table)
  infant mortality, 24
  internal migration, 15
  legitimate birth-rates, 19
  migration of foreigners to, 15
Say, J. B., 724
Scandinavia, effect of migration on growth of cities, 13
Scarborough, *164, 318*; 444
Scarlet fever, 628, 651
  deaths, 665 (table)
  epidemics, 652
Scarr, Archie, 774, 778, 782
Scharf, A., 244[2, 8, 9, 11]
Scharf, G., 463
Schinkel, K. F. von, 440
Schneewind, J., 902
Schnore, L. F., 52[29], 54[49], 55[60], 56[64], 98, 104[61, 68], 120[23]
Schoenwald, R. L., **669–92**, 684[3], 688[53], 690[69]
Schofield, S., 757[7]
School, attendance, 80, 421
  Board, 322, 862, 866
  middle-class attitudes, 63
  punishment, 65, 76
  violence, 62

School Board offices (London), *264*; 319
School Board visitors, 153
Schoyen, A. R., 189[34]
Schumpeter, J. A., 101[9]
Schwab, W. B., 190[45]
Scobie, J. R., 54[53]
Scotland, annual rate of urban concentration, 5 (table)
  church architecture, 443–4
  city slums, 400–1
  civic government, 395, 396
  common stairs of tenements, 406–7
  difficulty of applying statutes, 392–5
  'feuing' system, 404–5
  infant mortality, 649
  'making-down', 407
  migration of foreigners, 15
  mortality, 401
  overcrowding, 23, 403
  population, 8
  public health, 393–4
  sex ratio of population, 16
  typhus epidemic, 633
  urban tenements, 402–4
  variety of cities, 407–8
  Victorian city, **391–408**
Scott, C., 221
Scott, G., 434
Scott, General, *279*
Scott, Sir G. Gilbert, *190, 330*; 301, 316, 325, 327, 328, 330[27], 440, 443, 444
Scott, H. H., 665[81]
Scott, P., 208[9]
Scott, S., 462
Scott, Sir W., 440, 507, 509, 524, 534, 749
Scott, W., 98–9, 100, 104[69]
Scott, W. B., *330*; 456, 457
Scottish Scenery, 233
Scouloudi, Irene, 328[6]
Scripture Reader of St Mark's, The, 599
Sears, R. R., 684[11]
Seaside holidays, *162–6*
Seaton Delaval, 436
Secret Agent, The, 531, 535–6
Sedding, J. P., 326
Seething Lane (London), 316
Sefton Park (Liverpool), *171*; 433
Selby, C., 219
Select Committee (S.C.) on Orange Lodges (1834), 802
S.C. on Public Walks (1833), 338
S.C. on Railway Rates (1881), 290
S.C. on Town Holdings (1886–9), 339, 340, 344, 345

S.C., Joint, on Electric and Cable Railways (1892), 284–285
S.C. reports, 718
Selway, N. C., 186[5]
Senior, Nassau, 86, 721
Sennett, R., 385[50], 732[19]
Seoul, population, 9
Serbia, sex ratio of population in, 16
Series of Lithographical Drawings on the London and Birmingham Railway, A, 463
Sesame and Lilies, 886
Seven Dials, Dickens in, 538
  James Catnach's shop, *41*
  migrants, 373
  slums, 461, 586
  street life, *93*
Seven Lamps of Architecture, 503
Seventy Years a Showman, 192
Sewage Utilization Act (1865), 392
Sewers *see* Sanitation; Water Systems
Sexby, J. J., 156[69]
Seymour, R., *339*; 451, 565, 566
Seymour Price, J., 386[66]
Shadwell, 139, 539, 586
  pubs, 166
Shadwell, A., 23, 52[32, 35], 53[43], 187[16]
Shaftesbury, Lord, 297, 458, 726–7
  Children's Employment Commission (1861), 646
  Evangelicalism, 727
  Ragged Schools, 727
Shaftesbury Avenue (London), 319, 325, 367
Shakespeare, William, 541, 742
Shanghai, population, 9
Shannon, H. A., 50[12], 382[3], 416n., 427[3, 5, 6]
Shaw, A., 409[19]
Shaw, Capt. D., 588–9
Shaw, G. B., 380, 594
Shaw, J. H., 777, 782
Shaw, John, 347
Shaw, John, Jr, 348, 349, 350, 351, 356[60], 357[71]
Shaw, Nellie, 888[18]
Shaw, R. Norman, *266, 271, 277*; 317, 326, 327, 328, 330[29], 434, 445
  new architectural styles, 320–321, 323
Shaw, S., 126, 154[31]
Sheepshanks, J., 450
Sheffield, agricultural community near, 878–9

Alsop Field development, 341, 345
attraction of countryside, 341
attractions to house-builders, 345
church attendance, 855, 857
cost of railways, 289
cutlery industry, 342
employment, 134
failure of railway lines, 287
mortality, 20
music halls, 174
phthisis deaths, 647
population, 13
Primitive Methodism, 858
quantity and quality of housing, 335, 338–9, 342, 345
railway development, 286–9
railway hotel, 296
railway station, 280, 301
Reform election, 801
supervision of development, 338, 352, 353
terraced housing, 337, 342
Town Hall, *176*
working-class housing, 338, 342

Sheffield and Rotherham Independent, 337
Shelley, P. B., 523
Shephard, L., 209[16, 18]
Shepherd's Bush, 313
  station, 284
Sheppard, J., 179
Sherif, T., 662[45]
Sherwell, A., 171, 187[13, 17], 189[43], 600
Shields, 125
Shils, E., 276[31]
Shipley, 883
Shkarovsky-Raffé, A., 208[1]
Shops, *see* Street life
Shoreditch, Grecian Theatre, *60*
  overcrowding, 23, 404
Shrewsbury, compared to Sheffield, 337
Sibthorp, R. W., 839
Sickert, W. R., *361*; 468, 492
Sigerist, H. E., 661[22]
Signs of the Times, 498, 514
Silas Marner, 518, 524, 526, 527, 528–9
Silver, A., 732[19]
Simey, Margaret B., 103[43, 46, 51], 836[37]
Simey, T. S., 103[43, 46, 51]
Simmel, G., 122[69], 170, 875
Simmons, J., **277–310**; 307[36], 308[58], 309[88, 113], 332[97]
Simon, Sir J., 94, 398, 401, 604, 610, 613, 617, 619[2], 622[53], 623[62], 634, 636, 639, 651, 656, 658
  Reports to the Privy Council, 638–9, 644, 650
Simon, Kate, 336, 354[11]
Simpson, G., 120[29]
Sims, G. R., *16*; 155[51, 67], 160[231], 221, 588, 589, 613
Sinclair, 284
Singer, C., 691[74]
Singer, S. F., 57[72]
Sinks of London Laid Open, 727
Sion College, 319
Sisley, A., 465
Sjoberg, G., 690[66]
Skalweit, A., 53[38]
Sketches by Boz, 451, 547, 548, 554, 586, 698
Slater, P. E., 684[10]
Slum Silhouettes, 599
Slums, aggravation by demolition, 365
  animals in, *390*; 364
  charitable efforts to deal with, 367
  conditions, 91, *347*, 389–90, *393–4*, *398*; 266, 269–70, 333
  criminal classes, 364
  effects of clearance, 614, 634
  lack of definition, 363
  lack of public control, 366–7
  Mansion House Council on the Dwellings of the Poor, 606
  migrants, 372–3
  origins, 362–3
  in relation to suburbs, **360–381**, 901
  Reports, 610
  Victorian prosperity underpinned by, 361
  see also Housing; Suburbs; and individual cities
Small Dwellings Acquisition Act (1899), 392
Small House at Allington, The, 351
Smallpox, 628, 654
  death-rates, 654
  in London and Liverpool, 655
  vaccination, 655
Smart, W., 409[32]
Smelser, N. J., 676, 687[43, 46]
Smiles, S., 782, 783
Smirke, S., 314, 437, 440, 442, 444
Smith, Adam, 28, 53[45], 89, 724, 725
Smith, Albert, 472[38]
Smith, Alexander, 485
Smith, E., 620[13], 624[77], 626, 651
Smith, F., 689[59]
Smith, G., 143, 155[58], 156[90], 159[178]
Smith, H., 432
Smith, Sir Henry, 158[155]
Smith, J. Wales, 773
Smith, M., 338
Smith, Samuel, 182
Smith, T. L., 55[58]
Smith, T. Southwood, 270, 603, 635
Smith, W. H., 218
Smithfield Meat Market, 370, 540
Smithies, T. B., *29*
Smyth, C., 835[26]
Smyth, T., 861
Smollett, T., 520
Snow, J., 635, 636, 637
Snowden, P., 188[27]
Soane, Sir J., 314, 437, 439
Social science, contribution of medical officers, 617
  statistics in relation to, 91, 92
  urban problems encouraging growth of, 91
  see also Statistics
Socialist League, 198
Socialists, appeal of Tolstoy to, 880
  community experiments, 879
  romanticism, 886
  see also Morris, William
Society for the Encouragement of Arts, Manufactures and Commerce, 608
Society for the Suppression of Vice, 699
Society of Medical Officers of Health, 616
Soho, 843
  sleeping rough, 123
  Swiss in, 546
Solly, H., 54[50]
Solomon, A., 457–8
Solomon, S., *349*; 453, 458
Soloway, R., 856, 857, 869[10, 11]
Some Notice of Office Buildings in the City of London, 441
Somers Town, 538
Somerset, Lady Henry, 182
Somerville, A., 158[145]
Sommer, R., 685[20]
Son of the State, A, 599
Songs of Innocence and of Experience, 522
Sorby, J., 343, 355[36]
Sorrows of the Streets, The, 727
Soufflot, J. G., 431
Southampton, *130*
  cholera, 636
  railway, 277
Southend, 296
Southgate, 433

Index

Southwark, 538, 568, 597, 625, 845, 846
  the Borough, 586, 589
    Borough High Street, *116, 121*; 162
    King's Head, *25*
    Street, *265*
    tramps, 126
Souvenir, 899
Spain, rate of urban concentration, 41 (table)
Spark, F., 773
Spectator, 463, 576[5]
Spencer, Hon. G., 838
Spencer, H., 27, 682–3
Spendthrifts, 243
Spens, W. C., 393, 408[9]
Sperling, S. J., 689[63]
Spiers & Pond, 325
Spiller, J., 443
Spiritualism, 858
Spitalfields, licensed premises, 164 (map)
  Roman Catholic mission, 845
  tramps, 129
  weavers, 129
Spooner, I., 821
Spring, T., 31, 173
Spurgeon, C. H., 239, 240
Stafford, 818
  compared to Sheffield, 337
Stafford, J., 215
Staffordshire, effect of industry on appearance, 248
  infant mortality rates, 24
  migrants from, 137
  typhoid fever, 640
Staley, A., 473[52]
Standard, 697
Standard Theatre (London), 212, 224
Stanfield, W. C., 464
Stange, G. R., 475–94, 888[10]
Stansfeld, H., 782, 783
Stansfield, A., 760[44]
Stanton, A. S., 866
Starnthwaite, agricultural colony, 879
Starr, S., 467
State of the Law as regards the Abatement of Nuisances and the Protection of the Public Health in Scotland, The, 395
Statistics:
  appointment of Statistical Officer by L.C.C., 94
  contrasts in nineteenth- and twentieth-century information, 99
  difficulties in categorization, 98
  health improvements stemming from, 95–6
  main categories, 99
  methods of using, 92–3, 97
  official collections, 94
  qualitative evidence, 95
  quantitative evidence, 95, 97
  relation of qualitative to quantitative evidence, 97, 100
  in relation to economics, 93, 94
  of urban railways, 98
  value of quantitative analysis, 101
  Victorian attitude towards, 84–6, 97, 894
Stead, W. T., 697
Stedman Jones, G., 383[29]
Steel, Robert W., 119[6]
Steel, W. L., 308[86]
Steele, J. C., 664[65]
Steele, Sir R., 222
Stenger, E., 245[32]
Stephens, F. G., 486
Stephens, W. R. W., 785[13]
Stephenson, G., 186[4], 286, 287
Stephenson, Robert, 439, 456
  & Co., 293, 456
Stepney, fair, 133
  Society for the Relief of the Distressed, 591
  Union workhouse, 136
Stereoscope, development of, *87, 90*
  see also Photography
Stevens, W., 266
Stevenson, J. J., 320, 322, 326, 328
Stevenson, L. G., 690[63]
Stevenson, Dr, 616
Stevenson, R. L., 529
Stevenson, T. H. C., 53[44]
Stewart, Alexander P., 619[4], 630, 632, 661[17, 21, 23, 26], 662[40], 691[74]
Stewart, A. T. Q., 811[76]
Stewart, C. J., 409[24]
Stewart, John A., 599
Stewart, R., 469[7], 556[18]
Stieb, E. W., 689[63]
Stockholm, as primate city, 39
Stockport, 310, 855
Stockton, railway, 279
Stoke Newington, off-licensed premises, 168
Stoke-on-Trent, railway station, *253;* 280, 301
Stokes, F. G., 871[51]
Stolnitz, G. J., 55[55]
Stone, L. O., 54[53]
Stones of Venice, The, 315, 320, 442
Stourbridge, 138
Strand (London), 538, 696
  end of migration to, 418
  Law Courts, 320
  licensed premises, 165
  pub density, 163, 165
  refreshment-house licences, 168
  theatre, *59*
Strauss, A. L., 104[70]
Strawberry Hill, 433
Streatham, 442, 538
Street, G. E., 320, 327, 442, 444, 447[15]
Street-folk, *107–8*
  ballad-sellers, *41, 42*
  characteristics, 711, 716
  crossing sweepers, 150
  described by Mayhew, 708, 710, 715, 716, 717, 721, 722
  occupations, 710, 711, 719
  sandwich-board men, *17, 152, 380;* 136, 150
  sexual freedom of, 730
  traders, *86, 95, 99–105, 111, 134, 141;* 149, 192, 193, 194, 239, 400
  see also Street life
Street improvements, 427
  control by Metropolitan Board of Works, 318
  effect of railways, 302
  effect on slums, 365, 367
  Georgian influence, 313
  West End, *175;* 367
Street life, *92, 107, 109–13, 124–125;* 141
  markets, *141, 179, 204, 211*
  mobile circus, *19*
  musicians, *7, 18*
  processions, *226, 369, 373*
  scenes, *12, 28, 93, 118, 121–3, 175*
  shops, *142–6*
Street literature, **191–207,** 899
  anticipation of tabloid newspaper by, 206, 207, 899
  character of broadsides, *43–7;* 191–2
  circulation figures, 207
  cost of living revealed in, 197
  crime as subject-matter, *45, 47;* 198–9
  irrelevant nature of illustrations, 194
  as literature of urban working class, 191, 197, 201, 206
  in Manchester, 740
  political themes, 196–8
  polka as subject, 198, 202–4
  production, 199–200
  religious subject-matter, 196

Index

Royal Family as subject, 196, 206
subjects, 193
superficiality, 206
temperance encouraged in, 195
writers, 200
Stroud, Dorothy, 328[6]
Stuart, Grace, 686[30]
Sturge, J., 781
Sturt, G., 158[139, 149]
Suburbs, attractions to middle-classes, 369–70, 445, 874
class distinctions intensified by, 370, 418
effect of improved transport, 23, 29–30
entertainment, 35
growth, *338*; 13, 15, 298–301, 896
as image of the Picturesque, 433
in relation to slums, **359–81**
as remedy for problems of city density, 29
investment in building in, 376–9
middle-class character, 298
as response to cultural need, 334
slums created by, 360
to relieve slums, 291
see also Chalcots; Slums; and individual suburbs
Sue, E., 48, 219, 481, 659
Suffolk, migrants from, *20*; 143
Suicide, in cities, 117
Sullivan, B., 212
Sullivan, L. H., 442, 480
Summerson, Sir J., **311–32**; 329[12], 330[28], 332[102], 335, 337, 348, 354[9], 356[50], 433, 437, 442, 446[5]
Sumner, J. B., 836[36]
Sun, 714
Sun Pictures in Scotland, 230
Sunday Monitor, 698
Supple, B., 386[65]
Surbiton, 433
Surrey, migrants to, 133, 139
migration through, 117
Surrey Gardens, *155*
Surrey Theatre, *52*; 212, 220, 224
Sussex, hopping in, 138, 139
migrants, 133, 138, 144
Sussman, H. L., 888[3]
Sutcliffe, A., 428[24]
Sutcliffe, F., 235
Sutherland Edwards, H., 733[27]
Swain, C., 743, 757[11], 758[18]
Swainson, H., 445
Swan, Guida, 870[41, 42]

Swan, W., 862, 863, 867
Swann, J., 761[50]
Swansea, 295
railway at, 278
Swanson, G. E., 686[33]
Sweden, marriage rates of urban population, 18
rate of urban concentration, 41 (table)
rates of increase in population, 14
rural fertility, 18
vital statistics, 35 (table)
Sweeney, J. L., 470[13]
Swift, J., 266, 476
Swindon, employment on railway, 294
locomotive works, 294
railway, 292, 293
Swinstead, J. Howard, 158[163]
Switzerland, annual rate of urban concentration, 5 (table)
infant mortality, 25
Sybil, 882
Sydenham, 233
Sydney, as agglomeration of suburbs, 29
primacy, 38
sex ratio of population, 16
Sykes, J., 309[91], 623[74]
Symondson, A., 869[23]
Symons, A., 490–2, 494[29]

T

Taeuber, Irene, 9n., 14, 50[14]
Taine, H., 173, 188[22], 371, 452, 657, 693, 704[1]
Tait's Edinburgh Magazine, 698
Talbot, W. H. F., 229, 230–1, 234, 244[7]
Tales of Mean Streets, 591, 597, 598, 599
Tancred, 535, 549
Tarn, J. N., 189[43], 383[24]
Tate, H., 323
Taunt, H. W., 240, 241, 245[29]
Tavistock Square (London), 353
Tawney, R. H., 898
Taylor, A. J., 310[123]
Taylor, G. R., 57[72], 685[16]
Taylor, J., 446[8]
Taylor, M. W., 640, 652
Taylor, N., 332[103], **431–47**, 446[2]
Taylor, R. V., 787[47]
Taylor, Tom, 216, 222, 223, 458
Taylor, T. P., 213, 217
Taylor, W. C., 49[4]
Taylor, W. J., 704[2]
Telford, T., 438, 440, 441

Temkin, O., 689[63], 690[64]
Temperance movement, *34*, *39*, *40*; 161
activities, 181–5
entertainments started by, 175
location of temperance societies, 852[92]
in relation to party politics, 162
role compared with pubs, 184–5
Roman Catholic encouragement, 847–9
as subject of melodrama, 213, 217–18
as subject of paintings, 451
as subject of street ballads, 195
support of women, 174
among the working class, 183
Temple Bar, *66*; 365
station, 283
Ten Hours (Amendment) Act (1860), 113
Tennent, E., 790–3
Tennyson, Alfred, 477–8, 553
Tess of the D'Urbervilles, 523
Tewkesbury, effect of railway, 292–3
Thackeray, W. M., 374, 385[51], 518, 523, 539, 580[39], 714–15, 716
Thames, River, *136–8*
pollution, *407*; 637–8, 680
see also Embankment
Thames Street (London), 316
Theatres, distribution in London, 214 (map)
division between East End and West End of London, 224
growth, 211
increasing respectability, 224
life in, depicted in poetry, 492
prostitutes, 694, 695
stage realism, 218–22
technical advances, 218
for urban proletariat, 212–13, 224
see also Drama
Theatre Royal (London), *57*
Theatrical Journal, 695
Thernstrom, S., 385[50], 732[19]
They That Walk in Darkness, 599
Thiepval, memorial at, *337*; 446
Tholfsen, T. R., 835[31], 836[38]
Thomas, D. B., 245[32]
Thomas, K., 705[21]
Thomas, William, 438
Thomas, W. Luson, 459, 580[37], 581[44]
Thomis, M. J., 785[2, 8]
Thompson, D. M., 835[27], 868[4]

465

Index

Thompson, E. P., 88[21], 102[26], 516[11], 576[3], 687[40], 712, 732[17], 733[20], 856, 868[7]
Thompson, F., *308*; 301, 439
Thompson, P., **59–80**, 329[13], 447[15]
Thompson, W., 53[39]
Thompson, W. S., 51[29]
Thomson, Alexander 'Greek', *328*; 442, 443
Thomson, H. C., 619[8]
Thomson, J., 742, 877
Thomson, Patricia, 705[21]
Thomson, W., 860
Thorndike, Lynn, 690[68]
Thorne, G., 187[12]
Thorne, J., 300
Thorne, Sir R. Thorne, 662[29,32,35,48], 665[84], 666[91]
Thorne, W., 156[83], 160[223]
Threadneedle Street (London), 258
Three Brides, The, 640, 650, 665[78]
Tientsin, population, 9
Tillett, Ben, 156[77]
Tillotson, Kathleen, 102[24]
Tillyard, F., 121[40]
Timbs, J., 188[22], 189[31]
Times, The, 297, 454, 546, 560, 570, 576, 577[6], 845
Tinker, C. B., 536[1]
Tissot, J. T., 455, 470[20]
Tite, Sir W., 301
To London Town, 598
Tobias, J. J., 276[29], 735[79]
Toft, J., 787[57]
Tokyo-Yokohama, child-woman ratio, 36
  infant mortality, 36
  population, 9
Tolstoy, L., 880, 889[37]
Tom Jones, 520
Tomahawk, 560, 576[5]
Tomlin, E. W. F., 557[20]
Tomlinson, W. W., 308[69]
Tomorrow, 886
Tönnies, F., 110–11, 120[26,31]
Topley, W. W., 246[32]
Torrens, W. M. McC., 616
Torrens Acts *see* Artizans' and Labourers' Dwellings Acts
Tory Party, 185
  Radicalism, 817
  traditions, 174
  working men's drink club promoted by, 185
  see also Leeds
Tothill Fields, 126
Tottenham, 538
  medical officer, 604
Tottie, T., 777, 782
Toulouse, 281

Toulouse-Lautrec, 465
Towards Democracy, 879
Tower Bridge, 320
Town councils, social composition, 98
Town-planning, abandonment of Georgian approach, 341
  industrial villages as example of, 881–7
  railway stations not used in, 301
  reasons for development, 107
Toynbee, Arnold, 593
Toynbee Hall, 592–4
Trafalgar Square, 231
Traill, H. D., 874, 875, 888[6]
Tramcars, *127*
  in Cardiff, 295
  effect on housing, 23, 30, 338
  kept out of central London, 284
  in Manchester, *177*; 252, 253, 255, 259
  routes, 163
Tramps, *15*
  and police, *14*; 129
  sleeping places, *9, 392*; 125–6, 127, 128–9
  see also Migrants
Transport, as remedy for problems of city density, 29–30
  bicycles, *157*
  canals, *194*; 250
  creation of jobs by extension, 371
  horse-trams, *127, 227*
  horses, *109–11, 132*; 250
  pubs as centre for, 162
  review of system, 285
  river, 165
  role of coaching inn, 162
  see also Railways; Tramcars
Travellers' Club, 313, 432
Travis & Mangnall, *327*
Tremenheere, H. S., 85, 101[5]
Trevelyan, C. E., 159[174]
Trevor, J., 864, 870[48]
Trew, A., 799
Trinity Square, station, 283
Trollope, A., 374, 385[51], 518, 527, 528, 535
Trowbridge, 444
Trudgill, E., **693–705**
True Womanhood, 551
Truro, 301
Tsuru, S., 57[73]
Tuberculosis, 628
  death-rates, 641–2
  Departmental Committee on (1888), 648
  effect of environmental factors, 643, 644, 646
  problem of diagnosing, 641
  treatment, 647
  in urban communities, 642
Tudur Jones, R., 833[1]
Tuer, A. W., 579[23]
Tunbridge Wells, church architecture, 444
  railway architecture, 301
Tunsiri, V. I., 785
Tupper, J. L., 452
Turnbull, J., 664[64]
Turner, J. M. W., *356–7*; 450, 464, 465
  Ruskin's attitude to, 504–7, 508
Turnor, R., 691[73]
Turpin, Dick, 179
Turvey, R., 104[66]
Twickenham, 538
Tyneside, effect of industry on appearance, 248
  high incidence of tuberculosis, 643
Typhoid, 628, 631, 632
  incidence in towns, 639
  mortality rates, 635
  reports, 634
  theories of causes, 635, 640–1
Typhus, 628, 630–4
  decline, 633
  ignorance of causes, 631, 632, 905
  records, 633
  in relation to socio-economic circumstances, 633–4

U

Uddington, 392
Uncommercial Traveller, The, 543, 553, 554, 586
Underwood, E. A., 619[5]
Underwood, G. A., 443
Unemployment, 134
Unholy Matrimony, 599
United Kingdom Alliance, 181, 182
United States, agglomeration of population, 9
  annual rate of urban concentration, 5 (table)
  concentration of population, 4, 10, 11 (table)
  evangelical revival, 195
  fertility, 43
  growth of cities, 8
  industrialization of labour force and product, 11 (table)
  interest in urban problems of low-income countries, 41
  internal migration, 15
  international migration, 14

466

Index

level of urbanization, 35 (table)
marriage-rates of urban population, 17
mobility of female population, 15
pollution, 43-4
scarlet fever, 652
sex ratio of urban population, 16
size of family units, 19
size of households, 25
tuberculosis, 642
urban population, 7
urbanization and per capita gross domestic product, 38
vital statistics, 35 (table)
writers' attitude to city, 480-1
Universities Settlement Association, 592
Unto this Last, 89, 502
Urban politics, **763-84**
Urban transformation, demographic processes in, 11-12
effects of, 28
meaning of urban-industrial transformation, 45
metropolitan phase, 12
manifestations of industrial, 40
nineteenth-century, 4-10
sex and age structure of population in later phases of, 17
see also Cities; Mortality; Population
Urbanization:
causes, 34
and control of death-rates, 965
correlation with city-size distribution, 39
correlation with non-agricultural labour force, 38
definition of, 49[8]
as function of population growth, 895
gross domestic product in relation to, 37-8
impact, 107-8
as index of rising material standards, 896
lesson of, 906
level and rate of world, 33 (table)
nineteenth-century bias against, 873
as social progress, 97, 105, 904
of society, 4
selective character, 42
and violence, 801
as world process, 894
see also Cities; Mortality; Population

Urbanize, meaning of, 5
Uruguay, rate of urban concentration in, 41 (table)
U.S.S.R., rate of urban concentration, 41 (table)
urbanization and per capita gross domestic product 38
urbanization, 32, 33 (table)
vital statistics, 35 (table)
Utilitarianism, 875, 902
Uxbridge, medical officer, 604

V

Vaccination Act (1840), 654; (1867), 654; (1898, 1907), 655
Vaccination Extension Act (1853), 654
Van den Berg, J. H., 684[5], 686[33]
Van Dyke, Catherine, 555[1]
Van Ghent, Dorothy, 542, 555[9]
Van Gogh, V., 465, 466, 474[61], 581[44]
Vanbrugh, Sir J., *299*; 431, 433, 436, 437, 441, 442
Vanderkiste, R. W., 850[19]
Vanity Fair, 523
Vauban, S. Le P. de, 436
Vaughan Williams, R., 156[72]
Vaux, J. H., 382[6]
Venezuela, migration of foreigners to, 37
rate of urban concentration, 41 (table)
Venice, 228
Ruskin's attitude, 504
Vere, L. G., 849[5], 850[16]
Verity, T., 324, 325
Verlaine, P., 482, 492
Vernon, R., 450
Vetch, Captain, 419
Vicinus, Martha, **739-61**, 757[10]
Victoria, Queen, *82*; 233, 235, 636
Victoria station (London), 281, 283
Victoria Theatre (London), 212, 213
Victorian Cities, 85, 98, 100, 101n.
Vienna, crude mortality, 21 (table)
infant mortality, 24
migration of foreigners, 15
population, 9
Vietnam, urbanization and per capita income, 37
Villette, 532, 535
Vincent, J. R., 98, 104[67], 785[2]
Vittoria, 535

Vizetelly, H., 563, 577[9,16]
Von Archenholz, J., 693, 704[1]
Von Herkomer, H., 452, 459, 568, 578[26], 580[37,440]
Von Jurashchek, F., 52[38]
Von Raumer, F., 693, 704[1]
Von Ungarn-Sternberg, R. 56[66]
Voysey, architect, 442

W

Wade, Mason, 556[18]
Wade, R. C., 246[32]
Wadsworth, A. P., 576[1]
Wagstaffe, W. W., 658, 667[104]
Wakefield, C. M., 835[24]
Wakefield, G., 887
Wales, annual rate of urban concentration, 5 (table)
concentration of population, 11 (table)
effect of industry on appearance, 248
increases in town and country population, 14, 15
industrialization of labour force and product, 11 (table)
internal migration, 15
marriage-rates of urban population, 16-17
migration to towns, 13
mortality, 20, 24
Nonconformity, 906
overcrowding, 23
population, 6, 8
sex and age structure of population, 16-17
size of households, 25
see also England and Wales; Great Britain
Walford, C., 51[21]
Walker, D., 442, 446[11]
Walker, R. B., 190[46], 833[3]
Walker, W. S., 393, 394
Wallington Hall, *330*; 456
Wallis, H., 473[59]
Walpole, Horace, 341, 431, 433, 442
Walters, E., *183, 189, 191*; 441
Walton, W. S., 619[5]
Walworth, 379, 865
Wandsworth, 130, 609
Wanstead Flats, fair, 133
Wappäus, J. E., 18, 19, 51[27]
Wapping, 586, 842
Ward, B., 850[15,17]
Ward, D., 52[35]
Ward, F. J., 330[46]
Ward, W., 851[51]
Ward, W. H., *312*
Wardell, J., 785[4]
Wardlaw, R., 705[9]

467

Index

Ware, employment of migrants, 144
  medical officer, 604
Warne, C., 304
Warren, J. G. H., 308[68]
Warsaw, population, 9
Warwick, 818
  migration through, 117
Washington, D.C., mortality rates, 27
Water systems, conflict over introduction of, 771
  effects of improvements, 10
  importance, 21–2, 638
  lack of provision, 106–7, 266, 269, 271, 339
  pride of improvements, 85
  provision of public baths, 146
Waterhouse, A., *219*, 202, 273, 275; 301, 321, 324, 326, 327, 328, 443
Wateringbury, 139
Waterloo Road (London), pubs, 163
Waterloo station (London), 282
Waterlow, Sir S., 568
Watson, Sheriff J., 400
Watts, G. F., *353*; 458, 459
Waugh, E., 742, 750–2, 753, 756, 760[45]
Way of all Flesh, The, 532
Weatherbee, Mary, 536[14]
Weales' Rudimentary Series, 431–432
Webb, Beatrice (née Potter), 93, 103[45], 155[42], 178, 189[33], 386[77], 589
Webb, P., 320, 326, 328, 445
Webb, S., 155[42], 156[83], 159[208], 160[217, 226], 187[14], 386[77]
Weber, A. F., 5, 7, 8, 9n, 13, 14, 19, 20, 26, 27, 30, 41, 48[2], 50[12, 14, 16, 17, 19], 51[20, 24, 26, 29], 53[39, 41, 44], 54[48], 55[61], 112, 120[36], 121[48], 122[70], 417, 427[2, 4, 6]
Webster, Mary, 579[23]
Webster, T., 455
Wedderburn, A., 187[16], 310[118], 470[17], 494[22], 556[11], 888[1]
Wedgwood, T., 228, 229
Wedmore, F., 465[60]
Weir, A., 869[21]
Welch, H. D., 308[83]
Welch, P. J., 834[19]
Weldon, W., 94
Wellek, R., 888[9]
Wellesley, Marchioness of, 848
Wellington, railway station, 293
Wellington, Duke of, as subject of street ballad, 198, 201–2

Wells, A. F., 732
Wells, H. G., 374, 385[51], 874, 885, 888[5], 889[34]
Wells, R., 445
Welsh, A., 557[20]
Welton, T. A., 52[37], 119[8], 120[22, 37]
Welwyn Garden City, 887
Wesley Bready, J., 735[65]
West, C., 653
West Bromwich, English-born population, 118
West Ham, as boom town, 125
West Hartlepool, docks, *129*
West India Dock Road, pubs, 166
West London Theatre, 224
Westergaard, H., 51[26]
Westminster, Duke of, temperance policy of, 166, 181
Westminster, medical officer, 609
  off-licensed premises, 168
  Palace Hotel, 318, 324
  Palace of, 314, 327
  Pye Street, 365
  'rookeries', 403
  St Martin's Northern Schools, *332*
  station, 284
  tramps, 126
  Victoria Street, *319*; 317, 318, 365
Westminster Abbey, 304
Westminster Cathedral, 445
Westminster Chambers, 318
Westminster Review, 576[5], 744
Weston, E., 174
Westward Ho!, 535
Weylland, Rev. J. M., 171, 179, 189[34]
Weylland, S. C., 187[16]
Whalton Manor, 445
Wheatley, F., 454, 565
Wheeler, J., 305[4], 757[11]
Whipple, G. C., 55[56]
Whistler, J. A. McN., *358*; 455, 467–8
Whitby, *23*; 147, 235–6
White, B. D., 785[2], 787[62]
White, J., 354[6], 355[26]
White, L. A., 686[25]
White, Morton and Lucia, 480, 493[6]
White, R. W., 685[22], 686[34]
White, W., 444, 447[15]
White Franklin, A., 666[95]
Whitechapel, Bull's Eye, 3
  Dickens in, 538
  migrants, 373, 418
  poverty, 596
  Public Library, 594
  pubs, 182

  soup kitchens, 123
  slums, 461, 586
  station, 282
  theatre, 213
Whitechapel Road, pubs, 164, 165
Whitehall, *119*
  Banqueting Hall, *68*
  scarcity of pubs, 166
Whitehall Court, 319, 323
Whiteing, R., 600
Whitford, F. J., 811[61], 812[87]
Whiting, J. W. M., 684[11], 685[22]
Whitley, W. T., 473[50]
Whitman, W., 480, 481
Whittaker, T., 183
Whittle, P. A., 156[75]
Wickham, E. R., 833[3]
Wickwar, W. H., 576[2]
Widowers' Houses, 380
Wiener, J. H., 576[2]
Wigmore Street (London), 324
Wilberforce, S., 861, 870[30]
Wild, J. W., *332*; 314, 442
Wilde, Oscar, 530
Wilkie, D., 452, 454, 473[46]
Wilkinson, E., 156[92]
Wilkinson, R., *52*, 57
Wilkinson, T. O., 56[67]
Wilkinson, T. T., 750, 757[5, 9], 759[39]
Wilkinson, T. W., 130, 155[51, 66], 158[163]
Will o' the Wisp, 377; 560, 569
Willan, R., 642, 662[49]
Willcox, W. F., 50[17], 51[20]
Willersley Castle, 438
Willesden, sleeping rough in, 128
  station, 283
Willett, W., 336
Williams, F. S., 153[4], 307[52]
Williams, G., 190[46], 785[2]
Williams, L. J., 308[75, 78]
Williams, R., 159[211]
Williams, Raymond, 509, 516[8, 10]
Williams, Robert, 382[7]
Williams, T. Desmond, 120[17]
Williamson, Dr J., 777, 782
Willmott, P., 56[66], 187[13]
Wilson, A., 746
Wilson, G. B., 187[9, 17]
Wilson, G. W., 233
Wilson, H., 323
Wilson, J. M., 310[119]
Wilson, Mona, 158[159]
Wilson, R. G., 785[5]
Wimbledon, 538
Wimperis, J. T., 324
Winckelmann, J. J., 431
Wine on the Lees, 599
Winfield, R. W., 831
Wingo, L., 56[65]

468

Winskill, P. T., 188[24], 189[43]
Winter, A., 386[71]
Winter, D. G., 692[75]
Wirth, L., 102[22], 875, 888[8]
Wiseman, Cardinal, 382[9], 838, 844, 845
Wohl, A. S., 353[13], 354[17], 383[26], **603–24**, 622[46,50]
Woking, 445
Wolff, K., 888[8]
Wolff, M. W., **559–81, 893–907**
Wolverhampton, compared to Sheffield, 337
  disease, 638
  effect of railways, 293
  infant mortality, 650
  Primitive Methodism, 858
Woman in White, The, 531
Women, in agriculture, 651
  in factories, 639, 650, 806
  growth of opportunities, 115–116
  incidence of puerperal sepsis, 656
  as migrants, 74, 113–14, 137
  precarious economic situation, 657
  predominance over men in numbers, 116
  in telephone service, 148
  see also Population; Prostitution
Wood, S., 301, 302, 432
Woodbury, W., 239
Woodcock, H., 155[67]
Woods, R. A., 600
Woodward, B., 443
Woodward, E. L., 121[43], 188[23]
Woolcombe, W., 642, 662[49]
Woolf, Virginia, 493

Woolwich, *299*; 126, 436, 862, 870[40]
  migration to, 140
Worcester, 292, 818
  railway, 278
Worcestershire, migrants to, 138
Wordsworth, W., 85, 477, 487, 522, 523, 546, 547, 556[15], 876, 902, 904
Working classes, *see* Classes, working
Workmen's Compensation Act (1897), 646
Workmen's fares, 284, 289, 298, 419
Worthing, 99
Worthington, T., *190, 218*
Workhouse, *see* Poor Law
Workshop Act (1867), 646
Wren, Sir Christopher, *66*; 432–3
Wright, C. D., 53[41]
Wright, D. G., 787[64]
Wright, E., 102[13]
Wright, F. L., 442, 480
Wright, G. Payling, 664[53]
Wright, J., 438, 450
Wright, L., 690[66,68]
Wright, T., 135, 157[110], 581[47], 859–60, 861, 862, 867, 869[24,25], 870[31,37], 871[60]
Wrigley, A., 760[40], 761[61]
Wrigley, E. A., 49[11], 51[26], 104[64], 428[19]
Wrong, D. H., 53[44], 56[66]
Wroot, H. E., 209[23]
Wuthering Heights, 518, 524, 527
Wyatt, J., 432, 443
Wyatt, M. D., *255*; 314, 439–40
Wyld, W., *341*; 465
Wyllie, W. L., 468
Wynn, W., 349, 356[59]

Y

Yarmouth, 140
  migrants to, 143
  herring sale, *8*
Yates, May, 761[61]
Yaziki, T., 56[67]
Yeats, W. B., 492
Yeo, Eileen, 732[17]
Yonge, Charlotte, 640, 650, 662[46], 665[78]
York, poverty, 96
  pubs, 176–8
  railway station, 301, 302
  structure of city, 247
  women drinkers, 168
Yorke, G. M., 832
Yorke, H., 787[61]
Yorke, H. A., 307[38]
Yorkshire, effects of industry on appearance, 247
  migrants to, 133
  Poor Law administration, 771
Young, G. M., 120[39], 469[6], 684[3]
Young, M., 56[66], 187[13]
Young, Fr W., 845
Youngson, A. J., 395
Yudkin, J., 103[34]
Yule, G. U., 94

Z

Zangwill, I., 374, 599
Zemer, H., 470[20]
Zimmerman, G. D., 813[131]
Zollschau, G. K., 122[65]
Zoond, Vera, 620[14,18]
Zurich, 281
Zweig, F., 190[44]

Index by Marie Forsyth